9-76-83

COMBINATORIAL ENUMERATION

WILEY-INTERSCIENCE
SERIES IN DISCRETE MATHEMATICS

Graham, Rothschild, and Spencer
RAMSEY THEORY

Tucker
APPLIED COMBINATORICS

Pless
INTRODUCTION TO THE THEORY OF ERROR-CORRECTING CODES

Nemirovsky and Yudin
PROBLEM COMPLEXITY AND METHOD EFFICIENCY IN OPTIMIZATION
(Translated by E.R. Dawson)

Goulden and Jackson
COMBINATORIAL ENUMERATION

COMBINATORIAL ENUMERATION

I. P. GOULDEN AND D. M. JACKSON

Department of Combinatorics and Optimization
University of Waterloo
Ontario, Canada

With a Foreword by Gian-Carlo Rota

A Wiley-Interscience Publication

JOHN WILEY & SONS

New York • Chichester • Brisbane • Toronto • Singapore

Library of Congress Cataloging in Publication Data

Goulden, I. P.
Combinatorial enumeration.

(Wiley-Interscience series in discrete mathematics, ISSN 0277-2698)
"A Wiley-Interscience publication."
Includes bibliographical references and index.
1. Combinatorial enumeration problems. I. Jackson, D. M. II. Title. III. Series.
QA164.8.G68 1983 511'.62 82-20101
ISBN 0-471-86654-7

Printed in the United States of America

10 9 8 7 6 5 4 3 2 1

For

Claire, Cyntha, Harriet, Jennifer,

our parents,

and Professor G. Matthews

Foreword

The progress of mathematics can be viewed as a movement from the infinite to the finite. At the start, the possibilities of a theory, for example, the theory of enumeration, appear to be boundless. Rules for the enumeration of sets subject to various conditions, or combinatorial objects as they are often called, appear to obey an indefinite variety of recursions, and seem to lead to a welter of generating functions. We are at first led to suspect that the class of objects with a common property that may be enumerated is indeed infinite and unclassifiable.

As cases file upon cases, however, patterns begin to emerge. Freakish instances are quietly discarded; impossible problems are recognized as such, and what is left organizes itself along a few general criteria.

We would like these criteria to eventually boil down to one, but by and large we must be content with a small finite number.

And so with the theory of enumeration, as Jackson and Goulden show in this book. There are two basic patterns, ordinary generating functions and exponential generating functions, the first counting unlabeled or linearly ordered objects, the second counting labeled objects. The various combinatorial interpretations of the Lagrange inversion formula give the deepest results in enumeration. The test case is the enumeration of permutations subject to various geometric conditions. The still largely mysterious q-analogs arise from adding an extra parameter to the enumeration of permutations.

Lastly, there is the connection between circular enumeration and exponential generating functions; this, as well as the other topics, is developed thoroughly and with a wealth of examples by Goulden and Jackson. Their book will be required reading from now on by any worker in combinatorics.

Gian-Carlo Rota

Cambridge, Massachusetts
April 1983

Preface

The theory of enumeration has developed rapidly during the past century, with the increasing awareness of the importance of discrete structures. Work on its mathematical foundations has been inspired by MacMahon's "Combinatory Analysis," published in 1915, and Rota's series entitled "On the Foundations of Combinatorial Theory," begun in 1964. Our objectives in writing this book are to give a unified account of a generating function approach to this area and to give a reasonably complete collection of representative results. We have illustrated the theory with a range of examples to reveal something of its generality and subtlety. The book is written not only for the combinatorial theorist but also for the mathematician, the physicist, and the computer scientist, in whose fields problems of this type occur. Hitherto, much of the material included here has been available only in the research journals.

The general principle behind our account is a very simple one. First, combinatorial arguments are used to derive bijections (decompositions) between sets of discrete structures, and these are then reduced to functional relationships between formal power series by associating generating functions with sets. The type of the generating function (whether **ordinary** (Chapter 2) or **exponential** (Chapter 3)) depends on the decomposition. In manipulating generating functions, we appeal to results from analysis and linear algebra that are developed from the ring of formal power series and Laurent series in Chapter 1.

Among the structures considered are permutations, sequences, integer partitions, trees, maps, plane partitions, and lattice paths. The examples following each decomposition have been selected to illustrate the variety of the enumerative results that can be obtained from a single decomposition. Many of these can be derived separately, and more quickly, by methods peculiar to the particular problem. However, such methods may be hard to discover without knowing the results in advance and tend to give less insight into the relationships between problems.

The exercises are organized as a compendium of supplementary results whose solutions are given in detail to encourage readers to probe further. They contain additional decompositions, further development and generalization of the ideas presented in the text, and a gradual evolution of the technical details, both combinatorial and algebraic.

We have not attempted to give a complete bibliography of the field; instead, we have confined ourselves to references that either relate closely to specific points in the material or that are more detailed accounts of particular topics. In the interest of brevity, these two types of references are not distinguished in the notes and references following each section.

Inevitably, it has been necessary to exclude a number of important enumerative areas. The main exclusions are incidence algebras, ring-theoretic methods, theory of chromatic polynomials, asymptotics, root systems, and graphical enumeration, each of which warrants separate treatment.

We have benefited both directly and indirectly from conversations with friends and colleagues. In particular we wish to thank D. Ž. Djoković, P. Flajolet, I. M. Gessel, M. Guy, J. Lawrence, A. Mandel, R. P. Stanley, and N. Wormald. One of us (D.M.J.) would like to express a debt of gratitude to the late Dr. J. C. P. Miller for his encouragement of this project, and the Department of Pure Mathematics and Mathematical Statistics (University of Cambridge), the Computer Laboratory (University of Cambridge), and the Institute National de Recherche en Informatique et en Automatique (Paris), for their hospitality during several summers of uninterrupted work. Finally, we are grateful to Mrs. Susan Embro and Mrs. Sandy Tamowski for so skillfully executing the long and, at times, trying task of typing our manuscript and to H. D. L. Night for preparing the illustrations.

I. P. Goulden
D. M. Jackson

Waterloo, Ontario
April 1983

Contents

Notation

For graph-theoretic definitions see Bondy and Murty (1976). In general, sans serif capitals denote sets and boldface letters denote matrices or vectors. In the text [a.b.c.] denotes exercise c in section b of chapter a, and a.b.c. denotes paragraph c in section b of chapter a.

N	$\{0, 1, 2, \ldots\}$
N_+	$\{1, 2, \ldots\}$
N_n	$\{1, \ldots, n\}$
$\lvert\mathsf{S}\rvert$	cardinality of the set S
S^*	free monoid on S
S^+	$\mathsf{S}^* - \{\varepsilon\}$, where ε is the empty string
$\mathbb{Q}$	the set of rationals
$\mathbb{Z}$	the set of all integers
$\mathsf{M}_{m,n}(\mathsf{R})$	the set of all $m \times n$ matrices with elements in R
$\mathsf{M}_n(\mathsf{R})$	$\mathsf{M}_{n,n}(\mathsf{R})$
$\lfloor \lambda \rfloor$	integer part of λ
δ_{ij}	Kronecker delta: $\delta_{ij} = \begin{Bmatrix} 1 & \text{if } i = j \\ 0 & \text{if } i \neq j \end{Bmatrix}$
$\mapsto$	elementwise action of a mapping
$\mathbf{x}^{\mathbf{i}}$	$x_1^{i_1} x_2^{i_2} \ldots x_n^{i_n}$ where $\mathbf{x} = (x_1, \ldots, x_n)$, $\mathbf{i} = (i_1, \ldots, i_n)$
$\mathbf{i}!$	$i_1! \cdots i_n!$
$\begin{bmatrix} m \\ \mathbf{i} \end{bmatrix}$	multinomial coefficient $m!/\mathbf{i}!$ where $i_1 + \cdots + i_n = m$
$f \circ \boldsymbol{\gamma}$ (umbral)	$\sum_{k \geqslant 0} f_k \gamma_k$ where $f = 1 + f_1 x + f_2 x^2 + \cdots$ and $\boldsymbol{\gamma} = (\gamma_0, \gamma_1, \gamma_2, \ldots)$
$[b_{ij}]_{m \times n}$	the $m \times n$ matrix whose (i, j)-element is b_{ij}
$\lVert a_{ij} \rVert_{m \times m}$	determinant of the $m \times m$ matrix whose (i, j)-element is a_{ij}
$\operatorname{cof}_{ij}\mathbf{A}$	cofactor of the (i, j)-element of $\mathbf{A}$
$\operatorname{diag}(\mathbf{x})$	the diagonal matrix with x_i in row i
$\mathbf{B}[\alpha \mid \beta]$	the submatrix of $\mathbf{B}$ with row and column labels in $\alpha \subseteq \mathsf{N}_m$, $\beta \subseteq \mathsf{N}_n$, respectively.

$\mathbf{B}(\alpha\|\beta)$	$\mathbf{B}[\mathsf{N}_m - \alpha\|\mathsf{N}_n - \beta]$
$[\mathbf{B}\|\mathbf{b}]_i$	the matrix obtained by replacing column i of $\mathbf{B}$ by the column vector $\mathbf{b}$, where $\mathbf{b}$ has m elements
adj $\mathbf{A}$	the adjoint of $\mathbf{A} \in \mathsf{M}_m(\mathsf{R})$
$\mathbf{K}!$	$\prod_{\substack{1 \leqslant i \leqslant m \\ 1 \leqslant j \leqslant n}} k_{ij}!$ where $\mathbf{K} = [k_{ij}]m \times n$
$\mathbf{A}^{\mathbf{K}}$	$\prod_{\substack{1 \leqslant i \leqslant m \\ 1 \leqslant j \leqslant n}} a_{ij}^{k_{ij}}$ where $\mathbf{A} = [a_{ij}]m \times n$
$\mathbf{J}_{m,n}$	$[1]_{m,n}$
$\mathbf{J}_n$	$\mathbf{J}_{n,n}$

COMBINATORIAL ENUMERATION

"Si l'on conçoit une fonction A, d'une variable t, développée dans une série ascendante par rapport aux puissances de cette variable, le coefficient de l'une quelconque de ces puissances sera une fonction de l'exposant ou indice de cette puissance. A est ce que je nomme *fonction génératrice* de ce coefficient ou de la fonction de l'indice."

Laplace, 1795
(*Oeuvres Complètes*)

1

Mathematical Preliminaries

1.1. THE RING OF FORMAL POWER SERIES

1.1.1. Formal Power Series

Let R be a ring with unity. The **characteristic** of R is the smallest positive integer n such that $na = a + \cdots + a = 0$ for all $a \in \mathsf{R}$. R is of characteristic zero if $na = 0$ implies $n = 0$ or $a = 0$, where $n \in \mathbb{Z}$ and $a \in \mathsf{R}$. Let $\mathbf{x} = \{x_1, x_2, \ldots\}$ be a set of commutative **indeterminates**, and let

$$\mathsf{R}[[\mathbf{x}]] = \left\{ \sum_{\mathbf{i} \geqslant \mathbf{0}} c_{\mathbf{i}} \mathbf{x}^{\mathbf{i}} | c_{\mathbf{i}} \in \mathsf{R}, \mathbf{i} \geqslant \mathbf{0} \right\}.$$

The elements of this set are called **(formal) power series** and $\mathbf{x}^{\mathbf{i}}$ is a **monomial**. Let $+$ and $\cdot$ denote addition and multiplication of power series. Then $(\mathsf{R}[[\mathbf{x}]], +, \cdot)$ is a ring with unity $\mathbf{x}^{\mathbf{0}}$. This ring, usually denoted by $\mathsf{R}[[\mathbf{x}]]$ for brevity, has no **zero divisors** if and only if R itself has no zero divisors. If $\mathbf{y} \subseteq \mathbf{x}$, then $\mathsf{R}[[\mathbf{x}]]$ has as subrings $\mathsf{R}[[\mathbf{y}]]$, and R itself, the latter by the identification $r \mapsto r\mathbf{x}^{\mathbf{0}}$ for $r \in \mathsf{R}$.

1.1.2. The Coefficient Operator

Let $f(\mathbf{x}) = \sum_{\mathbf{i} \geqslant \mathbf{0}} c_{\mathbf{i}} \mathbf{x}^{\mathbf{i}} \in \mathsf{R}[[\mathbf{x}]]$. Then $c_{\mathbf{i}}$ is called the **coefficient** of $\mathbf{x}^{\mathbf{i}}$ in $f(\mathbf{x})$, and R is called the **coefficient ring** of $\mathsf{R}[[\mathbf{x}]]$. Let

$$[\mathbf{x}^{\mathbf{i}}] : \mathsf{R}[[\mathbf{x}]] \to \mathsf{R} : f \mapsto c_{\mathbf{i}}.$$

Then $[\mathbf{x}^{\mathbf{i}}]$ is called a **coefficient operator** on $\mathsf{R}[[\mathbf{x}]]$. The **constant term** in f is $[\mathbf{x}^{\mathbf{0}}]f$, often denoted by $f(\mathbf{0})$. Let

$$\mathsf{R}[[\mathbf{x}]]_0 = \{ f \in \mathsf{R}[[\mathbf{x}]] | [\mathbf{x}^{\mathbf{0}}] f = 0 \}$$

and

$$\mathsf{R}[[\mathbf{x}]]_1 = \{ f \in \mathsf{R}[[\mathbf{x}]] | (f(\mathbf{0}))^{-1} \text{ exists} \}.$$

These two subsets of $\mathsf{R}[[\mathbf{x}]]$ are of considerable importance to us, as we shall see shortly. A **polynomial** is an element of $\mathsf{R}[[\mathbf{x}]]$ with a finite number of nonzero coefficients. $\mathsf{R}[\mathbf{x}]$ denotes the set of all polynomials in $\mathsf{R}[[\mathbf{x}]]$.

Suppose that $\mathbf{x} = \mathbf{y} \,\dot\cup\, \mathbf{z}$, a partition of $\mathbf{x}$. Each power series in $\mathsf{R}[[\mathbf{x}]]$ may be regarded uniquely as a power series in $\mathbf{z}$ with coefficients in $\mathsf{R}[[\mathbf{y}]]$, so $\mathsf{R}[[\mathbf{x}]] \cong (\mathsf{R}[[\mathbf{y}]])[[\mathbf{z}]]$. Accordingly, the coefficient of $\mathbf{z}^{\mathbf{i}}$ in $f(\mathbf{x})$ depends on the coefficient ring. In general, we adopt the convention that $[\mathbf{z}^{\mathbf{i}}]f(\mathbf{x}) = [\mathbf{z}^{\mathbf{i}}]F(\mathbf{x})$ where $F(\mathbf{x}) \in (\mathsf{R}[[\mathbf{y}]])[[\mathbf{z}]]$ is the image of $f \in \mathsf{R}[[\mathbf{y}, \mathbf{z}]]$ under the natural isomorphism. For example, under this convention

$$[x^n] \sum_{i,j \geqslant 0} x^i y^j = \sum_{j \geqslant 0} y^j.$$

1.1.3. Infinite Sums and Products

The set $\mathsf{F} = \{f_j \in \mathsf{R}[[\mathbf{x}]], j \geqslant 0\}$ is called a **summable** family if each monomial in $\mathsf{R}[[\mathbf{x}]]$ occurs (with nonzero coefficient) in a finite number of elements of F. For a summable family F, product is distributive over addition, addition is associative and commutative, and the infinite product $\prod_{j \geqslant 0}(1 + f_j)$, for $f_j \in \mathsf{R}[[\mathbf{x}]]_0$, is defined by analogy with the finite case as follows. Each monomial $w \in \mathsf{R}[[\mathbf{x}]]_0$ has a finite number of ordered factorizations into monomials in $\mathsf{R}[[\mathbf{x}]]_0$. Since F is summable, only finitely many f_j have a nonzero coefficient for any particular monomial factor of w. The infinite product is obtained by summing over all ordered factorizations of monomials in $\mathsf{R}[[\mathbf{x}]]_0$, a sum that is well defined.

For example, $\prod_{i \geqslant 1}(1 + x_i)$, $\prod_{i \geqslant 1}(1 + z^i)$, $\prod_{i \geqslant 1}(1 + x_i + y_i^3)$ are all defined. However, no interpretation is given to $\prod_{i \geqslant 1} x_i$ or $\prod_{i \geqslant 1}(x_i + y_i)$, although $\prod_{i=1}^n x_i$ and $\prod_{i=1}^n (x_i + y_i)$ are defined for $n < \infty$.

1.1.4. Compositional and Multiplicative Inverses

Let $f = \sum_{\mathbf{i}} c_{\mathbf{i}} \mathbf{x}^{\mathbf{i}} \in \mathsf{R}[[\mathbf{x}]]$ and $\mathbf{g} = (g_1, g_2, \ldots)$, where $g_j \in \mathsf{R}[[\mathbf{x}]]$ for $j \geqslant 1$. The **composition** of f and $\mathbf{g}$ (obtained by **substitution** of $\mathbf{g}$ for $\mathbf{x}$) is $\sum_{\mathbf{i}} c_{\mathbf{i}} \mathbf{g}^{\mathbf{i}}$ and is denoted by $f(\mathbf{g}(\mathbf{x}))$ or $f(\mathbf{x})|_{\mathbf{x}=\mathbf{g}}$. We say that the composition (or, equivalently, substitution) is **admissible** if $\{c_{\mathbf{i}} \mathbf{g}^{\mathbf{i}} | \mathbf{i} \geqslant \mathbf{0}\}$ is a summable family. Composition is associative and is distributive over addition and multiplication provided all intermediate compositions are admissible.

1. If $\mathbf{x}$ is finite and if $g_j \in \mathsf{R}[[\mathbf{x}]]_0$ for all $j \geqslant 1$, then $f(\mathbf{g}(\mathbf{x}))$ is admissible for every $f \in \mathsf{R}[[\mathbf{x}]]$.

If $\mathsf{R}[[\mathbf{y}, \mathbf{z}]] \cong (\mathsf{R}[\mathbf{y}])[[\mathbf{z}]]$, then $\mathbf{y}$ is said to be a **bounded set** of variables for $f \in \mathsf{R}[[\mathbf{y}, \mathbf{z}]]$.

2. If $\mathbf{y}$ is a bounded set for $f \in \mathsf{R}[[\mathbf{x}]]$, where $\mathbf{y} \subseteq \mathbf{x}$, then the substitution of $\mathbf{g}$ for $\mathbf{y}$ in f is admissible for every $\mathbf{g}$.

The most frequent use of the latter result is the substitution of elements of R, usually 1, for entries of **y**. The substitution of **0** for **y** is always admissible.

If $f = xg$ where $g \in \mathsf{R}[[x]]_1$, then there is a unique power series $f^{[-1]}(x) \in \mathsf{R}[[x]]_0$, called the **compositional inverse** of f, such that $f(f^{[-1]}(x)) = f^{[-1]}(f(x)) = x$. To see this it suffices to compare coefficients on both sides of $f(f^{[-1]}(x)) = x$, yielding a set of recurrence equations for the coefficients of $f^{[-1]}(x)$ in terms of those of $f(x)$. An explicit series expansion for $f^{[-1]}(x)$ is obtained by means of the Lagrange theorem and is given following **1.2.4.**

If $f \in \mathsf{R}[[\mathbf{x}]]_1$, then there is a unique power series $f^{-1}(\mathbf{x}) \in \mathsf{R}[[\mathbf{x}]]_1$, called the **multiplicative inverse** of f, such that $f^{-1}(\mathbf{x})f(\mathbf{x}) = f(\mathbf{x})f^{-1}(\mathbf{x}) = 1$. Moreover, $f^{-1}(\mathbf{x})$ exists if and only if $f(\mathbf{x}) \in \mathsf{R}[[\mathbf{x}]]_1$ and is given by $f^{-1}(\mathbf{x}) = f^{-1}(\mathbf{0})\Sigma_{i\geqslant 0}\{1 - f^{-1}(\mathbf{0})f(\mathbf{x})\}^i$ when R is commutative. The multiplicative inverse is also denoted by $f(\mathbf{x})^{-1}$.

1.1.5. The Formal Derivative and Integral

If $f(x) = \Sigma_{i\geqslant 0}\, c_i x^i \in \mathsf{R}[[x]]$, then the **formal derivative** of $f(x)$ **with respect to** x is

1. $$D_x f(x) = \sum_{i\geqslant 0} (i+1)c_{i+1}x^i.$$

This is also denoted by $f'(x)$ and by df/dx. If $f \in \mathsf{R}[[\mathbf{x}]]$ and $\mathbf{y} = \mathbf{x} - \{x\}$, then the preceding definition is extended to this case by regarding f as an element of $(\mathsf{R}[[\mathbf{y}]])[[x]]$. In this case D_x is denoted by $\partial/\partial x$. The product rule, chain rule, and **Leibniz's theorem** hold for the formal derivative and are proved by comparing coefficients. For example, if $f(x) = \Sigma_{i\geqslant 0}\, c_i x^i$ and $g(x) = \Sigma_{j\geqslant 0}\, d_j x^j$, then

$$\begin{aligned} D_x(f(x)g(x)) &= \sum_{i,\,j\geqslant 0} (i+j)c_i d_j x^{i+j-1} \\ &= \sum_{i,\,j\geqslant 0} i c_i d_j x^{i-1+j} + \sum_{i,\,j\geqslant 0} j d_j c_i x^{j-1+i} \\ &= g(x)D_x f(x) + f(x)D_x g(x). \end{aligned}$$

This is the **product rule**. The **chain rule** is established first for $f(x) = x^n$, $n \geqslant 1$, by induction on n and by the product rule, and then by linear extension for any power series $f(x)$. If $f(0) = 1$, we prove that $D_x f^{-n}(x) = -nf^{-n-1}(x)D_x f(x)$, for $n \geqslant 1$, by taking the formal derivative of both sides of the equation $f^{-n}(x)f^n(x) = 1$, applying the product rule.

The operators D_{x_i} and D_{x_j} commute on $\mathsf{R}[[\mathbf{x}]]$. If R has characteristic zero, then

2. $$[\mathbf{x}^\mathbf{i}]f(\mathbf{x}) = \frac{1}{\mathbf{i}!}\frac{\partial^\mathbf{i} f}{\partial \mathbf{x}^\mathbf{i}}\bigg|_{\mathbf{x}=\mathbf{0}} \qquad \text{for } f \in \mathsf{R}[[\mathbf{x}]].$$

This is Taylor's theorem, and it gives a differential representation for the coefficient operator.

If $f, g \in \mathsf{R}[[x]]$ and R has characteristic zero, then

3. $D_x f = D_x g$ and $f(0) = g(0) \Leftrightarrow f(x) = g(x)$.

This result can be used to solve formal differential equations uniquely.

If $f(x) = \sum_{i \geqslant 0} c_i x^i \in \mathsf{R}[[x]]$, then the **formal integral** of $f(x)$ with respect to x is

4. $$I_x f(x) = \sum_{i \geqslant 1} i^{-1} c_{i-1} x^i,$$

where R has characteristic zero. This is also denoted by $\int_0^x f(t)\,dt$, and it is extended to the multivariate case by means of the natural isomorphism $\mathsf{R}[[\mathbf{x}]] \cong (\mathsf{R}[[\mathbf{y}]])[[x]]$, where $\mathbf{y} = \mathbf{x} - \{x\}$. The integration by parts rule holds for the formal integral operator, and the operators I_{x_i} and I_{x_j} commute on $\mathsf{R}[[\mathbf{x}]]$.

If $\{f_j \in \mathsf{R}[[x]], j \geqslant 0\}$ is a summable family and if $\alpha_j \in \mathsf{R}$ for $j \geqslant 0$, then

5. $$D_x\left(\sum_{j \geqslant 0} \alpha_j f_j\right) = \sum_{j \geqslant 0} \alpha_j D_x f_j,$$

and if R has characteristic zero, then

6. $$I_x\left\{\sum_{j \geqslant 0} \alpha_j f_j\right\} = \sum_{j \geqslant 0} \alpha_j I_x f_j.$$

Also, for $f(x) \in \mathsf{R}[[x]]$,

7. $D_x(I_x f(x)) = f(x),\ I_x(D_x f(x)) = f(x) - f(0)$.

1.1.6. The Logarithmic, Exponential, and Binomial Power Series

Certain power series will be used with great frequency. These are now defined, and we assume that R is a commutative ring with unity, with characteristic 0 and containing the rationals.

Let x be an indeterminate. Then the **exponential series** is

1. $$\exp x = \sum_{j \geqslant 0} \frac{x^j}{j!} \in \mathsf{R}[[x]],$$

also denoted by e^x. The **logarithmic series** is

2. $$\log(1 - x)^{-1} = \sum_{j \geqslant 1} \frac{x^j}{j} \in \mathsf{R}[[x]].$$

3. If y is an indeterminate, then the **binomial series** is

$$(1+x)^y = \sum_{j\geq 0} y(y-1)\cdots(y-j+1)\frac{x^j}{j!}$$

$$= \sum_{j\geq 0}\binom{y}{j}x^j \in (\mathsf{R}[y])[[x]].$$

If $f \in \mathsf{R}[[x]]_0$, then by **1.1.4(1)**, the compositions of these functions with f are admissible so $\exp f$, $\log(1+f)$ and $(1+f)^y$ are defined. Since y is bounded for $(1+x)^y$, we can substitute any element of $\mathsf{R}[[x]]$ for y. These power series, when defined, have the same properties as the analogous analytic functions. The proofs of some of these properties follow.

First, we immediately have $D_x \exp x = \exp x$, $D_x \log(1-x)^{-1} = (1-x)^{-1}$ and $D_x(1+x)^y = y(1+x)^{y-1}$ by applying the formal derivative term by term. The application of the chain rule to these is immediate except possibly in the case of the logarithm. In this case we make the substitution $x = 1 - f^{-1}$, where $f(0) = 1$. Thus $D_x \log f(x) = f(x) D_x\{1 - f^{-1}(x)\}$, and

4. $D_x \log f(x) = f^{-1}(x) D_x f(x)$.

By the chain rule, $D_x \log(\exp x) = (\exp x)^{-1}\exp x = 1 = D_x x$, so

5. $\log(\exp x) = x$, from **1.1.5(3)**,

since both $\log(\exp x)$ and x have derivative 1 and constant term 0. Similarly, using the product and chain rules,

$$D_x(1-x)\exp\log(1-x)^{-1}$$
$$= -\exp\log(1-x)^{-1} + (1-x)(1-x)^{-1}\exp\log(1-x)^{-1} = 0,$$

so that

$$(1-x)\exp\log(1-x)^{-1} = 1,$$

and

6. $\exp\log(1-x)^{-1} = (1-x)^{-1}$.

Again, this is because both $(1-x)\exp\log(1-x)^{-1}$ and 1 have derivative 0 and constant term 1.

Now we consider properties of the binomial series. It is easy to prove for positive integers n, by induction, that

7. $(1+x)^n = \prod_{i=1}^{n}(1+x) = \sum_{j\geq 0}\binom{n}{j}x^j,$

and the positioning of y as an exponent in the binomial series is justified for $y = n$. This is the **binomial theorem** for positive integers. Thus, for positive integers m and n, $[x^k]$ can be applied to the binomial series expansion of the identity $(1 + x)^n(1 + x)^m = (1 + x)^{n+m}$, giving the **Vandermonde convolution**

8. $$\sum_{i=0}^{k} \binom{n}{i}\binom{m}{k-i} = \binom{n+m}{k}.$$

But if $f(x)$ is a polynomial in x of degree k, and the equation $f(x) = 0$ has more than k roots, then $f(x) = 0$ identically. Thus the polynomial $\binom{y+z}{k} - \sum_{i=0}^{k}\binom{y}{i}\binom{z}{k-i}$ in indeterminates y and z must be identically 0, since it has an infinite number of roots, namely, all positive integers. Accordingly we have the binomial series identity

9. $$(1 + x)^y(1 + x)^z = (1 + x)^{y+z}.$$

Substitution of $-y$ for z yields $(1 + x)^y(1 + x)^{-y} = (1 + x)^0 = 1$, so

10. $$\{(1 + x)^y\}^{-1} = (1 + x)^{-y}.$$

This allows us to prove that

11. $$\log(1 + x)^y = y\log(1 + x)$$

by the following differential argument:

$$\begin{aligned} D_x\log(1 + x)^y &= (1 + x)^{-y}y(1 + x)^{y-1} \\ &= y(1 + x)^{-1} = D_x y\log(1 + x), \end{aligned}$$

and

$$\log(1 + 0)^y = 0 = y\log(1 + 0).$$

Combining these results gives

$$\begin{aligned} \{(1 + x)^y\}^z &= \exp\log\{(1 + x)^y\}^z = \exp z\log(1 + x)^y \\ &= \exp zy\log(1 + x) = \exp\log(1 + x)^{yz}, \end{aligned}$$

so

12. $$\{(1 + x)^y\}^z = (1 + x)^{yz}.$$

Finally, by the binomial theorem

$$\exp(x+y)=\sum_{n\geq 0}\frac{(x+y)^n}{n!}=\sum_{n\geq 0}\sum_{i=0}^{n}\frac{x^i}{i!}\frac{y^{n-i}}{(n-i)!},\text{ so}$$

13. $\exp(x+y)=(\exp x)(\exp y)$.

The substitution of $-x$ for y yields $\exp(0)=(\exp x)(\exp -x)$, and we have

14. $(\exp x)^{-1}=\exp(-x)$.

By making the substitution $x=f$, for $f\in \mathsf{R}[[x]]_0$ and $y=g$, for $g\in\mathsf{R}[[x]]$, in the preceding results, we obtain many of the results that are familiar to us in terms of the corresponding analytic functions. The only results that do not hold for the formal power series are those that correspond to making inadmissible substitutions. For example, it is not the case that $\exp(\log x)=x$, since $\log x$ does not exist as a formal power series.

1.1.7. Circular and Hyperbolic Power Series

Let

$$\sin x=\sum_{n\geq 0}(-1)^n\frac{x^{2n+1}}{(2n+1)!},\qquad \cos x=\sum_{n\geq 0}(-1)^n\frac{x^{2n}}{(2n)!},$$

$$\sinh x=\sum_{n\geq 0}\frac{x^{2n+1}}{(2n+1)!},\qquad \cosh x=\sum_{n\geq 0}\frac{x^{2n}}{(2n)!}.$$

These are the power series analogues of the corresponding trigonometric functions. We note that $\exp ix=\cos x+i\sin x$ and $\exp x=\cosh x+\sinh x$ where $i^2=-1$, from which

$$\sin x=(1/2i)(e^{ix}-e^{-ix}),\qquad \cos x=\tfrac{1}{2}(e^{ix}+e^{-ix}),$$

$$\sinh x=\tfrac{1}{2}(e^{x}-e^{-x}),\qquad \cosh x=\tfrac{1}{2}(e^{x}+e^{-x}).$$

All the usual trigonometric identities may be deduced from these by using properties of the exponential power series. For example,

$$\begin{aligned}\sin^2 x+\cos^2 x&=\{(1/2i)(e^{ix}-e^{-ix})\}^2+\{\tfrac{1}{2}(e^{ix}+e^{-ix})\}^2\\&=-\tfrac{1}{4}\{e^{2ix}-2e^{ix-ix}+e^{-2ix}\}+\tfrac{1}{4}\{e^{2ix}+2e^{ix-ix}+e^{-2ix}\}\\&=1.\end{aligned}$$

The power series $\tan x, \cot x, \sec x, \operatorname{cosec} x$ are defined in terms of the power series $\sin x$ and $\cos x$ following their definitions as functions of a real variable. The corresponding hyperbolic power series are defined in an analogous way.

1.1.8. Formal Differential Equations

Many classical methods for solving differential equations are still valid in $\mathsf{R}[[\mathbf{x}]]$. However, care must be exercised at each step to ensure that expressions obtained by invoking this connection exist in $\mathsf{R}[[\mathbf{x}]]$. As an example consider the differential equation

1. $D_x f = af^2 + bf$ with the boundary condition $f(0) = \alpha$

for $f \in \mathsf{R}[[x]]$, where $a, b, \alpha \in \mathsf{R}$ and a^{-1}, b^{-1} exist. This equation is a **Riccati** equation, a kind that occurs naturally in certain types of enumerative problems (see [**3.3.46**]).

We linearize the equation as follows. Let $h \in \mathsf{R}[[x]]$ be such that $h(0) = 1$ and $f = -a^{-1}D_x \log h$. It follows that h satisfies the differential equation $D_x\{D_x h\} = bD_x h$.

Since $f(0) = -a^{-1}h^{-1}(0)(D_x h(x))|_{x=0}$, then $(D_x h(x))|_{x=0} = -a\alpha$. But $D_x\{-a\alpha e^{bx}\} = b\{-a\alpha e^{bx}\}$ and $-a\alpha e^{bx}|_{x=0} = -a\alpha$. Then by **1.1.5(3)** we have $D_x h = -a\alpha e^{bx}$. Applying I_x to both sides we have

$$h(x) - h(0) = I_x(D_x h), \qquad \text{from } \mathbf{1.1.5(7)}$$

$$= -a\alpha I_x(e^{bx}) = -b^{-1}a\alpha\{e^{bx} - 1\},$$

from which $h(x) - 1 = -a\alpha b^{-1}\{e^{bx} - 1\}$, since $h(0) = 1$. Thus $h(x) = 1 - a\alpha b^{-1}\{e^{bx} - 1\}$.

We have obtained a unique power series $h(x)$ that satisfies the linearized form of the original equation, and $f(x)$ is uniquely determined by $f(x) = -a^{-1}h^{-1}(x)\, D_x h(x)$. It follows that

2. $$f(x) = \frac{\alpha e^{bx}}{1 - a\alpha b^{-1}\{e^{bx} - 1\}}.$$

The preceding procedure is identical to the one used when $f(x)$ is a function of a real variable and the solution we have obtained, of course, is the same as the one obtained by classical means. This is always the case when the classical solution is analytic at the origin. However, there are differential equations each of which has a solution in the ring of formal power series but no solution that is a function analytic at the origin (i.e., the power series has zero radius of convergence when viewed as a function). For example,

$$x^2 D_x g + (x - 1)g + 1 = 0$$

is satisfied by

$$g(x) = \sum_{n \geqslant 0} n! x^n \in \mathbb{Q}[[x]].$$

As a function, $g(x)$ is not analytic at the origin.

1.1.9. Roots of a Power Series

Occasionally, we need to solve polynomial equations for power series, for example, in finding nth roots. Let R be a commutative ring with characteristic 0 and with no zero divisors. We wish to determine $f \in \mathsf{R}[[\mathbf{x}]]$ such that

1. $f^n(\mathbf{x}) = g(\mathbf{x})$ where $g(\mathbf{0}) = \alpha^n$, $\alpha \in \mathsf{R}$, α^{-1} exists, with the initial condition $f(\mathbf{0}) = \alpha$.

Then the unique such power series is

2. $$f(\mathbf{x}) = \alpha(\alpha^{-n} g(\mathbf{x}))^{1/n} = \alpha \sum_{i \geqslant 0} \binom{1/n}{i} (\alpha^{-n} g(\mathbf{x}) - 1)^i,$$

since $\alpha^{-n} g(\mathbf{x}) - 1 \in \mathsf{R}[[\mathbf{x}]]_0$. This is a solution to **(1)** since

$$f^n(\mathbf{x}) = \alpha^n \{(\alpha^{-n} g(\mathbf{x}))^{1/n}\}^n = \alpha^n (\alpha^{-n} g(\mathbf{x}))^1 = g(\mathbf{x}),$$

from **1.1.6(12)**, and since $f(\mathbf{0}) = \alpha$. To establish uniqueness, suppose that f and h are both solutions to **(1)**, so that

$$0 = f^n(\mathbf{x}) - h^n(\mathbf{x}) = (f - h)(f^{n-1} + f^{n-2}h + \cdots + fh^{n-2} + h^{n-1}).$$

Since R, and therefore $\mathsf{R}[[\mathbf{x}]]$, has no zero divisors, then either $f - h = 0$ or $f^{n-1} + f^{n-2}h + \cdots + h^{n-1} = 0$. But

$$(f^{n-1} + f^{n-2}h + \cdots + h^{n-1})|_{\mathbf{x}=\mathbf{0}} = n\alpha^{n-1} \neq 0$$

since $\alpha \neq 0$ and R has characteristic 0 and no zero divisors. Thus $f = h$ and **(2)** is the unique solution to **(1)**.

This result is used most frequently when $f(\mathbf{x})$ satisfies a quadratic equation with a given initial condition.

1.1.10. Matrices Over the Ring of Formal Power Series

We regard matrices over $\mathsf{R}[[\mathbf{x}]]$ as power series whose coefficients are $n \times n$ matrices over R. Thus let ψ denote the isomorphism (called the tensor product map)

$$\psi : \mathsf{M}_n(\mathsf{R}[[\mathbf{x}]]) \to (\mathsf{M}_n(\mathsf{R}))[[\mathbf{x}]] : [f_{ij}(\mathbf{x})]_{n\times n} \mapsto \sum_{\mathbf{i}} \mathbf{A}_{\mathbf{i}}\mathbf{x}^{\mathbf{i}}$$

where $[\mathbf{A}_{\mathbf{i}}]_{ij} = [\mathbf{x}^{\mathbf{i}}]f_{ij}(\mathbf{x})$. Thus $n \times n$ matrices with elements that are power series may be treated as power series with coefficients in $\mathsf{M}_n(\mathsf{R})$, a ring with unity. Accordingly, finite sums and products in $\mathsf{M}_n(\mathsf{R}[[\mathbf{x}]])$ are precisely those calculated in $(\mathsf{M}_n(\mathsf{R}))[[\mathbf{x}]]$.

If $f \in \mathsf{R}[[x]]$ and $\mathbf{A} \in \mathsf{M}_n(\mathsf{R}[[\mathbf{x}]]_0)$, let $f(x) = \sum_{j\geqslant 0} f_j x^j$ and regard $f_j \in \mathsf{R}$ as the element $f_j\mathbf{I}_n \in \mathsf{M}_n(\mathsf{R})$. The substitution $f(\mathbf{A})$ under this convention is always admissible since $\{x\}$ is a finite set of indeterminates. This follows from a suitable modification of **1.1.4(1)**. In particular,

1. $(\mathbf{I} - \mathbf{A})^{-1} = \sum_{i\geqslant 0} \mathbf{A}^i.$
2. $\log(\mathbf{I} - \mathbf{A})^{-1} = \sum_{i\geqslant 1} i^{-1}\mathbf{A}^i.$
3. $\exp(\mathbf{A}) = \sum_{i\geqslant 0} (i!)^{-1}\mathbf{A}^i.$

are all summable. The **trace, determinant**, and **permanent** of $\mathbf{A}$ are defined since these are polynomial functions of the elements of $\mathbf{A}$, assuming that R is commutative. Moreover, for $\mathbf{A} \in \mathsf{M}_n(\mathsf{R}[[\mathbf{x}]]_0)$,

$$(\mathbf{I} - \mathbf{A})^{-1} = \operatorname{adj}(\mathbf{I} - \mathbf{A})/\det(\mathbf{I} - \mathbf{A}),$$

and $(\mathbf{I} - \mathbf{A})\mathbf{c} = \mathbf{B}$ can be solved by Cramer's rule for the vector $\mathbf{c}$.

We now state three results for determinants used in later sections. If $\mathbf{A}, \mathbf{B} \in \mathsf{M}_n(\mathsf{R}[[\mathbf{x}]])$, then

4. $\det(\mathbf{A} + \mathbf{B}) = \sum_{\substack{\alpha,\beta\subseteq \mathsf{N}_n \\ |\alpha| = |\beta|}} (-1)^{\sigma(\alpha,\beta)} \det \mathbf{A}[\alpha|\beta] \det \mathbf{B}(\alpha|\beta).$

where $\sigma(\alpha, \beta) = \sum_{i\in\alpha} i + \sum_{j\in\beta} j$. The proof of this result is quite straightforward. The Laplace expansion, and the cofactor expansion as a special case, are immediate corollaries.

If $\mathbf{A}, \mathbf{B}^T \in \mathsf{M}_{n,k}(\mathsf{R}[[\mathbf{x}]])$, then

5. $|\mathbf{I}_n + \mathbf{AB}| = |\mathbf{I}_k + \mathbf{BA}|.$

The proof is left as an exercise [**1.1.14**]. In the special case $k = 1$ this result yields $|\mathbf{I} + \mathbf{A}| = 1 + \operatorname{trace} \mathbf{A}$, where $\mathbf{A} \in \mathsf{M}_n(\mathsf{R}[[\mathbf{x}]])$ and has rank equal to 1.

Finally, if $\mathbf{A} \in \mathsf{M}_n(\mathsf{R}[[\mathbf{x}]]_0)$, then

6. $\operatorname{trace}\log(\mathbf{I}-\mathbf{A})^{-1} = \log\det(\mathbf{I}-\mathbf{A})^{-1}$.

The proof involves properties of the formal integral operator and proceeds as follows.

$$\begin{aligned}
\operatorname{trace}\log(\mathbf{I}-\mathbf{A})^{-1} &= \operatorname{trace}\sum_{j\geq 1} j^{-1}\mathbf{A}^j \\
&= \int_0^1 \operatorname{trace}\sum_{i\geq 0} y^i\mathbf{A}^{i+1}\,dy \\
&= \int_0^1 \operatorname{trace}\mathbf{A}(\mathbf{I}-y\mathbf{A})^{-1}\,dy \\
&= \int_0^1 \operatorname{trace}\{\mathbf{A}\operatorname{adj}(\mathbf{I}-y\mathbf{A})\}|\mathbf{I}-y\mathbf{A}|^{-1}\,dy \\
&= \int_0^1 \left\{\sum_{i=1}^n\sum_{j=1}^n a_{ij}\operatorname{cof}_{ij}(\mathbf{I}-y\mathbf{A})\right\}|\mathbf{I}-y\mathbf{A}|^{-1}\,dy \\
&= \int_0^1 -\{D_y|\mathbf{I}-y\mathbf{A}|\}|\mathbf{I}-y\mathbf{A}|^{-1}\,dy \\
&= -\log|\mathbf{I}-y\mathbf{A}|_0^1 = \log|\mathbf{I}-\mathbf{A}|^{-1}.
\end{aligned}$$

By making the admissible substitution of $\mathbf{B}$ for $\log(\mathbf{I}-\mathbf{A})^{-1}$, where $\mathbf{B} \in \mathsf{M}_n(\mathsf{R}[[\mathbf{x}]]_0)$, and applying exp to both sides of (**6**), we obtain

7. $\exp\operatorname{trace}\mathbf{B} = \det\exp\mathbf{B}$,

an identity due to Jacobi.

1.1.11. Formal Laurent Series

Let

1. $\mathsf{R}((\mathbf{x})) = \left\{\sum_{\mathbf{i}} c_{\mathbf{i}}\mathbf{x}^{\mathbf{i}} \,|\, c_{\mathbf{i}} \in \mathsf{R}, \operatorname{card}\{\mathbf{i}\,|\,c_{\mathbf{i}} \neq 0, \mathbf{i} \not\geq \mathbf{0}\} < \infty\right\}.$

This is the set of **(formal) Laurent series**, and it forms a ring with respect to addition and multiplication of power series. Again $c_{\mathbf{i}}$ is a coefficient, and $\mathbf{x}^{\mathbf{i}}$ is a monomial. If $f \in \mathsf{R}((\mathbf{x}))$, then the **valuation** of f is defined to be

$$\operatorname{val}(f) = \begin{cases} \mathbf{k} & \text{if } f(\mathbf{x}) = \mathbf{x}^{\mathbf{k}}g(\mathbf{x}) \text{ and } g(\mathbf{x}) \in \mathsf{R}[[\mathbf{x}]]_1 \\ \infty & \text{otherwise.} \end{cases}$$

For a Laurent series f, the multiplicative inverse exists iff $\mathrm{val}(f) < \infty$. If $\mathrm{val}(f) = \mathbf{k}$, then $f = \mathbf{x}^{\mathbf{k}} g$ where $g \in \mathsf{R}[[\mathbf{x}]]_1$, and we define

2. $f^{-1} = \mathbf{x}^{-\mathbf{k}} g^{-1}$.

Thus the power series $x(1-x)^{-1}$ has the multiplicative inverse $x^{-1}(1-x)$ in the ring of formal Laurent series, although it has no inverse in the ring of formal power series.

The formal derivative and integral are defined as for formal power series, with the difference that there is no $f \in \mathsf{R}((\mathbf{x}))$ such that $D_x f = x^{-1}$. This fact is exploited in Section 1.2, where we prove the Lagrange theorem for implicit functions.

NOTES AND REFERENCES

Our account of the ring of formal power series is based on Mandel (private communication); for other approaches see Henrici (1974), Niven (1969), and Tutte (1975). Muir (1960) is a useful source of information on determinants. For further identities see Gould (1972) and Riordan (1968).

1.1.8 Reid (1972); [**1.1.5**] Stanton and Sprott (1962), Riordan (1968); [**1.1.6**] Breach et al. (1976); [**1.1.8**] Riordan (1958); [**1.1.12**] Chihara (1978); [**1.1.13**] Carlitz (1962); [**1.1.15**] Sherman and Morrison (1949); [**1.1.18**] Hadamard (1892).

EXERCISES

1.1.1. Let $f(x), g(x) \in \mathsf{R}[[x]]$, where $f(0) = g(0) = 0$. If $\{g'(0)\}^{-1}$ exists, show that

$$\left.\frac{f(x)}{g(x)}\right|_{x=0} = \left.\frac{f'(x)}{g'(x)}\right|_{x=0},$$

which is **L'Hôpital's rule.**

1.1.2. Show that (a) $\binom{-n}{k} = \binom{n+k-1}{k}(-1)^k$, (b) $\binom{-\frac{1}{2}}{k} = \binom{2k}{k}(-4)^{-k}$.

1.1.3. Show that

$$\sum_{i=a}^{b-1} ix^i = x^a(1-x)^{-2}\{a + (1-a)x - bx^{b-a} + (b-1)x^{b-a+1}\}.$$

1.1.4. Suppose that $a_0, a_1, \ldots$ and $b_0, b_1, \ldots$ are such that $b_n = \sum_{i=0}^{n} \binom{n}{i} a_i$ for $n \geqslant 0$. Show that $a_n = \sum_{i=0}^{n} (-1)^{n-i} \binom{n}{i} b_i$ for $n \geqslant 0$, by considering the formal power series $A(t) = \sum_{n \geqslant 0} a_n t^n/n!$ and $B(t) = \sum_{n \geqslant 0} b_n t^n/n!$. These relations between the a_i's and b_i's are called an **inverse pair**.

1.1.5. Prove the inverse pairs

(a) $a_n = \sum_{k=0}^{n} \binom{p-k}{p-n} b_k, \qquad b_n = \sum_{k=0}^{n} (-1)^{n-k} \binom{p-k}{p-n} a_k, \qquad n = 0, \ldots, p,$

(b) $a_n = \sum_{k=0}^{n} \binom{2k}{k} b_{n-k}, \qquad b_n = \sum_{k=0}^{n} (1-2k)^{-1} \binom{2k}{k} a_{n-k}, \qquad n \geq 0.$

1.1.6. Show that $\sum_{m=0}^{n} \sum_{k=0}^{m} \binom{n}{m} \binom{m}{k}^3 = \sum_{m=0}^{n} \binom{2m}{m} \binom{n}{m}^2$.

1.1.7. Show that $\sum_{k=0}^{n} \binom{n}{k} \cos(k\theta) = 2^n \cos^n(\frac{1}{2}\theta) \cos(\frac{1}{2}n\theta)$.

1.1.8. Let $c_n = \sum_{i \geq 0} \binom{n+il}{m+ik}$. Show that

$$\sum_{n \geq 0} c_n t^n = t^m (1-t)^{k-m-1} \{(1-t)^k - t^{k-l}\}^{-1}.$$

1.1.9. (a) Let $\omega = e^{2\pi i/m}$. Show that

$$\sum_{k=0}^{m-1} \omega^{nk} = \begin{cases} m & \text{if } m \mid n \\ 0 & \text{otherwise.} \end{cases}$$

(b) Let $f(x) = \sum_{j \geq 0} f_j x^j$. Show that

$$\sum_{k \equiv l (\mathrm{mod}\ m)} f_k x^k = \frac{1}{m} \sum_{j=0}^{m-1} f(x\omega^j) \omega^{-lj}.$$

This is called **multisection** of a series. Series **bisection** corresponds to $m = 2$.

1.1.10. Show that

(a) $\sum_{k=0}^{n} \binom{3n}{3k} = \frac{1}{3}(2^n + 2(-1)^n)$,

(b) $\sum_{k \equiv l (\mathrm{mod}\ m)} \binom{n}{k} = \dfrac{2^n}{m} \sum_{j=0}^{m-1} \cos^n\left(\dfrac{\pi j}{m}\right) \cos \dfrac{(n-2l)\pi j}{m}$.

1.1.11. (a) Let $J_{m,k} = \int_{-1}^{1} (1+x)^m (1-x)^k \, dx$. Show that

$$J_{m,k} = \frac{2^{m+k+1}}{m+k+1} \binom{m+k}{k}^{-1} \qquad \text{for } m, k \geq 0.$$

(b) Hence show that

$$\sum_{k=0}^{n} \binom{n}{k}^{-1} = \frac{n+1}{2^n} \sum_{j \geq 0} \frac{1}{2j+1} \binom{n+1}{2j+1}.$$

1.1.12. Suppose that the polynomials $M_0(x), M_1(x), \ldots$ satisfy the recurrence $M_{k+1} = xM_k - k^2 M_{k-1}$, $k \geq 0$, where $M_{-1} = 0$, $M_0 = 1$. Show that

$$\sum_{k \geq 0} M_k(x) \frac{u^k}{k!} = (1+u^2)^{-1/2} \exp(x \tan^{[-1]}(u)).$$

The polynomials $M_k(x)$ are called **Meixner polynomials**.

1.1.13. Let $\mathbf{t} = (t_1,\ldots,t_k)$ and $\mathbf{n} = (n_1,\ldots,n_k)$. Show that

$$\sum_{\mathbf{n}\geqslant \mathbf{1}} \min(\mathbf{n})\mathbf{t}^{\mathbf{n}} = (1 - t_1 \cdots t_k)^{-1} \prod_{i=1}^{k} t_i(1-t_i)^{-1}.$$

1.1.14. Prove **1.1.10(5)**, that

$$|\mathbf{I}_n + \mathbf{AB}| = |\mathbf{I}_k + \mathbf{BA}| \quad \text{for} \quad \mathbf{A}, \mathbf{B}^T \in \mathsf{M}_{n,k}(\mathsf{R}[[\mathbf{x}]]).$$

1.1.15. Suppose that $\mathbf{M} \in \mathsf{M}_n(\mathsf{R}[[\mathbf{x}]])$, with rank($\mathbf{M}$) = 1 and trace($\mathbf{M}$) $\neq -1$. Show that

$$(\mathbf{I}_n + \mathbf{M})^{-1} = \mathbf{I}_n - (1 + \operatorname{trace}\mathbf{M})^{-1}\mathbf{M}.$$

This is a special case of the **Sherman–Morrison formula.**

1.1.16. Let $[\mathbf{M}]_{ij} = 1 - \delta_{i1}\delta_{1j}$ for $i, j = 1,\ldots, n+1$. Show that

$$[\mathbf{M}^k]_{ij} = \begin{cases} \displaystyle\sum_{l=1}^{k-1} n^l \binom{l-1}{k-l-1}, & i = j = 1, k > 1 \\ \displaystyle\sum_{l=1}^{k-1} n^l \binom{l}{k-l-1}, & i = 1, j > 1 \text{ or } i > 1, j = 1, k > 1 \\ \displaystyle\sum_{l=0}^{k-1} n^l \binom{l+1}{k-l-1}, & i, j > 1, k \geqslant 1. \end{cases}$$

1.1.17. An $n \times n$ **circulant** is a matrix $[a_{ij}]_{n\times n}$ for which $a_{ij} = a_{kl}$ whenever $i - j \equiv (k - l) \bmod n$. Let circ$(x_1,\ldots,x_n)$ denote the $n \times n$ circulant with first row $(x_1,\ldots, x_n)$. Show that

$$|\operatorname{circ}(x_1,\ldots, x_n)| = \prod_{k=0}^{n-1} \left(\sum_{j=0}^{n-1} \omega^{kj} x_j \right),$$

where $\omega = e^{2i\pi/n}$.

1.1.18. Let $a(x) \in \mathsf{R}[[x]]_1$ and $a(0) = 1$. Prove that

(a) $\|[x^{p+j-i}]a(x)\|_{q\times q} = (-1)^{pq} \|[x^{q+j-i}]a^{-1}(x)\|_{p\times p}$,

(b) $[x^q]a^{-1}(x) = (-1)^q \|[x^{1+j-i}]a(x)\|_{q\times q}$.

These determinants are called **Hänkel determinants**. Part (a) is **Hadamard's theorem.**

1.2. THE LAGRANGE THEOREM FOR IMPLICIT FUNCTIONS

The Lagrange theorem is an important result for solving functional equations that arise in combinatorial enumeration. We prove the theorem for the ring of formal power series, although an analogous result holds for analytic functions. The proof given here is an algebraic one, and a purely combinatorial proof appears in Chapter 5.

We begin by considering the properties of the operator $[x^{-1}]$ on $\mathsf{R}((x))$ where R is commutative. This operator is of central importance since we may express the extraction of any coefficient in terms of it. Thus

$$[x^n]f(x) = [x^{-1}]x^{-(n+1)}f(x),$$

where $f \in \mathsf{R}((x))$. The operator $[x^{-1}]$ may be regarded as a **formal residue operator**, since $[x^{-1}]f(x)$ is the formal residue of the Laurent series $f(x)$ at $x = 0$. For this reason, $[x^{-1}]f(x)$ may be denoted by $\operatorname{Res} f(x)$ if preferred.

The fundamental observation about the formal residue operator is that the term x^{-1} cannot arise as the derivative of x^n for any n, a point noted at the end of **1.1.11**.

1.2.1. Proposition

Let $f, g \in \mathsf{R}(x))$. *Then*

1. $[x^{-1}]f'(x) = 0$, and **2.** $[x^{-1}]f'(x)g(x) = -[x^{-1}]f(x)g'(x)$. □

The next result permits a change of variable during the calculation of formal residues. It is the main result of this section, from which all of the others may be deduced.

1.2.2. Theorem (Residue Composition)

Let $f(x), r(x) \in \mathsf{R}((x))$ *and let* $\operatorname{val}(r) = \alpha > 0$. *Then*

$$\alpha[x^{-1}]f(x) = [z^{-1}]f(r(z))r'(z).$$

Proof: We first prove the result for $f(x) = x^n$, where n is an integer.

If $n \neq -1$, then $[z^{-1}]r^n(z)r'(z) = (n+1)^{-1}[z^{-1}](d/dz)r^{n+1}(z) = 0$ by (**1**) of Proposition 1.2.1, since $r^{n+1}(z) \in \mathsf{R}((z))$.

If $n = -1$, then $[z^{-1}]r^n(z)r'(z) = [z^{-1}]r'(z)r^{-1}(z)$. But $r(z) = \beta z^\alpha h(z)$ where $h(z) \in \mathsf{R}[[z]]_1$, $h(0) = 1$, $\beta \neq 0$, since $\operatorname{val}(r) = \alpha$. Now $h^{-1}(z)$ and $\log h(z)$ exist, so $[z^{-1}]r'(z)r^{-1}(z) = [z^{-1}]\{\alpha z^{-1} + h'(z)h^{-1}(z)\} = \alpha + [z^{-1}](d/dz)(\log h(z)) = \alpha$ since $\log h(z) \in \mathsf{R}[[z]]$.

It follows that, for all integers n, $[z^{-1}]r^n(z)r'(z) = \alpha\delta_{n,-1} = \alpha[x^{-1}]x^n$. Let $f(x) = \Sigma_{n\geqslant k} a_n x^n$ where $\text{val}(f) = k < \infty$. Since $\text{val}(r) > 0$, then $f(r(z))$ exists. Thus

$$\alpha[x^{-1}]f(x) = [z^{-1}]\sum_{n\geqslant k} a_n r^n(z)r'(z) = [z^{-1}]f(r(z))r'(z). \qquad \square$$

As an example of the use of the residue composition theorem we prove a binomial identity.

1.2.3. An Identity (by Residue Composition)

We wish to obtain the sum

$$S = \sum_{k=0}^{n} \binom{2n+1}{2k+1}\binom{j+k}{2n}.$$

Bisecting $(1+x)^{2n+1}$ yields

$$\sum_{k=0}^{n} \binom{2n+1}{2k+1} x^{2k} = \frac{1}{2x}\{(1+x)^{2n+1} - (1-x)^{2n+1}\} = f(x), \text{ say}.$$

Thus

$$S = [y^{2n}](1+y)^j \sum_{k=0}^{n} (1+y)^k [x^{2k}]f(x)$$

$$= [y^{-1}]y^{-(2n+1)}(1+y)^j f\left((1+y)^{1/2}\right) \qquad \text{since } f(x) \in \mathsf{R}[[x^2]].$$

To evaluate this coefficient change variables by setting $y = z^2(z^2-2)$. Thus $\text{val}(y(z)) = 2$ since $z^2 - 2 \in \mathsf{R}[[z]]_1$, and $(1+y)^{1/2} = 1 - z^2$. By the residue composition theorem

$$S = [z^{-1}](z^2-1)^{2j}\left\{\frac{1}{(z^2-2)^{2n+1}} - \frac{1}{z^{4n+2}}\right\}z$$

$$= [z^{-1}](z^2-1)^{2j} z^{-(4n+1)} \qquad \text{since } (z^2-2)^{-(2n+1)} \in \mathsf{R}[[z]]$$

$$= [z^{4n}](z^2-1)^{2j} = \binom{2j}{2n}$$

so

$$\sum_{k=0}^{n} \binom{2n+1}{2k+1}\binom{j+k}{2n} = \binom{2j}{2n}. \qquad \square$$

At first sight the substitution $y = z^2(z^2 - 2)$ used in the preceding example might seem unnecessarily complicated since the substitution $y = z^2 - 1$ appears to be more reasonable. However, in the latter case $\alpha = 0$, which is disallowed by the residue composition theorem.

1.2.4. Theorem (Lagrange)

Let $\phi(\lambda) \in \mathsf{R}[[\lambda]]_1$. Then there exists a unique formal power series $w(t) \in \mathsf{R}[[t]]_0$ such that $w = t\phi(w)$. Moreover

1. *If $f(\lambda) \in \mathsf{R}((\lambda))$, then*

$$[t^n]f(w) = \begin{cases} \frac{1}{n}[\lambda^{n-1}]\{f'(\lambda)\phi^n(\lambda)\} & \text{for } n \neq 0, n \geqslant \text{val}(f) \\ [\lambda^0]f(\lambda) + [\lambda^{-1}]f'(\lambda)\log(\phi(\lambda)\phi^{-1}(0)) & \text{for } n = 0. \end{cases}$$

2. *If $F(\lambda) \in \mathsf{R}[[\lambda]]$, then*

$$\sum_{n\geqslant 0} c_n t^n = F(w)\{1 - t\phi'(w)\}^{-1}, \quad \text{where } c_n = [\lambda^n]F(\lambda)\phi^n(\lambda).$$

Proof: Let $\Phi(w) = w/\phi(w)$ so $\Phi(w) = t$ and $\text{val}(\Phi) = 1$, since $\phi(\lambda) \in \mathsf{R}[[\lambda]]_1$. Thus $\Phi^{[-1]}(\lambda)$ exists, and $w = \Phi^{[-1]}(t)$ is the unique solution of $w = t\phi(w)$.

1. For any integer n

$$[t^n]f(w) = [t^{-1}]t^{-(n+1)}f(\Phi^{[-1]}(t)) = [w^{-1}]\Phi^{-(n+1)}(w)f(w)\Phi'(w), \quad (1)$$

where the second equality is from the residue composition theorem with the change of variable $t = \Phi(w)$.

If $n \neq 0$, then from (1)

$$\begin{aligned} [t^n]f(w) &= -\frac{1}{n}[w^{-1}]f(w)(\Phi^{-n}(w))' \\ &= \frac{1}{n}[w^{-1}]f'(w)\Phi^{-n}(w) \qquad \text{by Proposition 1.2.1(2)} \\ &= \frac{1}{n}[w^{n-1}]f'(w)\phi^n(w). \end{aligned}$$

If $n = 0$, then differentiating $\Phi(w)$ by parts we have from (1)

$$\begin{aligned} [t^n]f(w) &= [w^0]f(w) - [w^{-1}]f(w)\phi'(w)\phi^{-1}(w) \\ &= [w^0]f(w) + [w^{-1}]f'(w)\log(\phi(w)\phi^{-1}(0)) \end{aligned}$$

by Proposition 1.2.1(2) and **1.1.5(7)**, and the result follows.

2. Let $f(w) = \int_0^w F(\lambda)\phi^{-1}(\lambda)\,d\lambda$, so $f(w) \in \mathsf{R}[[w]]$ since $F(\lambda) \in \mathsf{R}[[\lambda]]$ and $\phi(\lambda) \in \mathsf{R}[[\lambda]]_1$. Thus, by **(1)**

$$f(w) = f(0) + \sum_{n \geqslant 1} \frac{1}{n} t^n [\lambda^{n-1}] \phi^n(\lambda) f'(\lambda).$$

Differentiating both sides by t gives, after rearrangement,

$$f'(w)\frac{dw}{dt} = \sum_{n \geqslant 0} t^n [\lambda^n] \phi^{n+1}(\lambda) f'(\lambda).$$

But differentiating $w = t\phi(w)$ we get $dw/dt = \phi(w)\{1 - t\phi'(w)\}^{-1}$ and the result follows since $f'(w)\phi(w) = F(w)$. □

This is a formal version of the classical Lagrange–Bürmann theorem and can be used to calculate $f(w)$, where $w(t)$ is the formal power series solution of the functional equation $w = t\phi(w)$. Note that when $f(w) = w$ we obtain the result

$$\Phi^{[-1]}(t) = [\lambda^{-1}] \log\{1 - t\Phi^{-1}(\lambda)\}^{-1},$$

where $\mathrm{val}(\Phi) = 1$.

1.2.5. A Functional Equation

Suppose that $w(t)$ satisfies the functional equation $w = te^w$. We determine $w(t)$, $w^{-1}(t)$, and $w^{-2}(t)$ by the Lagrange theorem, which may be applied since $e^\lambda \in \mathsf{R}[[\lambda]]_1$.

1. Let $f(\lambda) = \lambda$. Then $\mathrm{val}(f) = 1$ so

$$w(t) = [\lambda^0]\lambda + [\lambda^{-1}]\log e^\lambda + \sum_{n \geqslant 1} \frac{1}{n} t^n [\lambda^{n-1}] e^{n\lambda} = \sum_{n \geqslant 1} n^{n-1} \frac{t^n}{n!}.$$

2. Let $f(\lambda) = \lambda^{-1}$. Then $\mathrm{val}(f) = -1$ so $w^{-1}(t) = -\sum_{n \geqslant -1} n^n \dfrac{t^n}{(n+1)!}$.

3. Let $f(\lambda) = \lambda^{-2}$. Then $\mathrm{val}(f) = -2$ so $w^{-2}(t) = -2 \sum_{\substack{n \geqslant -2 \\ n \neq 0}} n^{n+1} \dfrac{t^n}{(n+2)!}$.

□

Calculating $w^{-1}(t)$ from **(1)** by expanding $t^{-1}\{\sum_{n \geqslant 0}(n+1)^n t^n/(n+1)!\}^{-1}$ as a Laurent series in t (see **1.1.11(2)**) leads to a more cumbersome expression than the one given in **(2)** above. We may therefore appreciate the usefulness of the power series $f(\lambda)$ of the Lagrange theorem.

As an example of the use of the alternative form, (**2**), of the Lagrange theorem, we derive the generating function for the central trinomial numbers.

1.2.6. The Central Trinomial Numbers

The central trinomial numbers are defined by $c_n = [\lambda^n](1 + \lambda + \lambda^2)^n$. Thus $c_n = [\lambda^n]F(\lambda)\phi^n(\lambda)$, where $F(\lambda) = 1$ and $\phi(\lambda) = 1 + \lambda + \lambda^2 \in \mathsf{R}[[\lambda]]_1$. Then (**2**) of the Lagrange theorem gives $c(t) = \sum_{n\geqslant 0} c_n t^n = \{1 - t(1 + 2w)\}^{-1}$, where w satisfies $w = t(1 + w + w^2)$. The unique $w(t) \in \mathsf{R}[[t]]_0$ that satisfies this quadratic functional equation is $w = \{-(t-1) - (1 - 2t - 3t^2)^{1/2}\}/2t$ so $c(t) = (1 - 2t - 3t^2)^{-1/2}$. □

The Lagrange theorem may also be used to expand a series $f(x) \in \mathsf{R}[[x]]$ in terms of another series $\psi(x) \in \mathsf{R}[[x]]$ where val $\psi = 1$. This is done by setting $\psi(x) = x\phi^{-1}(x)$, where $\phi(x) \in \mathsf{R}[[x]]_1$, and by letting $\psi(x) = y$, so $x = y\phi(x)$. The Lagrange theorem then gives $f(x)$ as a power series in y.

1.2.7. Abel's Extension of the Binomial Theorem

To expand $e^{\alpha x}$ in terms of xe^{-x}, for an arbitrary indeterminate α, let $y = xe^{-x}$, so $x = ye^x$. Since $e^x \in \mathsf{R}[[x]]_1$ and $\text{val}(e^{\alpha\lambda}) = 0$, the Lagrange theorem gives

$$e^{\alpha x} = [\lambda^0]e^{\alpha\lambda} + [\lambda^{-1}](\alpha e^{\alpha\lambda}\cdot\lambda) + \sum_{n\geqslant 1}\frac{1}{n}y^n[\lambda^{n-1}]\alpha e^{(\alpha+n)\lambda}$$

$$= \sum_{n\geqslant 0}\frac{1}{n!}\alpha(\alpha+n)^{n-1}x^n e^{-nx}.$$

Let β be an indeterminate. Expanding $e^{\alpha x}$, $e^{\beta x}$, and $e^{(\alpha+\beta)x}$ in terms of $y = xe^{-x}$ by the above, and applying $[y^n]$ to $e^{\alpha x}e^{\beta x} = e^{(\alpha+\beta)x}$ gives, for $n \geqslant 0$,

$$\alpha\beta\sum_{k=0}^{n}\binom{n}{k}(\alpha+k)^{k-1}(\beta+n-k)^{n-k-1} = (\alpha+\beta)(\alpha+\beta+n)^{n-1}. \quad \square$$

Most of these ideas may be extended to the multivariate case by introducing an obvious generalization of the univariate formal residue operator. As a result we obtain multivariate versions of the residue composition theorem and the Lagrange theorem.

1.2.8. Theorem (Multivariate Residue Composition)

Let $f(\mathbf{x}), g_1(\mathbf{x}), \ldots, g_m(\mathbf{x}) \in \mathsf{R}((\mathbf{x}))$, *and let* $\text{val}(g_i) = (p_{i1}, \ldots, p_{im}) \geqslant \mathbf{0}$ *be finite where* $p_{i1} + \cdots + p_{im} > 0$ *for* $1 \leqslant i \leqslant m$. *Then*

$$|\mathbf{P}|\cdot[\mathbf{x}^{-1}]f(\mathbf{x}) = [\mathbf{z}^{-1}]\{f(\mathbf{g}(\mathbf{z}))\cdot J(\mathbf{g})\}$$

where $\mathbf{g} = (g_1, \ldots, g_m)$, $\mathbf{z} = (z_1, \ldots, z_m)$, $J(\mathbf{g}(\mathbf{z})) = \left\| \dfrac{\partial g_i}{\partial z_j} \right\|_{m \times m}$ *and* $\mathbf{P} = [p_{ij}]_{m \times m}$.

Proof: Since $\operatorname{val}(g_i) \geqslant \mathbf{0}$, and $p_{i1} + \cdots + p_{im} > 0$ for $1 \leqslant i \leqslant m$, then $f(\mathbf{g}(\mathbf{x}))$ exists in $\mathsf{R}((\mathbf{x}))$. Let $f(\mathbf{x}) = \sum_{\mathbf{k}} c(\mathbf{k})\mathbf{x}^{\mathbf{k}} \in \mathsf{R}((\mathbf{x}))$, and D_j denote $\partial/\partial z_j$, so

$$\begin{aligned}
[\mathbf{z}^{-1}]\{f(\mathbf{g}(\mathbf{z})) \cdot J(\mathbf{g})\} &= [\mathbf{z}^{-1}] \sum_{\mathbf{k}} c(\mathbf{k}) \mathbf{g}^{\mathbf{k}}(\mathbf{z}) \| D_j g_i(\mathbf{z}) \| \\
&= \sum_{\mathbf{k}} c(\mathbf{k}) [\mathbf{z}^{-1}] G_{\mathbf{k}}(\mathbf{z}), \quad \text{where } G_{\mathbf{k}}(\mathbf{z}) = \| g_i^{k_i}(\mathbf{z}) D_j g_i(\mathbf{z}) \|.
\end{aligned}$$

Since $\operatorname{val} g_i(\mathbf{z}) = (p_{i1}, \ldots, p_{im}) = \mathbf{p}_i$, let $g_i(\mathbf{z}) = \beta_i \mathbf{z}^{\mathbf{p}_i} h_i(\mathbf{z})$, where $h_i(\mathbf{z}) \in \mathsf{R}[[\mathbf{z}]]_1$, $h_i(\mathbf{0}) = 1$, and $\beta_i \neq 0$ for $i = 1, \ldots, m$. Thus

$$g_i^{k_i}(\mathbf{z}) D_j g_i(\mathbf{z}) = \begin{cases} D_j \left\{ \dfrac{1}{k_i + 1} g_i^{k_i + 1}(\mathbf{z}) \right\} & \text{if } k_i \neq -1 \\[2ex] \dfrac{p_{ij}}{z_j} + D_j \log h_i(\mathbf{z}) & \text{if } k_i = -1 \end{cases}$$

so that $G_{\mathbf{k}}(\mathbf{z}) = \left\| \dfrac{p_{ij}}{z_j} \delta_{k_i, -1} + \sum_{\mathbf{l}_i} \dfrac{l_{ij}}{z_j} a_i(\mathbf{l}_i, k_i) \mathbf{z}^{\mathbf{l}_i} \right\|$, where

$$\sum_{\mathbf{l}_i} a_i(\mathbf{l}_i, k_i) \mathbf{z}^{\mathbf{l}_i} = \begin{cases} \dfrac{1}{k_i + 1} g_i^{k_i + 1}(\mathbf{z}), & k_i \neq -1 \\[2ex] \log h_i(\mathbf{z}), & k_i = -1 \end{cases}$$

and $\mathbf{l}_i = (l_{i1}, \ldots, l_{im})$. Using multilinearity on each row of $G_{\mathbf{k}}(\mathbf{z})$, we get $G_{\mathbf{k}}(\mathbf{z}) = \sum_{\alpha \subseteq \mathsf{N}_m} \| A_{ij}(\alpha) \|$, where

$$A_{ij}(\alpha) = \begin{cases} \dfrac{p_{ij}}{z_j} \delta_{k_i, -1} & \text{if } i \in \mathsf{N}_m - \alpha \\[2ex] \sum_{\mathbf{l}_i} \dfrac{l_{ij}}{z_j} a_i(\mathbf{l}_i, k_i) \mathbf{z}^{\mathbf{l}_i} & \text{if } i \in \alpha. \end{cases}$$

Again using multilinearity to expand on row i, for each $i \in \alpha$,

$$G_{\mathbf{k}}(\mathbf{z}) = \sum_{t=0}^{m} \sum_{\alpha = \{\alpha_1, \ldots, \alpha_t\} \subseteq \mathsf{N}_m} \sum_{\mathbf{l}_{\alpha_1}} \cdots \sum_{\mathbf{l}_{\alpha_t}} \mathbf{z}^{\mathbf{l}_{\alpha_1} + \cdots + \mathbf{l}_{\alpha_t} - \mathbf{1}} \| B_{ij} \| \prod_{s=1}^{t} a_{\alpha_s}(\mathbf{l}_{\alpha_s}, k_{\alpha_s}),$$

where

$$B_{ij} = \begin{cases} p_{ij}\delta_{k_i,-1} & \text{if } i \in \mathsf{N}_m - \alpha \\ l_{ij} & \text{if } i \in \alpha. \end{cases}$$

Thus for a nonzero contribution to $[\mathbf{z}^{-1}]G_{\mathbf{k}}(\mathbf{z})$, we must have $\mathbf{l}_{\alpha_1} + \cdots + \mathbf{l}_{\alpha_t} = \mathbf{0}$. But in this case $\|B_{ij}\| = 0$, by the Laplace expansion, unless $t = 0$. If $t = 0$, then $\|B_{ij}\| = \|p_{ij}\| \prod_{s=1}^{m} \delta_{k_s,-1} = [\mathbf{z}^{-1}]G_{\mathbf{k}}(\mathbf{z})$ and the result follows. □

Comparing this result with the univariate case we note that $\text{val}(f)$ has been replaced by $|\mathbf{P}|$, a determinant formed from the multivariate valuations of $g_1, \ldots, g_m$, and that $r'(x)$ has been replaced by the Jacobian $J(\mathbf{g})$ of $g_1(\mathbf{z}), \ldots, g_m(\mathbf{z})$ with respect to $z_1, \ldots, z_m$.

1.2.9. Theorem (Multivariate Lagrange)

Let $f(\boldsymbol{\lambda}) \in \mathsf{R}((\boldsymbol{\lambda}))$ and $\phi_1(\boldsymbol{\lambda}), \ldots, \phi_m(\boldsymbol{\lambda}) \in \mathsf{R}[[\boldsymbol{\lambda}]]_1$ where $\boldsymbol{\lambda} = (\lambda_1, \ldots, \lambda_m)$. Suppose that $w_i = t_i\phi_i(\mathbf{w})$ for $i = 1, \ldots, m$, where $\mathbf{w} = (w_1, \ldots, w_m)$. Let $\boldsymbol{\phi} = (\phi_1, \ldots, \phi_m)$ and $\mathbf{t} = (t_1, \ldots, t_m)$. Then

1. $$f(\mathbf{w}(\mathbf{t})) = \sum_{\mathbf{k}} \mathbf{t}^{\mathbf{k}}[\boldsymbol{\lambda}^{\mathbf{k}}]\left\{ f(\boldsymbol{\lambda})\boldsymbol{\phi}^{\mathbf{k}}(\boldsymbol{\lambda}) \left\| \delta_{ij} - \frac{\lambda_j}{\phi_i(\boldsymbol{\lambda})} \frac{\partial \phi_i(\boldsymbol{\lambda})}{\partial \lambda_j} \right\| \right\}.$$

2. *If $F(\boldsymbol{\lambda}) \in \mathsf{R}[[\boldsymbol{\lambda}]]$, then*

$$\left. \frac{F(\mathbf{w})}{\left\| \delta_{ij} - t_i \dfrac{\partial}{\partial w_j} \phi_i(\mathbf{w}) \right\|} \right|_{\mathbf{w}=\mathbf{w}(\mathbf{t})} = \sum_{\mathbf{k} \geqslant \mathbf{0}} \mathbf{t}^{\mathbf{k}}[\boldsymbol{\lambda}^{\mathbf{k}}]F(\boldsymbol{\lambda})\boldsymbol{\phi}^{\mathbf{k}}(\boldsymbol{\lambda}).$$

Proof: **1.** By the argument used in the univariate case, there exists a unique series $w_i(\mathbf{t}) \in \mathsf{R}[[\mathbf{t}]]_0$ such that $w_i = t_i\phi_i(\mathbf{w})$, for $i = 1, \ldots, m$. Now $[\mathbf{t}^{\mathbf{k}}]f(\mathbf{w}) = [\mathbf{t}^{-1}]\mathbf{t}^{-(\mathbf{k}+1)}f(\mathbf{w})$ and substituting $t_i = w_i\phi_i^{-1}(\mathbf{w})$ we have, by the multivariate residue composition theorem,

$$[\mathbf{t}^{\mathbf{k}}]f(\mathbf{w}) = [\mathbf{w}^{-1}]f(\mathbf{w})\boldsymbol{\phi}^{\mathbf{k}+1}\mathbf{w}^{-(\mathbf{k}+1)}J(\mathbf{t}),$$

since $\|[\text{val}(t_i(\mathbf{w}))]\|_{m\times m} = |\mathbf{I}_m| = 1$. But

$$J(\mathbf{t}) = \|D_j t_i\| = \|\delta_{ij}\phi_i^{-1}(\mathbf{w}) - w_i\phi_i^{-2}(\mathbf{w}) \cdot D_j\phi_i(\mathbf{w})\|$$
$$= \boldsymbol{\phi}^{-1}(\mathbf{w})\|\delta_{ij} - w_j\phi_i^{-1}(\mathbf{w})D_j\phi_i(\mathbf{w})\|$$

by multiplying row i by w_i^{-1} and column j by w_j for $1 \leqslant i, j \leqslant m$. Thus

$$[\mathbf{t}^{\mathbf{k}}]f(\mathbf{w}) = [\mathbf{w}^{\mathbf{k}}]f(\mathbf{w})\boldsymbol{\phi}^{\mathbf{k}}(\mathbf{w}) \left\| \delta_{ij} - w_j\phi_i^{-1}(\mathbf{w}) \frac{\partial \phi_i(\mathbf{w})}{\partial w_j} \right\|.$$

2. Let $f(\mathbf{w}) = F(\mathbf{w})\left\|\delta_{ij} - w_i\phi_i^{-1}(\mathbf{w})\frac{\partial}{\partial w_j}\phi_i(\mathbf{w})\right\|^{-1}$. Then $f(\mathbf{w}) \in \mathsf{R}[[\mathbf{w}]]$. The result follows immediately from (**1**), since

$$\left\|\delta_{ij} - w_j\phi_i^{-1}(\mathbf{w})\frac{\partial \phi_i(\mathbf{w})}{\partial w_j}\right\| = \left\|\delta_{ij} - w_i\phi_i^{-1}(\mathbf{w})\frac{\partial \phi_i(\mathbf{w})}{\partial w_j}\right\| \neq 0,$$

and $t_i = w_i\phi_i^{-1}(\mathbf{w})$. □

The specialization of (**1**) of the preceding result to (**1**) of the univariate case is not immediate since further simplification of

$$[\lambda^k]f(\lambda)\phi^k(\lambda)\{1 - x\phi^{-1}(x)\phi'(x)\}$$

is possible. The specialization of (**2**) of the multivariate case to (**2**) of the univariate case is, however, immediate. We may regard $w_i = t_i\phi_i(\mathbf{w})$ for $1 \leqslant i \leqslant m$ as a system of simultaneous functional equations for the power series $w_1(\mathbf{t}), \ldots, w_m(\mathbf{t}) \in \mathsf{R}[[\mathbf{t}]]_0$ which the multivariate Lagrange theorem solves. The following is a typical example, for $m = 2$.

1.2.10. A Functional Equation in Two Variables

We wish to express $uv(1 - u - v)$ as a power series in x and y, where u and v satisfy

$$\left.\begin{aligned} u &= x(1-v)^{-2} \\ v &= y(1-u)^{-2}. \end{aligned}\right\}$$

Since $(1-u)^{-2}, (1-v)^{-2} \in \mathsf{R}[[u, v]]_1$, the multivariate Lagrange theorem can be applied to obtain

$$(1-u)^{-r}(1-v)^{-s} = \sum_{i,j\geqslant 0} x^i y^j [\lambda^i\mu^j](1-\lambda)^{-r-2j}(1-\mu)^{-s-2i}\Delta,$$

where Δ is the determinant given by $\Delta = 1 - 4xy(1-\lambda)^{-3}(1-\mu)^{-3}$. Thus

$$(1-u)^{-r}(1-v)^{-s}$$
$$= \sum_{i,j\geqslant 0} \frac{x^i}{i!}\frac{y^j}{j!}\frac{(2j+i+r-1)!}{(2j+r)!}\frac{(2i+j+s-1)!}{(2i+s)!}(2js + 2ir + rs).$$

But $uv(1 - u - v)$

$$= xy\{(1-u)^{-1}(1-v)^{-2} + (1-u)^{-2}(1-v)^{-1} - (1-u)^{-2}(1-v)^{-2}\},$$

so

$$uv(1-u-v) = \sum_{i,j\geqslant 1} \frac{x^i}{i!}\frac{y^j}{j!}\frac{(2j+i-2)!}{(2i-1)!}\frac{(2i+j-2)!}{(2j-1)!}. \qquad \square$$

We use the second form of the multivariate Lagrange theorem to prove a celebrated result of MacMahon.

1.2.11. Theorem (MacMahon Master Theorem)

Let $\mathbf{A} = [a_{ij}]_{m\times m}$ *and let* $\mathbf{X} = \operatorname{diag}(x_1,\ldots,x_m)$. *Then*

$$[\mathbf{x}^{\mathbf{k}}]\prod_{i=1}^{m}(a_{i1}x_1 + \cdots + a_{im}x_m)^{k_i} = [\mathbf{x}^{\mathbf{k}}]|\mathbf{I} - \mathbf{XA}|^{-1},$$

where $\mathbf{k} = (k_1,\ldots,k_m)$.

Proof: Let $F(\boldsymbol{\lambda}) = 1$ and $\phi_i(\boldsymbol{\lambda}) = a_{i1}\lambda_1 + \cdots + a_{im}\lambda_m$. Then, by the second form of the multivariate Lagrange theorem, we have

$$[\boldsymbol{\lambda}^{\mathbf{k}}]\prod_{i=1}^{m}(a_{i1}\lambda_1 + \cdots + a_{im}\lambda_m)^{k_i} = [\mathbf{x}^{\mathbf{k}}]\left\|\delta_{ij} - x_i\frac{\partial\phi_i}{\partial x_j}\right\|^{-1}$$

and the result follows, since $\partial\phi_i/\partial x_j = a_{ij}$. $\square$

In the light of this proof, the MacMahon master theorem may be regarded as a specialization of the multivariate Lagrange theorem to the case when the functions $\phi_i(\boldsymbol{\lambda})$ are linear. As an example of the use of the MacMahon master theorem we consider a binomial identity.

1.2.12. Dixon's Identity

To evaluate the sum $S = \sum_{k=0}^{n}(-1)^k\binom{n}{k}^3$, first express S as a coefficient in a series. Since each summand is the product of three binomial coefficients with upper index n, consider the expression

$$\left(1-\frac{x}{y}\right)^n\left(1-\frac{y}{z}\right)^n\left(1-\frac{z}{x}\right)^n$$

$$= \sum_{0\leqslant i,j,k\leqslant n}\binom{n}{i}\binom{n}{j}\binom{n}{k}(-1)^{i+j+k}x^{i-k}y^{j-i}z^{k-j}.$$

To force the lower indices in the binomial coefficients to be equal, we apply the operator $[x^0y^0z^0]$. Thus

$$S = [x^0y^0z^0]\left(1 - \frac{x}{y}\right)^n\left(1 - \frac{y}{z}\right)^n\left(1 - \frac{z}{x}\right)^n$$

$$= [x^ny^nz^n](y-x)^n(z-y)^n(x-z)^n$$

$$= [x^ny^nz^n]|\mathbf{I} - \mathbf{XM}|^{-1},$$

by the MacMahon master theorem, where $\mathbf{X} = \operatorname{diag}(x, y, z)$ and

$$\mathbf{M} = \begin{bmatrix} 0 & -1 & 1 \\ 1 & 0 & -1 \\ -1 & 1 & 0 \end{bmatrix}.$$

It follows that

$$S = [x^ny^nz^n](1 + xy + yz + zx)^{-1} = \sum_{r,s,t \geqslant 0} (-1)^{r+s+t}\begin{bmatrix} r+s+t \\ r, s, t \end{bmatrix},$$

where the summation is over all (r, s, t) for which $t + r = r + s = s + t = n$. Thus $r = s = t = \frac{1}{2}n$ and r, s, t are integers, so

$$S = \begin{cases} (-1)^m(3m)!(m!)^{-3} & \text{if } n = 2m \\ 0 & \text{otherwise.} \end{cases}$$ □

In **1.2.12** we could have applied the MacMahon master theorem with other choices of $\mathbf{A}$ besides $\mathbf{M}$. Each of these is obtained by permuting the rows of $\mathbf{M}$, and one such choice is

$$\mathbf{B} = \begin{bmatrix} -1 & 1 & 0 \\ 0 & -1 & 1 \\ 1 & 0 & -1 \end{bmatrix} = \begin{bmatrix} 0 & 0 & 1 \\ 1 & 0 & 0 \\ 0 & 1 & 0 \end{bmatrix}\mathbf{M}.$$

For this matrix, in the notation of **1.2.12**, we have

$$S = [x^ny^nz^n]|\mathbf{I} - \mathbf{XB}|^{-1} = [x^ny^nz^n]\{1 + (x+y+z) + (xy + yz + zx)\}^{-1}.$$

This is a very much less convenient series with which to work. It should be noted that there is no paradox since

$$[x^ny^nz^n]\{1 + xy + yz + zx\}^{-1}$$

$$= [x^ny^nz^n]\{1 + (x+y+z) + (xy+yz+zx)\}^{-1}$$

for $n \geqslant 0$ does not imply that the two series are equal as formal power series (they are clearly not).

We conclude this section with a special form of the multivariate Lagrange theorem that will be of importance in the combinatorial material of the later chapters.

1.2.13. Corollary (Lagrange Theorem for Monomials)

Let $\phi_i \in \mathsf{R}[[\boldsymbol{\lambda}]]_1$ for $i = 1,\ldots,m$. Suppose that $w_i = t_i\phi_i(\mathbf{w})$, for $1 \leqslant i \leqslant m$, where $\phi_i(\mathbf{w})$ is independent of w_j for each (i, j) in a fixed subset S of N_m^2. Then, for non-negative integers $r_1,\ldots,r_m$ and positive integers $(k_1,\ldots, k_m)$ where $\mathbf{r} = (r_1,\ldots, r_m)$ and $\mathbf{k} = (k_1,\ldots, k_m)$ we have

$$[\mathbf{t}^{\mathbf{k}}]\mathbf{w}^{\mathbf{r}} = (k_1 \cdots k_m)^{-1}\sum_{\boldsymbol{\mu}}\|\delta_{ij}k_i - \mu_{ij}\|\prod_{i=1}^{m}\{[w_1^{\mu_{i1}} \cdots w_m^{\mu_{im}}]\phi_i^{k_i}\}.$$

The summation is over all matrices $\boldsymbol{\mu} = [\mu_{ij}]_{m\times m}$ of non-negative integers such that $\sum_{i=1}^{m}\mu_{ij} = k_j - r_j$ for $1 \leqslant j \leqslant m$ and $\mu_{ij} = 0$ for each $(i, j) \in \mathsf{S}$.

Proof: From the multivariate Lagrange theorem

$$[\mathbf{t}^{\mathbf{k}}]\mathbf{w}^{\mathbf{r}} = [\mathbf{w}^{\mathbf{k}-\mathbf{r}}]\boldsymbol{\phi}^{\mathbf{k}}\left\|\delta_{ij} - \frac{w_j}{\phi_i(\mathbf{w})}\frac{\partial\phi_i}{\partial w_j}\right\| = [\mathbf{w}^{\mathbf{k}-\mathbf{r}}]\left\|\phi_i^{k_i}\delta_{ij} - \frac{w_j}{k_i}\frac{\partial}{\partial w_j}\phi_i^{k_i}\right\|$$

$$= \left\{\frac{1}{(\mathbf{k}-\mathbf{r})!}\frac{\partial^{\mathbf{k}-\mathbf{r}}}{\partial\mathbf{w}^{\mathbf{k}-\mathbf{r}}}\left\|\phi_i^{k_i}\delta_{ij} - \frac{w_j}{k_i}\frac{\partial}{\partial w_j}\phi_i^{k_i}\right\|\right\}\Bigg|_{\mathbf{w}=\mathbf{0}}.$$

Let $E_i^{(\boldsymbol{\mu})} = D_{w_1}^{\mu_{i1}} \cdots D_{w_m}^{\mu_{im}}$. Then, by differentiating the determinant by columns,

$$[\mathbf{t}^{\mathbf{k}}]\mathbf{w}^{\mathbf{r}} = \left\{\sum_{\boldsymbol{\mu}}\frac{1}{\boldsymbol{\mu}!}\left\|\delta_{ij}\left(E_i^{(\boldsymbol{\mu})}\phi_i^{k_i}\right) - k_i^{-1}\left(E_i^{(\boldsymbol{\mu})}w_j\frac{\partial}{\partial w_j}\phi_i^{k_i}\right)\right\|\right\}\Bigg|_{\mathbf{w}=\mathbf{0}}$$

where the summation is over $\boldsymbol{\mu} = [\mu_{ij}]_{m\times m}$ such that $\sum_{i=1}^{m}\mu_{ij} = k_j - r_j$. But, by Leibniz's theorem

$$k_i^{-1}\left(E_i^{(\boldsymbol{\mu})}w_j\frac{\partial}{\partial w_j}\phi_i^{k_i}\right)\Bigg|_{\mathbf{w}=\mathbf{0}} = k_i^{-1}\mu_{ij}\left(E_i^{(\boldsymbol{\mu})}\phi_i^{k_i}\right)\Big|_{\mathbf{w}=\mathbf{0}},$$

where $\mu_{ij} = 0$ if $(i, j) \in \mathsf{S}$. Thus

$$[\mathbf{t}^{\mathbf{k}}]\mathbf{w}^{\mathbf{r}} = \sum_{\boldsymbol{\mu}}\frac{1}{\boldsymbol{\mu}!}\|\delta_{ij} - k_i^{-1}\mu_{ij}\|\prod_{i=1}^{m}\{\left(E_i^{(\boldsymbol{\mu})}\phi_i^{k_i}\right)\}\Bigg|_{\mathbf{w}=\mathbf{0}}$$

and the result follows. □

It is important to notice that this corollary gives an expression for $[\mathbf{t}^{\mathbf{k}}]\mathbf{w}^{\mathbf{r}}$ in terms of determinants of matrices over the integers. Such determinants occur in a number of classical results. The matrix-tree theorem and the BEST (de Bruijn–van Aardenne Ehrenfest–Stone–Tutte) theorem are examples of this phenomenon.

NOTES AND REFERENCES

1.2.2 Jacobi (1830); **1.2.3** Gessel (private communication); **1.2.4** Henrici (1964), Jacobi (1830), Whittaker and Watson (1927); **1.2.6** Pólya and Szegö (1964); **1.2.7** Riordan (1968); **1.2.8** Jacobi (1830); **1.2.9** Garsia and Joni (1977), Good (1960), Hofbauer (1979), Jacobi (1830), Tutte (1975); **1.2.10** Brown and Tutte (1964); **1.2.11** Good (1962a), MacMahon (1915); **1.2.12** Dixon (1891); **1.2.13** Goulden and Jackson (1981a).

[**1.2.5**] Jabotinsky (1953); [**1.2.6**] Mullin and Stanton (1969); [**1.2.7, 8**] Riordan (1968); [**1.2.9**] Watson (1952); [**1.2.10**] Henrici (1974); [**1.2.11**] Pólya and Szegö (1964); [**1.2.12**] Halphen (1879); [**1.2.13**] Good (1962b); [**1.2.14**] Good (1976).

EXERCISES

1.2.1. Show that

$$\sum_{n\geqslant k}\binom{2n}{n-k}t^n=(1-4t)^{-1/2}\left\{\frac{1-2t-(1-4t)^{1/2}}{2t}\right\}^k.$$

1.2.2. Show that $(1-x)^{-1}e^{\alpha x}=\sum_{m\geqslant 0}\dfrac{(m+\alpha)^m}{m!}(xe^{-x})^m$.

1.2.3. Show that the solution to the functional equation

$$w=t+xw^2(1-w)^{-1}$$

is

$$w=\sum_{i>k\geqslant 0}\frac{t^iy^k}{i+k}\binom{i+k}{k}\binom{i-2}{i-k-1}.$$

1.2.4. Show that the solution to the functional equation $w=a+t\phi(w)$, where $\phi(\lambda+a)\in \mathsf{R}[[\lambda]]_1$, is given by

$$w=a+\sum_{n\geqslant 1}(t^n/n)[\lambda^{n-1}]\phi^n(a+\lambda).$$

1.2.5. Let $\lambda^{-1}\Phi(\lambda) \in \mathsf{R}[[\lambda]]_1$ and $[\mathbf{A}]_{ij} = [\lambda^i]\Phi^j(\lambda)$ for $i, j \geqslant 1$. Show that

$$[\mathbf{A}^{-1}]_{ij} = ij^{-1}[\lambda^{-j}]\Phi^{-i}(\lambda), \qquad i, j \geqslant 1.$$

1.2.6. Show that

(a) $\binom{an}{n-1} = \sum_{i=1}^{n} \frac{1}{i}\binom{ai}{i-1}\binom{a(n-i)}{n-i}$,

(b) $(n-1)n^{n-2} = \sum_{k=1}^{n-1}\binom{n-1}{k-1}k^{k-1}(n-k)^{n-k-1}$.

1.2.7. Prove the inverse pair

$$a_n = \sum_{k\geqslant 0}\binom{n}{k}b_{n-ck},$$

$$b_n = \sum_{k\geqslant 0}(-1)^k\left\{\binom{n-k(c-1)-1}{k} - c\binom{n-k(c-1)-1}{k-1}\right\}a_{n-ck},$$

for $n \geqslant 0$, where c is an integer. This is called a **Chebyshev** inverse pair.

1.2.8. Prove the inverse pair

$$a_n = \sum_{k\geqslant 0}\binom{n}{k}\alpha(\alpha+n-k)^{n-k-1}b_k,$$

$$b_n = \sum_{k\geqslant 0}(-1)^{n-k}\binom{n}{k}\alpha(\alpha-n+k)^{n-k-1}a_k,$$

for $n \geqslant 0$. This is called an **Abel** inverse pair.

1.2.9. Let $f(x) = \sum_{n\geqslant 0} f_n x^n \in \mathsf{R}[[x]]$, and let

$$J_n(x) = \sum_{i\geqslant 0}\frac{(-1)^i}{i!(n+i)!}\left(\frac{x}{2}\right)^{n+2i},$$

a **Bessel function**. Show that $f(x) = \sum_{n\geqslant 0} a_n J_n(x)$ where $a_0 = f_0$ and

$$a_n = n\sum_{j\geqslant 0} 2^{n-2j}\frac{(n-j-1)!}{j!}f_{n-2j}, \qquad n > 0.$$

1.2.10. Show that

$$\left\{\frac{d^{n-1}}{dx^{n-1}}\left(\frac{x-1}{x^a-1}\right)^n\right\}\Bigg|_{x=1} = \frac{1}{a}\left(\frac{1}{a}-1\right)\cdots\left(\frac{1}{a}-n+1\right).$$

1.2.11. Let $P_n(h) = \dfrac{1}{2^n n!}\dfrac{d^n}{dh^n}(h^2 - 1)^n$, $n \geqslant 0$, be the **Legendre polynomials.** Show that $\Sigma_{n\geqslant 0} P_n(h) t^n = (1 - 2ht + t^2)^{-1/2}$.

1.2.12. Let $f^{(n)}(x)$ denote the nth derivative of $f(x)$. Show that

$$\frac{d^n}{dx^n}\{x^{n-1} f(x^{-1})\} = (-1)^n x^{-n-1} f^{(n)}(x^{-1}).$$

1.2.13. Show that

(a) $$\sum_i (-1)^i \binom{m+n}{m+i}\binom{n+k}{n+i}\binom{k+m}{k+i} = \frac{(m+n+k)!}{m!n!k!},$$

(b) $$\sum_i \binom{m+n}{m+i}\binom{n+k}{n+i}\binom{k+m}{k+i}$$

$$= \sum_{j\geqslant 0} (-1)^j \frac{(m+n+k-j)!(u-1)^{2j} u^{-j}}{(2j)!(m-j)!(n-j)!(k-j)!}$$

1.2.14. Let $\overline{m}_j = \Sigma_{i=1}^n m_i a_{ij}$, $u_i = x_i \exp\{-\Sigma_{j=1}^n a_{ij} x_j\}$,

$$\mathbf{X} = \operatorname{diag}(x_1, \ldots, x_n) \text{ and } \mathbf{A} = [a_{ij}]_{n\times n}.$$

Show that

$$\sum_{m_1,\ldots,m_n \geqslant 0} \overline{m}_1^{m_1} \cdots \overline{m}_n^{m_n} \frac{u_1^{m_1} \cdots u_n^{m_n}}{m_1! \cdots m_n!} = |\mathbf{I} - \mathbf{XA}|^{-1}.$$

2

The Combinatorics of the Ordinary Generating Function

2.1. INTRODUCTION

Throughout this book we use the term **configuration** to denote almost any mathematical object—for example, a permutation, a tree, a map, a non-negative integer matrix, a linear transformation on a vector space over GF(q), or a function on a finite set.

The **ordinary generating function** for a set S of distinct structures, or **combinatorial configurations**, with respect to a weight function ω, which records designated enumerative information, is

$$\sum_{\sigma \in \mathsf{S}} x^{\omega(\sigma)}$$

where x is an indeterminate and $\omega(\sigma)$ is a non-negative integer. In this chapter we consider the combinatorial significance of this function. This may be distilled into five main results called the elementary **counting lemmas**. Many of the methods and ideas encountered here are generalized in subsequent chapters, and accordingly this chapter serves as an important point of departure for much of the remaining work. The following paragraphs outline the chapter as a whole and the role served by each section in presenting some of the basic enumerative ideas.

2.1.1. The Elementary Counting Lemmas

Section 2.2 gives the main counting lemmas. If F and G are sets of distinct configurations with ordinary generating functions f and g, respectively, then under reasonable conditions these lemmas express the generating functions for

1. $\mathsf{F} \mathbin{\dot\cup} \mathsf{G}$ **disjoint union**
2. $\mathsf{F} \times \mathsf{G}$ **Cartesian product**
3. $\mathsf{F} \circ \mathsf{G}$ **composition**
4. F' **derivative** with respect to a subconfiguration

as $f + g$ (sum), fg (product), $f \circ g$ (composition) and f' (derivative), respectively. These basic operations on sets are called the **elementary set operations**. The remaining lemma is the Principle of Inclusion and Exclusion.

2.1.2. Decompositions and Weight Functions

The elementary counting lemmas serve as the interface between the **combinatorics** of a set of configurations and the generating function algebra. The combinatorics is expressed in terms of a **decomposition**, which is a bijection whose domain and codomain are defined in terms of the set to be enumerated, and possibly other sets, by means of only those set operations listed in **2.1.1**. Clearly for these operations to hold, a decomposition must be **weight preserving**, in the sense that it preserves designated enumerative information. In fact, for operations **(2)** and **(3)** to hold, it is necessary that the enumerative information be **additively** preserved across the decomposition. It is a comment on the power of combinatorial theory that it is possible to find useful weight-preserving decompositions while being confined to the four elementary set operations.

In any decomposition we may recognize quite easily which enumerative information is preserved. Designated enumerative information, exposed in the decomposition, is recorded by **marking** it with an indeterminate. Some decompositions give an immediate expression for a generating function. These are called **direct**. Others necessitate a multiplicative or compositional inversion, and these are called **indirect** decompositions. The remaining decompose a set in terms of itself and are therefore called **recursive** decompositions.

2.1.3. Direct and Indirect Decompositions, Combinatorial Marking, and Multivariate Generating Functions

Section 2.3 gives a number of preliminary examples of the use of decompositions. For most of these we obtain a generating function and demonstrate how the general coefficient may be extracted, as well as showing how recurrence equations for these coefficients may be obtained. All the decompositions in this section are direct, and involve disjoint union and Cartesian product.

Section 2.4 deals with the introductory aspects of sequence enumeration and makes extensive use of multivariate generating functions. The elementary counting lemmas extend immediately to this case. We give instances of the use of set composition and indirect decompositions. Throughout this section systematic use is made of combinatorial marking, mentioned in **2.1.2**, for recording enumerative information. One of the results is specialized to permutations, giving an instance of an **exponential** generating function. This is considered in detail in Chapter 3. The general theory and further applications of sequence enumeration are given in Chapter 4. Section 2.4 also contains some brief comments on **multiplicatively** preserved weights and the use of the **Dirichlet** generating function in this situation. Analogous elementary counting lemmas exist for this generating function, although this is not pursued since the multiplicative case may be formally transformed into the additive case.

2.1.4. A Classical Application of Enumerative Arguments

Sections 2.5 and 2.6 are concerned with the application of enumerative arguments to obtain q-identities. All the decompositions are direct. Section 2.5 deals with the enumeration of integer partitions, and the decompositions are obtained by considering the Ferrers graph representation of a partition. Section 2.6 obtains related identities, by considering inversions in bimodal permutations. The main result is called the ***q*-analogue** of the binomial theorem. These sections relate to the **Eulerian** generating function, a topic explored more fully in Chapter 4.

2.1.5. Recursive and "At-Least" Decompositions

Section 2.7 is concerned with planted plane trees and various applications of the fundamental recursive decomposition for this set. The Lagrange theorem is used extensively in the solution of the functional equations induced by this decomposition, which additively preserves degree, chromatic, and height information. The general notion of height in a configuration and the connection between this and continued fractions is considered in Section 5.2. The section also includes some applications of the differential of a set.

Section 2.8 concerns the enumeration of sequences with respect to an arbitrary, but fixed, set of distinguished substrings. Although this class includes some problems treated by composition in Section 2.4, it also contains problems that are not tractable by those methods. A general theorem for enumerating sequences with respect to distinguished substrings is given. This is obtained by an "at-least" decomposition and the Principle of Inclusion and Exclusion.

Section 2.9 is devoted to rooted planar maps. The decompositions for these configurations are recursive and often lead to functional equations requiring an additional stage in their solution before the Lagrange theorem may be applied. This is called the **quadratic method**.

2.2. THE ELEMENTARY COUNTING LEMMAS

This section deals with the elementary counting lemmas for the ordinary generating function. For notational simplicity alone, the exposition is confined to the univariate case, since extension to the multivariate case is natural. Enumeration is concerned with counting distinguishable configurations so the following definition is needed.

2.2.1. Definition (Distinguishability)

Let S *be a set of configurations, and let* E *be an equivalence relation on* S. *Then*

1. σ_1 *is* **indistinguishable** *from* σ_2 *iff* $\sigma_1 \mathsf{E} \sigma_2$.
2. $\sigma_1, \sigma_2 \in \mathsf{S}$ *are said to be* **distinct** *iff they are distinguishable.* □

To count the number of configurations in a set S that possess a particular number of specified characteristics it is necessary to introduce a weight function.

2.2.2. Definition (Weight Function, Weight)

Let S *be a set of configurations and let* $\omega : \mathsf{S} \to \{0, 1, 2, \dots\}$. *Then*

1. ω *is called a* **weight function** *on* S.
2. *For* $\sigma \in \mathsf{S}$, $\omega(\sigma)$ *is called the* **weight** *of* σ. □

We may now state a problem that is general enough to include the majority of enumeration problems.

2.2.3. Remark (The General Enumerative Problem)

Let ω *be a weight function on a set* S *of distinct configurations. Then the general enumerative problem is as follows*:

$$\textit{Find} \quad |\{\sigma \in \mathsf{S} \,|\, \omega(\sigma) = n\}|.$$

In other words, find the number of distinct configurations of weight n. □

Clearly, in any particular case, the initial task in problem solving is to identify the set S, the weight function ω, and the equivalence relation E associated with indistinguishability. Usually we are concerned with situations where E is the identity. The following example gives instances of these in various problems.

2.2.4. Example (Enumerative Problems)

1. Find the number of sequences of prescribed length which are formed with the symbols 0, 1, and 2. Thus (a) $\mathsf{S} = \{0, 1, 2\}^*$, (b) E is the identity relation on S, (c) For $\sigma \in \mathsf{S}$, $\omega(\sigma)$ is the length of σ.
2. Find the number of subsets of N_n whose elements have a prescribed sum. Thus (a) $\mathsf{S} = 2^{\mathsf{N}_n}$, (b) E is the identity relation on S, (c) For $\sigma = \{\sigma_1, \dots, \sigma_k\} \in 2^{\mathsf{N}_n}$, $\omega(\sigma) = \sigma_1 + \cdots + \sigma_k$. □

For the purposes of enumerating the elements of S with respect to the weight function ω, it is natural to consider the following formal power series.

2.2.5. Definition (Ordinary Generating Function)

The **generating function** *of a set* S *of distinct configurations that is* **ordinary** *with respect to the weight function* ω *on* S *is*

$$\sum_{\sigma \in \mathsf{S}} x^{\omega(\sigma)}$$

and is denoted by $\Phi_{\mathsf{S}}^{(\omega)}(x)$. *We say that* x **marks** ω, *and that* $\Phi_{\mathsf{S}}^{(\omega)}(x)$ **enumerates S with respect to** ω. □

When ω counts the number of occurrences of a particular subconfiguration of σ, x is said to mark this subconfiguration.

2.2.6. Remark (*s*-Objects)

When certain subconfigurations are designated for enumerative purposes in this way, we call them ***s*-objects**. *If* $\sigma \in \mathsf{S}$, *then* $\omega_s(\sigma)$ *denotes the number of s-objects in* S. □

The next example gives the ordinary generating function corresponding to a simple enumeration problem.

2.2.7. Example (a Generating Function)

Let S be the set of sequences over $\{0, 1, 2\}$ and let $\omega(\sigma)$ be the length of $\sigma \in \mathsf{S}$. Clearly, there are exactly three choices for each of the n positions in a sequence of length n, so there are $c_n = 3^n$ sequences with $\omega(\sigma) = n$. It follows from Definition 2.2.5 that

$$\Phi_{\mathsf{S}}^{(\omega)}(x) = \sum_{\sigma \in \mathsf{S}} x^{\omega(\sigma)} = \sum_{n \geqslant 0} x^n \sum_{\substack{\sigma \in \mathsf{S} \\ \omega(\sigma) = n}} 1.$$

But $\sum_{\substack{\sigma \in \mathsf{S} \\ \omega(\sigma) = n}} 1 = c_n$, and we have already established that this is equal to 3^n. Thus

$$\Phi_{\mathsf{S}}^{(\omega)}(x) = \sum_{n \geqslant 0} 3^n x^n = (1 - 3x)^{-1}.$$ □

The generating function $\Phi_{\mathsf{S}}^{(\omega)}(x)$ was deduced from $\{c_n | n \geqslant 0\}$ by calculating $\sum_{n \geqslant 0} c_n x^n$. However, this is a reversal of what is soon to become our usual strategy. Our intention is to develop a theory of enumeration that will enable us first to derive $\Phi_{\mathsf{S}}^{(\omega)}(x)$, and then to deduce c_n by applying $[x^n]$ to this formal power series. This idea is formalized in the following proposition.

2.2.8. Proposition

Let S *be a set of distinct configurations, and let* ω *be a weight function on* S. *Then*

$$[x^k]\Phi_{\mathsf{S}}^{(\omega)}(x) = |\{\sigma \in \mathsf{S} \mid \omega(\sigma) = k\}|$$

(*the number of configurations in* S *of weight* k). □

This proposition explains, in part, the reasons for introducing the ordinary generating function. It provides us with a strategy for solving a combinatorial problem by looking for a superset B in which the given configurations are recognized as those having a prescribed weight. Clearly, any such B will be suitable, providing, of course, that the generating function of B may be determined. In casting a particular problem in the general form stated in **2.2.3**, we select the set S to be the set B for which this task is the most convenient. This selection of S is usually obvious from a combinatorial point of view.

We now define the notion of a decomposition. This is essentially a combinatorial device, in the sense that it is concerned with the combinatorial structure of sets of configurations.

2.2.9. Definition (Decomposition, ω-Preserving)

Let S *and* T *be sets of distinct configurations.*

1. *If there exists a bijection* $\Omega : \mathsf{S} \to \mathsf{T}$, *then we say that* Ω *is a* **decomposition** *of* S *and we write* $\mathsf{S} \xrightarrow{\sim} \mathsf{T}$.
2. *If there exists a weight function* ω' *on* T *such that* $\omega(\sigma) = \omega'(\Omega(\sigma))$ *for all* $\sigma \in \mathsf{S}$, *then* Ω *is called an* ω-**preserving decomposition** *of* S. *We write* $\mathsf{S} \underset{\omega}{\xrightarrow{\sim}} \mathsf{T}$. □

The following proposition is now immediate.

2.2.10. Proposition

If $\mathsf{S} \underset{\omega}{\xrightarrow{\sim}} \mathsf{T}$ *where* S *and* T *are sets of distinct configurations and* ω *is a weight function on* S, *then there exists a weight function* ω' *on* T *such that*

$$\Phi_{\mathsf{S}}^{(\omega)}(x) = \Phi_{\mathsf{T}}^{(\omega')}(x).$$ □

The strategy for determining $\Phi_{\mathsf{S}}^{(\omega)}(x)$ is, of course, to select an ω-preserving decomposition Ω of S and a weight function ω' on T such that it is in fact easier to determine the generating function $\Phi_{\mathsf{T}}^{(\omega')}(x)$. This is a crucial point and is developed further in the discussion that follows **2.2.16**. Example 2.2.15 contains the generating function for the set whose decomposition is given below.

2.2.11. Example (Terquem Problem: Decomposition)

Problem. Find the number $c(n, k)$ of subsets $\{\sigma_1, \ldots, \sigma_k\}$ of N_n such that $\sigma_i \equiv i \bmod 2$ and $\sigma_1 < \cdots < \sigma_k$.

The Set of Configurations. The set of configurations to be enumerated is

$$\mathsf{S}_k = \{\{\sigma_1, \ldots, \sigma_k\}_n | \sigma_1 < \cdots < \sigma_k \leqslant n; n \geqslant 0\},$$

where we have used the subscript k in S_k to emphasize that k is fixed for the problem. The configuration $\{\rho_1, \ldots, \rho_k\}_n \in \mathsf{S}_k$ is indistinguishable from $\{\sigma_1, \ldots, \sigma_k\}_m \in \mathsf{S}_k$ if and only if $m = n$, and $\sigma_i = \rho_i$ for $i = 1, \ldots, k$ (so the equivalence relation for indistinguishability is the identity).

If the weight function is

$$\omega : \mathsf{S}_k \to \{0, 1, \ldots\} : \{\sigma_1, \ldots, \sigma_k\}_n \mapsto n,$$

then $c(n, k)$ is precisely the number of elements in S_k with weight equal to n. In other words,

$$c(n, k) = [x^n]\Phi_{\mathsf{S}_k}^{(\omega)}(x).$$

The Decomposition. Let O denote the set of all positive odd integers, and let

$$\Omega : \mathsf{S}_k \to \mathsf{O}^k \times \mathsf{N} : \{\sigma_1, \ldots, \sigma_k\}_n \mapsto (\alpha_1, \ldots, \alpha_{k+1}),$$

where $\alpha_i = \sigma_i - \sigma_{i-1}$ for $1 \leqslant i \leqslant k+1$, in which $\sigma_0 = 0$ and $\sigma_{k+1} = n$.

Now $\sigma_j = \alpha_1 + \cdots + \alpha_j$ for $1 \leqslant j \leqslant k$ and $n = \alpha_1 + \cdots + \alpha_{k+1}$, so Ω is bijective. It follows from Definition 2.2.9 that Ω is a decomposition of S_k.

Let the weight function ω' be

$$\omega' : \mathsf{O}^k \times \mathsf{N} \to \mathsf{N} : (\alpha_1, \ldots, \alpha_{k+1}) \mapsto \alpha_1 + \cdots + \alpha_{k+1}.$$

Then, clearly, for $(\alpha_1, \ldots, \alpha_{k+1}) = \Omega(\sigma)$ we have

$$\omega(\sigma) = n = \alpha_1 + \cdots + \alpha_{k+1} = \omega'(\Omega(\sigma)) \qquad \text{for all } \sigma \in \mathsf{S}_k.$$

Thus, from Definition 2.2.9, Ω is an ω-preserving decomposition of S_k, so

$$\Phi_{\mathsf{S}_k}^{(\omega)}(x) = \Phi_{\mathsf{O}^k \times \mathsf{N}}^{(\omega')}(x)$$

from Proposition 2.2.10. Finally, by Proposition 2.2.8, we conclude

$$c(n, k) = [x^n]\Phi^{(\omega')}_{\mathsf{O}^k \times \mathsf{N}}(x). \qquad \square$$

It is reasonable to ask whether $\Phi^{(\omega')}_{\mathsf{O}^k \times \mathsf{N}}(x)$ can be obtained in terms of the generating function for O and N alone. Indeed, it is reasonable to pose this question in general and ask whether there are other set-theoretic operations, besides Cartesian product, that are useful in connection with combinatorial decompositions. In fact, we shall see next that there are four basic set operations that can be used, namely, disjoint union, Cartesian product, composition, and differentiation. Of these, the latter two probably will be unfamiliar, but they are defined presently. Associated with these operations are four elementary counting lemmas, which give the corresponding generating function as the sum, product, composition, and derivative of the generating functions for the sets included in the decomposition.

2.2.12. Lemma (Sum)

Let ω be a weight function on a set S of distinct configurations. Let A, B be disjoint subsets of S. Then

$$\Phi^{(\omega)}_{\mathsf{A} \,\dot\cup\, \mathsf{B}}(x) = \Phi^{(\omega)}_{\mathsf{A}}(x) + \Phi^{(\omega)}_{\mathsf{B}}(x).$$

Proof: From Definition 2.2.5

$$\Phi^{(\omega)}_{\mathsf{A} \,\dot\cup\, \mathsf{B}}(x) = \sum_{\sigma \in \mathsf{A} \,\dot\cup\, \mathsf{B}} x^{\omega(\sigma)} = \sum_{\sigma \in \mathsf{A}} x^{\omega(\sigma)} + \sum_{\sigma \in \mathsf{B}} x^{\omega(\sigma)} = \Phi^{(\omega)}_{\mathsf{A}}(x) + \Phi^{(\omega)}_{\mathsf{B}}(x),$$

since the union is disjoint. $\square$

2.2.13. Example (Sum)

Let A be the set of all sequences over $\{0, 1, 2\}$, and let $\mathsf{A}_0, \mathsf{A}_1$ be the subsets of sequences in A of even and odd length respectively. Let $\omega(\sigma)$ denote the length of $\sigma \in \mathsf{A}$. Clearly $\mathsf{A} = \mathsf{A}_0 \,\dot\cup\, \mathsf{A}_1$, so by the sum lemma

$$\Phi^{(\omega)}_{\mathsf{A}}(x) = \Phi^{(\omega)}_{\mathsf{A}_0}(x) + \Phi^{(\omega)}_{\mathsf{A}_1}(x). \qquad \square$$

An analogous result may be obtained for the Cartesian product.

2.2.14. Lemma (Product)

Let α, β, ω be weight functions on the sets $\mathsf{A}, \mathsf{B}, \mathsf{A} \times \mathsf{B}$ of distinct configurations. If

$$\omega((a, b)) = \alpha(a) + \beta(b) \qquad \textit{for all } (a, b) \in \mathsf{A} \times \mathsf{B},$$

then

$$\Phi^{(\omega)}_{\mathsf{A}\times\mathsf{B}}(x) = \Phi^{(\alpha)}_{\mathsf{A}}(x)\Phi^{(\beta)}_{\mathsf{B}}(x).$$

Proof: From Definition 2.2.5

$$\Phi^{(\omega)}_{\mathsf{A}\times\mathsf{B}}(x) = \sum_{(a,b)\in\mathsf{A}\times\mathsf{B}} x^{\omega(a,b)} = \sum_{a\in\mathsf{A}} x^{\alpha(a)} \sum_{b\in\mathsf{B}} x^{\beta(b)} = \Phi^{(\alpha)}_{\mathsf{A}}(x)\Phi^{(\beta)}_{\mathsf{B}}(x). \quad \square$$

As an example of the use of the product lemma, we return to Example 2.2.11 and complete its solution.

2.2.15. Example (Terquem Problem: Generating Function)

Let $c(n, k)$ be the number of subsets $\{\sigma_1, \dots, \sigma_k\}_n$ of N_n with $1 \leqslant \sigma_1 < \cdots < \sigma_k$ and $\sigma_i \equiv i \bmod 2$ for $1 \leqslant i \leqslant k$. From Example 2.2.11

$$c(n, k) = [x^n]\Phi^{(\omega')}_{\mathsf{O}^k\times\mathsf{N}}(x),$$

where $\omega' : \mathsf{O}^k \times \mathsf{N} \to \mathsf{N} : (\alpha_1, \dots, \alpha_{k+1}) \mapsto \alpha_1 + \cdots + \alpha_{k+1}$. Let ψ, ϕ be weight functions defined on O and N, respectively, by $\psi(i) = i$ for $i \in \mathsf{O}$ and $\phi(j) = j$ for $j \in \mathsf{N}$. Then

$$\omega'(\alpha_1, \dots, \alpha_{k+1}) = \psi(\alpha_1) + \cdots + \psi(\alpha_k) + \phi(\alpha_{k+1}).$$

Thus, from the product lemma,

$$\Phi^{(\omega')}_{\mathsf{O}^k\times\mathsf{N}}(x) = \left(\Phi^{(\psi)}_{\mathsf{O}}(x)\right)^k \Phi^{(\phi)}_{\mathsf{N}}(x).$$

The generating functions for O and N with respect to ψ and ϕ, respectively, may be evaluated without difficulty as follows:

$$\Phi^{(\psi)}_{\mathsf{O}}(x) = \sum_{\sigma\in\mathsf{O}} x^{\psi(\sigma)} = \sum_{j\geqslant 0} x^{2j+1} = x(1 - x^2)^{-1},$$

$$\Phi^{(\phi)}_{\mathsf{N}}(x) = \sum_{\sigma\in\mathsf{N}} x^{\phi(\sigma)} = \sum_{j\geqslant 0} x^{j} = (1 - x)^{-1}.$$

It follows that

$$\Phi^{(\omega')}_{\mathsf{O}^k\times\mathsf{N}}(x) = x^k(1 - x)^{-1}(1 - x^2)^{-k},$$

so

$$c(n, k) = [x^{n-k}](1 - x)^{-1}(1 - x^2)^{-k}.$$

To extract this coefficient we could, of course, expand both $(1-x)^{-1}$ and $(1-x^2)^{-k}$ as power series to obtain a double sum. One of these summations may be avoided by noting that $(1-x)^{-1} = (1+x)(1-x^2)^{-1}$. Thus

$$c(n,k) = [x^{n-k}] \sum_{i=0}^{1} \sum_{r=0}^{\infty} \binom{k+r}{k} x^{2r+i}$$

so, for a nonzero contribution to $c(n,k)$, $n-k = 2r+i$ where $0 \leqslant i < 2$. Thus $r = \lfloor \frac{1}{2}(n-k) \rfloor$ so $c(n,k) = \binom{\lfloor \frac{1}{2}(n+k) \rfloor}{k}$. □

As a check, note that there are four permissible subsets of N_6 of size 3. These are $\{1,2,3\}$, $\{1,2,5\}$, $\{1,4,5\}$, and $\{3,4,5\}$. Moreover, $c(6,3) = \binom{\lfloor 9/2 \rfloor}{3} = 4$.

The sum and product lemmas may be more familiar in a probabilistic context, in which they are known as the rule of sum for mutually exclusive events, and the rule of product for independent events. It is important to recognize combinatorially the linear relation that occurs between the weight functions in the statement of the product lemma. The following definition is therefore introduced.

2.2.16. Definition (Additively Weight-Preserving Decomposition)

Let $\mathsf{A}, \mathsf{B}, \mathsf{C}$ *be sets of distinct configurations. Let* $\Omega : \mathsf{C} \underset{\omega}{\tilde{\to}} \mathsf{A} \times \mathsf{B}$ *be an* ω*-preserving decomposition. If*

$$\omega(c) = \alpha(a) + \beta(b) \qquad \textit{for all } c \in \mathsf{C}, \textit{ with } \Omega(c) = (a,b),$$

where α, β *are weight functions on* A, B, *respectively, then* Ω *is called an* **additively** *weight-preserving decomposition of* C. □

For almost all the configurations we treat, it is a simple matter to establish the weight preservation, additivity, and bijectivity of the decompositions. In fact, each of these properties may usually be recognized at a combinatorial level.

We now return to the point, raised following Proposition 2.2.10, about the usefulness of the ω-preserving decomposition $\Omega : \mathsf{S} \tilde{\to} \mathsf{T}$ in the task of enumerating S. It is necessary to seek a decomposition that is additively weight preserving and such that T is expressed in terms of disjoint unions and products of sets whose generating functions are accessible. An example of this is provided by **2.2.11** and **2.2.15**. Moreover, in casting a particular problem, in the first place, into the general form stated in **2.2.3**, we can choose S to be any set in which the given configurations are precisely those having a prescribed weight. In fact, S is selected as the set such that there is a T and a decomposition Ω with the preceding properties.

By construction, a generating function contains the desired enumerative information, so we now raise the question of the practicability of obtaining it by coefficient extraction. To assist us in this we have at our disposal Taylor's theorem, Lagrange's theorem, and, of course, methods for deriving recurrence equations. For multivariate generating functions that are also symmetric, there are other techniques that may be used, and these are described in Section 3.5.

In Example 2.2.11 use was made of weight-preserving decompositions. There is, however, no mechanical procedure for deriving such decompositions in general. For the more common configurations such as sequences, trees, distributions, maps, and {0, 1}-matrices, appropriate decompositions are usually easy to find. For the more complex configurations, such as plane partitions, decompositions are more difficult to find. However, once a decomposition has been found, typically it is weight-preserving for several weight functions. In other words, a decomposition may be used for a *class of problems*, rather than merely for an individual problem.

We have seen in Example 2.2.11 that the decomposition of S_k into O and N enables us to determine $\Phi_{\mathsf{S}_k}^{(\omega)}(x)$ directly in terms of known generating functions $\Phi_{\mathsf{O}}^{(\psi)}(x)$ and $\Phi_{\mathsf{N}}^{(\phi)}(x)$ by means of the elementary counting lemmas. For this reason such decompositions are called direct. In general, the codomain of a decomposition is a formula involving disjoint unions and Cartesian products of a number of distinct sets $\mathsf{B}_1, \ldots, \mathsf{B}_m$. The ways in which a set may appear in a decomposition are classified below.

2.2.17. Remark (Direct, Indirect, Recursive Decompositions)

A decomposition of A *with codomain involving the distinct sets* $\mathsf{B}_1, \ldots, \mathsf{B}_m$ *alone is called*

1. **Direct** *for* A *if there is no* i *such that* $\mathsf{B}_i = \mathsf{A}$.
2. **Indirect** *for all* B_j *such that* $\mathsf{B}_j \neq \mathsf{A}$.
3. **Recursive** *for* A *if there is an* i *such that* $\mathsf{B}_i = \mathsf{A}$. □

An indirect decomposition is given in the next example.

2.2.18. Example (an Indirect Decomposition)

We wish to determine the number c_n of {0, 1}-sequences of length n that begin with a 1.

Let S be the set of all such sequences for $n \geqslant 0$, and let the weight $\omega(\sigma)$ of a {0, 1}-sequence σ be the length of σ. The set $\{0, 1\}^*$ is obtained from S by prefixing each sequence in S, in turn, by each sequence in 0^*. Thus

$$\{0, 1\}^* \underset{\omega}{\widetilde{\rightarrow}} 0^* \times \mathsf{S}$$

is clearly an *additively* ω-preserving decomposition Ω of $\{0, 1\}^*$, since $\omega(c) = \omega(a) + \omega(b)$ for $\Omega(c) = (a, b)$. Of course, Ω is indirect for S. From the sum and product lemmas

$$\Phi_{\{0,1\}^*}^{(\omega)}(x) = \Phi_{0^*}^{(\omega)}(x)\Phi_{\mathsf{S}}^{(\omega)}(x).$$

But

$$\Phi_{\{0,1\}^*}^{(\omega)}(x) = \sum_{k\geqslant 0} \Phi_{\{0,1\}^k}^{(\omega)}(x) = \sum_{k\geqslant 0} \left(\Phi_{\{0,1\}}^{(\omega)}(x)\right)^k$$

$$= \sum_{k\geqslant 0} \left\{\Phi_{\{0\}}^{(\omega)}(x) + \Phi_{\{1\}}^{(\omega)}(x)\right\}^k = \sum_{k\geqslant 0} (2x)^k.$$

Thus $\Phi_{\{0,1\}^*}^{(\omega)}(x) = (1 - 2x)^{-1}$, and, similarly, $\Phi_{0^*}^{(\omega)}(x) = (1 - x)^{-1}$,

so $$(1 - 2x)^{-1} = (1 - x)^{-1}\Phi_{\mathsf{S}}^{(\omega)}(x).$$

It follows that $\Phi_{\mathsf{S}}^{(\omega)}(x) = (1 - x)(1 - 2x)^{-1}$, whence $c_n = 2^{n-1}$. □

Combinatorially, we may expect indirect decompositions to occur when a problem to be solved is combined with a second to give a third problem. Such decompositions entail an inversion, or an inversive step, before the generating function may be obtained. In the previous example this step amounted to determining the multiplicative inverse of the function $(1 - x)^{-1}$. In general, however, the situation may be considerably more complicated. We complete this discussion with an example of a recursive decomposition that gives an alternative treatment of Example 2.2.7.

2.2.19. Example (a Recursive Decomposition)

We wish to determine the number c_n of sequences of length n over the alphabet $\{0, 1, 2\}$.

Let S be the set of all such sequences of length $n \geqslant 0$. Clearly, any sequence in S, except the empty sequence ε, may be obtained by prefixing the sequences of S by 0, 1, or 2. Thus

$$\mathsf{S} \,\tilde{\rightarrow}\, \{\varepsilon\} \cup \{0, 1, 2\} \times \mathsf{S}$$

is an additively ω-preserving recursive decomposition of S, where $\omega(\sigma)$ is the length of $\sigma \in \mathsf{S}$. From the sum and product lemmas

$$\Phi_{\mathsf{S}}^{(\omega)}(x) = \Phi_{\{\varepsilon\}}^{(\omega)}(x) + \Phi_{\{0,1,2\}}^{(\omega)}(x)\Phi_{\mathsf{S}}^{(\omega)}(x).$$

But $$\Phi_{\{0,1,2\}}^{(\omega)}(x) = \Phi_{\{0\}}^{(\omega)}(x) + \Phi_{\{1\}}^{(\omega)}(x) + \Phi_{\{2\}}^{(\omega)}(x) = 3x,$$

whence $$\Phi_{\mathsf{S}}^{(\omega)}(x) = 1 + 3x\Phi_{\mathsf{S}}^{(\omega)}(x).$$

Thus $\Phi_{\mathsf{S}}^{(\omega)}(x) = (1 - 3x)^{-1}$ so $c_n = 3^n$. □

In general, a recursive decomposition leads to a functional equation for the required generating function. If this equation satisfies the conditions under which the Lagrange theorem may be applied, we can usually obtain an explicit power series solution. In the remaining cases little more can be done, apart from determining a recurrence equation for the coefficients in the power series solution.

We now equip ourselves with a combinatorial operation arising from the successive use of disjoint unions and Cartesian products. This new operation is called **composition.** It is carried out by replacing certain subconfigurations, the s-objects of Remark 2.2.6, in a first set of distinct configurations by those of a second set, in all possible ways, to obtain a third set of distinct configurations.

2.2.20. Definition (Composition)

Let S, D, *and* C *be sets of distinct combinatorial configurations. Then* C *is called a* **composition** *of* S **with** D *if there exist s-objects and an operation called* **replacement** *such that*

1. *Replacement of each s-object in any* $\sigma \in \mathsf{S}$ *by an element of* D *gives an element of* C.
2. *Each element of* C *is uniquely constructed in this way.*

We write $\mathsf{C} = \mathsf{S} \circ \mathsf{D}$. □

It is important to note that S and D may have many compositions, depending on the specific realization of the replacement operation and the selection of s-objects.

2.2.21. Example (Composition)

Let S be the set of all $\{0, 1\}$-sequences with no blocks of adjacent 0's, and let D be the set $\{0, 00, 000, \dots\} = 0(0^*)$. Then

1. Substitution of each 0 in any sequence in S by an element in D gives an element of $\{0, 1\}^*$.
2. Each element of $\{0, 1\}^*$ is uniquely constructed in this way. For example $0011100010 \in \{0, 1\}^*$ may only be obtained from $0111010 \in \mathsf{S}$. This is done by replacing the first zero by $00 \in \mathsf{D}$, the second by $000 \in \mathsf{D}$, and the third by $0 \in \mathsf{D}$.

3. An s-object is a 0 occurring in elements of S.

It follows that $\{0, 1\}^* = \mathsf{S} \circ \mathsf{D}$. This is, of course, an indirect decomposition for S. □

We may now give the following elementary counting lemma associated with composition.

2.2.22. Lemma (Composition)

Let ω', ω be weight functions on the sets $\mathsf{S} \circ \mathsf{D}, \mathsf{D}$ of distinct configurations. If $\sigma' \in \mathsf{S} \circ \mathsf{D}$ is obtained from $\sigma \in \mathsf{S}$, with $\omega_s(\sigma) = k$ and $\alpha = (\alpha_1, \ldots, \alpha_k) \in \mathsf{D}^k$, where

$$\omega'(\sigma') = \omega(\alpha_1) + \cdots + \omega(\alpha_k),$$

then

$$\Phi_{\mathsf{S} \circ \mathsf{D}}^{(\omega')}(x) = \Phi_{\mathsf{S}}^{(\omega_s)}\left(\Phi_{\mathsf{D}}^{(\omega)}(x)\right)$$

provided the substitution is admissible.

Proof: Now

$$\begin{aligned}
\Phi_{\mathsf{S} \circ \mathsf{D}}^{(\omega')}(x) &= \sum_{\sigma' \in \mathsf{S} \circ \mathsf{D}} x^{\omega'(\sigma')}, && \text{from Definition 2.2.5} \\
&= \sum_{k \geqslant 0} \sum_{\substack{\sigma' = (\sigma, \alpha) \in \mathsf{S} \times \mathsf{D}^k \\ \omega_s(\sigma) = k}} x^{\omega'(\sigma')}, && \text{by the sum lemma} \\
&= \sum_{k \geqslant 0} \sum_{\substack{(\sigma, \alpha_1, \ldots, \alpha_k) \in \mathsf{S} \times \mathsf{D}^k \\ \omega_s(\sigma) = k}} x^{\omega(\alpha_1) + \cdots + \omega(\alpha_k)} \\
&= \sum_{k \geqslant 0} \sum_{\substack{\sigma \in \mathsf{S} \\ \omega_s(\sigma) = k}} \left\{ \sum_{\beta \in \mathsf{D}} x^{\omega(\beta)} \right\}^k \\
&= \sum_{\sigma \in \mathsf{S}} \left\{ \sum_{\beta \in \mathsf{D}} x^{\omega(\beta)} \right\}^{\omega_s(\sigma)} && \text{and the result follows.}
\end{aligned}$$

□

A weight-preserving decomposition $\Omega : \mathsf{C} \xrightarrow{\sim} \mathsf{S} \circ \mathsf{D}$ is said to be additive when the linear relation for weights, given in the composition lemma, holds. This is essentially Definition 2.2.16 for disjoint unions of Cartesian products. The admissibility (**1.1.4**) of the composition is usually easy to check since, in almost all combinatorial situations where composition is natural, we have $\Phi_{\mathsf{D}}^{(\omega)}(0) = 0$.

Example 2.2.21 can now be concluded with the aid of this lemma.

2.2.23. Example (Composition)

We wish to determine the number c_n of $\{0, 1\}$-sequences of length n with no blocks of adjacent 0's.

Let S be the set of all such sequences for $n \geqslant 0$. For $\sigma \in \{0, 1\}^*$ let $\omega(\sigma) = (i, j)$ be the weight of σ, where σ has i 1's and j 0's. If $\mathsf{A} \subseteq \{0, 1\}^*$ let

$$\Phi_{\mathsf{A}}^{(\omega)}(x, y) = \sum_{\sigma \in \mathsf{A}} x^i y^j.$$

This is the first instance of an ordinary generating function in more than one variable. Of course, each of the counting lemmas holds in each variable of an ordinary generating function. Clearly, $c_n = [x^n]\Phi_{\mathsf{S}}^{(\omega)}(x, x)$.

From Example 2.2.21 and the composition lemma

$$\Phi_{\{0,1\}^*}^{(\omega)}(x, y) = \Phi_{\mathsf{S}}^{(\omega)}\left(\Phi_{0(0^*)}^{(\omega)}(x, y), y\right).$$

It remains to determine the generating functions for $\{0, 1\}^*$ and $0(0^*)$ with respect to ω. By the sum and product lemmas we have immediately that $\Phi_{\{0,1\}^*}^{(\omega)}(x, y) = \sum_{k \geqslant 0} \Phi_{\{0,1\}^k}^{(\omega)}(x, y) = (1 - x - y)^{-1}$ and $\Phi_{0(0^*)}^{(\omega)} = x(1 - x)^{-1}$. Then $(1 - x - y)^{-1} = \Phi_{\mathsf{S}}^{(\omega)}(x(1 - x)^{-1}, y)$.

To obtain $\Phi_{\mathsf{S}}^{(\omega)}(x, y)$, set $u = x(1 - x)^{-1}$, from which $x = u(1 + u)^{-1}$, and it follows that

$$\Phi_{\mathsf{S}}^{(\omega)}(u, y) = \left\{1 - u(1 + u)^{-1} - y\right\}^{-1}.$$

Thus $\Phi_{\mathsf{S}}^{(\omega)}(x, x) = (1 + x)(1 - x - x^2)^{-1}$.

We elect to obtain a recurrence equation for c_n although, equally well, partial fractions or Taylor's theorem could be used to expand $(1 - x - x^2)^{-1}$ explicitly. Thus

$$(1 - x - x^2) \sum_{i \geqslant 0} c_i x^i = 1 + x.$$

Extracting the coefficient of x^{n+2} gives

$$c_{n+2} = c_{n+1} + c_n \qquad \text{for } n \geqslant 0.$$

For the constant and linear terms, $c_0 = 1$ and $c_1 - c_0 = 1$, so $c_1 = 2$. The number c_n is called the $(n + 1)$-st **Fibonacci number** F_{n+1}. □

Unlike the product lemma, the composition lemma has been stated in terms of a combinatorial operation, called *replacement*, for producing a configuration that realizes the elements of the composition. It would have been possible to have defined the composition $\mathsf{S} \circ \mathsf{A}$ of two sets of distinct configurations without reference to this operation. For example,

$$\mathsf{S} \circ \mathsf{A} = \dot{\bigcup_{k \geqslant 0}} \mathsf{S}_k \times \mathsf{A}^k,$$

where $\mathsf{S}_k = \{\sigma \in \mathsf{S} \mid \omega_s(\sigma) = k\}$, suffices. The interpretation of the elements of $\mathsf{S} \circ \mathsf{A}$, in this case, is involved in establishing the decomposition $\mathsf{T} \tilde{\rightarrow} \mathsf{S} \circ \mathsf{A}$ for a set T of distinct configurations. Precisely the same work is involved, in the product lemma, in checking $\mathsf{T} \tilde{\rightarrow} \mathsf{S} \times \mathsf{A}$, for example.

The reason for treating composition differently, by including a realization of the elements of $\mathsf{S} \circ \mathsf{A}$ through the operation of replacement, is that, combinatorially, it is natural to avoid any reference to the parameter k and the subset S_k. Instead, we consider the substitution of an arbitrary element of A for an arbitrary s-object in an arbitrary $\sigma \in \mathsf{S}$, and establish that the desired configuration is uniquely constructed in this way.

The final combinatorial operation we consider is differentiation of a set with respect to an s-object.

2.2.24. Definition (s-Derivative)

Let S *be a set of distinct configurations with* s*-objects. Then the* s*-derivative of* S*, denoted by* S' *or* $(d/ds)\mathsf{S}$*, is defined by*

$$S' = \dot{\bigcup_{\sigma \in \mathsf{S}}} \{\sigma\} \times \mathsf{N}_{\omega_s(\sigma)}. \qquad \square$$

Combinatorially, it is useful to regard $(\sigma, i) \in \mathsf{S}'$ as a configuration σ whose ith s-object is distinguished. This is done by placing ˜ over the distinguished s-object.

2.2.25. Example (s-Derivative of a Set of Permutations)

Let P_2 be the set of permutations on N_2. Let the ith s-object of $\sigma_1\sigma_2 \in \mathsf{P}_2$ be σ_i for $i = 1, 2$. Then $(d/ds)\mathsf{P}_2 = \{(12, 1), (12, 2), (21, 1), (21, 2)\}$ or, equivalently, $\{\tilde{1}2, 1\tilde{2}, \tilde{2}1, 2\tilde{1}\}$. $\square$

The following elementary counting lemma gives the generating function for the s-derivative of S in terms of the generating function for S.

2.2.26. Lemma (Differentiation)

Let S *be a set of distinct configurations with s-objects, and let* $\omega_s(\sigma)$ *denote the number of s-objects in* $\sigma \in$ S. *Then*

$$\Phi_{\mathsf{S}'}^{(\omega_s)}(x) = x\frac{d}{dx}\Phi_{\mathsf{S}}^{(\omega_s)}(x)$$

where $\omega_s(\sigma, i) = \omega_s(\sigma)$, *and* x *marks an s-object.*

Proof: Let $\Phi_{\mathsf{S}}^{(\omega_s)}(x) = \Sigma_{n\geqslant 0}\, a_n x^n$ and $\Phi_{\mathsf{S}'}^{(\omega_s)}(x) = \Sigma_{n\geqslant 0}\, b_n x^n$. Then, by Definition 2.2.24, we have $b_n = na_n$ and the result follows. □

The following example may be treated in several ways and gives a convenient illustration of the s-derivative.

2.2.27. Example (s-Derivative)

We wish to find the number c_p of sequences of length p over N_n. Let $\mathsf{S} = \mathsf{N}_n^*$. An s-object is any element of a sequence in S. To construct an element ρ of $(\mathsf{S} - \{\varepsilon\})'$, where ε denotes the empty sequence, note that ρ may be expressed uniquely in the form $\alpha\tilde{k}\beta$ with $\alpha, \beta \in \mathsf{S}$, and $k \in \mathsf{N}_n$, where $\tilde{k}$ is the distinguished s-object of ρ. It follows that

$$(\mathsf{S} - \{\varepsilon\})' \stackrel{\sim}{\rightarrow} \mathsf{S} \times \mathsf{N}_n \times \mathsf{S}$$

is an additively length-preserving decomposition. Of course, in this case $\omega_s(\sigma)$ is precisely the length of σ.

Thus, by the product and differentiation lemma,

$$x\frac{d}{dx}\{S(x) - 1\} = nxS^2(x)$$

where $S(x)$ is the generating function for S with respect to ω_s. The initial condition is $S(0) = 1$ since there is one sequence of length zero over N_n, namely, the empty sequence ε. Applying the integral operator I_x, defined in **1.1.5(4)**, to this equation we have

$$I_x\frac{d}{dx}S^{-1}(x) = -I_x n, \qquad \text{so} \qquad S^{-1}(x) - S^{-1}(0) = -nx$$

from **1.1.5(7)**. It follows that $S(x) = (1 - nx)^{-1}$ and that $c_p = n^p$. □

We have equipped ourselves with the operation of distinguishing an s-object. This is the most limited of the combinatorial operations, but it is of value since S′ may sometimes be easier to decompose than S in terms of union, product, and composition.

In principle, our strategy for enumerating sets of combinatorial configurations is straightforward. First, we look for a weight-preserving decomposition of a set into associated sets by some combination of the four basic operations of disjoint union, product, composition, and differentiation. Second, this decomposition is used to derive a functional relationship between the generating functions of the sets employed in the decomposition. The transition from a decomposition to a functional relationship is achieved by applying the elementary counting lemmas associated with each of these operations. These lemmas translate the four basic set operations of disjoint unions, product, composition, and differentiation into sum, product, composition, and differentiation of generating functions.

In view of these observations, a functional relationship derived from a decomposition, in the simplest case, can be an explicit expression for the required generating function. In the more complex cases it will be a functional equation, a differential equation, or indeed systems of these, in the multivariate case.

We now consider a generating function that may be used in enumerating a finite set S of distinct configurations with respect to P, a set of distinct nonempty subsets $P_1, \ldots, P_q$ of S. We refer to P_i as the ith **property** and define two generating functions that arise in connection with S and P.

2.2.28. Definition ("Exact," "At-Least" Generating Functions)

For $\alpha = \{\alpha_1, \ldots, \alpha_k\} \subseteq N_q$, *let the number of elements in* S *that belong to the subsets* $P_{\alpha_1}, \ldots, P_{\alpha_k}$ *and*

1. *No other* P_i*'s be* $e(\alpha)$.
2. *Possibly some other* P_i*'s be* $n(\alpha)$.

Then

3. *The* **"exact" generating function** *for* S *with respect to* P *is*

$$E(x) = \sum_{k \geqslant 0} e_k x^k \text{ where } e_k = \sum_{\substack{\alpha \\ |\alpha| = k}} e(\alpha).$$

4. *The* **"at-least" generating function** *for* S *with respect to* P *is*

$$N(x) = \sum_{k \geqslant 0} n_k x^k \text{ where } n_k = \sum_{\substack{\alpha \\ |\alpha| = k}} n(\alpha).$$

□

Since e_k is the number of elements of S that belong to exactly k members of P, we may write

$$E(x) = \Phi_{\mathsf{S}}^{(\omega)}(x)$$

where, for $\sigma \in \mathsf{S}$, $\omega(\sigma)$ is the number of members of P to which σ belongs. On the other hand, each element of S that belongs to exactly m members of P is counted exactly $\binom{m}{k}$ times in n_k, once for each of the k-subsets of these particular m properties. These observations justify the use of the terms "exact" and "at least" in the above definition. The following elementary counting lemma gives a connection between the "exact" and "at-least" generating functions of S with respect to P.

2.2.29. Lemma (the Principle of Inclusion and Exclusion)

Let $E(x)$, $N(x)$ be the "exact" and "at-least" generating functions of a finite set S *with respect to a set* P *of q properties. Then*

1. $E(x) = N(x-1)$,
2. $e_0 = \sum_{m=0}^{q}(-1)^m n_m$.

Proof: Clearly $E(x) = \sum_{\sigma \in \mathsf{S}} x^{\omega(\sigma)}$ where $\omega(\sigma)$, for $\sigma \in \mathsf{S}$, is the number of members of P to which σ belongs. We may construct $N(x)$, following Definition 2.2.28, by taking all k-subsets of the properties and counting, with weight k, those objects $\sigma \in \mathsf{S}$ possessing at least this subset of k properties. However, this is equivalent to considering each element $\sigma \in \mathsf{S}$ and recording it $2^{\omega(\sigma)}$ times, once with weight k for each k-subset of the $\omega(\sigma)$ properties which σ possesses. Thus

$$N(x) = \sum_{\sigma \in \mathsf{S}} \sum_{\alpha \subseteq \mathsf{N}_{\omega(\sigma)}} x^{|\alpha|} = \sum_{\sigma \in \mathsf{S}} (1+x)^{\omega(\sigma)}.$$

Accordingly, $N(x) = E(1+x)$ and **(1)** follows. Note that the substitution of $1+x$ into $E(x)$ is admissible **(1.1.4)** since S is a finite set and ω has an upper bound of q.

For **(2)** note, from **(1)**, that $e_0 = [x^0]N(x-1)$ and the result follows. □

There are many combinatorial situations in which it is a simpler task to determine n_k, or equivalently $N(x)$, than $E(x)$. These are the situations in which we are able to construct configurations that have a designated set of k properties, and perhaps some others. It is the fact that the remaining properties do not "interfere" with the calculation that is important.

2.2.30. Example (Derangements)

Let d_m be the number of permutations on N_m that have no fixed points. This number is called the **derangement number**.

Let S be the set of permutations on m symbols, and let P_i denote the subset of these in which i is a fixed point for $i = 1, \ldots, m$. Accordingly, for this set of properties, $d_m = e_0$. To obtain n_k we determine the number of permutations which have k specified properties, and then sum this over all subsets of k properties. There are $(m-k)!$ permutations that have at least k specified properties, and there are $\binom{m}{k}$ subsets of k properties. Thus $n_k = \binom{m}{k}(m-k)!$ so

$$d_m = m! \sum_{k=0}^{m} \frac{(-1)^k}{k!}$$

by (**2**) of the Principle of Inclusion and Exclusion. □

In the preceding example, n_k was determined by explicit calculations without first constructing $N(x)$. In fact, we obtain additional power in the use of the Principle of Inclusion and Exclusion by determining $N(x)$ itself as a consequence of the elementary counting lemmas.

NOTES AND REFERENCES

Other approaches have been given by Bender and Goldman (1971), Doubilet (1972), Doubilet, Rota, and Stanley (1972), Henle (1972), Joyal (1981), and MacMahon (1915).

2.2.11 Terquem (1839); **2.2.30** Montmort (1708).

2.3. PRELIMINARY EXAMPLES

We consider now the application of the elementary counting lemmas to a number of simple configurations. In each case the decompositions are stated separately so that the purely combinatorial part of the argument is clearly visible. We complete the discussion of each example by extracting the general terms from generating functions or by deriving recurrence equations. Most of the problems considered concern the enumeration of subsets, multisets, and compositions.

2.3.1. Decomposition (Subsets)

Let S *be the set of all subsets of* N_n. *Then*

$$\mathsf{S} \xrightarrow{\sim} \{0,1\}^n : \sigma \mapsto (j_1, \ldots, j_n),$$

where

$$j_i = \begin{cases} 1 & \text{if } i \in \sigma \\ 0 & \text{if } i \notin \sigma. \end{cases}$$ □

Under this decomposition, $\{1, 3\}$, considered as a subset of N_3, is encoded as $(1, 0, 1)$. The sum of the components of this 3-tuple is two, the size of the set which it encodes. Indeed, it is clear that the size of a subset is preserved by the decomposition and is recoverable in this way in general. Considered as a subset of N_4, $\{1, 3\}$ is encoded as $(1, 0, 1, 0)$. To enumerate the number of r-subsets of an n-set we proceed as follows.

2.3.2. Subsets

Let $c_n(r)$ be the number of r-subsets of N_n. Then

$$c_n(r) = [x^r]\Phi_{\mathsf{S}}^{(\omega)}(x),$$

where S is the set of all subsets of N_n and $\omega(\sigma) = |\sigma|$ for $\sigma \in \mathsf{S}$. Let Ω denote Decomposition 2.3.1. Then $\omega(\sigma) = j_1 + \cdots + j_n$, where $\Omega(\sigma) = (j_1, \ldots, j_n)$. It follows that Ω is an additively ω-preserving decomposition of S. In the notation of Definition 2.2.9, ω' is, of course, defined by $\omega'(i) = i$, for $i \in \{0, 1\}$. Thus, by the sum and product lemmas

$$\Phi_{\mathsf{S}}^{(\omega)}(x) = \left\{ \sum_{i \in \{0,1\}} x^i \right\}^n = (1 + x)^n,$$

so $c_n(r) = [x^r](1 + x)^n = \binom{n}{r}$. □

We use $(1 + x)^n$ as an example of the application of a common technique for obtaining recurrence equations.

2.3.3. Recurrence Equation for Binomial Coefficients

Let $f_n(x) = (1 + x)^n$. Differentiating with respect to x we have $f_n'(x) = nf_{n-1}(x)$. Comparing the coefficients of x^r gives

$$(r + 1)c_n(r + 1) = nc_{n-1}(r) \qquad \text{for } r \geqslant 0,$$

since $f_n(x) = \sum_{r=0}^{n} c_n(r)x^r$. This is a recurrence equation for $c_n(r)$. Replace n by $n - j$ and r by $r - j$ and take the product from $j = 0$ to r to get

$$\frac{c_n(r + 1)}{c_{n-r-1}(0)} = \binom{n}{r + 1}.$$

But $c_j(0) = 1$, so $c_n(r) = \binom{n}{r}$.

Another recurrence equation for $c_n(r)$ may be obtained by noting that $(1+x)^n = (1+x)(1+x)^{n-1}$, and by comparing coefficients of x^r on both sides to give

$$c_n(r) = c_{n-1}(r) + c_{n-1}(r-1) \qquad \text{for } n \geqslant 1, r \geqslant 0.$$

where $c_n(-1) = 0$, $c_n(0) = 1$, for $n \geqslant 0$. □

The total number of subsets of N_n is, of course, $c_n(0) + c_n(1) + \cdots + c_n(n)$. This may be calculated by noting that $f_n(x) = c_n(0) + c_n(1)x + \cdots + c_n(n)x^n$, so $f_n(1)$ is the required number. This substitution is admissible, by **1.1.4**, since $f_n(x)$ is bounded in x. But $f_n(x) = (1+x)^n$, so $f_n(1) = 2^n$. The total number of subsets of N_n is therefore 2^n.

It is clear that Decomposition 2.3.1 may be extended from sets to multisets, simply by recording the frequency of occurrence of each element. We therefore have the following decomposition.

2.3.4. Decomposition (Multisets)

Let S *be the set of all multisets of* N_n. *Then*

$$\mathsf{S} \tilde{\rightarrow} \mathsf{N}^n : \sigma \mapsto (j_1, \ldots, j_n),$$

where j_i *is the number of occurrences of* i *in* σ. □

For example, $\{1,1,1,2\}$, considered as a multiset of N_2, corresponds to $(3,1)$ under this decomposition, and the sum of the elements in this 2-tuple is 4, the number of elements in the multiset $\{1,1,1,2\}$. This relationship, of course, holds in general. Considered as a multiset of N_3, $\{1,1,1,2\}$ corresponds to $(3,1,0)$.

2.3.5. Multisets

Let $d_n(r)$ be the number of multisets of N_n of size r. Then

$$d_n(r) = [x^r]\Phi_{\mathsf{S}}^{(\omega)}(x),$$

where S is the set of all multisets of N_n and $\omega(\sigma) = |\sigma|$, the number of elements in the multiset, for $\sigma \in \mathsf{S}$. Now $\omega(\sigma) = j_1 + \cdots + j_n$, where $(j_1, \ldots, j_n) = \Omega(\sigma)$ and Ω denotes Decomposition 2.3.4. By the sum and product lemmas

$$\Phi_{\mathsf{S}}^{(\omega)}(x) = \left\{ \sum_{i \in \mathsf{N}} x^i \right\}^n = (1-x)^{-n}.$$

It follows that

$$d_n(r) = [x^r](1-x)^{-n} = \binom{n+r-1}{r}. \qquad \square$$

It is easy to see that the number of multisets of N_n is not finite, because each element of N_n may be selected an arbitrary number of times. But $(1-x)^{-n} = d_n(0) + d_n(1)x + d_n(2)x^2 + \cdots$. This is not bounded in x so the substitution $x = 1$ is not admissible.

2.3.6. Definition (Composition of an Integer)

For $r > 0$ let $\alpha = (\alpha_1, \ldots, \alpha_r)$, where $\alpha_1, \ldots, \alpha_r > 0$ and $\alpha_1 + \cdots + \alpha_r = n$. Then α is called a **composition** *of n with r* **parts**. *The empty vector ε is a composition of* 0. □

There are eight compositions of 4. These are given in Table 2.3.1.

For compositions there is a particularly simple decomposition, derivable immediately from Definition 2.3.6, which we use several times, with different weight functions.

2.3.7. Decomposition (Compositions of an Integer)

Let S be the set of all compositions. Then

$$\mathsf{S} = \dot{\bigcup_{k \geqslant 0}} \mathsf{N}_+^k. \qquad \square$$

To enumerate compositions of an integer proceed as follows.

2.3.8. Compositions of an Integer

Let S be the set of all compositions, and let $c(n)$ be the number of compositions of n. Then

$$c(n) = [x^n]\Phi_{\mathsf{S}}^{(\omega)}(x),$$

Table 2.3.1 Compositions of 4

Number of Parts	Compositions
1	(4)
2	(1, 3), (3, 1), (2, 2)
3	(2, 1, 1), (1, 2, 1), (1, 1, 2)
4	(1, 1, 1, 1)

where $\omega(\alpha) = \alpha_1 + \cdots + \alpha_l$ for $\alpha = (\alpha_1, \ldots, \alpha_l) \in \mathsf{S}$. Thus, by the sum and product lemmas,

$$\Phi_{\mathsf{S}}^{(\omega)}(x) = \sum_{k \geq 0} \left\{ \sum_{i \in \mathsf{N}_+} x^i \right\}^k = (1-x)(1-2x)^{-1} = 1 + x(1-2x)^{-1},$$

so $c(n) = [x^{n-1}](1-2x)^{-1} = 2^{n-1}$, for $n \geq 1$. □

As a partial check, note that $c(4) = 8$, in agreement with Table 2.3.1.

By recognizing relationships between generating functions, we may be able to establish the existence of correspondences between the sets of configurations enumerated by the functions. Although such correspondences occur infrequently in practice, it is nevertheless worthwhile checking to see whether a relationship may be found. The following is a correspondence we are able to derive from the limited amount of material that has been developed already. It is an immediate one, but it suffices to demonstrate the point.

2.3.9. Correspondence (Multisets–Compositions)

There is a bijection between the sets of

1. *k-part compositions of n.*
2. *Multisets of size $n-k$ in N_k.*

Proof: From Decomposition 2.3.7, the number of k-part compositions of n is

$$[x^n](x + x^2 + \cdots)^k = [x^{n-k}](1-x)^{-k}.$$

But from **2.3.5**, $[x^{n-k}](1-x)^{-k}$ is the number of multisets of size $n-k$ in N_k. Since these two numbers are equal, there is a bijection between the corresponding sets. □

Although a bijection exists between the two sets, it is entirely another matter to obtain a combinatorial description of it. In the present case, however, we can see that the composition $(i_1, \ldots, i_k)$ of n corresponds to the multiset $\{\alpha_1, \ldots, \alpha_{n-k}\}$ of N_k, where there are $i_j - 1$ occurrences of j in the multiset, for $1 \leq j \leq n-k$. Thus, for example, the composition $(1, 3, 1)$ of 5 corresponds to the multiset $\{2, 2\}$ of N_3.

Decomposition 2.3.7 may, of course, be used to enumerate compositions of n with respect to their number of parts, by giving us the necessary structural properties. It remains to find a weight function suitable for this more refined enumeration.

Having checked that a decomposition is additively weight-preserving, we explicitly mimic the accumulation of weights by marking objects whose weights are to be preserved in an appropriate way with indeterminates. Clearly, we are doing no more than using the additively weight-preserving properties that have already been established, but in a way that is combinatorially directed. The shift in point of view is merely one of emphasis, but it is combinatorially important. This is demonstrated in the enumeration of compositions with respect to parts.

2.3.10. Compositions and Parts

Let $c(n, k)$ be the number of compositions of n with exactly k parts. We use Decomposition 2.3.7.

The generating function for any position in a composition is

$$\sum_{i \in \mathsf{N}_+} x^i y = xy(1-x)^{-1},$$

where x marks the size of a part and y marks the occurrence of a part. Thus, by the product lemma, $c(n, k) = [x^n y^k] h(x, y)$, where

$$h(x, y) = \sum_{k \geqslant 0} \left(xy(1-x)^{-1} \right)^k = 1 + xy\{1 - (1+y)x\}^{-1}.$$

It follows immediately that $c(n, k) = \binom{n-1}{k-1}$. □

The less refined enumeration, given in **2.3.8**, is recovered by setting $y = 1$ in $h(x, y)$.

The use of a differential method for obtaining recurrence equations has already been demonstrated in **2.3.3**. When a generating function is a quotient of polynomials we may again use the simple technique given in **2.2.23**. This is illustrated next.

2.3.11. Recurrence Equation for Compositions and Parts

The ordinary generating function for the number $c(n, k)$ of compositions of n with exactly k parts is

$$\sum_{k, n \geqslant 0} c(n, k) x^n y^k = (1-x)\{1 - (1+y)x\}^{-1}.$$

Thus $(1 - x - xy)\Sigma_{k, n \geqslant 0}\, c(n, k) x^n y^k = 1 - x$. Applying $[x^p y^q]$ to both sides gives the recurrence equation

$$c(p, q) - c(p-1, q) - c(p-1, q-1) = 0 \qquad \text{if } p, q \geqslant 1,$$

with initial conditions $c(p, 0) = c(0, q) = 0$ for $p, q > 0$ and $c(0, 0) = 1$. □

Combinatorial identities may often be obtained by extracting the general coefficient from a formal power series in two different ways. In the next example we obtain the generating function for a certain set of compositions and then derive an identity by expanding this function in two different ways.

2.3.12. An Identity from Compositions

Let $c(n, l)$ be the number of compositions of n with exactly l even parts. Let x mark the size of a part and y a part of even size. The generating function for each part is therefore $x + yx^2 + x^3 + yx^4 + x^5 + \cdots$. Thus from Decomposition 2.3.7

$$c(n, l) = [x^n y^l] f(x, y),$$

where

$$f(x, y) = \sum_{k \geqslant 0} (x + yx^2 + x^3 + yx^4 + x^5 + \cdots)^k$$

$$= (1 - x^2)\{1 - x - (1 + y)x^2\}^{-1},$$

so

$$f(x, y) = (1 + x)\{1 - x^2(1 + y)(1 - x)^{-1}\}^{-1}$$

$$= \{1 - x(1 + xy)(1 - x^2)^{-1}\}^{-1}.$$

Applying first the coefficient operator $[y^l]$ and then $[x^n]$ to these two expressions gives

$$\sum_{k \geqslant 0} \binom{k}{l} \left\{ \binom{n - k - 1}{k - 1} + \binom{n - k - 2}{k - 1} \right\}$$

$$= \sum_{k=0}^{\lfloor (n-l)/2 \rfloor} \binom{n - l - 2k}{l} \binom{n - l - k - 1}{k}.$$

Of course, both are expressions for $c(n, l)$. □

We now return to the enumeration of subsets of a set.

2.3.13. Definition (Succession in a Set)

Let α be a subset of N. *A* **succession** *in α is a subset of the form $\{i, i + 1\}$.* □

To enumerate subsets with respect to successions we have the following decomposition, which is a modification of Decomposition 2.2.11.

2.3.14. Decomposition (Subsets)

Let S_k be the set of all k-subsets of N_n for $n \geqslant 0$. Then

$$\mathsf{S}_k \tilde{\rightarrow} \mathsf{N}_+^k \times \mathsf{N} : \{\sigma_1, \ldots, \sigma_k\}_n \mapsto (j_1, \ldots, j_{k+1}),$$

where $\sigma_1 < \sigma_2 < \cdots < \sigma_k$, $\sigma_i - \sigma_{i-1} = j_i$ for $i = 1, \ldots, k+1$, $\sigma_0 = 0$, and $\sigma_{k+1} = n$. □

To enumerate k-subsets of N_n with respect to successions, observe that a succession in $\{\sigma_1, \ldots, \sigma_k\}_n$ is indicated by the occurrence of a 1 in $(j_2, \ldots, j_k)$. Decomposition 2.3.1 is inadequate for the present purpose because it does not preserve information about successions conveniently.

2.3.15. Subsets with Successions

Let $c(m, n, k)$ be the number of k-subsets σ of N_n with exactly m successions. Then

$$n = j_1 + \cdots + j_k \qquad \text{and} \qquad m = \text{number of 1's in } (j_2, \ldots, j_k)$$

where $\Omega(\sigma) = (j_1, \ldots, j_{k+1})$ and Ω denotes Decomposition 2.3.14.

Let x mark the size of the set whose k-subsets are to be enumerated, and let y mark the occurrence of a succession. Then $c(m, n, k) = [x^n y^m] f(x, y)$, where, by the product lemma,

$$\begin{aligned} f(x, y) &= (x + x^2 + x^3 + \cdots)(yx + x^2 + x^3 + \cdots)^{k-1} \\ &\quad \times (1 + x + x^2 + \cdots) \\ &= x^k (1 - x)^{-2} \{y + x(1 - x)^{-1}\}^{k-1}. \end{aligned}$$

The extraction of the coefficient of $x^n y^m$ is routine, yielding

$$c(m, n, k) = \binom{k-1}{m}\binom{n-k+1}{k-m}.$$ □

The number of 2-subsets of N_5 with no successions is therefore $\binom{4}{2} = 6$. These 2-subsets are $\{1,3\}$, $\{1,4\}$, $\{1,5\}$, $\{2,4\}$, $\{2,5\}$, and $\{3,5\}$. We next consider the enumeration of Skolem subsets of N_n, defined as follows.

2.3.16. Definition (Skolem Subset)

Let $\sigma = \{\sigma_1, \ldots, \sigma_k\}$ satisfy the following conditions:

1. $0 = \sigma_0 < \sigma_1 < \sigma_2 < \cdots < \sigma_k \leqslant \sigma_{k+1} = n$.
2. $\sigma_i - \sigma_{i-1} \equiv 1 \bmod p$, *for $i = 1, \ldots, k$.*

Then σ is called a **Skolem k-subset** *of N_n with* **index** *p.* □

Skolem subsets have the following decomposition, obtained immediately from Definition 2.3.16.

2.3.17. Decomposition (Skolem Subsets)

Let S_k *be the set of all Skolem k-subsets of index p. Then*

$$\mathsf{S}_k \tilde{\to} \{i \in \mathsf{N} \mid i \equiv 1 \bmod p\}^k \times \mathsf{N} : \{\sigma_1, \ldots, \sigma_k\}_n \mapsto (j_1, \ldots, j_{k+1}),$$

where $\sigma_1 < \cdots < \sigma_k$, $\sigma_{i+1} - \sigma_i = j_{i+1}$ *for* $i = 0, \ldots, k$ *and* $\sigma_0 = 0$, $\sigma_{k+1} = n$. □

We now use this decomposition in the enumeration of Skolem k-sets of index p.

2.3.18. The Skolem Problem

Let $c(n, k, p)$ be the number of Skolem k-subsets of N_n with index p. If $\sigma \in \mathsf{S}_k$, then σ is a Skolem k-subset of N_n, where $n = j_1 + \cdots + j_{k+1}$, $(j_1, \ldots, j_{k+1}) = \Omega(\sigma)$, and Ω is given in Decomposition 2.3.17. Let x mark the size of the set of which the Skolem k-sets are subsets. Then, from the product lemma, we have $c(n, k, p) = [x^n] f(x)$, where

$$f(x) = \left\{ \sum_{i \equiv 1 (\bmod p)} x^i \right\}^k \sum_{i \in \mathsf{N}} x^i,$$

so $f(x) = x^k (1 - x^p)^{-k} (1 - x)^{-1}$.

To extract the coefficient in closed form, note that $f(x) = x^k (1 - x^p)^{-(k+1)} (1 + x + x^2 + \cdots + x^{p-1})$, so

$$c(n, k, p) = \binom{\lfloor \{n + (p-1)k\}/p \rfloor}{k}. \qquad \square$$

The Terquem problem (**2.2.15**) corresponds to the case $p = 2$. The final subset problem deals with a slightly modified set of successions.

2.3.19. Definition (Circular Succession)

A **circular succession** *on* N_n *is a succession on* N_n *or* $\{1, n\}$. □

Decomposition 2.3.14 does not allow us to recognize the circular succession $\{1, n\}$ on N_n as we do the other circular successions. The following decomposition, however, does.

2.3.20. Decomposition (Subsets)

Let S_k *be the set of all k-subsets of* N_+. *Let the s-objects of* $\{\sigma_1,\ldots,\sigma_k\}_n$ *be the elements of* N_n *and let* $\sigma_0 = \sigma_k$. *Then*

$$\mathsf{S}_k \times \mathsf{N}_k \stackrel{\sim}{\rightarrow} \frac{d}{ds}\mathsf{N}_+^k : (\{\sigma_1,\ldots,\sigma_k\}_n, i) \mapsto ((\alpha_1,\ldots,\alpha_k), j),$$

where $\sigma_1 < \cdots < \sigma_k$,

1. $j = \sigma_i$
2. $\alpha_r = \sigma_{(i+r)\bmod k} - \sigma_{(i+r-1)\bmod k} + n\delta_{r,k-i+1}$, *for* $r = 1,\ldots,k$.

Proof: Note that

$$\sigma_r = \begin{cases} j - n + \alpha_1 + \alpha_2 + \cdots + \alpha_{k-i+r} & \text{for } 1 \leqslant r \leqslant i \\ j + \alpha_1 + \alpha_2 + \cdots + \alpha_{r-i} & \text{for } i + 1 \leqslant r \leqslant k. \end{cases}$$

Thus the mapping is invertible and the result follows. □

The decomposition works as follows. The differences between adjacent elements in $\{\sigma_1,\ldots,\sigma_k\}_n \in \mathsf{S}_k$ together with $\sigma_k - \sigma_1$, in cyclic order starting from some distinguished position $i \in \mathsf{N}_k$, give $(\alpha_1,\ldots,\alpha_k)$. To recover $\{\sigma_1,\ldots,\sigma_k\}_n$ from these differences we need to supply the initial element $\sigma_i \in \mathsf{N}_n$ to which the differences may then be added. We treat this element as a distinguished s-object and to obtain $(d/ds)\mathsf{S}_k$ we must allow this element to be each $j \in \mathsf{N}_n$ in turn.

2.3.21. Example (Circular Successions)

Consider the subset $\sigma = \{1, 5, 6, 8, 11\}$ of N_{11}, denoted by $\{1, 5, 6, 8, 11\}_{11} \in \mathsf{S}_5$. Suppose that the third element of σ, namely, $\sigma_3 \in \mathsf{N}_{11}$, is distinguished. We denote $\sigma \subseteq \mathsf{N}_{11}$ with the third element distinguished by $(\{1, 5, 6, 8, 11\}_{11}, 3) \in \mathsf{S}_5 \times \mathsf{N}_5$. Thus $i = 3, j = 6$ from (**1**), and $k = 5$, in the notation of Decomposition 2.3.20.

The image of this element of $\mathsf{S}_5 \times \mathsf{N}_5$ under the decomposition is $((\alpha_1, \alpha_2, \alpha_3, \alpha_4, \alpha_5), 6)$ where, from (**2**),

$$\alpha_1 = \sigma_4 - \sigma_3 = 2, \qquad \alpha_4 = \sigma_2 - \sigma_1 = 4$$

$$\alpha_2 = \sigma_5 - \sigma_4 = 3, \qquad \alpha_5 = \sigma_3 - \sigma_2 = 1$$

$$\alpha_3 = \sigma_1 - \sigma_5 + 11 = 1.$$

Thus $(\{1,5,6,8,11\}_{11}, 3)$ corresponds to $((2,3,1,4,1), 6)$ under the decomposition. It should be noted that the number of circular successions in σ is equal to the number of 1's in $(\alpha_1, \alpha_2, \alpha_3, \alpha_4, \alpha_5)$, so the decomposition preserves the number of circular successions additively.

To reverse this decomposition, note that $n = \alpha_1 + \cdots + \alpha_5 = 11$, and

$$\begin{aligned}
\sigma_4 &= 6 + \alpha_1 &&= 8 \\
\sigma_5 &= 6 + \alpha_1 + \alpha_2 &&= 11 \\
\sigma_1 &= 6 - 11 + \alpha_1 + \cdots + \alpha_3 &&= 1 \\
\sigma_2 &= 6 - 11 + \alpha_1 + \cdots + \alpha_4 &&= 5 \\
\sigma_3 &= 6 - 11 + \alpha_1 + \cdots + \alpha_5 &&= 6.
\end{aligned}$$

Moreover, $6 = \sigma_i$ from (**1**), so $i = 3$. Thus

$$(\{\sigma_1, \ldots, \sigma_5\}_{11}, i) = (\{1,5,6,8,11\}_{11}, 3). \qquad \square$$

We now enumerate subsets of N_n with respect to circular successions.

2.3.22. Subsets and Circular Successions

Let $c(m, n, k)$ be the number of k-subsets of N_n with exactly m circular successions. Now the number of circular successions in $\sigma = \{\sigma_1, \ldots, \sigma_k\}_n$ is equal to the number of 1's in $\alpha = (\alpha_1, \ldots, \alpha_k)$, where $\Omega(\sigma, i) = (\alpha, j)$, and Ω is Decomposition 2.3.20. Also n may be recovered from α since $n = \alpha_1 + \cdots + \alpha_k$, so Ω additively preserves the size n of the superset of $\{\sigma_1, \ldots, \sigma_k\}$ and the number of circular successions. Thus

$$c(m, n, k) = [x^n y^m] f_k(x, y),$$

where, by the differentiation lemma,

$$k f_k(x, y) = x \frac{\partial}{\partial x} (yx + x^2 + x^3 + \cdots)^k.$$

We note that x marks s-objects and y marks circular successions. Since $[x^n] x \dfrac{\partial}{\partial x} = n[x^n]$ on the ring of formal power series then $kc(m, n, k) = n[x^n y^m]\{yx + x^2(1 - x)^{-1}\}^k$, so

$$c(m, n, k) = \frac{n}{k} \binom{k}{m} \binom{n - k - 1}{k - m - 1}. \qquad \square$$

NOTES AND REFERENCES

2.3.15 Kaplansky (1943); **2.3.18** Netto (1927); **2.3.22** Kaplansky (1943); [**2.3.5**] Ostrowski (1929); [**2.3.6**] Liu (1968); [**2.3.9**] Lagrange (1963); [**2.3.10**] Moser and Abramson (1969a); [**2.3.11**] Goulden and Jackson (1978), Moser and Abramson (1969b); [**2.3.12**] Tanny (1975); [**2.3.13**] Moser and Abramson (1969b); [**2.3.14**] Read (1983); [**2.3.15**] Klarner (1967).

EXERCISES

2.3.1. (a) Find a recurrence for the number of compositions into parts each congruent to $l \pmod m$ for any fixed l and m such that $1 \leqslant l \leqslant m$.

(b) Show that the number of compositions of n into odd parts is F_n, the nth Fibonacci number.

2.3.2. Show that the number of (ordered) ways of obtaining a total of n in k rolls of a six-sided die is

$$\sum_{i \geqslant 0} (-1)^i \binom{k}{i} \binom{n-6i-1}{k-1}.$$

2.3.3. (a) Show that the number of solutions in non-negative integers to $x_1 + \cdots + x_k = n$ is $\binom{n+k-1}{n}$.

(b) Show that the number of solutions to $x_1 + \cdots + x_k = n$ with $a_i \leqslant x_i < b_i$ for $i = 1, \ldots, k$ is

$$[x^n] x^{a_1 + \cdots + a_k} (1-x)^{-k} \prod_{i=1}^{k} (1 - x^{b_i - a_i}).$$

2.3.4. (a) Show that the number of solutions in non-negative integers to $x_1 + \cdots + x_k + 2x_{k+1} + \cdots + 2x_{2k} = n$ is

$$\sum_{i \geqslant 0} \binom{k+i-1}{k-1} \binom{n+k-2i}{k-1},$$

and hence establish the identity

$$\sum_{i \geqslant 0} \binom{2k+i-1}{i} \binom{n+k-i-1}{k-1} (-1)^{n-i}$$

$$= \sum_{i \geqslant 0} \binom{k+i-1}{k-1} \binom{n+k-2i-1}{k-1}.$$

2.3.5. (a) Show that each non-negative integer less than $\prod_{i=1}^{k} a_i$ has a unique representation of the form

$$x_0 + \sum_{j=1}^{k-1} x_j \prod_{i=1}^{j} a_i, \quad \text{where } 0 \leqslant x_j \leqslant a_{j+1} - 1,$$

for arbitrary $k, a_1, \ldots, a_k \geqslant 1$.

(b) Show that every non-negative integer has a unique representation as a number base l for any $l \geqslant 2$.

(c) Show that every non-negative integer has a unique representation of the form $\sum_{i \geqslant 1} i! x_{i-1}$ for $0 \leqslant x_{i-1} \leqslant i$.

(d) For fixed n and l, show that each non-negative integer less than $n!/(n-l)!$ has a unique representation of the form

$$\sum_{i=0}^{l-1} \frac{(n-l+i)!}{(n-l)!} x_i, \quad \text{where } 0 \leqslant x_i \leqslant n - l + i.$$

2.3.6. Let

$$S_{k,m} = \sum_{m > x_k > \cdots > x_1 \geqslant 0} z^{\binom{x_1}{1} + \cdots + \binom{x_k}{k}}.$$

Show by induction on k that $S_{k,m} = (1 - z^{\binom{m}{k}})/(1 - z)$ and hence deduce that each non-negative integer has a unique representation of the form

$$\binom{x_1}{1} + \binom{x_2}{2} + \cdots + \binom{x_k}{k}, \quad \text{where } 0 \leqslant x_1 < x_2 < \cdots < x_k,$$

for each $k \geqslant 1$.

2.3.7. (a) In a population of $n(2k+1)$ people, p are "alphas" and the rest are "betas." These people are arranged into n "ridings" of $2k+1$ each. Show that the number of such arrangements for which the alphas are in the majority in q ridings is

$$[x^p u^q] \left\{ \sum_{i=0}^{k} \binom{2k+1}{i} x^i + u \sum_{i=k+1}^{2k+1} \binom{2k+1}{i} x^i \right\}^n.$$

(b) Show that the expected number of ridings in which the alphas have a majority is

$$n \binom{n(2k+1)}{p}^{-1} \sum_{i=k+1}^{2k+1} \binom{2k+1}{i} \binom{(n-1)(2k+1)}{p-i}.$$

2.3.8. By using Decomposition 2.3.14, show that the number of k-subsets of N_n with m circular successions is

$$\frac{n}{k}\binom{k}{m}\binom{n-k-1}{k-m-1}.$$

(See also **2.3.22.**)

2.3.9. (a) Show that the number of subsets $\sigma_1 < \cdots < \sigma_k$ of N_n with $a \leqslant \sigma_j - \sigma_{j-1} < b$ for $j = 2,\ldots, k$ and $b > a > 0$ is

$$\sum_{i\geqslant 0}(-1)^i\binom{k-1}{i}\binom{n-(a-1)(k-1)-i(b-a)}{k}.$$

(b) Show that the number of subsets $\sigma_1 < \cdots < \sigma_k$ of N_n with $a \leqslant \sigma_j - \sigma_{j-1} < b$ for $j = 2,\ldots, k$ and $a \leqslant \sigma_1 - \sigma_k (\text{mod } n) < b$ is

$$\frac{n}{k}\sum_{i\geqslant 0}(-1)^i\binom{k}{i}\binom{n-(a-1)k-i(b-a)-1}{k-1}.$$

2.3.10. Show that the number of subsets $\sigma_1 < \cdots < \sigma_k$ of N_n with $a \leqslant \sigma_j - \sigma_{j-1} < b$ for $j = 2,\ldots, n$ and $n - c < \sigma_k - \sigma_1 \leqslant n - d$ is

$$[x^{n-d-a(k-1)}](1-x^{b-a})^{k-1}(1-x)^{-(k+1)}$$
$$\times\{d-(d-1)x-cx^{c-d}+(c-1)x^{c-d+1}\}.$$

2.3.11. An **$(\mathbf{l}, \mathbf{q})$-subset** of N_n is a subset $\sigma = \{\sigma_1,\ldots, \sigma_k\}_n$ of N_n such that $\sigma_1 < \sigma_2 < \cdots < \sigma_k$, $\sigma_1 \equiv (1 + l_1) \pmod{q_1}$ and $\sigma_j - \sigma_{j-1} \equiv (1 + l_j) \pmod{q_j}$ for $j = 2, 3,\ldots, k$.

(a) Show that the number of $(\mathbf{l}, \mathbf{q})$-subsets of N_n is

$$[x^n]x^{l_1+\cdots+l_k+k}(1-x)^{-1}\prod_{i=1}^{k}(1-x^{q_i})^{-1}.$$

(b) Show that the number of $(\mathbf{l}, \mathbf{q})$-subsets of N_n with $q_i = p$ for $i = 1,\ldots, k$ is

$$\binom{\lfloor\{n-(l_1+\cdots+l_k)+(p-1)k\}/p\rfloor}{k}.$$

2.3.12. Show that the number of subsets $\sigma = (\sigma_1,\ldots, \sigma_k)$ of N_n, with $k = (\alpha + \beta)l + j$, such that $\sigma_1 < \cdots < \sigma_k$, the first α elements have the same parity, the next β have the opposite parity, and so on, alternating parities between successive blocks of size $\alpha, \beta, \alpha, \beta, \ldots,$ is

$$\binom{\lfloor \frac{1}{2}(n+1)\rfloor + (k-j)/(\alpha+\beta)}{k} + \binom{\lfloor \frac{1}{2}n\rfloor + (k-j)/(\alpha+\beta)}{k}$$

if $0 < j \leqslant \alpha$,

$$\binom{\lfloor \frac{1}{2}n\rfloor + 1 + (k-j)/(\alpha+\beta)}{k} + \binom{\lfloor \frac{1}{2}(n+1)\rfloor + (k-j)/(\alpha+\beta)}{k}$$

if $\alpha < j \leqslant \alpha + \beta$.

Subsets with this property are called **alternating** subsets.

2.3.13. Show that the number of subsets $\sigma = \{\sigma_1, \ldots, \sigma_k\}$ of N_n such that $\sigma_1 < \cdots < \sigma_k$ and $\sigma_j \equiv (\sigma_{j-1} + 1 + l_j) \bmod p$, for $j = 2, \ldots, k$ is

$$\frac{kv + up + k}{u + k}\binom{u+k}{k},$$

where $u = \lfloor (n - k - l_2 - \cdots - l_k)/p \rfloor$, $v = n - k - l_2 - \cdots - l_k - pu$.

2.3.14. Show that the number of ways of covering a $2 \times n$ chessboard with m dominos (1×2 rectangles) and k 1×1 squares is

$$[x^m y^k](1-x)\{(1-x-x^2)(1-x) - (1+x)y^2\}^{-1},$$

where $2m + k = 2n$.

2.3.15. A **polyomino** is a finite union of unit squares in the plane such that the vertices of the squares have integer coordinates, and such that it is connected with no finite cut set. Two polyominoes are indistinguishable if there is a translation which transforms one into the other. Show that the number of polyominoes with n squares, such that each column is a contiguous line of squares is

$$\tfrac{1}{16}[x^n](5 - 13x + 7x^2)(1 - 5x + 7x^2 - 4x^3)^{-1}, \qquad n > 1.$$

2.4. SEQUENCES

In this section we begin a more systematic study of sequences. A number of sequence problems have already been considered, and we now develop some further decompositions.

2.4.1. Definition (Substring, Subsequence, Block)

Let $\sigma = \sigma_1 \cdots \sigma_l \in \mathsf{N}_n^*$. *Then* σ *is called a* **sequence** *over* N_n.

1. *If* $\sigma = \alpha\beta\gamma$, *then* β *is called a* **substring** *of* σ.
2. *If* $\{i_1, \ldots, i_m\} \subseteq \mathsf{N}_l$, *then* $\sigma_{i_1}\sigma_{i_2} \cdots \sigma_{i_m}$ *is called a* **subsequence** *of* σ.
3. k^j *is called a* **block** *of length* j, *for* $j \geqslant 1$ *and* $k = 1, \ldots, n$. □

To obtain decompositions we consider how a sequence may be decomposed into maximal subconfigurations.

2.4.2. Definition (Maximal Block)

A block in a sequence is said to be **maximal** *if it is not properly contained in a block.* □

Any $\{0, 1\}$-sequence may be decomposed uniquely in terms of its 0's. For example, 01011000111101 may be expressed in the form

$$(1^0)0(1)0(1^2)0(1^0)0(1^0)0(1^4)0(1).$$

This gives the following decomposition.

2.4.3. Decomposition ($\{0, 1\}$-Sequences)

$\{0, 1\}^* = (1^*0)^*1^*$. □

We now enumerate $\{0, 1\}$-sequences with respect to maximal blocks of 1's.

2.4.4. $\{0, 1\}$-Sequences and Maximal Blocks of 1's

Let $c(n, k)$ be the number of sequences in $\{0, 1\}^*$ of length n with exactly k maximal blocks of 1's. Let the weight of such a sequence σ be $\omega(\sigma) = (n, k)$. Let x mark an element of $\{0, 1\}$ and y mark a maximal block of 1's. Then

$$\Phi^{(\omega)}_{\{0,1\}^*}(x, y) = \sum_{n, k \geqslant 0} c(n, k) x^n y^k.$$

Thus, by the sum and product lemmas and Decomposition 2.4.3

$$\Phi^{(\omega)}_{\{0,1\}^*}(x, y) = \Phi^{(\omega)}_{(1^*0)^*}(x, y)\Phi^{(\omega)}_{1^*}(x, y) = \left\{1 - \Phi^{(\omega)}_{1^*0}(x, y)\right\}^{-1}\Phi^{(\omega)}_{1^*}(x, y)$$

$$= \left\{1 - \Phi^{(\omega)}_{1^*}(x, y)\Phi^{(\omega)}_{0}(x, y)\right\}^{-1}\Phi^{(\omega)}_{1^*}(x, y).$$

But $\Phi_{1^*}^{(\omega)}(x, y) = \sum_{k=0}^{\infty} \Phi_{1^k}^{(\omega)}(x, y) = 1 + xy + x^2y + \cdots$, since 1^k corresponds to a single maximal block of 1's if $k > 0$. It follows that $\Phi_{1^*}^{(\omega)}(x, y) = 1 + xy(1 - x)^{-1}$. Since $\Phi_0^{(\omega)}(x, y) = x$ we have

$$\Phi_{\{0,1\}^*}^{(\omega)}(x, y) = \{1 + (y - 1)x\}\{1 - 2x - (y - 1)x^2\}^{-1}$$

so

$$c(n, k) = \binom{n+1}{2k}. \qquad \square$$

Note that $c(4,2) = \binom{5}{4} = 5$. The five $\{0, 1\}$-sequences of length 4 with two maximal blocks of 1's are 1011, 1101, 0101, 1010, and 1001.

To retain information about maximal blocks of 0's as well we notice that every $\{0, 1\}$-sequence may be uniquely decomposed in terms of maximal blocks of 0's alternating with maximal blocks of 1's. For example, 01011000111101 is decomposed into

$$(0^1 1^1)(0^1 1^2)(0^3 1^4)(0^1 1^1).$$

This gives the following decomposition.

2.4.5. Decomposition ($\{0, 1\}$-Sequences)

$\{0, 1\}^* = 1^*(00^*11^*)^*0^*.$ $\square$

The next example demonstrates the use of this decomposition.

2.4.6. $\{0, 1\}$-Sequences and Maximal Blocks of 0's and 1's

Let $c(n)$ be the number of $\{0, 1\}$-sequences of length n in which a maximal block of 0's of odd length is never followed by a maximal block of 1's of odd length. Let the weight of such a sequence σ be $\omega(\sigma) = n$.

Let S be the set of all such sequences for $n \geqslant 0$. Consider $\{0, 1\}^*$. By Decomposition 2.4.5, this is equal to $1^*(00^*11^*)^*0^*$. Now 00^*11^* gives all sequences in S consisting of a maximal block of 0's followed by a maximal block of 1's. From this set we must excise the set $0(0^2)^*1(1^2)^*$ of illegal sequences. Thus

$$\mathsf{S} = 1^*\left(00^*11^* - 0(0^2)^*1(1^2)^*\right)^*0^*.$$

Let x mark an element of $\{0, 1\}$. Then $\Phi_{\mathsf{S}}^{(\omega)}(x) = \sum_{n \geqslant 0} c(n)x^n$. From the sum and product lemmas

$$\begin{aligned}\Phi_{\mathsf{S}}^{(\omega)}(x) &= \Phi_{1^*}^{(\omega)}(x)\left\{1 - \Phi_{00^*}^{(\omega)}(x)\Phi_{11^*}^{(\omega)}(x) + \Phi_{0(0^2)^*}^{(\omega)}(x)\Phi_{1(1^2)^*}^{(\omega)}(x)\right\}^{-1}\Phi_{0^*}^{(\omega)}(x)\\ &= (1 + x)^2\{1 - 2x^2(1 + x)\}^{-1},\end{aligned}$$

since $$\Phi^{(\omega)}_{00*}(x) = \Phi^{(\omega)}_{11*}(x) = x(1-x)^{-1}$$

and $$\Phi^{(\omega)}_{0(0^2)*}(x) = \Phi^{(\omega)}_{1(1^2)*}(x) = x(1-x^2)^{-1}.$$

Thus $$c(n) = \sum_{k\geqslant 0} 2^k \binom{k+2}{n-2k}.$$ □

In principle, this method generalizes to larger alphabets, but it is then necessary to use a decomposition that adds a level of recursion for each symbol of the alphabet. This is awkward and may be avoided by the methods of Chapter 4. In certain simpler cases, however, we can enumerate N_n^* and in doing so can begin to make systematic use of multivariate generating functions. Throughout this section the symbol $i \in \mathsf{N}_n$ is marked by the indeterminate x_i, and $\mathbf{x}$ denotes $(x,\ldots, x_n)$.

2.4.7. Definition (Type of a Sequence)

Let $\sigma \in \mathsf{N}_n^*$ have i_j occurrences of j, for $1 \leqslant j \leqslant n$. Then $\tau(\sigma) = (i_1,\ldots, i_n)$ is called the **type** *of σ.* □

Since permutations of N_n have type $(1,\ldots, 1)$, we may specialize results for sequences to results for permutations by applying the operator $[x_1 \cdots x_n]$ to the generating function for sequences. The first example of this is given in **2.4.21**.

2.4.8. Sequences and Type

Let $c(\mathbf{i})$ be the number of sequences over N_n with type $\mathbf{i} = (i_1,\ldots, i_n)$. Let the weight of $\sigma \in \mathsf{N}_n^*$ be $\omega(\sigma) = \tau(\sigma)$. Thus $c(\mathbf{i}) = [\mathbf{x}^{\mathbf{i}}]\Phi^{(\omega)}_{\mathsf{N}_n^*}(\mathbf{x})$. But $\Phi^{(\omega)}_{\mathsf{N}_n^*}(\mathbf{x}) = \{1 - \Phi^{(\omega)}_{\mathsf{N}_n}(\mathbf{x})\}^{-1} = \{1 - (x_1 + \cdots + x_n)\}^{-1}$ by the sum and product lemmas, so

$$c(\mathbf{i}) = \left[\begin{matrix} i_1 + \cdots + i_n \\ i_1,\ldots, i_n \end{matrix}\right].$$ □

Consider first the enumeration of sequences with restrictions placed on type.

2.4.9. Sequences and Type Restrictions

Let $c(n)$ be the number of sequences over N_4 of length n with an even number of 1's.

To determine $c(n)$, let $c(i_1,\ldots, i_4)$ be the number of such sequences of type $(i_1,\ldots, i_4)$, and let

$$G(x_1,\ldots, x_4) = \sum_{i_1,\ldots,i_4\geqslant 0} c(i_1,\ldots, i_4)x_1^{i_1} \cdots x_4^{i_4}$$

where x_j marks the occurrence of a j in a sequence, for $1 \leqslant j \leqslant 4$. The required number $c(n)$ is therefore given by $c(n) = [x^n]G(x,\ldots, x)$, where x marks the occurrence of any element in N_4. Let $F(x_1,\ldots, x_4)$ be the generating function for N_4^* with respect to type. Then

$$G(x_1,\ldots, x_4) = \tfrac{1}{2}\{F(x_1, x_2, x_3, x_4) + F(-x_1, x_2, x_3, x_4)\},$$

by bisection of series. From **2.4.8**, we have $F = \{1 - (x_1 + \cdots + x_4)\}^{-1}$. Thus $c(n) = [x^n]G(x,\ldots, x) = \frac{1}{2}[x^n]\{(1 - 4x)^{-1} + (1 - 2x)^{-1}\}$ whence

$$c(n) = 2^{n-1}(1 + 2^n). \qquad \square$$

A composition of an integer n may be regarded as a sequence over N_n whose elements sum to n. We take advantage of this in the next example.

2.4.10. Compositions and Part Restrictions

Let $d(n)$ be the number of compositions of n with parts 1, 2, 3, or 4 and with an even number of 1's. To determine $d(n)$, note that the sequence $\sigma = \sigma_1 \cdots \sigma_l \in \mathsf{N}_4^*$ of type $(i_1,\ldots, i_4)$ is a composition of $i_1 + 2i_2 + 3i_3 + 4i_4$. Thus

$$d(n) = [x^n] \sum_{i_1,\ldots,i_4\geqslant 0} c(i_1,\ldots, i_4)(x)^{i_1}(x^2)^{i_2}(x^3)^{i_3}(x^4)^{i_4},$$

In the notation of **2.4.9** it follows that

$$\begin{aligned} d(n) &= [x^n]G(x, x^2, x^3, x^4) \\ &= \tfrac{1}{2}[x^n](1 - x - x^2 - x^3 - x^4)^{-1} + \tfrac{1}{2}[x^n](1 + x - x^2 - x^3 - x^4)^{-1}. \end{aligned}$$

To obtain a recurrence equation for $d(n)$, let

$$\sum_{n\geqslant 0} a(n)x^n = (1 - x - x^2 - x^3 - x^4)^{-1}$$

and

$$\sum_{n\geqslant 0} b(n)x^n = (1 + x - x^2 - x^3 - x^4)^{-1},$$

so $$d(n) = \tfrac{1}{2}\{a(n) + b(n)\} \qquad \text{for } n \geqslant 0.$$

On comparing coefficients of x^n we see that $a(n)$ satisfies the recurrence equation

$$a(n+4) = a(n+3) + a(n+2) + a(n+1) + a(n) \qquad \text{for } n \geqslant 0$$

where $a(0) = a(1) = 1$, $a(2) = 2$, $a(3) = 4$. Similarly, $b(n)$ satisfies the recurrence equation

$$b(n+4) = -b(n+3) + b(n+2) + b(n+1) + b(n) \qquad \text{for } n \geqslant 0,$$

where $b(0) = 1$, $b(1) = -1$, $b(2) = 2$, $b(3) = -2$. □

The following remark is now obvious.

2.4.11. Remark (Sequences and Compositions)

Let $\Phi(x_1, \ldots)$ be the generating function for a set S of sequences with respect to type. Then the number of compositions of m in S is $[x^m]\Phi(x, x^2, \ldots)$. □

Sequence enumeration methods may also be used to obtain some elementary results on ordered factorizations of integers. This is perhaps surprising, since factorizations are multiplicative objects associated with integers. Note that 12 has a unique prime factorization, but eight ordered factorizations, namely, (12), (2, 6), (6, 2), (2, 2, 3), (2, 3, 2), (3, 2, 2), (3, 4), and (4, 3).

2.4.12. Ordered Factorizations

Let $c_m(i_1, i_2, \ldots)$ be the number of ordered factorizations of $N = p_1^{i_1} p_2^{i_2} \cdots$ with exactly m factors (each larger than 1), where $1 < p_1 < p_2 < \cdots$ are the consecutive primes.

Let F be the set of integers greater than 1. Then, since each positive integer has a unique prime factorization,

$$\mathsf{F} \xrightarrow{\sim} p_1^* p_2^* \cdots - \varepsilon : N \mapsto p_1^{i_1} p_2^{i_2} \ldots,$$

so the generating function for the unique prime factorizations of elements of F is

$$\Phi(x_{p_1}, x_{p_2}, \ldots) = \left\{ \prod_{i \geqslant 1} \left(1 - x_{p_i}\right)^{-1} \right\} - 1.$$

But each ordered factorization of a positive integer may be represented uniquely by an element of F^*. Thus, letting u mark a factor, we have

$$c_m(i_1, i_2, \ldots) = \left[u^m x_{p_1}^{i_1} x_{p_2}^{i_2} \cdots\right](1 - u\Phi)^{-1}.$$

This is readily reduced to

$$\left[u^m x_{p_1}^{i_1} x_{p_2}^{i_2} \cdots\right] \sum_{k \geq 0} u^k (1+u)^{-(1+k)} \prod_{i \geq 1} \left(1 - x_{p_i}\right)^{-k}.$$

Accordingly,

$$c_m(i_1, i_2, \ldots) = \sum_{k \geq 0} (-1)^{m-k} \binom{m}{k} \prod_{j \geq 1} \binom{k + i_j - 1}{i_j}. \qquad \square$$

Just as a composition may be viewed as a sequence over N_+, so may an ordered factorization be viewed as a sequence over $\mathsf{N}_+ - \{1\}$. The simple transformation of the sequence generating function for compositions (see Remark 2.4.11) has an analogue for factorizations. This is considered at the end of this section.

We consider next sequences over N_n with conditions applied to pairs of adjacent elements. Although more general conditions will be considered later (Chapter 4), the following definition is sufficient for present purposes.

2.4.13. Definition (Rise, Level, Fall)

A substring ij of $\sigma \in \mathsf{N}_n^*$ is a

1. **Rise** *if $i < j$.*
2. **Level** *if $i = j$.*
3. **Fall** *if $i > j$.* □

The remainder of this section is concerned with the enumeration of sequences when conditions are imposed on the number of rises, levels, and falls. The following kind of sequence is important in later decompositions.

2.4.14. Definition (Smirnov Sequence)

A **Smirnov sequence** *is a sequence with no levels.* □

We have the following indirect decomposition for Smirnov sequences.

2.4.15. Decomposition (Smirnov Sequences)

Let D be the set of all Smirnov sequences in N_n^*. Then

$$\mathsf{N}_n^* \;\tilde{\to}\; \mathsf{D} \circ \{1(1^*) \times \cdots \times n(n^*)\},$$

where $\circ$ *denotes the composition with respect to* s_1-,..., s_n*-objects, and* $i \in \mathsf{N}_n$ *is an* s_i*-object,* $i = 1,\ldots, n$. □

In other words, any sequence in N_n^* may be constructed uniquely by taking a sequence in D, with distinct adjacent elements, and replacing each element by a block of that element.

As an example of this decomposition, consider the sequence 441555444. This is obtained from $4154 \in \mathsf{D}$ by replacing the first 4 from the left by 44, 1 by 1, 5 by 555, and the last 4 by 444. Note that 4455444 is not generated from 4154, since 1 cannot be replaced by the empty string.

2.4.16. The Smirnov Problem

Let $S(\mathbf{x})$ and $D(\mathbf{x})$ be the generating functions of N_n^* and D, respectively, with respect to type.

The generating function for $i(i^*)$ with respect to type is $x_i(1 - x_i)^{-1}$. Thus, from Decomposition 2.4.15 and the composition lemma,

$$S(x_1,\ldots, x_n) = D\left(x_1(1 - x_1)^{-1},\ldots, x_n(1 - x_n)^{-1}\right).$$

To obtain D we let $y_i = x_i(1 - x_i)^{-1}$ so $x_i = y_i(1 + y_i)^{-1}$ for $1 \leqslant i \leqslant n$. It follows that

$$D(y_1,\ldots, y_n) = S\left(y_1(1 + y_1)^{-1},\ldots, y_n(1 + y_n)^{-1}\right).$$

Thus, from **2.4.8**,

$$D(x_1,\ldots, x_n) = \left\{1 - \sum_{i=1}^{n} x_i(1 + x_i)^{-1}\right\}^{-1}$$

and the number of Smirnov sequences of type **i** is $[\mathbf{x}^{\mathbf{i}}]D(\mathbf{x})$. □

To enumerate sequences with respect to levels and type we may again use Decomposition 2.4.15, but this time as a **direct** decomposition for N_n^*.

2.4.17. Sequences and Levels

Let $c_l(\mathbf{i})$ be the number of sequences in N_n^* of type **i** with l levels.

Let y mark levels. Then the generating function for $i(i^*)$ with respect to type and levels is

$$x_i + yx_i^2 + y^2x_i^3 + \cdots = x_i(1 - yx_i)^{-1},$$

since $i^k \in i^*$ contains $k - 1$ levels. But in Decomposition 2.4.15 the only levels in elements of N_n^* are those internal to elements of $i(i^*)$. Thus, from Decomposition 2.4.15 and the composition lemma,

$$c_l(\mathbf{i}) = [y^l \mathbf{x}^{\mathbf{i}}] D\left(x_1(1 - yx_1)^{-1}, \ldots, x_n(1 - yx_n)^{-1}\right).$$

Thus, from **2.4.16**,

$$c_l(\mathbf{i}) = [y^l \mathbf{x}^{\mathbf{i}}] \left\{1 - \sum_{i=1}^{n} x_i \{1 + (1 - y)x_i\}^{-1}\right\}^{-1}. \qquad \square$$

Clearly, further information may be preserved by Decomposition 2.4.15, such as the number of maximal blocks of length j for $j \geqslant 1$. Again, we use Decomposition 2.4.15 as a direct decomposition for N_n^*.

2.4.18. Sequences and Maximal Blocks

Let $c(\mathbf{i}, \mathbf{k})$ be the number of sequences in N_n^* of type $\mathbf{i} = (i_1, \ldots, i_n)$, and with k_j maximal blocks of length j, for $j \geqslant 1$, where $\mathbf{k} = (k_1, k_2, \ldots)$. By **2.4.16** the generating function for Smirnov sequences in N_n^* with respect to type is

$$D(\mathbf{x}) = \left\{1 - \sum_{i=1}^{n} x_i (1 + x_i)^{-1}\right\}^{-1}.$$

Let f_j be an indeterminate marking a maximal block of length j. Thus the generating function for $i(i^*)$ with respect to type and maximal block size is

$$f_1 x_i + f_2 x_i^2 + f_3 x_i^3 + \cdots.$$

Thus, by Decomposition 2.4.15 and the composition lemma,

$$c(\mathbf{i}, \mathbf{k}) = [\mathbf{x}^{\mathbf{i}} \mathbf{f}^{\mathbf{k}}] G(\mathbf{x}, \mathbf{f}),$$

where $G(\mathbf{x}, \mathbf{f}) = D(f_1 x_1 + f_2 x_1^2 + \cdots, \cdots, f_1 x_n + f_2 x_n^2 + \cdots)$ and $\mathbf{f} = (f_1, f_2, \ldots)$. Let $f(x) = 1 + f_1 x + f_2 x^2 + \cdots$. Then

$$G(\mathbf{x}, \mathbf{f}) = \left\{1 + \sum_{j=1}^{\infty} F_j s_j\right\}^{-1}$$

where $f^{-1}(x) = 1 + F_1 x + F_2 x^2 + \cdots$ and $s_j = x_1^j + \cdots + x_n^j$ for $j \geqslant 1$. To simplify the expression for G, let $\mathbf{s} = (1, s_1, \ldots)$. Then

$$c(\mathbf{i}, \mathbf{k}) = [\mathbf{x}^{\mathbf{i}} \mathbf{f}^{\mathbf{k}}] (f^{-1} \circ \mathbf{s})^{-1},$$

where $f^{-1} \circ \mathbf{s}$ is the umbral composition of $f^{-1}(x)$ with $\mathbf{s}$. $\square$

This result is a special case of a more general theorem, called the maximal string decomposition theorem, considered in Chapter 4. Note that $f(x)$ is the generating function for maximal blocks, in a fixed arbitrary symbol, with respect to length. Also s_j is the generating function for maximal blocks of length j and is, of course, a power sum symmetric function.

We now use an indirect decomposition to enumerate N_n^+ with respect to type and falls. This is called the **Simon Newcomb problem.**

2.4.19. Decomposition (Sequences and Falls)

Let D_n *denote the set of sequences in* N_n^+ *with no falls. Then*

$$(\mathsf{D}_n 0)^+ \tilde{\rightarrow} \{\mathsf{N}_n^+ \circ (\{0\} \times \{\varepsilon, 0\})\}0,$$

where the composition is with respect to s_1*- and* s_2*-objects. An* s_1*-object is a fall and an* s_2*-object is a nonfall.*

Proof: The replacement operation (of Definition 2.2.20) for s_1-objects and s_2-objects is defined as follows.

$$ij \circ \{0\} = i0j \qquad \text{if } ij \text{ is a fall}$$

$$ij \circ \{\varepsilon, 0\} = \{ij, i0j\} \qquad \text{if } ij \text{ is a nonfall.}$$

Accordingly, $\{\mathsf{N}_n^+ \circ (\{0\} \times \{\varepsilon, 0\})\}0$ is the set of all sequences in $\{0, \dots, n\}^+$ in which nonempty, nondecreasing strings in N_n^+ precede the first 0 and separate consecutive 0's, and which end in 0. But this is simply $(\mathsf{D}_n 0)^+$. □

For example,

$$\{1322 \circ (0 \times \{\varepsilon, 0\})\}0 = \{130220, 1030220, 1302020, 10302020\}.$$

Note that

$$130220 \in (\mathsf{D}_3 0)^2;\ 1030220, 1302020 \in (\mathsf{D}_3 0)^3;\ 10302020 \in (\mathsf{D}_3 0)^4.$$

We may now determine the generating function for sequences with respect to type and falls.

2.4.20. The Simon Newcomb Problem

Let $c(m, \mathbf{i})$ denote the number of sequences in N_n^+ of type $\mathbf{i} = (i_1, \dots, i_n)$ and with m falls.

Let z mark a fall and let $d(\sigma)$ denote the number of falls in $\sigma \in \mathsf{N}_n^+$. Then

$$c(m, \mathbf{i}) = [\mathbf{x}^{\mathbf{i}} z^m]\Gamma(\mathbf{x}, z), \qquad \text{where } \Gamma(\mathbf{x}, z) = \sum_{\sigma \in \mathsf{N}_n^+} \mathbf{x}^{\tau(\sigma)} z^{d(\sigma)}$$

is the generating function for N_n^+ with respect to type and falls. Let y mark a nonfall. Then the generating function for N_n^+ with respect to type, falls, and nonfalls is

$$G(\mathbf{x}, z, y) = \sum_{\sigma \in \mathsf{N}_n^+} \mathbf{x}^{\tau(\sigma)} y^{v(\sigma)} z^{d(\sigma)} = y^{-1}\Gamma(y\mathbf{x}, y^{-1}z),$$

since $v(\sigma) = (i_1 + \cdots + i_n) - d(\sigma) - 1$ is, of course, precisely the number of nonfalls in σ.

Let $D_n(\mathbf{x}, z)$ be the generating function for D_n with respect to type and falls. We now enumerate $(\mathsf{D}_n 0)^+$ with t marking 0 and x_i marking i for $i = 1,\ldots, n$. Thus, from Decomposition 2.4.19,

$$tD_n(\mathbf{x}, t)\{1 - tD_n(\mathbf{x}, t)\}^{-1} = tG(\mathbf{x}, t, 1 + t),$$

since the generating functions for $\{0\}$ and $\{\varepsilon, 0\}$ are t and $1 + t$, respectively. From the preceding relationship between G and Γ we therefore obtain

$$t(1 + t)^{-1}\Gamma\left((1 + t)\mathbf{x}, t(1 + t)^{-1}\right) = tD_n(\mathbf{x}, t)\{1 - tD_n(\mathbf{x}, t)\}^{-1}.$$

Setting $(1 + t)\mathbf{x} = \mathbf{y}$ and $t(1 + t)^{-1} = f$, so that $t = f(1 - f)^{-1}$ and $\mathbf{x} = (1 - f)\mathbf{y}$, yields

$$\Gamma(\mathbf{y}, f) = D_n\left((1 - f)\mathbf{y}, f(1 - f)^{-1}\right)\left\{1 - f - fD_n\left((1 - f)\mathbf{y}, f(1 - f)^{-1}\right)\right\}^{-1}.$$

But $\mathsf{D}_n \cup \{\varepsilon\} = 1^*2^* \cdots n^*$, since the elements of D_n have no falls, whence $1 + D_n(\mathbf{x}, t) = \prod_{i=1}^n (1 - x_i)^{-1}$ by the product lemma. Thus

$$\Gamma(\mathbf{y}, f) = \left(\prod_{i=1}^{n}\{1 - (1 - f)y_i\}^{-1} - 1\right)\left(1 - f\prod_{i=1}^{n}\{1 - (1 - f)y_i\}^{-1}\right)^{-1}. \quad \square$$

We now return briefly to the question of specializing results for sequences to results for the subset of these that are permutations.

2.4.21. Permutations and Falls

Let $c(n, i)$ be the number of permutations on N_n with i falls. Then, from **2.4.20**,

$$c(n, i) = [f^i x_1 \cdots x_n]\Gamma(\mathbf{x}, f).$$

The coefficients that are linear in x_i are unaffected by setting $x_i^2 = 0$, $i = 1,\ldots, n$. In this case $\prod_{i=1}^n \{1 - (1 - f)x_i\}^{-1} = e^{(1-f)x}$, where $x = x_1 + \cdots + x_n$. It follows that

$$c(n, i) = [f^i x_1 \cdots x_n]P(f, x),$$

where $P(f, x) = (e^x - e^{fx})(e^{fx} - fe^x)^{-1} = \Sigma_{k \geq 0} p_k(f)x^k$, say. Now $[x_1 \cdots x_n]x^k = n!\delta_{k,n}$, so

$$c(n, i) = [f^i]n!p_n(f) = \left[f^i \frac{x^n}{n!}\right] P(f, x)$$
$$= \left[f^i \frac{x^n}{n!}\right] (e^x - e^{fx})(e^{fx} - fe^x)^{-1}.$$

We note that $c(n, i)$ is called an **Eulerian number**. □

The generating function obtained in **2.4.21** is said to be exponential in x. The elementary counting lemmas associated with the exponential generating function are considered in Chapter 3.

The final decomposition is based on subsequences, as opposed to substrings.

2.4.22. Definition (Sequence with Strictly Increasing Support)

A sequence in N_n^* *with* **strictly increasing support** *is a sequence in which* $12 \cdots n$ *occurs at least once as a subsequence.* □

For example, the sequences in N_2^* of type $(2, 2)$ with strictly increasing support are 1122, 1221, 1212, 2121, and 2112. The illegal sequence of type $(2, 2)$ is 2211.

2.4.23. Decomposition (Sequences with Strictly Increasing Support)

Let S_n *denote the set of sequences in* N_n^* *with strictly increasing support. Then*

$$\mathsf{S}_n = (\mathsf{N}_n - \{1\})^* 1 (\mathsf{N}_n - \{2\})^* 2 \cdots (\mathsf{N}_n - \{n\})^* n \mathsf{N}_n^*.$$

Proof: If $\sigma \in \mathsf{S}$ we may locate the unique subsequence $12 \cdots n$ that has each element as close as possible to the beginning of the sequence. Look for the first occurrence of 1 from the left. For $i = 1, \ldots, n - 1$, look for the first occurrence of $i + 1$ after the previously located i. Consider the substring α between i and $i + 1$ determined by this algorithm. By construction, α is any sequence in N_n^* with no $i + 1$'s. Thus $\alpha \in (\mathsf{N}_n - \{i + 1\})^*$. Clearly, any sequence in N_n^* may follow the symbol n located by the algorithm. The result follows. □

2.4.24. Sequences of Type (2,..., 2) with Strictly Increasing Support

Let $c_2(n)$ be the number of sequences in N_n^* of type $(2, \ldots, 2)$ with strictly increasing support.

From Decomposition 2.4.23 and the product lemma we have $c_2(n) = [x_1^2 \cdots x_n^2]F(x_1, \ldots, x_n)$, where

$$F(x_1, \ldots, x_n) = (x_1 \cdots x_n)(1 - x)^{-1} \prod_{j=1}^{n} (1 - x + x_j)^{-1}$$

and $x = x_1 + \cdots + x_n$. Thus

$$c_2(n) = [x_1 \cdots x_n](1-x)^{-(n+1)} \prod_{j=1}^{n} \left\{1 + (1-x)^{-1} x_j\right\}^{-1}.$$

Now set $x_1^2 = \cdots = x_n^2 = 0$ since we are concerned only with linear terms in $x_1, \ldots, x_n$. Thus

$$\begin{aligned} c_2(n) &= [x_1 \cdots x_n](1-x)^{-(n+1)} \exp\{-x(1-x)^{-1}\} \\ &= [x_1 \cdots x_n] \sum_{k=0}^{\infty} \frac{(-1)^k}{k!} x^k (1-x)^{-(n+k+1)} \\ &= \sum_{k,l \geq 0} \frac{(-1)^k}{k!} \binom{n+k+l}{l} [x_1 \cdots x_n] x^{k+l} \\ &= (2n)! \sum_{k=0}^{n} \frac{(-1)^k}{(n+k)!} \binom{n}{k}. \end{aligned}$$ □

It is an easy matter with this formula to check that $c_2(2) = 5$, in agreement with the set of permissible sequences in S_2 already given.

We now return to the enumeration of ordered factorizations, considered in **2.4.12**, and introduce another type of generating function to deal with the situation.

2.4.25. Definition (Dirichlet Generating Function)

The generating function for a set S *of distinct configurations that is* **Dirichlet** *with respect to the weight function* ω *on* S *is*

$$\Delta_{\mathsf{S}}^{(\omega)}(x) = \sum_{\sigma \in \mathsf{S}} (\omega(\sigma))^{-x}.$$ □

We have the following product lemma for Dirichlet generating functions.

2.4.26. Lemma (Product)

Let α, β, ω *be weight functions on the sets*, $\mathsf{A}, \mathsf{B}, \mathsf{A} \times \mathsf{B}$ *of distinct configurations. If*

$$\omega(a, b) = \alpha(a)\beta(b) \qquad \textit{for all } (a, b) \in \mathsf{A} \times \mathsf{B}$$

then $\Delta_{\mathsf{A}\times\mathsf{B}}^{(\omega)}(x) = \Delta_{\mathsf{A}}^{(\alpha)}(x)\, \Delta_{\mathsf{B}}^{(\beta)}(x)$.

Proof: From Definition 2.4.25

$$\Delta^{(\omega)}_{\mathsf{A}\times\mathsf{B}}(x) = \sum_{(a,b)\in\mathsf{A}\times\mathsf{B}} (\alpha(a)\beta(b))^{-x} = \sum_{a\in\mathsf{A}} (\alpha(a))^{-x} \sum_{b\in\mathsf{B}} (\beta(b))^{-x}$$

$$= \Delta^{(\alpha)}_{\mathsf{A}}(x)\,\Delta^{(\beta)}_{\mathsf{B}}(x).$$

□

Similarly, there is a sum lemma for Dirichlet generating functions. It is clear from Lemma 2.4.26 that the Dirichlet generating function is used with multiplicatively weight-preserving decompositions, defined as follows.

2.4.27. Definition (Multiplicatively Weight-Preserving Decomposition)

Let $\mathsf{A}, \mathsf{B}, \mathsf{C}$ *be sets of distinct configurations. Let* $\Omega : \mathsf{C} \xrightarrow{\sim} \mathsf{A} \times \mathsf{B}$ *be an* ω*-preserving decomposition. If*

$$\omega(c) = \alpha(a)\beta(b) \qquad \text{for all } c \in \mathsf{C}, \text{ with } \Omega(c) = (a, b),$$

where α, β *are weight functions on* A, B, *respectively, then* Ω *is called a* **multiplicatively** *weight-preserving decomposition of* C. □

2.4.28. Ordered Factorizations and Dirichlet Generating Functions

We use the decomposition of **2.4.12**. As before let F denote the set of all integers greater than 1. Thus each ordered factorization of a positive integer may be represented uniquely in F^*. For $\sigma = \sigma_1 \cdots \sigma_m \in \mathsf{F}^*$, let $\omega(\sigma) = (\omega'(\sigma_1) \cdots \omega'(\sigma_m), m)$, where $\omega'(\sigma_i) = \sigma_i$. Let $d_m(n)$ be the number of ordered factorizations of n with m factors. Then from Definition 2.4.25 and the sum and product lemmas for ordinary and Dirichlet generating functions

$$d_m(n) = [u^m n^{-x}]\,\Delta^{(\omega)}_{\mathsf{F}^*}(x)$$

$$= [u^m n^{-x}]\{1 - u\,\Delta^{(\omega')}_{\mathsf{F}}(x)\}^{-1}.$$

But $\Delta^{(\omega')}_{\mathsf{F}}(x) + 1 = \sum_{q=1}^{\infty} q^{-x} = \zeta(x)$ where $\zeta(x)$ is the **Riemann zeta function.** Thus

$$d_m(n) = [u^m n^{-x}]\{1 + u - u\zeta(x)\}^{-1}.$$

Note that the total number of ordered factorizations of n is obtained by setting $u = 1$, to give $[n^{-x}]\{2 - \zeta(x)\}^{-1}$. □

We do not pursue the idea of Dirichlet generating functions further for the following reason.

2.4.29. Remark (Sequences and Ordered Factorizations)

Let S *be a subset of the set of sequences on* $\mathsf{N}_+ - \{1\}$, *and let the ordinary generating function for* S *be* $f(x_2, x_3, \ldots)$. *Then the number of ordered factorizations of* n *in* S *is* $[n^{-x}]f(2^{-x}, 3^{-x}, \ldots)$. □

NOTES AND REFERENCES

For further information on Dirichlet generating functions see Hardy and Wright (1938).

2.4.12 MacMahon (1891); **2.4.16** MacMahon (1915), Smirnov, Sarmanov, and Zaharov (1966); **2.4.17** Carlitz (1972), David and Barton (1962); **2.4.19, 20** Gessel (1977); **2.4.20** Dillon and Roselle (1969); **2.4.21** Riordan (1958); **2.4.24** Horton (private communication); [**2.4.4**] Feller (1950); [**2.4.6**] Lovász (1979); [**2.4.9, 10**] Liu (1968); [**2.4.13**] Carlitz (1979); [**2.4.17**] Carlitz (1972); [**2.4.18**] Carlitz (1977); [**2.4.19**] Comtet (1974); [**2.4.20**] Mullin (1964a); [**2.4.21–23**] Goulden and Jackson (1981); [**2.4.22**] Good (1965); [**2.4.23**] Hutchinson and Wilf (1975).

EXERCISES

2.4.1. (a) Show that the number of $\{0, 1\}$-sequences of length n with m maximal blocks is $2\binom{n-1}{m-1}$.

(b) Show that the expected number of maximal blocks in a $\{0, 1\}$-sequence of length n is $\frac{1}{2}(n+1)$.

2.4.2. Show that the number of $\{0, 1\}$-sequences of length n with r pairs of adjacent 1's and no adjacent 0's is

$$[x^n y^r](1 + x)(1 + (1 - y)x)(1 - yx - x^2)^{-1}.$$

2.4.3. Show that the number of $\{0, 1\}$-sequences with all maximal blocks of 1's having even size and all maximal blocks of 0's having odd size is

$$\sum_{j \geqslant 0} \binom{n - j + 1}{3j - n + 1}.$$

2.4.4. Toss a fair coin until r consecutive heads appear. Show that the probability of stopping after exactly $k \geqslant r$ tosses is

$$[x^{k-r}](2^{r+1} - 2^{r+1}x + x^{r+1})^{-1}(2 - x).$$

2.4.5. By using Decompositions 2.3.1 and 2.4.5 show that the number of k-subsets of N_n with m successions is $\binom{k-1}{m}\binom{n-k+1}{k-m}$. (See also **2.3.15.**)

2.4.6. Show that the number of sequences on N_4 of length n in which 1, 2 are not adjacent is

$$\frac{5+\sqrt{17}}{2\sqrt{17}}\left\{\frac{3+\sqrt{17}}{2}\right\}^n - \frac{5-\sqrt{17}}{2\sqrt{17}}\left\{\frac{3-\sqrt{17}}{2}\right\}^n.$$

2.4.7. Show that the number of sequences of length n on N_{k+2} with an even number of 1's and an odd number of 2's is

$$\tfrac{1}{4}\left((k+2)^n - (k-2)^n\right).$$

2.4.8. Show that the number of sequences of length l on N_{2n} in which no even number can be adjacent to itself is

$$[x^l](1+x)/\left(1-(2n-1)x-nx^2\right).$$

2.4.9. Show that the number of arrangements of two of each of n letters such that no pair of identical letters is adjacent is $\sum_{k=0}^{n}(-1)^k 2^{k-n}\binom{n}{k}(2n-k)!$

2.4.10. Show that the number of arrangements of five copies of each element of N_n such that each element is next to another copy of itself is $\sum_{i=0}^{n}(-1)^i\binom{n}{i}(2n-i)!$

2.4.11. Show that the number of sequences of length l on N_n

(a) in which all maximal blocks have length at least k is

$$[x^l]\frac{1-x+x^k}{1-x-(n-1)x^k},$$

(b) in which no maximal block has length at least k is

$$[x^l]\frac{1-x^k}{1-nx+(n-1)x^k}.$$

2.4.12. Show that the number of sequences on N_n of type $\mathbf{i}$ with j_l blocks of l's of length k_l, for $l = 1,\ldots, n$ is

$$[\mathbf{x}^{\mathbf{i}}\mathbf{y}^{\mathbf{j}}]\left\{1-\sum_{l=1}^{n}\frac{x_l(1-y_lx_l)-x_l^{k_l}(1-y_l)}{1-y_lx_l-x_l^{k_l}(1-y_l)}\right\}^{-1}$$

where $\mathbf{y} = (y_1,\ldots, y_n)$ and $\mathbf{j} = (j_1,\ldots, j_n)$.

2.4.13. Show that the number of compositions of n in which adjacent parts are distinct mod p is

$$[x^n]\left\{1 - \sum_{i=1}^{p} \frac{x^i}{1 + x^i - x^p}\right\}^{-1}.$$

2.4.14. Show that the number of sequences of length l on N_n with k **successions** (a substring of the form $(i, i + 1)$) is

$$[x^l t^{l-k}]\left\{1 - t\frac{nx - (n+1)x^2(1-t) + x^{n+2}(1-t)^{n+1}}{(1 - x(1-t))^2}\right\}^{-1}.$$

2.4.15. (a) Show that the number of sequences of length l on N_n with k **circular successions** (a succession or $(n, 1)$) is $n\binom{l-1}{k}(n-1)^{l-k-1}$.

(b) Give a direct argument to establish the preceding result.

2.4.16. Derive the generating function for sequences in N_n^* with respect to levels and type, given in **2.4.17**, by the Principle of Inclusion and Exclusion.

2.4.17. (a) Show that the number of sequences in N_n^+ with i rises, j levels, and k falls, and of type $\mathbf{m}$ is

$$[\mathbf{x}^{\mathbf{m}} r^i l^j f^k]\frac{\prod_{i=1}^{n}(1 + (r - l)x_i) - \prod_{i=1}^{n}(1 + (f - l)x_i)}{r\prod_{i=1}^{n}(1 + (f - l)x_i) - f\prod_{i=1}^{n}(1 + (r - l)x_i)}.$$

(b) Show that the number of Smirnov sequences of type $\mathbf{i}$, with k rises, is

$$[\mathbf{x}^{\mathbf{i}} r^k]\frac{\prod_{i=1}^{n}(1 + rx_i) - \prod_{i=1}^{n}(1 + x_i)}{r\prod_{i=1}^{n}(1 + x_i) - \prod_{i=1}^{n}(1 + rx_i)}.$$

(c) Use part (b) to obtain the generating function for Smirnov sequences given in **2.4.16**.

2.4.18. (a) Show that the number of compositions of m with parts $\leqslant n$, i rises, j levels, and k falls is given by

$$[r^i l^j f^k x^m]\frac{\prod_{i=1}^{n}(1 + (r - l)x^i) - \prod_{i=1}^{n}(1 + (f - l)x^i)}{r\prod_{i=1}^{n}(1 + (f - l)x^i) - f\prod_{i=1}^{n}(1 + (r - l)x^i)}.$$

(b) Show that the number of ordered factorizations of m with factors $\leqslant n$, i rises, j levels, and k falls is given by

$$[r^i l^j r^k m^{-x}]\frac{\prod_{i=2}^{n}(1+(r-l)i^{-x})-\prod_{i=2}^{n}(1+(f-l)i^{-x})}{r\prod_{i=2}^{n}(1+(f-l)i^{-x})-f\prod_{i=2}^{n}(1+(r-l)i^{-x})}.$$

2.4.19. A permutation $\sigma_1 \cdots \sigma_n$ on N_n, $n \geqslant 1$, is called **indecomposable** if there exists no $m < n$ for which $\sigma_1 \cdots \sigma_m$ is a permutation on N_m. Show that the number of indecomposable permutations on N_n for $n \geqslant 1$ is

$$[t^n]\Big(\sum_{i\geqslant 0} i!t^i\Big)^{-1}.$$

2.4.20. A **monic** polynomial of degree n over GF(p) is a polynomial $x^n + \sum_{i=0}^{n-1} f_i x^i$ where $f_0, \ldots, f_{n-1} \in \mathrm{GF}(p)$. An **irreducible** monic polynomial over GF(p) has no monic polynomial of degree $\geqslant 1$ as a proper factor.

(a) If there are m_i distinct irreducible monic polynomials of degree i over GF(p), show that, for primes p,

$$\prod_{i\geqslant 1}(1-z^i)^{-m_i} = (1-pz)^{-1}.$$

(b) Deduce from part (a) that $m_i \geqslant 1$ for all $i \geqslant 1$, so GF(p^i) exists for all primes p and all $i \geqslant 1$.

2.4.21. Let $c_{\alpha,l}(\mathbf{k},\mathbf{M})$ be the number of sequences on N_n of type $\mathbf{k}$, which begin with α and end with l, and have m_{ij} occurrences of the substring ij, for $1 \leqslant i, j \leqslant n$, where $\mathbf{M} = [m_{ij}]_{n\times n}$.

(a) Show that $c_{\alpha,l}(\mathbf{k},\mathbf{M}) = [\mathbf{x}^{\mathbf{k}}\mathbf{A}^{\mathbf{M}}]f_{\alpha,l}(\mathbf{x},\mathbf{A})$ where $\mathbf{A} = [a_{ij}]_{n\times n}$ and

$$f_{i,l} = x_i\left\{\delta_{il} + \sum_{j=1}^{n} a_{ij} f_{j,l}\right\} \qquad \text{for } i = 1, \ldots, n.$$

(b) Deduce from part (a) that

$$c_{\alpha,l}(\mathbf{k},\mathbf{M}) = \frac{(\mathbf{k}-\mathbf{1})!}{\mathbf{M}!}\|\delta_{ij}k_i - m_{ij}\|_{n\times n}$$

where $$\sum_{i=1}^{n} m_{ij} = k_j - \delta_{\alpha j}, \qquad \sum_{j=1}^{n} m_{ij} = k_i - \delta_{il}.$$

2.4.22. Let $h(\mathbf{D})$ be the number of directed Hamiltonian cycles in a digraph on n vertices with adjacency matrix $\mathbf{D} = [d_{ij}]_{n\times n}$.

(a) Show that $h(\mathbf{D}) = [\mathbf{x}^{1+\boldsymbol{\delta}_l}]f_{l,l}(\mathbf{x}, \mathbf{D})$ where $f_{l,l}$ is as given in [**2.4.21**(a)], and $\boldsymbol{\delta}_l = (\delta_{l1},\ldots,\delta_{ln})$, for any $l = 1,\ldots,n$.

(b) Deduce from part (a) that

$$h(\mathbf{D}) = \sum_{\beta}(-1)^{|\beta|}(\det \mathbf{D}[\beta|\beta])(\operatorname{per} \mathbf{D}(\beta|\beta))$$

where the summation is over all $\beta \subseteq \mathsf{N}_n - \{l\}$ for any $l \in \mathsf{N}_n$.

2.4.23. Let $e(\mathbf{D})$ be the number of directed closed Euler trails in a digraph with adjacency matrix $\mathbf{D}$, and let the in-degree and out-degree of vertex i be equal to b_i, for $i = 1,\ldots,n$. Show that, for any $l \in \mathsf{N}_n$,

$$e(\mathbf{D}) = (\mathbf{b} - \mathbf{1})!\, cof_{ll}\left[\delta_{ij}b_i - d_{ij}\right]_{n\times n}.$$

2.5. PARTITIONS OF AN INTEGER

In this section enumerative arguments are used to derive a number of classical identities associated with partitions of integers. Surprisingly, all the decompositions used are direct. Further results are obtained in Section 2.6 by more elaborate methods.

2.5.1. Definition (Partition)

Let n be a non-negative integer and let $\alpha = (\alpha_1,\ldots,\alpha_k)$ be integers such that $\alpha_1 \geqslant \alpha_2 \geqslant \cdots \geqslant \alpha_k > 0$ and $\alpha_1 + \cdots + \alpha_k = n$, for $k \geqslant 1$. Then α is a **partition** *of n with k* **parts**. *The empty partition is denoted by ε.* □

The eleven partitions of 6 are given in Table 2.5.1. Partitions are to be distinguished from compositions, given by Definition 2.3.6.

The following decomposition gives the obvious relationship between partitions and sequences.

2.5.2. Decomposition (Partitions)

Let Π denote the set of all partitions. Then $\Pi \xrightarrow{\sim} (1^*)(2^*)\cdots$. □

We may now determine the number of partitions of n.

Table 2.5.1 Partitions of 6

1-part:	(6)	4-part:	(3, 1, 1, 1), (2, 2, 1, 1)
2-part:	(5, 1), (4, 2), (3, 3)	5-part:	(2, 1, 1, 1, 1)
3-part:	(4, 1, 1), (3, 2, 1), (2, 2, 2)	6-part:	(1, 1, 1, 1, 1, 1).

2.5.3. The Number of Partitions

Let $p(n)$, the **partition number**, denote the number of partitions of n, and let q mark the size of a part. Then, from Decomposition 2.5.2

$$p(n) = [q^n](1-q)^{-1}(1-q^2)^{-1}\ldots,$$

since $(1-q^k)^{-1}$ is the generating function for k^*. □

This expression is nothing more than an encoding of the problem, but it can be used to check that there are indeed eleven partitions of 6 (see Table 2.5.1).

2.5.4. Example (Calculation of $p(6)$)

From **2.5.3**, $p(6) = [q^6](1-q)^{-1}(1-q^2)^{-1}\cdots$. To extract this coefficient we use a number of shortcuts for discarding terms with degree larger than 6.

$$\begin{aligned}
p(6) &= [q^6](1-q)^{-1}(1-q^2)^{-1}\cdots(1-q^6)^{-1}\\
&= [q^6](1-q)^{-1}\cdots(1-q^5)^{-1}(1+q^6)\\
&= 1 + [q^6](1-q)^{-1}\cdots(1-q^4)^{-1}(1+q^5)\\
&= 2 + [q^6](1-q)^{-1}\cdots(1-q^3)^{-1}(1+q^4)\\
&= 4 + [q^6](1-q)^{-1}(1-q^2)^{-1}(1+q^3+q^6)\\
&= 5 + ([q^3]+[q^6])(1-q)^{-1}(1-q^2)^{-1} = 11.
\end{aligned}$$

This is in agreement with Table 2.5.1. □

Decomposition 2.5.2 may be used to count certain subsets of partitions as follows. These sets are used in later decompositions.

2.5.5. Distinct Parts

Let D denote the set of partitions with distinct parts. Then

$$\mathsf{D} \xrightarrow{\sim} (\varepsilon \cup 1)(\varepsilon \cup 2)\ldots,$$

so the number of partitions of n with distinct parts is $[q^n](1+q)(1+q^2)\cdots$.

□

2.5.6. Largest Part Exactly m

Let M_m denote the set of partitions with largest part exactly m. Then

$$\mathsf{M}_m \stackrel{\sim}{\rightarrow} (1^*)(2^*)\cdots(m^*)m,$$

so the number of partitions of n with largest part exactly m is

$$[q^n](1-q)^{-1}(1-q^2)^{-1}\cdots(1-q^m)^{-1}q^m$$
$$= [q^{n-m}](1-q)^{-1}(1-q^2)^{-1}\cdots(1-q^m)^{-1}. \qquad \square$$

The first operation on partitions we consider is called conjugation.

2.5.7. Definition (Conjugate)

Let $\alpha = (\alpha_1, \ldots, \alpha_k)$ be a partition of n. Let β_j be the number of parts of α of size at least j, for $j = 1, \ldots, \alpha_1 = k'$. Then the partition $\tilde{\alpha} = (\beta_1, \ldots, \beta_{k'})$ of n is called the **conjugate** *of α. If $\alpha = \tilde{\alpha}$, then α is* **self-conjugate**. □

The conjugates of (4, 1, 1), (3, 2, 1), and (2, 2, 2) are (3, 1, 1, 1), (3, 2, 1), and (3, 3). Note that (3, 2, 1) is a self-conjugate partition of 6.

The following decomposition is obtained immediately by conjugation, since the number of parts in α is equal to the largest part of $\tilde{\alpha}$.

2.5.8. Decomposition (Partitions with Given Largest Part)

Let P_m be the set of partitions with exactly m parts, and M_m the set of partitions with largest part exactly m. Then

$$\mathsf{M}_m \stackrel{\sim}{\rightarrow} \mathsf{P}_m. \qquad \square$$

For example, there are three three-part partitions of 6, namely, (4, 1, 1), (3, 2, 1), and (2, 2, 2). There are three partitions of 6 with largest part three, namely, (3, 3), (3, 2, 1), and (3, 1, 1, 1).

2.5.9. All Partitions—Euler's Theorem

From Decomposition 2.5.2 the number of partitions of n with exactly m parts is

$$[t^m q^n](1-tq)^{-1}(1-tq^2)^{-1}\ldots,$$

so the number of partitions of n with at most m parts is

$$[t^m q^n](1-t)^{-1}(1-tq)^{-1}(1-tq^2)^{-1}\cdots.$$

On the other hand, by Decomposition 2.5.8 this is also equal to the number of partitions of n with largest part at most m. By Decomposition 2.5.2 this is

$$[q^n](1-q)^{-1}\cdots(1-q^m)^{-1}.$$

But this is equal to $[t^m q^n]\{1+\sum_{k\geq 1} t^k(1-q)^{-1}\cdots(1-q^k)^{-1}\}$. Equating the two expressions we have the following identity due to Euler.

$$\prod_{i=0}^{\infty}(1-tq^i)^{-1} = 1 + \sum_{k\geq 1} t^k \prod_{i=1}^{k}(1-q^i)^{-1}. \qquad \square$$

Further decompositions may be obtained by representing a partition by its Ferrers graph.

2.5.10. Definition (Ferrers Graph, Durfee Square)

Let $\alpha = (\alpha_1, \ldots, \alpha_k)$ be a partition. Then

1. *The* **Ferrers graph** *$F(\alpha)$ of α consists of k rows of equally spaced dots such that the jth row has α_j dots and the leading dots in each row appear in the left-most column.*
2. *The* **Durfee square** *of $F(\alpha)$ is the largest $m \times m$ square in $F(\alpha)$ that contains m^2 dots and includes the leading dot in the first row.* $\square$

The $m \times m$ Durfee square is denoted by D_m. Figure 2.5.1 gives the Ferrers graphs and Durfee squares of the three-part partitions of 6.

The conjugate of a partition α may be obtained as the set of column sums of dots in $F(\alpha)$. Thus the conjugate of (4, 1, 1) is (3, 1, 1, 1).

In decomposing Ferrers graphs we use the following combinatorial operation.

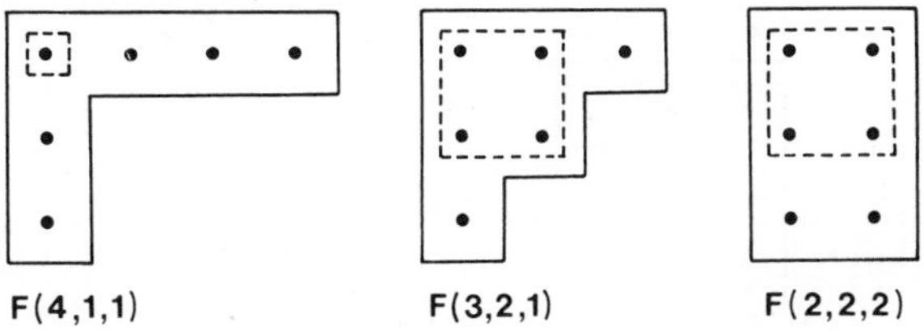

Figure 2.5.1. Ferrers graphs for three-part partitions of 6.

2.5.11. Definition (Abutment)

Let $\alpha = (i_1, \ldots, i_p)$ *and* $\beta = (j_1, \ldots, j_q)$ *be partitions. Then*

1. *The* **row abutment** *of* α *and* β *is the partition* $(i_1 + j_1, \ldots, i_r + j_r)$, *where*

$$r = \max(p, q), \quad i_k = 0 \text{ for } k > p \quad \text{and} \quad j_k = 0 \text{ for } k > q.$$

2. *The* **column abutment** *of* α *and* β *is the partition whose conjugate is the row abutment of* $\tilde{\alpha}$ *and* $\tilde{\beta}$. □

Thus (4, 1, 1) is formed by row abutting (1) and (3), and column abutting the result with (1, 1). These terms are, of course, motivated by operations on the rows and columns of Ferrers graphs, and are illustrated in Figure 2.5.1.

2.5.12. Decomposition (All Partitions)

Let Q_m *denote the set of all partitions with largest part at most* m. *Let* R_m *denote the set of partitions with at most* m *parts. If* Π *is the set of all partitions, then*

$$\Pi \stackrel{\sim}{\rightarrow} \dot{\bigcup_{m \geqslant 0}} \{D_m\} \times \mathsf{Q}_m \times \mathsf{R}_m.$$

Proof: Row-abut an element of R_m with D_m to form a Ferrers graph. Column-abut this with an element of Q_m, to form the Ferrers graph of a partition in Π. This is reversible because of the uniqueness of the Durfee square. □

The decomposition is illustrated in Figure 2.5.2, in which the partition

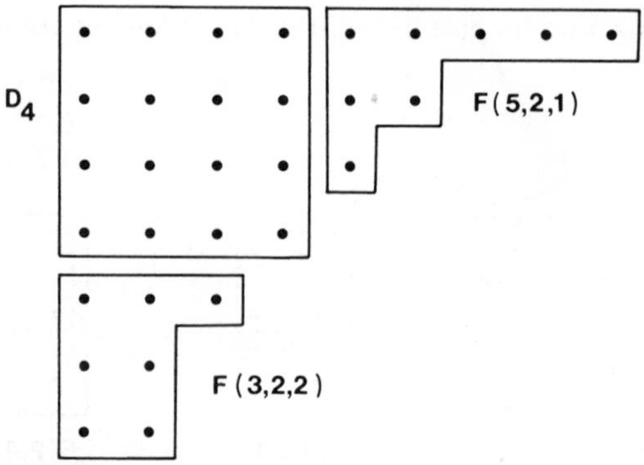

Figure 2.5.2. Decomposition of (9, 6, 5, 4, 3, 2, 2).

(9, 6, 5, 4, 3, 2, 2) is decomposed into

$$(D_4, (3,2,2), (5,2,1)) \in \{D_4\} \times \mathsf{Q}_4 \times \mathsf{R}_4.$$

The next result is almost immediate.

2.5.13. All Partitions—q-Analogue of Kummer's Theorem

We obtain an identity by enumerating the set Π of all partitions with respect to their number of parts. Let z mark the occurrence of a part. Then the generating function for Π is $\prod_{k\geq 1}(1 - zq^k)^{-1}$, by Decomposition 2.5.2.

Another expression for this is obtained by means of Decomposition 2.5.12. The generating function for D_m is $q^{(m^2)}z^m$ since it is an m-part partition of m^2. The generating function for R_m is $\prod_{i=1}^m(1 - q^i)^{-1}$, by Decomposition 2.5.8. This generating function does not contain z since the element of R_m does not contribute additional parts in Decomposition 2.5.12. The generating function for Q_m is $\prod_{i=1}^m(1 - zq^i)^{-1}$, by Decomposition 2.5.2. Thus, from Decomposition 2.5.12

$$\prod_{i\geq 1}(1 - zq^i)^{-1} = 1 + \sum_{m\geq 1} z^m q^{(m^2)} \prod_{i=1}^{m} (1 - q^i)^{-1}(1 - zq^i)^{-1}.$$

Equivalently, replacing z by zq^{-1},

$$\prod_{i\geq 0}(1 - zq^i)^{-1} = 1 + \sum_{m\geq 1} z^m q^{m(m-1)} \prod_{i=1}^{m} (1 - q^i)^{-1}(1 - zq^{i-1})^{-1}.$$

This is a q-analogue of Kummer's theorem. □

We obtain another classical result by specialization.

2.5.14. All Partitions—Euler's Theorem

In **2.5.13**, let $z = q$. Thus

$$\prod_{i\geq 1}(1 - q^i)^{-1} = 1 + \sum_{m\geq 1} q^{(m^2)}(1 - q)^{-2} \cdots (1 - q^m)^{-2}. \qquad \square$$

This result gives another way of calculating $p(n)$, and is illustrated for $p(6)$, previously considered in **2.5.4.**,

$$\begin{aligned} p(6) &= [q^6]q(1 - q)^{-2} + [q^6]q^4(1 - q)^{-2}(1 - q^2)^{-2} \\ &= [q^5](1 - q)^{-2} + [q^2](1 - q)^{-2}(1 + 2q^2) \\ &= 6 + [q^2](1 - q)^{-2} + 2 \end{aligned}$$

whence $p(6) = 11$.

So far we have considered the set of all partitions as a source of identities. We now consider partitions with distinct parts.

2.5.15. Definition (Maximal Triangle)

Let F be a Ferrers graph of a partition. Let T_k be the triangle formed with the first $k - i + 1$ dots of row i of F, for $i = 1, \ldots, k$, for some k. The **maximal triangle** *of F is the triangle T_k for which k is maximum.* □

This leads to the following decomposition for the set of partitions with distinct parts.

2.5.16. Decomposition (Partitions with Distinct Parts)

Let D denote the set of partitions with distinct parts, and let R_m denote the set of partitions with at most m parts. Then

$$\mathsf{D} \xrightarrow{\sim} \dot{\bigcup_{m \geqslant 0}} \{T_m\} \times \mathsf{R}_m.$$

Proof: Row-abut an element of R_m to T_m. The result is a Ferrers graph on m rows, no two of which have an equal number of dots. This Ferrers graph therefore represents a partition in D. The construction is reversible because of the uniqueness of the maximal triangle. □

The decomposition is illustrated in Figure 2.5.3, where $(5, 3, 2)$ is decomposed into $(T_3, (2, 1, 1))$, and can be used to obtain an identity.

2.5.17. Partitions with Distinct Parts—An Identity

An identity is obtained by enumerating partitions with distinct parts with respect to the number of parts. Let z mark the occurrence of a part. From **2.5.5** this generating function is $\prod_{k \geqslant 1}(1 + zq^k)$. On the other hand, we may use Decomposition 2.5.16. The generating function for $\{T_m\}$ is $q^{m(m+1)/2}z^m$ since

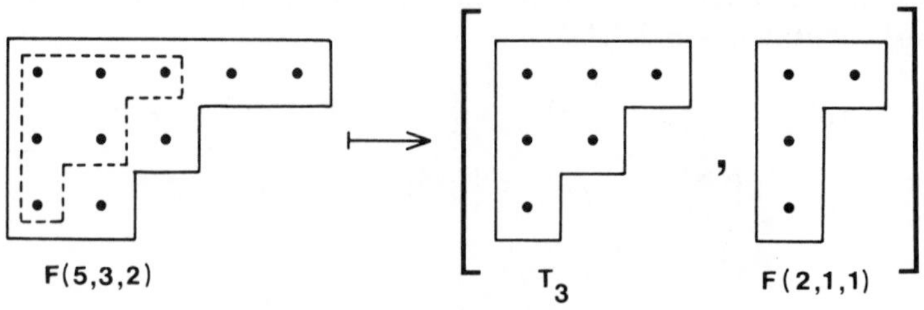

Figure 2.5.3. Decomposition of $(5, 3, 2)$.

T_m contains exactly $\frac{1}{2}m(m+1)$ dots and exactly m rows. The generating function for R_m is $\prod_{i=1}^{m}(1-q^i)^{-1}$ and is independent of z since elements of R_m do not contribute additional rows, under the decomposition. Thus by Decomposition 2.5.16, and the sum and product lemmas

$$\prod_{i\geqslant 1}(1+zq^i) = 1 + \sum_{m\geqslant 1} z^m q^{\binom{m+1}{2}} \prod_{i=1}^{m}(1-q^i)^{-1}.$$

Replace z by zq^{-1} to give the following identity

$$\prod_{i\geqslant 0}(1+zq^i) = 1 + \sum_{m\geqslant 1} z^m q^{\binom{m}{2}} \prod_{i=1}^{m}(1-q^i)^{-1}. \qquad \square$$

Again, by specializing we may obtain another result due to Euler.

2.5.18. Partitions with Distinct Parts—Euler's Theorem

In **2.5.17** set $z = q$ in the last equation to obtain

$$\prod_{i\geqslant 1}(1+q^i) = 1 + \sum_{m\geqslant 1} q^{\binom{m+1}{2}} \prod_{i=1}^{m}(1-q^i)^{-1}. \qquad \square$$

Another identity may be obtained from self-conjugate partitions.

2.5.19. Decomposition (Self-conjugate Partitions)

Let C *denote the set of self-conjugate partitions. Let* $\mathsf{R}_m^{(2)}$ *denote the set of partitions with at most m parts, each of which is even. Then*

$$\mathsf{C} \xrightarrow{\sim} \dot{\bigcup_{m\geqslant 0}} \{D_m\} \times \mathsf{R}_m^{(2)}.$$

Proof: Let $\alpha \in \mathsf{R}_m$, the set of partitions with at most m parts. Then the column abutment of $\tilde{\alpha}$ with the row abutment of D_m and α represents a self-conjugate partition. The construction is reversible. Now the row abutment of α and α is in $\mathsf{R}_m^{(2)}$, and any element in $\mathsf{R}_m^{(2)}$ may be obtained from a unique α in this way. $\square$

For example, $(7,6,4,3,2,2,1)$ is a self-conjugate partition, whose Durfee square is D_3. It is formed by row-abutting $(4,3,1) \in \mathsf{R}_3$ and column-abutting the conjugate of $(4,3,1)$ to D_3. The row abutment of $(4,3,1)$ with itself is $(8,6,2) \in \mathsf{R}_3^{(2)}$, so $(7,6,4,3,2,2,1)$ corresponds to $(D_3,(8,6,2))$ under this decomposition. This is illustrated in Figure 2.5.4.

The following is another decomposition for self-conjugate partitions.

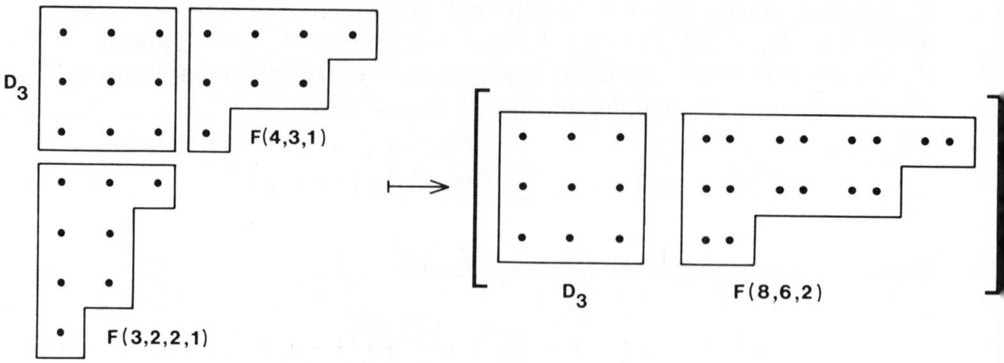

Figure 2.5.4. Decomposition of (7, 6, 4, 3, 2, 2, 1).

2.5.20. Decomposition (Self-conjugate Partitions)

Let $\mathsf{D}^{(1)}$ *denote the set of partitions with odd distinct parts and* C *denote the set of self-conjugate partitions. Then*

$$\mathsf{C} \xrightarrow{\sim} \mathsf{D}^{(1)}.$$

Proof: Let $\alpha \in \mathsf{C}$ have the Durfee square D_m. From α we construct a partition $\beta = (\beta_1, \ldots, \beta_m)$ as follows. Let β_1 be the sum of the dots along the first row and down the first column of $F(\alpha)$. Delete these dots to obtain a partition $\alpha^{(1)}$, with Durfee square D_{m-1}. This partition is again self-conjugate. Repeat this process to obtain $\beta_2, \ldots, \beta_m$ and $\alpha^{(2)}, \ldots, \alpha^{(m-1)}$. Clearly, $(\beta_1, \ldots, \beta_m)$ is a partition. Moreover, $\beta_1, \ldots, \beta_m$ are distinct odd numbers since $\alpha, \alpha^{(1)}, \ldots, \alpha^{(m-1)}$ are self-conjugate. Thus $\beta \in \mathsf{D}^{(1)}$. This construction is reversible, and the result follows. □

The construction is illustrated in Figure 2.5.5, in which $(7, 6, 4, 3, 2, 2, 1) \in \mathsf{C}$ corresponds to $(13, 9, 3) \in \mathsf{D}^{(1)}$.

We now use these two decompositions to obtain another identity.

2.5.21. Self-conjugate Partitions—An Identity

Let G be the generating function for self-conjugate partitions with respect to the size of the Durfee squares. The size is marked by z. Then from Decomposition 2.5.19

$$G = 1 + \sum_{m \geqslant 1} z^m q^{(m^2)} \prod_{i=1}^{m} \left(1 - q^{2i}\right)^{-1},$$

since $z^m q^{(m^2)}$ is the generating function for D_m, and $\Pi_{i=1}^m (1-q^{2i})^{-1}$ is the generating function for $\mathsf{R}_m^{(2)}$.

On the other hand, from Decomposition 2.5.20 and **2.5.5**

$$G = \prod_{i \geqslant 0} (1 + zq^{2i+1}),$$

since each part is odd, and the parts are distinct. Equating the two expressions for G gives the identity

$$\prod_{i\geqslant 0}\left(1 + zq^{2i+1}\right) = 1 + \sum_{m\geqslant 1} z^m q^{(m^2)} \prod_{i=1}^{m} \left(1 - q^{2i}\right)^{-1}. \qquad \square$$

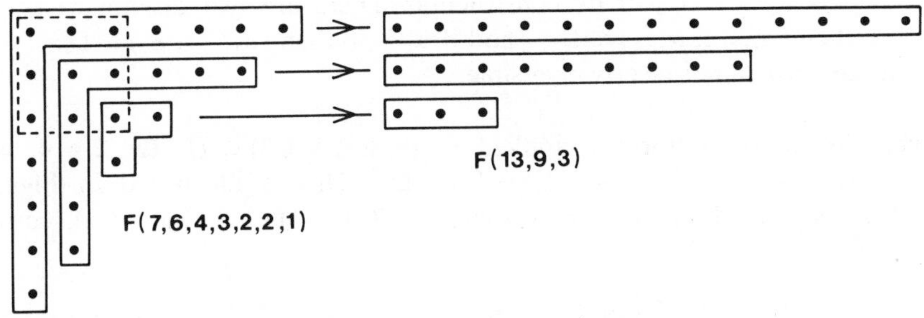

Figure 2.5.5. Decomposition of (7, 6, 4, 3, 2, 2, 1).

Note that **2.5.21** may be derived algebraically by replacing q by q^2 and then z by zq in **2.5.17**. By specialization we may obtain another identity due to Euler.

2.5.22. Self-conjugate Partitions—Euler's Theorem

Setting $z = 1$ in **2.5.21** we have

$$\prod_{i\geqslant 0}\left(1 + q^{2i+1}\right) = 1 + \sum_{m\geqslant 1} q^{(m^2)} \prod_{i=1}^{m} \left(1 - q^{2i}\right)^{-1}. \qquad \square$$

The final decomposition is more complex than the previous ones and is used to obtain two classical identities.

2.5.23. Decomposition (Sylvester: All Partitions)

Let D_i denote the set of partitions with exactly i parts, each distinct. Then, for any $k \geqslant 0$

$$\Pi \times \{T_k\} \overset{\rightarrow}{\rightarrow} \dot{\bigcup_{j\geqslant 0}} \mathsf{D}_{k+j} \times \left(\mathsf{D}_j \,\dot{\cup}\, \mathsf{D}_{j-1}\right)$$

with the convention that $\mathsf{D}_0 \,\dot{\cup}\, \mathsf{D}_{-1} = \mathsf{D}_0$.

Proof: Consider an arbitrary element $(\gamma, T_k) \in \Pi \times \{T_k\}$, where $\gamma = (\gamma_1, \ldots, \gamma_r)$ for some $r \geqslant 0$. Adjoin T_k to the top of $F(\gamma)$ to form a configuration θ. This is illustrated in Figure 2.5.6(i), in which the Ferrers graphs are indicated by their outline alone.

Now decompose θ uniquely into an ordered pair, (β, α), of partitions by the following construction, which is indicated in Figure 2.5.6(ii). Continue the hypotenuse h of T_k through $F(\gamma)$, separating θ into two subconfigurations of dots, namely, those below h and those above h.

The dots below h lie in $k + j$ columns. This uniquely determines j, where $j \geqslant 0$, and a partition $\beta = (\beta_1, \ldots, \beta_{k+j})$ in which there are β_i dots in column i of this subconfiguration. Moreover, $\beta_1 > \cdots > \beta_{k+j} > 0$ so $\beta \in \mathsf{D}_{k+j}$.

The dots above h lie in j rows, with α_i dots in row i for $i = 1, \ldots, j$. But $\alpha_1 > \cdots > \alpha_{j-1} > \alpha_j \geqslant 0$ by construction. There are two disjoint cases. If $\alpha_j = 0$, then $(\alpha_1, \ldots, \alpha_{j-1}) \in \mathsf{D}_{j-1}$ and if $\alpha_j \neq 0$, then $(\alpha_1, \ldots, \alpha_j) \in \mathsf{D}_j$.

The construction is clearly reversible. □

When this decomposition is applied to $(9, 8, 8, 4, 3, 3) \in \Pi$, for $k = 4$, we obtain $((10, 9, 8, 5, 3, 2, 1), (4, 2, 1)) \in \mathsf{D}_7 \times \mathsf{D}_3$. This is illustrated in Figure 2.5.7. As a second example, note that $(9, 8, 7, 4, 3, 3) \in \Pi$, for $k = 4$, corre-

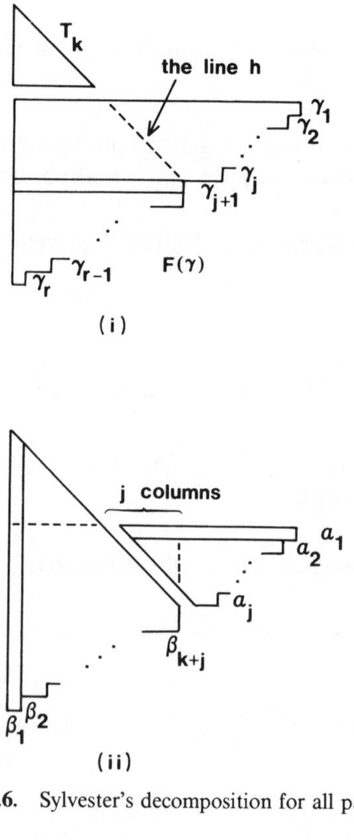

Figure 2.5.6. Sylvester's decomposition for all partitions.

sponds to $((10, 9, 8, 5, 3, 2, 1), (4, 2)) \in \mathsf{D}_7 \times \mathsf{D}_2$. This is obtained by deleting the encircled dot in Figure 2.5.7. These two examples illustrate the two cases $\mathsf{D}_{k+j} \times \mathsf{D}_j$ and $\mathsf{D}_{k+j} \times \mathsf{D}_{j-1}$, for $k = 4$ and $j = 3$, in the codomain of the decomposition.

We now use this decomposition to obtain an identity due to Jacobi.

2.5.24. The Jacobi Triple Product Identity

Let q mark the size of a part. The generating function of $\Pi \times \{T_k\}$ is therefore

$$q^{\binom{k+1}{2}} \prod_{m \geqslant 1} (1 - q^m)^{-1}.$$

On the other hand, the generating function for $\mathsf{D}_{k+j} \times (\mathsf{D}_j \cup \mathsf{D}_{j-1})$ is

$$[w^{k+j}] \prod_{m \geqslant 1} (1 + wq^m) \Big\{ [u^j] \prod_{r \geqslant 1} (1 + uq^r) + [u^{j-1}] \prod_{r \geqslant 1} (1 + uq^r) \Big\},$$

since D_i has distinct parts. But this is equal to

$$[w^{k+j} u^j](1 + u) \prod_{m \geqslant 1} (1 + wq^m)(1 + uq^m)$$

$$= [w^{k+j} u^j] \prod_{m \geqslant 1} (1 + wq^m)(1 + uq^{m-1}).$$

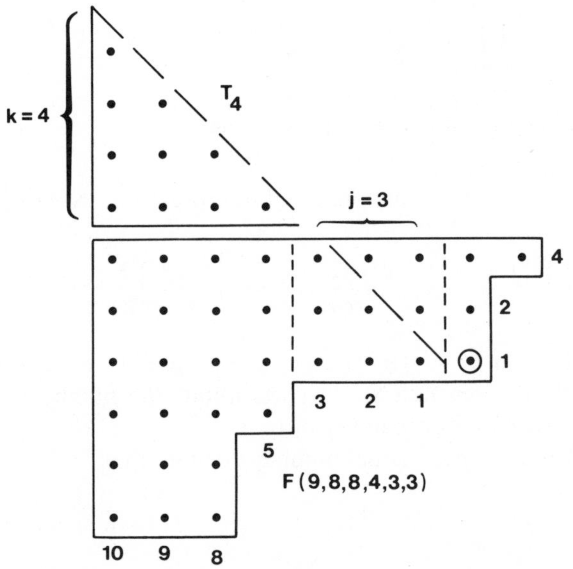

Figure 2.5.7. Decomposition of $(9, 8, 8, 4, 3, 3)$ for $k = 4$.

From Decomposition 2.5.23

$$q^{\binom{k+1}{2}}\prod_{m\geq 1}(1-q^m)^{-1}=\sum_{j\geq 0}[w^{k+j}u^j]\prod_{m\geq 1}(1+wq^m)(1+uq^{m-1})$$

so $[w^k]\Pi_{m\geq 1}(1+wq^m)(1+w^{-1}q^{m-1})=q^{\binom{k+1}{2}}\Pi_{m\geq 1}(1-q^m)^{-1}$. Summing over k, where the terms for negative k are given by symmetry, gives

$$\prod_{m\geq 1}(1+wq^m)(1+w^{-1}q^{m-1})(1-q^m)=\sum_{k=-\infty}^{\infty}w^k q^{\binom{k+1}{2}}.$$

Replacing q by q^2 and then setting $wq=y$ we obtain

$$\prod_{m\geq 1}(1+q^{2m-1}y)(1+q^{2m-1}y^{-1})(1-q^{2m})=\sum_{k=-\infty}^{\infty}y^k q^{(k^2)}.$$

This result is known as the Jacobi triple product identity. □

The final identity, due to Euler, is obtained by specializing this result.

2.5.25. Euler's Theorem for Pentagonal Numbers

In the Jacobi triple product identity, replace q by $q^{3/2}$ and y by $-q^{-1/2}$. This is admissible since the power series are bounded in y. Then

$$\prod_{m\geq 1}(1-q^m)=\sum_{m=-\infty}^{\infty}(-1)^m q^{m(3m-1)/2}.$$

This is Euler's theorem for pentagonal numbers. □

The result may be given a combinatorial interpretation. Note that

$$\prod_{m\geq 1}(1-q^m)=\left\{\sum_{p\text{ even}}-\sum_{p\text{ odd}}\right\}[z^p]\prod_{m\geq 1}(1+zq^m).$$

Let $[q^n]\Pi_{m\geq 1}(1-q^m)=\nu(n)$ so $\nu(n)$ is the number of distinct part partitions of n with an even number of parts minus the number of distinct part partitions of n with an odd number of parts.

It follows from the pentagonal number theorem that

$$\nu(n)=\begin{cases}(-1)^m & \text{if } n=\tfrac{1}{2}m(3m-1)\text{ for some integer } m\\ 0 & \text{otherwise.}\end{cases}$$

NOTES AND REFERENCES

For a complete treatment of partition theory see Andrews (1976), in which complete references for the textual material can be found; earlier treatments of interest are Hardy and Wright (1938) and MacMahon (1915).

[**2.5.1**] Euler (1748), Glaisher (1883); [**2.5.2**] Andrews (1976); [**2.5.3**] Euler (1748); [**2.5.4**] Andrews (1967), MacMahon (1915); [**2.5.5**] Subbarao (1971); [**2.5.6**] Andrews (1967), MacMahon (1915); [**2.5.7**] Ramanujan (1927); [**2.5.8**] Stanley (private communication); [**2.5.12**] MacMahon (1886); [**2.5.13**] Franklin (1881); [**2.5.14**] Ramanujan (1927); [**2.5.15**] Euler (1748); [**2.5.16–17**] Cauchy (1893); [**2.5.19**] Bressoud (1980), Rogers and Ramanujan (1919).

EXERCISES

2.5.1. (a) Prove that the number of partitions of n with distinct parts equals the number of partitions of n with only odd parts.

(b) Find an explicit bijection between the sets given in (a).

(c) Show that the number of partitions of n into parts not divisible by m is equal to the number of partitions of n with no part repeated m times or more.

2.5.2. Prove that the number of partitions of n in which only the odd parts can be repeated is the same as the number of partitions of n with no part repeated more than three times.

2.5.3. Show that the absolute value of the difference between the number of partitions of n with an even number of parts and the number of partitions of n with an odd number of parts is equal to the number of partitions of n into distinct odd parts.

2.5.4. Show that the number of partitions of n in which consecutive integers do not both appear as parts is equal to the number of partitions of n in which no part appears exactly once.

2.5.5. (a) Show that the number of partitions of n in which each part appears 2, 3, or 5 times equals the number of partitions of n into parts congruent to 2, 3, 6, 9, or 10 mod 12.

(b) Let a, b be positive integers with different numbers of 2's in their prime factorizations. Show that the number of partitions of n in which each part appears a, b or $a + b$ times equals the number of partitions of n into parts congruent to $a \bmod 2a$ or $b \bmod 2b$.

2.5.6. Show that the number of partitions of n in which no part appears exactly once is equal to the number of partitions of n with no parts congruent to 1 or 5 mod 6.

2.5.7. Show that the number of partitions of n with unique smallest part and largest part at most twice the smallest part is equal to the number of partitions of n in which the largest part is odd and the smallest part is larger than one-half of the largest part.

2.5.8. Show that the sum of the number of 1's over all partitions of n is equal to the sum of the number of different parts over all partitions of n.

2.5.9. By enumerating partitions with respect to number of parts and size of largest part, show directly that

$$\sum_{m\geqslant 1} \frac{q^m z w^m}{(1-zq)\cdots(1-zq^m)}$$

$$= \sum_{m\geqslant 1} \frac{q^{(m^2)} z^m w^m}{(1-zq)\cdots(1-zq^m)(1-wq)\cdots(1-wq^m)}.$$

2.5.10. By enumerating partitions with distinct parts with respect to the number of parts and the size of the largest parts, show directly that

$$\sum_{m\geqslant 1} q^m z w^m (1+zq)\cdots(1+zq^{m-1}) = \sum_{m\geqslant 1} \frac{q^{\binom{m+1}{2}} z^m w^m}{(1-wq)\cdots(1-wq^m)}.$$

2.5.11. Prove that

$$\left(\sum_{k\geqslant 0} \frac{x^{2k+1}}{(1-q)(1-q^2)\cdots(1-q^{2k+1})}\right)\left(\sum_{k\geqslant 0} \frac{x^{2k}}{(1-q)\cdots(1-q^{2k})}\right)^{-1}$$

$$= \left(\sum_{k\geqslant 0} \frac{x^{2k+1}q^{\binom{2k+1}{2}}}{(1-q)(1-q^2)\cdots(1-q^{2k+1})}\right)\left(\sum_{k\geqslant 0} \frac{x^{2k}q^{\binom{2k}{2}}}{(1-q)\cdots(1-q^{2k})}\right)^{-1}.$$

2.5.12. A partition of n is called **perfect** if all numbers from 1 to $n-1$ can be written uniquely as the sum of a subset of elements of the partition. Show that the number of perfect partitions of n is equal to the number of ordered factorizations of $n+1$ by finding a bijection between these sets.

2.5.13. If $\alpha = (\alpha_1, \ldots, \alpha_r)$ is a partition with distinct parts, so that $\alpha_1 > \cdots > \alpha_r > 0$, let $s(\alpha) = \alpha_r$, and $d(\alpha)$ be the largest k for which $\alpha_k = \alpha_1 - k + 1$. Give a decomposition for partitions α with distinct parts, which involves $s(\alpha)$ and $d(\alpha)$ and yields Euler's pentagonal number theorem **(2.5.25)** immediately.

2.5.14. (a) From the Jacobi triple product identity **(2.5.24)**, deduce that

$$\prod_{i\geqslant 1}(1-q^i)^3 = \sum_{k\geqslant 0}(-1)^k(2k+1)q^{\binom{k+1}{2}}.$$

(b) Combine Euler's pentagonal number theorem with part (a) to obtain an expansion for $F(q) = q\prod_{i\geqslant 1}(1-q^i)^4$. Hence, by noting that

$$q\prod_{i\geqslant 1}(1-q^i)^{-1} = F(q)\prod_{i\geqslant 1}(1-q^i)^{-5},$$

deduce that $p(5n+4) \equiv 0 \pmod 5$.

(c) Prove that $p(7n+5) \equiv 0 \pmod 7$ by considering

$$G(q) = q^2\prod_{i\geqslant 1}(1-q^i)^6 = q^2\left\{\sum_{i\geqslant 0}(-1)^i(2i+1)q^{\binom{i+1}{2}}\right\}^2,$$

and proceeding as in part (b).

2.5.15. Let $F(t) = \prod_{i\geqslant 0}(1-tq^i)^{-1} = \sum_{i\geqslant 0} c(i)t^i$.

(a) Show that $F(tq) = (1-t)F(t)$.

(b) Show that $c(n) = \prod_{i=1}^{n}(1-q^i)^{-1}$, as in **2.5.9**, by solving the recurrence equation for $c(n)$ that follows from (a). This method for obtaining the coefficients in the power series expansion of a product is known as **Euler's device**.

2.5.16. Show that

$$\prod_{i\geqslant 0}(1-atq^i)(1-tq^i)^{-1} = 1 + \sum_{n\geqslant 1} t^n\prod_{i=1}^{n}(1-aq^{i-1})(1-q^i)^{-1}$$

by Euler's device.

2.5.17. Show that

$$\prod_{i=0}^{n}(1-tq^i)^{-1} = 1 + \sum_{k\geqslant 1} t^k\prod_{i=1}^{k}(1-q^{n+i})(1-q^i)^{-1}$$

by Euler's device. Hence the generating function for partitions with at most k parts and largest part at most n is $\prod_{i=1}^{k}(1-q^{n+i})(1-q^i)^{-1}$.

2.5.18. Deduce Jacobi's triple product identity by Euler's device.

2.5.19. (a) Deduce from the Jacobi triple product identity that

(i) $$\prod_{m\geqslant 1}(1-q^{5m-3})(1-q^{5m-2})(1-q^{5m}) = \sum_{k=-\infty}^{\infty}(-1)^k q^{k(5k+1)/2}.$$

(ii) $$\prod_{m\geqslant 1}(1-q^{5m-4})(1-q^{5m-1})(1-q^{5m}) = \sum_{k=-\infty}^{\infty}(-1)^k q^{k(5k+3)/2}.$$

(b) Show that

$$\prod_{i\geqslant 1}(1-xq^i)^{-1}\left\{1+\sum_{n\geqslant 1}(-1)^n x^{2n}q^{\frac{1}{2}n(5n-1)}(1-xq^{2n})\right.$$

$$\left.\times\frac{(1-xq)\cdots(1-xq^{n-1})}{(1-q)\cdots(1-q^n)}\right\}=1+\sum_{k\geqslant 1}\frac{x^k q^{(k^2)}}{(1-q)\cdots(1-q^k)}.$$

(c) Deduce from (a) and (b) that

$$\text{(i)}\quad \prod_{m\geqslant 0}(1-q^{5m+1})^{-1}(1-q^{5m+4})^{-1}=1+\sum_{k\geqslant 1}q^{(k^2)}\prod_{i=1}^{k}(1-q^i)^{-1},$$

$$\text{(ii)}\quad \prod_{m\geqslant 0}(1-q^{5m+2})^{-1}(1-q^{5m+3})^{-1}=1+\sum_{k\geqslant 1}q^{k^2+k}\prod_{i=1}^{k}(1-q^i)^{-1}.$$

These are the **Rogers–Ramanujan identities.**

2.6. INVERSIONS IN PERMUTATIONS AND q-IDENTITIES

Section 2.5 dealt with the development of partition identities by means of Ferrers graphs, Durfee squares, and maximal triangles. The identities involved the expansion of infinite products. We now consider the combinatorial development of comparable identities for finite products, but base the development on permutations enumerated with respect to inversions, rather than on partitions of integers.

2.6.1. Definition (Inversion)

Let $\sigma=\sigma_1\cdots\sigma_k$ be a sequence over N_n. Then an **inversion** *in σ is a pair (σ_i,σ_j) with $i<j$ and $\sigma_i>\sigma_j$. The number of inversions in σ is denoted by $I(\sigma)$.* □

If $\sigma=45231$, then $(5,3)$ is an inversion since $\sigma_2=5>3=\sigma_4$. The other inversions are $(5,2)$, $(5,1)$, $(4,2)$, $(4,3)$, $(4,1)$, $(3,1)$, and $(2,1)$. So $I(\sigma)=8$.

The number of inversions in a permutation is determined by the following algorithm.

2.6.2. Algorithm (Inversion)

Let $\sigma=\sigma_1\cdots\sigma_n$ be a permutation on N_n, and let j be defined by $\sigma_j=n$. Then

$$I(\sigma)=n-j+I(\hat{\sigma})$$

where $\hat{\sigma}$ is the permutation on N_{n-1} obtained from σ by deleting n.

Proof: Let $\hat{k}_n$ be the number of inversions in σ not involving n, and let k_n be the number of inversions in σ which do involve n. Then $I(\sigma) = k_n + \hat{k}_n$. But $\hat{k}_n = I(\hat{\sigma})$. Moreover, (n, σ_l) is an inversion when $j < l \leqslant n$. There are exactly $n - j$ of these, so $k_n = n - j$ and the result follows. □

Thus $I(45231) = 3 + I(4231) = 6 + I(231) = 7 + I(21) = 8 + I(1) = 8$. The eight inversions are listed above. The following result is immediate.

2.6.3. Proposition

Let $\sigma_1 \cdots \sigma_k$ be a permutation of the elements of $\alpha \subset \mathsf{N}_+$ where $|\alpha| = k$. Let $\phi : \alpha \to \mathsf{N}_+$ be order preserving. Then

$$I(\phi(\sigma_1) \cdots \phi(\sigma_k)) = I(\sigma_1 \cdots \sigma_k).$$

Proof: Let $i < j$. Then (σ_i, σ_j) is an inversion if and only if $(\phi(\sigma_i), \phi(\sigma_j))$ is an inversion. □

We may use the inversion algorithm to enumerate permutations with respect to inversions.

2.6.4. Lemma (Inversions)

The number of permutations in N_n with exactly r inversions is $[q^r]n!_q$, where $n!_q = \prod_{i=1}^{n}(1 - q^i)/(1 - q)$.

Proof: Let $f_n(q)$ be the generating function for the set of permutations on N_n with respect to inversions. Each permutation on N_{n+1} may be obtained uniquely by inserting $n + 1$ into one of the $n - 1$ positions between elements of σ, or at the beginning or end of σ, for some unique permutation σ on N_n. The set of contributions from the insertion of $n + 1$ into σ to the number of inversions is $\{0, 1, \ldots, n\}$, by the inversion algorithm. Let q mark an inversion. Then

$$f_{n+1}(q) = (1 + q + q^2 + \cdots + q^n) f_n(q) = \frac{1 - q^{n+1}}{1 - q} f_n(q).$$

This is a recurrence equation for $f_n(q)$ with initial condition $f_1(q) = 1$. The result follows. □

Consider, for example, the generating function $3!_q$ for permutations on N_3 with respect to inversions. Now $3!_q = (1 - q)(1 - q^2)(1 - q^3)(1 - q)^{-3} = 1 + 2q + 2q^2 + q^3$. Thus there is one permutation (123) with no inversions, and there are two permutations (132, 213) with one, two permutations (231, 312) with two, and one permutation (321) with three.

It should be noted that $n!_q = n!$ when $q = 1$, since $n!_q = \prod_{i=0}^{n-1} \sum_{j=0}^{i} q^j$ from Lemma 2.6.4.

Consider a permutation σ on N_n and a vector $\mathbf{i} = (i_1, \ldots, i_k)$ where $i_1 + \cdots + i_k = n$ and $i_1, \ldots, i_k \geqslant 0$. Let π_1 be the subset of N_n whose elements are the first i_1 elements of σ, π_2 be the subset formed by the next i_2, and so on, until π_k is the subset formed by the last i_k elements. Then $\Pi = (\pi_1, \ldots, \pi_k)$ is called the **ordered partition** of σ of type $\mathbf{i}$. For example, in 45231 the ordered partition of type (3, 2) is ({2, 4, 5}, {1, 3}).

We may now divide the inversions of σ into two sets induced by the ordered partition Π of type $\mathbf{i}$. If an inversion of σ involves elements in two distinct blocks of Π, then it is a **between-set inversions of type i**. If an inversion involves elements in only one set, it is called a **within-set inversion of type i**.

The following example illustrates these two sets of inversions.

2.6.5. Example (Between-Set and Within-Set Inversions)

Consider the permutation $\sigma = 45231$ which has ({2, 4, 5}, {1, 3}) as its ordered partition of type (3, 2). The three within-set inversions of type (3, 2) in σ are (4, 2), (5, 2), and (3, 1). The five between-set inversions of type (3, 2) in σ are (2, 1), (4, 1), (4, 3), (5, 1) and (5, 3). The sum of the numbers of between-set and within-set inversions is 8, the number of inversions in 45231. □

The following lemma enables us to enumerate certain configurations with respect to within-set and between-set inversions.

2.6.6. Lemma (Between-Set and Within-Set Generating Functions)

1. *Let Π be a fixed ordered partition of N_n of type $\mathbf{i} = (i_1, \ldots, i_k)$. There are*

$$[q^r](i_1!_q \cdots i_k!_q)$$

permutations with partition Π having r within-set inversions of type $\mathbf{i}$.

2. *There are*

$$[q^r]\begin{bmatrix} n \\ \mathbf{i} \end{bmatrix}_q, \qquad \text{where } \begin{bmatrix} n \\ \mathbf{i} \end{bmatrix}_q = n!_q/(i_1!_q \cdots i_k!_q),$$

ordered partitions of N_n of type $\mathbf{i}$ having r between-set inversions.

Proof:

1. Let P_α be the set of permutations on $\alpha \subseteq \mathsf{N}_n$. Let $\phi : \mathsf{N}_{|\alpha|} \to \alpha : j \mapsto i$, where i is the jth smallest element in α. Thus ϕ is an order-preserving bijection. Thus the generating function for P_α with respect to inversions is $\sum_{\sigma' \in \mathsf{P}_\alpha} q^{I(\sigma')} = \sum_{\sigma \in \mathsf{P}_{|\alpha|}} q^{I(\phi(\sigma))}$, where P_i is the set of all permutations on N_i. But, by Proposition 2.6.3, this is equal to

$$\sum_{\sigma \in \mathsf{P}_{|\alpha|}} q^{I(\sigma)} = |\alpha|!_q \qquad \text{by Lemma 2.6.4.}$$

Let S_Π be the set of all permutations with ordered partition $\Pi = (\pi_1, \ldots, \pi_k)$ of type **i**. Then $\mathsf{S}_\Pi \tilde{\to} \mathsf{P}_{\pi_1} \times \cdots \times \mathsf{P}_{\pi_k} : \rho_1 \cdots \rho_k \mapsto (\rho_1, \ldots, \rho_k)$. We now apply the product lemma to this decomposition to obtain the generating function for S_Π with respect to within-set inversions. This is $|\pi_1|!_q \cdots |\pi_k|!_q$ by the preceding argument, and the result follows.

2. Let $B(q)$ be the generating function for the number of ordered partitions of N_n of type **i**, with respect to between-set inversions. We enumerate P_n with respect to inversions in two different ways. First, from Lemma 2.6.4, the generating function for P_n is $n!_q$. Second, P_n may be obtained by arranging the elements of each block of an ordered partition of type **i** in all possible ways, for each partition of type **i**. Thus by the product lemma the generating function for P_n is $B(q)(i_1!_q \cdots i_k!_q)$ by **(1)**, since the number of inversions is the sum of the numbers of between-set and within-set inversions. Thus

$$n!_q = B(q)(i_1!_q \cdots i_k!_q),$$

and the result follows. □

The generating function $\begin{bmatrix} n \\ i, n-i \end{bmatrix}_q$, which we denote by $\binom{n}{i}_q$, is called a **Gaussian coefficient**, or a **q-binomial coefficient**. Further instances of its use are given in Chapter 4.

As an example of the use of Lemma 2.6.6(2) consider the enumeration of ordered partitions of N_4 of type (2, 2) with respect to between-set inversions. Table 2.6.1 gives a list of all such ordered partitions in the left-hand column. It should be recollected that the blocks are ordered among themselves, so $(\{1,2\},\{3,4\})$ and $(\{3,4\},\{1,2\})$ are distinct. The direct calculation shown in Table 2.6.1 indicates that the between-set generating function is $1 + q + 2q^2 + q^3 + q^4$. Thus there is one ordered partition of type (2, 2) with no between-set inversions, one with one, two with two, one with three, and one with four.

On the other hand, by Lemma 2.6.6(2) the generating function is

$$\binom{4}{2}_q = \frac{4!_q}{2!_q 2!_q} = \frac{(1-q)(1-q^2)(1-q^3)(1-q^4)}{(1-q)^2(1-q^2)^2}$$

$$= (1+q+q^2)(1+q^2) = 1 + q + 2q^2 + q^3 + q^4.$$

This agrees with Table 2.6.1.

The q-binomial coefficient has many properties analogous to those of the binomial coefficient. The following gives one such property.

Table 2.6.1 Ordered Partitions of N_4 of Type (2, 2)

Ordered Partition	Between-Set Inversions	Number of Between-Set Inversions	Contribution
$(\{1,2\},\{3,4\})$	$\varnothing$	0	1
$(\{1,3\},\{2,4\})$	$(3,2)$	1	q
$(\{1,4\},\{2,3\})$	$(4,2),(4,3)$	2	q^2
$(\{2,3\},\{1,4\})$	$(2,1),(3,1)$	2	q^2
$(\{2,4\},\{1,3\})$	$(2,1),(4,1),(4,3)$	3	q^3
$(\{3,4\},\{1,2\})$	$(3,1),(3,2)(4,1),(4,2)$	4	q^4
			$1+q+2q^2+q^3+q^4$

2.6.7. Recurrence Equation for q-Binomial Coefficients

Consider the set S_k of ordered partitions of N_n of type $(k, n-k)$. The generating function g_k of S_k with respect to between-set inversions may be obtained in two different ways. First, from Lemma 2.6.6, we have immediately that $g_k = \binom{n}{k}_q$. As a second method, we consider the element n. There are two cases.

Case 1. n is in the k-set. In this case (n, j) is a between-set inversion for each j in the $(n-k)$-set. This contributes q^{n-k} to the generating function. By Lemma 2.6.6, the remaining $k-1$ elements in the k-set contribute $\binom{n-1}{k-1}_q$ to the generating function. Thus, by the product lemma, the contribution to g_k is $q^{n-k}\binom{n-1}{k-1}_q$.

Case 2. n is in the $(n-k)$-set. In this case, n contributes no between-set inversions. By Lemma 2.6.6 the remaining $n-k-1$ elements in the $(n-k)$-set contribute $\binom{n-1}{k}_q$ to g_k. Thus, by the sum lemma

$$\binom{n}{k}_q = \binom{n-1}{k}_q + q^{n-k}\binom{n-1}{k-1}_q.$$

This is, of course, a q-analogue of the familiar identity $\binom{n}{k} = \binom{n-1}{k} + \binom{n-1}{k-1}$, which may be obtained by setting $q = 1$. □

We now introduce two subconfigurations that serve the same purpose here that Ferrers graphs, Durfee squares, and maximal triangles did in Section 2.5.

2.6.8. Definition (Increasing, Decreasing, Cup-, Cap-permutations)

1. *Let $(\alpha)_<$ and $(\alpha)_>$ denote the strings formed by arranging the elements of $\alpha \subseteq \mathsf{N}_n$ in increasing and decreasing order respectively. Then $(\alpha)_<$ is called*

the **increasing permutation** *on* α *and* $(\alpha)_>$ *is called the* **decreasing permutation** *on* α.

2. *If* (α, β) *is an ordered partition of* N_n *of type* (i, j), *then* $(\alpha)_>(\beta)_<$ *is called a* **cup-permutation of shape** (i, j), *and* $(\alpha)_<(\beta)_>$ *is called a* **cap-permutation of shape** (i, j). □

For example (256)(7431) is a cap-permutation of shape (3, 4). The cap-permutation (2567)(431) has shape (4, 3). These are different cap-permutations, although they are identical as permutations. The next result enumerates cap-permutations and cup-permutations with respect to inversions.

2.6.9. Lemma (Cup- and Cap-permutations)

There are $[z^k x^{n-k} q^j] Q_n(z, x)$ *(respectively,* $[z^k x^{n-k} q^j] Q_n(x, z)$*) cup-permutations (respectively, cap-permutations) on* N_n *of shape* $(k, n-k)$ *with* j *inversions, where*

1. $Q_n(z, x) = \prod_{l=0}^{n-1} (x + q^l z).$
2. $Q_n(z, x) = \sum_{l=0}^{n} \binom{n}{l}_q q^{\binom{l}{2}} z^l x^{n-l}.$

Proof:

1. We construct all cup-permutations on N_n by considering the effect of attaching n to the left-hand end or the right-hand end of a cup-permutation in N_{n-1}. Insertion of n on the left contributes an additional $n - 1$ inversions, by the inversion algorithm, and one additional element in the left-hand set. Let z mark an element in the left-hand set. Then the contribution from this insertion is zq^{n-1}. On the other hand, insertion on the right contributes no additional inversions, and one additional element in the right-hand set. Let x mark an element in the right-hand set. Then the contribution from this insertion is x.

The contribution from the insertion of n is therefore $x + zq^{n-1}$. It follows by the product lemma that $Q_n(z, x)$ satisfies the recurrence equation

$$Q_n(z, x) = (x + zq^{n-1}) Q_{n-1}(z, x) \qquad \text{for } n \geqslant 2,$$

with initial condition $Q_1(z, x) = x + z$. Thus (**1**) follows for cup-permutations.

2. Alternatively we may construct all cup-permutations by generating all cup-permutations with shape $(k, n-k)$ for $k = 0, \ldots, n$. Accordingly, consider the ordered partitions of type $(k, n-k)$. The generating function of these partitions with respect to between-set inversions is $\binom{n}{k}_q$, by Lemma 2.6.6. On the k-set of an ordered partition of type $(k, n-k)$ construct a decreasing permutation. By the inversion algorithm, the number of inversions in this permutation is $1 + 2 + \cdots + (k-1) = \binom{k}{2}$. The permutation has k elements so its contribution to the generating function is $z^k q^{\binom{k}{2}}$, since elements of the k-set are marked by z. Finally, construct an increasing permutation on the $(n-k)$-set. This contributes no inversions. Its contribution to the generating function is therefore x^{n-k}, since elements of the $(n-k)$-set are marked by x.

It follows by the product lemma that the generating function for cup-permutations, with shape $(k, n-k)$, with respect to inversions is

$$\binom{n}{k}_q q^{\binom{k}{2}} z^k x^{n-k}.$$

Thus, by the sum lemma,

$$Q_n(z, x) = \sum_{k=0}^{n} \binom{n}{k}_q q^{\binom{k}{2}} z^k x^{n-k}.$$

For cap-permutations, the argument for **(2)** is similar. Finally, **(1)** for cap-permutations follows from **(1)** for cup-permutations and from **(2)**. □

The preceding lemma may be used to obtain a more general result that may be specialized in several ways to obtain partition identities. The result is obtained by considering permutations of the form $(\alpha_1)_>(\alpha_2)_<(\alpha_3)_>(\alpha_4)_<$, called **bimodal** permutations.

2.6.10. Theorem (Bimodal Permutation)

If $Q_n(x, y) = \prod_{i=0}^{n-1}(y + q^i x)$, *then*

$$\sum_{k=0}^{n} \binom{n}{k}_q Q_k(x, y) Q_{n-k}(w, z) = \sum_{k=0}^{n} \binom{n}{k}_q Q_k(w, y) Q_{n-k}(x, z).$$

Proof: Consider (bimodal) permutations of the form $(\alpha_1)_>(\alpha_2)_<(\alpha_3)_>(\alpha_4)_<$ where $(\alpha_1, \ldots, \alpha_4)$ is an ordered partition of N_n. Let x, y, w, z mark elements in $\alpha_1, \ldots, \alpha_4$ respectively. We enumerate these permutations in two different ways.

First, note that $(\alpha_1)_>(\alpha_2)_<$ and $(\alpha_3)_>(\alpha_4)_<$ are cup-permutations on the blocks π_1 and π_2, respectively, of an ordered partition (π_1, π_2) of N_n of type $(k, n-k)$ for some k, where $0 \leqslant k \leqslant n$. From Lemma 2.6.9, the generating functions for the cup-permutations with respect to within-set inversions are $Q_k(x, y)$ and $Q_{n-k}(w, z)$, for each such partition. The generating function for ordered partitions of type $(k, n-k)$ with respect to between-set inversions is $\binom{n}{k}_q$, from Lemma 2.6.6. Thus the generating function for bimodal permutations on N_n with respect to inversions is

$$\sum_{k=0}^{n} \binom{n}{k}_q Q_k(x, y) Q_{n-k}(w, z).$$

Next, note that $(\alpha_1)_>$ is a decreasing permutation, $(\alpha_2)_<(\alpha_3)_>$ is a cap-permutation, and $(\alpha_4)_<$ is an increasing permutation on the blocks π_1, π_2, and π_3, respectively, of an ordered partition (π_1, π_2, π_3) of N_n of type (i, k, l) for some $i, k, l \geqslant 0$ with $i + k + l = n$. The generating function for $\{(\alpha_1)_>\}$ with respect to within-set inversions is $x^i q^{\binom{i}{2}}$. The generating function for $\{(\alpha_4)_<\}$ with respect to within-set inversions is z^l, since there are no inversions in an increasing permutation. The generating function for cap-permutations $\{(\alpha_2)_< (\alpha_3)_>\}$ on the k-set π_2 with respect to within-set inversions is $Q_k(w, y)$, since y marks elements of α_2 and w marks elements of α_3. Finally, the generating function for ordered partitions of type (i, k, l) with respect to between-set inversions is $\begin{bmatrix} n \\ i, k, l \end{bmatrix}_q$, by Lemma 2.6.6. Thus, by the product lemma, the contribution to the generating function for the bimodal permutations from ordered partitions of type (i, k, l) is

$$\begin{bmatrix} n \\ i, k, l \end{bmatrix}_q x^i z^l q^{\binom{i}{2}} Q_k(w, y).$$

The generating function for bimodal permutations on N_n is therefore

$$\sum_{i+k+l=n} \begin{bmatrix} n \\ i, k, l \end{bmatrix}_q x^i z^l q^{\binom{i}{2}} Q_k(w, y)$$

$$= \sum_{k=0}^{n} \binom{n}{k}_q Q_k(w, y) \sum_{i=0}^{n-k} \binom{n-k}{i}_q q^{\binom{i}{2}} x^i z^{n-k-i}.$$

But, by Lemma 2.6.9, this is equal to

$$\sum_{k=0}^{n} \binom{n}{k}_q Q_k(w, y) Q_{n-k}(x, z).$$

The result follows by equating the two expressions for the generating function for bimodal permutations. □

2.6.11. Corollary (q-Analogue of the Binomial Theorem)

$$Q_n(-x, z) = \sum_{k=0}^{n} \binom{n}{k}_q Q_k(-x, y) Q_{n-k}(-y, z).$$

Proof: Replace x by $-x$ and set $w = -y$ in Theorem 2.6.10. The result follows, since $Q_i(-y, y) = \delta_{i,0}$. □

We now obtain some q-binomial identities involving finite products.

2.6.12. Three Finite Product Identities

1. $$\prod_{i=0}^{n-1}(z-q^i)=\sum_{k=0}^{n}(-1)^k\binom{n}{k}_q q^{\binom{k}{2}}z^{n-k}.$$

2. $$z^n=\sum_{k=0}^{n}\binom{n}{k}_q\prod_{i=0}^{k-1}(z-q^i).$$

3. $$\prod_{i=0}^{n-1}(x-q^iz)=\sum_{k=0}^{n}(-1)^k\binom{n}{k}_q q^{\binom{k}{2}}\left\{\prod_{i=0}^{k-1}(z-q^i)\right\}\left\{\prod_{l=0}^{n-k-1}(x-q^{k+l})\right\}.$$

Proof:

1. Put $y=0$, $x=1$ in the q-analogue of the binomial theorem.
2. Put $y=1$, $x=0$ in the q-analogue of the binomial theorem.
3. In the q-analogue of the binomial theorem replace z by zq^{n-1} and y by q^{n-1} to transform the left-hand side to α and the right-hand side to β.

Now $\alpha=Q_n(-x,zq^{n-1})$, so

$$(-1)^n q^{-\binom{n}{2}}\alpha=\prod_{i=0}^{n-1}(x-zq^i).$$

On the other hand,

$$(-1)^n q^{-\binom{n}{2}}\beta$$

$$=(-1)^n q^{-\binom{n}{2}}\sum_{k=0}^{n}\binom{n}{k}_q Q_k(-x,q^{n-1})Q_{n-k}(-q^{n-1},zq^{n-1})$$

$$=(-1)^n q^{-\binom{n}{2}}\sum_{k=0}^{n}\binom{n}{k}_q\left\{\prod_{i=0}^{k-1}(q^{n-1}-xq^i)\right\}\left\{\prod_{j=0}^{n-k-1}(zq^{n-1}-q^{n-1+j})\right\}$$

$$=\sum_{k=0}^{n}(-1)^k\binom{n}{k}_q q^{\binom{k}{2}}\prod_{i=0}^{k-1}(z-q^i)\prod_{l=0}^{n-k-1}(x-q^{k+l}),$$

after routine simplification, replacing k by $n-k$. But $\alpha=\beta$, and the result follows. □

To obtain identities involving infinite products, we need to give an interpretation of $\lim_{n\to\infty}\binom{n}{k}_q$. This is done by combinatorial means in the next result.

2.6.13. Proposition

$$\lim_{n\to\infty}\binom{n}{k}_q=\prod_{j=1}^{k}(1-q^j)^{-1}.$$

Proof: Let $\alpha_1, \ldots, \alpha_k \geqslant 1$ be such that $\alpha_1 + \cdots + \alpha_k \leqslant n$. Consider the ordered partition of N_n of type $(k, n-k)$ in which $\{\alpha_1, \alpha_1 + \alpha_2, \ldots, \alpha_1 + \cdots + \alpha_k\}$ is the k-set. The number of between-set inversions due to the element $\alpha_1 + \cdots + \alpha_i$ is $\alpha_1 + \cdots + \alpha_i - i$. The total number of between-set inversions for this partition is therefore

$$\sum_{i=1}^{k} (\alpha_1 + \cdots + \alpha_i - i) = \sum_{j=1}^{k} (k - j + 1)(\alpha_j - 1).$$

The generating function for the set of ordered partitions with respect to between-set inversions is therefore

$$\sum_{\substack{\alpha_1, \ldots, \alpha_k \geqslant 1 \\ \alpha_1 + \cdots + \alpha_k \leqslant n}} q^{(\alpha_k - 1) + 2(\alpha_{k-1} - 1) + \cdots + k(\alpha_1 - 1)}.$$

But, from **2.6.6**, the generating function for the set of ordered partitions with respect to between-set inversions is $\binom{n}{k}_q$. Thus

$$\lim_{n \to \infty} \binom{n}{k}_q = \sum_{\alpha_1, \ldots, \alpha_k \geqslant 1} q^{(\alpha_k - 1) + 2(\alpha_{k-1} - 1) + \cdots + k(\alpha_1 - 1)} = \prod_{i=1}^{k} (1 - q^i)^{-1}. \quad \square$$

We conclude this section with a number of identities involving infinite products. Two of these have been given in Section 2.5, but they are included here for completeness.

2.6.14. Four Infinite Product Identities

1. $\displaystyle\prod_{i \geqslant 0} (1 + tq^i) = 1 + \sum_{k \geqslant 1} q^{\binom{k}{2}} t^k \prod_{i=1}^{k} (1 - q^i)^{-1}.$

2. $\displaystyle\prod_{i \geqslant 0} (1 - tq^i)^{-1} = 1 + \sum_{k \geqslant 1} t^k \prod_{i=1}^{k} (1 - q^i)^{-1}.$

3. $\displaystyle\prod_{i \geqslant 0} (1 - atq^i)(1 - tq^i)^{-1} = 1 + \sum_{k \geqslant 1} t^k \prod_{i=1}^{k} (1 - aq^{i-1})(1 - q^i)^{-1}.$

4. $\displaystyle\prod_{i \geqslant 0} (1 - zq^i)$
$$= 1 + \sum_{k \geqslant 1} (-1)^k q^{\binom{k}{2}} \left\{ \prod_{i=1}^{k} (z - q^{i-1})(1 - q^i)^{-1} \right\} \prod_{l \geqslant 0} (1 - q^{k+l}).$$

Proof: In the q-analogue of the binomial theorem, we make the following substitutions and let n tend to infinity using Proposition 2.6.13.

1. $z = 1, y = 0, x = -t.$
2. $z = 1, y = t, x = 0.$
3. $z = 1, y = t, x = at.$
4. In **2.6.12(3)** set $x = 1$. $\square$

NOTES AND REFERENCES

This section is based on Goulden and Jackson (1982a); for an alternative approach see Goldman and Rota (1970).

2.6.4 Rodrigues (1839); **2.6.6** Gessel (1977); **2.6.7** Gauss (1863); **2.6.11** Goldman and Rota (1970); **2.6.12** Cauchy (1893), Goldman and Rota (1970); **2.6.14** Euler (1748), Cauchy (1893), Goldman and Rota (1970); [**2.6.1**] Rényi (1962); [**2.6.5**] Hermite (1891); [**2.6.6**] Carlitz (1956), Szegö (1926), Rogers (1893a, b); [**2.6.7**] Foata (1968), MacMahon (1915); [**2.6.8**] MacMahon (1915); [**2.6.9**] Sylvester (1882); [**2.6.11**] Pólya (1970); [**2.6.12–16**] Goldman and Rota (1970).

EXERCISES

2.6.1. An element j in a permutation σ on N_n is defined to be **outstanding** if $\sigma_j > \sigma_i$ for all $1 \leqslant i < j$. The element 1 is outstanding for all σ. Showing that the number of permutations on N_n with k outstanding elements is $[q^k]\prod_{i=0}^{n-1}(i+q)$.

2.6.2. (a) Prove combinatorially that

$$\binom{n}{k}_q \binom{k}{m}_q = \binom{n}{m}_q \binom{n-m}{k-m}_q.$$

(b) Deduce that

$$\sum_{k=m}^{n} \binom{n}{k}_q \binom{k}{m}_q (-1)^{k-m} q^{\binom{k-m}{2}} = \delta_{n,m}.$$

2.6.3. Prove combinatorially that

(a) $$\binom{n}{k}_q = q^k \binom{n-1}{k}_q + \binom{n-1}{k-1}_q.$$

(b) $$\binom{n+m+1}{m+1}_q = \sum_{k=0}^{n} q^k \binom{k+m}{m}_q.$$

(c) $$\binom{a+b}{n}_q = \sum_{k=0}^{n} q^{(a-k)(n-k)} \binom{a}{k}_q \binom{b}{n-k}_q.$$

This is the **q-Vandermonde identity**.

2.6.4. Show that

$$\sum_{k=0}^{n} \binom{2n}{2k}_q q^{\binom{2k}{2}} = \prod_{i=1}^{2n-1} (1+q^i).$$

2.6.5. (a) By considering Cauchy's identity **2.6.12(1)**, show that

$$\prod_{i=1}^{n}(1+xq^{2i-1})(1+x^{-1}q^{2i-1}) = \sum_{k=-n}^{n} q^{(k^2)}x^k\binom{2n}{n+k}_{q^2}.$$

(b) Deduce the Jacobi triple product identity (**2.5.24**) from part (a).

2.6.6. (a) Prove that

$$\left\{\sum_{k\geqslant 0}\frac{t^k}{k!_q}\right\}^{-1} = \sum_{k\geqslant 0}(-1)^k q^{\binom{k}{2}}\frac{t^k}{k!_q}.$$

(b) Prove the inverse pair

$$a_n = \sum_{k=0}^{n}\binom{n}{k}_q b_k \qquad \text{for } n\geqslant 0,$$

$$b_n = \sum_{k=0}^{n}\binom{n}{k}_q(-1)^{n-k}q^{\binom{n-k}{2}}a_k \qquad \text{for } n\geqslant 0.$$

(c) Let

$$F_n(x) = \sum_{j=0}^{n}\binom{n}{j}_q x^j, \qquad n\geqslant 0,$$

the **Rogers–Szegö polynomials**. Prove that

$$\sum_{n\geqslant 0}\frac{F_n(x)t^n}{(1-q)\cdots(1-q^n)} = \prod_{i\geqslant 0}(1-q^i t)^{-1}(1-q^i tx)^{-1}.$$

(d) Deduce from (c) that

$$\sum_{j=0}^{n}\binom{n}{j}_q(-1)^j = \begin{cases}\prod_{i=1}^{m}(1-q^{2i-1}) & \text{if } n=2m\\ 0 & \text{otherwise.}\end{cases}$$

2.6.7. Define the **major index** of a permutation $\sigma=\sigma_1\cdots\sigma_n$ on N_n to be

$$m(\sigma) = \sum_{\substack{i=1\\ \sigma_{1+i}<\sigma_i}}^{n-1} i,$$

and let $f(\sigma)$ denote the number of falls in σ. A new permutation $\hat{\sigma}$ on N_{n+1} is formed by inserting the element $n + 1$ between two adjacent elements of σ or at either end of σ.

(a) If $n + 1$ is inserted in front of σ, show that $m(\hat{\sigma}) = m(\sigma) + f(\sigma) + 1$.

(b) If $n + 1$ is inserted in the ith fall from the right of σ, show that $m(\hat{\sigma}) = m(\sigma) + i$.

(c) If $n + 1$ is inserted in the ith rise from the left of σ, show that $m(\hat{\sigma}) = m(\sigma) + f(\sigma) + i + 1$.

(d) If $n + 1$ is inserted at the right end of σ, then show that $m(\hat{\sigma}) = m(\sigma)$.

(e) Deduce that the number of permutations σ on N_n with $m(\sigma) = k$ is $[q^k]n!_q$. We note that if $a(\sigma) = k$, for all permutations σ on N_n, exactly $[q^k]n!_q$ times, then $a(\sigma)$ is called a **MacMahonian** (or Mahonian) **statistic**.

2.6.8. (a) Prove that the number of $\{0, 1\}$-sequences with k 1's, $(n - k)$ 0's, and m inversions is $[q^m]\binom{n}{k}_q$.

(b) By counting $\{0, 1\}$-sequences, prove that

$$\binom{n}{k}_q = q^k\binom{n-1}{k}_q + \binom{n-1}{k-1}_q.$$

(c) By counting $\{0, 1\}$-sequences, prove the q-Vandermonde identity [**2.6.3**(c)].

2.6.9. (a) Deduce, from the proof of Proposition 2.6.13, that the number of partitions of n into at most i parts, each at most m, is $[q^n]\binom{m+i}{m}_q$.

(b) Show that the number of partitions of n into $i + 1$ parts, with largest part $m + 1$ is $[q^n]q^{m+i+1}\binom{m+i}{m}_q$.

2.6.10. By counting partitions, give combinatorial proofs for

(a) $$\binom{n}{k}_q = q^k\binom{n-1}{k}_q + \binom{n-1}{k-1}_q.$$

(b) $$\prod_{i=1}^{n}(1 + q^i x) = \sum_{k=0}^{n} q^{\binom{k+1}{2}}\binom{n}{k}_q x^k.$$

(c) $$\sum_{i=1}^{n} w^i q^i x(1 + qx)\cdots(1 + q^{i-1}x)$$

$$= \sum_{k=1}^{n} w^k q^{\binom{k+1}{2}} x^k \sum_{l=0}^{n-k} w^l q^l \binom{k+l-1}{l}_q.$$

2.6.11. Consider all sequences of k 0's and $(n - k)$ 1's and an associated **path** from $(0,0)$ to $(k, n - k)$ in which we increase the x-coordinate by 1 for each 0 and the y-coordinate by 1 for each 1 in the sequence. Show that the number of such paths for which the area enclosed by the path, the x-axis, and the line $x = k$ is m is given by $[q^m]\binom{n}{k}_q$.

2.6.12. Let $V_n(q)$ denote the vector space of dimension n over GF(q), for q a prime power.

(a) Show that there are $(q^n - 1)(q^n - q) \cdots (q^n - q^{m-1})$ m-tuples of linearly independent vectors in $V_n(q)$.

(b) Deduce that there are $\binom{n}{m}_q$ subspaces of $V_n(q)$ with dimension m.

2.6.13. By considering subspaces of a vector space, prove that

$$\binom{n}{m}_q = q^m \binom{n-1}{m}_q + \binom{n-1}{m-1}_q.$$

2.6.14. By considering the n-dimensional subspaces of $V_{a+b}(q)$ that intersect the a-dimensional subspace $V_a(q)$ in a k-dimensional space, prove the q-Vandermonde identity [**2.6.3(c)**].

2.6.15. Consider linear transformations from $V_n(q)$ into a space Z with $z > q^n$ vectors. By determining the number whose nullity is $n - k$, prove Cauchy's identity **2.6.12(2)**.

2.6.16. Consider vector spaces $V_n(q)$, X, Y, Z such that $X \subset Y \subset Z$ and $\dim X > n = \dim V_n(q)$. Suppose that X, Y, Z contain x, y, z vectors, respectively. By considering all one-to-one linear transformations $f: V_n \to Z$ such that $f(V_n) \cap X$ and $f(V_n) \cap Y$ have dimensions 0 and k, respectively, prove the q-analogue of the binomial theorem (**2.6.11**).

2.7. PLANTED PLANE TREES

We now turn our attention to the enumeration of certain types of trees. These are important configurations in combinatorics and its applications and are of interest at this point since they involve the use of recursive decompositions. Although decompositions have been used in the earlier sections, these have been either direct or indirect and there has not yet been a situation in which a recursive decomposition has been essential. We shall use the Lagrange theorem for constructing power series solutions for the functional equations that arise.

A **rooted tree** is a tree in which a single vertex, called the **root vertex**, is distinguished. A **rooted plane** tree is a rooted tree embedded in the plane. Two such trees are isomorphic if one may be transformed into the other by continuous operations in the plane. For example, the trees (i) and (iii), given in Figure 2.7.1 are isomorphic, while (i) and (ii) are not. We adopt the convention that the root vertex is distinguished with a circle.

This section deals with rooted plane trees whose roots are monovalent. Such trees are called **planted plane trees**. The planted plane tree consisting of a single edge is called the **trivial** planted plane tree and is denoted by ε throughout the section. A **cubic tree** is a tree whose vertices are either monovalent or trivalent. To decompose a planted plane tree we need the following definition.

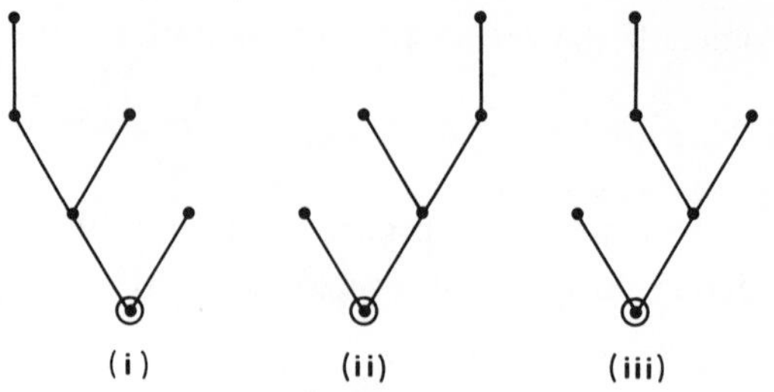

Figure 2.7.1.

2.7.1. Definition (Branch, Branch List)

Let the trivial tree uv_0 be rooted at u. For $k \geqslant 1$, let $t_1, \ldots, t_k \in \mathsf{P}$, the set of planted plane trees, with root vertices $v_1, \ldots, v_k$. Let t be the tree obtained by identifying the vertices $v_0, \ldots, v_k$ as a single vertex v, and rooting the result at u so that $t_1, \ldots, t_k$, uv_0 are encountered sequentially in a clockwise circulation of v. Then t_j is called the jth **branch** *of t. We call $(t_1, \ldots, t_k)$ the* **branch list** *of t and denote it by $\Lambda(t)$. Furthermore, $\Lambda(\varepsilon) = \varnothing$.* □

Figure 2.7.2 gives a planted plane cubic tree and its two branches, t_1 and t_2, in clockwise order.

We begin with the enumeration of planted plane cubic trees, to illustrate the use of a simple recursive decomposition.

2.7.2. Decomposition (Planted Plane Cubic Trees)

Let C be the set of planted plane cubic trees. Then

$$\mathsf{C} - \{\varepsilon\} \xrightarrow{\sim} \{\varepsilon\} \times \mathsf{C}^2 : t \mapsto (\varepsilon, \Lambda(t)).$$

Proof: The trivial tree ε is excluded from the domain, since $\Lambda(\varepsilon)$ is empty. The result follows directly from Definition 2.7.1. □

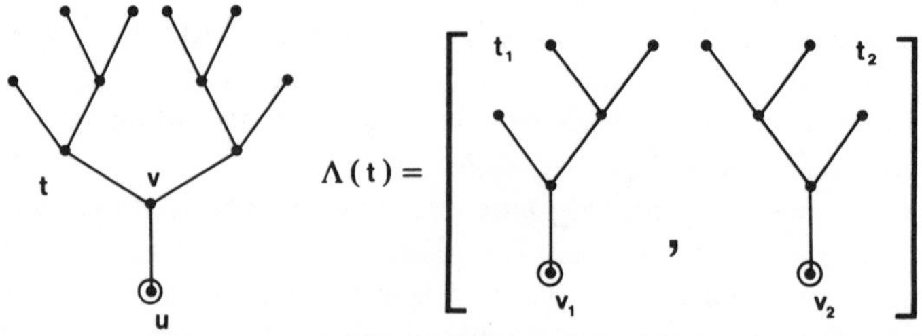

Figure 2.7.2. The branches of a planted plane tree.

The elements of C may be enumerated as follows.

2.7.3. Planted Plane Cubic Trees and Nonroot Monovalent Vertices

We determine the number $c(n)$ of planted plane cubic trees with n nonroot monovalent vertices.

Let $C(x)$ be the generating function for C with respect to nonroot monovalent vertices, marked by x. The generating function for $\{\varepsilon\} \times \mathsf{C}^2$ is C^2, since no additional monovalent vertices are inserted. The generating function for the set $\{\varepsilon\}$ is x, since ε contains one nonroot monovalent vertex. Thus, from Decomposition 2.7.2 and the sum and product lemmas,

$$C^2 - C + x = 0,$$

with the condition that $C(0) = 0$ since $c(0) = 0$. This functional equation may be solved by rewriting it as $C = x(1 - C)^{-1}$. Thus, by the Lagrange theorem,

$$C = \sum_{n \geq 1} \frac{x^n}{n} [\lambda^{n-1}](1 - \lambda)^{-n} = \sum_{n \geq 0} \frac{x^{n+1}}{n+1} \binom{2n}{n},$$

whence

$$c(n+1) = \frac{1}{n+1} \binom{2n}{n}.$$

This is called a **Catalan number**. □

The idea behind Decomposition 2.7.2 may be exploited more generally in the following decomposition for the set of planted plane trees.

2.7.4. Decomposition (Branch)

Let P *be the set of planted plane trees. Then*

$$\mathsf{P} - \{\varepsilon\} \tilde{\rightarrow} \dot{\bigcup_{k \geq 1}} \{\varepsilon\} \times \mathsf{P}^k : t \mapsto \big(\varepsilon, (t_1, \ldots, t_p)\big)$$

where $p + 1$ *is the degree of the vertex adjacent to the root of* t *and* $\Lambda(t) = (t_1, \ldots, t_p)$.

Proof: Direct from Definition 2.7.1. □

As an immediate application of this decomposition, consider the enumeration of planted plane trees with respect to nonroot vertices.

2.7.5. Planted Plane Trees and Nonroot Vertices

Let P be the set of planted plane trees and let $p(n)$ be the number of such trees on n nonroot vertices. Let $P(x)$ be the generating function for P with respect to nonroot vertices, marked by x. From the branch decomposition

$$P - x = x\left(P + P^2 + \cdots\right).$$

It follows that $P^2 - P + x = 0$ with the condition that $P(0) = 0$, since $p(0) = 0$. Thus, from **2.7.3**

$$P = \sum_{n \geqslant 0} \frac{x^{n+1}}{n+1}\binom{2n}{n}. \qquad \square$$

Note that $p(n) = c(n)$, from **2.7.3** and **2.7.5**, so there is a weight-preserving bijection between the sets P and C. However, it is a reasonable, but separate, question to ask for a combinatorial description of it. This is given in [**2.7.9**].

A more substantial instance of the use of the branch decomposition occurs when we wish to preserve information about degrees of vertices.

2.7.6. Definition (Degree Sequence)

Let $\mathbf{i} = (i_1, i_2, \ldots)$ *be such that* i_j *is the number of nonroot vertices of degree* j *in a planted plane tree* t, *for* $j \geqslant 1$. *Then* $\mathbf{i}$ *is called the* **degree sequence** *of* t. $\square$

2.7.7. Planted Plane Trees and Degree Sequence

Let $c(\mathbf{i})$ be the number of trees in the set P of planted plane trees with degree sequence $\mathbf{i}$.

Let t mark a nonroot vertex of any degree, and let x_j mark a nonroot vertex of degree j for $j \geqslant 1$. Let

$$F(t, \mathbf{x}) = \sum_{\mathbf{i} \geqslant \mathbf{0}} c(\mathbf{i}) \mathbf{x}^{\mathbf{i}} t^{i_1 + i_2 + \cdots}.$$

Then, from the branch decomposition,

$$F = t \sum_{k \geqslant 0} x_{k+1} F^k,$$

since $\{\varepsilon\} \times \mathsf{P}^k$ contributes an additional vertex, which is of degree $k+1$. This is a functional equation for F. By the Lagrange theorem

$$c(i_1, \ldots, i_n) = \frac{1}{n}\left[x_1^{i_1} \cdots x_n^{i_n}\right]\left[\lambda^{n-1}\right]\left(x_1 + \lambda x_2 + \lambda^2 x_3 + \cdots\right)^n.$$

Thus

$$c(i_1,\ldots,i_n) = \begin{cases} \dfrac{1}{n}\left[\begin{matrix} n \\ i_1,\ldots,i_n \end{matrix}\right] & \text{if } i_1 + i_2 + \cdots + i_n = n \text{ and} \\ & i_1 + 2i_2 + \cdots + ni_n = 2n - 1 \\ 0 & \text{otherwise.} \end{cases}$$ □

When less extensive information about the degrees of vertices is required, it is more convenient to return to the original decomposition than to specialize the more general result given in **2.7.7**. This point is illustrated next.

2.7.8. Planted Plane Trees and Bivalent Vertices

Let $b(m, n)$ be the number of planted plane trees with m bivalent vertices and n nonroot vertices.

Let $B(x, y)$ be the generating function for P with respect to nonroot vertices, marked by x, and bivalent vertices, marked by y. From the branch decomposition

$$B = x\{1 + yB + B^2 + B^3 + \cdots\}$$

since $\{\varepsilon\} \times \mathsf{P}^k$ contributes an additional bivalent vertex only when $k = 1$. Thus $B(x, y)$ satisfies the functional equation

$$B = x\{(1 - B)^{-1} + (y - 1)B\}.$$

By the Lagrange theorem,

$$B(x, y) = \sum_{n \geq 1} \frac{x^n}{n}[\lambda^{n-1}]\{(1 - \lambda)^{-1} + (y - 1)\lambda\}^n.$$

Thus, after routine manipulation,

$$b(m, n) = \frac{1}{n} \sum_{k \geq 1} \binom{n}{k}\binom{2k-2}{k-1}\binom{n-k}{m}(-1)^{n-k-m}.$$

But

$$\binom{n}{k}\binom{n-k}{m} = \binom{n}{m}\binom{n-m}{k},$$

so

$$b(m, n) = (-1)^{n+m}\frac{1}{n}\binom{n}{m}\sum_{k \geq 1}(-1)^k\binom{n-m}{k}\binom{2k-2}{k-1}.$$ □

When $m = 0$, there are no bivalent vertices, and trees with this property are called **homeomorphically irreducible**. The number of such trees on n nonroot vertices is therefore

$$(-1)^n \frac{1}{n} \sum_{k \geq 1} (-1)^k \binom{n}{k} \binom{2k-2}{k-1}.$$

By **2.7.8**, $B(x, y)$ satisfies the functional equation

$$B = x\{(1 - B)^{-1} + (y - 1)B\},$$

whence $B^2 - B + x\{1 - x(y-1)\}^{-1} = 0$. It follows from **2.7.5** that $B(x, y) = P(x\{1 - x(y-1)\}^{-1})$, where $P(x)$ is the generating function for the set of planted trees with respect to nonroot vertices. A combinatorial explanation of this relationship is given by the following decomposition.

2.7.9. Decomposition (Planted Plane Trees)

Let H *be the set of homeomorphically irreducible trees in the set* P *of planted plane trees. Then*

$$\mathsf{P} \xrightarrow{\sim} \mathsf{H} \circ \{0(0^*)\}$$

where $\circ$ *denotes composition with respect to nonroot vertices.*

Proof: Let v denote an arbitrary nonroot vertex of a tree $h \in \mathsf{H}$. Direct each edge of h toward the root vertex of h. Let e denote the unique edge incident with v and directed from v toward the root of h. Let $v \circ 0^{p+1}$ denote the subdivision of e by exactly p vertices.

Clearly, all trees in P may be obtained uniquely in this way. The result follows. □

This decomposition is very closely related to Decomposition 2.4.15, for Smirnov sequences. The decomposition is an indirect one for H and is illustrated in Figure 2.7.3.

2.7.10. Planted Plane Trees and Bivalent Vertices, by Composition

We wish to determine the generating function $B(x, y)$ for P with respect to bivalent vertices, marked by y, and nonroot vertices, marked by x. Let $H(x)$ be the generating function for the set H of homeomorphically irreducible trees in P with respect to nonroot vertices. Then from Decomposition 2.7.9 and the composition lemma

$$P(x) = H\left(x(1 - x)^{-1}\right) \quad \text{and} \quad B(x, y) = H\left(x(1 - xy)^{-1}\right).$$

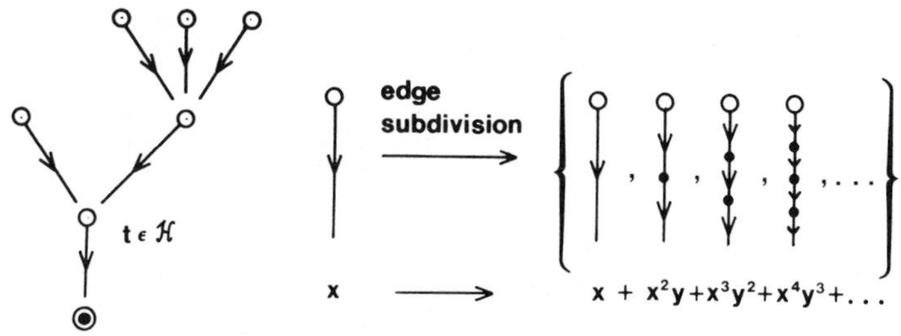

Figure 2.7.3. A homeomorphically irreducible tree and edge subdivision.

These are functional equations for $H(x)$, $B(x, y)$ in terms of $P(x)$. It follows that

$$B(x, y) = P\left(x\{1 + x(1 - y)\}^{-1}\right), \qquad \text{since } \left\{x(1 - x)^{-1}\right\}^{[-1]} = x(1 + x)^{-1}.$$

This is the relationship between $B(x, y)$ and $P(x)$ that was noted in the comments following **2.7.8.** □

In fact, Decomposition 2.7.9 preserves information that is not conveniently available from the recursion in the branch decomposition. As an example, we consider planted plane trees in which there are no **isolated** bivalent vertices, where an isolated bivalent vertex is a bivalent vertex that is not adjacent to another bivalent vertex.

2.7.11. Planted Plane Trees with No Isolated Bivalent Vertices

Let $c(n)$ denote the number of planted plane trees with no isolated bivalent vertices and n nonroot vertices. Let x mark a nonroot vertex. Then, by Decomposition 2.7.9,

$$P(x) = H\left(x(1 - x)^{-1}\right)$$

where $H(x)$ is the generating function for the set of homeomorphically irreducible trees in P with respect to nonroot vertices. Moreover,

$$\sum_{n \geqslant 0} c(n)x^n = H\left(x + x^3 + x^4 + \cdots\right),$$

since edge subdivision by a single vertex gives an isolated bivalent vertex. It follows that

$$c(n) = [x^n]P\left(x(1 - x + x^2)(1 - x^2 + x^3)^{-1}\right),$$

where $P(x) = \Sigma_{m\geqslant 0}\dfrac{x^{m+1}}{m+1}\dbinom{2m}{m}$ from **2.7.5**. Thus

$$c(n) = [x^n] \sum_{m\geqslant 0} \frac{1}{m+1}\binom{2m}{m}x^{m+1}\left(\frac{1+x^3}{1-x^2}\right)^{m+1}\left\{1+\frac{x(1+x^3)}{1-x^2}\right\}^{-(m+1)}.$$

We may, of course, obtain an expression for $c(n)$ from this that involves a triple summation of products of binomial coefficients. □

The branch decomposition may also be used to retain chromatic information. The following definition is needed.

2.7.12. Definition (2-Chromatic Tree)

A **2-chromatic** *tree is a tree whose vertices are colored with one of two colors so that adjacent vertices have different colors.* □

For planted plane 2-chromatic trees, we have the following corollary of the branch decomposition.

2.7.13. Decomposition (Planted Plane 2-Chromatic Trees)

Let $\mathsf{P}_1, \mathsf{P}_2$ *be the set of planted plane* 2-*chromatic trees with root colors* 1 *and* 2, *respectively. Then*

$$\mathsf{P}_1 - \{\varepsilon_1\} \,\tilde{\rightarrow}\, \dot{\bigcup_{k\geqslant 1}} \{\varepsilon_1\} \times \mathsf{P}_2^k : t \mapsto (\varepsilon_1, \Lambda(t))$$

$$\mathsf{P}_2 - \{\varepsilon_2\} \,\tilde{\rightarrow}\, \dot{\bigcup_{k\geqslant 1}} \{\varepsilon_2\} \times \mathsf{P}_1^k : t \mapsto (\varepsilon_2, \Lambda(t)),$$

where ε_1 *and* ε_2 *are the trivial trees in* P_1 *and* P_2, *respectively.* □

The decomposition is illustrated in Figure 2.7.4 and is used to enumerate P_1.

2.7.14. Planted Plane 2-Chromatic Trees

Let $c_i(m, n)$ be the number of planted plane 2-chromatic trees on m nonroot vertices of color 1, n nonroot vertices of color 2, and having root color i, where $i = 1$ or 2. Let x_i mark nonroot vertices of color i, for $i = 1, 2$. Let

$$h_i(x_1, x_2) = \sum_{m,n\geqslant 0} c_i(m, n)x_1^m x_2^n \qquad \text{for } i = 1, 2.$$

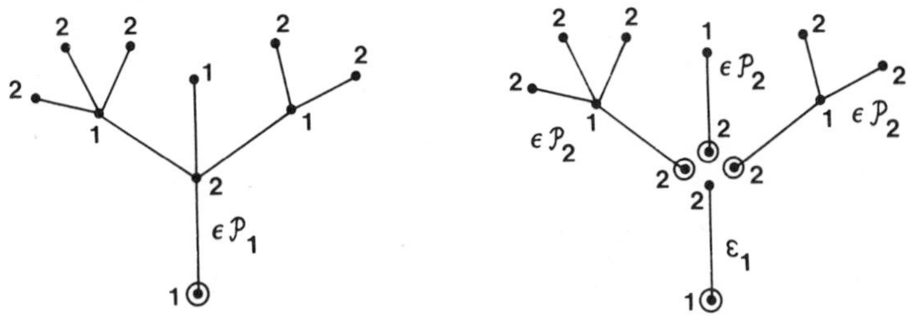

Figure 2.7.4. Branch decomposition of a 2-chromatic tree.

Then, from Decomposition 2.7.13,

$$h_1 = x_2\{1 + h_2 + h_2^2 + \cdots\} \qquad \text{and} \qquad h_2 = x_1\{1 + h_1 + h_1^2 + \cdots\}.$$

Thus h_1, h_2 satisfy the system of simultaneous functional equations

$$\left.\begin{aligned} h_1 &= x_2(1 - h_2)^{-1} \\ h_2 &= x_1(1 - h_1)^{-1}. \end{aligned}\right\}$$

We eliminate h_2 to obtain $h_1 = x_2\{1 - x_1(1 - h_1)^{-1}\}^{-1}$, so by the Lagrange theorem

$$c_1(m, n) = \frac{1}{n}[x_1^m][\lambda^{n-1}]\{1 - x_1(1 - \lambda)^{-1}\}^{-n},$$

whence

$$c_1(m, n) = \frac{n + m - 1}{mn}\binom{n + m - 2}{m - 1}^2$$

for $m, n \geqslant 1$, and $c_1(0, 1) = 1$. □

The branch decomposition may be used to enumerate planted plane trees with respect to height. However, we have to consider the action of the decomposition in greater detail.

2.7.15. Definition (Height of a Vertex)

Let v be a nonroot vertex in a planted plane tree t. The **height** *of v in t is one less than the number of edges in the unique path from v to the root of t.* □

We wish to find the number $c_h(n, d)$ of vertices of degree d and height h in the set of all planted plane trees on n nonroot vertices. When $n = 4$ there are $\frac{1}{4}\binom{6}{3} = 5$ planted plane trees, from **2.7.5**. These are given in Figure 2.7.5 together with the values of $c_h(4, d)$ for $0 \leqslant h \leqslant 3$.

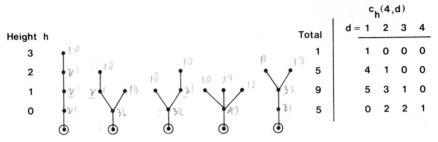

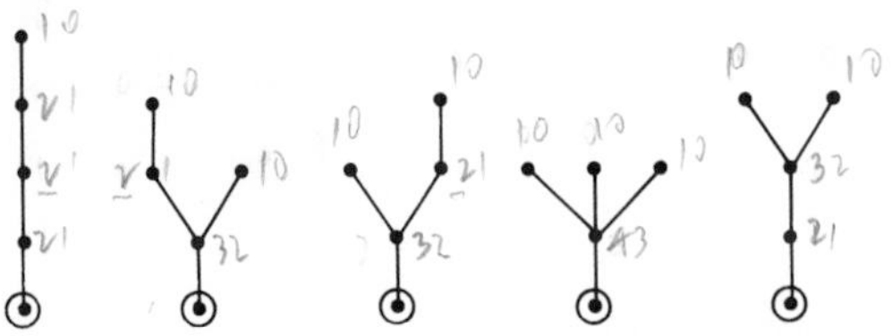

Height h	Total	$c_h(4,d)$ d = 1	2	3	4
3	1	1	0	0	0
2	5	4	1	0	0
1	9	5	3	1	0
0	5	0	2	2	1

Figure 2.7.5. Number of vertices with given height and degree in planted plane trees on four nonroot vertices.

Let the s-objects of trees in P be nonroot vertices of degree d at height h. Then each tree in P with k such vertices is counted k times in $d\mathsf{P}/ds$. Thus, by the differentiation lemma,

$$c_h(n, d) = [x^n] z \frac{\partial}{\partial z} P_{h,d}(x, z)|_{z=1},$$

where $P_{h,d}(x, z)$ is the generating function for P with respect to nonroot vertices marked by x and s-objects marked by z.

We determine $c_h(n, d)$ in two different ways. In **2.7.16** we apply the branch decomposition to P to obtain an expression for $P_{h,d}(x, z)$ and then differentiate with respect to z to obtain $c_h(n, d)$. In **2.7.17**, on the other hand, we give a decomposition for $d\mathsf{P}/ds$ itself, and obtain $c_h(n, d)$ immediately as a consequence.

2.7.16. Vertices of Given Degree and Height in Planted Plane Trees (First Method)

From the preceding,

$$c_h(n, d) = [x^n] z \frac{\partial}{\partial z} P_{h,d}(x, z)|_{z=1}.$$

Moreover, from the branch decomposition,

$$P_{i,d}(x, z) = \phi(P_{i-1,d}(x, z)) \qquad \text{for } i \geqslant 1, \text{ where } \phi(\lambda) = x(1-\lambda)^{-1}.$$

The initial condition is, again by the branch decomposition,

$$\begin{aligned} P_{0,d}(x, z) &= x\{1 + P(x) + \cdots + P^{d-2}(x) + zP^{d-1}(x) + P^d(x) + \cdots\} \\ &= \psi(P(x)), \qquad \text{where } \psi(\lambda) = x\{(1-\lambda)^{-1} + (z-1)\lambda^{d-1}\} \end{aligned}$$

and, of course, $P(x)$ is the generating function for P with respect to nonroot vertices and, from **2.7.5**, satisfies

$$P(x) = x\{1 - P(x)\}^{-1}.$$

It follows that

$$P_{h,d}(x, z) = \phi^{[h]}(\psi(P(x))).$$

Thus, by the chain rule,

$$z\frac{\partial}{\partial z}P_{h,d}(x, z) = z\frac{\partial}{\partial z}\{\phi^{[h]}(\psi(P(x)))\}$$

$$= z\phi'(\phi^{[h-1]}(\psi(P(x))))\frac{\partial}{\partial z}\{\phi^{[h-1]}(\psi(P(x)))\}.$$

Iterating, we obtain

$$z\frac{\partial}{\partial z}P_{h,d}(x, z) = z\left\{\prod_{i=1}^{h}\phi'(\phi^{[h-i]}(\psi(P(x))))\right\}\frac{\partial}{\partial z}\{\psi(P(x))\}.$$

But $(\partial/\partial z)\{\psi(P(x))\} = xP^{d-1}(x)$, $\psi(t)|_{z=1} = \phi(t)$, and $\phi^{[k]}(P(x)) = P(x)$ for all $k \geqslant 0$. Accordingly,

$$z\frac{\partial}{\partial z}P_{h,d}(x, z)|_{z=1} = \left\{\prod_{i=1}^{h}\phi'(\phi^{[h-i+1]}(P(x)))\right\}xP^{d-1}(x)$$

$$= \{\phi'(P(x))\}^{h}xP^{d-1}(x).$$

However, $\phi'(t) = x(1-t)^{-2}$, so $\phi'(P(x)) = x^{-1}P^2(x)$, whence

$$z\frac{\partial}{\partial z}P_{h,d}(x, z)|_{z=1} = x^{1-h}P^{2h+d-1}(x)$$

so $c_h(n, d) = [x^{n+h-1}]P^{2h+d-1}(x)$. It follows from the Lagrange theorem that

$$c_h(n, d) = \frac{1}{n+h-1}[\lambda^{n+h-2}](2h+d-1)\lambda^{2h+d-2}(1-\lambda)^{-(n+h-1)}$$

$$= \frac{2h+d-1}{n+h-1}[\lambda^{n-d-h}](1-\lambda)^{-(n+h-1)}$$

whence $$c_h(n, d) = \frac{2h+d-1}{2n-d-1}\binom{2n-d-1}{n+h-1}.$$ □

As a check, note that

$$c_h(4, 2) = \tfrac{1}{5}(2h+1)\binom{5}{3+h},$$

so $c_0(4,2) = 2$, $c_1(4,2) = 3$, and $c_2(4,2) = 1$, in agreement with Figure 2.7.5.

In fact, $d\mathsf{P}/ds$ itself may be constructed combinatorially by deleting the (unique) path from the distinguished vertex of a tree in $d\mathsf{P}/ds$ to its root. This gives the following decomposition.

2.7.17. Decomposition (Planted Plane Trees with a Single Distinguished Vertex of Degree *d* and Height *h*)

Let $\mathsf{Q} = \dot{\cup}_{k \geq 0} \mathsf{P}^k$, *and let* π_i *denote a path of vertex length i. Then*

$$\frac{d\mathsf{P}}{ds} \xrightarrow{\sim} \mathsf{P}^{d-1} \times \mathsf{Q}^{2h} \times \{\pi_{h+2}\}$$

where the differentiation is with respect to s-objects, which are vertices of degree d at height h. □

The decomposition is illustrated in Figure 2.7.6. The distinguished vertex is v, and the path π_{h+2} is $uv_0 \cdots v_{h-1}v$, where u is the root vertex. The order of the

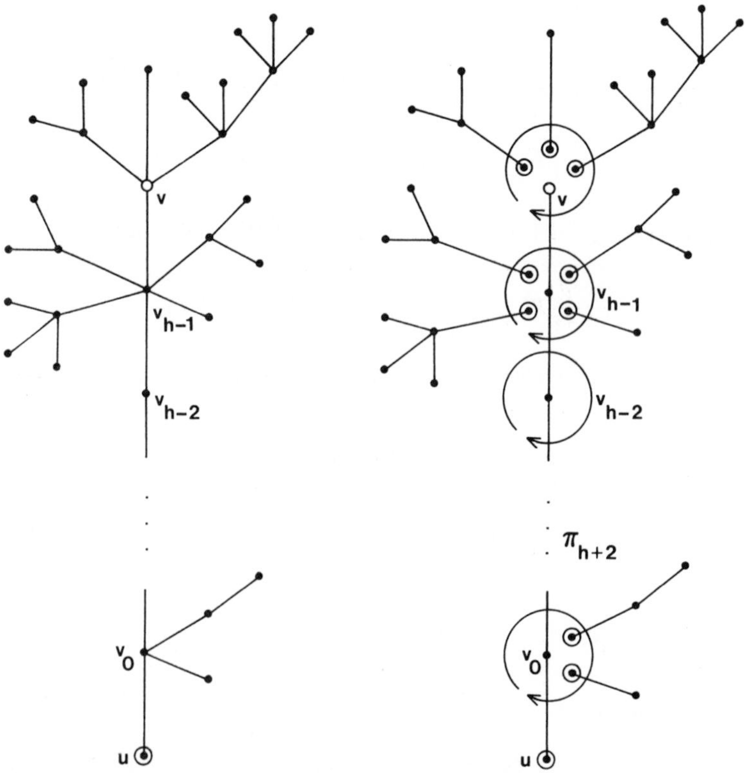

Figure 2.7.6. Decomposition of a tree in P with a single distinguished vertex v of degree 4 and height h.

planted plane trees at each v_j and v is specified in a clockwise direction from the path.

2.7.18. Vertices of Given Degree and Height in Planted Plane Trees (Second Method)

Let x mark nonroot vertices in planted plane trees. We know that

$$c_h(n, d) = [x^n] z \frac{\partial}{\partial z} P_{h,d}(x, z)|_{z=1}.$$

Then from Decomposition 2.7.17,

$$\sum_{n \geq 0} c_h(n, d) x^n = P^{d-1}(1 - P)^{-2h} x^{h+1},$$

since the generating functions for π_{h+2} and Q are x^{h+1} and $(1 - P)^{-1}$, respectively, and the generating function for P is $P(x)$. But from **2.7.5**, P satisfies the functional equation $P = x(1 - P)^{-1}$, whence

$$c_h(n, d) = [x^{n+h-1}] P^{2h+d-1}(x).$$

This agrees with **2.7.16**. □

There are certain aspects of the branch decomposition of P that may be used to obtain other decompositions. The recursive step in this decomposition consists of identifying the roots of a set of trees in P as a single vertex. Thus at this stage we are able to record the degree of this vertex as it is constructed.

We now have some insight into a way in which we might have derived the branch decomposition as a means of enumerating P with respect to degree sequence. We seek a recursive decomposition that constructs a vertex. This vertex must be uniquely specifiable in order to obtain a decomposition that is bijective. The most obvious vertex that is uniquely specifiable is the (unique) vertex adjacent to the root. Consider now the deletion of this vertex. The tree is decomposed into its branches, and the original tree is of course uniquely recoverable from these. This is precisely the statement of the branch decomposition.

2.7.19. Remark (Finding Decompositions)

To find a decomposition for the enumeration of a set S *of distinct configurations with respect to an s-object*

1. *Decide whether there is a uniquely specifiable s-object in an arbitrary configuration* σ *in* S.
2. *Delete this s-object from* σ.
3. *If* σ *decomposes into a number of disjoint configurations that may be uniquely recombined to form* σ, *then we have found a decomposition for* S. □

To demonstrate how Remark 2.7.19 may sometimes be used to find decompositions, consider the enumeration of planted plane trees with respect to left-most paths.

2.7.20. Definition (Left-most Path)

Direct the edges of $t \in \mathsf{P}$ away from the root. The **left-most** *path π of t is a directed path, from the root to a monovalent vertex, containing the left-most edge in t at each vertex in π.* □

To obtain a decomposition for P that records information about the left-most path, we try the strategy described in Remark 2.7.19. When the left-most path is deleted, the tree decomposes into an ordered collection of planted plane trees, giving the following decomposition.

2.7.21. Decomposition (Left-most Path)

Let $\mathsf{Q} = \dot{\cup}_{j \geqslant 0} \mathsf{P}^j$ and let π_k denote a path with k vertices, rooted at a terminal vertex. Then

$$\mathsf{P} \xrightarrow{\sim} \dot{\bigcup_{k \geqslant 0}} \mathsf{Q}^k \times \{\pi_{k+2}\}. \qquad \square$$

This decomposition is illustrated in Figure 2.7.7, where we have also indicated the left-most paths of the constituent trees in Q.

When each of the left-most paths is marked at each of the recursive steps, we obtain the **left-most path covering** of a planted plane tree. This is exhibited in Figure 2.7.8 for the tree given in Figure 2.7.7. It contains one maximal left-most path of length 2, three of length 3, two of length 4, and one of length 5. Note that a tree can have no left-most paths of length 1. We now make use of Decomposition 2.7.21 in the enumeration of planted plane trees with respect to degree sequence and lengths of left-most paths.

2.7.22. Planted Plane Trees, Left-most Paths, and Degree Sequence

Let $\mathbf{m} = (m_1, m_2, \ldots)$, $\mathbf{i} = (i_2, i_3, \ldots)$, and let $c(\mathbf{m}, \mathbf{i})$ be the number of planted plane trees with m_j nonroot vertices of degree $j \geqslant 1$, and i_j left-most paths of length $j \geqslant 2$.

The generating function for P in which x_i marks a nonroot vertex of degree i and f_i a left-most path of length i is

$$h(\mathbf{x}, \mathbf{f}) = \sum_{\mathbf{m}, \mathbf{i}} c(\mathbf{m}, \mathbf{i}) \mathbf{x}^{\mathbf{m}} \mathbf{f}^{\mathbf{i}}$$

where $\mathbf{x} = (x_1, \ldots)$ and $\mathbf{f} = (f_2, f_3, \ldots)$. We use Decomposition 2.7.21 and

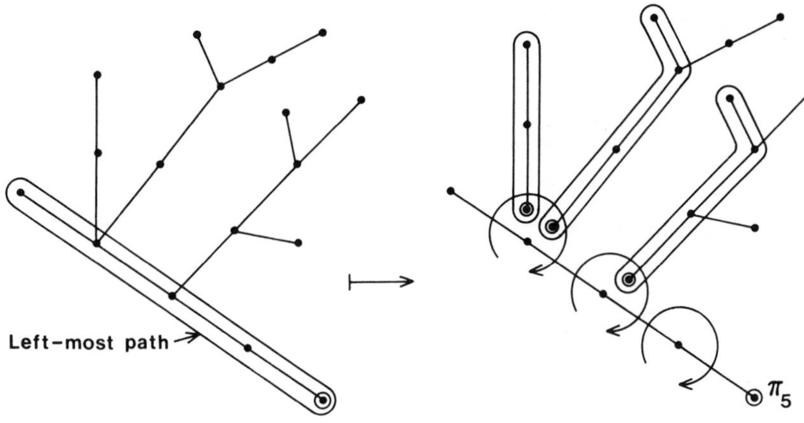

Figure 2.7.7. Left-most path decomposition.

consider first the term $\{\pi_{k+2}\} \times \mathsf{Q}^k$, where $\mathsf{Q} = \dot{\cup}_{j \geq 0} \mathsf{P}^j$. Each occurrence of P^j involves the construction of exactly one vertex, which has degree $j + 2$. Thus the generating function for P^j is $x_{j+2}h^j$, so the generating function for Q is $\sum_{j \geq 0} x_{j+2}h^j$. No trees are attached to the nonroot end of π_{k+2}, so the generating function for this monovalent vertex is x_1. The generating function for π_{k+2} is f_{k+2}, since it has length $k + 2$, and the vertices it contains have already been included in the generating function for Q^k. The generating

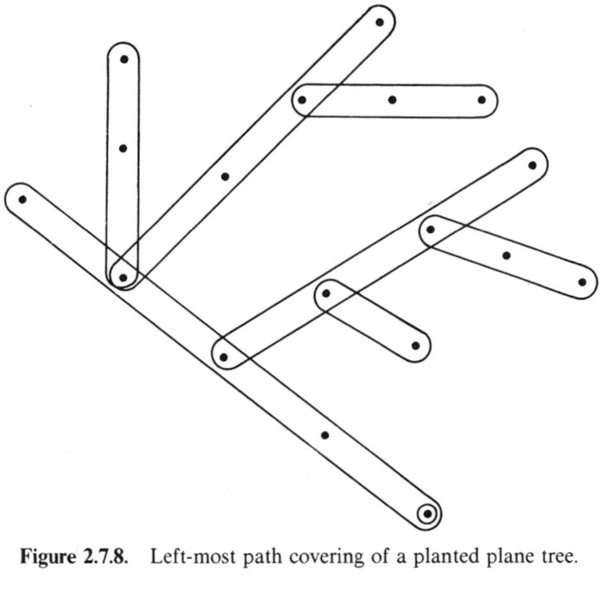

Figure 2.7.8. Left-most path covering of a planted plane tree.

function for $\{\pi_{k+2}\} \times \mathsf{Q}^k$ is therefore

$$x_1 f_{k+2}\Big(\sum_{i\geq 0} x_{i+2}h^i\Big)^k$$

so, by Decomposition 2.7.21,

$$h = x_1 \sum_{k\geq 2} f_k\Big\{\sum_{i\geq 0} x_{i+2}h^i\Big\}^{k-2}.$$

This is a functional equation for h, which may be solved by the Lagrange theorem as follows.

$$h = \sum_{n\geq 1} \frac{1}{n} x_1^n [\lambda^{n-1}]\Big\{\sum_{k\geq 2} f_k\Big\{\sum_{i\geq 0} x_{i+2}\lambda^i\Big\}^{k-2}\Big\}^n,$$

whence

$$[\mathbf{f}^{\mathbf{i}}]h = \sum_{n\geq 1} \frac{1}{n} x_1^n \begin{bmatrix} n \\ \mathbf{i} \end{bmatrix} [\lambda^{n-1}]\Big\{\sum_{j\geq 0} x_{j+2}\lambda^j\Big\}^{\Sigma_{l\geq 2}(l-2)i_l}.$$

It follows that

$$c(\mathbf{m},\mathbf{i}) = \frac{1}{m_1}\begin{bmatrix} m_1 \\ i_2, i_3, \ldots \end{bmatrix}\begin{bmatrix} m_2 + m_3 + \cdots \\ m_2, m_3, \ldots \end{bmatrix},$$

where $\sum_{j\geq 2} i_j = 1 + \sum_{l\geq 2}(l-2)m_l = m_1$ and $\sum_{l\geq 2} m_l = \sum_{j\geq 2}(j-2)i_j$.

□

NOTES AND REFERENCES

2.7.3 Euler (1758); **2.7.4, 5** Good (1965); **2.7.7, 14** Tutte (1964); **2.7.16** Dershowitz and Zaks (1980), Flajolet (private communication); **2.7.18** Flajolet (private communication); [**2.7.6**] Knuth (1968a); [**2.7.9**] de Bruijn and Morselt (1967), Harary, Prins, and Tutte (1964), Klarner (1970); [**2.7.12**] Catalan (1838), Etherington (1937), Schröder (1970); [**2.7.14**] Cayley (1890); [**2.7.15**] Gordon and Torkington (1980); [**2.7.16**] Jackson and Goulden (1981a).

EXERCISES

2.7.1. (a) Show that the number of planted plane trees on $mk + 1$ nonroot vertices, each of which has degree congruent to 1 (mod k), is

$$\frac{1}{mk+1}\binom{m(k+1)}{m}.$$

(b) Show that the number of planted plane trees with $2n$ nonroot vertices, $2m + 1$ of which have odd degrees, is

$$\frac{1}{2n+m}\binom{2n}{2m+1}\binom{2n+m}{m}.$$

2.7.2. Show that the number of planted plane trees having n nonroot vertices
(a) with k of degree $s + 1$ is

$$\frac{1}{n}\sum_{i=k}^{n-1}\binom{n}{i}\binom{2n-2-(s+1)i}{n-i-1}\binom{i}{k}(-1)^{i-k},$$

(b) with k of degree $s + 1$ and m of degree $t + 1$, where $s \neq t$, is

$$\frac{1}{n}\sum_{i+j+l=n}\begin{bmatrix} n \\ i, j, l\end{bmatrix}\binom{j}{k}\binom{l}{m}\binom{n-2-sj-lt+i}{i-1}(-1)^{j+l-k-m}.$$

2.7.3. Prove that the number of planted plane trees on n nonroot vertices, in which the vertex adjacent to the root has

(a) degree m is $\dfrac{m-1}{n-1}\dbinom{2n-m-2}{n-2}$,

(b) odd degree is $\dfrac{2}{n-1}\displaystyle\sum_{i=0}^{n-3}(-1)^i(i+1)\binom{2n-3-i}{n}$, for $n > 1$.

2.7.4. Prove that the number of planted plane trees on n nonroot vertices, in which every bivalent vertex is isolated, is

$$\sum_{m=0}^{n-1}\sum_{i=0}^{n-m-1}\frac{(-1)^i}{m+1}\binom{2m}{m}\binom{m+i}{i}\binom{m+i+1}{n-m-i-1}.$$

2.7.5. Prove that the number of vertices at height h in planted plane trees on n nonroot vertices is

$$\frac{2h+1}{n+h}\binom{2n-2}{n-h-1}.$$

2.7.6. Let h_n be the average value of the sum of heights of the trivalent vertices in the (equiprobable) planted plane cubic trees with n trivalent vertices. Show that

$$h_n = (n+1)4^n\binom{2n}{n}^{-1} - (3n+1) \qquad \text{for } n \geqslant 0.$$

2.7.7. Show that the number of planted plane trees with i_j maximal left-most paths of length j (and thus $n = i_2 + i_3 + \cdots$ nonroot monovalent vertices and $m = \sum_{k \geqslant 2}(k-2)i_k$ other nonroot vertices) is

$$\frac{(n-1)!}{\mathbf{i}!}\binom{m+n-2}{n-1}.$$

2.7.8. Let $P(x)$ be the generating function for planted plane trees with respect to nonroot vertices. Use the Principle of Inclusion and Exclusion to prove that the number of planted plane trees on n nonroot vertices, m of which are bivalent, is given by $[x^n y^m]P(x\{1-(y-1)x\}^{-1})$, as in **2.7.10.**

2.7.9. (a) Let C be the set of planted plane cubic trees. Establish a decomposition $\mathsf{C} - \{\varepsilon\} \mathrel{\tilde{\to}} \dot{\cup}_{k \geqslant 1}\{\pi_{k+2}\} \times \mathsf{C}^k$, where π_{k+2} is a path with vertex length $k+2$.

(b) Let P be the set of planted plane trees. Establish a decomposition $\mathsf{P} - \{\varepsilon\} \mathrel{\tilde{\to}} \mathsf{P}^2$.

(c) Find an explicit one-to-one correspondence between the elements of P with n nonroot vertices and the elements of C with n nonroot monovalent vertices.

2.7.10. (a) Let P_1 be the set of planted plane 2-chromatic trees with root vertex of color 1. Establish the decomposition $\mathsf{P}_1 - \{\varepsilon_1\} \to \dot{\cup}_{k \geqslant 1}\{\varepsilon_2\} \times \mathsf{P}_1^{k+1}$, where ε_i is the trivial planted plane tree with root color i, for $i = 1, 2$. Hence establish a one-to-one correspondence between 2-chromatic planted plane trees with root vertex of color 1, m nonroot vertices of color 1, and n nonroot vertices of color 2, and planted plane trees with n monovalent nonroot vertices and m vertices with degree greater than 1.

(b) Find an explicit bijection between the sets given in (a).

2.7.11. If we differentiate the equation $P = x(1-P)^{-1}$, given in **2.7.5**, then we obtain $xP' = P + P^2P'$. Give a decomposition for P, the set of planted plane trees, that leads immediately to this differential equation.

2.7.12. (a) Let $b(n)$ be the number of ways of computing the product $A_1 \cdots A_n$ in a nonassociative, noncommutative algebra. This is equivalent to "**bracketing**" $A_1 \cdots A_n$ so that each pair of brackets contains exactly two objects. If $B(x) = \sum_{n \geqslant 1} b(n)x^n$, show that $B(x) = x + B^2(x)$, where we let $b(1) = 1$, and hence that

$$b(n) = \frac{1}{n}\binom{2n-2}{n-1},$$

as in **2.7.3.**

(b) Prove that the number of ways of bracketing n objects so that each pair of brackets contains at least two objects is

$$\frac{1}{n}\sum_{i=0}^{n-1}(-1)^i 2^{n-1-i}\binom{n}{i}\binom{2n-2-i}{n-1}.$$

(c) If $c(n)$ is the number of ways of computing the product $A_1 \cdots A_n$ in a nonassociative, commutative algebra, show that $C(x) = \sum_{n \geqslant 1} c(n)x^n$ satisfies the functional equation $C(x) = x + \frac{1}{2}\{C(x)^2 + C(x^2)\}$, where $c(1) = 1$.

2.7.13. Let $c(n)$ denote the number of sequences $\sigma = \sigma_1 \cdots \sigma_{2n}$ of n 1's and n (-1)'s such that the **partial sums** $s_j = \sigma_1 + \cdots + \sigma_j$ are positive for

$$1 \leqslant j \leqslant 2n - 1. \text{ If } C(x) = \sum_{n \geqslant 1} c(n)x^n,$$

show that $C(x) = x\{1 - C(x)\}^{-1}$, so that

$$c(n) = \frac{1}{n}\binom{2n-2}{n-1},$$

as in **2.7.3**.

2.7.14. In the plane, show that the number of ways of drawing nonintersecting diagonals in a convex n-gon with a single distinguished edge to yield i_j polygons with j sides, for $j \geqslant 3$, is

$$\frac{(n-2+i_3+i_4+\cdots)!}{(n-1)!i_3!i_4!\cdots}.$$

The configurations are called rooted **dissections** of the n-gon.

2.7.15. Let t be a planted plane tree with degree sequence **i**. Suppose that the edges of t are cut in half, leaving a collection of i_j vertices with j incident "half-edges," called **half-edge structures** of degree j, for $j \geqslant 1$, and the root vertex with an incident half-edge. Derive **2.7.7** directly by determining the number of ways of reconstituting the preceding collection of half-edge structures to yield a planted plane tree.

2.7.16. A ***k*-chromatic** tree is a tree whose vertices are colored with one of k colors so that adjacent vertices have different colors. The **chromatic partition** of a planted plane k-chromatic tree is $\mathbf{L} = [l_{ij}]_{k\times\infty}$, where the tree has l_{ij} nonroot vertices of color i and degree j. Let $c_r(\mathbf{L}, \mathbf{n})$ be the number of planted plane k-chromatic trees with root color r, chromatic partition $\mathbf{L}$, and $n_i = \sum_{j \geqslant 1} l_{ij}$ nonroot vertices of color i, where $\mathbf{n} = (n_1, \ldots, n_k)$.

(a) Show that $c_r(\mathbf{L}, \mathbf{n}) = [\mathbf{W}^{\mathbf{L}}\mathbf{x}^{\mathbf{n}}]f_r(\mathbf{W}, \mathbf{x})$, where $\mathbf{W} = [w_{ij}]_{k\times\infty}$, $\mathbf{x} = (x_1, \ldots, x_k)$, and

$$f_i = \sum_{\substack{j=1\\ j\neq i}}^{k} x_j g_j(f_j), \qquad g_j(\lambda) = \sum_{m \geqslant 1} w_{jm}\lambda^{m-1}.$$

(b) Show that $c_r(\mathbf{L}, \mathbf{n}) = [\mathbf{W}^{\mathbf{L}}\mathbf{x}^{\mathbf{n}}](y - y_r)$, where $y_i = x_i g_i(y - y_i)$ for $i = 1, \ldots, k$, and $y = y_1 + \cdots + y_k$.

(c) Show that $c_r(\mathbf{L}, \mathbf{n}) = (\mathbf{L}!)^{-1} \sum_{i \geq 0} (N - i)! p_i$, where

$$\sum_{i \geq 0} p_i x^i = \sum_{j=1}^{k} \sum_{m \geq 0} \binom{n_j - 1 + \delta_{rj}}{m} \binom{q_j}{m} m! (-x)^m$$

and

$$q_i = \sum_{j \geq 1} (j - 1) l_{ij}, \qquad \text{for } i = 1, \ldots, k.$$

2.8. SEQUENCES WITH DISTINGUISHED SUBSTRINGS

In this section we consider the general problem of enumerating sequences over N_n that have a specified number of substrings from a prescribed set A of distinguished substrings. Many problems may be expressed in terms of enumerating N_n^* with respect to some set A. Consider, for example, the problem of counting sequences over N_n with exactly i strictly increasing substrings of length k, where k is fixed. The sequence 23572451, for example, has exactly three increasing substrings of length $p = 3$, namely, 235, 357, and 245. The set of distinguished substrings in this case is $\{i_1 \cdots i_k | 1 \leq i_1 < \cdots < i_k \leq n\}$.

The obvious difficulty with this general class of problems is that the sequences in A may, in general, overlap among themselves. For example, suppose that $\mathsf{A} = \{5721, 7215, 2572\}$. then 2572, 5721, and 7215 occur as mutually overlapping substrings of 257215. We assume that $\mathsf{A} = \{A_1, \ldots, A_p\}$ has the property that there is no pair $A, B \in \mathsf{A}$ such that A is a substring of B. In this case A is called a **reduced set**.

2.8.1. Definition (A-Type of a Sequence)

Let $\mathsf{A} = \{A_1, \ldots, A_p\}$ *be a reduced set of sequences over* N_n. *If* $\sigma \in \mathsf{N}_n^*$ *has* m_i *occurrences of the substring* A_i, *then* $\kappa(\sigma) = (m_1, \ldots, m_p)$ *is called the* **A-type** *of* σ. □

Our general purpose is to enumerate sequences in N_n^* with respect to type and to A-type by an "at-least" decomposition (see Definition 2.2.28). The following definition is needed.

2.8.2. Definition (k-Cluster)

A **k-cluster** $(k \geq 1)$ *on* N_n *with respect to* A *is a triple* $(\sigma_1 \cdots \sigma_r, A_{i_1} \cdots A_{i_k}, (l_1, \ldots, l_k)) \in \mathsf{N}_n^r \times \mathsf{A}^k \times \mathsf{N}_r^k$, *for some* $r \geq 1$, *satisfying the following condi-*

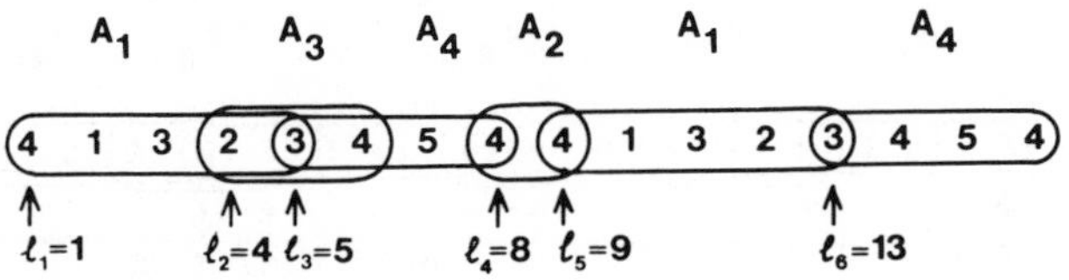

Figure 2.8.1. The 6-cluster $(\sigma, A_1A_3A_4A_2A_1A_4, (1,4,5,8,9,13))$.

tions, in which r_j *is the length of* A_{i_j} *for* $j = 1,\dots, k$

1. $\sigma_{l_j}\sigma_{l_j+1}\cdots\sigma_{l_j+r_j-1} = A_{i_j}, j = 1,\dots, k$ *(*A_{i_j} *starts in position* l_j*).*
2. $0 < l_{j+1} - l_j < r_j, j = 1,\dots, k-1$ *(*$A_{i_j}, A_{i_{j+1}}$ *overlap in* σ*).*
3. $r = l_k + r_k - 1$ *and* $l_1 = 1$ *(*A_{i_1} *contains* σ_1*;* A_{i_k} *contains* σ_r*).*

The set of all k-clusters on N_n *with respect to* A*, for all* $k \geqslant 1$*, is denoted by* $\mathsf{D}(\mathsf{A})$*.* □

Figure 2.8.1 gives a 6-cluster on N_5 with respect to $\{41323, 44, 234, 3454\}$, in which $\sigma = 4132345441323454$. The preceding conditions on a k-cluster ensure, in such a diagram, that each element of σ belongs to at least one element of A and that adjacent pairs of elements in $A_{i_1}\cdots A_{i_k}$ overlap in σ. The position in σ of the first element of the jth distinguished substring (shown circled in Figure 2.8.1) in the cluster is denoted by l_j. Figure 2.8.2 gives another 6-cluster with the same sequence σ.

With the set $\mathsf{D}(\mathsf{A})$ of clusters we associate the following function.

2.8.3. Definition (Cluster Generating Function)

Let $\mathbf{x} = (x_1,\dots, x_n)$ *and* $\mathbf{y} = (y_1,\dots, y_p)$ *be indeterminates. Then the cluster generating function,* $C(\mathbf{x},\mathbf{y})$*, is defined by*

$$C(\mathbf{x},\mathbf{y}) = \sum_{(\mu_1,\mu_2,\mu_3)\in \mathsf{D}(\mathsf{A})} \mathbf{x}^{\tau(\mu_1)}\mathbf{y}^{\alpha(\mu_2)},$$

where $\tau(\mu_1)$ *is the type of* μ_1 *(see Definition 2.4.7),* $\alpha(\mu_2) = (j_1,\dots, j_p)$*, and* j_i *is the number of occurrences of* A_i *in* μ_2 *for* $i = 1,\dots, p$*.* □

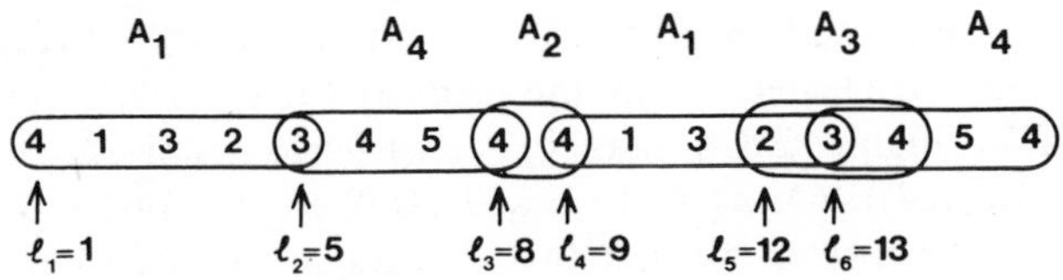

Figure 2.8.2. The 6-cluster $(\sigma, A_1A_4A_2A_1A_3A_4, (1,5,8,9,12,13))$.

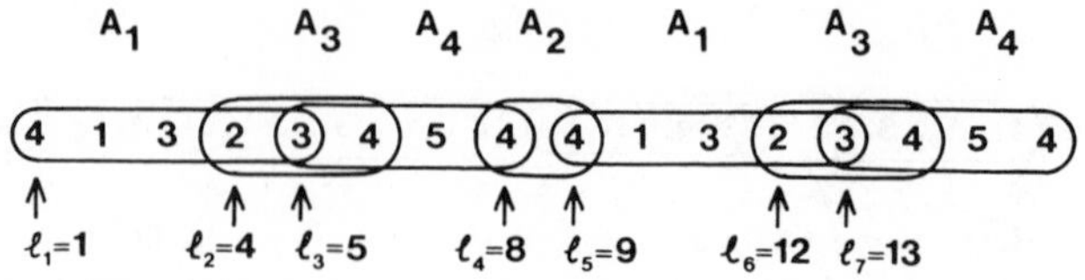

Figure 2.8.3. The 7-cluster $(\sigma,\ A_1A_3A_4A_2A_1A_3A_4, (1,4,5,8,9,12,13))$.

It should be noted that $\alpha(\mu_2)$ in the preceding definition is less than or equal to $\kappa(\sigma)$, the A-type of σ. For example, for the 6-clusters given in Figures 2.8.1 and 2.8.2 we have $\alpha(\mu_2) = (2,1,1,2)$. However, an inspection of σ indicates that its A-type is $\kappa(\sigma) = (2,1,2,2)$. The unique 7-cluster on σ with $\alpha(\mu_2) = (2,1,2,2)$ is given in Figure 2.8.3. All other clusters on σ are constructed by choosing μ_2 as a subsequence $A_{i_1} \cdots A_{i_k}$, of $A_1A_3A_4A_2A_1A_3A_4$, consisting of overlapping substrings that cover σ.

To obtain a relationship between the generating function for N_n^* with respect to A, and the cluster generating function, the following definitions are needed.

2.8.4. Definition

Let E(A) *be the union of*

$$\{(\sigma, \varepsilon, \varnothing) | \sigma \in \mathsf{N}_n^*\}$$

and

$$\left(\sigma_1 \cdots \sigma_r, A_{i_1} \cdots A_{i_k}, (l_1, \ldots, l_k)\right) \in \mathsf{N}_n^r \times \mathsf{A}^k \times \mathsf{N}_r^k \text{ for all } k, r \geq 1,$$

where

1. $\sigma_{l_j} \cdots \sigma_{l_j + r_j - 1} = A_{i_j}, j = 1, \ldots, k.$
2. $1 \leq l_1 < \cdots < l_k \leq r - r_k + 1$, *where* r_j *is the length of* A_{i_j} *for* $j = 1, \ldots, k$. □

The elements of E(A) may be represented, like those of D(A), by circling the A_{i_j} in σ that begins at position l_j for $1 \leq j \leq k$. In this case, however, the conditions do not require that each element of σ is contained in at least one oval, nor that adjacent ovals overlap. Figure 2.8.4 gives an element of E(A).

The set E(A) may be constructed by selecting elements of D(A), ordering them linearly, and inserting arbitrary elements from N_n^* between, before, or after these selected clusters. Thus the element $(\sigma, A_1A_3A_4A_3, (1,4,8,12))$ of E(A), given in Figure 2.8.4, may be constructed from the three clusters $(413234, A_1A_3, (1,4)), (44, A_4, (1)), (234, A_3, (1))$, in this order, with the substring 5 inserted between the first two clusters, 13 inserted between the last two, and 54 after the last cluster. The inserted elements may, of course,

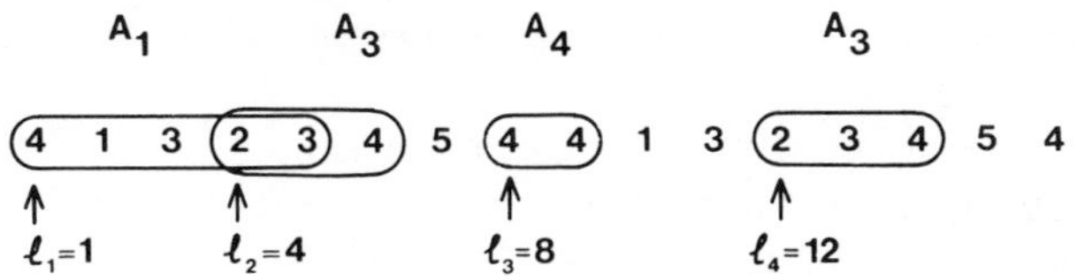

Figure 2.8.4. The element $(\sigma, A_1A_3A_4A_3, (1,4,8,12)) \in \mathsf{E}(\mathsf{A})$.

introduce additional distinguished substrings. Moreover, there may be distinguished substrings in the clusters that are not recorded by the cluster generating function. In fact, E(A) contains all sequences σ of A-type **m** exactly $2^{m_1+\cdots+m_p}$ times, once for each subset of the distinguished substrings that are recorded in the μ_2's of clusters. This construction is therefore "at-least" with respect to A-type, and we have the following result.

2.8.5. Proposition

The generating function for N_n^* *with respect to* A *which is "at least" with respect to* A*-type is*

$$\Psi(\mathbf{x},\mathbf{y}) = \sum_{(\mu_1,\mu_2,\mu_3)\in \mathsf{E}(\mathsf{A})} \mathbf{x}^{\tau(\mu_1)}\mathbf{y}^{\alpha(\mu_2)}. \qquad \square$$

The next result gives the connection between the cluster generating function and the generating function for N_n^* with respect to type and A-type.

2.8.6. Theorem (Distinguished Substring)

The number of sequences over N_n *of type* **k** *and* A*-type* **m** *is* $[\mathbf{x}^{\mathbf{k}}\mathbf{y}^{\mathbf{m}}]\Phi(\mathbf{x},\mathbf{y})$, *where*

$$\Phi(\mathbf{x},\mathbf{y}) = \{1-(x_1+\cdots+x_n) - C(\mathbf{x},\mathbf{y}-\mathbf{1})\}^{-1}$$

and where $C(\mathbf{x},\mathbf{y})$ *is the cluster generating function for the reduced set,* A, *of distinguished sequences.*

Proof: By the Principle of Inclusion and Exclusion,

$$\Phi(\mathbf{x},\mathbf{y}) = \Psi(\mathbf{x},\mathbf{y}-\mathbf{1}).$$

But we have the immediate decomposition

$$\mathsf{E}(\mathsf{A}) \mathrel{\tilde{\to}} \{\mathsf{D}(\mathsf{A}) \cup \mathsf{N}_n\}^*.$$

Thus $\Psi(\mathbf{x},\mathbf{y}) = \{1-(x_1+\cdots+x_n) - C(\mathbf{x},\mathbf{y})\}^{-1}$ and the result follows. $\square$

It often happens that the cluster generating function may be determined by combinatorial means. We now give two instances of this.

2.8.7. Sequences with No pth Powers of Strings of Length k

We wish to determine the number $c(l, q)$ of sequences in N_n^* of length l that contain q substrings of the form w^p, where w is any sequence in N_n^* of length k. For this purpose k and p are fixed positive integers.

Clearly, $\mathsf{B}_p = \{w^p | w \in \mathsf{N}_n^*, \ |w| = k\}$ is a reduced set of distinguished substrings. To obtain the cluster generating function, first construct all 2-clusters and then extend this construction to clusters in general. Let $W_1 = w_1^p \in \mathsf{B}_p$ and $W_2 = w_2^p \in \mathsf{B}_p$, and let x mark the length of a sequence. The generating function for B_p with respect to length is

$$\phi(x) = n^k x^{pk},$$

since $|\mathsf{B}_p| = n^k$ and each sequence has length pk.

Now consider the ways in which W_1 and W_2 may overlap to form a cluster. There are two cases.

Case 1. The last block, w_1, of W_1 and the first block, w_2, of W_2 overlap in a substring β with length between 1 and $k - 1$, as shown in Figure 2.8.5. Let $w_2 = \beta\gamma$, where γ has length j. There are n^j choices for γ, so, since β is determined by w_1, there are n^j choices for w_2, and W_2 is then determined. The length of the nonoverlapping portion of W_2 is $k(p-1)+j$, where $1 \leqslant j \leqslant k-1$ since β is nonempty. Thus the generating function for the nonoverlapping portion of W_2 is

$$\sum_{1 \leqslant j \leqslant k-1} n^j x^{k(p-1)+j}.$$

Case 2. W_1 and W_2 overlap in a substring of length greater than or equal to k, as shown in Figure 2.8.6. Then w_2 is fully determined from w_1. If j is the length of the nonoverlapping portion of W_2, then $1 \leqslant j \leqslant k(p-1)$, so the

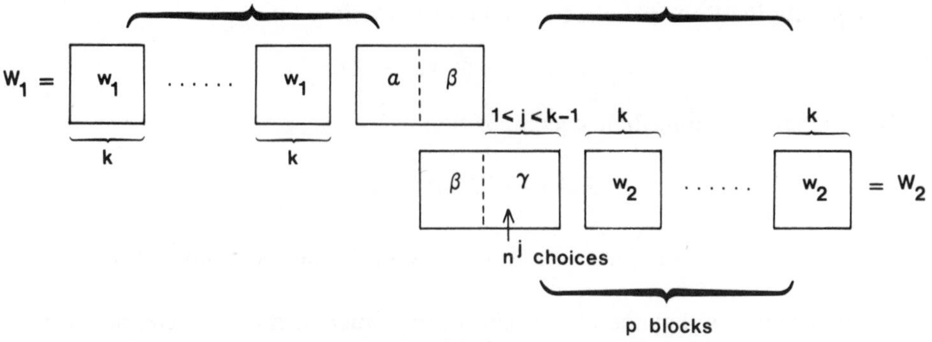

Figure 2.8.5. Construction of a 2-cluster.

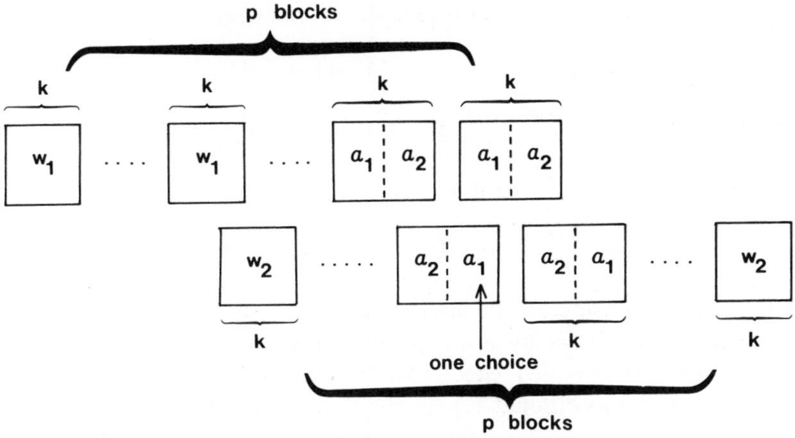

Figure 2.8.6. Construction of a 2-cluster.

generating function for this portion is

$$\sum_{1 \leqslant j \leqslant k(p-1)} x^j.$$

Combining these two cases, we have obtained the generating function $f(x)$ for the nonoverlapping substring of W_2. This is

$$f(x) = \sum_{1 \leqslant j \leqslant k-1} n^j x^{k(p-1)+j} + \sum_{1 \leqslant j \leqslant k(p-1)} x^j$$

$$= \{x - x^{k(p-1)}\}(1-x)^{-1} + \{x^{k(p-1)} - n^k x^{kp}\}(1-nx)^{-1}.$$

For $(r+1)$-clusters, there are $r+1$ sequences $W_1, \ldots, W_{r+1}$. Among these there are r pairwise overlaps between W_i and W_{i+1} for $1 \leqslant i \leqslant r$, and each such overlap is enumerated by $f(x)$. The generating function for the set B_p of all permissible W_1 is $\phi(x)$. Thus, the number of $(r+1)$-clusters of length m is

$$[x^m]\phi(x)f^r(x).$$

Let $C(x, y)$ be the cluster generating function and let y mark a distinguished substring. Then

$$C(x, y) = \sum_{r \geqslant 0} y^{r+1}\phi(x)f^r(x) = y\phi(x)\{1 - yf(x)\}^{-1}.$$

It follows from Theorem 2.8.6 that

$$c(l, q) = [x^l y^q]\{1 - nx - C(x, y-1)\}^{-1}.$$

After routine manipulation $c(l, q) = [x^l y^q]FG^{-1}$, where

$$F = (1-x)(1-nx) - (y-1)\{x(1-x^{k(p-1)}) - nx^2(1-x^{k(p-1)-1}) - n^k x^{kp}(1-x)\}$$

$$G = (1-nx)\{(1-x)(1-nx) - (y-1)\big(x(1-x)^{k(p-1)} - nx^2(1-x^{k(p-1)-1})\big)\}. \qquad \square$$

We have already considered, in **2.4.18**, the enumeration of sequences with respect to maximal blocks by a compositional argument using Smirnov sequences. In enumerating sequences with respect to **strictly increasing substrings**, a purely compositional argument of this sort does not work. By a strictly increasing substring we mean a sequence $j_1 \cdots j_p$ with $1 \leqslant j_1 < \cdots < j_p \leqslant n$. However, we are able to deal with both this subconfiguration and blocks by means of distinguished subsequences.

2.8.8. Sequences and Strictly Increasing Substrings

We determine the number $c(\mathbf{k}, i)$ of sequences in N_n^* of type $\mathbf{k}$ with exactly i strictly increasing substrings of length p, where p is fixed.

Let A be the set of all strictly increasing substrings of length p in N_n^*. Clearly, because p is fixed, A is a reduced set of distinguished substrings. We may construct any k-cluster (μ_1, μ_2, μ_3) of length m on A as follows.

Let μ_1 be a strictly increasing sequence in N_n^* of length at least p. Let $\mu_3 = (l_1, \ldots, l_k)$ be a set of starting positions for elements of A in μ_1, and let $l_{i+1} - l_i = d_i$ for $i = 1, \ldots, k-1$. Then $l_1 = 1$, $(d_1, \ldots, d_{k-1})$ is an arbitrary element of N_{p-1}^{k-1}, and $d_1 + \cdots + d_{k-1} + p = |\mu_1|$, the length of μ_1. The conditions ensure that the distinguished substrings overlap and cover μ_1. Thus the number of choices of μ_3 for k-clusters on any increasing sequence μ_1 of length m is

$$[x^m]x^p(x + \cdots + x^{p-1})^{k-1},$$

and μ_2 is uniquely determined by μ_1, μ_3.

Let y mark distinguished substrings. Since the increasing sequences of length m are enumerated by

$$\gamma_m = [x^m]\prod_{i=1}^{n}(1 + xx_i),$$

then

$$C(x, y-1) = \sum_{m \geqslant p}\sum_{k \geqslant 1}\gamma_m (y-1)^k [x^m] x^p (x - x^p)^{k-1}(1-x)^{-(k-1)}$$

$$= F \circ \gamma$$

where $F(x) = (y-1)x^p\{1-(y-1)(x-x^p)(1-x)^{-1}\}^{-1}$, $\boldsymbol{\gamma} = (\gamma_0, \gamma_1, \cdots)$ and $\circ$ denotes umbral composition. Thus, from Theorem 2.8.6,

$$c(\mathbf{k}, i) = [\mathbf{x}^{\mathbf{k}} y^i]\{1-(x_1+\cdots+x_n) - F\circ\boldsymbol{\gamma}\}^{-1}. \qquad \square$$

Finally, we consider the determination of the cluster generating function for an arbitrary reduced set $\mathsf{A} = \{A_1, \ldots, A_p\}$ of distinguished sequences. In this case advantage cannot be taken of the special properties of A which were used in **2.8.7** and **2.8.8**. Instead, it is necessary to know in complete detail how the elements of A overlap in pairs. This information is encoded in the following matrix.

2.8.9. Definition (Connector Matrix)

Let $\mathsf{A} = \{A_1, \ldots, A_p\}$ *be a reduced set. Let* $\mathbf{V}$ *be the* $p \times p$ *matrix* $[v_{ij}]_{p\times p}$, *where*

$$v_{ij} = \sum_{\alpha} \mathbf{x}^{\tau(\alpha)},$$

the sum being taken over all nonempty strings α *such that* $A_i = \alpha\beta$ *and* $A_j = \beta\gamma$, *for some nonempty strings* β *and* γ. *Then* $\mathbf{V}$ *is called the* **connector matrix** *for* A.

$\square$

For example, let $A_1 = 1213$, $A_2 = 13121$, and $A_3 = 333$. Then the connector matrix for $\{A_1, A_2, A_3\}$ is

$$\begin{bmatrix} 0 & x_1x_2 & x_1^2x_2 \\ x_1x_3 + x_1^2x_2x_3 & x_1^2x_2x_3 & 0 \\ 0 & 0 & x_3 + x_3^2 \end{bmatrix}$$

The following lemma gives an expression for the cluster generating function in terms of the connector matrix.

2.8.10. Lemma (Cluster Generating Function for an Arbitrary Set)

The cluster generating function for a reduced set $\mathsf{A} = \{A_1, \ldots, A_p\}$ *with connector matrix* $\mathbf{V}$ *is*

$$C(\mathbf{x}, \mathbf{y}) = \operatorname{trace}(\mathbf{I} - \mathbf{YV})^{-1}\mathbf{YLJ},$$

where $\mathbf{Y} = \operatorname{diag}(y_1, \ldots, y_p)$, $\mathbf{L} = \operatorname{diag}(\mathbf{x}^{\tau(A_1)}, \ldots, \mathbf{x}^{\tau(A_p)})$ *and* $\mathbf{J} = [1]_{p\times p}$.

Proof: Let $f_i(\mathbf{x}, \mathbf{y})$ be the cluster generating function for clusters with an initial segment A_i. We decompose the set of clusters enumerated by f_i by prefixing those enumerated by f_j with A_i, taking account of all possible overlaps. The generating function for this is $y_i v_{ij} f_j$. This, however, excludes the sequence consisting of A_i alone, which is enumerated by $y_i \mathbf{x}^{\tau(A_i)}$. Thus

$$C(\mathbf{x}, \mathbf{y}) = f_1 + \cdots + f_p,$$

where

$$f_i = y_i \mathbf{x}^{\tau(A_i)} + \sum_{j=1}^{p} y_i v_{ij} f_j \qquad \text{for } i = 1, \ldots, p.$$

This is a system of equations for $f_1, \ldots, f_p$, which may be rewritten in the form

$$\mathbf{f}^T = \mathbf{L}\mathbf{y}^T + \mathbf{Y}\mathbf{V}\mathbf{f}^T,$$

where $\mathbf{f} = (f_1, \ldots, f_p)$ and $\mathbf{y} = (y_1, \ldots, y_p)$. It follows that

$$\mathbf{f}^T = (\mathbf{I} - \mathbf{Y}\mathbf{V})^{-1}\mathbf{L}\mathbf{y}^T$$

whence $C(\mathbf{x}, \mathbf{y}) = \operatorname{trace}(\mathbf{I} - \mathbf{Y}\mathbf{V})^{-1}\mathbf{L}\mathbf{Y}\mathbf{J}$. □

We conclude this section with an example of the use of this lemma in enumerating {0, 1}-sequences with no occurrences of certain designated substrings. This is difficult by the methods of Section 2.4.

2.8.11. Example

Let $c(n)$ be the number of {0, 1}-sequences of length n with no occurrences of the substrings $A_1 = 10101101$ and $A_2 = 1110101$.

The connector matrix with respect to length is, by Definition 2.8.9

$$\mathbf{V} = \begin{bmatrix} x^5 + x^7 & x^7 \\ x^2 + x^4 + x^6 & x^6 \end{bmatrix}. \qquad \text{Also } \mathbf{L} = \begin{bmatrix} x^8 & 0 \\ 0 & x^7 \end{bmatrix}.$$

From Lemma 2.8.10 and Theorem 2.8.6 we have, setting $\mathbf{y} = \mathbf{0}$,

$$\begin{aligned} c(n) &= [x^n]\{1 - 2x + \operatorname{trace}(\mathbf{I} + \mathbf{V})^{-1}\mathbf{L}\mathbf{J}\}^{-1} \\ &= [x^n]\{1 - 2x + x^7(1 + x - x^3)(1 + x^5 + x^6 + x^7 - x^9)^{-1}\}^{-1}. \end{aligned}$$

We may use this to obtain a linear recurrence equation for $c(n)$ in the usual way. □

NOTES AND REFERENCES

This section is based on Goulden and Jackson (1979).

2.8.6 Kim, Putcha, and Roush (1977), Zeilberger (1981); **2.8.8** Jackson and Aleliunas (1977); [**2.8.8**] Guibas and Odlyzko (1981a, 1981b).

EXERCISES

2.8.1. Show that the number of $\{0,1\}$-sequences of length n with no occurrences of 0100100 is $[x^n](1+x^3+x^6)(1-2x+x^3-2x^4+x^6-x^7)^{-1}$.

2.8.2. Show that the number of sequences of length l over N_n with no occurrences of 22122 is

$$\sum_{i,j,k\geqslant 0}(-1)^{i+j}n^{k-i}\binom{k}{i}\binom{i+j-1}{j}\binom{j}{l-k-4i-3j}.$$

2.8.3. Show that the number of k-subsets of N_n with m successions (**2.3.13**) is $\binom{k-1}{m}\binom{n-k+1}{k-m}$, as in **2.3.15**.

2.8.4. Show that the number of $\{0,1\}$-sequences of length l with no occurrences of 11101011 or 101111 is

$$[x^l](1+x^5+x^6-x^8-x^9-x^{10})(1-2x+x^5-2x^7+x^9+x^{11})^{-1}.$$

2.8.5. Show that the number of sequences of type $\mathbf{i}$ in N_n^*, with m_j blocks (**2.4.1**) of k j's, for $j=1,\ldots,n$ is

$$[\mathbf{y}^{\mathbf{m}}\mathbf{x}^{\mathbf{i}}]\left\{1-\sum_{j=1}^{n}\left(x_j(1-x_jy_j)+(y_j-1)x_j^k\right)\left(1-y_jx_j+(y_j-1)x_j^k\right)^{-1}\right\}^{-1}.$$

2.8.6. (a) Derive [**2.4.14**] by means of distinguished substrings.
(b) Derive [**2.4.15**] by means of distinguished substrings.

2.8.7. Derive [**2.4.17**(a)] by means of distinguished substrings.

2.8.8. Prove that the number of sequences over N_n of length l with no substrings in the reduced set $\mathsf{A}=\{A_1,\ldots,A_p\}$ is

$$[x^l]\left\{1-nx+\operatorname{trace}\left(\chi(x^{-1})\right)^{-1}\mathbf{J}\right\}^{-1}$$

where $[\chi(x)]_{ij}=\Sigma_\beta x^{|\beta|}$, in which the sum is over all nonempty strings β such that $A_i=\alpha\beta$, $A_j=\beta\gamma$, for $\alpha,\gamma\in\mathsf{N}_n^*$, where $i,j=1,\ldots,p$. The matrix $\chi(x)$ is called the **correlation matrix**.

2.8.9. Prove that the number of unrooted circular permutations on N_n with A-type m is

$$[\mathbf{y}^{\mathbf{m}}\mathbf{x}^{\mathbf{1}}]\Big(\log\{1-(x_1+\cdots+x_n)-C(\mathbf{x},\mathbf{y}-\mathbf{1})\}^{-1}$$
$$+\operatorname{trace}\log\big(\mathbf{I}-(\mathbf{Y}-\mathbf{I})\mathbf{V}^{-1}\big)\Big),$$

where a distinguished substring is allowed to overlap circularly with itself.

2.9. ROOTED PLANAR MAPS AND THE QUADRATIC METHOD

In this section we consider the enumeration of combinatorial configurations called rooted planar maps. Unlike the situation in the enumeration of planted plane trees, the Lagrange theorem, in general, cannot be applied directly to the functional equations encountered here. The equations typically involve two variables, two unknown functions, and are at most quadratic in the unknown functions. However, a strategy called the **quadratic method** may be employed to determine a solution to these equations. This method necessitates deriving a pair of equations to which the Lagrange theorem can be readily applied, and is analogous to separation of variables for differential equations.

2.9.1. The Quadratic Method

We wish to determine formal power series $f(x, y)$ and $h(y)$ such that

$$(g_1 f + g_2)^2 = g_3,$$

where $g_i \equiv g_i(x, y) \equiv G_i(x, y, h(y))$, and $G_i(x, y, z)$ is a known function of x, y, z for $i = 1, 2, 3$. Now let $D(x, y) = (g_1 f + g_2)^2 = g_3$ and suppose that $\alpha \equiv \alpha(y)$ is such that $D(x, y)|_{x=\alpha} = 0$, where, of course, we must check that this substitution is admissible in any particular application. Then $(g_1 f + g_2)|_{x=\alpha} = 0$ and so $(\partial/\partial x)D(x, y)|_{x=\alpha} = 0$. This gives the following pair of simultaneous equations, involving α, y and $h(y)$,

$$\left.\begin{aligned} g_3|_{x=\alpha} &= 0 \qquad (1)\\ \left.\frac{\partial g_3}{\partial x}\right|_{x=\alpha} &= 0 \qquad (2)\end{aligned}\right\}.$$

1. *To obtain* $h(y)$: We could try to eliminate α between (1) and (2), but in the examples encountered in the enumeration of planar maps, it is not possible to carry this out. Instead, eliminate h to give an equation involving α and y. This often expresses y in terms of α, and we determine $h(y)$ by the Lagrange theorem.

2. *To obtain* $f(x, y)$: In many combinatorial situations $h(y)$ is the function of interest, so $f(x, y)$ need not be calculated. If necessary, however, $f(x, y)$ can be determined by solving the equation

$$g_1 f + g_2 = \pm (g_3)^{1/2},$$

where the sign is determined by initial conditions. It is again often convenient to carry out this calculation in terms of x and α and to determine $f(x, y)$ from the Lagrange theorem. □

In combinatorial applications $h(y)$ is in general defined in terms of $f(x, y)$. Thus in the examples given in this section, $h(y) = [x^2]f(x, y)$ or $h(y) = f(1, y)$. The following is a definition of a planar map.

2.9.2. Definition (Planar Map)

A **planar map** *M is a connected nonnull graph G, with loops and multiple edges allowed, embedded in the plane.* □

A planar map M therefore separates the plane into a number of regions, called **faces**, that are disjoint and connected. A single face, called the **outer face**, is unbounded and contains the point at infinity (if G is a tree, then M has no other face). All other faces are called **inner faces**. The edges and vertices of G are the **edges** and **vertices** of M.

An example of a planar map is given in Figure 2.9.1, with various vertices and edges named for future reference.

The **degree of a vertex** in M is the number of edges incident with the vertex, with loops counted twice. The **degree of a face** of M is the number of edges incident with the face. If G is a single vertex, then the degree of the outer face is 0, and M is called the **vertex map**. An **isthmus**, which is an edge whose deletion disconnects the graph G, is counted twice in the degree of any face with which it is incident. If a map M can be expressed as the union of two maps, each with at least one edge and whose intersection is a single vertex, then that vertex is called a **cut-vertex** of M.

For example, in Figure 2.9.1 the degree of vertex v_5 is 4, of vertex v_7 is 4, and of the outer face is 10. The isthmuses are edges e_4, e_5, and e_6. The cut-vertices are v_1, v_3, v_5, and v_7.

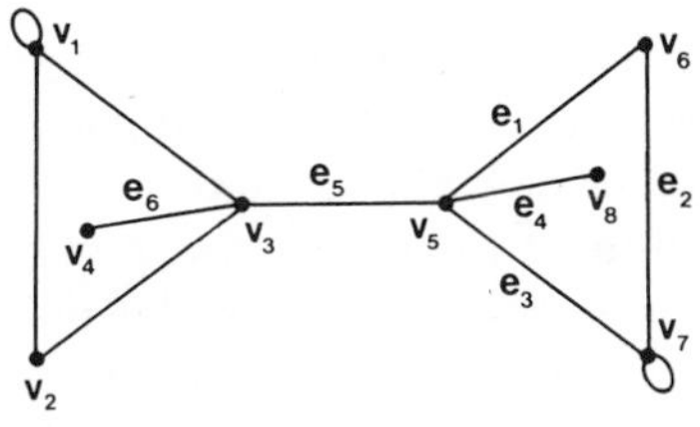

Figure 2.9.1. A planar map.

Since the plane is orientable, there is a natural cyclic sequence of edges that are encountered as each vertex is circulated in a specific direction. Similarly, there is a natural cyclic sequence of edges associated with each face, as the perimeter of the face is traversed in a given direction. For example, if we circulate v_5 of Figure 2.9.1 in a counterclockwise direction beginning with e_4, we obtain the sequence of edges $e_4e_1e_5e_3$. If we traverse the face incident with v_5, v_6, v_7, v_8 in a counterclockwise direction beginning with e_1, then we obtain the sequence of edges $e_1e_4e_4e_3e_2$.

On several occasions it will be convenient to use counting arguments based on the following results, whose proofs are given in [**2.9.1**].

2.9.3. Proposition (Euler's Polyhedral Formula)

Let n_1, n_2, and n_3 be the number of vertices, edges, and faces, respectively, in an arbitrary planar map M. Then

1. $n_1 - n_2 + n_3 = 2$.
2. $\Sigma_u \text{degree}(u) = \Sigma_f \text{degree}(f) = 2n_2$, *where the first summation is over the set of vertices in M and the second summation is over the set of faces in M.* □

For example, in the planar map in Figure 2.9.1 $n_1 = 8$, $n_2 = 11$, $n_3 = 5$ and **Euler's polyhedral formula** (Proposition 2.9.3(1)) is verified in this case. The sum of the degrees of the vertices in this map is $4 + 2 + 4 + 1 + 4 + 2 + 4 + 1 = 22 = 2n_2$ and Proposition 2.9.3(2) is verified for vertices.

Throughout this section we shall be interested only in rooted planar maps, which are defined now.

2.9.4. Definition (Rooted Planar Map, Root Edge, Root Face, Root Vertex)

A **rooted planar map** *is a planar map in which a single edge in the outer face is directed in a counterclockwise direction. The outer face is called the* **root face**, *the directed edge the* **root edge**, *and the vertex which is incident with the tail of the root edge is called the* **root vertex**. □

Suppose that M_1, M_2 and M_3 are rooted planar maps obtained from the planar map in Figure 2.9.1 by selecting as root edges $\overrightarrow{v_2v_3}$, $\overrightarrow{v_6v_5}$, and $\overrightarrow{v_5v_7}$, respectively. Then M_1 and M_3 are distinct rooted planar maps, but M_1 and M_2 are indistinguishable as rooted planar maps.

Let the rooted map consisting of a single directed edge be denoted by δ and of a single directed loop by l. The vertex map, denoted by v, is a rooted planar map with no root edge, but just a single root vertex. A planar map with no cut-vertices is called a **nonseparable planar map** or **2-connected planar map**. Thus δ, l, and v are nonseparable rooted planar maps, and l is the only nonseparable rooted planar map that contains a loop. A planar map with no isthmuses is called a **2-edge-connected planar map**. Thus l and v are 2-edge-connected rooted planar maps, but δ is not.

Throughout this section we shall be concerned with the enumeration of the set of rooted planar maps itself, as well as various subsets of this set. These subsets are nonseparable and 2-edge-connected rooted planar maps and **rooted near-triangulations**. The enumerations are presented in increasing order of the complexity involved in the accompanying set decomposition.

The first set of rooted planar maps that we enumerate is the set of rooted near-triangulations, denoted by T. The elements of T are nonseparable rooted planar maps in which every nonroot face has degree 3. Further, we adopt the convention that δ is the only element of T with no nonroot faces. By considering the effect of removing the root edge and uniquely rooting the remaining planar map, we have the following decomposition for T.

2.9.5. Decomposition (Rooted Near-triangulation; Root Edge)

Let S *be the set of rooted near-triangulations with root face of degree* 2. *Then*

$$\mathsf{T} \stackrel{\sim}{\to} \{\delta\} \,\dot\cup\, \{\delta\} \times \mathsf{T}^2 \,\dot\cup\, \{\delta\} \times (\mathsf{T} - \mathsf{S}).$$

Proof: Consider the root-edge r of an arbitrary element T of $\mathsf{T} - \{\delta\}$, directed from root-vertex u_1 to u_2. Now r lies on a nonroot face that has degree 3. Let u_3 denote the vertex on this face that is not incident with r. There are two disjoint cases to consider.

Case 1. If u_3 lies on the root face of T, as shown in Figure 2.9.2(i), then when r is removed, u_3 is a cut-vertex of the resulting map and T decomposes into an ordered pair of rooted near-triangulations T_1 and T_2, with a unique common vertex. The root edges of T_1 and T_2 are defined to be $\overrightarrow{u_1u_3}$ and $\overrightarrow{u_3u_2}$, respectively.

Case 2. If u_3 does not lie on the root face of T, as shown in Figure 2.9.2(ii), then when r is removed we obtain the rooted near-triangulation T_1, whose root edge is defined to be $\overrightarrow{u_1u_3}$. Since the degree of the root face of T_1 is one larger than the degree of the root face of T and no loops are allowed in T, the root face of T_1 cannot have degree 2.

The construction of Case 1 and Case 2 are reversible and the result follows. □

This decomposition is now used to deduce a functional equation for the generating function for rooted near-triangulations with respect to number of faces and number of edges in the outer face. The solution to this functional equation is obtained by applying the quadratic method.

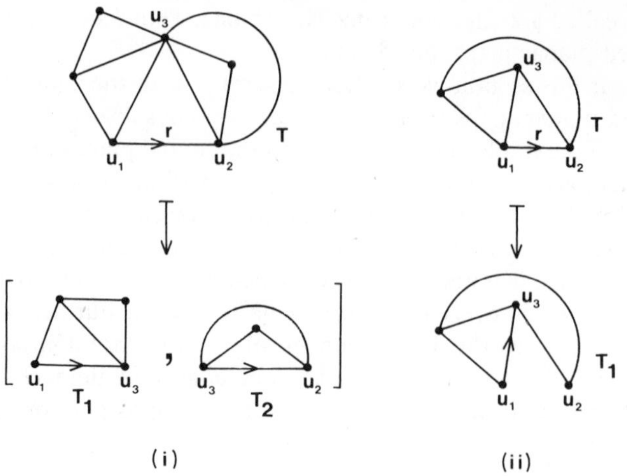

Figure 2.9.2. The root-edge decomposition for rooted near-triangulations.

2.9.6. Rooted Near-triangulations and Inner Faces

Let $T(x, y)$ be the ordinary generating function for T in which x marks edges in the outer face and y marks inner faces. Similarly, let $S(y)$ be the generating function for S in which y marks inner faces. Then, from the root-edge decomposition for rooted near-triangulations,

$$T(x, y) = x^2 + x^{-1}yT^2(x, y) + x^{-1}y\big(T(x, y) - x^2S(y)\big),$$

since by removing the root edge we introduce one additional edge to the outer face and lose one inner face.

To obtain $S(y)$ we follow **2.9.1(1)** and apply the quadratic method. First, complete the square in T to obtain

$$(2yT + y - x)^2 = 4y^2x^2S + (y - x)^2 - 4yx^3 = D(x, y), \tag{1}$$

and then introduce $\alpha \equiv \alpha(y)$, where $D|_{x=\alpha} = (\partial/\partial x)D|_{x=\alpha} = 0$, giving the simultaneous equations

$$4y^2\alpha^2S + (y - \alpha)^2 - 4y\alpha^3 = 0 \tag{2}$$

$$8y^2\alpha S + 2(\alpha - y) - 12y\alpha^2 = 0. \tag{3}$$

Note that the substitution $x = \alpha(y)$ is admissible since D is bounded in x. Eliminating S between (2) and (3), we obtain $(y - \alpha)^2 - \alpha(\alpha - y) + 2y\alpha^3 = 0$, so

$$y = \alpha(1 - 2\alpha^2). \tag{4}$$

Substituting (4) in (2),

$$S = \left(4\alpha^4(1 - 2\alpha^2) - \left(\alpha(1 - 2\alpha^2) - \alpha\right)^2\right)/4\alpha^4(1 - 2\alpha^2)^2,$$

so

$$S = (1 - 3\alpha^2)(1 - 2\alpha^2)^{-2}. \tag{5}$$

Thus we have succeeded in expressing $S(y)$ in terms of α, which in turn satisfies the functional equation

$$\alpha = y(1 - 2\alpha^2)^{-1}$$

from (4). Accordingly, we may apply the Lagrange theorem to obtain

$$\begin{aligned}
[y^{2n}]S(y) &= \frac{1}{2n}[\lambda^{2n-1}]\left\{\frac{d}{d\lambda}\frac{1 - 3\lambda^2}{(1 - 2\lambda^2)^2}\right\}\left\{(1 - 2\lambda^2)^{-1}\right\}^{2n} \\
&= \frac{1}{2n}[\lambda^{2n-1}]2\lambda(1 - 6\lambda^2)(1 - 2\lambda^2)^{-2n-3} \\
&= \frac{2^{n-1}}{n}\left\{\binom{3n+1}{n-1} - 3\binom{3n}{n-2}\right\} = \frac{(3n)!2^{n+1}}{n!(2n+2)!}.
\end{aligned}$$

This is the number of rooted near-triangulations with $2n$ inner faces and root face of degree 2. There are none with an odd number of inner faces. □

For example, there are four rooted near-triangulations with four inner faces and root face of degree 2, in agreement with the preceding result for $n = 2$. These are displayed in Figure 2.9.3.

We now determine $T(x, y)$ by following **(2)** of the quadratic method.

2.9.7. Rooted Near-triangulations, Inner Faces, and Degree of Outer Face

To determine $T(x, y)$, we observe first that, from **2.9.6**, D can be expressed in terms of α and x by substituting (4) and (5) in (1), to give

$$\begin{aligned}
D &= 4x^2\alpha^2(1 - 3\alpha^2) + \left(\alpha(1 - 2\alpha^2) - x\right)^2 - 4x^3\alpha(1 - 2\alpha^2) \\
&= (1 - 2\alpha^2)(x - \alpha)^2\{1 - 2\alpha^2 - 4\alpha x\}.
\end{aligned}$$

Thus

$$D^{1/2} = (1 - 2\alpha^2)(x - \alpha)\left\{1 - 4\alpha x(1 - 2\alpha^2)^{-1}\right\}^{1/2}$$

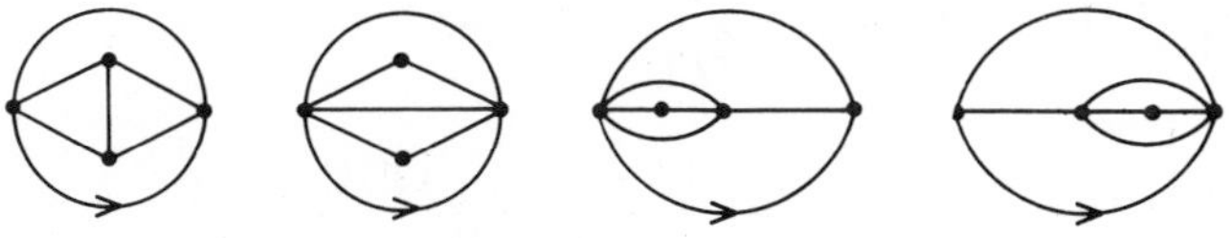

Figure 2.9.3. Rooted near-triangulations with four inner faces and root face of degree 2.

and solving for T in the original quadratic functional equation by **(2)** of the quadratic method, we have

$$T = \frac{x-y}{2y} \pm \frac{D^{1/2}}{2y} = \frac{\alpha x}{1-2\alpha^2} + \frac{x-\alpha}{2\alpha}\left\{1 - \left(1 - 4\alpha x(1-2\alpha^2)^{-1}\right)^{1/2}\right\},$$

where the negative root has been selected since the positive root contributes negative powers of y. (These are not allowed for combinatorial reasons.) Thus

$$T(x, y) = \frac{\alpha x}{1-2\alpha^2} + \frac{x-\alpha}{\alpha} \sum_{n \geqslant 0} \frac{\alpha^{n+1} x^{n+1} (2n)!}{(1-2\alpha^2)^{n+1} n!(n+1)!}$$

so

$$[x^n]T(x, y) = \frac{(2n-4)!}{(n-2)!(n-1)!} \frac{\alpha^{n-2}}{(1-2\alpha^2)^{n-1}} - \frac{(2n-2)!}{(n-1)!n!} \frac{\alpha^n}{(1-2\alpha^2)^n}, \qquad n \geqslant 1,$$

a power series in y. But $\alpha = y(1-2\alpha^2)^{-1}$ from (4) of **2.9.6**, so

$$[x^n]T(x, y) = \frac{(2n-4)!}{(n-2)!(n-1)!} \frac{y^{n-2}}{(1-2\alpha^2)^{2n-3}} - \frac{(2n-2)!}{(n-1)!n!} \frac{y^n}{(1-2\alpha^2)^{2n}}.$$

If a near-triangulation has n edges in the outer face, then, from Proposition 2.9.3, it must have $n + 2j$ inner faces for some $j \geqslant -1$. Thus we need to determine

$$\begin{aligned}[x^n y^{n+2j}]T(x, y) &= \frac{(2n-4)!}{(n-2)!(n-1)!}[y^{2j+2}](1-2\alpha^2)^{-(2n-3)} \\ &\quad - \frac{(2n-2)!}{(n-1)!n!}[y^{2j}](1-2\alpha^2)^{-2n}.\end{aligned}$$

Applying the Lagrange theorem with $\alpha = y(1-2\alpha^2)^{-1}$, from **2.9.6**, we have

$$\begin{aligned}[y^{2m}](1-2\alpha^2)^{-k} &= \frac{1}{2m}[\lambda^{2m-1}]4k\lambda(1-2\lambda^2)^{-k-1-2m} \\ &= \frac{k(k+3m-1)!2^m}{m!(k+2m)!}.\end{aligned}$$

Substituting, we have, for $j \geqslant 1$, $n \geqslant 2$,

$$\begin{aligned}[x^n y^{n+2j}]T(x, y) &= \frac{(2n-4)!}{(n-2)!(n-1)!} \frac{(2n+3j-1)!(2n-3)2^{j+1}}{(j+1)!(2n+2j-1)!} \\ &\quad - \frac{(2n-2)!}{(n-1)!n!} \frac{(2n+3j-1)!}{j!(2n+2j)!}(2n)2^j \\ &= \frac{2^{j+2}(2n+3j-1)!(2n-3)!}{(j+1)!(2n+2j)!((n-2)!)^2}.\end{aligned}$$

If $j = 0, -1$, this formula also holds. This is the number of rooted near-triangulations with $n + 2j$ inner faces and n edges on the outer face. □

Note that we need not have carried out the explicit calculation of $h(y)$ in **2.9.6**, since it is contained as the case $n = 2$ in $T(x, y)$.

Next, we consider the set of all rooted planar maps, denoted by M. In **2.9.8** we give four simultaneous decompositions involving M and three other sets of rooted planar maps, R, L_1, and L_2. R is the subset of M consisting of rooted planar maps that have a root edge that is not an isthmus. Note that l is in R but δ and v are not. L_1 consists of those maps in R with root-face degree equal to 1 and L_2 consists of those maps in R in which the inner face incident with the root edge has degree equal to 1. In the subsequent calculation of the generating function for M, the generating functions for R, L_1, and L_2 are eliminated between the simultaneous functional equations deduced from these decompositions. Thus R, L_1, and L_2 play the role of auxiliary sets of maps and are introduced because the four decompositions that involve M, R, L_1, and L_2 are substantially simpler than the single recursive decomposition that can be given for M.

In one of the decompositions we delete the root edge, and introduce a root edge into the resulting map in a special way, by the **root-shift** operator Δ. For a map M in $\mathsf{R} - \mathsf{L}_1$, let $\Delta(M)$ be the unique map obtained as follows. Suppose that the root-edge r of M is directed from vertex u_1 to u_2 and that u_3 is the vertex that immediately follows u_2 as the outer face is traversed in a counterclockwise direction. If an edge r_1 is added to M, directed from u_1 to u_3 and drawn counterclockwise in the outer face of M, and the edge r is deleted, then the resulting map, rooted on edge r_1, is $\Delta(M)$. Two examples of this operation are illustrated in Figure 2.9.4. Note that M and $\Delta(M)$ have the same number of edges, and that the root-face degree of $\Delta(M)$ is one less than the root-face degree of M.

2.9.8. Decomposition (Rooted Planar Map; Root Edge)

1. $\mathsf{M} - \{v\} \stackrel{\sim}{\rightarrow} \{\delta\} \times \mathsf{M}^2 \,\dot\cup\, \mathsf{R}$. **3.** $\mathsf{L}_1 \stackrel{\sim}{\rightarrow} \{l\} \times \mathsf{M}$.

2. $\mathsf{R} - \mathsf{L}_1 \stackrel{\sim}{\rightarrow} \mathsf{R} - \mathsf{L}_2 : M \mapsto \Delta(M)$. **4.** $\mathsf{L}_2 \stackrel{\sim}{\rightarrow} \{l\} \times \mathsf{M}$.

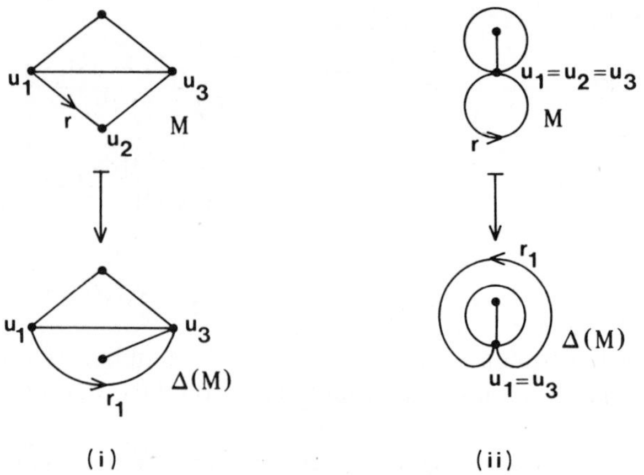

Figure 2.9.4. The action of the root-shift operator Δ.

Proof: **1**. Consider an arbitrary map M in $\mathsf{M} - \{v\}$. Then if the root edge is not an isthmus we have $M \in \mathsf{R}$. Otherwise consider the root-edge r, directed from vertex u_1 to u_2. If r is removed, then M decomposes into an ordered pair of rooted planar maps, M_1 and M_2, with root-vertices u_1 and u_2, respectively. The root-edge r_2 of M_2 is defined to be the edge that follows $\overrightarrow{u_1u_2}$ as we traverse the outer face of M, and the root-edge r_1 of M_1 is defined to be the edge that follows $\overrightarrow{u_2u_1}$ as we traverse the outer face of M. The construction is reversible and is illustrated in Figure 2.9.5(i).

2. This decomposition follows immediately from the discussion following the definition of the root-shift operator Δ.

3. Consider an arbitrary map in L_1. If the root vertex of this map is u_1, and r_1 is the edge following the root-edge r in the counterclockwise circulation of u_1, then the map, with root edge chosen to be r_1, obtained by deleting r, is an arbitrary element of M. This construction is reversible and is illustrated in Figure 2.9.5(ii). If the original map is the loop l, then we obtain the vertex map v by these means.

4. For an arbitrary map in L_2, remove the root-edge loop r, and root the resulting planar map on the edge, r_1, which immediately follows r as the root face is traversed in a counterclockwise direction. As a result we obtain a unique element of M, illustrated in Figure 2.9.5(iii). If the map is l, then we obtain v by this procedure. □

This decomposition is used to determine the number of rooted planar maps on n edges.

2.9.9. Rooted Planar Maps

Let $M(x, y)$, $R(x, y)$, $L_1(x, y)$, and $L_2(x, y)$ be the generating functions for M, R, L_1, and L_2, respectively, in which y marks edges and x marks edges in

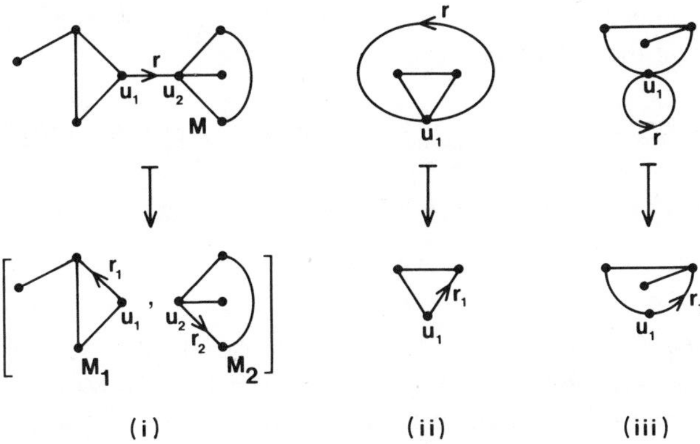

Figure 2.9.5. Root-edge decomposition for rooted planar maps.

the outer face. Then, from Decomposition 2.9.8,

$$M(x, y) - 1 = x^2yM^2(x, y) + R(x, y), \tag{1}$$

$$R(x, y) - L_1(x, y) = x\{R(x, y) - L_2(x, y)\}, \tag{2}$$

and $$L_1(x, y) = xyM(1, y), \qquad L_2(x, y) = xyM(x, y).$$

Eliminating L_1 and L_2 from (2), we have

$$R(x, y) = xy(1 - x)^{-1}\{M(1, y) - xM(x, y)\}.$$

To calculate $h(y) \equiv M(1, y)$, we eliminate $R(x, y)$ between the previous equation and (1) to give

$$(1 - x)\{M(x, y) - 1\} = x^2y(1 - x)M^2(x, y) + xy\{M(1, y) - xM(x, y)\}.$$

This is a quadratic functional equation for $M(x, y)$. To obtain $h(y)$ we follow the quadratic method, first completing the square, to get

$$\begin{aligned}&\{2x^2y(1 - x)M(x, y) - 1 + x - x^2y\}^2 \\ &= (1 - x + x^2y)^2 - 4x^2y(1 - x)^2 - 4x^3y^2(1 - x)h(y) \\ &= D(x, y).\end{aligned} \tag{3}$$

We again introduce $\alpha \equiv \alpha(y)$ where $D|_{x=\alpha} = (\partial/\partial x)D|_{x=\alpha} = 0$. This substitution is admissible because $M(x, y)$ is bounded in x. This follows, since, from Proposition 2.9.3, the number of edges in the outer face is bounded above by twice the number of edges in the map. In fact, it is more convenient to consider the equivalent equations $(1-x)^{-1}D|_{x=\alpha} = (\partial/\partial x)(1-x)^{-1}D|_{x=\alpha} = 0$, which give the simultaneous equations

$$1-\alpha+2\alpha^2 y+\alpha^4 y^2(1-\alpha)^{-1}-4\alpha^2 y(1-\alpha)-4\alpha^3 y^2 h(y)=0 \quad (4)$$

$$-1-4\alpha y+12\alpha^2 y+4\alpha^3 y^2(1-\alpha)^{-1}+\alpha^4 y^2(1-\alpha)^{-2}-12\alpha^2 y^2 h(y)=0. \quad (5)$$

Note that $\alpha(0)=1$ from (3), so $(1-\alpha)^{-1}$ does not exist as a power series. However, $\alpha'(0)=1$ from (3), by comparing coefficients, so $(1-\alpha)^{-1}y$ does exist, and (4) and (5) are well formed. Eliminating $h(y)$ between (4) and (5), we have

$$3-2\alpha-2\alpha^2 y-\alpha^4 y^2(1-\alpha)^{-2}=0. \quad (6)$$

Solving this quadratic for y gives

$$\alpha^2 y = -(1-\alpha)^2-(1-\alpha)(2-\alpha),$$

where the negative root is selected, since $\alpha(0)=\alpha'(0)=1$. Thus y is given in terms of α by

$$y=-\alpha^{-2}(1-\alpha)(3-2\alpha), \quad (7)$$

and substituting this in (4), we obtain

$$h=\alpha(4-3\alpha)(3-2\alpha)^{-2}. \quad (8)$$

Now let $\beta=1-\alpha^{-1}$ in (7) and (8), giving the following simultaneous functional equations

$$\left.\begin{aligned} \beta &= y(1-3\beta)^{-1} \\ h &= (1-4\beta)(1-3\beta)^{-2} \end{aligned}\right\}.$$

We determine $h(y)$ by the Lagrange theorem, to obtain, for $n \geqslant 1$,

$$\begin{aligned}
[y^n]h(y) &= \frac{1}{n}[\lambda^{n-1}]\left\{\frac{d}{d\lambda}(1-4\lambda)(1-3\lambda)^{-2}\right\}\{(1-3\lambda)^{-1}\}^n \\
&= \frac{2}{n}[\lambda^{n-1}](1-6\lambda)(1-3\lambda)^{-n-3} \\
&= \frac{2(2n)!3^n}{n!(n+2)!}.
\end{aligned}$$

This is the number of rooted planar maps on n edges, for $n \geqslant 0$. □

In the preceding solution, the equation $y = -\alpha^{-2}(1-\alpha)(3-2\alpha)$ is not suitable for solution by the Lagrange theorem. Thus we must find a substitution for a new parameter β that yields a suitable form. There are many such substitutions, but $\beta = 1 - \alpha^{-1}$, by trial and error, yields the most compact solution when the Lagrange theorem is applied.

Although it is theoretically possible to calculate $M(x, y)$ by **(2)** of the quadratic method, in practice no convenient expression has been obtained for $[x^n y^m]M(x, y)$, the number of rooted planar maps with n edges in the outer face and m edges.

We now turn to the enumeration of C, the set of 2-edge-connected rooted planar maps. Let H be the subset of such maps for which the map obtained by the removal of the root edge is also 2-edge-connected. In **2.9.10** we give a decomposition for C in terms of H. The latter is an auxiliary set, which is eliminated by means of decompositions obtained by suitably restricting **2.9.8**.

2.9.10. Decomposition (2-Edge-Connected Rooted Planar Map)

$$\mathsf{C} - \mathsf{H} - \{v\} \mathrel{\tilde{\rightarrow}} \dot{\bigcup_{k \geqslant 2}} \mathsf{H}^k.$$

Proof: Consider an arbitrary map M in $\mathsf{C} - \mathsf{H} - \{v\}$. Then the map obtained by deleting the root edge r of M contains k isthmuses, for some $k \geqslant 1$. These isthumuses must lie on the outer face in M. Let $r_1, \ldots, r_k$ denote the isthmuses, directed counterclockwise, which are encountered, in that order, as the edges of the outer face are traversed starting at the root edge r. Suppose that r_i is directed from vertex u_i to w_i, $i = 1, \ldots, k$, and r is directed from u_{k+1} to w_0. If the edges $r, r_1, \ldots, r_k$ are removed from M, we are left with an ordered collection of 2-edge-connected maps $M_1, \ldots, M_{k+1}$. Now let $\hat{M}_i$ be the unique rooted map obtained from M_i by adding a root edge, directed from u_i to w_{i-1}, and drawn in the inner face of M that is incident with r, for $i = 1, \ldots, k+1$. If $u_i = w_{i-1}$, the root edge is a loop. This decomposition is illustrated in Figure 2.9.6. Then $(\hat{M}_1, \ldots, \hat{M}_{k+1})$ is in H^{k+1}, since the removal of the root edge of $\hat{M}_i$ leaves the 2-edge-connected graph M_i. The procedure is reversible, and the

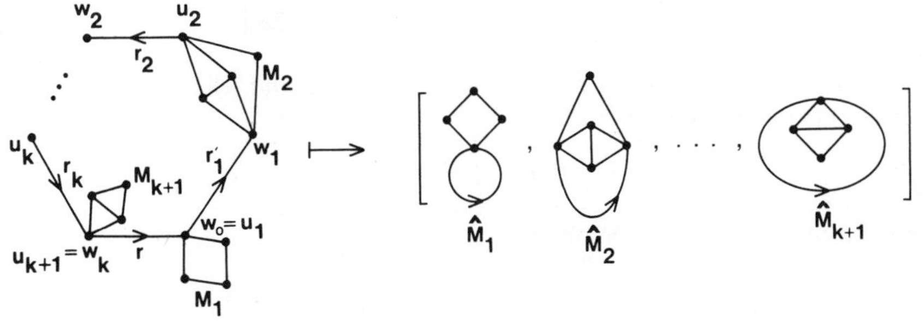

Figure 2.9.6. Decomposition of a two-edge-connected map.

sum of the degrees of the outer faces of $\hat{M}_1, \ldots, \hat{M}_{k+1}$ is equal to the degree of the outer face of M. The result follows, since k can be any positive integer. □

This decomposition is now used to deduce the number of 2-edge-connected rooted planar maps with a given number of edges.

2.9.11. 2-Edge-Connected Rooted Planar Maps

Let $C(x, y)$ and $H(x, y)$ be the generating functions for C and H, respectively, in which x marks edges in the outer face and y marks edges. Then, from Decomposition 2.9.10,

$$C(x, y) - H(x, y) - 1 = H^2(x, y)\{1 - H(x, y)\}^{-1}.$$

Consider the three decompositions obtained by restricting Decompositions 2.9.8(2), (3), and (4) to the sets of all 2-edge-connected maps in $\mathsf{R} - \mathsf{L}_1$, L_1, and L_2, respectively. The ranges of the resulting decompositions are the sets of all 2-edge-connected maps in $\mathsf{R} - \mathsf{L}_2$ and $\{l\} \times \mathsf{C}$, $\{l\} \times \mathsf{C}$, respectively. Applying the sum and product lemmas to these decompositions and eliminating the generating functions for the 2-edge-connected maps in L_1 and L_2, we get

$$H(x, y) - xyC(1, y) = x\{H(x, y) - xyC(x, y)\}.$$

Our aim is to determine $b(y) \equiv C(1, y)$, so, eliminating $H(x, y)$, we have

$$(1 - x)C(x, y) - xyC(x, y)\{b(y) - xC(x, y)\} = 1 - x.$$

Following the quadratic method we complete the square to get

$$(2x^2yC + \beta)^2 = \beta^2 + 4x^2y(1 - x) = D(x, y)$$

where $\beta \equiv \beta(x, y) = 1 - x - xyb(y)$. We now make the admissible substitution $x = \alpha(y)$ such that $D|_{x=\alpha} = (\partial/\partial x)D|_{x=\alpha} = 0$. These two equations are

$$\beta^2(\alpha, y) + 4\alpha^2y(1 - \alpha) = 0 \tag{1}$$

$$2\beta(\alpha, y)(-1 - yb(y)) - 12\alpha^2y + 8\alpha y = 0. \tag{2}$$

Replacing $\alpha(-1 - yb(y))$ by $\beta(\alpha, y) - 1$ in (2), and eliminating $\beta^2(\alpha, y)$ between (1) and (2), gives

$$\beta(\alpha, y) = -2\alpha^3y. \tag{3}$$

Substituting (3) into (1) we obtain

$$y = \alpha^{-4}(\alpha - 1), \tag{4}$$

and using $\beta(\alpha, y) = 1 - \alpha - \alpha y b(y)$ and (3) and (4), we have

$$yb(y) = \alpha^{-2}(1-\alpha)(\alpha-2). \tag{5}$$

Now let $\alpha - 1 = \zeta$ so (4) and (5) yield the following system of simultaneous functional equations:

$$\left.\begin{aligned} \zeta &= y(1+\zeta)^4 \\ yb(y) &= \zeta(1-\zeta)(1+\zeta)^{-2} \end{aligned}\right\}.$$

By the Lagrange theorem, for $n \geqslant 0$,

$$\begin{aligned}[y^n]C(1, y) &= [y^n]b(y) = [y^{n+1}]yb(y) \\ &= \frac{1}{n+1}[\lambda^n]\left\{\frac{d}{d\lambda}\lambda(1-\lambda)(1+\lambda)^{-2}\right\}(1+\lambda)^{4(n+1)} \\ &= \frac{1}{n+1}[\lambda^n](1-3\lambda)(1+\lambda)^{4n+1} = \frac{2(4n+1)!}{(n+1)!(3n+2)!}.\end{aligned}$$

This is the number of 2-edge-connected rooted planar maps on n edges, for $n \geqslant 0$. □

Finally, we consider the enumeration of the set P of rooted nonseparable maps. The indirect decomposition for P involves a composition with the set M of all rooted planar maps. This decomposition is an example of a **change of connectivity**, and other examples are given as exercises.

2.9.12. Decomposition (Nonseparable Rooted Planar Map)

$$\mathsf{M} \xrightarrow{\sim} \mathsf{P} \circ (\mathsf{M}^2)$$

where the composition is with respect to s-objects, which are edges.

Proof: Consider an arbitrary map M in $\mathsf{P} - \{v\}$. Each element of $\mathsf{M} - \{v\}$ can be uniquely obtained by the following operation in each face of M. As a face, say f, is traversed in a counterclockwise direction beginning at an arbitrary vertex v_1, we encounter a sequence $v_1, e_1, v_2, e_2, \ldots, v_k, e_k$, for some $k \geqslant 1$, consisting alternately of vertices and edges. Then, for $i = 1, \ldots, k$, we insert an arbitrary rooted planar map $M_i \in \mathsf{M}$ in f by identifying the root vertex of M_i with the vertex v_i. If this procedure is carried out in all faces of M simultaneously, then we obtain a map M', a unique element of $\mathsf{M} - \{v\}$, whose root edge is defined to be the root edge of M. The operation of this construction in a single face f of M is illustrated in Figure 2.9.7. Note that in

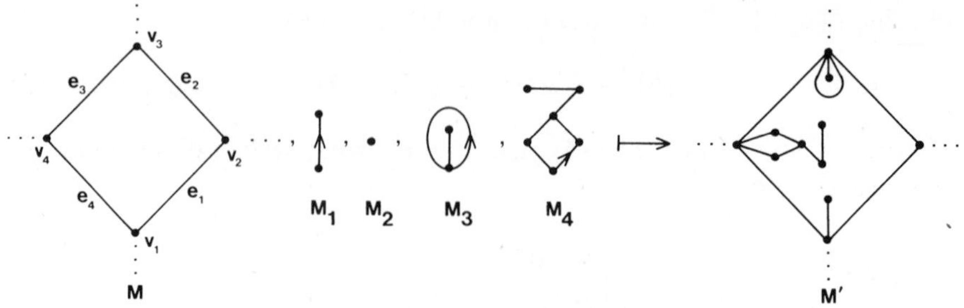

Figure 2.9.7. Construction for rooted planar maps from nonseparable rooted planar maps.

the example illustrated we have $M_2 = v$, so no map is inserted in f at v_2, since the vertex map v is identified with v_2 in the construction. Finally, since each edge is incident with two faces, or is encountered twice in a single face if it is an isthmus, we have associated two elements of M uniquely with each edge of M. Moreover, the number of edges in M' is equal to the number of edges in M plus the sum of the numbers of edges in the inserted maps. Thus the number of edges is additively preserved in this decomposition. Note that the defining graph of M is the maximal nonseparable subgraph of the defining graph of M' that contains the root edge of M'. Thus M can be uniquely recovered from M' and the construction is reversible. □

We now use this decomposition to enumerate nonseparable rooted planar maps. Since the decomposition involves the set M of all rooted planar maps, this enumeration is only possible because the generating function for M is known from **2.9.9**.

2.9.13. Nonseparable Rooted Planar Maps

Let $h(y)$ and $P(y)$ be the generating functions for M, the set of all rooted planar maps, and P, the set of nonseparable rooted planar maps, respectively, in which y marks edges. We wish to obtain $P(y)$. From Decomposition 2.9.12 and the composition lemma for ordinary generating functions

$$h(y) = P(yh^2(y)). \tag{1}$$

But, from **2.9.9**, $h(y)$ satisfies the simultaneous equations

$$\left.\begin{aligned} y &= \alpha(1-3\alpha) \\ h &= (1-4\alpha)(1-3\alpha)^{-2}, \end{aligned}\right\} \tag{2}$$

called **parametric equations** for h and y. To determine the power series P, change variables to $x = yh^2(y)$ and eliminate h and y between (2) and (1).

Thus a system of parametric equations for P and x is

$$\left.\begin{aligned} x &= \alpha(1-4\alpha)^2(1-3\alpha)^{-3} \\ P &= (1-4\alpha)(1-3\alpha)^{-2}. \end{aligned}\right\} \tag{3}$$

Finally, setting $\theta = \alpha(1-3\alpha)^{-1}$, so $(1-3\alpha)^{-1} = 1+3\theta$ and $(1-4\alpha)(1-3\alpha)^{-1} = 1-\theta$, we have, from (3),

$$\left.\begin{aligned} \theta &= x(1-\theta)^{-2} \\ P &= (1-\theta)(1+3\theta). \end{aligned}\right\} \tag{4}$$

Applying the Lagrange theorem to these equations yields, for $n \geqslant 1$,

$$\begin{aligned} [x^n]P(x) &= \frac{1}{n}[\lambda^{n-1}]\left\{\frac{d}{d\lambda}(1-\lambda)(1+3\lambda)\right\}\left\{(1-\lambda)^{-2}\right\}^n \\ &= \frac{1}{n}[\lambda^{n-1}](2-6\lambda)(1-\lambda)^{-2n} \\ &= \frac{2(3n-3)!}{n!(2n-1)!}. \end{aligned}$$

This is the number of nonseparable rooted planar maps with n edges, for $n \geqslant 1$. □

There are six nonseparable rooted planar maps with four edges, in agreement with the preceding result for $n = 4$. These are illustrated in Figure 2.9.8.

In the preceding solution it should be noted that the Lagrange theorem can be applied directly to the simultaneous functional equations given in (3). However, the solution is then a binomial convolution, which is avoided by the change of parameter $\theta = \alpha(1-3\alpha)^{-1}$, giving the equivalent equations (4). Change of variable and change of parameter are used extensively throughout the exercises.

The four examples that have been discussed give an idea of the types of decompositions that can be obtained for rooted planar maps. Each involves the use of the Lagrange theorem in a nontrivial way. In particular, the quadratic method has been applied in three of the cases, and by carefully manipulating a pair of equations we are able to determine the solution in an implicit form suitable for the Lagrange theorem. The successful application of the quadratic

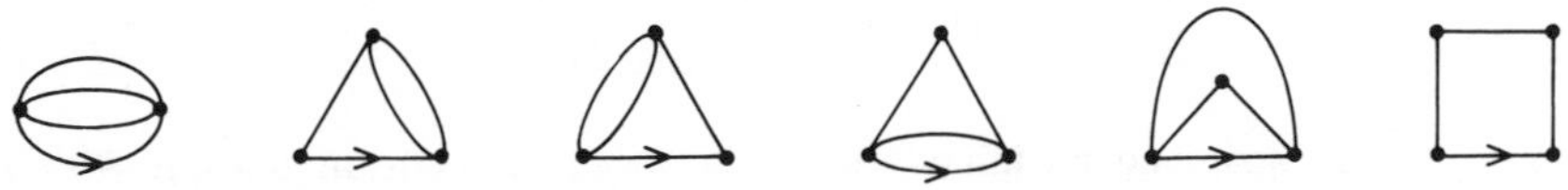

Figure 2.9.8. Nonseparable rooted planar maps with four edges.

method, given an initial equation for which it is suitable, relies very greatly on our ability to find the particular transformations that yield this implicit form. These have been different in each instance, and the solutions can only serve as examples of the means for determining generating functions satisfying the functional equations of **2.9.1**.

NOTES AND REFERENCES

For a connection to chromatic theory and the four-color theorem see Tutte (1973b); for enumeration on surfaces of higher genus see Walsh and Lehman (1972a, 1972b, 1975); for unrooted planar maps see Wormald (1981a, b).

2.9.1 Tutte (1973a); **2.9.3** Biggs, Lloyd, and Wilson (1976); **2.9.6, 7** Mullin (1964b); **2.9.9** Tutte (1963); **2.9.11** Wormald (private communication); **2.9.13** Tutte (1963); [**2.9.2**] Tutte (1962); [**2.9.3**] Brown (1964); [**2.9.4**] Mullin (1965); [**2.9.5**] Tutte (1962); [**2.9.6**] Wormald (private communication); [**2.9.7**] Brown (1963); [**2.9.8**] Brown and Tutte (1964); [**2.9.9**] Tutte (1963).

EXERCISES

2.9.1. Prove Proposition 2.9.3.

2.9.2. (a) A **planar triangulation** is a rooted near-triangulation in which the outer face has degree 3. Let $S(y)$ be the generating function, which was determined in **2.9.6**, for rooted near-triangulations with outer face of degree 2, where y marks inner faces. Show that $S(y) - 1$ is the generating function for planar triangulations, where y marks all faces.

(b) A **strict triangulation** is a planar triangulation with no multiple edges. If $T(y)$ is the generating function for strict triangulations with respect to faces, show that $S(y) - 1 = T(yS^{3/2}(y))$.

(c) Use the parametric equations for $S(y)$ given in **2.9.6** to show that

$$T(t) = \sum_{n \geq 1} \frac{2(4n-3)!}{n!(3n-1)!} t^{2n}.$$

(d) A **simple triangulation** is a strict triangulation with no separating triangle. A **separating triangle** is a triangle that has at least one vertex inside it and at least one outside it. (For example, $v_1 v_2 v_3$ in Figure 2.9.1 of the text.) Show that the number of simple triangulations with $2n$ faces, for $n > 1$, is

$$\frac{1}{n} \sum_{i=0}^{n-1} (-1)^{n-i} \binom{2n+i-1}{i} \binom{2n-2i}{2}.$$

2.9.3. (a) A **strict near-triangulation** is a rooted near-triangulation with no multiple edges. Let $B(x, y)$ be the generating function for strict near-triangulations, with x marking edges in the outer face and y marking inner faces. Show that $T(x, y) = B(xS^{1/2}(y), yS^{3/2}(y))$, where $T(x, y)$ and $S(y)$, given in **2.9.7** and **2.9.6**, are generating functions for rooted near-triangulations with respect to edges in the outer face and inner faces, and rooted near-triangulations with outer face of degree 2, with respect to inner faces.

(b) Use the results of **2.9.7** to show that the number of strict near-triangulations with n edges in the outer face and $n + 2j$ inner faces is

$$\frac{2(2n-3)!(2n+4j-1)!}{(n-1)!(n-3)!(j+1)!(2n+3j)!} \qquad \text{for } n \geqslant 3, j \geqslant 0,$$

and

$$\frac{(2n-4)!}{(n-1)!(n-2)!} \qquad \text{for } n \geqslant 2, j = -1.$$

2.9.4. (a) A **simple near-triangulation** is a strict near-triangulation with no separating triangle. Let $U(x, y)$ be the generating function for simple near-triangulations with x marking edges in the outer face and y marking inner faces. Show that $B(x, y) - x^3y = U(x, y^{-1}T(y)) - x^3y^{-1}T(y)$, where $B(x, y)$ and $T(y)$, given in [**2.9.3**(a)] and [**2.9.2**(b)] respectively, are generating functions for strict near-triangulations with respect to edges in the outer face and inner faces, and strict triangulations with respect to all faces.

(b) Use the results of [**2.9.3**(b)] to show that the number of simple near-triangulations with n edges in the outer face and $n + 2j$ inner faces is

$$\frac{(2n-4)!(n+3j-1)!}{(j+1)!(n-1)!(n-4)!(n+2j)!}, \qquad n \geqslant 4, j \geqslant 0.$$

2.9.5. (a) A **2-separator** of a planar map M is a pair $\{u, v\}$ of distinct vertices such that M can be expressed as the union of two planar maps whose intersection is the pair $\{u, v\}$ of vertices. A **strong near-triangulation** is a strict near-triangulation with no 2-separators. The single edge δ is not a strong near-triangulation. Let $A(x, y)$ be the generating function for strong near-triangulations with x marking edges in the outer face and y marking inner faces. Find a decomposition that leads to the equation

$$B(z, w) - z^2 = z^2B^{-1}(z, w)A(z^{-1}B(z, w), w),$$

where $B(z, w)$ is the generating function for strict near-triangulations, given in [**2.9.3**(b)].

(b) From part (a) and the results for strict near-triangulations, show that the number of strong near-triangulations with n edges in the outer face and $n + 2j$ inner faces is

$$\sum_{i \geqslant 1} \frac{(-1)^{n-i}(n+i-4)!(4j+2n+i-3)!}{(i-1)!i!(n-i-1)!(j+1)!(3j+2n+i-2)!} f(n, i, j)$$

where

$$f(n, i, j) = (n - i - 1)(2n + i - 6)(3j + 2n + i - 2)$$
$$+ (n + i - 3)(2n + i - 2)(j + 1).$$

(c) Find a decomposition that leads directly to the functional equation $(x^2 - A(x, y))(A(x, y) - x^3y + x^2T(y)) = xyA(x, y)$, where $T(y)$ is the generating function for strict triangulations with respect to all faces, determined in [**2.9.2**(c)].

2.9.6. (a) Let $b(y)$ and $P(y)$ be the generating functions for 2-edge-connected and nonseparable rooted planar maps, respectively, where y marks edges. Show that $b(y) = P(yb^2(y))$.

(b) Use the parametric equations for 2-edge-connected maps, given in **2.9.11**, to determine the number (given in **2.9.13**) of nonseparable rooted planar maps with n edges.

2.9.7. Let $P(x, y)$ be the ordinary generating function for nonseparable rooted planar maps with y marking edges and x marking edges incident with the outer face. Let $Q(x, y)$ be the generating function for those nonseparable rooted planar maps that, when the root edge is deleted, yield a nonseparable planar map. By finding decompositions similar to **2.9.8** and **2.9.10**, show that

(a) $Q(x, y) - xy - x^2y(P(1, y) - 1 - y) = x\{Q(x, y) - xy - y(P(x, y) - 1 - xy)\}$.

(b) $P(x, y) - x^2y - 1 = xy\{1 - (xy)^{-1}(Q(x, y) - xy)\}^{-1}$.

(c) Deduce a quadratic equation for $P(x, y)$ and use (**1**) of the quadratic method to determine the number (given in **2.9.13**) of nonseparable rooted planar maps with n edges.

(d) Use (**2**) of the quadratic method to show that the number of nonseparable rooted planar maps with n edges incident with the outer face, and m edges, is

$$\frac{n}{(2m - n)!} \sum_{j=n}^{\min(m,2n)} \frac{(2j - n)(3n - 2j - 1)(j - 2)!(3m - n - j - 1)!}{(j - n + 1)!(j - n)!(2n - j)!(m - j)!},$$

for $m \geqslant n \geqslant 2$.

2.9.8. (a) Let $yzP(x, y, z)$ be the ordinary generating function for nonseparable rooted planar maps with x marking faces and z marking vertices. From the decompositions used in [**2.9.7**(a), (b)], deduce that

$$(P(x, y, z) - x^2z - 1)(P(x, y, z) - xP(1, y, z)) = xy(1 - x).$$

(b) Use the quadratic method to show that the number of nonseparable rooted planar maps with $n+1$ faces and $m+1$ vertices is

$$\frac{(2n+m-2)!(2m+n-2)!}{n!m!(2n-1)!(2m-1)!} \qquad \text{for } m, n \geqslant 1.$$

2.9.9. (a) Let $P(y)$ be the generating function for nonseparable rooted planar maps with respect to edges, and $Q(y)$ be the generating function for those nonseparable rooted planar maps which, when the root edge is deleted, yield a nonseparable planar map. From [**2.9.7**(b)] show that

$$Q(y) = 2y - y^2(P(y) - y - 1)^{-1}.$$

(b) Let $D(y)$ be the generating function with respect to edges, for those nonseparable rooted planar maps which, when the root edge is deleted, yield a map for which the pair of vertices incident with the root edge is not a 2-separator. The map δ is included, but l and v are not. Show that $D(y) = Q(y)$.

(c) A **3-connected** rooted planar map is a nonseparable rooted planar map with no 2-separators. Let $E(y)$ be the generating function with respect to edges for 3-connected rooted planar maps with more than three edges, which are called **rooted *c*-nets**. Show that

$$(y^{-1}F(y))^{-1}E(y^{-1}F(y)) = 2 - y - 2(1 + y^{-1}F(y))^{-1} - y^{-1}F(y),$$

where $F(y) = P(y) - 2y - 1$.

(d) Using the parametric equation for $P(y)$ given in **2.9.13**, deduce from part (c) that the number of rooted c-nets with n edges is

$$2(-1)^n + \frac{1}{6(n-1)!}\sum_{j\geqslant 0}(-1)^j\frac{(j+4)!(2n-j-6)!}{j!(n-j-3)!}, \qquad n \geqslant 4.$$

3

The Combinatorics of the Exponential Generating Function

3.1. INTRODUCTION

The material of Chapter 2 may be extended by introducing another class of combinatorial configurations called **s-tagged** (or, more briefly, **tagged**) **configurations**. These may be thought of as configurations in the sense of Chapter 2, but now the n s-objects carry, as **tags**, distinct integers between 1 and n. For example, a labeled graph may be regarded as a tagged configuration whose tagged objects are vertices. A matrix may be regarded as a tagged configuration with two distinct types of tagged objects, namely, rows and columns, tagged by row and column labels, respectively.

To deal with tagged configurations, we introduce the **exponential generating function** for a set S of distinct s-tagged configurations. This is defined to be

$$\sum_{\sigma \in \mathsf{S}} \frac{x^{\omega_s(\sigma)}}{\omega_s(\sigma)!}$$

where $\omega_s(\sigma)$, the number of s-objects in σ, is called the **tag weight** of σ. In this chapter we consider the combinatorial significance of this power series.

3.1.1. The Elementary Counting Lemmas

The main counting lemmas for the exponential generating function are given in Section 3.2. If F and G are sets of distinct s-tagged configurations, with exponential generating functions f and g, respectively, then these lemmas give the exponential generating function for

1. $\mathsf{F} \,\dot{\cup}\, \mathsf{G}$ disjoint union
2. $\mathsf{F} * \mathsf{G}$ $*$-product

3. $\mathsf{F}\circledast\mathsf{G}$ $\quad$ $*$-composition
4. $\dfrac{d\mathsf{F}}{ds}$ $\quad$ $*$-derivative

as $f + g$ (sum), fg (product), $f \circ g$ (composition), and f' (derivative), respectively. These lemmas serve as the interface between the **combinatorics** of a set of tagged configurations and the algebra of exponential generating functions. The combinatorics is expressed in terms of decompositions utilizing the preceding four expressions. The decompositions are required to be additively tag-weight preserving and may again be classified as direct, indirect, or recursive.

3.1.2. The $*$-Product and $*$-Composition

To deal with tagged configurations we introduce another combinatorial product, called the **$*$-product**. Intuitively, a tagged configuration with n s-objects may be regarded as a configuration that has been "built" on a set of tagged s-objects with **tag-set** N_n. The $*$-product is the operation that decomposes a tagged configuration into two configurations, one built on tagged s-objects with tag-set α and the other on tagged s-objects with tag-set β, where $\{\alpha, \beta\}$ is a partition of N_n.

Section 3.3 deals with some applications of the elementary counting lemmas to the enumeration of trees and functions. Some of the decompositions are related to those given in Chapter 2 for the corresponding untagged configurations, whereas others have no such counterparts. We continue to emphasize the fact that often a configuration may be decomposed in several different ways, depending on the enumerative information that is to be preserved. As a particular example of this, we give direct, indirect, and recursive decompositions for derangements, complementing the "at-least" decomposition given in **2.2.30**.

Section 3.4 contains a detailed study of two rather more complicated configurations. The purpose of this is to demonstrate the enumerative power now at our disposal and to do this in a context necessitating the use of a decomposition that, in turn, requires several subdecompositions. The two configurations are proper 2-covers of a set, and homeomorphically irreducible simple labeled graphs. For the latter an "at-least" construction is used.

3.1.3. The $*$-Derivative

It is often convenient to be able to **distinguish** a tagged s-object in a configuration or to **delete** a specific tagged s-object from a configuration. Both of these operations may be realized by the operation called **$*$-differentiation**. Section 3.4 contains examples of the determination of differential equations by distinguishing a tagged s-object. These examples include a differential decomposition for proper 2-covers. A differential decomposition for simple 3-regular labeled graphs is given in Section 3.5.

3.1.4. The Γ-Series

Section 3.5 gives a general method for deriving recurrence equations for coefficients in symmetric multivariate generating functions. For a symmetric multivariate generating function $T(\mathbf{t})$, where $\mathbf{t} = (t_1, t_2, \ldots)$ we derive a function $\Gamma(T)$ of $\mathbf{y} = (y_1, y_2, \ldots)$ with the property that

$$[\mathbf{t}^{\mathbf{i}}]T = \left[\frac{\mathbf{y}^{\mathbf{j}}}{\mathbf{j}!}\right]\Gamma(T),$$

where $\mathbf{j} = (j_1, j_2, \ldots)$, $\mathbf{i} = (i_1, i_2, \ldots)$, and $\mathbf{i}$ contains j_k k's, for $k \geqslant 1$. The function $\Gamma(T)$ is called the **Γ-series** of $T(\mathbf{t})$. This transformation is particularly useful when $\mathbf{i}$ is such that $i_l = 0$ for $l > p$ and $i_l = m$ for $1 \leqslant l \leqslant p$. In this case $j_l = p$ if $l = m$ and $j_l = 0$ if $l \neq m$, so we require the coefficient of $y_m^p/p!$ in $\Gamma(T)$. This is called the coefficient of the **regular term** of degree m in T.

The use of Γ-series is demonstrated in connection with simple 3-regular labeled graphs and nonnegative integer square matrices with row and column sums equal to 2.

3.2. THE ELEMENTARY COUNTING LEMMAS

This section contains the elementary counting lemmas associated with the exponential generating function. Again, for simplicity, the exposition is confined to the univariate case. We begin by defining a **tagged configuration.**

3.2.1. Definition (s-Tagged Configuration, Tag Set, Tag Weight)

1. *An* **s-tagged** *configuration σ on N_k is a configuration with k s-objects, such that assigned to each s-object is a distinct element of N_k, called its* **tag**.
2. *If $\alpha = \{\alpha_1, \ldots, \alpha_k\} \subseteq \mathsf{N}_+$, where $\alpha_1 < \cdots < \alpha_k$, then σ_α denotes the configuration, on α, obtained by replacing the tag i by α_i for $i = 1, \ldots, k$. If Π is a set of s-tagged configurations then*

$$\hat{\Pi} = \{\sigma_\alpha | \sigma \in \Pi, \alpha \subset \mathsf{N}_+\}.$$

3. *α is called the* **tag set** *of σ_α and is denoted by $\Omega_s(\sigma_\alpha)$, and $|\alpha|$ is called the* **tag weight** *of σ and σ_α, denoted by $\omega_s(\sigma)$, $\omega_s(\sigma_\alpha)$, respectively. If $\omega_s(\sigma) = 0$, then σ is an s-tagged configuration and $\sigma_\varnothing = \sigma \in \hat{\Pi}$.* □

To show that a configuration is s-tagged, we must identify the tagged s-objects. This is illustrated in the next example.

3.2.2. Example (s-Tagged Configurations)

1. Let D be the set of derangements **(2.2.30)** $\sigma = \sigma_1\sigma_2\sigma_3\sigma_4$ on N_4. The s-objects are the four positions in the derangement, and the s-tag assigned to position i is σ_i, for $i = 1,\ldots,4$. Thus the tag set of σ is N_4, and D is a set of s-tagged configurations. If $\sigma = 2413 \in \mathsf{D}$ and $\alpha = \{3,5,6,8\}$, then

$$\sigma_\alpha = 2413_{\{3,5,6,8\}} = 5836 \in \hat{\mathsf{D}}.$$

2. Let $\mathsf{U} = \{\varnothing, \mathsf{N}_1, \mathsf{N}_2, \ldots\}$. The tagged s-objects of $\sigma = \{1,\ldots,k\} \in \mathsf{U}$ are the k elements of σ. Thus U is a set of s-tagged configurations. If $\sigma = \{1,2,3\} \in \mathsf{U}$ and $\alpha = \{4,7,9\}$, then

$$\sigma_\alpha = \{1,2,3\}_{\{4,7,9\}} = \{4,7,9\} \in \hat{\mathsf{U}}. \qquad \square$$

The set U in the second of the two preceding examples will occur frequently in subsequent decompositions, since $\hat{\mathsf{U}}$ is the power set of N_+. We now associate a generating function with a set of tagged configurations as follows.

3.2.3. Definition (Exponential Generating Function)

Let Π be a set of s-tagged configurations. Then the generating function for Π that is **exponential** *with respect to ω_s is*

$$\sum_{\sigma\in\Pi} \frac{x^{\omega_s(\sigma)}}{\omega_s(\sigma)!}$$

and is denoted by $\Psi_\Pi^{(s)}(x)$ (or $\Psi_\Pi(x)$, when the context permits). We say that x **marks ω_s (exponentially).** $\square$

The next result recovers enumerative information from a generating function, and is an immediate consequence of Definition 3.2.3.

3.2.4. Proposition

The number of s-tagged configurations in Π with tag weight k is $\left[\dfrac{x^k}{k!}\right]\Psi_\Pi^{(s)}(x)$. $\square$

To use decompositions we need the following result, which is an immediate consequence of the definitions.

3.2.5. Proposition

Let A, B *be sets of s-tagged and t-tagged configurations, respectively. If a given decomposition*

$$\mathsf{A} \xrightarrow{\sim} \mathsf{B}$$

is tag-weight preserving, then

$$\Psi_{\mathsf{A}}^{(s)}(x) = \Psi_{\mathsf{B}}^{(t)}(x). \qquad \square$$

The following counting lemma is an immediate consequence of the definitions.

3.2.6. Lemma (Sum)

Let A, B *be disjoint subsets of a set* Π *of s-tagged configurations. Then*

$$\Psi_{\mathsf{A} \cup \mathsf{B}}^{(s)}(x) = \Psi_{\mathsf{A}}^{(s)}(x) + \Psi_{\mathsf{B}}^{(s)}(x). \qquad \square$$

The sum lemma is now applied to derive the exponential generating functions for certain sets that will be used frequently.

3.2.7. Example (Exponential Generating Functions)

1. Consider the set U defined in Example 3.2.2(2). Then by Definition 3.2.3

$$\Psi_{\mathsf{U}}^{(s)}(x) = \Psi_{\{\varnothing\}}^{(s)}(x) + \sum_{k \geqslant 1} \Psi_{\{\mathsf{N}_k\}}^{(s)}(x)$$

$$= \sum_{k \geqslant 0} \frac{x^k}{k!}$$

whence $\Psi_{\mathsf{U}}^{(s)}(x) = e^x$.

2. Let P be the set of all permutations on β for each $\beta \in \mathsf{U}$, and let P_k be the set of all permutations on N_k. Then, by the sum lemma,

$$\Psi_{\mathsf{P}}^{(s)}(x) = \sum_{k \geqslant 0} \Psi_{\mathsf{P}_k}^{(s)}(x).$$

However, each element of P_k has tag weight equal to k, and $|\mathsf{P}_k| = k!$, so

$$\Psi_{\mathsf{P}_k}^{(s)}(x) = x^k$$

whence $\Psi_{\mathsf{P}}^{(s)}(x) = (1 - x)^{-1}$. $\square$

The functions e^x and $(1 - x)^{-1}$ occur often in connection with tagged configurations because of the importance, in decompositions, of the sets U, P of unordered and ordered subsets of N_+. The next example gives an instance in which both functions appear.

3.2.8. Example (Derangements)

From Example 3.2.2(1), the set D of derangements is a set of s-tagged configurations. Thus

$$\Psi_{\mathsf{D}}^{(s)}(x) = \sum_{n \geqslant 0} d_n \frac{x^n}{n!}$$

where d_n, the derangement number, is known from **2.2.30** to be

$$\frac{d_n}{n!} = 1 - \frac{1}{1!} + \frac{1}{2!} - \cdots + \frac{(-1)^n}{n!}.$$

It follows that $\Psi_{\mathsf{D}}^{(s)}(x) = e^{-x}(1-x)^{-1}$. □

From Examples 3.2.7 and 3.2.8,

$$\Psi_{\mathsf{U}}^{(s)}(x)\Psi_{\mathsf{D}}^{(s)}(x) = \Psi_{\mathsf{P}}^{(s)}(x).$$

For ordinary generating functions, such a multiplicative relationship can be deduced directly from a Cartesian product decomposition for (untagged) configurations. It is reasonable to ask if there exists an analogous combinatorial product for tagged configurations which implies a multiplicative relationship for exponential generating functions. The $*$-product, which is introduced next, is such a combinatorial product. We return to the derangement problem in Section 3.3.

3.2.9. Definition ($*$-Product)

Let A, B *be sets of s-tagged configurations. Then the* **$*$-product** *of* A *with* B *is*

$$\mathsf{A} * \mathsf{B} = \dot{\bigcup_{k \geq 0}} \{(a, b) \in \hat{\mathsf{A}} \times \hat{\mathsf{B}} \mid \Omega_s((a, b)) = \mathsf{N}_k\}$$

where the tag set of (a, b) *is defined to be*

$$\Omega_s((a, b)) = \Omega_s(a) \,\dot{\cup}\, \Omega_s(b).$$ □

It is important to note that the tag weight of (a, b) is equal to $\omega_s(a) + \omega_s(b)$. This is an immediate consequence of Definition 3.2.9.

3.2.10. Example ($*$-Product)

Let the tagged s-objects of a permutation $\sigma_1\sigma_2$ on N_2 be σ_1, σ_2 and let $\mathsf{A} = \{12\}$ and $\mathsf{B} = \{21\}$. Then

$$\mathsf{A} * \mathsf{B} = \{(12, 43), (13, 42), (14, 32), (23, 41), (24, 31), (34, 21)\}.$$ □

Associated with the $*$-product is the following counting lemma.

3.2.11. Lemma ($*$-Product)

Let A, B *be sets of s-tagged configurations. Then*

$$\Psi_{\mathsf{A}*\mathsf{B}}^{(s)}(x) = \Psi_{\mathsf{A}}^{(s)}(x)\Psi_{\mathsf{B}}^{(s)}(x).$$

Proof: From Definition 3.2.3

$$\Psi^{(s)}_{A*B}(x) = \sum_{(a,b)\in A*B} x^{\omega_s((a,b))}/\omega_s((a,b))!.$$

Now, from Definition 3.2.1(2),

$$|\{a \in \hat{A} | \Omega_s(a) = \alpha\}| = |\{a \in \hat{A} | \Omega_s(a) = \beta\}| \quad \text{if } |\alpha| = |\beta|,$$

where $\alpha, \beta \subset N_+$. Thus

$$\Psi^{(s)}_{A*B}(x) = \sum_{k\geqslant 0} \sum_{\substack{a\in A \\ \omega_s(a)+\omega_s(b)=k}} \sum_{b\in B} \frac{x^k}{k!} \frac{k!}{\omega_s(a)!\omega_s(b)!},$$

since $\omega_s((a,b)) = \omega_s(a) + \omega_s(b)$. It follows that

$$\begin{aligned}\Psi^{(s)}_{A*B}(x) &= \sum_{a\in A} \sum_{b\in B} \frac{x^{\omega_s(a)+\omega_s(b)}}{\omega_s(a)!\omega_s(b)!} \\ &= \Psi^{(s)}_{A}(x)\Psi^{(s)}_{B}(x).\end{aligned}$$

□

As an example of the use of this lemma we return to **3.2.10**.

3.2.12. Example (a Permutation Problem)

Let c be the number of permutations $k_1k_2k_3k_4$ on N_4 such that $k_1 < k_2$ and $k_3 > k_4$. Let S denote this set. Now $k_1k_2k_3k_4$ has tag set $\{k_1, k_2, k_3, k_4\}$, so S is a set of s-tagged configurations.

In the notation of Example 3.2.10,

$$S \tilde{\to} A*B : k_1k_2k_3k_4 \mapsto (k_1k_2, k_3k_4).$$

Clearly, this is tag-weight preserving, since the tag weight of $k_1k_2k_3k_4$ is 4, which is equal to $\omega_s(k_1k_2) + \omega_s(k_3k_4)$. From the $*$-product lemma,

$$\Psi^{(s)}_{S}(x) = \Psi^{(s)}_{A}(x)\Psi^{(s)}_{B}(x).$$

But $A = \{12\}$ and $B = \{21\}$, so, by Definition 3.2.3, we have

$$\Psi^{(s)}_{A}(x) = \Psi^{(s)}_{B}(x) = \frac{1}{2!}x^2,$$

and
$$\Psi_{\mathsf{S}}^{(s)}(x) = \left(\frac{x^2}{2}\right)^2.$$

From Proposition 3.2.4 it follows that

$$c = \left[\frac{x^4}{4!}\right]\Psi_{\mathsf{S}}^{(s)}(x), \quad \text{whence } c = \binom{4}{2} = 6.$$

The set of permissible permutations is $\mathsf{S} = \{1243, 1342, 1432, 2341, 2431, 3421\}$. □

The next example is a more substantial one, which uses a more complicated decomposition. The decomposition is direct.

3.2.13. Example (a Sequence Problem)

Let $c_l(p)$ be the number of sequences over N_3 of length l in which no symbol occurs exactly p times. Let S denote the set of such sequences for $l = 0, 1, 2, \ldots$. The objects of $\sigma_1 \cdots \sigma_l \in \mathsf{S}$ are $\sigma_1, \ldots, \sigma_l$ and the tag of σ_i is i. Thus S is a set of tagged configurations. Let $\mathsf{A} = \mathsf{U} - \{\mathsf{N}_p\}$, where U is the set of tagged objects defined in Example 3.2.2(2). Now

$$\mathsf{S} \mathrel{\tilde{\to}} \mathsf{A} * \mathsf{A} * \mathsf{A} : \sigma_1 \ldots \sigma_l \mapsto (\{i \mid \sigma_i = 1\}, \{i \mid \sigma_i = 2\}, \{i \mid \sigma_i = 3\})$$

is a tag-weight preserving decomposition. Thus, by Proposition 3.2.5 and the $*$-product lemma, $\Psi_{\mathsf{S}}(x) = \{\Psi_{\mathsf{A}}(x)\}^3$. But

$$\Psi_{\mathsf{A}}(x) = \Psi_{\mathsf{U}}^{(x)} - \frac{x^p}{p!}, \qquad \text{by the sum lemma}$$

$$= e^x - \frac{x^p}{p!}, \qquad \text{by Example 3.2.7(1).}$$

It follows from Proposition 3.2.4 that $c_l(p) = \left[\frac{x^l}{l!}\right]\left(e^x - \frac{x^p}{p!}\right)^3$, so

$$c_l(p) = 3^l - 3\binom{l}{p}2^{l-p} + 3\binom{2p}{p}\binom{l}{2p} - \frac{l!}{p!^3}\delta_{3p,l}.$$ □

The next combinatorial operation is the composition of two tagged configurations.

3.2.14. Definition ($*$-Composition with Respect to s-Objects)

Let S be a set of s-tagged configurations and let D be a set of t-tagged configurations in which each configuration has a nonempty set of t-objects. Then a *set* C *of t-tagged configurations is called a* $*$**-composition** *of* S *with* D *if there exists an operation, denoted by R and called* **replacement**, *for which the following conditions hold.*

1. *Let* $\sigma \in \mathsf{S}$ *where* $\Omega_s(\sigma) = \mathsf{N}_k$ *for some* $k \geqslant 0$. *Let* $(\alpha_1, \ldots, \alpha_k)_\prec$ *be a partition of* N_n, *for some* $n \geqslant 0$, *with nonempty blocks such that* $\alpha_1 \prec \cdots \prec \alpha_k$ ($\prec$ *denotes the ordering of blocks in increasing order of their smallest elements*). *Then the replacement of the s-object in* σ, *tagged with* i, *by* $d_i \in \hat{\mathsf{D}}$ *on* α_i, *for* $i = 1, \ldots, k$ *gives an element of* C. *We denote this element of* C *by* $R(\sigma, d_1, \ldots, d_k)$.
2. *Each element of* C *is uniquely constructed in this way and we write* $\mathsf{C} = \mathsf{S} \circledast_s \mathsf{D}$ (*or* $\mathsf{S} \circledast \mathsf{D}$, *when the context permits*). □

As an example, we give a decomposition for the set of all partitions with k blocks, each of which is nonempty.

3.2.15. Example ($*$-Composition)

Let $\mathsf{U}_+ = \mathsf{U} - \{\varepsilon\}$, where U is the set of t-tagged configurations defined in Example 3.2.2(2). Let Π_k be the set of all partitions of N_n (whose elements are tagged t-objects), with k blocks, each nonempty, for all $n \geqslant 0$. Let the elements of N_k be s-tagged objects. Then we define replacement by

$$R(\mathsf{N}_k, \alpha_1, \ldots, \alpha_k) = \{\alpha_1, \ldots, \alpha_k\}$$

and it follows immediately that $\Pi_k = \mathsf{N}_k \circledast_s \mathsf{U}_+$, so the set of all partitions with no empty blocks is $\mathsf{U} \circledast_s \mathsf{U}_+$. □

Next we give the counting lemma associated with $*$-composition.

3.2.16. Lemma ($*$-Composition)

Let S *be a set of s-tagged configurations, and let* T *be a set of t-tagged configurations with nonempty sets of t-objects. Then*

$$\Psi^{(t)}_{\mathsf{S} \circledast_s \mathsf{T}}(x) = \Psi^{(s)}_{\mathsf{S}}\left(\Psi^{(t)}_{\mathsf{T}}(x)\right).$$

Proof: From Definitions 3.2.3 and 3.2.14

$$\Psi^{(t)}_{\mathsf{S} \circledast_s \mathsf{T}} = \sum_{k \geqslant 0} \sum_{\substack{\sigma \in \mathsf{S} \\ \omega_s(\sigma) = k}} \sum_{c} x^{\omega_t(c)}/\omega_t(c)!,$$

where the sum is over all $c = (d_1, \ldots, d_k) \in \hat{\mathsf{T}} \times \cdots \times \hat{\mathsf{T}}$ such that $(\Omega_t(d_1), \ldots, \Omega_t(d_k))_\prec$ is a partition of N_n for some $n \geqslant 0$. Since the blocks

$\Omega_t(d_i)$ are nonempty

$$\Psi^{(t)}_{S \circledast_s T}(x) = \sum_{k \geq 0} \sum_{\substack{\sigma \in S \\ \omega_s(\sigma)=k}} \frac{1}{k!} \sum_{c=(d_1,\ldots,d_k) \in T *_t \cdots *_t T} x^{\omega_t(c)}/\omega_t(c)!$$

$$= \sum_{k \geq 0} \sum_{\substack{\sigma \in S \\ \omega_s(\sigma)=k}} \frac{1}{k!} \{\Psi^{(t)}_T(x)\}^k \qquad \text{by the } *\text{-product lemma,}$$

since $\omega_t(c) = \omega_t(d_1) + \cdots + \omega_t(d_k)$ from Definition 3.2.9. Thus

$$\Psi^{(t)}_{S \circledast_s T}(x) = \sum_{\sigma \in S} \{\Psi^{(t)}_T(x)\}^{\omega_s(\sigma)}/\omega_s(\sigma)! = \Psi^{(s)}_S(\Psi^{(t)}_T(x)). \qquad \square$$

We now complete Example 3.2.15.

3.2.17. Partitions of a Set

Let $p_{n,k}$ be the number of partitions of N_n with k blocks, each nonempty. From Example 3.2.15 and the $*$-composition lemma

$$p_{n,k} = \left[\frac{x^n}{n!}\right] \Psi_{\mathsf{N}_k}(\Psi_{\mathsf{U}_+}(x)).$$

But from Example 3.2.7(1) we have $\Psi_{\mathsf{N}_k}(x) = x^k/k!$ and $\Psi_{\mathsf{U}_+}(x) = e^x - 1$. It follows that $p_{n,k} = [x^n/n!](e^x - 1)^k/k!$ whence

$$p_{n,k} = \frac{1}{k!} \sum_{j=0}^{k} (-1)^{k-j} \binom{k}{j} j^n. \qquad \square$$

The number $p_{n,k}$ is called a **Stirling number** of the **second kind**. By the sum lemma, the number of partitions of N_n with no empty blocks is $p_{n,0} + \cdots + p_{n,n}$, a **Bell exponential number**.

We conclude this section with a discussion of a combinatorial operation, called tagged differentiation.

3.2.18. Definition ($*$-Differentiation)

Let A, B *be sets of s-tagged configurations. If* $\omega_s(a) > 0$ *for all* $a \in \mathsf{A}$, *and there exists a bijection* $f: \mathsf{A} \to \mathsf{B}$ *such that*

$$\omega_s(f(a)) = \omega_s(a) - 1 \qquad \textit{for all } a \in \mathsf{A},$$

then we say that B *is the* **∗-derivative** *of* A, *and write*

$$\frac{d}{ds}\mathsf{A} = \mathsf{B}. \qquad \square$$

The following counting lemma is an immediate consequence.

3.2.19. Lemma (∗-Differentiation)

Let S *be a set of s-tagged configurations with nonempty tag-sets. Then*

$$\Psi^{(s)}_{(d/ds)\mathsf{S}}(x) = \frac{d}{dx}\Psi^{(s)}_{\mathsf{S}}(x).$$

Proof: By Definition 3.2.3,

$$\begin{aligned}\Psi^{(s)}_{(d/ds)\mathsf{S}}(x) &= \sum_{\sigma\in(d/ds)\mathsf{S}} x^{\omega_s(\sigma)}/\omega_s(\sigma)! \\ &= \sum_{\sigma\in\mathsf{S}} x^{\omega_s(\sigma)-1}/\{\omega_s(\sigma)-1\}! \qquad \text{by Definition 3.2.18} \\ &= \frac{d}{dx}\Psi^{(s)}_{\mathsf{S}}(x).\end{aligned} \qquad \square$$

In any particular instance the usefulness of the ∗-differentiation lemma is decided by our ability to obtain a bijection, denoted by f in Definition 3.2.18, with the required property. It should be noted that a set does not necessarily have a unique ∗-derivative because of the choice of the bijection. In practice, the following construction is often used.

3.2.20. Remark (a Construction for the ∗-Derivative)

We often realize tagged differentiation by

1. *Selecting a single tag in each s-tagged configuration. This may be done, for example, by selecting* (a) *the* **smallest** *tag or* (b) *the* **largest** *tag.*
2. *Treating the selected tag in a special way. This may be done, for example, by* (a) **marking** *the selected tag with a marker indicating that the corresponding s-object had the selected tag or* (b) **deleting** *the selected tag.* □

As an example of the use of the ∗-differentiation lemma we consider the classical problem of alternating permutations.

3.2.21. Definition (Alternating Permutation)

A permutation $\sigma_1 \cdots \sigma_n$ *on* N_n, *for* $n \geqslant 1$, *is said to be* **alternating** *if* $\sigma_1 < \sigma_2 > \sigma_3 < \cdots$. *The empty string* ε *is an alternating permutation of length zero.* □

This problem gives an instance in which $*$-differentiation may be realized by **deletion** of the **largest** tag.

3.2.22. André's Problem

Let a_n be the number of alternating permutations on N_n. The elements of the permutations are the tagged objects. Let P, Q denote, respectively, the sets of alternating permutations of odd and even length. If p_n and q_n denote the numbers of permutations of length n in P and Q, respectively, then $a_n = p_n + q_n$.

Let $\sigma = \alpha n \beta \in \mathsf{P} - \{1\}$ have length n, so n is the largest tag. Then the decomposition

$$\frac{d}{ds}(\mathsf{P} - \{1\}) \stackrel{\sim}{\to} \mathsf{P} * \mathsf{P} : \alpha n \beta \mapsto (\alpha, \beta)$$

is tag-weight preserving. Similarly, the decomposition

$$\frac{d}{ds}(\mathsf{Q} - \{\varepsilon\}) \stackrel{\sim}{\to} \mathsf{P} * \mathsf{Q} : \alpha n \beta \mapsto (\alpha, \beta)$$

is tag-weight preserving.

Let $p(x) = \Psi_{\mathsf{P}}(x)$ and $q(x) = \Psi_{\mathsf{Q}}(x)$. Then, from the sum, $*$-product, and $*$-differentiation lemmas,

$$\frac{d}{dx}(p - x) = p^2.$$

$$\frac{d}{dx}(q - 1) = pq.$$

There is one alternating permutation of length zero in Q, namely, the empty string, and no alternating permutation of length zero in P. Thus the initial conditions are $p(0) = 0$, $q(0) = 1$. Integrating these equations yields $p(x) = \tan x$ and $q(x) = \sec x$, so

$$a_n = \left[\frac{x^n}{n!}\right](\sec x + \tan x). \qquad \square$$

The preceding decompositions are recursive for P and for Q.

NOTES AND REFERENCES

Other approaches have been given by Bender and Goldman (1971), Cartier and Foata (1969), Doubilet (1972), Doubilet, Rota, and Stanley (1972), Foata

(1974), Foata and Schützenberger (1970), Henle (1972), Joyal (1981), and Mullin and Rota (1970).

3.2.17 Rota (1964b); **3.2.22** André (1881).

3.3. TREES AND CYCLES IN PERMUTATIONS AND FUNCTIONS

We now consider the application of the elementary counting lemmas to two types of tagged configurations, namely, cycles and trees. Our purpose is to show that the $*$-product, $*$-composition, and $*$-derivative of tagged configurations are indeed the precise combinatorial operations that adapt the apparatus of weight-preserving decompositions to the context of tagged objects. We begin by settling the question, raised in the discussion following **3.2.8**, concerning a decomposition for derangements. Throughout this section let P denote the set of all permutations on the tag-sets $\mathsf{N}_1, \mathsf{N}_2, \ldots$, together with the empty string ε. Let D be the subset of P consisting of derangements (see **2.2.30** for definition). As before, let U denote the set $\{\varnothing, \mathsf{N}_1, \mathsf{N}_2, \ldots\}$.

Derangements have the following indirect decomposition, which is obtained immediately from the definition.

3.3.1. Decomposition (Derangements)

Let $\mathsf{J} = \{\varepsilon\} \,\dot{\cup}\, \{1 \cdots n \mid n \geqslant 1\}$. *Then*

$$\mathsf{P} \overset{\sim}{\rightarrow} \mathsf{J} * \mathsf{D} : \sigma_1 \cdots \sigma_n \mapsto \left(\sigma_{i_1} \cdots \sigma_{i_r}, \sigma_{j_1} \cdots \sigma_{j_{n-r}}\right),$$

where $\sigma_{i_l} = i_l$ *for* $l = 1, \ldots, r$, $\sigma_{j_k} \neq j_k$ *for* $k = 1, \ldots, n - r$ *and where* $i_1 < \cdots < i_r, j_1 < \cdots < j_{n-r}$. □

Under this decomposition, for example, $7132564 \mapsto (356, 7124) \in \hat{\mathsf{J}} \times \hat{\mathsf{D}}$, since $356 = 123_{\{3,5,6\}}$ and $7124 = 4123_{\{1,2,4,7\}}$, where $123 \in \mathsf{J}$ and $4123 \in \mathsf{D}$.

3.3.2. Derangements (Indirect Decomposition)

Decomposition 3.3.1 is tag-weight preserving, so by the $*$-product lemma, $\Psi_{\mathsf{P}}(x) = \Psi_{\mathsf{J}}(x)\Psi_{\mathsf{D}}(x)$. It follows immediately that $\Psi_{\mathsf{D}}(x) = e^{-x}(1-x)^{-1}$, since $\Psi_{\mathsf{J}}(x) = e^x$ and $\Psi_{\mathsf{P}}(x) = (1-x)^{-1}$ from Example 3.2.7. Thus

$$d_n = \left[\frac{x^n}{n!}\right]\Psi_{\mathsf{D}}(x) = n!\left(1 - \frac{1}{1!} + \frac{1}{2!} - \cdots + \frac{(-1)^n}{n!}\right). \qquad \square$$

Decomposition 3.3.1 is indirect for D, so it is necessary to obtain the (multiplicative) inverse of $\Psi_{\mathsf{J}}(x)$. This corresponds in **2.2.30** to the use of the Principle of Inclusion and Exclusion, which is a combinatorial inversion.

A recurrence equation for the general coefficient in the generating function for D is now obtained by equating coefficients in a differential equation.

3.3.3. A Recurrence Equation for the Derangement Number

From **3.3.2** $(1-x)D(x) = e^{-x}$ where $D(x) = \sum_{n \geqslant 0} d_n x^n/n!$ and d_n is the derangement number for N_n. Differentiating and eliminating e^{-x} gives

$$(1-x)D(x) = -\frac{d}{dx}\{(1-x)D(x)\}, \qquad \text{so } (x-1)D' + xD = 0.$$

Applying $[x^n/n!]$ to both sides yields the recurrence equation

$$d_{n+1} = n(d_n + d_{n-1})$$

with initial conditions $d_0 = 1, d_{-1} = 0$. □

A derangement is a permutation with no 1-cycles. We therefore turn our attention to the enumeration of permutations, in general, with respect to cycles of each length.

3.3.4. Definition (Circular Permutation)

A **circular permutation** *is a string* $\sigma = i_1 \cdots i_n$, $n \geqslant 1$, *of the numbers* $1, \ldots, n$, *in some order, in which cyclic rotations of* σ *are indistinguishable from each other.* □

We adopt the convention that $i_1 = 1$ in a circular permutation $i_1 \cdots i_n$. For example, the distinct circular permutations on N_3 are 123 and 132. Throughout this section the set of all circular permutations on tag-sets $\mathsf{N}_1, \mathsf{N}_2, \ldots$ is denoted by C.

3.3.5. Decomposition (Cycle; for Permutations)

$$\mathsf{P} \tilde{\to} \mathsf{U} \circledast \mathsf{C} : \sigma \mapsto \{c_1, \ldots, c_r\},$$

where $\{c_1, \ldots, c_r\}$ *is the set of disjoint cycles of* σ.

Proof: Each permutation is uniquely expressible as a commutative product of (nonempty) disjoint cycles, whose elements are, of course, tagged objects. □

As an example of this decomposition, consider $\sigma = 5\ 10\ 3\ 6\ 1\ 8\ 7\ 4\ 2\ 9 \in \mathsf{P}$. Now σ is uniquely expressible as

$$\pi = \begin{pmatrix} 1 & 2 & 3 & 4 & 5 & 6 & 7 & 8 & 9 & 10 \\ 5 & 10 & 3 & 6 & 1 & 8 & 7 & 4 & 2 & 9 \end{pmatrix}$$

$$= \begin{pmatrix} 1 & 5 \\ 5 & 1 \end{pmatrix} \begin{pmatrix} 2 & 10 & 9 \\ 10 & 9 & 2 \end{pmatrix} \begin{pmatrix} 3 \\ 3 \end{pmatrix} \begin{pmatrix} 4 & 6 & 8 \\ 6 & 8 & 4 \end{pmatrix} \begin{pmatrix} 7 \\ 7 \end{pmatrix},$$

the disjoint cycle representation of π. Thus, with the convention for representing circular sequences,

$$\sigma \mapsto \{1\,5, 2\,10\,9, 3, 4\,6\,8, 7\}$$

under the decomposition. In fact, this is equal to

$$\{12_{\{1,5\}}, 132_{\{2,9,10\}}, 1_{\{3\}}, 123_{\{4,6,8\}}, 1_{\{7\}}\}.$$

We give two examples of the use of this decomposition.

3.3.6. Derangements (Direct Decomposition)

Restricting the cycle decomposition to D, we have $\mathsf{D} \xrightarrow{\sim} \mathsf{U} \circledast (\mathsf{C} - \{1\})$, since derangements have no 1-cycles. The decomposition is tag-weight preserving so, from the $*$-composition lemma,

$$D(x) = \exp\{C(x) - x\},$$

where $C(x) = \Psi_{\mathsf{C}}(x)$.

Again, from the cycle decomposition and $*$-composition lemma,

$$(1 - x)^{-1} = \exp C(x)$$

so eliminating $C(x)$ we have

$$D(x) = e^{-x}(1 - x)^{-1}. \qquad \square$$

Of course, $C(x) = \Psi_{\mathsf{C}}(x)$ could have been obtained directly, by noting that there are exactly $(n-1)!$ circular permutations on N_n, so $C(x) = x + \frac{1}{2}x^2 + \frac{1}{3}x^3 + \cdots = \log(1 - x)^{-1}$.

As a second example, we consider a configuration where it is necessary to use a generating function that is exponential in one variable and ordinary in another.

3.3.7. Involutions

An **involution** is a permutation g such that g^2 is an identity permutation. The number of involutions on N_n is called an **involution number** and is denoted here by c_n. Let I denote the set of all involutions on the tag sets $\mathsf{N}_1, \mathsf{N}_2, \ldots,$ together with ε, the empty string.

To obtain the number $c_{n,k}$ of involutions on N_n with exactly k cycles, the cycle decomposition is restricted to I. Since involutions contain only 1-cycles or 2-cycles,

$$\mathsf{I} \xrightarrow{\sim} \mathsf{U} \circledast \{1, 12\}.$$

This is tag-weight preserving, and additively λ-preserving, where $\lambda(\sigma)$ is the number of cycles in $\sigma \in \mathsf{I}$. Let x mark tag weight and y mark λ. Thus, from the decomposition for I, by the $*$-composition lemma,

$$\sum_{n,k\geqslant 0} c_{n,k}\frac{x^n}{n!}y^k = \exp\{y(x+\tfrac{1}{2}x^2)\}.$$

In particular, $c_n = [x^n/n!]\exp(x+\frac{1}{2}x^2)$. □

Note that the generating function $f(x) = \exp(x+\frac{1}{2}x^2)$ satisfies the differential equation $f'(x) = (1+x)f(x)$ with the boundary condition $f(0) = 1$. Applying $[x^n/n!]$ to both sides, we see that c_n satisfies the recurrence equation

$$c_{n+1} = c_n + nc_{n-1}$$

with initial conditions $c_0 = 1$, $c_{-1} = 0$.

If $\exp\{y(x+\frac{1}{2}x^2)\} = \sum_{n\geqslant 0} H_n(y)x^n/n!$, then $H_n(y)$ is called a **Hermite polynomial**. It will be seen, particularly in Chapter 5, that other special functions also admit combinatorial interpretations.

We consider next the enumeration of rooted labeled trees. It will, in fact, be possible to make use of the ideas behind the decompositions for the corresponding unlabeled configurations, given in Chapter 2. There are two reasons for considering trees now. First, they may be used to demonstrate the use of product, composition, and derivative of tagged configurations in a context that is more complicated than the one just considered. Second, results on cycles and trees may be combined to obtain results on the enumeration of functions.

3.3.8. Definition (Labeled Tree)

1. *A* **labeled tree** *is a vertex-tagged tree.*
2. *Two labeled trees t_1, t_2 are* **indistinguishable** *if there exists a bijection from the vertex set of t_1 to the vertex set of t_2 that preserves adjacency and tags.* □

Figure 3.3.1 gives three rooted labeled trees, in which the root is represented by a circled vertex. Trees (i) and (ii) are indistinguishable, while (iii) is distinguishable from (i) and (ii).

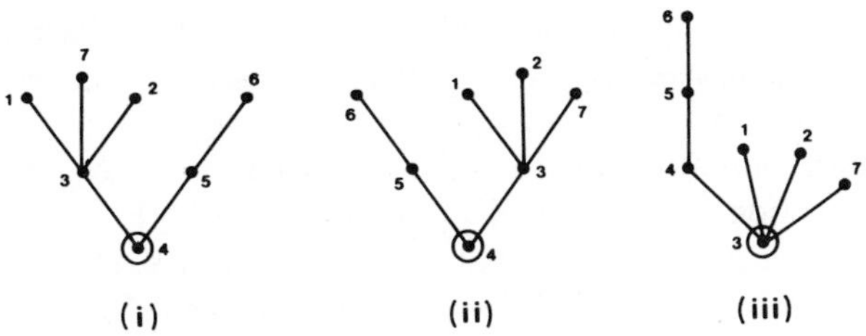

Figure 3.3.1. Labeled trees.

Let T be the set of rooted labeled trees, and let the tagged objects of $t \in \mathsf{T}$ be the labeled vertices. Let $t_1, \ldots, t_p \in \hat{\mathsf{T}}$ be the trees obtained from $t \in \mathsf{T}$ by deleting its root vertex and incident edges, for some $p \geqslant 0$. The root vertices of $t_1, \ldots, t_p$ are the vertices adjacent to the root vertex in t. If $\Lambda(t) = (t_1, \ldots, t_p)$, where the t_i are arranged in increasing order of their smallest tags, then, following Definition 2.7.1, $\Lambda(t)$ is called the **labeled branch list** of t. The following recursive decomposition for T is now immediate.

3.3.9. Decomposition (Labeled Branch)

Let v denote the vertex tree in T. *Then*

$$\mathsf{T} \tilde{\rightarrow} \{v\} * (\mathsf{U} \circledast \mathsf{T}) : \sigma \mapsto (v, \Lambda(\sigma)). \qquad \square$$

The labeled branch decomposition may be used as extensively, for labeled trees, as the branch decomposition **(2.7.4)** was for unlabeled trees. This point is demonstrated in the following example. At the end of this section this decomposition is used in conjunction with the multivariate Lagrange theorem to derive the matrix-tree theorem.

3.3.10. Rooted Labeled Trees

The number $c(n)$ of rooted labeled trees on n vertices is $[x^n/n!]\Psi_{\mathsf{T}}(x)$, where, from the labeled branch decomposition, the $*$-product, and the $*$-composition lemmas

$$T = xe^T,$$

in which $T(x) = \Psi_{\mathsf{T}}(x)$. To see this, note that $\Psi_{\{v\}}(x) = x$, since v has a single vertex, and that $\Psi_{\mathsf{U}}(x) = e^x$. By the Lagrange theorem,

$$T(x) = \sum_{n \geqslant 1} \frac{1}{n} x^n [\lambda^{n-1}] e^{n\lambda} = \sum_{n \geqslant 1} n^{n-1} \frac{x^n}{n!},$$

so

$$c(n) = n^{n-1}. \qquad \square$$

Of course, since each labeled rooted tree may be rooted in n distinct ways, we may immediately deduce Cayley's result that there are n^{n-2} labeled trees on n vertices.

We now turn our attention to functions from N_n to N_n. Consider, for example, the function from N_{13} to N_{13} given in Table 3.3.1.

The obvious calculations show that $f^{[m]}(i) \in \{3, 5, 8, 9, 11\}$ for all $i \in \mathsf{N}_{13}$, provided $m \geqslant 2$. Moreover, $f^{[6k+j]}(x) = f^{[j]}(x)$, provided $j \geqslant 2$. Thus, after repeated applications of f, the sequence of function values eventually becomes

Table 3.3.1. A Function $f: \mathsf{N}_{13} \to \mathsf{N}_{13}$

$f(1) = 11$	$f(5) = 8$	$f(9) = 3$	$f(13) = 10$
$f(2) = 5$	$f(6) = 9$	$f(10) = 11$	
$f(3) = 9$	$f(7) = 5$	$f(11) = 5$	
$f(4) = 10$	$f(8) = 11$	$f(12) = 11$	

periodic, with period 6, on the subset $\{3, 5, 8, 9, 11\}$ of N_{13}. The elements of this subset are called **recurrent** elements of f. The remaining elements of N_{13} are called the **transient** elements of f.

Our purpose now is to examine a decomposition of the set F of all functions from N_n to N_n, for all $n \geqslant 1$. To obtain a decomposition for recording recurrent and transient elements, we consider the following representation of $f: \mathsf{N}_n \to \mathsf{N}_n$. Construct a directed graph on the vertex set N_n with the property that $\overrightarrow{ij}$ is in the edge set if and only if $f(i) = j$. The resulting graph is called the **functional digraph** for f. Figure 3.3.2 gives the functional digraph for the function defined in Table 3.3.1.

It is clear that the recurrent elements of f are precisely those which lie on a directed cycle and that the period is the lcm of the cycle lengths.

3.3.11. Proposition

A functional digraph is a labeled digraph, each of whose components is a directed cycle, possibly of length equal to 1, *incident with trees such that each tree has exactly one vertex in common with the cycle. The edges of each tree are directed toward the (unique) vertex on the cycle associated with it.* □

The next decomposition for F in terms of T is an immediate consequence of Proposition 3.3.11, so its element action is omitted. For this purpose the edges of $t \in \mathsf{T}$ are directed toward its root.

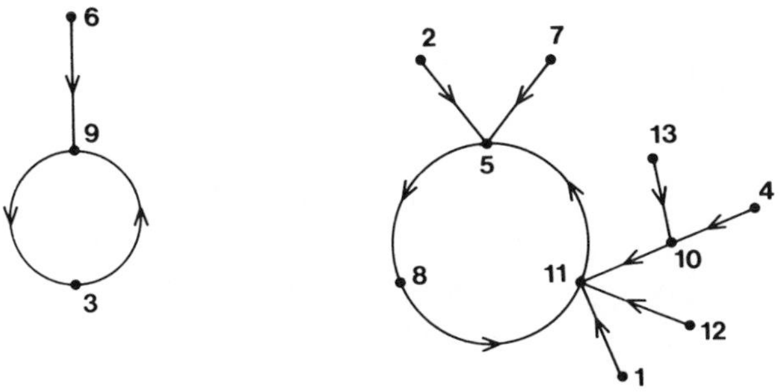

Figure 3.3.2. The functional digraph for Table 3.3.1.

3.3.12. Decomposition (Cycle; for Functions)

$$\mathsf{F} \xrightarrow{\sim} \mathsf{U} \circledast \mathsf{C} \circledast \mathsf{T}. \qquad \square$$

The following result is an application of this decomposition to the enumeration of functions with respect to cycles of given length in the corresponding functional digraphs.

3.3.13. Functions and Cycle Type

Let $c_n(\mathbf{i})$ be the number of functions f from N_n to N_n with i_j cycles of length j, where $\mathbf{i} = (i_1, i_2, \dots)$ is called the **cycle type** of f.

Let t_j mark cycles of length j. The cycle decomposition for functions is tag-weight preserving and additively preserving for cycle type. Letting $\Psi_{\mathsf{T}}(x) = T(x)$, we have, by the $*$-composition lemma,

$$c_n(\mathbf{i}) = \left[\mathbf{t}^{\mathbf{i}} \frac{x^n}{n!}\right] \exp\left\{ \sum_{k \geq 1} \frac{1}{k} t_k T^k(x) \right\},$$

since $\Psi_{\mathsf{C}} = t_1 x + \frac{1}{2} t_2 x^2 + \cdots$. But $T = xe^T$, so by the Lagrange theorem, after routine manipulation,

$$c_n(\mathbf{i}) = \left((1^{i_1} i_1!)(2^{i_2} i_2!) \cdots \right)^{-1} n^{n-q} q! \binom{n-1}{q-1}$$

for $q = i_1 + 2i_2 + 3i_3 + \cdots \leq n$. $\square$

The second application involves a more delicate use of the Lagrange theorem.

3.3.14. Expectation of the Number of Cycles of Given Length in Functions

Let $c(r, n)$ be the expectation of the number of cycles of length exactly r in the set of all functions from N_n to N_n, with the hypothesis that the n^n functions are equiprobable.

Let $d_k(r, n)$ be the number of functions from N_n to N_n with exactly k cycles of length r. Then from the cycle decomposition for functions

$$\sum_{n, k \geq 0} d_k(r, n) \frac{x^n}{n!} z^k = \exp\left\{ T + \tfrac{1}{2} T^2 + \cdots + \frac{1}{r-1} T^{r-1} \right.$$

$$\left. + \frac{z}{r} T^r + \frac{1}{r+1} T^{r+1} + \cdots \right\},$$

so
$$c(r, n) = \frac{1}{n^n} \sum_{k \geq 0} k d_k(r, n) = \frac{1}{n^n} \left[\frac{x^n}{n!}\right] \left\{ \frac{\partial}{\partial z} \sum_{n, k \geq 0} d_k(r, n) \frac{x^n}{n!} z^k \right\}\Bigg|_{z=1}$$

$$= \frac{1}{n^n} \left[\frac{x^n}{n!}\right] \frac{1}{r} T^r (1 - T)^{-1}.$$

To obtain further simplification before appealing to the Lagrange theorem, differentiate $T = xe^T$ with respect to x and eliminate e^T, giving $xT' = T(1-T)^{-1}$. Accordingly, $T^r(1-T)^{-1} = x(T^r/r)'$. It follows that

$$c(r,n) = r^{-2}n^{-n}n![x^{n-1}](T^r)' = r^{-2}n^{-n+1}n![x^n]T^r.$$

Applying the Lagrange theorem to this simplified expression gives

$$c(r,n) = (r-1)!\binom{n}{r}n^{-r}. \qquad \square$$

Attention is drawn to the fact that the expression for $c(r,n)$ would have been considerably more complicated, if we had not used the functional equation to simplify $T^r(1-T)^{-1}$. The cycle decomposition for functions is now used to count idempotent functions, of index m, from N_n to N_n. These are functions f for which $f^{[m+1]} = f$.

3.3.15. Idempotent Functions

Let K_m be the set of idempotent functions of index m in F. Let T_1 be the set of rooted labeled trees with edges directed toward the root and whose nonroot monovalent vertices are adjacent to the root. Restricting the cycle decomposition to K_m, we have

$$\mathsf{K}_m \xrightarrow{\sim} \mathsf{U} \circledast \left(\dot{\bigcup_{k|m}} \mathsf{C}_k \right) \circledast \mathsf{T}_1,$$

where C_k is the set of circular permutations of length k. To see this, first note that the transient elements of $f \in \mathsf{K}_m$ appear in the functional digraph of f as tags of monovalent vertices adjacent to an element on a cycle. This ensures that $f(i)$ is a recurrent element for all $i \in \mathsf{N}_n$. The recurrent elements are tags of vertices on cycles, each of whose length divides m, so $f^{[m]}(f(i)) = f(i)$ for all $i \in \mathsf{N}_n$.

It follows from the $*$-composition lemma that the number $c(n,m)$ of functions in K_m from N_n to N_n is

$$c(n,m) = \left[\frac{x^n}{n!}\right] \exp \sum_{k|m} \frac{1}{k}\{\Psi_{\mathsf{T}_1}(x)\}^k,$$

since the decomposition is tag-weight preserving. For T_1 we have the decomposition

$$\mathsf{T}_1 \xrightarrow{\sim} \{1\} * \mathsf{U} : \sigma \mapsto (i, \alpha)$$

where i is the tag of the root of σ, and $\alpha = \mathsf{N}_{k+1} - \{i\} \in \hat{\mathsf{N}}_k$, for some k, is the set of tags of nonroot monovalent vertices of σ. Since this is tag-weight preserving, $\Psi_{\mathsf{T}_1}(x) = xe^x$ from the $*$-product lemma, whence

$$c(n, m) = \left[\frac{x^n}{n!}\right] \exp \sum_{k \mid m} \frac{1}{k}(xe^x)^k. \qquad \square$$

In deriving recurrence equations in this section we have seen that many of the generating functions satisfy differential equations. The $*$-differentiation lemma is now used to obtain some of these equations combinatorially. More extensive uses are given in Section 3.4. We begin with a recursive decomposition for permutations with respect to their cycle structure. Let $\pi \in \mathsf{P} - \{\varepsilon\}$, and delete the largest tag in π. This can be done in another way by locating the cycle c in π that contains the largest tag and then deleting the largest tag in c. Deleting the elements in c from π, we obtain a sequence in $\hat{\mathsf{P}}$. The following decomposition is obtained by these means.

3.3.16. Decomposition (Recursive Cycle; for Permutations)

$$\frac{d}{ds}(\mathsf{P} - \{\varepsilon\}) \mathrel{\tilde{\rightarrow}} \frac{d\mathsf{C}}{ds} * \mathsf{P}. \qquad \square$$

As an application of this decomposition we consider derangements and involutions.

3.3.17. Derangements and Involutions (Recursive Decomposition)

Restricting the recursive cycle decomposition for permutations to derangements yields

$$\frac{d\mathsf{D}}{ds} \mathrel{\tilde{\rightarrow}} \left(\frac{d}{ds}(\mathsf{C} - \{1\})\right) * \mathsf{D}.$$

Since this is tag-weight preserving, it follows from the $*$-differentiation lemma and the $*$-product lemma that

$$\frac{d}{dx} D = \left\{\frac{d}{dx}\{\log(1 - x)^{-1} - x\}\right\} D$$

where $D(x) = \Psi_{\mathsf{D}}(x)$ and, of course, $\Psi_{\mathsf{C}}(x) = \log(1 - x)^{-1}$ as before (**3.3.6**). Thus

$$(x - 1)D' + xD = 0,$$

in agreement with **3.3.3**. On the other hand, restricting the recursive cycle decomposition to the set I of involutions gives

$$\frac{d\mathsf{I}}{ds} \mathrel{\tilde{\rightarrow}} \left(\frac{d}{ds}\{1, 12\}\right) * \mathsf{I}$$

since involutions contain only 1- or 2-cycles. Moreover, this decomposition is tag-weight preserving, so

$$f' = \left(\frac{d}{dx}\left(x + \frac{x^2}{2}\right)\right)f = (1 + x)f,$$

in agreement with **3.3.7**. □

In certain circumstances it is convenient to incorporate the derivative of a tagged set into another combinatorial operation. The latter is a tagged analogue of the derivative for untagged sets and is given in the next remark.

3.3.18. Remark (Distinguished Tagged *s*-Objects)

Let S *be a set of s-tagged configurations, each with a nonempty tag set. Let* $(sd/ds)\mathsf{S}$ *denote the set obtained by distinguishing each s-object, in turn, of each* $\sigma \in \mathsf{S}$. *Then*

$$\Psi_{(sd/ds)\mathsf{S}}(x) = x\frac{d}{dx}\Psi_{\mathsf{S}}(x). \qquad \square$$

Note that $(sd/ds)\mathsf{S}$ is a particular combinatorial interpretation of $\{1\} * (d/ds)\mathsf{S}$, in which the element $i = 1_i \in \{1\}$ is the tag of the *s*-object that is distinguished in $(d/ds)\mathsf{S}$. For example, if $\mathsf{S} = \{12, 21\}$, the set of permutations on the tag-set N_2, then $(sd/ds)\mathsf{S} = \{1\tilde{2}, \tilde{1}2, \tilde{2}1, 2\tilde{1}\}$, where a distinguished tagged object is marked with a tilde.

As an example of the use of this modification of the $*$-differentiation lemma, we obtain a correspondence between two sets of configurations by applying the strategy given in Remark 2.7.19.

3.3.19. Correspondence (Functions from N_n to N_n-Rooted Labeled Trees)

There is a bijection between

1. *The set of all functions from* N_n *to* N_n.
2. *The set of all rooted labeled trees on n vertices with a single distinguished vertex.*

Proof: Now consider $t \in (sd/ds)\mathsf{T}$. There is a unique path from the root vertex of t to the distinguished vertex of t. Each vertex of this path is the root of a tree in $\hat{\mathsf{T}}$ that is edge-disjoint with the path. Thus t corresponds uniquely to an element of $\mathsf{T} * \cdots * \mathsf{T}$ (k times, for some $k \geqslant 1$, the number of vertices in the path). If the root vertex and distinguished vertex are identical, then, of course, $k = 1$. Thus $(sd/ds)\mathsf{T} \xrightarrow{\sim} \mathsf{P} \circledast \mathsf{T}$ is tag-weight preserving.

But from the cycle decomposition for functions (**3.3.12**) and permutations (**3.3.5**)

$$\mathsf{F} \overset{\sim}{\to} \mathsf{U} \circledast \mathsf{C} \circledast \mathsf{T} \overset{\sim}{\to} \mathsf{P} \circledast \mathsf{T}$$

is tag-weight preserving. Thus $(sd/ds)\mathsf{T} \overset{\sim}{\to} \mathsf{F}$ is tag-weight preserving, and the result follows. □

Figure 3.3.3 illustrates this correspondence for the tree $t \in (sd/ds)\mathsf{T}$ given in (i). The distinguished vertex is 4, so the unique path from the root is 13 2 9 21 4. The trees in $\hat{\mathsf{T}}$ rooted at 13, 2, 9, 21, 4 with edges disjoint from the path are listed in (ii). Now the sequence 13 2 9 21 4 = $41352_{\langle 2,4,9,13,21\rangle} \in \hat{\mathsf{P}}$ is encoded as the permutation $\begin{pmatrix} 2 & 4 & 9 & 13 & 21 \\ 13 & 2 & 9 & 21 & 4 \end{pmatrix}$, whose disjoint cycle representation is $\begin{pmatrix} 2 & 13 & 21 & 4 \\ 13 & 21 & 4 & 2 \end{pmatrix}\begin{pmatrix} 9 \\ 9 \end{pmatrix}$. These cycles and their incident trees are given in (iii) and constitute the functional digraph of a function from N_{21} to N_{21}. The correspondence may be regarded as a tagged analogue of Decomposition 2.7.17 and is now used to obtain Cayley's result on labeled rooted trees.

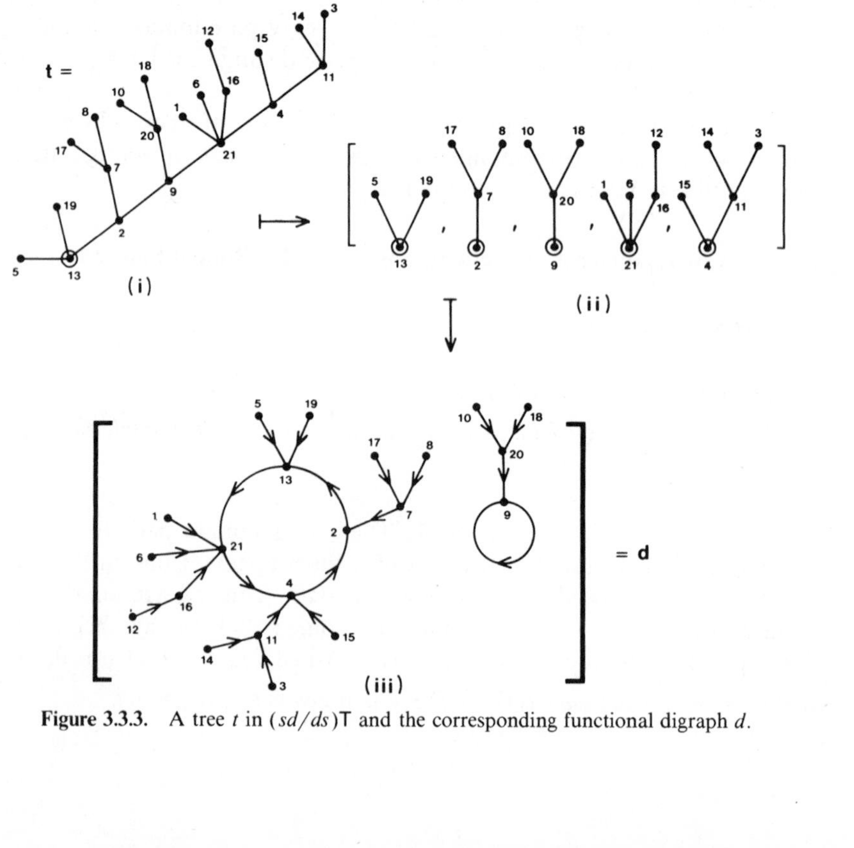

Figure 3.3.3. A tree t in $(sd/ds)\mathsf{T}$ and the corresponding functional digraph d.

3.3.20. Rooted Labeled Trees (Direct Decomposition)

Let c_n denote the number of rooted labeled trees on n vertices. Then, from Correspondence 3.3.19, we have $nc_n = n^n$, since n^n is the number of functions from N_n to N_n. Thus $c_n = n^{n-1}$, in agreement with **3.3.10**. □

Further examples of differential decompositions and the use of Remark 3.3.18 are given in Section 3.4. We conclude this section by returning to the enumeration of trees. In particular, we consider the spanning trees of a graph.

3.3.21. Definition (Spanning Tree)

Let g be a graph with vertex set $\mathsf{V}(g)$ and edge set $\mathsf{E}(g)$. Let $\mathsf{V}(t), \mathsf{E}(t)$ be the vertex set and edge set of a tree t. Then t is said to be a **spanning tree** *of g if*

$$\mathsf{V}(t) = \mathsf{V}(g), \quad \textit{and} \quad \mathsf{E}(t) \subseteq \mathsf{E}(g). \qquad \square$$

In enumerating the spanning trees of a labeled graph we need to preserve information about pairs of adjacent vertices. This is done by coloring the vertices of the graph. Let t be a k-colored rooted labeled tree with d_{ij} edges from vertices of color i to vertices of color j, where the edges are directed away from the root for this purpose. The matrix $\mathbf{D} = [d_{ij}]_{k \times k}$ is called the **edge partition** of t. The next lemma enables us to enumerate with respect to edge partition and is proved by means of the labeled branch decomposition, which preserves color information.

3.3.22. Lemma (Edge-Weighted Tree)

Let $\theta_c(\mathbf{D})$ be the number of k-colored labeled rooted trees with (a) root color c, (b) edge partition $\mathbf{D}$, (c) $n_j = \delta_{cj} + (d_{1j} + \cdots + d_{kj})$ vertices of color j for $j = 1, \ldots, n$, (d) $N = n_1 + \cdots + n_k$ vertices. If $\mathbf{n} = (n_1, \ldots, n_k)$, $\mathbf{x} = (x_1, \ldots, x_k)$ and $\mathbf{A} = [a_{ij}]_{k \times k}$ then

$$\theta_c(\mathbf{D}) = \left[\mathbf{A}^{\mathbf{D}} \mathbf{x}^{\mathbf{n}} \frac{z^N}{N!}\right] f_c(\mathbf{A}, \mathbf{x}, z),$$

where $f_1, \ldots, f_k$ satisfy the following system of functional equations:

$$f_i = z x_i \exp(a_{i1} f_1 + \cdots + a_{ik} f_k) \qquad \textit{for } i = 1, \ldots, k.$$

Proof: Let a_{ij} be an indeterminate marking an edge from a vertex of color i to a vertex of color j, where the edge is directed away from the root for this purpose. Let x_i mark a vertex of color i and z mark a vertex. Let T_i denote the set of all k-colored rooted labeled trees with root color i and let ν_i denote the vertex tree in T_i. Then from the labeled branch decomposition

$$\mathsf{T}_i \xrightarrow{\sim} \{\nu_i\} * \left(\mathsf{U} \circledast \{\mathsf{T}_1 \,\dot\cup\, \mathsf{T}_2 \,\dot\cup\, \cdots \,\dot\cup\, \mathsf{T}_k\}\right).$$

The result follows from the elementary counting lemmas. □

The following theorem enumerates the rooted spanning trees of a labeled graph. It is called the matrix-tree theorem, and the form derived here is a general one in which the edges are directed, and marked with an indeterminate.

3.3.23. Theorem (Matrix-Tree)

The number of trees, rooted at c, on the vertex set $\{1,\ldots,k\}$, *with* m_{ij} *occurrences of the edge* $\vec{ij}$ *(directed away from the root) is*

$$[\mathbf{A}^{\mathbf{M}}]\operatorname{cof}_{cc}\left[\delta_{ij}\alpha_i - a_{ij}\right]_{k\times k},$$

where $\mathbf{M} = [m_{ij}]_{k\times k}$ *(and is, of course, a* $(0,1)$*-matrix) and* $\alpha_j = a_{1j} + \cdots + a_{kj}$ *for* $j = 1,\ldots,k$.

Proof: Let $F_i(\mathbf{A},\mathbf{x}) = f_i(\mathbf{A},\mathbf{x},1)$. Among the trees enumerated by $f_c(\mathbf{A},\mathbf{x},z)$ are those with exactly one vertex of color i for $i = 1,\ldots,k$. These trees are, of course, labeled, and this may be done in $k!$ ways. The generating function for such trees when the labeling is ignored is $[z^k x_1 \ldots x_k/k!]f_c/k!$. But this is equal to $[\mathbf{x}^{\mathbf{1}}]F_c$. Thus, the generating function for the number of labeled trees on the vertex set $\{1,\ldots,k\}$, rooted at c, with the edge $\vec{ij}$ (directed away from the root for this purpose) marked by a_{ij} is $[\mathbf{x}^{\mathbf{1}}]F_c$.

Now from Lemma 3.3.22 $F_1,\ldots,F_k$ satisfy the system of functional equations $F_i = x_i \exp \Phi_i$ where $\Phi_i = a_{i1}F_1 + \cdots + a_{ik}F_k$ for $i = 1,\ldots,k$. Thus, from the multivariate Lagrange theorem **(1.2.9)**,

$$[\mathbf{x}^{\mathbf{1}}]F_c = [\mathbf{F}^{\mathbf{1}}]F_c(\exp\{\Phi_1 + \cdots + \Phi_k\})\|\delta_{ij} - a_{ij}F_j\|.$$

But $\Phi_1 + \cdots + \Phi_k = \alpha_1 F_1 + \cdots + \alpha_k F_k$, so $\exp\{\Phi_1 + \cdots + \Phi_k\} = \prod_{j=1}^{k} e^{\alpha_j F_j}$, and

$$[\mathbf{x}^{\mathbf{1}}]F_c = [\mathbf{F}^{\mathbf{1}}]F_c\|\delta_{ij}e^{\alpha_j F_j} - a_{ij}F_j e^{\alpha_j F_j}\| = [\mathbf{F}^{\mathbf{1}}]F_c\|\delta_{ij}(1 + \alpha_j F_j) - a_{ij}F_j\|$$

by discarding terms of degree 2 and higher in F_j, for $j = 1,\ldots,k$. However, F_j occurs only in column j so, for a nonzero contribution to the coefficient, we retain only the linear terms in F_j in column j, for $j \neq c$. This gives

$$[\mathbf{x}^{\mathbf{1}}]F_c = [\mathbf{F}^{\mathbf{1}}]F_c\|\delta_{ij}(\delta_{cj} + \alpha_j F_j) - a_{ij}F_j\|.$$

Expressing this as the sum of two determinants that differ in column c alone yields

$$[\mathbf{x}^1]F_c = [\mathbf{F}^1]F_c\|\delta_{ij}\alpha_j F_j - a_{ij}F_j\| + [\mathbf{F}^1]F_c\|(\delta_{ij}\alpha_j F_j - a_{ij}F_j)(1 - \delta_{cj}) + \delta_{ij}\delta_{cj}\|.$$

However,

$$[\mathbf{F}^1]F_c\|\delta_{ij}\alpha_j F_j - a_{ij}F_j\| = [\mathbf{F}^1]F_c(F_1 \dots F_k)\|\delta_{ij}\alpha_j - a_{ij}\| = 0,$$

since the matrix has column sums equal to zero. Expanding the other matrix by column c gives

$$[\mathbf{x}^1]F_c = [\mathbf{F}^1]F_c \operatorname{cof}_{cc}\left[(\delta_{ij}\alpha_j - a_{ij})F_j\right]_{k\times k} = \operatorname{cof}_{cc}\left[\delta_{ij}\alpha_j - a_{ij}\right]_{k\times k}.$$

This completes the proof. □

By setting $k = 1$, $a_{11} = 1$, $z = 1$, and $x_1 = x$ in Lemma 3.3.22, we recover the result, given in **3.3.10**, that the number of rooted labeled trees on n vertices is $[x^n/n!]T(x)$, where $T(x)$ satisfies the equation $T = xe^T$. At the functional equation level, this makes a striking connection between two classical, and apparently unrelated, results, namely, Cayley's result and the matrix-tree theorem.

The next corollary gives the more familiar form of the matrix-tree theorem, in which the matrix of indeterminates is replaced by the adjacency matrix of a graph. First, we consider directed graphs and enumerate the rooted spanning trees whose edges are each directed away from the root or are each directed toward the root. The former is called an **out-directed spanning arborescence** and the latter is an **in-directed spanning arborescence**.

3.3.24. In-Directed and Out-Directed Spanning Arborescences

We determine the number of out-directed spanning arborescences, rooted at c, of a directed graph on the vertex set $\{1,\dots,k\}$, with adjacency matrix $\Lambda = [\lambda_{ij}]_{k\times k}$, where λ_{ij} gives the number of edges directed from vertex i to vertex j. Let

$$\operatorname{cof}_{cc}\left[\delta_{ij}\alpha_i - a_{ij}\right]_{k\times k} = \sum_{\mathbf{M}} \tau_c(\mathbf{M})\mathbf{A}^{\mathbf{M}}$$

where, from the matrix-tree theorem, $\tau_c(\mathbf{M})$ is the number of labeled trees rooted at c, on the vertex set $\{1,\dots,k\}$, with m_{ij} occurrences ($0 \leqslant m_{ij} \leqslant 1$) of the edge $\overrightarrow{ij}$ directed away from the root. For an edge $\overrightarrow{ij}$ that is included in such a tree (i.e., an edge for which $m_{ij} = 1$), there are exactly λ_{ij} choices for the corresponding edge in a spanning tree of a graph with adjacency matrix Λ. The

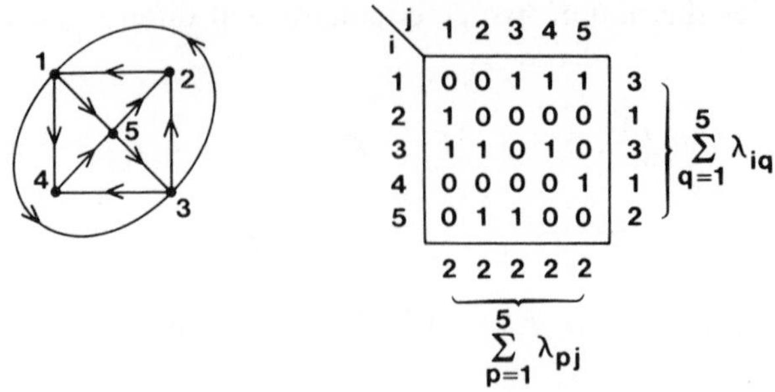

Figure 3.3.4. A directed graph and its adjacency matrix.

required number is therefore

$$\sum_{\mathbf{M}} \tau_c(\mathbf{M})\Lambda^{\mathbf{M}} = \operatorname{cof}_{cc}\left[\delta_{ij}(\lambda_{1j} + \cdots + \lambda_{kj}) - \lambda_{ij}\right]_{k\times k}.$$

For in-directed spanning arborescences the number is

$$\operatorname{cof}_{cc}\left[\delta_{ij}(\lambda_{i1} + \cdots + \lambda_{ik}) - \lambda_{ij}\right]_{k\times k}.$$

This is obtained by transposing Λ and applying the preceding argument. □

As an example of the use of this corollary consider the directed graph given in Figure 3.3.4 together with its adjacency matrix and row and column sums. From **3.3.24** the number of in-directed spanning arborescences rooted on the vertex with label 1 is

$$\operatorname{cof}_{11}\begin{bmatrix} 3 & 0 & -1 & -1 & -1 \\ -1 & 1 & 0 & 0 & 0 \\ -1 & -1 & 3 & -1 & 0 \\ 0 & 0 & 0 & 1 & -1 \\ 0 & -1 & -1 & 0 & 2 \end{bmatrix} = 5.$$

These five in-directed spanning arborescences are displayed in Figure 3.3.5.

For undirected graphs, the following specialization of the matrix-tree theorem is immediately obtained.

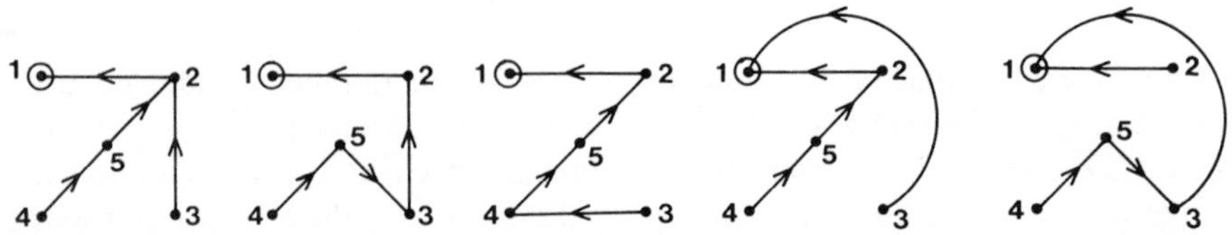

Figure 3.3.5. The five in-directed spanning arborescences of the graph given in Figure 3.3.4.

3.3.25. Matrix-Tree Theorem for Undirected Graphs

The number of spanning trees of a graph g on the vertex set $\{1,\ldots, k\}$ with adjacency matrix $\Lambda = [\lambda_{ij}]_{k\times k}$ is

$$\operatorname{cof}_{cc}\left[\delta_{ij}(\lambda_{1j} + \cdots + \lambda_{kj}) - \lambda_{ij}\right]_{k\times k}, \qquad \text{for any } c \text{ where } 1 \leqslant c \leqslant k,$$

and λ_{ij} is the number of edges (undirected) between vertex i and vertex j. To see this, construct a directed graph g' with λ_{ij} edges directed from vertex i to vertex j, and $\lambda_{ji} = \lambda_{ij}$ edges directed from vertex j to vertex i. Then the out-directed spanning arborescences of g' rooted at any fixed vertex c are in one-to-one correspondence with the spanning trees of g. □

As an example of the use of this result we again determine the number of labeled trees on n vertices.

3.3.26. Labeled Trees

The number of labeled trees on n vertices is equal to the number of spanning trees of the complete graph on n vertices. The adjacency matrix for this graph is $\mathbf{J}_n - \mathbf{I}_n$. Thus, by **3.3.25**, the number of spanning trees is

$$\begin{aligned}\operatorname{cof}_{11}(n\mathbf{I}_n - \mathbf{J}_n) &= n^{n-1}|\mathbf{I}_{n-1} - n^{-1}\mathbf{J}_{n-1}| \\ &= n^{n-1}\{1 - n^{-1}\operatorname{trace}\mathbf{J}_{n-1}\}, \qquad \text{from } \mathbf{1.1.10(5)} \\ &= n^{n-1}\{1 - n^{-1}(n-1)\} = n^{n-2}.\end{aligned}$$

There are thus n^{n-2} labeled trees on n vertices and n^{n-1} rooted labeled trees on n vertices, in agreement with **3.3.10**. □

NOTES AND REFERENCES

Cycles. **3.3.7**, [**3.3.11**] Pólya (1937); [**3.3.14**] Chowla, Herstein, and Scott (1952); [**3.3.15**] Blum (1974); [**3.3.16**] Pólya and Szegö (1927); [**3.3.17**] Foata and Schützenberger (1970), David and Barton (1962); [**3.3.19**] Lovász (1979); [**3.3.20**] Bognár et al. (1970).

Trees. **3.3.10** Cayley (1889); [**3.3.21**] Clarke (1958); [**3.3.22**] Meir and Moon (1968); [**3.3.23**] Moon (1964); for further work on labeled trees see Moon (1970).

Functions. [**3.3.2**] Tomescu (1975), Lovász (1979); [**3.3.28**] Harris (1960); [**3.3.32**] Rubin and Sitgreaves (1954); [**3.3.33**] Foata and Riordan (1974).

Matrix-Tree Theorem. **3.3.23–25** Brooks et al. (1940), Good (1965), Tutte (1948); **3.3.26**, [**3.3.39**] Weinberg (1958); [**3.3.37**] Moon (1970); [**3.3.38**] Lovász

(1979); **[3.3.41]** Sedláček (1969); **[3.3.42]** Goulden and Jackson (1982c), Knuth (1968b); **[3.3.43]** de Bruijn and van Aardenne-Ehrenfest (1951); **[3.3.44]** Rényi (1970).

[3.3.1] Comtet (1974); **[3.3.4]** Gilbert (1956); **[3.3.5]**, **[3.3.7, 8]** Farrell (1979); **[3.3.9]** Fishburn (1979); **[3.3.10]** Schröder (1870); **[3.3.45]** Gessel and Stanley (1978); **[3.3.46]** Carlitz and Scoville (1974); **[3.3.48, 49]** Gessel (1980b), Mallows and Riordan (1968).

EXERCISES

3.3.1. Let s_n be the number of ways of writing a sum of order n (e.g., $s_2 = 3$, since the ways of writing are $\Sigma_{i_1}\Sigma_{i_2}, \Sigma_{i_2}\Sigma_{i_1}, \Sigma_{\{i_1, i_2\}}$). Show that

$$s_n = \left[\frac{x^n}{n!}\right](2 - e^x)^{-1}.$$

3.3.2. (a) Let F_n be the set of all functions from N_m to N_n for all $m \geqslant 0$. If the tagged objects of $f \in \mathsf{F}_n$ are the elements of the domain of f, find a decomposition

$$\mathsf{F}_n \tilde{\to} \underbrace{\mathsf{U} * \cdots * \mathsf{U}}_{n \text{ times}} : f \mapsto (\alpha_1, \ldots, \alpha_n)$$

which is tag-weight preserving, where the tagged objects of $\mathsf{N}_k \in \mathsf{U}$ are the elements of N_k.

(b) Show that the number of functions from N_m to N_n is n^m.

(c) Show that the number of bijections from N_m to N_n is $n!\delta_{n,m}$.

(d) Show that the number of injections from N_m to N_n is $m!\binom{n}{m}$.

(e) Show that the number of surjections from N_m to N_n is

$$\sum_{k=0}^{n} (-1)^{n-k}\binom{n}{k}k^m,$$

a Stirling number of the second kind.

(f) Show that the number of functions from N_m to N_n such that for each $j \in \mathsf{N}_r$, $f(i) = j$ for some $i \in \mathsf{N}_n$, is

$$\sum_{k=0}^{r} (-1)^k\binom{r}{k}(n-k)^m.$$

(g) Show that the number of functions $f: \mathsf{N}_n \to \mathsf{N}_n$ such that if f takes i as a value then it takes on each value less than or equal to i is

$$[x^n/n!](2 - e^x)^{-1}.$$

3.3.3. (a) Let $S(x, y)$ be the exponential generating function for surjections from N_m, consisting of tagged s-objects marked by x, to N_n, a set of tagged t-objects marked by y, for $n, m \geqslant 0$. Show that $S(x, y) = \exp\{y(e^x - 1)\}$ and hence that

$$\frac{\partial}{\partial x} S = yS + y\frac{\partial}{\partial y} S.$$

(b) Give a differential decomposition for surjections that allows us to deduce this equation immediately.

3.3.4. Let G be a set of graphs whose tagged objects are labeled vertices. Suppose that the connected components of G are chosen from $\hat{\mathsf{B}}$, where B is a set of connected graphs whose labeled vertices are tagged objects. If $G(x)$ and $B(x)$ are the exponential generating functions for G and B with x marking tagged objects, show that $G(x) = \exp B(x)$. This is called the **logarithmic connection** for labeled graph enumeration.

3.3.5. (a) Show that the number of ways of covering the complete graph on n labeled vertices K_n with complete graphs is $[x^n/n!]\exp(e^x - 1)$.

(b) Show that the number of ways of covering K_n with complete bipartite graphs (at least one vertex in each part of the bipartition) is

$$[x^n/n!]\exp\tfrac{1}{2}(e^x - 1)^2.$$

3.3.6. (a) Show that the number of ways of covering K_n with paths of vertex-length 1 or more is

$$\left[\frac{x^n}{n!}\right]\exp\left\{\frac{x(2 - x)}{2(1 - x)}\right\}.$$

(b) Show that the number of ways of covering K_n with cycles of length 3 or more is

$$\left[\frac{x^n}{n!}\right](1 - x)^{-1/2}\exp\left(-\frac{x}{2} - \frac{x^2}{4}\right).$$

3.3.7. A **star graph** on k vertices contains $k - 1$ monovalent vertices joined to a vertex of degree $k - 1$, for $k \geqslant 3$, a single edge for $k = 2$, or a single vertex for $k = 1$. Show that the number of ways to cover K_n with star graphs is

$$[x^n/n!]\exp\{x(e^x - \tfrac{1}{2}x)\}.$$

3.3.8. Show that the number of ways of covering the complete bipartite graph $K_{n,m}$ with complete bipartite graphs containing at least one vertex in each part of the bipartition is

$$[x^n y^m/n!m!]\exp\{(e^x - 1)(e^y - 1)\}.$$

3.3.9. Let R be a binary relation on N_n, and define

$$aR = \{(x, y) \in R \mid (y, x) \notin R\}$$

and $cR = \mathsf{N}_n^2 - R$.

(a) R is a **generalized weak order** if caR is transitive. Show that there is a one-to-one correspondence between these relations and an ordered partition of N_n in which each block has an arbitrary symmetric relation. Hence show that the number of generalized weak orders on N_n is

$$\left[\frac{x^n}{n!}\right]\left\{1 - \sum_{i \geqslant 1} 2^{\binom{i+1}{2}} \frac{x^i}{i!}\right\}^{-1}.$$

(b) Show that the number of transitive generalized weak orders on N_n is $[x^n/n!]\{2 - e^x \exp(e^x - 1)\}^{-1}$.

(c) Show that the number of asymmetric generalized weak orders on N_n is $[x^n/n!](2 - e^x)^{-1}$.

3.3.10. For some integer $a \geqslant 2$, we "chain" a finite set S by forming a new set whose elements are a subset α of S, of size a, and the individual elements of $\mathsf{S} - \alpha$. A **regular chain** is formed by successive chaining, to a single set, starting with N_n, with order of chaining not mattering.

(a) If c_n is the number of regular chains on N_n, and $\Sigma_{n \geqslant 1} c_n t^n/n! = C(t)$, show that $C(t) = t + C^a(t)/a!$.

(b) Hence show that

$$c_n = \begin{cases} a!^{-b}(n-1)!\dbinom{ab}{n-1} & \text{if } b = \dfrac{n-1}{a-1} \text{ is an integer} \\ 0 & \text{otherwise.} \end{cases}$$

3.3.11. Show that the number of permutations on N_n with i_j cycles of length j, for $j \geqslant 1$, is

$$\left[\frac{x^n}{n!} t_1^{i_1} t_2^{i_2} \cdots\right] P(x, t_1, t_2, \ldots) = n!\left\{\prod_{j \geqslant 1} i_j! j^{i_j}\right\}^{-1},$$

where $n = i_1 + 2i_2 + \cdots$, and $P(x, t_1, t_2, \ldots) = \exp \Sigma_{i \geqslant 1} t_i \dfrac{x^i}{i}$.

3.3.12. (a) Show that the number of permutations in which every cycle length is divisible by d is

$$\frac{n!}{m! d^m} \prod_{i=0}^{m-1} (id + 1) \qquad \text{if } n = md.$$

(b) Show that the number of permutations in which no cycle length is divisible by d is

$$n! \sum_{i=0}^{\lfloor n/d \rfloor} \binom{1/d}{i} (-1)^i.$$

3.3.13. Show that the number of permutations on N_n with k odd cycles and j even cycles is

$$\left[u^k v^j \frac{x^n}{n!}\right] (1+x)^{(u-v)/2} (1-x)^{-(u+v)/2}.$$

3.3.14. Show that the number of permutations σ on N_n such that σ^m is the identity permutation on N_n is

$$\left[\frac{x^n}{n!}\right] \exp \sum_{d|m} d^{-1} x^d.$$

3.3.15. A **square permutation** $\rho \in \mathsf{P}$ is such that $\rho = \sigma^2$ for at least one $\sigma \in \mathsf{P}$. Show that the number of square permutations on N_n is given by

$$\left[\frac{x^n}{n!}\right] (1+x)^{1/2} (1-x)^{-1/2} \prod_{i \geqslant 1} \cosh\left(\frac{x^{2i}}{2i}\right).$$

3.3.16. (a) Let $\mathbf{A} = [a_{ij}]_{n \times n}$ be a symmetric matrix of indeterminates. Show that the number of distinct monomials in the expansion of $|\mathbf{A}|$ is

$$b_n = \left[\frac{x^n}{n!}\right] (1-x)^{-1/2} \exp\left(\frac{x}{2} + \frac{x^2}{4}\right).$$

(b) Show that $b_{n+1} = (n+1)b_n - \binom{n}{2} b_{n-2}$, $n \geqslant 0$, and $b_0 = 1$, $b_i = 0$, $i < 0$.

3.3.17. (a) Show that the number of permutations on N_n with k cycles is given by

$$\left[u^k \frac{x^n}{n!}\right] (1-x)^{-u},$$

a **Stirling number of the first kind.**

(b) An **upper record** in a permutation σ on N_n is an element $i \in \mathsf{N}_n$ such that $\sigma_i > \sigma_j$ for all $j < i$. Show that the number of permutations on N_n with k upper records is

$$\left[u^k \frac{x^n}{n!}\right] (1-x)^{-u}.$$

(c) Find a one-to-one correspondence between permutations on N_n with k cycles and permutations on N_n with k upper records. For the permutation $\sigma = 539614287$ with four cycles, give the corresponding permutation with four upper records.

3.3.18. (a) Let $G(u, x)$ be the generating function for permutations where x marks the size of the permutation and u marks the number of cycles. Show, from [**3.3.17**], that G satisfies

$$\frac{\partial}{\partial x} G = x \frac{\partial}{\partial x} G + uG.$$

(b) Give a differential decomposition that allows us to deduce this differential equation directly.

3.3.19. Show that the number of permutations on N_n in which the cycle containing n has length m, is $(n-1)!$, for any $m = 1, \ldots, n$.

3.3.20. (a) Show that the number of permutations on N_{k+m} in which every cycle contains at least one element of N_k is $k(k+m-1)!$

(b) Show that the number of permutations on N_{k+m} in which every cycle contains at least one element from N_k and at least one element from $\mathsf{N}_{k+m} - \mathsf{N}_k$ is $km(k+m-2)!$

(c) Show that the number of permutations on N_{k+m} in which every cycle contains an odd number of elements from N_k and an even number of elements from $\mathsf{N}_{k+m} - \mathsf{N}_k$ is

$$\left[\frac{x^k}{k!}\frac{y^m}{m!}\right]\left\{\frac{(1+x)^2 - y^2}{(1-x)^2 - y^2}\right\}^{1/4}.$$

3.3.21. (a) Show that the number of labeled rooted trees on n vertices with root vertex of degree m is

$$n\binom{n-2}{m-1}(n-1)^{n-m-1}.$$

(b) Show that the number of labeled rooted trees in which each vertex has odd degree is

$$n2^{-n}\sum_{k=0}^{n}\binom{n}{k}(n-2k)^{n-2}.$$

3.3.22. Show that the number of labeled rooted trees in which none of the n vertices has degree d is

$$n(n-2)! \sum_{k=0}^{\left\lfloor \frac{n-2}{d-1} \right\rfloor} \binom{n}{k} \frac{(-1)^k}{\{(d-1)!\}^k} \frac{(n-k)^{n-2-k(d-1)}}{(n-2-k(d-1))!}.$$

When $d = 2$ these trees are **homeomorphically irreducible**.

3.3.23. Prove that the number of labeled rooted trees with **degree sequence** $(i_1, \ldots)$ (i.e., i_j vertices are of degree j, for $j \geqslant 1$), is

$$n(n-2)!n!\left\{\prod_{j\geqslant 1} i_j!(j-1)!^{i_j}\right\}^{-1} \qquad \text{for } n \geqslant 2,$$

where $i_1 + i_2 + \cdots = n$, $i_1 + 2i_2 + 3i_3 + \cdots = 2n - 2$.

3.3.24. (a) By differentiating $T = xe^T$, show that the generating function $T(x)$ for labeled rooted trees satisfies the differential equation

$$x\frac{d}{dx}T = T + T\left(x\frac{d}{dx}T\right).$$

(b) Give a differential decomposition that allows us to deduce this equation directly.

3.3.25. (a) In a labeled rooted tree, let the **height** of a vertex be the edge length of the unique path from that vertex to the root. Let $A_{h,d}(x, z)$ be the generating function for labeled rooted trees in which labeled vertices (tagged t-objects) are marked by x, and those of degree d at height h (s-objects) are marked by z. Show that

$$A_{h,d}(x, z) = x \exp A_{h-1,d}(x, z), \qquad h \geqslant 2,$$

$$A_{1,d}(x, z) = x\{\exp A_{0,d-1}(x, z)\},$$

$$A_{0,d-1}(x, z) = x\left\{e^{T(x)} + (z-1)\frac{T^{d-1}(x)}{(d-1)!}\right\}, \qquad T(x) = xe^{T(x)}.$$

Hence show that the total number of vertices of degree d at height h in all labeled rooted trees on n vertices is

$$t_n(h, d) = (h + d - 1)\frac{n!(n-1)^{n-h-d-1}}{(d-1)!(n-h-d)!}.$$

(b) Determine $t_n(h, d)$ by finding a decomposition for $(d/ds)\mathsf{T}$ in terms of T, the set of labeled rooted trees.

3.3.26. (a) Let a planted labeled tree be a rooted labeled tree in which the root vertex is joined to an unlabeled (untagged) vertex of degree 1. If $H(x)$ is the generating function for homeomorphically irreducible planted labeled trees, with x marking tagged vertices, show that $H = x(e^H - H)$, and hence that

$$x\frac{d}{dx}H = H + x(1+x)H\frac{dH}{dx} - x^2\frac{dH}{dx}.$$

(b) Find a differential decomposition that leads to this differential equation immediately.

3.3.27. Show that the number of functions $f\colon \mathsf{N}_n \to \mathsf{N}_n$ such that all cycles have length k is

$$(n-1)!\sum_{i\geqslant 0}\frac{k^{-i}}{i!}\frac{n^{n-(i+1)k}}{(n-(i+1)k)!}.$$

3.3.28. Show that the number of functions $f\colon \mathsf{N}_n \to \mathsf{N}_n$ with exactly k cycles is given by

$$\sum_{i=k-1}^{n-1}\binom{n-1}{i}n^{n-1-i}[z^{k-1}](z+1)\dots(z+i).$$

3.3.29. Show that the number of functions $f\colon \mathsf{N}_n \to \mathsf{N}_n$ in which all cycles have even length is

$$(n-1)!\sum_{i=0}^{\left\lfloor\frac{n-2}{2}\right\rfloor}\frac{n^{n-2-2i}}{(n-2-2i)!}\frac{(i+1)}{2^{2i+1}}\binom{2i+2}{i+1}.$$

3.3.30. Show that the mth factorial moment of the number of cycles of length r in functions $f\colon \mathsf{N}_n \to \mathsf{N}_n$ is

$$(rm)!r^{-m}n^{-rm}\binom{n}{rm}.$$

3.3.31. (a) Show that the number of functions $f\colon \mathsf{N}_n \to \mathsf{N}_n$ such that $f^{[m+k]} = f^{[k]}$ is given by

$$\left[\frac{x^n}{n!}\right]\exp\left\{\sum_{d\mid m} d^{-1}T_k^d\right\}$$

where T_k is the exponential generating function for labeled rooted trees in which monovalent vertices have height at most k.

(b) Show that $T_k = x\exp(T_{k-1})$, $k \geqslant 1$, $T_0 = x$.

3.3.32. Show that the number of functions $f\colon \mathsf{N}_n \to \mathsf{N}_n$ with k recurrent elements is $k!\binom{n-1}{k-1}n^{n-k}$.

3.3.33. Show that the number of functions $f\colon \mathsf{N}_n \to \mathsf{N}_n$ with no cycles of length greater than 1 is $(n+1)^{n-1}$.

3.3.34. Consider n lines in general position, so that no three are concurrent. A **frame** consists of n of the $\binom{n}{2}$ points of intersection, such that no three of the n points lies on the same line. Show that the number of frames on n lines is

$$\left[\frac{x^n}{n!}\right](1-x)^{-1/2}\exp\left(-\frac{x}{2}-\frac{x^2}{4}\right).$$

3.3.35. Show that both the expected value and variance of the number of cycles of length 1 in a random permutation on N_n are equal to 1, for any $n \geqslant 2$.

3.3.36. (a) Show that the number of sequences on N_n of length m and containing k different objects is

$$\binom{n}{k}\left[\frac{x^m}{m!}\right](e^x-1)^k.$$

(b) Show that the average number of different objects in sequences of length m on N_n is $n\left(1-\left(1-\frac{1}{n}\right)^m\right)$, where the n^m sequences are regarded as equiprobable.

3.3.37. Show that the number of spanning trees of the complete graph K_n with i edges in the subgraph K_m is

$$\binom{m-1}{i}m^i(n-m)^{m-1-i}n^{n-m-1}.$$

3.3.38. Show that the number of spanning trees of the complete bipartite graph $K_{n,m}$ is $n^{m-1}m^{n-1}$.

3.3.39. Show that the number of spanning trees of the graph obtained by removing m edges, all incident with a single vertex, from K_n is

$$(n-m-1)(n-1)^{m-1}n^{n-m-2}.$$

3.3.40. The **wheel** W_n on $n+1$ vertices is a graph consisting of a cycle on n vertices, each of which is connected to a single vertex not on the cycle. Let t_n be the number of spanning trees of W_n. Using the result [**1.1.17**] for the determinant of a circulant, show that, for $n \geqslant 3$,

(a) $t_n = 2^{-n}\{(3+\sqrt{5})^n + (3-\sqrt{5})^n\} - 2$

(b) $t_n = \prod_{k=1}^{n-1}(1+4\sin^2(k\pi/n))$.

3.3.41. Let $g-e$ be the graph obtained from a graph g by deleting the edge e. We obtain $g\cdot e$ by deleting e and identifying its incident vertices. Suppose that $t(g)$ is the number of spanning trees of g.

(a) Show that $t(g)=t(g-e)+t(g\cdot e)$, for any e in g. This is called the **deletion–contraction algorithm**.

(b) If e is the only edge joining vertices i and j in g, and g_e^λ is the graph constructed by repeating e λ times, show that

$$t(g_e^\lambda) = \lambda t(g) - (\lambda - 1)t(g - e).$$

(c) If W_n is the wheel defined in [**3.3.40**], show by applying parts (a) and (b) that $t(W_{n+3}) = 4t(W_{n+2}) - 4t(W_{n+1}) + t(W_n)$, for $n \geqslant 1$ and $t(W_1) = 1$, $t(W_2) = 5$, $t(W_3) = 16$.

(d) The **ladder**, denoted by L_n, is the graph obtained by joining the corresponding vertices of two paths on n vertices, as follows.

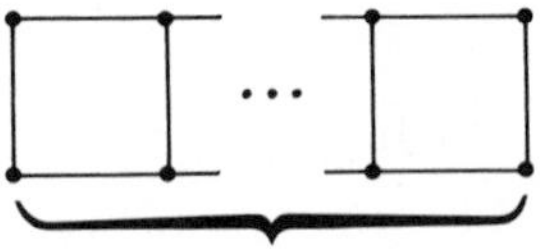

Show that $t(L_{n+2}) = 4t(L_{n+1}) - t(L_n)$, for $n \geqslant 1$, and $t(L_1) = 1$, $t(L_2) = 4$.

(e) Hence deduce that

$$t(L_n) = \frac{1}{2\sqrt{3}}\{(2+\sqrt{3})^n - (2-\sqrt{3})^n\}, \qquad n \geqslant 1.$$

3.3.42. (a) Show that the number of rooted labeled k-colored trees with root color c and edge partition $\mathbf{D}$ is

$$\theta_c(\mathbf{D}) = N!\mathbf{n}^{\mathbf{q}-\mathbf{1}}(\mathbf{D}!)^{-1}\operatorname{cof}_{cc}\left[\delta_{ij}n_i - d_{ij}\right]_{k\times k},$$

where $n_i = \delta_{ic} + (d_{1i} + \cdots + d_{ki})$, $q_i = d_{i1} + \cdots + d_{ik}$, for $i = 1,\ldots,k$, $\mathbf{n} = (n_1,\ldots,n_k)$, $\mathbf{q} = (q_1,\ldots,q_k)$, and $N = n_1 + \cdots + n_k$.

(b) Let $\lambda_c(\mathbf{n},\mathbf{T})$ be the number of labeled k-colored trees rooted at a vertex of color c, with n_i nonroot vertices of color i, for $i = 1,\ldots,k$, and edges from a vertex of color i to a vertex of color j (directed away from the root) allowed if $T_{ij} = 1$ and not allowed if $T_{ij} = 0$. Show that

$$\lambda_c(\mathbf{n},\mathbf{T}) = N!(\mathbf{n}!)^{-1}\operatorname{cof}_{cc}\left[\delta_{ij}\left(\sum_{l=1}^{k} n_l T_{lj}\right) - n_i T_{ij}\right]_{k\times k} \prod_{j=1}^{k}\left(\sum_{i=1}^{k} n_i T_{ij}\right)^{n_j - 1}.$$

3.3.43. Let g be a digraph on vertex set $\{1,\ldots,k\}$ such that

$$\text{in-degree}(i) = \text{out-degree}(i) = d_i, \text{ for } i = 1,\ldots,k.$$

Suppose that g has α directed closed Euler trails and β_c in-directed spanning arborescences rooted at c. By comparing [**2.4.23**] with the matrix-tree theorem, deduce that $\alpha = (\mathbf{d}-\mathbf{1})!\beta_c$ for any $c = 1,\ldots, k$, where $\mathbf{d} = (d_1,\ldots, d_k)$. This is the **de Bruijn–van Aardenne Ehrenfest–Smith–Tutte (BEST) theorem.**

3.3.44. Let $T(x_1,\ldots, x_n)$ be the generating function for labeled trees on vertices $\{1,\ldots, n\}$ in which x_i is an ordinary marker for the degree of vertex i. Show that $T(x_1,\ldots, x_n) = x_1\ldots x_n(x_1 + \cdots + x_n)^{n-2}$.

3.3.45. Let B be the set of sequences $\sigma_1\ldots\sigma_{2n}$ on N_n of type $(2,\ldots, 2)$, for $n \geqslant 0$, such that if $i<j<k$ and $\sigma_i = \sigma_k$ then $\sigma_j > \sigma_i$. Let $B(x, f)$ be the generating function for B, where f is an ordinary marker for falls (**2.4.13**) and x is an exponential marker for alphabet size.

(a) Show that $(\partial/\partial x)(B-1) = (f(B-1)+1)^2 B$ by means of a differential decomposition.

(b) Show that the number of sequences in B on N_k with m falls is $[f^m](1-f)^{2k+1}\Sigma_{n\geqslant 1} f^{n-1}S(n+k, n)$, where $S(i, j) = [x^i/i!](e^x-1)^j/j!$ is a Stirling number of the second kind.

3.3.46. (a) Let $g(x, f)$ be the generating function for permutations on N_n, for $n \geqslant 1$, in which x is an exponential marker for length and f is an ordinary marker for falls (**2.4.13**). Give a differential decomposition which yields the equation $(\partial/\partial x)(g-x) = fg^2 + (1+f)g$, where $g(0, f) = 0$.

(b) Solve this Riccati equation (**1.1.8**) to show that the number of permutations on N_n with k falls is

$$\left[f^k\frac{x^n}{n!}\right](e^x - e^{fx})(e^{fx} - fe^x)^{-1},$$

in agreement with **2.4.21**.

(c) For a permutation $\sigma_1\ldots\sigma_n$ we say that σ_i is a **double rise** if $\sigma_{i-1} < \sigma_i < \sigma_{i+1}$, a **double fall** if $\sigma_{i-1} > \sigma_i > \sigma_{i+1}$, a **modified maximum** if $\sigma_{i-1} < \sigma_i > \sigma_{i+1}$, and a **modified minimum** if $\sigma_{i-1} > \sigma_i < \sigma_{i+1}$, for $i = 1,\ldots, n$, where $\sigma_0 = \sigma_{n+1} = 0$. Show that the number of permutations on N_n with i_1 modified minima, i_2 modified maxima, i_3 double rises, and i_4 double falls is

$$\left[u_1^{i_1}u_2^{i_2-1}u_3^{i_3}u_4^{i_4}\frac{x^n}{n!}\right]\frac{e^{\alpha_2 x} - e^{\alpha_1 x}}{\alpha_2 e^{\alpha_1 x} - \alpha_1 e^{\alpha_2 x}},$$

where $\alpha_1\alpha_2 = u_1u_2$, $\alpha_1 + \alpha_2 = u_3 + u_4$.

3.3.47. (a) Let A_n and B_n be the sets of alternating sequences σ on N_n, of odd and even lengths, respectively (i.e. $\sigma_1 < \sigma_2 \geqslant \sigma_3 < \sigma_4 \geqslant \cdots$). If $A_n(x_1,\ldots, x_n)$ and $B_n(x_1,\ldots, x_n)$ are ordinary generating functions with respect to sequence type for A_n and B_n, and $\nabla C_n = C_n - C_{n-1}$ for any sequence $\{C_0, C_1,\ldots\}$, show that

$$\nabla A_n = x_n(A_{n-1}A_n + 1), \qquad n \geqslant 1,$$

$$\nabla B_n = x_n A_{n-1} B_n, \qquad n \geqslant 1,$$

where $B_n(\mathbf{0}) = 1$, $A_n(\mathbf{0}) = 0$, $n \geqslant 0$.

(b) Show that

$$\nabla C_n D_n = C_n \nabla D_n + D_{n-1} \nabla C_n$$

$$\nabla\left(\frac{1}{C_n}\right) = \frac{-\nabla C_n}{C_n C_{n-1}}.$$

∇ is called the **backward difference operator.**

(c) Let $A_n = -(\nabla H_{n+1})/x_n H_n$ where $H_n(\mathbf{0}) = 1$. Show that $H_n(\mathbf{x}) = \Sigma_{k\geqslant 0}(-1)^k \gamma_n(2k)$, and hence

$$A_n(\mathbf{x}) = \frac{\Sigma_{k\geqslant 0}(-1)^k \gamma_n(2k+1)}{\Sigma_{k\geqslant 0}(-1)^k \gamma_n(2k)},$$

where $\gamma_n(i) = [z^i]\Pi_{j=1}^n(1 + zx_j) = \Sigma_{1\leqslant \alpha_1 < \cdots < \alpha_i \leqslant n} x_{\alpha_1} \cdots x_{\alpha_i}$.

(d) Show that

$$B_n(\mathbf{x}) = \left\{\sum_{k\geqslant 0}(-1)^k \gamma_n(2k)\right\}^{-1}.$$

3.3.48. Let A be the set of labeled trees rooted at the vertex labeled 1. If $1 < i < j$, then (i, j) is a **tree inversion** in $t \in \mathsf{A}$ whenever the vertex labeled j lies on the unique path in t from the root vertex to the vertex labeled i. Let B be the set of trees in A with root vertex of degree 1, and let $A(x, q)$, $B(x, q)$ be the generating functions for A, B, respectively, where x is an exponential marker for vertices and q is an ordinary marker for tree inversions.

(a) Show that

(i) $(\partial/\partial x)A(x, q) = \exp\{(\partial/\partial x)B(x, q)\}$ and

(ii) $(\partial/\partial x)B(x, q) - q^{-1}A(xq, q) = q^{-1}\{(\partial/\partial x)B(x, q) - A(x, q)\}$.

(b) Deduce from part (a) that

$$A(x, q) = (q-1)\log\left\{\sum_{n\geqslant 0}(q-1)^{-n} q^{\binom{n}{2}} \frac{x^n}{n!}\right\}.$$

3.3.49. Let $A(x, q) = \Sigma_{n\geqslant 1} A_n(q)x^n/n!$ be the generating function for labeled trees given in [**3.3.48**(b)]. Prove that

(a) $A_n(q)$ is a polynomial of degree $\binom{n-1}{2}$ in q.

(b) $A_n(0) = (n-1)!$.

(c) $A_n(1) = n^{n-2}$.

(d) $\Sigma_{n\geqslant 0} A_{n+1}(-1)x^n/n! = \sec x + \tan x$. (Thus $A_{n+1}(-1) =$ number of alternating permutations on N_n, from **3.2.22**.)

3.4. 2-COVERS OF A SET AND HOMEOMORPHICALLY IRREDUCIBLE LABELED GRAPHS

This section deals with only two tagged configurations, namely, 2-covers of a set and homeomorphically irreducible labeled graphs. Each of these, particularly the second, needs a succession of decompositions for certain tagged subconfigurations, and they are therefore considerably more complex than the configurations considered in Section 3.3. Our purpose is to present a full and detailed study of these configurations, to demonstrate the power now at our disposal. Among other things, we show that reasonable computational procedures may be derived for obtaining numerical information about these configurations.

It will be necessary on occasion to use configurations that have more than one set of tagged objects. The counting lemmas extend naturally to the multivariate case and are used without further comment. As a preliminary, however, we consider briefly the problem of determining the number of $\{0,1\}$-matrices that do not contain any rows or columns of 0's. To obtain a decomposition we need the following sets. Let $\mathsf{Z}, \mathsf{B}, \mathsf{M}$ denote the sets of $\{0\}$-matrices, $\{0,1\}$-matrices with no rows or columns of zeros, and $\{0,1\}$-matrices, respectively, on the set R of row indices, as s-tags, and the set C of column indices, as t-tags, for all $(\mathsf{R},\mathsf{C}) \in \mathsf{U} \times \mathsf{U}$. Throughout this section $\mathsf{U} = \{\varnothing, \mathsf{N}_1, \mathsf{N}_2, \dots\}$, P_k is the set of all permutations on N_k and P is the set of all permutations including the empty string ε. For $\mathbf{M} \in \mathsf{M}$, let $\alpha(\mathbf{M}), \beta(\mathbf{M})$ denote, respectively, the sets of indices of the rows and columns of 0's. Let $\alpha'(\mathbf{M}), \beta'(\mathbf{M})$ denote, respectively, the sets of indices of rows and columns that do not consist of 0's alone.

We now state a decomposition for M. It is important to note the convention that $0 \times k$ and $k \times 0$ "matrices" exist in M and Z for all $k \geqslant 0$, and in B only for $k = 0$. Since the matrices are on s-objects and t-objects, the $*$-product is with respect to both s-objects and t-objects independently.

3.4.1. Decomposition ($\{0, 1\}$-Matrices)

$$\mathsf{M} \xrightarrow{\sim} \mathsf{Z} * \mathsf{B} : \mathbf{M} \mapsto (\mathbf{S}, \mathbf{T}),$$

where $\mathbf{S} \in \hat{\mathsf{Z}}$ *is on* $(\alpha(\mathbf{M}), \beta(\mathbf{M}))$ *and* $\mathbf{T} \in \hat{\mathsf{B}}$ *is on* $(\alpha'(\mathbf{M}), \beta'(\mathbf{M}))$.

Proof: Let $\mathbf{M}$ be $p \times q$. The mapping is invertible, since

$$[\mathbf{M}]_{ij} = 0 \quad \text{for all } (i, j) \in \mathsf{N}_p \times \mathsf{N}_q - \alpha(\mathbf{M}) \times \beta(\mathbf{M}) - \alpha'(\mathbf{M}) \times \beta'(\mathbf{M}).$$

The result follows. □

This decomposition is illustrated below.

3.4.2. Example

Let

$$\mathbf{M} = \begin{bmatrix} 1 & 0 & 1 & 0 & 0 & 0 \\ 0 & 0 & 0 & 0 & 0 & 0 \\ 1 & 0 & 0 & 0 & 0 & 1 \\ 0 & 0 & 0 & 0 & 0 & 0 \\ 1 & 0 & 1 & 0 & 0 & 0 \end{bmatrix}.$$

Rows 2 and 4, and columns 2, 4, and 5 consist of 0's alone. Thus $\alpha(\mathbf{M}) = \{2,4\}$ and $\beta(\mathbf{M}) = \{2,4,5\}$, so $\alpha'(\mathbf{M}) = \{1,3,5\}$ and $\beta'(\mathbf{M}) = \{1,3,6\}$. Let $\mathbf{A}$ be the submatrix of $\mathbf{M}$ with rows indices in $\{2,4\}$ and column indices in $\{2,4,5\}$. Let $\mathbf{B}$ be the submatrix of $\mathbf{M}$ with row indices in $\{1,3,5\}$ and column indices in $\{1,3,6\}$. Then

$$\mathbf{A} = \begin{bmatrix} 0 & 0 & 0 \\ 0 & 0 & 0 \end{bmatrix} \in \mathsf{Z} \quad \text{and} \quad \mathbf{B} = \begin{bmatrix} 1 & 1 & 0 \\ 1 & 0 & 1 \\ 1 & 1 & 0 \end{bmatrix} \in \mathsf{B}.$$

Moreover, if $\mathbf{M} \mapsto (\mathbf{S}, \mathbf{T})$ under Decomposition 3.4.1, then

$$\mathbf{S} = \mathbf{A}_{(\{2,4\},\{2,4,5\})} \in \hat{\mathsf{Z}} \quad \text{and} \quad \mathbf{T} = \mathbf{B}_{(\{1,3,5\},\{1,3,6\})} \in \hat{\mathsf{B}}. \qquad \square$$

We now use this decomposition to count $\{0,1\}$-matrices with no rows or columns of 0's.

3.4.3. {0, 1}-Matrices with No Rows or Columns of 0's

Let $c(m,n,p)$ be the number of $m \times n$ $\{0,1\}$-matrices with exactly p 1's and with no rows or columns of 0's.

The elements of M, B, and Z are tagged configurations on rows, called s-objects, and columns, called t-objects. Clearly, Decomposition 3.4.1 is both s-tag weight and t-tag weight preserving. Let $\omega(\sigma)$ denote the number of 1's in a $\{0,1\}$-matrix. Then Decomposition 3.4.1 is also additively ω-preserving. Let x, y, z mark rows, columns and 1's, respectively. Let $\chi_{\mathsf{M}}(x,y,z)$, $\chi_{\mathsf{Z}}(x,y,z)$, $\chi_{\mathsf{B}}(x,y,z)$ denote the generating functions of $\mathsf{M}, \mathsf{Z}, \mathsf{B}$, which are exponential with respect to s-tag weight and t-tag weight and ordinary with respect to ω. Then, from Decomposition 3.4.1 and the product lemmas,

$$\chi_{\mathsf{M}}(x,y,z) = \chi_{\mathsf{Z}}(x,y,z)\chi_{\mathsf{B}}(x,y,z).$$

But $\chi_{\mathsf{Z}} = \sum_{i,\, j \geqslant 0} \frac{x^i}{i!}\frac{y^j}{j!} = e^{x+y}$ and $\chi_{\mathsf{M}} = \sum_{i,\, j \geqslant 0} \frac{x^i}{i!}\frac{y^j}{j!}(1+z)^{ij}$,

since each of the ij cells in an $i \times j$ matrix may contain a 0 or a 1. Thus

$$\chi_{\mathsf{B}}(x, y, z) = e^{-x-y} \sum_{i,\, j \geqslant 0} \frac{x^i}{i!}\frac{y^j}{j!}(1+z)^{ij}.$$

It follows by routine calculation that

$$c(m, n, p) = \left[\frac{x^m}{m!}\frac{y^n}{n!}z^p\right]\chi_{\mathsf{B}}(x, y, z)$$

$$= (-1)^{m+n} \sum_{i,\, j \geqslant 0} (-1)^{i+j}\binom{m}{i}\binom{n}{j}\binom{ij}{p}. \qquad \square$$

We may use **3.4.3** to obtain a recurrence equation for $c(m, n, p)$. This is done, following Section 3.3, by determining a differential equation for the generating function for $c(m, n, p)$.

3.4.4. Recurrence Equation

Let $F(x, y, z) = \sum_{m,\, n,\, p \geqslant 0} c(m, n, p)x^m y^n z^p/m!n!$ where, from **3.4.3**,

$$e^{x+y}F = \sum_{m,\, n \geqslant 0} (1+z)^{mn}\frac{x^m}{m!}\frac{y^n}{n!}.$$

Applying $(1+z)\,\partial/\partial z$ to both sides gives

$$(1+z)\frac{\partial}{\partial z}e^{x+y}F = \sum_{m,\, n \geqslant 1} (1+z)^{mn}\frac{x^m}{(m-1)!}\frac{y^n}{(n-1)!}.$$

However, we may derive an expression for the right-hand side of this equation in another way, by differentiating $e^{x+y}F$ partially with respect to x and y. Thus

$$xy\frac{\partial^2}{\partial x\,\partial y}(e^{x+y}F) = \sum_{m,\, n \geqslant 1} (1+z)^{mn}\frac{x^m}{(m-1)!}\frac{y^n}{(n-1)!},$$

whence
$$(1+z)\frac{\partial F}{\partial z} = xy\left\{F + \frac{\partial F}{\partial x} + \frac{\partial F}{\partial y} + \frac{\partial^2 F}{\partial x\,\partial y}\right\}.$$

The linear recurrence equation for $c(m, n, p)$, obtained by applying $[x^m y^n z^p / m!n!]$ to both sides, is

$$\begin{aligned}(p+1)c(m,n,p+1) = -p\,c(m,n,p) &+ mn\,c(m-1,n-1,p)\\ &+ mn\,c(m,n-1,p)\\ &+ mn\,c(m-1,n,p) + mn\,c(m,n,p).\end{aligned}$$

The initial conditions are

$$c(m,n,0) = \delta_{m,0}\delta_{n,0} \qquad \text{for } m, n \geqslant 0$$

$$c(m,0,p) = \delta_{m,0}\delta_{p,0} \qquad \text{for } m, p \geqslant 0$$

$$c(0,n,p) = \delta_{n,0}\delta_{p,0} \qquad \text{for } n, p \geqslant 0$$

since $F(0, y, z) = F(x, 0, z) = F(x, y, 0) = 1$. □

In fact, the differential equation for $F(x, y, z)$ may be derived directly by means of a differential construction. This is a more subtle example of the use of $*$-differentiation than has been given in Section 3.3.

3.4.5. A Differential Decomposition for Matrices with No 0 Rows or Columns

Let V be the set of {0, 1}-matrices with at most one row of 0's and at most one column of 0's and with exactly one entry, distinguished as follows. The distinguished entry lies in a row of 0's and a column of 0's if these exist. If there is no such row, then the distinguished entry may lie in any row, and if there is no such column, then the distinguished entry may lie in any column. Let $V(x, y, z)$ be the generating function for V, which is exponential in x and y, marking rows and columns, respectively, and ordinary in z, marking 1's. We derive two different expressions for V, using differential decompositions.

Expression 1. Consider separately those elements of V in which the distinguished entry is a 1 and those in which it is a 0.

1. The matrices in V with a distinguished 1 are obtained by distinguishing a 1 in an element of B, so the contribution of this set to V is $z\partial F/\partial z$.
2. The matrices in V with a distinguished 0 are obtained by removing a 1 in an element of B and replacing it by a distinguished 0. The contribution of this set is $(\partial/\partial z)F$.

Thus $V(x, y, z) = (1 + z)(\partial/\partial z)F(x, y, z)$.

Expression 2. Distinguish an entry indirectly by distinguishing its row and its column. If there is a row of 0's, then it must be distinguished; otherwise any row can be distinguished, and the case is similar for columns. We consider separately elements of V according to the presence of a row or column of 0's.

1. The matrices in V with no row or column of 0's are obtained by distinguishing a row and a column in an element of B, so the contribution of this set to V is $x(\partial/\partial x)y(\partial/\partial y)F$.
2. We construct the elements of V with no row of 0's and a single column of 0's by adjoining a (distinguished) column of 0's to elements of B, with generating function yF and then distinguishing a row in this matrix, to give a contribution of $x(\partial/\partial x)yF$.
3. The contribution of elements of V with a single row of 0's and no column of 0's is $xy\ \partial F/\partial y$.
4. The contribution of elements of V with a row of 0's and a column of 0's is xyF.

Thus $V(x, y, z) = xy(1 + \partial/\partial x)(1 + \partial/\partial y)F$. Equating the two expressions for V, we obtain the differential equation of **3.4.4**. □

We now consider the enumeration of a configuration called a 2-cover. This is a more complicated structure than a {0, 1}-matrix with no rows or columns of 0's, in the sense that the decomposition that we employ uses more combinatorial operations.

3.4.6. Definition (Proper k-Cover, Restricted Proper k-Covers)

Let $\mathsf{B} = \{\mathsf{B}_1, \ldots, \mathsf{B}_r\}$ *be a set of distinct nonempty subsets of* N_n *such that each element of* N_n *appears in exactly* k *elements of* B. *Then* B *is called a* **proper** **k-cover** *of* N_n *of* **order** r, *and is called* **restricted** *if every intersection of* k B_i*'s contains at most one element.* □

If K_k denotes the set of {0, 1}-matrices with distinct nonzero rows and with columns that each sum to k, then it is clear that a proper k-cover of order r corresponds to exactly $r!$ matrices of K_k. These matrices are the ones obtained by permuting the r rows of the $r \times n$ matrix whose (i, j)-element is equal to 1 if $j \in \mathsf{B}_i$ and 0 otherwise. In enumerating proper 2-covers it is therefore sufficient to enumerate K_2, for which there is the following indirect decomposition. However, another approach is needed (see [**3.5.7**]) for proper covers in general.

3.4.7. Decomposition (Proper 2-Covers)

Let M_2 *denote the set of* {0, 1}*-matrices whose column sums are equal to* 2. *Let* $\mathsf{E}_2 \subseteq \mathsf{M}_2$ *denote the set of two-rowed matrices, on at least one column, whose column sums are equal to* 2. *Let* W *denote the set of "matrices" with no columns.*

Let the rows be tagged s-objects and the columns be tagged t-objects. Then

$$\mathsf{M}_2 \xrightarrow{\sim} \mathsf{W} * (\mathsf{U} \circledast \mathsf{E}_2) * \mathsf{K}_2,$$

where $*$-products and $*$-composition are taken with respect to s-objects and t-objects.

Proof: Any matrix in M_2 may be uniquely decomposed into the following: (1) … in $\hat{\mathsf{W}}$ whose s-tag set is the set of labels of zero-rows and whose … (2) an unordered collection of elements in $\hat{\mathsf{E}}_2$, each of whose …ls of rows that are identical and whose t-tag set is the …ch 1's appear in these rows; (3) an element of $\hat{\mathsf{M}}_2$ …rows and is therefore in $\hat{\mathsf{K}}_2$. The construction is clearly re… □

Consider, for example, the matrix given in Figure 3.4.1(i). Figure 3.4.1(ii) gives the matrix, obtained from (i) by row and column permutation, which exposes the corresponding elements of $\hat{\mathsf{W}}$, $\hat{\mathsf{E}}_2$, and $\hat{\mathsf{K}}_2$.

These are

$$[\ ]_{(\{5\},\varnothing)}, \begin{bmatrix} 1 \\ 1 \end{bmatrix}_{(\{2,8\},\{4\})}, \begin{bmatrix} 1 & 1 \\ 1 & 1 \end{bmatrix}_{(\{3,7\},\{2,6\})}, \begin{bmatrix} 0 & 1 & 1 \\ 1 & 0 & 1 \\ 1 & 1 & 0 \end{bmatrix}_{(\{1,4,6\},\{1,3,5\})}.$$

We now use the decomposition to determine the number of proper 2-covers.

3.4.8. Proper 2-Covers

Let x mark rows and y mark columns. Since Decomposition 3.4.7 is s-tag and t-tag weight preserving, we have

$$\Psi_{\mathsf{M}_2}(x, y) = e^x\{\exp \Psi_{\mathsf{E}_2}(x, y)\} K_2(x, y),$$

$$\text{(i)}\quad \begin{bmatrix} 0 & 0 & 1 & 0 & 1 & 0 \\ 0 & 0 & 0 & 1 & 0 & 0 \\ 0 & 1 & 0 & 0 & 0 & 1 \\ 1 & 0 & 0 & 0 & 1 & 0 \\ 0 & 0 & 0 & 0 & 0 & 0 \\ 1 & 0 & 1 & 0 & 0 & 0 \\ 0 & 1 & 0 & 0 & 0 & 1 \\ 0 & 0 & 0 & 1 & 0 & 0 \end{bmatrix}_{8\times 6} \qquad \text{(ii)}\quad \begin{array}{c} 5 \\ 2 \\ 8 \\ 3 \\ 7 \\ 1 \\ 4 \\ 6 \end{array} \begin{bmatrix} & & & & & \\ 1 & & & & \mathbf{0} & \\ 1 & & & & & \\ & 1 & 1 & & & \\ & 1 & 1 & & & \\ & & & 0 & 1 & 1 \\ & & & 1 & 0 & 1 \\ & \mathbf{0} & & 1 & 1 & 0 \end{bmatrix} \atop \begin{array}{cccccc} 4 & 2 & 6 & 1 & 3 & 5 \end{array}$$

Figure 3.4.1. A matrix in M_2.

where $K_2(x, y) = \Psi_{\mathsf{K}_2}(x, y)$. But there are $\binom{m}{2}^n$ matrices in M_2 which are $m \times n$ so

$$\Psi_{\mathsf{M}_2}(x, y) = \sum_{m, n \geq 0} \binom{m}{2}^n \frac{x^m}{m!} \frac{y^n}{n!} = \sum_{m \geq 0} \frac{x^m}{m!} \exp\binom{m}{2} y.$$

Moreover, $\Psi_{\mathsf{E}_2}(x, y) = \frac{1}{2}x^2(e^y - 1)$, since there is exactly one matrix, consisting entirely of 1's, on two rows and k columns, for each $k \geq 1$. It follows immediately that

$$K_2(x, y) = \exp\left\{ -x - \frac{x^2}{2}(e^y - 1) \right\} \sum_{m \geq 0} \frac{x^m}{m!} \exp\binom{m}{2} y.$$

The number of proper 2-covers of N_n of order m is therefore

$$c_2(m, n) = \frac{1}{m!}\left[\frac{x^m}{m!}\frac{y^n}{n!}\right] K_2(x, y) = \left[x^m \frac{y^n}{n!}\right] K_2(x, y). \qquad \square$$

Restricted proper k-covers correspond to matrices in K_k that have distinct columns, and we use this fact to derive a generating function for them as follows.

3.4.9. Restricted Proper 2-Covers

Let A_2 denote the subset of matrices in K_2 with distinct columns. Each matrix σ_1 in K_2 may be constructed from a unique matrix σ_2 in A_2 by replicating each column of σ_2 the required (nonzero) number of times and permuting them so that the submatrix of σ_2, formed by the left-most occurrences of each distinct column in σ_2, is the matrix σ_1. Recollect that a canonical ordering is required for $*$-composition. Thus

$$\mathsf{A}_2 \circledast_t (\mathsf{U} - \{\varnothing\}) \xrightarrow{\sim} \mathsf{K}_2$$

is s-tag and t-tag weight preserving. It follows from the $*$-composition lemma that

$$A_2(x, e^y - 1) = K_2(x, y),$$

where $A_2(x, y) = \Psi_{\mathsf{A}_2}^{(s,t)}(x, y)$, and $a(m, n) = [x^m y^n / n!]\Psi_{\mathsf{A}_2}^{(s,t)}(x, y)$ is the number of restricted proper 2-covers of N_n of order m. Thus

$$\begin{aligned} A_2(x, y) &= K_2(x, \log(1 + y)) \\ &= \exp\left\{ -x - \frac{x^2}{2} y \right\} \sum_{m \geq 0} \frac{x^m}{m!} (1 + y)^{\binom{m}{2}}. \qquad \square \end{aligned}$$

3.4.10. A Differential Equation for Restricted Proper 2-Covers

Let $H(x, y) = \exp(x + \frac{1}{2}x^2 y)$ and $F(x, y) = \sum_{m \geq 0} x^m (1+y)^{\binom{m}{2}}/m!$, so $HA_2 = F$ from **3.4.9**. We first obtain a differential equation for F, and from this derive one for A_2 by means of the relationship between A_2 and F. Noting immediately that F satisfies

$$(1+y)\frac{\partial F}{\partial y} = \tfrac{1}{2}x^2 \frac{\partial^2 F}{\partial x^2} \quad \text{gives} \quad (1+y)\frac{\partial}{\partial y}(HA_2) = \tfrac{1}{2}x^2 \frac{\partial^2}{\partial x^2}(HA_2).$$

But $\partial H/\partial y = \frac{1}{2}x^2 H$ and $\partial H/\partial x = (1 + xy)H$, so

$$(1+y)\frac{\partial A_2}{\partial y} = \tfrac{1}{2}x^2 \frac{\partial^2 A_2}{\partial x^2} + x^2(1+xy)\frac{\partial A_2}{\partial x} + \tfrac{1}{2}x^3 y(2+xy)A_2. \qquad \square$$

By applying $[x^m y^n/n!]$ to both sides of this equation, we may obtain a linear recurrence equation for $a(m, n)$. The details are straightforward and are omitted. The next result gives a combinatorial interpretation for the differential equation for A_2.

3.4.11. A Differential Decomposition for Restricted Proper 2-Covers

Let $\mathsf{A}_2^{(c)}$ be the set of matrices with two distinguished rows obtained from A_2 as follows. In each matrix of A_2 distinguish the rows where 1's appear in the last column, and then delete this column. Let x and y mark, respectively, rows and columns exponentially. The generating function for $\mathsf{A}_2^{(c)}$ is therefore $\partial A_2/\partial y$. We obtain another expression for this generating function by considering five cases for an arbitrary $g \in \mathsf{A}_2^{(c)}$.

Case 1. g is a restricted proper 2-cover with two distinguished rows and with no copy of the deleted column. The contribution from all restricted proper 2-covers with two distinguished rows is $\frac{1}{2}x^2 \partial^2 A_2/\partial x^2$. Those that contain a copy of the deleted column are enumerated by $y\partial A_2/\partial y$, in which the distinguished column is considered to be the copy of the deleted column. The two rows that are distinguished by this procedure are the rows that contain 1's in the distinguished column. The contribution from this case is therefore

$$\frac{x^2}{2!}\frac{\partial^2 A_2}{\partial x^2} - y\frac{\partial A_2}{\partial y}.$$

Case 2. g consists of an element $c \in \mathsf{A}_2$ with one distinguished row, and a single row of 0's. The row of 0's must also be distinguished, since the deleted column must contain a 1 in this row. The contribution from this case is thus $x^2 \partial A_2/\partial x$, since $x\partial A_2/\partial x$ enumerates elements c with a single distinguished row, and x enumerates the row of 0's.

Case 3. g consists of an element $c \in \mathsf{A}_2$ with one distinguished row, and a pair of repeated rows, one of which is distinguished. Since columns must be distinct, these rows are 0 except for a single 1 in each of them. These two 1's appear in the same column, which may be inserted anywhere into c to construct g. The contribution from this case is $x^3y\partial A_2/\partial x$ since $x(d/dx)x^2/2! = x^2$ enumerates the two identical rows, one of which is distinguished, y enumerates the column containing the 1's in these rows, and $x\partial A_2/\partial x$ enumerates the element c, which has one distinguished row.

Case 4. g consists of an element $c \in \mathsf{A}_2$, with an additional row of 0's, which is distinguished, and an additional pair of repeated rows, one of which is distinguished. As in previous cases, the zero row is enumerated by x, the pair of equal rows, one of which is distinguished, is enumerated by x^2, and the column containing the 1's in the repeated rows is enumerated by y. The contribution from c is A_2, and the total contribution from this case is therefore x^3yA_2.

Case 5. g consists of two pairs of identical rows, in which one of each pair is distinguished, and an element $c \in \mathsf{A}_2$. The contribution from each of these pairs of rows is thus x^2 and that from the pair of columns containing the 1's in these identical rows is $y^2/2!$. The contribution of this case is $\frac{1}{2}x^4y^2A_2$, since c is an arbitrary element of A_2, enumerated by A_2.

Combining these cases and equating the two expressions for the generating function of $\mathsf{A}_2^{(c)}$, we immediately obtain the differential equation for A_2 given in **3.4.10.** □

The second configuration that is examined in detail is the homeomorphically irreducible simple labeled graph. A graph is said to be **simple** if it contains no loops or multiple edges, and **homeomorphically irreducible** if it contains no bivalent vertices. For brevity, these graphs are called simple labeled ***h*-graphs**. Several decompositions will be needed for dealing with this configuration, and it is interesting to observe that it is apparently necessary to use an "at-least" construction.

The decomposition is "at-least" for bivalent vertices and is based on the following idea. We construct all simple labeled graphs in which j vertices, for some j, called **initial vertices**, can have any degree, while the additional vertices, called **noninitial**, must be bivalent. Noninitial vertices are classified as

1. Vertices subdividing edges between initial vertices.
2. Vertices subdividing loops at initial vertices.
3. Vertices on cycles (simple) containing only noninitial vertices.

Let A be the set of all labeled graphs constructed in this way. The elements of

A have (untagged) s-objects, namely, edges; tagged t-objects, namely, labeled vertices; and (untagged) u-objects, namely, noninitial bivalent vertices.

The following notation will be used in connection with the decomposition. The notation concerns the three classes previously listed. Let L be the set of all labeled graphs consisting of loops, at an initial vertex, which have been subdivided by two or more noninitial vertices. Such graphs are necessarily simple. Let C be the set of all labeled cycles with three or more vertices, each noninitial. These graphs are simple. Let M be the set of all labeled graphs consisting of a multiple edge between a pair of initial vertices, subdivided by noninitial vertices such that at most one edge has no noninitial vertices. These graphs are again simple.

Let $\varnothing$ denote the absence of an edge between two initial vertices, let ε denote an edge between two initial vertices, and let M_1 denote the set obtained by subdividing the edge between two initial vertices, with at least one noninitial vertex, in all possible ways.

3.4.12. Decomposition (Simple Labeled *h*-Graphs)

1. $\mathsf{A} \xrightarrow{\sim} \dot{\bigcup}_{k \geq 0} \mathsf{N}_k * (\mathsf{P}_k \circledast \mathsf{U} \circledast \mathsf{L}) * (\mathsf{P}_{\binom{k}{2}} \circledast \mathsf{M}) * (\mathsf{U} \circledast \mathsf{C})$.
2. $\mathsf{M} \xrightarrow{\sim} \{\varnothing, \varepsilon\} * (\mathsf{U} \circledast \mathsf{M}_1)$.

Proof: The components of the construction are the following.

For N_k: There are k initial vertices, and these form an unordered collection of tagged t-objects.

For $\mathsf{P}_k \circledast \mathsf{U} \circledast \mathsf{L}$: The set L consists of subdivided loops containing bivalent noninitial vertices, which are both (untagged) u-objects and tagged t-objects, and edges, which are s-objects. The loop has a unique untagged vertex that is designated as the root. An element of $\mathsf{U} \circledast \mathsf{L}$ is obtained from unordered collections of loops in L by identifying the root vertices of these loops as the root of this element. At each of the k initial vertices from N_k, now canonically ordered by their labeling, we place an element of $\mathsf{U} \circledast \mathsf{L}$ by identifying its root with the labeled initial vertex.

For $\mathsf{P}_{\binom{k}{2}} \circledast (\{\varnothing, \varepsilon\} * (\mathsf{U} \circledast \mathsf{M}_1))$: The set M_1 consists of all single edges, between a pair of untagged vertices, which are designated "left" and "right" for this purpose, subdivided by at least one noninitial vertex. The bivalent noninitial vertices are both (untagged) u-objects and tagged t-objects, and the edges are (untagged) s-objects. The single edge ε contains a single edge, an s-object. An element of $\mathsf{U} \circledast \mathsf{M}_1$ is obtained from an unordered collection of edges in M_1 by identifying the left vertices of these edges as a single vertex and the right vertices as another single vertex. The two vertices are designated as a left untagged vertex and a right untagged vertex for the resulting configuration.

The elements of $\{\varnothing, \varepsilon\} * (\mathsf{U} \circledast \mathsf{M}_1)$ are formed by connecting the left and right vertices of an element of $\mathsf{U} \circledast \mathsf{M}_1$ by at most one edge. The resulting set of configurations is precisely M. Finally, in the set of k initial vertices there are $\binom{k}{2}$unordered pairs of initial vertices. We associate an element of M with each pair by identifying the left and right vertices with, respectively, the smaller and larger labeled vertices in the pair.

For $\mathsf{U} \circledast \mathsf{C}$: Each element of this set consists of a graph, on noninitial vertices, whose components are cycles.

Each configuration in A is obtained exactly once by this construction. The result follows. □

Now let H be the set of simple labeled graphs, with edges as s-objects, labeled vertices as tagged t-objects, and bivalent vertices as u-objects. It is important to note that the graphs in H appear many times in A. If $h \in \mathsf{H}$ has m bivalent vertices, then h will appear 2^m times in A, once for each of the subsets of those of its bivalent vertices that are designated as u-objects in the element of A.

Let $H(x, y, z)$, $A(x, y, z)$ be the generating functions for H, A, respectively, with x marking s-objects, y marking (tagged) t-objects, and z marking u-objects, the latter being bivalent vertices for elements of H and noninitial vertices for elements of A.

3.4.13. Proposition

$$H(x, y, z) = A(x, y, z - 1).$$

Proof: The preceding argument shows that $A(x, y, z)$ is the generating function for H that is "at least" with respect to bivalent vertices and exact for edges and all labeled vertices. The result follows by the Principle of Inclusion and Exclusion, since bivalent vertices are regarded as untagged objects. □

The generating function for the number of simple labeled h-graphs is, of course, $H(x, y, 0)$ and is obtained by applying Decomposition 3.4.12 and Proposition 3.4.13.

3.4.14. Simple Labeled *h*-Graphs

We begin by deriving the generating functions for C, L, and M, since $A(x, y, z)$ may be determined from these by means of Decomposition 3.4.12, with the aid of the $*$-product and $*$-composition lemmas.

1. For C: There are $\frac{1}{2}(i-1)!$ ways to label a cycle on $i \geqslant 3$ vertices. Each vertex is a tagged t-object and a u-object. Such cycles contain i s-objects, namely, edges. Thus

$$\begin{aligned}\Psi_{\mathsf{C}}(x, y, z) &= \tfrac{1}{2}\{\tfrac{1}{3}(xyz)^3 + \tfrac{1}{4}(xyz)^4 + \cdots\} \\ &= \tfrac{1}{2}\log(1 - xyz)^{-1} - \tfrac{1}{2}\{xyz + \tfrac{1}{2}(xyz)^2\}.\end{aligned}$$

The condition $i \geqslant 3$ applies, since the graphs are simple.

2. For L: There are $\frac{1}{2}i!$ ways of labeling an element of L on $i \geqslant 2$ noninitial vertices, since they are rooted at an initial vertex and can be traversed in either direction. The condition $i \geqslant 2$ is required, since the graphs are simple. Each subdivided loop on $i \geqslant 2$ noninitial vertices contains $i + 1$ edges (s-objects). The noninitial vertices are both u-objects and tagged t-objects. Thus

$$\Psi_{\mathsf{L}}(x, y, z) = \tfrac{1}{2}(x^3y^2z^2 + x^4y^3z^3 + \cdots) = \tfrac{1}{2}x^3y^2z^2(1 - xyz)^{-1}.$$

3. For M: From Decomposition 3.4.12(2)

$$\Psi_{\mathsf{M}}(x, y, z) = \Psi_{\{\varnothing, \varepsilon\}}(x, y, z)\exp\Psi_{\mathsf{M}_1}(x, y, z),$$

by the $*$-product and $*$-composition lemmas. But neither $\varnothing$ nor ε has noninitial vertices, $\varnothing$ has no edges, and ε has one. Thus $\Psi_{\{\varnothing, \varepsilon\}}(x, y, z) = 1 + x$. For M_1 we note that any edge, from a left vertex to a right vertex, subdivided by i noninitial vertices, can be labeled in $i!$ ways. Such a subdivided edge contains $i + 1$ edges (s-objects) and each of the i noninitial vertices is both a tagged t-object and a u-object. Thus

$$\Psi_{\mathsf{M}_1}(x, y, z) = x^2yz + x^3y^2z^2 + \cdots = x^2yz(1 - xyz)^{-1},$$

so

$$\Psi_{\mathsf{M}}(x, y, z) = (1 + x)\exp\{x^2yz(1 - xyz)^{-1}\}.$$

To complete the argument we use Decomposition 3.4.12(1) to obtain

$$A(x, y, z) = \Psi_{\mathsf{A}}(x, y, z) = (\exp\Psi_{\mathsf{C}})\sum_{k \geqslant 0}\frac{y^k}{k!}\{\exp\Psi_{\mathsf{L}}\}^k\Psi_{\mathsf{M}}^{\binom{k}{2}}.$$

But $H(x, y, 0) = A(x, y, -1)$ from Proposition 3.4.13. It follows that

$$\begin{aligned}H(x, y, 0) &= (1 + xy)^{-1/2}\exp(\tfrac{1}{2}xy - \tfrac{1}{4}x^2y^2) \\ &\quad \times \sum_{j \geqslant 0}\frac{y^j}{j!}\left\{(1 + x)\exp\frac{-x^2y}{1 + xy}\right\}^{\binom{j}{2}}\left\{\exp\frac{\frac{1}{2}x^3y^2}{1 + xy}\right\}^j\end{aligned}$$

so the number of simple labeled h-graphs on p edges and q vertices is

$$h(p, q) = \left[x^p\frac{y^q}{q!}\right]H(x, y, 0).$$

□

We complete our examination of simple labeled h-graphs by deriving a recurrence equation. Although this appears to be a more complex task than that for restricted proper 2-covers (**3.4.10**), the same method applies.

3.4.15. A Recurrence Equation for Simple Labeled h-Graphs

Let h_p be the number of simple labeled h-graphs on p vertices. Setting $x = 1$ in **3.4.14**, we have $h_p = [y^p/p!]h(y)$, where $h(y) = \sum_{j \geq 0} y^j 2^{\binom{j}{2}} g_j(y)/j!$ and

$$g_j(y) = \exp\{-\tfrac{1}{2}\log(1+y) + \tfrac{1}{2}(1+j)y - \tfrac{1}{4}y^2 - \tfrac{1}{2}j^2 y(1+y)^{-1}\}.$$

Thus $2(1+y)^2 g_j' + \{(1+y)y^2 + j^2 - j(1+y)^2\}g_j = 0$. Letting $g_j(y) = \sum_{k \geq 0} a_k^{(j)} y^k$ and comparing coefficients we have

$$h_p = p! \sum_{j=0}^{p} \frac{1}{j!} 2^{\binom{j}{2}} a_{p-j}^{(j)}$$

where $2(k+1)a_{k+1}^{(j)} = (j - j^2 - 4k)a_k^{(j)} + 2(1+j-k)a_{k-1}^{(j)} + (j-1)a_{k-2}^{(j)} - a_{k-3}^{(j)}$ for $k \geq 0$, with initial conditions $a_0^{(j)} = 1$, $a_l^{(j)} = 0$ for $l < 0$. □

NOTES AND REFERENCES

3.4.8 Comtet (1968), Bender (1974); **3.4.14, 15**, **[3.4.11]** Jackson and Reilly (1975); **[3.4.1]** Everett and Stein (1973); **[3.4.4]** Comtet (1966); **[3.4.7]** Gilbert (1956); **[3.4.8]** Goulden and Vanstone (1983); **[3.4.9]** Harary and Palmer (1973); **[3.4.10]** Gessel and Wang (1979).

EXERCISES

3.4.1. Show that the number of $\{0, 1\}$-matrices with m rows, exactly k of which are empty (contain only 0's), n columns, exactly l of which are empty, and containing p 1's is

$$(-1)^{m+n+k+l} \sum_{i,j \geq 0} (-1)^{i+j} \binom{m}{i}\binom{n}{j}\binom{m-i}{k}\binom{n-j}{l}\binom{ij}{p}.$$

3.4.2. Show that the number of $\{0, 1\}$-matrices with column sum 2, m distinct, nonempty rows, and n columns, k of which are distinct, is

$$\left\{\sum_{l=0}^{k} \binom{k}{l} l^n (-1)^{k-l}\right\}\left\{m!(-1)^m \sum_{i,j \geq 0} \binom{\binom{i}{2}}{k-j} \frac{(-1)^{i+j}2^{-j}}{i!j!(m-i-2j)!}\right\}$$

3.4.3. Show that the number of $\{0,1\}$-matrices with column sum 2, m nonempty rows, l of which are distinct, and n columns, k of which are distinct, is

$$\left\{\sum_{t=0}^{k}\binom{k}{t}t^n(-1)^{k-t}\right\}m!(-1)^l\sum_{i,j\geqslant 0}\left(\!\!\binom{\binom{i}{2}}{k-j}\!\!\right)\binom{j}{l+j-m}\frac{(-1)^{i+j}2^{-j}}{i!j!(m-i-2j)!}.$$

3.4.4. A **cover** of N_n is a set of distinct nonempty subsets of N_n whose union is N_n. Show that the number of covers of N_n with i_j subsets of size j, for $j \geqslant 1$, is given by

$$\sum_{k=0}^{n}(-1)^{n-k}\binom{n}{k}\prod_{j\geqslant 1}\left(\!\!\binom{\binom{k}{j}}{i_j}\!\!\right).$$

3.4.5. Show that the number of $m \times k$ matrices over N_n with distinct rows and columns is

$$\sum_{i\geqslant 0}k!\binom{n^i}{k}s(m,i),$$

where $s(m,i)=[u^i x^m/m!](1+x)^u$ is a Stirling number of the first kind.

3.4.6. Let $c(m,k)$ be the number of $m \times k$ matrices over N_n with no rows or columns consisting entirely of i's, for $i = 1,\ldots,n$. Use an "at-least" construction to show that

$$c(m,k)=n^{mk}+n\sum_{i\geqslant 1}(-1)^i\left\{\binom{m}{i}n^{(m-i)k}+\binom{k}{i}n^{m(k-i)}\right\}$$

$$+\sum_{i\geqslant 1}\sum_{j\geqslant 1}(-1)^{i+j}\binom{m}{i}\binom{k}{j}n^{(m-i)(k-j)}.$$

3.4.7. (a) Show that the number of simple connected graphs on n labeled vertices and i edges is

$$\left[y^i\frac{x^n}{n!}\right]\log\left(\sum_{m\geqslant 0}\frac{x^m}{m!}(1+y)^{\binom{m}{2}}\right).$$

(b) Show that the number of simple graphs with k components on n labeled vertices and i edges is

$$\left[y^i\frac{x^n}{n!}\right]\frac{1}{k!}\left\{\log\left(\sum_{m\geqslant 0}\frac{x^m}{m!}(1+y)^{\binom{m}{2}}\right)\right\}^k.$$

3.4.8. (a) Show that the number of simple graphs on n labeled vertices with i connected components and k edges is

$$g_n(i,k) = \left[z^i y^k \frac{x^n}{n!}\right]\left(\sum_{m\geq 0} \frac{x^m}{m!}(1+y)^{\binom{m}{2}}\right)^z.$$

(b) Show that the number of solutions $(x_1, x_2, \ldots, x_k)$ in $GF(p)$, to the equation $m_1x_1 + \cdots + m_kx_k = 0$, for $k \geqslant 1$, is p^k if $m_1 = m_2 = \cdots = m_k = 0$, and p^{k-1} otherwise.

(c) Let $N(\mathbf{m})$ be the number of solutions $(x_1, \ldots, x_k)$, for distinct x_i, in $GF(p)$ to the equation $m_1x_1 + \cdots + m_nx_n = 0$, where $m_1 + \cdots + m_n = 0$ but no subset of the m_i's has a zero sum. Show, by the Principle of Inclusion and Exclusion, that

$$N(\mathbf{m}) = \sum_{k=0}^{\binom{n}{2}} \sum_{i=1}^{n} (-1)^k p^{i-1+\delta_{i1}} g_n(i,k).$$

(d) Deduce, from parts (a) and (c), that

$$N(\mathbf{m}) = (n-1)!\binom{p-1}{n-1} + (-1)^{n-1}(p-1)(n-1)!.$$

3.4.9. (a) A **block** is a connected simple graph on at least two vertices with no vertex whose removal disconnects the graph. Let $B(x)$ be the exponential generating function for labeled blocks with x marking the labeled vertices, and let $C(x)$ be the exponential generating function for all connected simple labeled graphs. Show that the number of connected simple graphs on n vertices, rooted at a distinguished vertex that is incident with k blocks, is

$$\left[\frac{x^n}{n!}\right]\frac{x}{k!}(B'(xC'(x)))^k.$$

(b) Show that $xC'(x) = x\exp\{B'(xC'(x))\}$, and hence that $B'(xC'(x)) = \log C'(x)$.

3.4.10. From [3.3.48(b)] we know that

$$\sum_{n\geqslant 1} \frac{x^n}{n!} q^{n-1} A_n(q+1) = \log\left\{\sum_{i\geqslant 0} \frac{x^i}{i!}(q+1)^{\binom{i}{2}}\right\}.$$

The right-hand side of this equation is the generating function for the set C of simple labeled connected graphs, with q marking edges and x marking vertices, from [3.4.7]. Derive a combinatorial decomposition relating labeled trees in A and the graphs in C and use it to obtain the preceding equation.

3.4.11. Show that the number of homeomorphically irreducible graphs on n labeled vertices and m edges is

$$\left[x^m \frac{y^n}{n!}\right](1+xy)^{-1/2}\exp\left(-\tfrac{1}{2}xy+\tfrac{1}{4}x^2y^2\right)$$
$$\times \sum_{j\geqslant 0} \frac{y^j}{j!}(1-x)^{-\binom{j+1}{2}} \exp\left\{\frac{-x^2y}{2(1+xy)}j^2 - \tfrac{1}{2}x^2yj\right\}.$$

3.4.12. Show that the number of homeomorphically irreducible graphs on n labeled vertices and m edges, but with no loops, is

$$\left[x^m \frac{y^n}{n!}\right](1+xy)^{-1/2}\exp\left(\tfrac{1}{2}xy+\tfrac{1}{4}x^2y^2\right)$$
$$\times \sum_{j\geqslant 0} \frac{y^j}{j!}(1-x)^{-\binom{j}{2}} \exp\left\{\frac{-x^2y}{2(1+xy)}j^2 - \tfrac{1}{2}x^2yj\right\}.$$

3.4.13. Show that the number of homeomorphically irreducible graphs on n labeled vertices and m edges, with no multiple edges but a single loop allowed at each vertex, is

$$\left[x^m \frac{y^n}{n!}\right](1+xy)^{-1/2}\exp\left(-\tfrac{1}{2}xy-\tfrac{1}{4}x^2y^2\right)$$
$$\times \sum_{j\geqslant 0} \frac{y^j}{j!}(1+x)^{\binom{j+1}{2}} \exp\left\{\frac{-x^2y}{2(1+xy)}j^2 + \frac{x^2}{2}yj\right\}.$$

3.4.14. If $f(n)$ is the number of homeomorphically irreducible graphs on n labeled vertices, with a single loop allowed at each vertex and no multiple edges, show that

$$f(n) = n!\sum_{j=0}^{n} 2^{\binom{j+1}{2}} b_j(n-j)/j!,$$

where

$$2(m+1)b_j(m+1) = (j-2-j^2-4m)b_j(m) + (2j-2-2m)b_j(m-1)$$
$$+(j-3)b_j(m-2) - b_j(m-3), \qquad m \geqslant 0$$

and $b_j(0)=1$, $b_j(i)=0$, for $i<0$.

3.4.15. Show that the number of $(m_1+m_2)\times(n_1+n_2)$ matrices over N in which m_i row sums and n_i column sums are equal to i, for $i=1,2$, is

$$\left[\frac{x_1^{m_1}x_2^{m_2}}{(m_1+m_2)!}\frac{y_1^{n_1}y_2^{n_2}}{(n_1+n_2)!}\right](1-x_2y_2)^{-1/2}$$
$$\times \exp\left\{\tfrac{1}{2}x_2y_2 + (1-x_2y_2)^{-1}\left(x_1y_1+\tfrac{1}{2}x_1^2y_2+\tfrac{1}{2}y_1^2x_2\right)\right\}.$$

3.5. COEFFICIENT EXTRACTION FOR SYMMETRIC FUNCTIONS

We now briefly turn our attention to the situation in which it is possible to express an enumerative result as a particular coefficient in a symmetric multivariate generating function. Often such a generating function is little more than an encoding of the original problem. The task that remains, which may well be a substantial one, is the extraction of the coefficient.

To fix ideas, and to illustrate certain points as the section progresses, we consider a specific combinatorial configuration, the simple 3-regular labeled graph.

3.5.1. Definition (p-Regular Graph)

A graph is said to be p-**regular** *if each vertex has degree* p. □

We now derive the ordinary generating function for simple labeled graphs.

3.5.2. The Ordinary Generating Function for Simple Labeled Graphs

Let t_j mark the degree of vertex j for $j = 1,\ldots, n$. The generating function for the pair of vertices $\{i, j\}$, where $i < j$, is $1 + t_i t_j$ so the generating function for all pairs of vertices is

$$\prod_{1 \leqslant i<j \leqslant n} (1 + t_i t_j).$$

Let $r_p(n)$ be the number of simple p-regular labeled graphs on n vertices. It follows immediately that

$$r_p(n) = [t_1^p \cdots t_n^p] \prod_{1 \leqslant i<j} (1 + t_i t_j). \qquad \square$$

The preceding argument may be modified slightly to account for loops and multiple edges. For a loop on vertex i, the generating function is t_i^2, since the loop contributes two to the degree of i. Since there may be any number of loops at i the contribution to the generating function is $1 + t_i^2 + t_i^4 + \cdots = (1 - t_i^2)^{-1}$. Similarly, for multiple edges between vertices i and j, the contribution to the generating function is $1 + t_i t_j + (t_i t_j)^2 + \cdots = (1 - t_i t_j)^{-1}$. The number of p-regular labeled graphs on n vertices is therefore

$$r_p'(n) = [t_1^p \cdots t_n^p] \prod_{1 \leqslant i \leqslant j} (1 - t_i t_j)^{-1}.$$

Our general task in this section is to calculate $[\mathbf{t}^{\mathbf{i}}]T(\mathbf{t})$, where $\mathbf{i} = (i_1,\ldots) \geqslant \mathbf{0}$ and $T(\mathbf{t})$ is a prescribed symmetric function in $\mathbf{t} = (t_1,\ldots)$. This is accomplished by constructing another function $g(\mathbf{y})$, where $\mathbf{y} = (y_1, \dots)$, with the property that

$$\left[\frac{\mathbf{y}^{\tau(\mathbf{i})}}{\tau(\mathbf{i})!}\right] g(\mathbf{y}) = [\mathbf{t}^{\mathbf{i}}]T(\mathbf{t}),$$

in which $\tau(\mathbf{i})$ is the type of $\mathbf{i}$ (see **2.4.7**). Clearly, the usefulness of this procedure resides in the possibility that $[\mathbf{y}^{\tau(\mathbf{i})}/\tau(\mathbf{i})!]g(\mathbf{y})$ may be easier to obtain than $[\mathbf{t}^{\mathbf{i}}]T(\mathbf{t})$. For example, we may reasonably expect this to be the case for 3-regular graphs, for which $\mathbf{i} = (3, \dots, 3, 0, \dots)$, so $\tau(\mathbf{i}) = (0, 0, n, 0, \dots)$.

3.5.3. Definition (Monomial Symmetric Function)

Let $\mathbf{t} = (t_1, t_2, \dots)$, $\mathbf{j} = (j_1, j_2, \dots)$, *and* $\mathbf{i} = (i_1, i_2, \dots)$. *Then*

$$A_{\mathbf{i}}(\mathbf{t}) = \sum_{\substack{\mathbf{j} \\ \tau(\mathbf{j}) = \mathbf{i}}} \mathbf{t}^{\mathbf{j}} = [\mathbf{x}^{\mathbf{i}}] \prod_{m \geqslant 1} \left(1 + x_1 t_m + x_2 t_m^2 + \cdots\right)$$

is called a **monomial symmetric function.** □

Let $T(\mathbf{t})$ be an arbitrary symmetric function in $\mathbf{t}$. Then T may be uniquely expressed in the form

$$T(\mathbf{t}) = \sum_{\mathbf{i} \geqslant \mathbf{0}} c(\mathbf{i}) A_{\mathbf{i}}(\mathbf{t}),$$

where the $c(\mathbf{i})$ are independent of $\mathbf{t}$. This observation allows us to construct a suitable function $g(\mathbf{y}) = (\Gamma(T))(\mathbf{y})$, called the **Γ-series** of $T(\mathbf{t})$, as follows.

3.5.4. Proposition

Let $T(\mathbf{t}) = \Sigma_{\mathbf{i}}\, c(\mathbf{i}) A_{\mathbf{i}}(\mathbf{t})$ *and let* $\Gamma(T) = \Sigma_{\mathbf{i}}\, c(\mathbf{i}) \mathbf{y}^{\mathbf{i}}/\mathbf{i}!$. *Then*

$$[\mathbf{t}^{\mathbf{j}}]T(\mathbf{t}) = \left[\mathbf{y}^{\tau(\mathbf{j})}/\tau(\mathbf{j})!\right] \Gamma(T).$$

Proof: $[\mathbf{t}^{\mathbf{j}}]T(\mathbf{t}) = \Sigma_{\mathbf{i}}\, c(\mathbf{i})[\mathbf{t}^{\mathbf{j}}]A_{\mathbf{i}}(\mathbf{t})$. But $[\mathbf{t}^{\mathbf{j}}]A_{\mathbf{i}}(\mathbf{t}) = \delta_{\tau(\mathbf{j}), \mathbf{i}}$, from Definition 3.5.3. Thus $[\mathbf{t}^{\mathbf{j}}]T = c(\tau(\mathbf{j}))$ and the result follows. □

To obtain $(\Gamma(T))(\mathbf{y})$, we note first that the generating functions $T(\mathbf{t})$ that arise in practice may usually be expressed quite straightforwardly in terms of the **power sum symmetric functions**

$$s_k(\mathbf{t}) = t_1^k + t_2^k + \cdots \qquad \text{for } k \geqslant 1.$$

This expression, $G(\mathbf{s})$, is unique, where $\mathbf{s} = (s_1, s_2, \dots)$. But the Γ-series theorem (**3.5.7**) will enable us to transform a system of differential equations for $G(\mathbf{s})$ into a system of differential equations for $\Gamma(T)$. Therefore, we wish to derive a system of differential equations for $G(\mathbf{s})$ in the indeterminates $s_1, s_2, \ldots$. We now illustrate these steps in the enumeration of simple 3-regular labeled graphs.

3.5.5. A Differential Equation for Simple Labeled Graphs in Terms of Power Sum Symmetric Functions

From **3.5.2**, the generating function for simple labeled graphs is

$$T(\mathbf{t}) = \prod_{1 \leqslant i<j} (1 + t_i t_j).$$

Thus

$$T(\mathbf{t}) = \exp\log \prod_{1 \leqslant i<j} (1 + t_i t_j) = \exp \sum_{k \geqslant 1} (-1)^{k-1} \sum_{1 \leqslant i<j} (t_i t_j)^k / k.$$

But

$$\sum_{1 \leqslant i<j} (t_i t_j)^k = \frac{1}{2}\left\{\left(\sum_{i \geqslant 1} t_i^k\right)^2 - \sum_{i \geqslant 1} t_i^{2k}\right\} = \tfrac{1}{2}\{s_k^2 - s_{2k}\},$$

so $T(\mathbf{t}) = G(\mathbf{s}(\mathbf{t}))$, where

$$G(\mathbf{s}) = \exp \sum_{k \geqslant 1} \frac{(-1)^{k-1}}{2k}\{s_k^2 - s_{2k}\}.$$

Clearly, $G(\mathbf{s})$ satisfies the following system of differential equations:

$$\left.\begin{aligned} \frac{\partial G(\mathbf{s})}{\partial s_{2k+1}} &= \frac{s_{2k+1}}{2k+1} G(\mathbf{s}) && \text{for } k \geqslant 0 \\ \frac{\partial G(\mathbf{s})}{\partial s_{2k}} &= \frac{1}{2k}\left((-1)^k - s_{2k}\right) G(\mathbf{s}) && \text{for } k \geqslant 1. \end{aligned}\right\}$$

For p-regular graphs we may, of course, set $s_j = 0$ for $j > p$. □

The device for determining $G(\mathbf{s})$ from $T(\mathbf{t})$ by considering $\exp\log T(\mathbf{t})$ is an important one and may be used to advantage when $T(\mathbf{t})$ is expressible as an iterated product. To derive a system of differential equations for $\Gamma(T)$ from a system for $G(\mathbf{s})$, we express $\Gamma(\partial G(\mathbf{s})/\partial s_k)$ and $\Gamma(s_k G(\mathbf{s}))$ in terms of $\Gamma(T)$ and its derivatives with respect to y_i for $i \geqslant 1$. The following proposition is required for this purpose. For notational convenience throughout we use $(\Gamma(f(\mathbf{s})))(\mathbf{y})$ and $(\Gamma(f(\mathbf{s}(\mathbf{t}))))(\mathbf{y})$ interchangeably.

3.5.6. Proposition

Let

$$E_n(\mathbf{x}) = [z^n]\log(1 + zx_1 + z^2x_2 + \cdots)$$
$$= \sum_{i_1+2i_2+\cdots=n} (-1)^{m-1}(m-1)!\mathbf{x}^{\mathbf{i}}/\mathbf{i}!$$

where $m = i_1 + i_2 + \cdots$, $\mathbf{i} = (i_1, i_2, \ldots)$ *and* $\mathbf{x} = (x_1, x_2, \ldots)$. *Then*

1. $\sum_{\mathbf{i}\geq \mathbf{0}} A_{\mathbf{i}}(\mathbf{t})\mathbf{x}^{\mathbf{i}} = \exp\left\{\sum_{n\geq 1} E_n(\mathbf{x})s_n(\mathbf{t})\right\}.$
2. $\frac{\partial}{\partial s_n}A_{\mathbf{i}}(\mathbf{t}) = \sum_{\substack{\mathbf{j}\\ \mathbf{0}\leq\mathbf{j}\leq\mathbf{i}}} A_{\mathbf{j}}(\mathbf{t})[\mathbf{x}^{\mathbf{i}-\mathbf{j}}]E_n(\mathbf{x}).$

Proof:

1. From Definition 3.5.3

$$\sum_{\mathbf{i}}\mathbf{x}^{\mathbf{i}}A_{\mathbf{i}}(\mathbf{t}) = \prod_{j\geq 1}\left(1 + x_1t_j + x_2t_j^2 + \cdots\right)$$
$$= \exp\sum_{j\geq 1}\log\left(1 + x_1t_j + x_2t_j^2 + \cdots\right) = \exp\sum_{n,\,j\geq 1} t_j^nE_n(\mathbf{x})$$
$$= \exp\sum_{n\geq 1}E_n(\mathbf{x})\sum_{j\geq 1}t_j^n,$$

and (**1**) follows.

2. From (**1**)

$$\sum_{\mathbf{i}\geq\mathbf{0}}\mathbf{x}^{\mathbf{i}}\frac{\partial}{\partial s_n}A_{\mathbf{i}}(\mathbf{t}) = \frac{\partial}{\partial s_n}\sum_{\mathbf{i}\geq\mathbf{0}}\mathbf{x}^{\mathbf{i}}A_{\mathbf{i}}(\mathbf{t}) = \frac{\partial}{\partial s_n}\exp\sum_{n\geq 1}E_n(\mathbf{x})s_n$$
$$= E_n(\mathbf{x})\exp\sum_{n\geq 1}E_n(\mathbf{x})s_n = E_n(\mathbf{x})\sum_{\mathbf{j}\geq\mathbf{0}}A_{\mathbf{j}}(\mathbf{t})\mathbf{x}^{\mathbf{j}},$$

again from (**1**). Thus

$$\frac{\partial}{\partial s_n}A_{\mathbf{i}}(\mathbf{t}) = [\mathbf{x}^{\mathbf{i}}]E_n(\mathbf{x})\sum_{\mathbf{j}\geq\mathbf{0}}A_{\mathbf{j}}(\mathbf{t})\mathbf{x}^{\mathbf{j}} = \sum_{\mathbf{j}\geq\mathbf{0}}A_{\mathbf{j}}(\mathbf{t})[\mathbf{x}^{\mathbf{i}-\mathbf{j}}]E_n(\mathbf{x})$$

and the result follows. □

We now state the result that enables us to derive a system of differential equations for $\Gamma(T)$ from a system of differential equations for $G(\mathbf{s})$.

3.5.7. Theorem (Γ-Series)

Let T be a symmetric function in $\mathbf{t}$, *let* $T(\mathbf{t}) = G(\mathbf{s}(\mathbf{t}))$, *and* $C_n(\mathbf{y}, \partial/\partial\mathbf{y}) = y_n + \sum_{i\geq 1} y_{n+i}\,\partial/\partial y_i$. *Then*

1. $\Gamma\left(\dfrac{\partial G}{\partial s_n}\right) = E_n\left(\dfrac{\partial}{\partial \mathbf{y}}\right)\Gamma(G),$
2. $\Gamma(s_n G) = C_n\left(\mathbf{y}, \dfrac{\partial}{\partial \mathbf{y}}\right)\Gamma(G),$

where $\partial/\partial\mathbf{y}$ denotes $(\partial/\partial y_1, \partial/\partial y_2, \dots)$.

Proof:

1. Let $T(\mathbf{t}) = \sum_{\mathbf{i}\geq \mathbf{0}} c(\mathbf{i})A_{\mathbf{i}}(\mathbf{t})$. Thus

$$\Gamma\left(\frac{\partial G}{\partial s_n}\right) = \sum_{\mathbf{i}\geq\mathbf{0}} c(\mathbf{i})\Gamma\left(\frac{\partial}{\partial s_n}A_{\mathbf{i}}(\mathbf{t})\right)$$

so, from Proposition 3.5.6,

$$\begin{aligned}\Gamma\left(\frac{\partial G}{\partial s_n}\right) &= \sum_{\mathbf{i}\geq\mathbf{0}} c(\mathbf{i})\sum_{\mathbf{j}\leq\mathbf{i}}\{[\mathbf{x}^{\mathbf{i}-\mathbf{j}}]E_n(\mathbf{x})\}\Gamma(A_{\mathbf{j}}(\mathbf{t}))\\ &= \sum_{\mathbf{i}\geq\mathbf{0}} c(\mathbf{i})\sum_{\mathbf{j}\leq\mathbf{i}}[\mathbf{x}^{\mathbf{i}-\mathbf{j}}]E_n(\mathbf{x})\mathbf{y}^{\mathbf{j}}/\mathbf{j}!\\ &= \sum_{\mathbf{i}\geq\mathbf{0}} c(\mathbf{i})\sum_{\mathbf{j}\leq\mathbf{i}}[\mathbf{x}^{\mathbf{i}-\mathbf{j}}]E_n(\mathbf{x})\frac{\partial^{\mathbf{i}-\mathbf{j}}}{\partial\mathbf{y}^{\mathbf{i}-\mathbf{j}}}(\mathbf{y}^{\mathbf{i}}/\mathbf{i}!).\end{aligned}$$

Now let $\mathbf{i}-\mathbf{j}=\mathbf{k}\geq\mathbf{0}$, since $\mathbf{i}\geq\mathbf{j}$. Thus

$$\begin{aligned}\Gamma\left(\frac{\partial G}{\partial s_n}\right) &= \sum_{\mathbf{i}\geq\mathbf{0}} c(\mathbf{i})\sum_{\mathbf{k}\geq\mathbf{0}}[\mathbf{x}^{\mathbf{k}}]E_n(\mathbf{x})\frac{\partial^{\mathbf{k}}}{\partial\mathbf{y}^{\mathbf{k}}}(\mathbf{y}^{\mathbf{i}}/\mathbf{i}!)\\ &= \sum_{\mathbf{k}\geq\mathbf{0}}\frac{\partial^{\mathbf{k}}}{\partial\mathbf{y}^{\mathbf{k}}}[\mathbf{x}^{\mathbf{k}}]E_n(\mathbf{x})\sum_{\mathbf{i}\geq\mathbf{0}}c(\mathbf{i})\mathbf{y}^{\mathbf{i}}/\mathbf{i}!\\ &= E_n\left(\frac{\partial}{\partial\mathbf{y}}\right)\Gamma(T),\end{aligned}$$

and (1) follows.

2. Let $\boldsymbol{\delta}_n = (0,\ldots,0,1,0,\ldots)$ where the 1 appears in the nth position. Thus, from Definition 3.5.3,

$$s_n A_{\mathbf{i}} = \left(\sum_{\tau(\mathbf{l})=\boldsymbol{\delta}_n} x_1^{l_1} x_2^{l_2} \cdots \right) \sum_{\tau(\mathbf{j})=\mathbf{i}} x_1^{j_1} x_2^{j_2} \cdots$$

$$= \sum_{\tau(\mathbf{l})=\boldsymbol{\delta}_n} \sum_{\tau(\mathbf{j})=\mathbf{i}} x_1^{l_1+j_1} x_2^{l_2+j_2} \cdots$$

where $\mathbf{l} = (l_1, l_2, \ldots)$ and $\mathbf{j} = (j_1, j_2, \ldots)$. Now the effect of $\mathbf{l}$ is to change a single kth power in $\mathbf{j}$ to a $(k+n)$th power, for some k, in all possible ways. Each of the resulting monomials may be obtained in $i_{k+n} + 1$ ways, so if $\mathbf{m} = (m_1, \ldots)$, then

$$s_n A_{\mathbf{i}} = \sum_{k \geqslant 1} (1 + i_{n+k}) \sum_{\tau(\mathbf{m})=\mathbf{i}+\boldsymbol{\delta}_{n+k}-\boldsymbol{\delta}_k} x_1^{m_1} x_2^{m_2} \cdots + (1 + i_n) \sum_{\tau(\mathbf{m})=\mathbf{i}+\boldsymbol{\delta}_n} x_1^{m_1} x_2^{m_2} \cdots,$$

and the result follows immediately. □

It follows from the Γ-series theorem that

$$\Gamma\left(s_1^{i_1} \cdots \frac{\partial^{j_1}}{\partial s_1^{j_1}} \cdots G \right) = C_1^{i_1}\left(\mathbf{y}, \frac{\partial}{\partial \mathbf{y}}\right) \cdots E_1^{j_1}\left(\frac{\partial}{\partial \mathbf{y}}\right) \cdots \Gamma(G).$$

Thus any differential equation for $G(\mathbf{s})$ may be translated, by means of the Γ-series theorem, into a differential equation for $\Gamma(G)$. We illustrate the use of the Γ-series theorem in the enumeration of simple 3-regular labeled graphs.

3.5.8. The Γ-Series for Simple 3-Regular Labeled Graphs

From Example 3.5.2 the number of simple 3-regular labeled graphs on n vertices is given by

$$r_3(n) = \left[t_1^3 \cdots t_n^3 \right] T(\mathbf{t}), \qquad \text{where } T(\mathbf{t}) = \prod_{1 \leqslant i < j} (1 + t_i t_j).$$

Let $T(\mathbf{t}) = G(\mathbf{s}(\mathbf{t}))$. Then, from Example 3.5.5, we know that $G(\mathbf{s})$, with $s_i = 0$ for $i \geqslant 4$, satisfies the system of differential equations

$$\left. \begin{aligned} \frac{\partial G}{\partial s_{2k+1}} &= \frac{s_{2k+1}}{2k+1} G && \text{for } k = 0, 1 \\ \frac{\partial G}{\partial s_{2k}} &= \frac{1}{2k}\left\{ (-1)^k - s_{2k} \right\} G && \text{for } k = 1. \end{aligned} \right\}$$

Applying Γ we have

$$\Gamma\left(\frac{\partial G}{\partial s_1}\right) = \Gamma(s_1 G),$$

$$\Gamma\left(\frac{\partial G}{\partial s_2}\right) = -\tfrac{1}{2}\{\Gamma(G) + \Gamma(s_2 G)\},$$

$$\Gamma\left(\frac{\partial G}{\partial s_3}\right) = \tfrac{1}{3}\Gamma(s_3 G).$$

From Proposition 3.5.6, $E_1 = x_1$, $E_2 = x_2 - \frac{1}{2}x_1^2$, $E_3 = x_3 - x_1 x_2 + \frac{1}{3}x_1^3$. It follows from the Γ-series theorem that $\Gamma(G)$ satisfies the system of equations

$$\left.\begin{aligned} \frac{\partial}{\partial y_1}\Gamma(G) &= y_1\Gamma(G) + y_2\frac{\partial}{\partial y_1}\Gamma(G) + y_3\frac{\partial}{\partial y_2}\Gamma(G) \\ 2\frac{\partial}{\partial y_2}\Gamma(G) - \frac{\partial^2}{\partial y_1^2}\Gamma(G) &= -\Gamma(G) - y_2\Gamma(G) - y_3\frac{\partial}{\partial y_1}\Gamma(G) \\ 3\frac{\partial \Gamma(G)}{\partial y_3} - 3\frac{\partial^2 \Gamma(G)}{\partial y_1\,\partial y_2} + \frac{\partial^3}{\partial y_1^3}\Gamma(G) &= y_3\Gamma(G), \end{aligned}\right\}$$

where, from Proposition 3.5.4, $r_3(n) = [y_3^n/n!]\Gamma(G)$. □

From this particular system of partial differential equations we can deduce a single ordinary differential equations for $(\Gamma(G))(0, 0, y_3)$ as follows.

3.5.9. A Recurrence Equation for Simple 3-Regular Labeled Graphs

Let $V(y_1, y_2, y_3)$ be the Γ-series, derived in **3.5.8**, for simple 3-regular labeled graphs. Let f_i denote $\partial f/\partial y_i$, where f is a function of y_1, y_2, y_3. Thus, from **3.5.8**,

$$r_3(n) = \left[\frac{y_3^n}{n!}\right] V(y_1, y_2, y_3),$$

where V satisfies the system of differential equations

$$\left.\begin{aligned} V_1 &= y_1 V + y_2 V_1 + y_3 V_2 && (1) \\ 2V_2 - V_{11} &= -(1 + y_2)V - y_3 V_1 && (2) \\ 3V_3 - 3V_{12} + V_{111} &= y_3 V. && (3) \end{aligned}\right\}$$

Our intention is to eliminate y_1 and y_2 to obtain a differential equation involving y_3 alone. Eliminating V_2 between (1) and (2), we have

$$y_3V_{11} - (y_3^2 + 2(1 - y_2))V_1 + (2y_1 - y_3(1 + y_2))V = 0.$$

Now V_{12} may be eliminated between (3) and the result of differentiating (1) partially with respect to y_1, giving

$$3y_3V_3 = -y_3V_{111} + 3(1 - y_2)V_{11} - 3y_1V_1 + (y_3^2 - 3)V.$$

We can now eliminate y_2 by letting $U(y_1, y_3) = V(y_1, 0, y_3)$ and setting $y_2 = 0$ in the last two equations to obtain

$$y_3U_{11} - (y_3^2 + 2)U_1 + (2y_1 - y_3)U = 0 \tag{4}$$

$$3y_3U_3 = -y_3U_{111} + 3U_{11} - 3y_1U_1 + (y_3^2 - 3)U.$$

Eliminating U_{111} between the latter and the result of applying $\partial/\partial y_1$ to (4), we have

$$3y_3U_3 = (1 - y_3^2)U_{11} - (y_1 + y_3)U_1 + (y_3^2 - 1)U.$$

Eliminating U_{11} between this and (4), we have

$$3y_3^2U_3 = (2 - 2y_3^2 - y_3^4 - y_1y_3)U_1 + 2y_1(y_3^2 - 1)U. \tag{5}$$

Set $y_1 = 0$ in (4) and (5) to give

$$y_3U_{11} - (y_3^2 + 2)U_1 - y_3U = 0, \tag{6}$$

$$3y_3^2U_3 = (2 - 2y_3^2 - y_3^4)U_1, \tag{7}$$

where the subscript i of U from this point onward denotes $\partial/\partial y_i$ at $y_1 = 0$, and U denotes $U(0, y_3)$. Differentiate (5) partially with respect to y_3 and set $y_1 = 0$. Thus

$$3y_3^2U_{33} + 6y_3U_3 + 4y_3(1 + y_3^2)U_1 = (2 - 2y_3^2 - y_3^4)U_{13}.$$

Differentiate (5) partially with respect to y_1 and set $y_1 = 0$. This gives

$$3y_3^2U_{13} = (2 - 2y_3^2 - y_3^4)U_{11} - y_3U_1 + 2(y_3^2 - 1)U.$$

Eliminating U_{13} between the last two equations, we have

$$9y_3^4U_{33} + 18y_3^3U_3 = 2(y_3^2 - 1)(2 - 2y_3^2 - y_3^4)U - y_3(2 + 10y_3^2 + 11y_3^4)U_1$$
$$+ (2 - 2y_3^2 - y_3^4)^2U_{11}.$$

Eliminate U_{11} between this and (6) to give

$$9y_3^5U_{33} + 18y_3^4U_3 + y_3^5(2 - 2y_3^2 - y_3^4)U$$
$$= \{(y_3^2 + 2)(2 - 2y_3^2 - y_3^4)^2 - y_3^2(2 + 10y_3^2 + 11y_3^4)\}U_1.$$

Finally, we eliminate U_1 between this and (7) to obtain

$$9y_3^4(2 - 2y_3^2 - y_3^4)U_{33} - 3y_3(8 - 26y_3^2 - 6y_3^4 + 3y_3^6 + 6y_3^8 + y_3^{10})U_3$$
$$+y_3^4(2 - 2y_3^2 - y_3^4)^2U = 0.$$

Let $W(x) = U(0, y_3)$, where $x = y_3^2$. Now

$$U_3 = 2y_3 \left.\frac{dW}{dx}\right|_{x=y_3^2} \quad \text{and} \quad U_{33} = 2\left.\frac{dW}{dx}\right|_{x=y_3^2} + 4y_3^2\left.\frac{d^2W}{dx^2}\right|_{x=y_3^2},$$

so W satisfies the differential equation

$$6x^2(2 - 2x - x^2)\frac{d^2W}{dx^2} - (8 - 32x + 6x^3 + 6x^4 + x^5)\frac{dW}{dx}$$
$$+\tfrac{1}{6}x(2 - 2x - x^2)^2W = 0,$$

where $[x^n/(2n)!]W(x) = r_3(2n)$.

Of course, since the sum of the degrees in a graph is even, there are no simple 3-regular labeled graphs on an odd number of vertices. By comparing coefficients we may obtain a linear recurrence equation for $r_3(2n)$. The details are routine and have been omitted. □

In certain cases there may be two sets of indeterminates, and the generating function may be symmetric in each set separately. The Γ-series theorem extends to this case in the obvious way. This situation is illustrated with the enumeration of integer matrices with prescribed line sum, where by **line sum** we mean both row and column sums.

3.5.10. Non-negative Integer Matrices with Line Sum Equal to 2

Let $l_k(m)$ be the number of $m \times m$ matrices over the non-negative integers with each line sum equal to k. We wish to obtain a recurrence equation for $l_2(m)$.

Let r_i and c_i be indeterminates marking the sum for row i and column i, respectively, for $i \geqslant 1$. Now if k is the (i, j)-element of a matrix, then it contributes k to the row sum of row i and k to the column sum of column j. Its contribution to the generating function for the (i, j)-cell is therefore $(r_i c_j)^k$, so the generating function for the (i, j)-cell is $1 + (r_i c_j) + (r_i c_j)^2 + \cdots = (1 - r_i c_j)^{-1}$, since k may be any non-negative integer. Thus, by the product lemma,

$$l_2(m) = \left[r_1^2 \cdots r_m^2 c_1^2 \cdots c_m^2\right] T(\mathbf{r}, \mathbf{c}),$$

where $T(\mathbf{r}, \mathbf{c}) = \prod_{i, j \geqslant 1} (1 - r_i c_j)^{-1}$ and $\mathbf{r} = (r_1, r_2, \dots)$, $\mathbf{c} = (c_1, c_2, \dots)$.

Now $T(\mathbf{r}, \mathbf{c})$ is a symmetric function in $\mathbf{r}$ and is a symmetric function in $\mathbf{c}$ so the Γ-series theorem is applicable.

Let $s_k(\mathbf{r}) = r_1^k + r_2^k + \cdots$ and $t_k(\mathbf{c}) = c_1^k + c_2^k + \cdots$ for $k \geqslant 1$, the power sum symmetric functions for $\mathbf{r}$ and for $\mathbf{c}$. We obtain $T(\mathbf{r}, \mathbf{c})$ in terms of $\mathbf{s}$ and $\mathbf{t}$, where $\mathbf{s} = (s_1(\mathbf{r}), s_2(\mathbf{r}), \dots)$ and $\mathbf{t} = (t_1(\mathbf{c}), t_2(\mathbf{c}), \dots)$ as follows.

$$T(\mathbf{r}, \mathbf{c}) = \exp \sum_{i, j \geqslant 1} \log(1 - r_i c_j)^{-1} = \exp \sum_{k \geqslant 1} \frac{1}{k} s_k t_k.$$

For matrices with line sum equal to 2 we may set $s_i = 0$, $t_i = 0$, for $i \geqslant 3$. Let $T(\mathbf{r}, \mathbf{c}) = G(s_1(\mathbf{r}), s_2(\mathbf{r}); t_1(\mathbf{c}), t_2(\mathbf{c}))$. Thus $G = \exp(s_1 t_1 + \frac{1}{2} s_2 t_2)$, which satisfies

$$\left.\begin{aligned} \frac{\partial G}{\partial s_1} &= t_1 G, & \frac{\partial G}{\partial t_1} &= s_1 G \\ \frac{\partial G}{\partial s_2} &= \tfrac{1}{2} t_2 G, & \frac{\partial G}{\partial t_2} &= \tfrac{1}{2} s_2 G. \end{aligned}\right\}$$

Let $\Gamma(G) = V(x_1, x_2; y_1, y_2)$. Then, from Proposition 3.5.4,

$$l_2(m) = \left[\frac{x_2^m y_2^m}{m! m!}\right] V(0, x_2; 0, y_2),$$

where x_2, y_2 mark rows and columns, respectively, whose sum is equal to 2. It follows from the Γ-series theorem that V satisfies the following system of differential equations.

$$\left.\begin{aligned} \frac{\partial V}{\partial x_1} &= y_1 V + y_2 \frac{\partial V}{\partial y_1} \quad (1), & \frac{\partial V}{\partial y_1} &= x_1 V + x_2 \frac{\partial V}{\partial x_1} \quad (1') \\ \frac{\partial V}{\partial x_2} - \frac{1}{2} \frac{\partial^2 V}{\partial x_1^2} &= \tfrac{1}{2} y_2 V \quad (2), & \frac{\partial V}{\partial y_2} - \frac{1}{2} \frac{\partial^2 V}{\partial y_1^2} &= \tfrac{1}{2} x_2 V. \quad (2') \end{aligned}\right\}$$

To obtain a recurrence equation for $\{l_2(n) | n \geqslant 0\}$, we derive a differential equation for $W(x_2, y_2) = V(0, x_2; 0, y_2)$ in the following way. From (1) and (1′)

$$\frac{\partial V}{\partial x_1} = (y_1 + y_2 x_1)(1 - x_2 y_2)^{-1} V. \tag{3}$$

Differentiating (2′) partially with respect to x_2, we have

$$2\frac{\partial^2 V}{\partial x_2\,\partial y_2} = \frac{\partial^3 V}{\partial y_1^2\,\partial x_2} + x_2\frac{\partial V}{\partial x_2} + V.$$

Eliminating $\partial V/\partial x_2$ from the right-hand side of this equation by means of (2), we have

$$4\frac{\partial^2 V}{\partial x_2\,\partial y_2} = \frac{\partial^4 V}{\partial x_1^2\,\partial y_1^2} + y_2\frac{\partial^2 V}{\partial y_1^2} + x_2\frac{\partial^2 V}{\partial x_1^2} + x_2 y_2 V + 2V. \tag{4}$$

It remains to express the right-hand side of this equation in terms of x_2 and y_2. To do this note that, from (3),

$$\left.\frac{\partial^2 V}{\partial x_1^2}\right|_{x_1=y_1=0} = y_2(1 - x_2 y_2)^{-1} W.$$

From (1′) and (3),

$$\left.\frac{\partial^4 V}{\partial x_1^2\,\partial y_1^2}\right|_{x_1=y_1=0} = 2(1 + x_2 y_2)(1 - x_2 y_2)^{-2} W,$$

and

$$\left.\frac{\partial^2 V}{\partial y_1^2}\right|_{x_1=y_1=0} = x_2(1 - x_2 y_2)^{-1} W.$$

Substituting these expressions into (4) and simplifying gives

$$(4 - 8x_2 y_2 + 4x_2^2 y_2^2)\frac{\partial^2 W}{\partial x_2\,\partial y_2} = (4 - 2x_2^2 y_2^2 + x_2^3 y_2^3) W.$$

But a matrix with line sum k must be square, so $W(x_2, y_2) = F(x_2 y_2)$, where $F(x) = \Sigma_{k\geqslant 0} l_2(k) x^k/(k!)^2$. Thus $F(x)$ satisfies the equation

$$4x(1-x)^2 F'' + 4(1-x)^2 F' - (4 - 2x^2 + x^3) F = 0,$$

where $F(0) = 1$. By inspection, this may be written

$$\left\{2x(1-x)\frac{d}{dx} + (2 + 2x - x^2)\right\} G(x) = 0,$$

where $G(x) = 2(1-x)F' - (2-x)F$. But the only formal power series $G(x)$ satisfying the preceding equation is $G(x) = 0$, so $F(x)$ satisfies

$$2(1-x)F' - (2-x)F = 0.$$

Then $\{l_2(m) | m \geqslant 0\}$ satisfies the recurrence equation

$$l_2(m+1) = (m+1)^2 l_2(m) - \tfrac{1}{2}m^2(m+1)l_2(m-1) \qquad \text{for } m \geqslant 0,$$

where $l_2(0) = 1$, $l_2(-1) = 0$. Moreover, the explicit solution for F is $F(x) = (1-x)^{-1/2}e^{x/2}$. □

For $m \times m$ matrices with line sum 2 we have been able to obtain an explicit expression for the generating function $F(x)$, both by an algebraic method (**3.5.10**) and by a combinatorial decomposition [**3.4.15**]. However, such a combinatorial decomposition is not evident for matrices with line sum 3, although the algebraic method can be extended to derive a recurrence equation for $\{l_3(m) | m \geqslant 0\}$ [**3.5.5**].

We conclude this section by using a differential decomposition to derive directly the system of equations for simple 3-regular labeled graphs, given in **3.5.8**.

3.5.11. Differential Decomposition for Simple 3-Regular Labeled Graphs

Let $V(y_1, y_2, y_3)$ be the generating function for the set V of simple labeled graphs with degrees at most 3, with tagged vertices of degree i (exponentially) marked with y_i, for $i = 1, 2, 3$. We derive three differential decompositions by distinguishing first one monovalent vertex, then two distinct monovalent vertices, and finally three distinct monovalent vertices in graphs in V.

Case 1. Distinguish exactly one monovalent vertex in each element in V. The generating function for this is $y_1 \partial V/\partial y_1$. We now derive this in another way, by considering three subcases.

1. The distinguished monovalent vertex is adjacent to a vertex of degree 1, forming a component consisting of a single edge joining two vertices. The generating function for this is $y_1^2 V$.

2. The distinguished monovalent vertex is adjacent to a vertex of degree 2. We may construct such graphs by distinguishing a monovalent vertex v and then connecting this by an edge to a new monovalent vertex u. Now u is the distinguished monovalent vertex adjacent to a bivalent vertex v. The generating function for this is $y_1 y_2 \partial V/\partial y_1$. We note that the operator $y_2 \partial/\partial y_1$ arises because a monovalent vertex is first distinguished and then connected to another vertex, making the former bivalent.

3. The distinguished monovalent vertex is adjacent to a vertex of degree 3. Following **(2)**, the generating function for this is $y_1 y_3 \partial V/\partial y_2$.

It follows that

$$\frac{\partial V}{\partial y_1} = y_1 V + y_2 \frac{\partial V}{\partial y_1} + y_3 \frac{\partial V}{\partial y_2}.$$

Case 2. Distinguish two distinct monovalent vertices in each element in V. The generating function for this is $\frac{1}{2} y_1^2 \partial^2 V/\partial y_1^2$. We now derive this in another way, by considering three subcases.

1. The two distinguished vertices are connected by a path of edge-length 1, forming a component consisting of one edge. The generating function for this is $\frac{1}{2} y_1^2 V$.

2. The two distinguished vertices are connected by a path of edge-length 2. There are two possibilities:

i. The path contains exactly one bivalent vertex. The generating function for this is $\frac{1}{2} y_1^2 y_2 V$, since $\frac{1}{2} y_1^2 y_2$ is the generating function for a component consisting of a path of edge-length 2.

ii. The path contains exactly one trivalent vertex. Such graphs may be obtained by joining a distinguished monovalent vertex in an element of V to two new monovalent vertices, which are themselves the distinguished monovalent vertices in the resulting graph. The generating function for this is $\frac{1}{2} y_1^2 y_3 \partial V/\partial y_1$.

3. We may obtain the remaining such graphs by deleting from a graph in V a vertex u of degree 2 connected to distinct vertices a and b, and connecting a distinguished isolated vertex a' to a and a distinguished isolated vertex b' to b. The vertices a' and b' are the distinguished monovalent vertices and are not connected by a path of edge-length 1 or 2. The generating function for this is $y_1^2 \partial V/\partial y_2$.

It follows that

$$\frac{\partial^2 V}{\partial y_1^2} = 2\frac{\partial V}{\partial y_2} + V + y_2 V + y_3 \frac{\partial V}{\partial y_1}.$$

Case 3. Distinguish three distinct monovalent vertices in each element in V. The generating function for this is $\frac{1}{6} y_1^3 \partial^3 V/\partial y_1^3$. We now derive this in another way, by considering three subcases.

1. Exactly two of the distinguished vertices are joined by a path of edge-length 1. We may construct such graphs by joining two isolated vertices, u and v, by an edge and by distinguishing one monovalent vertex w in a graph in V. The generating function for this is $\frac{1}{2} y_1^2 (y_1 \partial V/\partial y_1)$.

2. At least one pair of distinguished vertices are joined by a path of edge-length 2. There are three possibilities:

i. All three distinguished vertices are joined by paths of edge-length exactly 2. Thus the distinguished vertices are the monovalent vertices of a component whose remaining vertex has degree 3. The generating function for this is therefore $\frac{1}{6}y_1^3y_3V$, since the component may be adjoined to any element in V.

ii. Exactly two of the distinguished vertices are joined by a path of edge length equal to 2. There are two subcases: **a.** The path contains a bivalent vertex. Such graphs may be constructed from a path of edge-length 2 joining two distinguished vertices, and a graph in V with exactly one distinguished monovalent vertex. The generating function for this is $\frac{1}{2}y_1^2y_2(y_1\partial V/\partial y_1)$. **b.** The path contains a vertex of degree 3. We may construct such graphs by considering a path *uvw* of edge-length 2 and a graph in V with exactly two distinct distinguished monovalent vertices *a* and *b*, separated by more than one edge. The vertices *v* and *a* are now identified, and *u*, *w*, and *b* are the distinct distinguished vertices of the resulting graph. The generating function due to all graphs in V treated in this way is $\frac{1}{2}y_1^2(y_1y_3(\partial^2V/\partial y_1^2))$. But this set includes graphs in which two distinguished monovalent vertices are separated by a single edge, and hence form a component, enumerated by y_1^2, adjoined to an element of V, enumerated by V. When this is treated in the preceding manner, the generating function is $\frac{1}{2}y_1^2(y_3y_1V)$, so the contribution of this case is $\frac{1}{2!}y_1^3y_3((\partial^2V/\partial y_1^2) - V)$.

3. No pairs of the distinguished monovalent vertices are joined by paths of edge length 1 or 2. We may construct such graphs by deleting from a graph in V a vertex of degree 3 connected to vertices *a*, *b*, and *c* and connecting *a* to *a'*, *b* to *b'*, and *c* to *c'*, where *a'*, *b'*, *c'* are isolated vertices. In the resulting graph, *a'*, *b'*, *c'* are the distinguished vertices. The generating function for this is $y_1^3\partial V/\partial y_3$.

It follows that

$$\frac{\partial^3V}{\partial y_1^3} = 3(1+y_2)\frac{\partial V}{\partial y_1} + 3y_3\frac{\partial^2V}{\partial y_1^2} - 2y_3V + 6\frac{\partial V}{\partial y_3}$$

$$= \frac{\partial}{\partial y_1}\left\{3(1+y_2)V + 3y_3\frac{\partial V}{\partial y_1}\right\} - 2y_3V + 6\frac{\partial V}{\partial y_3}$$

$$= -3\frac{\partial}{\partial y_1}\left\{2\frac{\partial V}{\partial y_2} - \frac{\partial^2V}{\partial y_1^2}\right\} - 2y_3V + 6\frac{\partial V}{\partial y_3}$$

from Case 2. Thus

$$\frac{\partial^3V}{\partial y_1^3} + 3\frac{\partial V}{\partial y_3} = y_3V + 3\frac{\partial^2V}{\partial y_1\,\partial y_2}.$$

Collecting these results yields the following system of partial differential equations for V:

$$\left.\begin{aligned} \frac{\partial V}{\partial y_1} &= y_1 V + y_2 \frac{\partial V}{\partial y_1} + y_3 \frac{\partial V}{\partial y_2} \\ 2\frac{\partial V}{\partial y_2} - \frac{\partial^2 V}{\partial y_1^2} &= -V - y_2 V - y_3 \frac{\partial V}{\partial y_1} \\ 3\frac{\partial V}{\partial y_3} - 3\frac{\partial^2 V}{\partial y_1 \partial y_2} + \frac{\partial^3 V}{\partial y_1^3} &= y_3 V. \end{aligned}\right\}$$

This is the system that was derived in **3.5.8**. □

We note that the Γ-series theorem enables us to derive systems of recurrence equations for simple labeled graphs in general. This task would involve a formidable case analysis if a differential construction were used.

NOTES AND REFERENCES

This section is based on Goulden, Jackson, and Reilly (1983); some related differential operators are given by Hammond (1883) and MacMahon (1915); for a discussion of properties of linear recurrence equations with polynomial coefficients, see Stanley (1980).

3.5.9 Read (1960); **3.5.10** Anand, Dumir, and Gupta (1966); [**3.5.3–5**] Goulden, Jackson, and Reilly (1983); [**3.5.4**] Read and Wormald (1980); [**3.5.6**] Reilly (private communication); [**3.5.7, 8**] Devitt and Jackson (1982).

EXERCISES

3.5.1. From **2.4.16** the number of Smirnov sequences of type **i** is $c(\mathbf{i}) = [\mathbf{x}^{\mathbf{i}}]T(\mathbf{x})$, where $T(\mathbf{x}) = \{1 - \Sigma_{i\geqslant 1} x_i(1+x_i)^{-1}\}^{-1}$. Deduce from the Γ-series theorem that

$$c(\mathbf{i}) = \sum_{m\geqslant 0} m![x^m]\prod_{i\geqslant 1}\{(-1)^i L_i^{(-1)}(x)\}^{f_i},$$

where $L_i^{(\alpha)}(x) = (-1)^i \sum_{k\geqslant 0}\binom{-i-\alpha-1}{i-k}\frac{x^k}{k!}$ *is a* **Laguerre polynomial** and $\mathbf{f} = (f_1, f_2, \dots) = \tau(\mathbf{i})$ is the type of **i**.

3.5.2. (a) Show that the number of sequences in N_+^* of type $\mathbf{i}$ in which the maximal blocks all have length at least 2 is

$$[\mathbf{x}^{\mathbf{i}}]\left\{1+\sum_{k\geq 1}(-1)^k(s_{3k-1}+s_{3k})\right\}^{-1},$$

where $\mathbf{x}=(x_1,x_2,\dots)$, $\mathbf{i}=(i_1,i_2,\dots)$, $s_j=x_1^j+x_2^j+\cdots$.

(b) Show that the solution to part (a) can be expressed as

$$\sum_{m\geq 0} m![x^m]P_1^{f_1}(x)P_2^{f_2}(x)\cdots,$$

where

$$P_i(x)=\sum_{k,l\geq 0}\binom{-k}{l}\binom{k}{i-2k-3l}\frac{x^k}{k!},$$

and $(f_1,f_2,\dots)=\tau(\mathbf{i})$ is the type of $\mathbf{i}$.

3.5.3. Let $q_3(2n)$ be the number of 3-regular labeled graphs on $2n$ vertices and let $W(x)=\sum_{n\geq 0}q_3(2n)x^n/(2n)!$. Show that $W(x)$ satisfies the differential equation

$$6x^2(x^2-2x-2)\frac{d^2W}{dx^2}-(x^5-6x^4+6x^3+24x^2+16x-8)\frac{dW}{dx}$$
$$+\frac{1}{6}(x^5-10x^4+24x^3-4x^2-44x-48)W=0.$$

3.5.4. Let $R_4(x)=\sum_{n\geq 0}r_4(n)x^n/n!$, where $r_4(n)$ is the number of simple 4-regular labeled graphs on n vertices. Let $V(y_1,\dots,y_4)=\Gamma(T)(y_1,\dots,y_4,0,\dots)$, where $T=\prod_{1\leq i<j}(1+t_it_j)$ is the generating function for simple labeled graphs. Describe a scheme that allows us to deduce a single ordinary differential equation for R_4 from the system of partial differential equations for V obtained from the Γ-series theorem.

3.5.5. Let $F(x)=\sum_{m\geq 0}l_3(m)x^m/(m!)^2$, where $l_3(m)$ is the number of $m\times m$ non-negative integer matrices with line sum 3. Let

$$V=\Gamma(T)(x_1,x_2,x_3,0,\dots;y_1,y_2,y_3,0,\dots),$$

where $T=\prod_{i,j\geq 1}(1-r_ic_j)^{-1}$ is the generating function for non-negative integer matrices. Describe a scheme that allows us to deduce a single ordinary differential equation for F from the system of partial differential equations for V obtained from the Γ-series theorem.

3.5.6. (a) Show that the number of proper k-covers of N_n is given by

$$[x_1^k \cdots x_n^k] \prod_{i \geqslant 1} \prod_{\substack{\{\alpha_1, \ldots, \alpha_i\} \subseteq \mathsf{N}_+ \\ \alpha_1 < \cdots < \alpha_i}} (1 + x_{\alpha_1} \cdots x_{\alpha_i}).$$

(b) Show that the generating function in part (a) may be expressed in terms of power sums as

$$G(\mathbf{s}) = \exp\left\{ \sum_{m \geqslant 1} \frac{(-1)^{m-1}}{m} \left\{ \exp\left(\sum_{i \geqslant 1} \frac{(-1)^{i-1}}{i} s_{mi} \right) - 1 \right\} \right\}.$$

(c) Let $V(y_1, y_2) = \Gamma(G(\mathbf{s}))|_{y_3 = y_4 = \cdots = 0}$, for $G(\mathbf{s})$ given in part (b). Show that

$$\frac{\partial V}{\partial y_2} - \frac{1}{2} \frac{\partial^2 V}{\partial y_1^2} = -\frac{1}{2} \frac{\partial V}{\partial y_1} - \frac{1}{2} V \exp y_2,$$

where $[y_2^n/n!]V$ is the number of proper 2-covers of N_n.

3.5.7. Show that the number of proper k-covers of N_n of order t is given by

$$\left[x^t \frac{y^n}{n!} \right] \exp\left(\sum_{m=1}^{k} (-x)^m / m \right) \sum_{u_1, \ldots, u_k \geqslant 0} (-1)^{u_1 + \cdots + u_k}$$

$$\times \frac{(-x)^{u_1 + 2u_2 + \cdots + ku_k}}{u_1! \cdots u_k!} \frac{\exp(y[z^k](1+z)^{u_1} \cdots (1+z^k)^{u_k})}{2^{u_2} \cdots k^{u_k}}.$$

3.5.8. Show that the number of restricted proper 3-covers of N_n of order t is given by

$$\left[x^t \frac{y^n}{n!} \right] \exp\left(-x - \tfrac{1}{2}x^2 + \tfrac{1}{3}x^3 y\right) \sum_{r \geqslant 0} \frac{x^r}{r!} (1+y)^{\binom{r}{3}} \exp\left\{\tfrac{1}{2}x^2(1+y)^r\right\}.$$

4

The Combinatorics of Sequences

4.1. INTRODUCTION

This chapter describes a general theory for enumerating sequences on N_+ with respect to the block of a bipartition, $\{\pi_1, \pi_2\}$ of N_+^2, to which each adjacent pair of elements belongs. The ordered string of π_i's to which successive pairs of adjacent elements belong is called the **pattern** of a sequence. Some of the problems of Chapter 2 may be expressed as pattern problems. For example,

1. André's problem (**3.2.22**): permutations with pattern in $(< \ \geqslant)^* \cup (< \ \geqslant)^* <$, (here π_1 is the set of rises (**2.4.13**), denoted by $<$);
2. Simon Newcomb problem (**2.4.20**): sequences with pattern in $\{<, \geqslant\}^*$ (i.e., all sequences) with k rises;
3. Smirnov problem (**2.4.16**): sequences with pattern in $(\neq)^*$.

Unlike the earlier chapters, the present one contains no new decompositions. Indeed, the main ones used are

4. $\{\pi_1, \pi_2\}^* = (\pi_1^*\pi_2)^*\pi_1^*$.
5. $\{\pi_1, \pi_2\}^* = \pi_2^*(\pi_1\pi_1^*\pi_2\pi_2^*)^*\pi_1^*$.

for the set of patterns. These are precisely the decompositions that have been used for binary sequences in Section 2.4. In view of this, this is the least combinatorial of the chapters.

To compensate for the paucity of decompositions we develop in Section 4.3 a collection of procedures, called the **pattern algebra**, for calculating the **fundamental generating functions** Ψ_1 and Ψ_2 associated with a set of patterns. Both preserve combinatorial information about the pattern and, moreover, Ψ_1 is ordinary with respect to sequence type and Ψ_2, defined only for $\pi_1 = <$, is Eulerian [**2.6.6**] with respect to inversions (**2.6.1**). A sum lemma and a product lemma are obtained for Ψ_1 and Ψ_2 by considering, respectively, the union and concatenation of patterns. The main result, called the **elimination theorem**, is obtained by using these elementary counting lemmas in conjunction with the fact that $\pi_2 = \mathsf{N}_+^2 - \pi_1$ eliminates π_2 from a pattern. This theorem expresses Ψ_1 and Ψ_2 for arbitrary sets of patterns in terms of the enumerators for sequences, called **π_1-strings**, whose patterns lie in π_1^*.

As a preliminary application of the pattern algebra we use decomposition **4**, which exposes maximal π_1-strings, to obtain the **maximal string decomposition theorem**, for enumerating sequences with respect to the lengths of their maximal π_1-strings. This result is first stated in Section 4.2, without proof, where we show how it may be specialized in a variety of ways to obtain many results in sequence enumeration including **1**, **2**, **3**. Explicit results for sequences of type **1**, namely, permutations, are obtained immediately by means of **transformation lemmas** for applying $[\mathbf{x}^1]$ to Ψ_1 for three specific bipartitions.

In Section 4.4 we give methods for enumerating circular permutations with respect to pattern. The main result used for this purpose is the **logarithmic connection theorem**, and by means of this we obtain a circular analogue to the maximal string decomposition theorem. This enables us, for example, to extend many of the sequence results of Section 4.2 to **circular** (permutation) **problems** without difficulty.

We conclude the chapter by using the logarithmic connection theorem and the MacMahon master theorem to calculate the permanents of certain matrices that arise combinatorially.

4.2. THE MAXIMAL STRING DECOMPOSITION THEOREM

In enumerating sequences with respect to pattern, we begin with the simplest case, namely, enumerating sequences with respect to the lengths of their maximal π_1-strings.

4.2.1. Definition (π_1-String, Maximal π_1-String Type)

Let $\Pi = (\pi_1, \pi_2)$ *be an arbitrary bipartition of* N_+^2.

1. *If* $l \geqslant 2$ *and* $\sigma = \sigma_1 \cdots \sigma_l \in \mathsf{N}_+^+$ *is such that* $(\sigma_i, \sigma_{i+1}) \in \pi_1$ *for* $1 \leqslant i \leqslant l-1$, *then* σ *is called a* **π_1-string** *of* **length** l. *Any sequence of length* 1 *is a* π_1*-string of length* 1.
2. *If* $\alpha \in \mathsf{N}_+^+$ *has* j_i *maximal* π_1*-strings of length* i, *then,* $\kappa_{\pi_1}(\alpha) = (j_1, \ldots)$ *is called the* **maximal π_1-string type** *of* α. □

For example, if π_1 is the relation $<$, so π_1 is the set of rises, then $\rho = 2157881624734$ may be uniquely expressed as the concatenation (2)(1578)(8)(16)(247)(34) of its maximal $<$-strings. We call this the **maximal $<$-string decomposition** of ρ. To see that 1578 is a maximal $<$-string, note that it is a $<$-string, since $1 < 5 < 7 < 8$, and that neither of the substrings 21578, 15788 of ρ is a $<$-string, since 21 and 88 are not rises.

Let S be a set of sequences on N_+. We associate with S the generating function

$$\sum_{\sigma \in \mathsf{S}} \mathbf{x}^{\tau(\sigma)} \mathbf{f}^{\kappa_{\pi_1}(\sigma)}$$

with respect to type τ (**2.4.7**) and maximal π_1-string type κ, where $\mathbf{f} = (f_1, \ldots)$ and f_i is an indeterminate marking maximal π_1-strings of length i. For example, in this generating function the monomial corresponding to ρ is

$$\mathbf{x}^{\tau(\rho)} \mathbf{f}^{\kappa_<(\rho)} = \left(x_1^2 x_2^2 x_3 x_4^2 x_5 x_6 x_7^2 x_8^2\right)\left(f_1^2 f_2^2 f_3 f_4\right).$$

The purpose of this section is to give a theorem (**4.2.3**) for determining this generating function and to apply this theorem to a variety of sequence problems. The generating function involves the following combinatorially defined functions.

4.2.2. Definition (π_1-String Enumerator, Maximal π_1-String Length Enumerator)

1. *The* π_1-**string enumerator** *for strings of length* $k > 0$ *is*

$$\gamma_k(\pi_1) = \sum_{\sigma} \mathbf{x}^{\tau(\sigma)}, \qquad \gamma_0(\pi_1) = 1$$

where the sum is over all π_1*-strings of length* k*. We denote* $(\gamma_0(\pi_1), \gamma_1(\pi_1), \ldots)$ *by* $\boldsymbol{\gamma}(\pi_1)$.

2. *The* **maximal** π_1**-string length enumerator** *is*

$$F(x) = 1 + f_1 x + f_2 x^2 + \cdots. \qquad \square$$

We may now state the main counting theorem of this section. Its proof is given in **4.3.12**. Definitions of $k!_q$ and inversions are given in **2.6.4** and **2.6.1**, respectively.

4.2.3. Theorem (Maximal String Decomposition)

There are

1. $[\mathbf{x}^{\mathbf{i}}\mathbf{f}^{\mathbf{k}}](F^{-1} \circ \boldsymbol{\gamma}(\pi_1))^{-1}$ *sequences on* N_+ *of type* $\mathbf{i}$ *and maximal* π_1*-string type* $\mathbf{k}$.
2. $[q^i \mathbf{f}^{\mathbf{k}} x^n / n!_q](F^{-1} \circ \boldsymbol{\eta}_q)^{-1}$ *permutations on* N_n *with maximal* <*-string type* $\mathbf{k}$ *and* i *inversions, where*

$$\boldsymbol{\eta}_q = \left(1, \frac{x}{1!_q}, \frac{x^2}{2!_q}, \ldots\right).$$

The compositions are umbral. $\square$

Although we are mainly concerned with the general bipartition (π_1, π_2), certain particular bipartitions occur often in practice. Although some of these have been encountered already, they are restated here for convenience.

4.2.4. Definition (Rises, Successions, *c*-Successions)

1. $\{(i, j) \in \mathsf{N}_+^2 \mid i < j\}$ *is the set of* **rises**.
2. $\{(i, j) \in \mathsf{N}_+^2 \mid i + 1 = j\}$ *is the set of* **successions**.
3. $\{(i, j) \in \mathsf{N}_n^2 \mid (i + 1) \equiv j (\mathrm{mod}\ n)\}$ *is the set of c-***successions** *on* N_n.

These sets are denoted by $<$, $+$ *and* $\oplus$, *respectively.* □

For example, the decomposition of 12346923434 into its maximal $+$-strings is (1234)(6)(9)(234)(34).

Let $\Phi(\gamma(\pi_1))$ be the generating function for the set S of sequences, with respect to type and maximal π_1-string type. This form for the generating function is guaranteed by the maximal string decomposition theorem. The generating function for permutations on N_n in S with respect to maximal π_1-string type is therefore

$$[x_1 \cdots x_n]\Phi(\gamma(\pi_1)),$$

where $\Phi \in (\mathbb{Q}[\mathbf{f}])[[\mathbf{x}]]$. Thus results on permutations and their maximal π_1-string type are accessible to us, provided we are able to apply $[x_1 \cdots x_n]$. Although this is difficult in general, the following lemma, whose proof is [**4.2.1**], indicates that it may be carried out straightforwardly in the case of rises, successions, and c-successions.

4.2.5. Lemma ($<$-Transformation, $+$-Transformation, $\oplus$-Transformation)

Let $\Phi(y_0, y_1, \ldots)$ *be a formal multivariate power series in* $(y_0, y_1, \ldots)$. *Then*

1. $[x_1 \cdots x_n]\Phi(\gamma(<)) = \left[\dfrac{x^n}{n!}\right]\Phi\left(1, \dfrac{x}{1!}, \dfrac{x^2}{2!}, \ldots\right)$
2. $[x_1 \cdots x_n]\Phi(\gamma(+)) = [x^n] \sum_{k \geqslant 0} k![y^k]\Phi(1, xy, x^2y, \ldots)$
3. $[x_1 \cdots x_n]\Phi(\gamma(\oplus)) = [x^n] \sum_{k \geqslant 1} (k-1)!x[y^k]\dfrac{\partial}{\partial x}\Phi(1, xy, x^2y, \ldots).$ □

The preceding lemma gives explicit transformations from multivariate generating functions for sequences to univariate generating functions for permutations. We note that (**1**) of Lemma 4.2.5 gives an exponential generating function for permutations with respect to length.

As an example of the use of the maximal string decomposition theorem, consider first π_1-alternating sequences. These are a natural generalization of alternating permutations, which are given in Definition 3.2.21.

4.2.6. Definition (π_1-Alternating Sequence)

A sequence $\sigma = \sigma_1\sigma_2\cdots$ *over* N_+ *is said to be* π_1-**alternating** *if* $(\sigma_1, \sigma_2) \in \pi_1$, $(\sigma_2, \sigma_3) \in \pi_2$, $(\sigma_3, \sigma_4) \in \pi_1, \ldots$. □

A sequence of even length is π_1-alternating if and only if each of its maximal π_1-strings has length equal to 2, so the maximal string decomposition theorem may be applied. However, π_1-alternating sequences of odd length have a final maximal π_1-string of length equal to 1, and hence cannot be recognized by their unordered collection of maximal π_1-string lengths alone. A strengthened form of the theorem for handling final maximal π_1-strings is given as Theorem 4.2.19.

4.2.7. Corollary (π_1-Alternating Sequences of Even Length)

There are

$$[\mathbf{x}^{\mathbf{i}}]\left\{\sum_{k\geqslant 0}(-1)^k\gamma_{2k}(\pi_1)\right\}^{-1}$$

π_1*-alternating sequences of type* **i** *and even length.*

Proof: The maximal π_1-string length enumerator for π_1-alternating sequences of even length is $F(x) = 1 + x^2$, since such sequences have only maximal π_1-strings of length 2. The result follows immediately from the maximal string decomposition theorem. □

This result is now specialized, in turn, to particular bipartitions.

4.2.8. <-Alternating Permutations of Even Length

From Corollary 4.2.7 and the <-transformation lemma, there are

$$\left[\frac{x^n}{n!}\right]\left\{\sum_{k\geqslant 0}(-1)^k\frac{x^{2k}}{(2k)!}\right\}^{-1} = \left[\frac{x^n}{n!}\right]\sec x$$

alternating permutations on N_n of even length. □

This agrees with **3.2.22**, in which a differential decomposition was used.

4.2.9. +-Alternating Permutations of Even Length

From Corollary 4.2.7 and the +-transformation lemma, there are

$$[x^n]\sum_{k\geqslant 0}k![y^k]\{1 - x^2y(1 + x^2)^{-1}\}^{-1}$$

+-alternating permutations on N_n of even length. By routine manipulation this reduces to $\Sigma_{k=1}^{m}(-1)^{m-k}\binom{m-1}{k-1}k!$, where $n = 2m$. □

When $n = 4$, there is only one permissible permutation, namely, 3412, in agreement with **4.2.9**.

4.2.10. ⊕-Alternating Permutations of Even Length

From Corollary 4.2.7 and the ⊕-transformation lemma, there are

$$[x^n] \sum_{k \geq 1} (k-1)!x[y^k]\frac{\partial}{\partial x}\{1 - x^2y(1+x^2)^{-1}\}^{-1}$$

⊕-alternating permutations on N_n of even length. By routine manipulation this reduces to $2m\Sigma_{k=1}^{m}(k-1)!\binom{m-1}{k-1}(-1)^{m-k}$ where $n = 2m$. □

When $n = 6$ there are six permissible permutations, namely, 125634, 341256, 563412, 614523, 236145, 452361, in agreement with **4.2.10**.

4.2.11. <-Alternating Permutations of Even Length, Inversions

From the maximal string decomposition theorem and Corollary 4.2.7 there are

$$\left[q^i \frac{x^n}{n!_q}\right]\left\{\sum_{k \geq 0} (-1)^k \frac{x^{2k}}{(2k)!_q}\right\}^{-1}$$

<-alternating permutations on N_n, of even length, with i inversions. □

The generating function given in **4.2.11** is denoted by $\sec_q x$ and is called a ***q*-analogue of the secant**. From **4.2.11**, the number of <-alternating permutations on N_6 with exactly two inversions is

$$\left[q^2 \frac{x^6}{6!_q}\right]\left\{\sum_{k \geq 0} (-1)^k \frac{x^{2k}}{(2k)!_q}\right\}^{-1} = [q^2]\left\{1 - 2\frac{6!_q}{2!_q 4!_q} + \frac{6!_q}{(2!_q)^3}\right\} = 1.$$

The single permutation is 132546, whose two inversions are (3,2) and (5,4).

The next group of examples concerns the enumeration of sequences with respect to the number of occurrences of π_1 in their pattern.

4.2.12. Corollary (Sequences, Type, Occurrences of π_1)

The number of sequences of type **i** *with exactly p occurrences of pairs of adjacent elements in* π_1 *is*

$$[\mathbf{x}^{\mathbf{i}}u^p]\left\{1 - \sum_{j \geq 1} (u-1)^{j-1}\gamma_j(\pi_1)\right\}^{-1}.$$

Proof: Maximal π_1-strings of length $k \geqslant 1$ contain $k-1$ adjacent pairs in π_1, so $f_k = u^{k-1}$, where u marks the occurrence of π_1 in the pattern. The maximal π_1-string length enumerator is therefore $F(x) = \{1-(u-1)x\}(1-ux)^{-1}$. The result follows immediately from the maximal string decomposition theorem. □

A number of specializations of this result are now given. These have been obtained earlier by other methods.

4.2.13. Sequences and Rises (Simon Newcomb Problem)

We wish to determine the number of sequences of type **i** with k rises. Since $\gamma_j(<) = [x^j]\prod_{i\geqslant 1}(1+xx_i)$, then, from Corollary 4.2.12, this number is

$$[u^k \mathbf{x}^{\mathbf{i}}](u-1)\left\{u - \prod_{i\geqslant 1}\{1+(u-1)x_i\}\right\}^{-1}. \qquad \square$$

The preceding generating function has a constant term of 1, corresponding to the empty sequence. When this is subtracted we recover the generating function given in **2.4.20**, which was stated in terms of falls. The equivalence of the results is established by noting that rises may be transformed into falls, without changing sequence type, by reading a sequence from right to left, instead of left to right.

4.2.14. Sequences and Levels (Smirnov Problem)

Consider the bipartition $(=, \neq)$. The first block, π_1, corresponds to the set of levels (Definition 2.4.13). But, from Definition 4.2.2(1), $\gamma_j(=) = s_j$, the power sum symmetric function. It follows from Corollary 4.2.12, setting $u = 0$, that there are

$$[\mathbf{x}^{\mathbf{i}}]\left\{1 - \sum_{j\geqslant 1}(-1)^{j-1}s_j\right\}^{-1}$$

sequences of type **i** with no levels, in agreement with **2.4.16**. □

Corollary 4.2.12 may, of course, be specialized further by considering the enumeration of permutations with respect to rises and inversions, successions or c-successions. These and other specializations are given in the exercises.

The third corollary deals with the enumeration of sequences with respect to π_1-strings of length p. This result has been obtained for $\pi_1 = <$ in **2.8.8** by means of distinguished substrings.

4.2.15. Corollary (Sequences, π_1-Strings of Length p)

There are

$$[u^j\mathbf{x}^{\mathbf{i}}]\left\{\left((1-x)(1-ux)\{1-ux+(u-1)x^p\}^{-1}\right)\circ\gamma(\pi_1)\right\}^{-1}$$

sequences of type $\mathbf{i}$ *with exactly* j π_1*-strings of length* p.

Proof: A maximal π_1-string of length k contains $k-p+1$ π_1-strings of length p if $k \geqslant p$, and none otherwise. Thus $f_k = u^{k-p+1}$ if $k \geqslant p$, and 1 otherwise, where u marks π_1-strings of length p. Thus $F(x) = (1-x)^{-1}(1-ux)^{-1}\{1-ux+(u-1)x^p\}$, and the result follows from the maximal string decomposition theorem. □

The following specializations are now immediate.

4.2.16. Sequences with π_1-Strings of Length 3

It follows from Corollary 4.2.15 that there are

$$[u^j\mathbf{x}^{\mathbf{i}}]\left(\sum_{k\geqslant 0}(-1)^k(1-u)^k\sum_{l=0}^{k}\binom{k}{l}\{\gamma_{k+l}(\pi_1)-u\gamma_{k+l+1}(\pi_1)\}\right)^{-1}$$

sequences of type $\mathbf{i}$ with j π_1-strings of length 3. □

4.2.17. Permutations with <-Strings of Length 3, Inversions

It follows from **4.2.16** and the maximal string decomposition theorem that there are

$$\left[q^iu^j\frac{x^n}{n!_q}\right]\left(\sum_{k\geqslant 0}(-1)^k(1+u)^k\sum_{l=0}^{k}\binom{k}{l}\left\{\frac{x^{k+l}}{(k+l)!_q}-u\frac{x^{k+l+1}}{(k+l+1)!_q}\right\}\right)^{-1}$$

permutations on N_n with j <-strings of length 3, and i inversions. □

The final example in this group involves a multiplicative weight.

4.2.18. Permutations with Prescribed Product of Maximal <-String Lengths

Let $c(m,n)$ be the number of permutations on N_n in which the product of the maximal <-string lengths is m. Let s be an indeterminate. By Remark 2.4.29 we set $f_i = i^{-s}$ in the maximal string decomposition theorem, and from the <-transformation lemma

$$c(m,n) = \left[\frac{x^n}{n!}m^{-s}\right]\left(F^{-1}\circ\left\{1,\frac{x}{1!},\frac{x^2}{2!},\ldots\right\}\right)^{-1}$$

where $F(x) = 1 + x/1^s + x^2/2^s + x^3/3^s + \cdots$. The generating function for $c(m, n)$ is exponential in x and Dirichlet (Definition 2.4.25) in s. □

We have already noted, in connection with π_1-alternating sequences of odd length, the necessity of treating the final maximal π_1-string differently from the nonfinal maximal π_1-strings. The following strengthened form of the maximal string decomposition theorem, whose proof is given as [**4.3.1**], is used for this purpose.

4.2.19. Theorem (Maximal String Decomposition; Distinguished Final String)

Let $F(x) = 1 + f_1x + f_2x^2 + \cdots$ *and* $G(x) = g_1x + g_2x^2 + \cdots$. *Then*

1. *There are* $[\mathbf{x}^{\mathbf{i}}\mathbf{f}^{\mathbf{k}}g_m](GF^{-1}\circ\gamma(\pi_1))(F^{-1}\circ\gamma(\pi_1))^{-1}$ *sequences in* N_+^+ *of type* $\mathbf{i}$, *nonfinal maximal* π_1-*string type* $\mathbf{k}$, *and with a final maximal* π_1-*string of length* $m \geqslant 1$.
2. *There are* $[q^i g_m \mathbf{f}^{\mathbf{k}} x^n/n!_q](GF^{-1}\circ\eta_q)(F^{-1}\circ\eta_q)^{-1}$ *permutations on* N_n *with nonfinal maximal* $<$-*string type* $\mathbf{k}$, *a final maximal* $<$-*string of length* $m \geqslant 1$, *and* i *inversions.* (*As before* $\eta_q = (1, x/1!_q, x^2/2!_q, \ldots)$). □

The generating functions $F(x)$ and $G(x)$, which appear in this theorem, are the **nonfinal** and **final maximal π_1-string length enumerators**, respectively. $G(x)$ has no constant term because we have adopted the convention that every nonempty sequence has a final maximal string. To obtain Theorem 4.2.3, we set $g_i = f_i$ for $i \geqslant 1$ in Theorem 4.2.19, so $G = F - 1$. It follows immediately that

$$\left(GF^{-1}\circ\gamma(\pi_1)\right)\left(F^{-1}\circ\gamma(\pi_1)\right)^{-1} = \left(F^{-1}\circ\gamma(\pi_1)\right)^{-1} - 1.$$

This may be reconciled with Theorem 4.2.3 by observing that the empty sequence, whose generating function is 1, does not have a nonempty final string, and accordingly is not counted by Theorem 4.2.19. We may, without confusion, refer to both Theorem 4.2.3 and Theorem 4.2.19 as the maximal string decomposition theorem.

We conclude this section by using this theorem to complete the examination of π_1-alternating sequences.

4.2.20. Corollary (π_1-Alternating Sequences of Odd Length)

There are

$$[\mathbf{x}^{\mathbf{i}}]\left(\sum_{k\geqslant 0}(-1)^k\gamma_{2k+1}(\pi_1)\right)\left(\sum_{k\geqslant 0}(-1)^k\gamma_{2k}(\pi_1)\right)^{-1}$$

π_1-*alternating sequences of odd length and type* $\mathbf{i}$.

Proof: A π_1-alternating sequence of odd length has a final maximal π_1-string of length 1, so $G(x) = x$. The nonfinal maximal π_1-strings are of length 2, so $F(x) = 1 + x^2$. The result follows immediately from the maximal string decomposition theorem (**4.2.19**). □

We give two specializations of this result.

4.2.21. <-Alternating Permutations of Odd Length, with Inversions

From Corollary 4.2.20 and the maximal string decomposition theorem, there are

$$\left[q^i \frac{x^n}{n!_q}\right]\left(\sum_{k\geqslant 0}(-1)^k\frac{x^{2k+1}}{(2k+1)!_q}\right)\left(\sum_{k\geqslant 0}(-1)^k\frac{x^{2k}}{(2k)!_q}\right)^{-1}$$

<-alternating permutations of odd length, with i inversions. □

The generating function given in **4.2.21** is denoted by $\tan_q x$, and is called a ***q*-analogue of the tangent.**

4.2.22. <-Alternating Permutations of Odd Length

By the <-transformation lemma and Corollary 4.2.20, or from **4.2.21**, with $q = 1$, there are

$$\left[\frac{x^n}{n!}\right]\left(\sum_{k\geqslant 0}(-1)^k\frac{x^{2k+1}}{(2k+1)!}\right)\left(\sum_{k\geqslant 0}(-1)^k\frac{x^{2k}}{(2k)!}\right)^{-1} = \left[\frac{x^n}{n!}\right]\tan x$$

<-alternating permutations of odd length n, in agreement with **3.2.22**. □

NOTES AND REFERENCES

4.2.3 Gessel (1977), Jackson and Aleliunas (1979); **4.2.11** Gessel (1977), Stanley (1976); **4.2.19** Jackson and Aleliunas (1979); **4.2.21** Gessel (1977), Stanley (1976); [**4.2.5**] Riordan (1958); [**4.2.6**] Gessel (1977), Stanley (1976); [**4.2.9**] Carlitz (1973b); [**4.2.10, 11**] Carlitz (1978); [**4.2.12**] Carlitz and Scoville (1974); [**4.2.13**] Salié (1963); [**4.2.14**] Tanny (1976); [**4.2.15**] Stanley (1976); [**4.2.16**] Carlitz, Scoville, and Vaughan (1976); [**4.2.17**] Reilly and Tanny (1980).

EXERCISES

4.2.1. Prove Lemma 4.2.5.

4.2.2. (a) Let $\Phi(y_0, y_1, \ldots)$ be a formal multivariate power series in $(y_0, y_1, \ldots)$. Prove that

$$\left[x_1^2 \cdots x_n^2\right]\Phi(\gamma(<)) = 2^{-n}\sum_{j=0}^{n}\binom{n}{j}\left[\frac{x^j}{j!}\frac{z^{2n-2j}}{(2n-2j)!}\right]\Phi(t_0, t_1, \ldots),$$

where

$$t_i = \sum_{l \geqslant 0}(-1)^l\frac{x^l}{l!}\frac{z^{i-2l}}{(i-2l)!} = [u^i]\exp(zu - xu^2) \qquad \text{for } i \geqslant 0.$$

(b) Show that the number of $<$-alternating sequences containing two occurrences of i, for $i = 1, \ldots, n$, is

$$2^{-n}\sum_{k=0}^{n}\binom{n}{k}(-1)^{n-k}\left[\frac{z^{2k}}{(2k)!}\right]\sec z.$$

4.2.3. Show that the number of sequences of type $\mathbf{i}$ in which each maximal π_1-string has length k is $[\mathbf{x}^{\mathbf{i}}]\{\sum_{i\geqslant 0}(-1)^i\gamma_{ki}(\pi_1)\}^{-1}$.

4.2.4. (a) Use the maximal string decomposition theorem to show that the number of permutations on N_{nk} with m inversions, and in which each maximal $<$-string has length jk for some $j \geqslant 1$, is $[q^m](nk)!_q\{k!_q\}^{-n}$.

(b) Obtain the result of part (a) by the methods of Section 2.6.

4.2.5. Show that the number of permutations on N_n with k maximal π_1-strings is

(a) $\left[u^k\frac{x^n}{n!}\right](1-u)\{1-ue^{(1-u)x}\}^{-1}$ for $\pi_1 = <$.

(b) $[u^k]\sum_{i\geqslant 1} i!u^i(1-u)^{n-i}\binom{n-1}{i-1}$ for $\pi_1 = +$.

(c) $[u^k]n\sum_{i\geqslant 1}(i-1)!u^i(1-u)^{n-i}\binom{n-1}{i-1}$ for $\pi_1 = \oplus$.

4.2.6. (a) Show that the number of sequences of type $\mathbf{i}$ with k maximal $<$-strings is

$$[u^k\mathbf{x}^{\mathbf{i}}](1-u)\left\{1 - u\prod_{j\geqslant 1}\left(1+(1-u)x_j\right)\right\}^{-1}.$$

(b) Show that the number of compositions of m with n parts and k maximal $<$-strings is

$$[u^k y^m x^n](1-u)\left\{1 - u\prod_{j\geqslant 1}\left(1+(1-u)xy^j\right)\right\}^{-1}.$$

(c) Show that the number of permutations on N_n with m inversions and k maximal $<$-strings is

$$\left[u^k q^m \frac{x^n}{n!_q}\right](1-u)\left\{1 - u\prod_{i\geqslant 0}\left(1-(1-u)x(1-q)q^i\right)^{-1}\right\}^{-1}.$$

4.2.7. Show that the number of sequences of type **i**, with each maximal π_1-string of length $> p$ is

$$[\mathbf{x}^{\mathbf{i}}]\left\{\sum_{m\geqslant 0}\sum_{i=0}^{m}\binom{m}{i}(-1)^i\left(\gamma_{pi+m}(\pi_1)-\gamma_{pi+m+1}(\pi_1)\right)\right\}^{-1}.$$

4.2.8. (a) Show that the number of permutation on N_n with m inversions, k of whose maximal $<$-strings have length $\geqslant p$, is

$$\left[u^k q^m \frac{x^n}{n!_q}\right]\left\{\sum_{i\geqslant 0}(1-u)^i\left(\frac{x^{pi}}{(pi)!_q}-\frac{x^{pi+1}}{(pi+1)!_q}\right)\right\}^{-1}.$$

(b) Show that the number of permutations on N_n, k of whose maximal $<$-strings have length $\geqslant 2$, is

$$[u^k x^n/n!](\cosh zx - z^{-1}\sinh zx)^{-1}, \qquad \text{where} \quad z=(1-u)^{1/2}.$$

4.2.9. Show that the number of $<$-alternating sequences of type **i** is

$$[\mathbf{x}^{\mathbf{i}}]\{2-\rho\Pi_{j\geqslant 1}(1+\rho x_j)+\rho\Pi_{j\geqslant 1}(1-\rho x_j)\}$$
$$\times\{\Pi_{j\geqslant 1}(1+\rho x_j)+\Pi_{j\geqslant 1}(1-\rho x_j)\}^{-1},$$

where $\rho^2=-1$.

4.2.10. Show that the number of permutations on N_n with k maximal $<$-strings, for which the final maximal string has length $l\equiv(t-j)(\bmod t)$, and all others have length $l\equiv 0(\bmod t)$, is

$$\left[u^{k-1}\frac{x^n}{n!}\right]z^j\phi_t^{(t-j)}(xz)\{1-u\phi_t^{(0)}(xz)\}^{-1},$$

where $\phi_t^{(i)}(x)=\Sigma_{l\geqslant 0}x^{tl+i}/(tl+i)!$ is called an **Olivier function**, and $z=(1-u)^{1/t}$.

4.2.11. Show that the number of permutations on N_n with k maximal $<$-strings, for which the final maximal $<$-string has length $\geqslant s$, and all others have lengths $\geqslant t$, is

$$\left[u^k \frac{x^n}{n!}\right]\left\{\sum_{i=1}^{t} \frac{\alpha_i^{t-s} e^{\alpha_i x}}{\alpha_i^{t-1} u^{-1} - t}\right\}\left\{\sum_{i=1}^{t} \frac{e^{\alpha_i x}}{t - u^{-1}\alpha_i^{t-1}}\right\}^{-1}$$

where $\alpha_1, \ldots, \alpha_t$ are the roots of $z^t - z^{t-1} + u = 0$ and $t > s$.

4.2.12. A **maximum** in a permutation $\sigma_1 \cdots \sigma_n$ on N_n is an element σ_i such that $\sigma_{i-1} < \sigma_i > \sigma_{i+1}$, where $2 \leqslant i \leqslant n - 1$. Show that the number of permutations on N_n with k maxima and i rises is

$$\left[r^i u^k \frac{x^n}{n!}\right](e^{\alpha_1 x} - e^{\alpha_2 x})(\alpha_1 e^{\alpha_2 x} - \alpha_2 e^{\alpha_1 x})^{-1},$$

where $\alpha_1 + \alpha_2 = r + 1$, $\alpha_1 \alpha_2 = ru$.

4.2.13. Show that the number of permutations on N_n in which the final maximal $<$-string has even length and all other maximal $<$-strings have length 2 is $\frac{1}{2}c_n$, for $n \geqslant 0$, where $c_n = [x^n/n!](\cosh x)/(\cos x)$ is a **Salié number**.

4.2.14. Show that the number of permutations on N_n

(a) With j rises is

$$\left[u^j \frac{x^n}{n!}\right](u-1)\{u - e^{(u-1)x}\}^{-1} \quad \text{(see also } \mathbf{2.4.21}\text{)}.$$

(b) With j successions is

$$[u^j x^n] \sum_{k \geqslant 0} k! x^k (1 - x(u-1))^{-k}.$$

(c) With j c-successions is

$$[u^j x^n] \sum_{k \geqslant 1} k! x^k (1 - x(u-1))^{-(k+1)}.$$

4.2.15. (a) By considering sequences on the alphabet N_+^k, show that the number of k-tuples of sequences of the same length with j occurrences of π_1 in the same position, and in which sequence l has type $\mathbf{i}_l = (i_{l1}, i_{l2}, \ldots)$, is

$$[u^j \mathbf{x}_1^{\mathbf{i}_1} \cdots \mathbf{x}_k^{\mathbf{i}_k}]\left\{1 - \sum_{m \geqslant 1} (u-1)^{m-1} \gamma_m(\pi_1, \mathbf{x}_1) \cdots \gamma_m(\pi_1, \mathbf{x}_k)\right\}^{-1},$$

where $\mathbf{x}_l = (x_{l1}, x_{l2}, \ldots)$ and $\gamma_m(\pi_1, \mathbf{x}_l) = \Sigma_\sigma \mathbf{x}_l^{\tau(\sigma)}$ where the summation is over all π_1-strings of length m.

(b) Show that the number of k-tuples of permutations on N_n with j occurrences of

(i) A rise in the same position in every permutation is

$$\left[u^j \frac{x^n}{(n!)^k}\right](u-1)\left\{u - \sum_{m \geqslant 0} (u-1)^m \frac{x^m}{(m!)^k}\right\}^{-1}.$$

(ii) Successions in the same position is

$$[u^j x^n] \sum_{i \geqslant 0} (i!)^k x^i \{1 - (u-1)x\}^{-i}.$$

(iii) c-Successions in the same position is

$$[u^k x^n] n^k \sum_{i \geqslant 1} ((i-1)!)^k x^i \{1 - (u-1)x\}^{-i}.$$

4.2.16. Show that the number of pairs of sequences on N_+ of the same length, with types $\mathbf{i}_1$ and $\mathbf{i}_2$, and with t occurrences of a rise in the first sequence in the same position as a level in the second sequence is

$$[u^t \mathbf{x}_1^{\mathbf{i}_1} \mathbf{x}_2^{\mathbf{i}_2}] \left\{ 1 + (1-u)^{-1} \sum_{i \geqslant 1} \left\{ \prod_{l \geqslant 1} (1 + (u-1)x_{2i}x_{1l}) - 1 \right\} \right\}^{-1}.$$

4.2.17. Let $\mathsf{A} = \{a_1 + 1, \ldots, a_1 + i_1, a_2 + 1, \ldots, a_2 + i_2, \ldots, a_m + 1, \ldots, a_m + i_m\}$, where $i_1, \ldots, i_m \geqslant 1$, $a_1 \geqslant 0$, and $a_{j+1} > a_j + i_j$ for $j = 1, \ldots, m-1$. Show that the number of permutations of the elements of A with j successions is

$$\sum_{\mathbf{k} \geqslant \mathbf{0}} (-1)^{s(\mathbf{i}) - s(\mathbf{k}) - j} \binom{s(\mathbf{i}) - s(\mathbf{k})}{j} s(\mathbf{k})! \binom{i_1 - 1}{k_1 - 1} \cdots \binom{i_m - 1}{k_m - 1}$$

where $\mathbf{k} = (k_1, \ldots, k_m)$, $\mathbf{i} = (i_1, \ldots, i_m)$, $s(\mathbf{k}) = k_1 + \cdots + k_m$, and $s(\mathbf{i}) = i_1 + \cdots + i_m$.

4.3. THE PATTERN ALGEBRA

We have seen in Section 4.2 that it is possible to enumerate sequences and permutations with respect to maximal π_1-strings. Now consider the enumeration of sequences with more general conditions on the pattern.

4.3.1. Definition (Pattern)

Let $\Pi = (\pi_1, \pi_2)$ *be an arbitrary bipartition of* $\mathsf{N}_+^2 = \pi_0$, *and let* $\mu = \pi_{m_1} \cdots \pi_{m_{l-1}} \in \{\pi_0, \pi_1, \pi_2\}^* = \Omega_\Pi$, *where* $l \geqslant 2$.

1. $\sigma_1 \cdots \sigma_l \in \mathsf{N}_+^+$ *is said to have* **pattern** μ *if* $(\sigma_j, \sigma_{j+1}) \in \pi_{m_j}$ *for* $j = 1, \ldots, l - 1$, *and the pattern has* **length** $|\mu|$ *equal to* l. *Let* ε *be the null string in* Ω_Π. *The set of sequences whose pattern is* ε *is* N_+.

2. *$\langle \mu \rangle$ is the set of sequences over N_+ with pattern μ. If $\mu_1, \mu_2 \in \Omega_\Pi$, then we define $\langle \mu_1 \cup \mu_2 \rangle$ by*

$$\langle \mu_1 \cup \mu_2 \rangle = \langle \mu_1 \rangle \cup \langle \mu_2 \rangle.$$

3. *Let $\alpha \subset \Omega_\Pi$. Then $\langle\langle \alpha \rangle\rangle$ denotes the set of all sequences in $\langle \alpha \rangle$ that are permutations on N_k for some $k \geqslant 0$.*

For convenience we denote N_+^2 by ω. □

For example, $3642715 \in \langle\langle \pi_1\pi_2\omega^2\pi_2\pi_1 \rangle\rangle$, where $\pi_1 = <$, so 3642715 is a permutation with pattern $\pi_1\pi_2\omega^2\pi_2\pi_1$. It is important to note that a sequence may have more than one pattern. For example, $3642715 \in \langle\langle \pi_1\pi_2(\pi_2\pi_1)^2 \rangle\rangle$, so 3642715 is also a permutation with pattern $\pi_1\pi_2(\pi_2\pi_1)^2$ where $\pi_1 = <$. However, a sequence has a unique pattern in $\{\pi_1, \pi_2\}^*$.

To associate a generating function with patterns, we first consider the **set of encodings** defined as follows.

4.3.2. Definition (Incidence Matrix, Set of Encodings)

1. *The* **incidence matrix** *for the pattern μ of length l is $\mathbf{I}(\mu)$ where*

$$[\mathbf{I}(\mu)]_{ij} = \sum_{\substack{\sigma_1 \cdots \sigma_l \in \langle \mu \rangle \\ \sigma_1 = i,\, \sigma_l = j}} x_{\sigma_1} \cdots x_{\sigma_{l-1}}.$$

2. *The* **set of encodings** *of Ω_Π is*

$$\mathsf{E} = \left\{ \sum_{i \geqslant 1} c_i \mathbf{I}(\mu_i) | c_i \in \mathbb{Q}[\mathbf{y}] \right\}$$

where $\Omega_\Pi = \{\mu_i | i \geqslant 1\}$ and $\mathbf{y} = (y_1, \ldots)$ is a set of commutative indeterminates. We denote $\mathbf{I}(\pi_1)$ by $\mathbf{A}$ and $\mathbf{I}(\pi_2)$ by $\mathbf{B}$. □

The $c_i(\mathbf{y})$ may be regarded as polynomials that encode combinatorial information about the pattern μ_i. With each enumerative problem concerning sequences with prescribed patterns, there is associated an element of E. For example, the element of E corresponding to π_1-alternating sequences of odd length is

$$\sum_{\mu \in (\pi_1\pi_2)^*} \mathbf{I}(\mu).$$

It is now possible to define generating functions associated with elements of E that retain information about the type $\tau(\sigma)$ of a sequence σ and the number $I(\rho)$ of inversions of a permutation ρ, as well as information about the pattern, which is encoded by $c_i(\mathbf{y})$, where σ and ρ have pattern μ_i.

4.3.3. Definition (Fundamental Generating Functions)

Let $\mathbf{U} = \sum_{i \geq 1} c_i \mathbf{I}(\mu_i)$ *be an arbitrary element of* E. *The fundamental generating functions for* $\mathbf{U}$ *are*

1. $\Psi_1(\mathbf{U}) = \sum_{i \geq 1} c_i \sum_{\sigma \in \langle \mu_i \rangle} \mathbf{x}^{\tau(\sigma)}$,
2. $\Psi_2(\mathbf{U}) = \sum_{i \geq 1} c_i \sum_{\sigma \in \langle\langle \mu_i \rangle\rangle} \frac{x^{|\sigma|}}{|\sigma|!_q} q^{I(\sigma)}$,

where $|\sigma|$ *denotes the length of* σ. □

For example, $[\mathbf{x}^{\mathbf{i}}]\Psi_1(\mathbf{I}((\pi_1\pi_2)^*))$ is the number of π_1-alternating sequences of odd length and type $\mathbf{i}$, while $[q^k x^n/n!_q]\Psi_2(\mathbf{I}((\pi_1\pi_2)^*))$, where $\pi_1 = <$, is the number of $<$-alternating permutations of odd length n, and with k inversions. The determination and use of the fundamental generating functions $\Psi_1(\mathbf{U})$ and $\Psi_2(\mathbf{U})$ for $\mathbf{U} \in \mathsf{E}$ will occupy the remainder of the section. We begin by considering the incidence matrices for disjoint unions of sets of patterns and for products (i.e., concatenation) of patterns. The following proposition is immediate.

4.3.4. Proposition (Incidence Matrices for Union and Product)

If $\mu_1, \mu_2 \in \Omega_\Pi$, *then*

1. $\mathbf{I}(\mu_1 \cup \mu_2) = \mathbf{I}(\mu_1) + \mathbf{I}(\mu_2)$, *if* $\langle \mu_1 \rangle \cap \langle \mu_2 \rangle = \varnothing$,
2. $\mathbf{I}(\mu_1\mu_2) = \mathbf{I}(\mu_1)\mathbf{I}(\mu_2)$,
3. $\mathbf{I}(\pi_1) + \mathbf{I}(\pi_2) = \mathbf{XJ}$,

where $\mathbf{X} = \operatorname{diag}(\mathbf{x})$. *We denote* $\mathbf{XJ}$ *by* $\mathbf{W}$. □

The sum and product lemmas for the fundamental generating functions are now stated.

4.3.5. Lemma (Sum, Product for the Fundamental Generating Functions)

Let $\mathbf{U}, \mathbf{V} \in \mathsf{E}$. *Then for* $i = 1$ *and* 2 *we have*

1. $\Psi_i(\mathbf{U} + \mathbf{V}) = \Psi_i(\mathbf{U}) + \Psi_i(\mathbf{V})$.
2. (i) $\Psi_1(\mathbf{UWV}) = \Psi_1(\mathbf{U})\Psi_1(\mathbf{V})$.
   (ii) $\Psi_2(\mathbf{UWV}) = \Psi_2(\mathbf{U})\Psi_2(\mathbf{V})$ *for* $\pi_1 = <$.
3. $\Psi_i(c\mathbf{U}) = c\Psi_i(\mathbf{U})$ *for* $c \in \mathbb{Q}[\mathbf{y}]$.

Moreover, $\Psi_1(\mathbf{U}) = \operatorname{trace} \mathbf{UW}$.

Proof:

1. Immediate, from Definition 4.3.3.

2. **Case $i = 1$:** From Definition 4.3.2 we have immediately that $\Psi_1(\mathbf{U}) = \text{trace } \mathbf{UW}$. Thus $\Psi_1(\mathbf{UWV}) = \text{trace}(\mathbf{UWVW}) = (\text{trace } \mathbf{UW})(\text{trace } \mathbf{VW}) = \Psi_1(\mathbf{U})\Psi_1(\mathbf{V})$.

Case $i = 2$: Let $\mathbf{U} = \sum_{i \geq 1} c_i \mathbf{I}(\mu_i)$ and $\mathbf{V} = \sum_{j \geq 1} d_j \mathbf{I}(\mu_j)$. Then

$$\Psi_2(\mathbf{UWV}) = \sum_{i, j \geq 1} c_i d_j \Psi_2\big(\mathbf{I}(\mu_i)\mathbf{W}\mathbf{I}(\mu_j)\big) = \sum_{i, j \geq 1} c_i d_j \Psi_2\big(\mathbf{I}(\mu_1 \omega \mu_j)\big)$$

$$= \sum_{i, j \geq 1} c_i d_j \sum_{\sigma \in \langle\langle \mu_i \omega \mu_j \rangle\rangle} x^{|\sigma|} \frac{q^{I(\sigma)}}{|\sigma|!_q}.$$

But by Lemma 2.6.6 we have

$$\sum_{\sigma \in \langle\langle \mu_i \omega \mu_j \rangle\rangle} q^{I(\sigma)} = \binom{m_i + m_j}{m_i}_q \sum_{\sigma' \in \langle\langle \mu_i \rangle\rangle} q^{I(\sigma')} \sum_{\sigma'' \in \langle\langle \mu_j \rangle\rangle} q^{I(\sigma'')}$$

where $|\mu_i| = m_i$, $|\mu_j| = m_j$, and $|\mu_i \omega \mu_j| = m_i + m_j$. Thus

$$\Psi_2(\mathbf{UWV}) = \sum_{i, j \geq 1} c_i d_j \sum_{\substack{\sigma' \in \langle\langle \mu_i \rangle\rangle \\ \sigma'' \langle\langle \mu_j \rangle\rangle}} \frac{x^{|\sigma'| + |\sigma''|}}{(|\sigma'| + |\sigma''|)!_q} \binom{|\sigma'| + |\sigma''|}{|\sigma'|}_q q^{I(\sigma')} q^{I(\sigma'')}$$

$$= \left\{ \sum_{i \geq 1} c_i \sum_{\sigma' \in \langle\langle \mu_i \rangle\rangle} x^{|\sigma'|} \frac{q^{I(\sigma')}}{|\sigma'|!_q} \right\} \left\{ \sum_{j \geq 1} d_j \sum_{\sigma'' \in \langle\langle \mu_j \rangle\rangle} x^{|\sigma''|} \frac{q^{I(\sigma'')}}{|\sigma''|!_q} \right\}$$

and the result follows.

3. Immediate. □

No further use is made here of the fact that $\Psi_1(\mathbf{U}) = \text{trace}\,\mathbf{UW}$ for $\mathbf{U} \in \mathsf{E}$, but the ramifications of this result are considered in Sections 4.4 and 4.5, where it is of considerable importance. The following result gives the connection between the fundamental generating functions and π_1-string enumerators.

4.3.6. Proposition (π_1-String Enumerator)

Let $f(x) = c_0 + c_1 x + c_2 x^2 + \cdots$. Then

1. $\Psi_1(f(\mathbf{A})) = (xf) \circ \gamma(\pi_1)$.
2. $\Psi_2(f(\mathbf{I}(<))) = (xf) \circ \eta_q$.

Proof:

1.
$$\begin{aligned}\Psi_1(f(\mathbf{A})) &= \sum_{i\geqslant 0} c_i\Psi_1(\mathbf{A}^i) && \text{by Lemma 4.3.5(3)}\\ &= \sum_{i\geqslant 0} c_i\gamma_{i+1}(\pi_1) && \text{by Definition 4.3.3(1) and}\\ & && \text{Proposition 4.3.4(2)}\\ &= \left(\sum_{i\geqslant 0} c_i x^{i+1}\right)\circ \gamma(\pi_1)\end{aligned}$$

and the result follows.

2. Similar. □

The following is a preliminary application of these results.

4.3.7. π_1-Alternating Sequences of Odd Length

The set of all π_1-alternating sequences of odd length is $\langle(\pi_1\pi_2)^*\rangle$. We identify with this problem the element

$$\mathbf{U} = \sum_{\mu\in(\pi_1\pi_2)^*} \mathsf{I}(\mu) \qquad \in \mathsf{E},$$

so the number of π_1-alternating sequences of odd length and type $\mathbf{i}$ is $[\mathbf{x}^{\mathbf{i}}]\Psi_1(\mathbf{U})$. Now by Proposition 4.3.4

$$\mathbf{U} = \sum_{k\geqslant 0} \mathsf{I}\left((\pi_1\pi_2)^k\right) = \sum_{k\geqslant 0} (\mathsf{I}(\pi_1\pi_2))^k = (\mathbf{I} - \mathsf{I}(\pi_1)\mathsf{I}(\pi_2))^{-1} = (\mathbf{I} - \mathbf{AB})^{-1}.$$

Our aim now is to derive an expression for $\Psi_1(\mathbf{U})$ in terms of $\Psi_1(\mathbf{A}^{k-1}) = \gamma_k(\pi_1)$. Let $\mathbf{H} = \mathbf{AB}$, so eliminating $\mathbf{B}$ using $\mathbf{B} = \mathbf{W} - \mathbf{A}$,

$$\mathbf{I} - \mathbf{H} = \mathbf{I} - \mathbf{A}(\mathbf{W} - \mathbf{A}) = \mathbf{I} + \mathbf{A}^2 - \mathbf{AW} = \mathbf{I} - (\mathbf{Q} - \mathbf{L}_1\mathbf{W}\mathbf{R}_1),$$

where $\mathbf{Q} = -\mathbf{A}^2$, $\mathbf{L}_1 = -\mathbf{A}$, $\mathbf{R}_1 = \mathbf{I}$. Postmultiplying by $\mathbf{U} = (\mathbf{I} - \mathbf{H})^{-1}$ and premultiplying by $(\mathbf{I} - \mathbf{Q})^{-1}$ gives

$$(\mathbf{I} - \mathbf{Q})^{-1} = \mathbf{U} + (\mathbf{I} - \mathbf{Q})^{-1}\mathbf{L}_1\mathbf{W}\mathbf{R}_1\mathbf{U}.$$

Applying Ψ_l, using Lemma 4.3.5, we have

$$\Psi_l\left((\mathbf{I} + \mathbf{A}^2)^{-1}\right) = \Psi_l(\mathbf{U}) - \Psi_l\left((\mathbf{I} + \mathbf{A}^2)^{-1}\mathbf{A}\right)\Psi_l(\mathbf{U})$$

for $l = 1$, and for $l = 2$ if $\pi_1 = <$. This is a linear equation for $\Psi_l(\mathbf{U})$, which yields

$$\Psi_l(\mathbf{U}) = \left\{1 - \Psi_l\left((\mathbf{I} + \mathbf{A}^2)^{-1}\mathbf{A}\right)\right\}^{-1}\Psi_l\left((\mathbf{I} + \mathbf{A}^2)^{-1}\right) \qquad \text{for } l = 1, 2.$$

But, from Proposition 4.3.6,

$$\Psi_1\big((\mathbf{I}+\mathbf{A}^2)^{-1}\big) = x(1+x^2)^{-1}\circ\gamma(\pi_1) = \sum_{k\geqslant 0}(-1)^k\gamma_{2k+1}(\pi_1)$$

and

$$1-\Psi_1\big((\mathbf{I}+\mathbf{A}^2)^{-1}\mathbf{A}\big) = \sum_{k\geqslant 0}(-1)^k\gamma_{2k}(\pi_1).$$

Combining these results gives

$$\Psi_1(\mathbf{U}) = \left\{\sum_{k\geqslant 0}(-1)^k\gamma_{2k+1}(\pi_1)\right\}\left\{\sum_{k\geqslant 0}(-1)^k\gamma_{2k}(\pi_1)\right\}^{-1},$$

in agreement with Corollary 4.2.20. In a similar way, using Proposition 4.3.6(2),

$$\Psi_2(\mathbf{U}) = \left\{\sum_{k\geqslant 0}(-1)^k\frac{x^{2k+1}}{(2k+1)!_q}\right\}\left\{\sum_{k\geqslant 0}(-1)^k\frac{x^{2k}}{(2k)!_q}\right\}^{-1}. \qquad \square$$

In **4.3.7** we have determined $\Psi_l((\mathbf{I}-\mathbf{H})^{-1})$, where $\mathbf{H}=\mathbf{AB}$, by solving a linear equation induced by $\mathbf{H}=-\mathbf{A}^2+\mathbf{AW}$. In general, $\Psi_l(\mathbf{C}(\mathbf{I}-\mathbf{H})^{-1}\mathbf{D})$ is determined by solving a system of linear equations of order s induced by an expansion

$$\mathbf{H}=\mathbf{Q}-\sum_{i=1}^{s}\mathbf{L}_i\mathbf{W}\mathbf{R}_i,$$

called a **factored expansion** of $\mathbf{H}$.

4.3.8. Theorem (Elimination)

Let $\mathbf{Q},\mathbf{H},\mathbf{C},\mathbf{D},\mathbf{L}_i,\mathbf{R}_i\in\mathsf{E}$ *for* $i=1,\dots,s$, *and let* $\mathbf{L}_0=\mathbf{0}$, $\mathbf{R}_0=\mathbf{C}$. *If* $\mathbf{H}=\mathbf{Q}-\sum_{k=1}^{s}\mathbf{L}_k\mathbf{W}\mathbf{R}_k$, *then for* $l=1,2$ *we have*

$$\Psi_l\big(\mathbf{C}(\mathbf{I}-\mathbf{H})^{-1}\mathbf{D}\big) = |[\mathbf{M}^{(l)}:\mathbf{d}^{(l)}]_0|\cdot|\mathbf{M}^{(l)}|^{-1}$$

where

$$[\mathbf{M}^{(l)}]_{ij}=\delta_{ij}+\Psi_l\big(\mathbf{R}_i(\mathbf{I}-\mathbf{Q})^{-1}\mathbf{L}_j\big), \qquad 0\leqslant i,j\leqslant s,$$

$$d_i^{(l)}=\Psi_l\big(\mathbf{R}_i(\mathbf{I}-\mathbf{Q})^{-1}\mathbf{D}\big), \qquad 0\leqslant i\leqslant s,$$

and $\mathbf{d}^{(l)}=(d_0^{(l)},\dots,d_s^{(l)})^T$.

Proof: For $\mathbf{T} \in \mathsf{E}$ we premultiply both sides of

$$\mathbf{I} - \mathbf{H} = \mathbf{I} - \mathbf{Q} + \sum_{j=1}^{s} \mathbf{L}_j \mathbf{W} \mathbf{R}_j$$

by $\mathbf{T}(\mathbf{I} - \mathbf{Q})^{-1}$, postmultiply by $(\mathbf{I} - \mathbf{H})^{-1}\mathbf{D}$, and apply Ψ_l to the result to obtain

$$\Psi_l\big(\mathbf{T}(\mathbf{I} - \mathbf{Q})^{-1}\mathbf{D}\big) = \Psi_l\big(\mathbf{T}(\mathbf{I} - \mathbf{H})^{-1}\mathbf{D}\big) + \sum_{k=1}^{s} \Psi_l\big(\mathbf{T}(\mathbf{I} - \mathbf{Q})^{-1}\mathbf{L}_k\big)\Psi_l\big(\mathbf{R}_k(\mathbf{I} - \mathbf{H})^{-1}\mathbf{D}\big)$$

by the sum and product lemmas for Ψ_l. Setting $\mathbf{T} = \mathbf{C}, \mathbf{R}_1, \ldots, \mathbf{R}_s$ in turn we obtain an $(s+1) \times (s+1)$ system of linear equations for the functions $\xi_j^{(l)} = \Psi_l(\mathbf{R}_j(\mathbf{I} - \mathbf{H})^{-1}\mathbf{D})$, where $0 \leqslant j \leqslant s$. Now $\xi_0^{(l)} = \Psi_l(\mathbf{R}_0(\mathbf{I} - \mathbf{H})^{-1}\mathbf{D}) = \Psi_l(\mathbf{C}(\mathbf{I} - \mathbf{H})^{-1}\mathbf{D})$ is the desired function and the equations may be written in the form

$$\xi_i^{(l)} + \sum_{j=1}^{s} \Psi_l\big(\mathbf{R}_i(\mathbf{I} - \mathbf{Q})^{-1}\mathbf{L}_j\big)\xi_j^{(l)} = d_i^{(l)}, \qquad \text{where } 0 \leqslant i \leqslant s.$$

The result follows by Cramer's rule. □

The elimination theorem expresses the generating functions $\Psi_l(\mathbf{C}(\mathbf{I} - \mathbf{H})^{-1}\mathbf{D})$ in terms of the subproblems $\Psi_l(\mathbf{R}_i(\mathbf{I} - \mathbf{Q})^{-1}\mathbf{L}_j)$. These can easily be evaluated by Proposition 4.3.6 if $\mathbf{Q}$ depends only on $\mathbf{A}$, $\mathbf{B}$ having already been eliminated. This accounts for the name of the theorem. Now $\mathbf{R}_i$, $\mathbf{L}_j$, and $\mathbf{D}$ present no problems since they are of bounded length and can be expanded directly in terms of $\mathbf{A}$. Under these circumstances the elements of $\mathbf{M}^{(l)}$ and $\mathbf{d}^{(l)}$ may be expressed in terms of $\gamma_k(\pi_1)$.

4.3.9. Definition (Left-, Right-Expansions)

Let $\mathbf{H} \in \mathsf{E}$. *Suppose* $\mathbf{H}$ *may be expressed in the form*

$$\mathbf{H} = \mathbf{Q} - \sum_{k=1}^{s} \mathbf{L}_k \mathbf{W} \mathbf{R}_k,$$

where $\mathbf{Q}, \mathbf{L}_k, \mathbf{R}_k \in \mathsf{E}$ *for* $1 \leqslant k \leqslant s$. *If* $\mathbf{Q}, \mathbf{L}_k$ *(resp.* $\mathbf{R}_k$*) are independent of* $\mathbf{I}(\pi_2)$ *for* $1 \leqslant k \leqslant s$, *then the factored expansion is called a* **left-expansion (resp. right-expansion)** *of* $\mathbf{H}$. □

Left- and right-expansions are obtained by the following algorithm.

4.3.10. Algorithm (Factored Expansion)

Let $\mathbf{H} = \sum_{i=1}^{m} c_i \mathbf{I}(\mu_i)$, *where* $c_i \in \mathbb{Q}[\mathrm{y}]$, $\mu_i \in \Omega_\Pi$ *for* $1 \leqslant i \leqslant m$. *We say that* $\mathbf{A}_i = \mathbf{I}(\mu_i)$ *is* **left-reduced** *if it contains no* $\mathbf{W}$ *or if it contains no occurrence of* $\mathbf{B}$ *to the left of the left-most occurrence of a* $\mathbf{W}$.

If $\mathbf{A}_i$ *is not left-reduced, replace the left-most occurrence of a* $\mathbf{B}$ *by* $\mathbf{W} - \mathbf{A}$ *to produce two terms, at least one of which is left-reduced. Repeat this process until all terms are left-reduced. The process clearly terminates, giving a left-expansion. If "left" is replaced everywhere by "right," then we obtain a right-expansion.* □

We now state a general strategy for enumerating sequences with respect to pattern.

4.3.11. Remark (General Strategy)

Let $\mathsf{U} \subseteq \Omega_\Pi$ *be a set of patterns.*

1. *Derive the element* $\mathbf{U} = \mathbf{C}(\mathbf{I} - \mathbf{H})^{-1}\mathbf{D} \in \mathsf{E}$ *corresponding to* U. *This is done by the sum and product lemmas for incidence matrices* (*Proposition* 4.3.4).
2. *Use the factored expansion algorithm to obtain a left- (resp. right-) expansion of* $\mathbf{H}$ *of the form*

$$\mathbf{H} = \mathbf{Q} - \sum_{k=1}^{s} \mathbf{L}_k \mathbf{W} \mathbf{R}_k .$$

   As a result, $\mathbf{Q}$ *and* $\mathbf{L}_i$ (*resp.* $\mathbf{R}_i$) *are independent of* $\mathbf{B}$, *for* $1 \leqslant i \leqslant s$.
3. *Use the elimination theorem to express* $\Psi_l(\mathbf{U})$ *in terms of* $\Psi_l(\mathbf{R}_i(\mathbf{I} - \mathbf{Q})^{-1}\mathbf{L}_j)$ *for* $1 \leqslant i, j \leqslant s$.
4. *Solve the subproblems* $\Psi_l(\mathbf{R}_i(\mathbf{I} - \mathbf{Q})^{-1}\mathbf{L}_j)$ *by expressing* $\mathbf{R}_i(\mathbf{I} - \mathbf{Q})^{-1}\mathbf{L}_j$ *in terms of* $\mathbf{A}$ *alone, eliminating* $\mathbf{B}$*'s in* $\mathbf{R}_i$ *or* $\mathbf{L}_j$, *and simplify using*

$$\Psi_1(f(\mathbf{A})) = (xf) \circ \gamma(\pi_1)$$

$$\Psi_2(f(\mathbf{A})) = (xf) \circ \eta_q \qquad \text{if } \pi_1 = < . \qquad \square$$

As an application of the general strategy and the use of the elimination theorem, we now prove the maximal string decomposition theorem (**4.2.3**).

4.3.12. Proof of the Maximal String Decomposition Theorem

Every sequence in N_+^+ has a unique pattern in $\{\pi_1, \pi_2\}^*$. Now

$$\{\pi_1, \pi_2\}^* = (\pi_1^* \pi_2)^* \pi_1^*$$

is a decomposition that exposes the maximal π_1-strings. Let f_j be an indeterminate marking a maximal π_1-string of length j. Thus, following Remark 4.3.11, the element in E corresponding to the enumeration of N_+^+ with respect to the lengths of maximal π_1-strings is

$$\mathbf{U} = \sum_{k \geq 0} \left\{ \sum_{j \geq 0} f_{j+1} \mathbf{I}\left(\pi_1^j \pi_2\right)\right\}^k \sum_{i \geq 0} f_{i+1} \mathbf{I}\left(\pi_1^i\right),$$

since $\pi_1^j \pi_2$ indicates the occurrence of a maximal π_1-string of length $j + 1$. Thus

$$\mathbf{U} = \left\{\mathbf{I} - \sum_{j \geq 0} f_{j+1} \mathbf{I}\left(\pi_1^j \pi_2\right)\right\}^{-1} \sum_{i \geq 0} f_{i+1} \mathbf{I}\left(\pi_1^i\right).$$

Let $F(x) = 1 + f_1 x + f_2 x^2 + \cdots$, the maximal π_1-string length enumerator, and let $f(x) = f_1 + f_2 x + f_3 x^2 + \cdots$. Then $\mathbf{U} = \mathbf{C}(\mathbf{I} - \mathbf{H})^{-1}\mathbf{D}$, where $\mathbf{C} = \mathbf{I}$, $\mathbf{H} = f(\mathbf{A})\mathbf{B}$, and $\mathbf{D} = f(\mathbf{A})$. Now $\mathbf{H} = \mathbf{Q} - \mathbf{L}_1\mathbf{W}\mathbf{R}_1$ where $\mathbf{Q} = -\mathbf{A}f(\mathbf{A})$, $\mathbf{L}_1 = -f(\mathbf{A})$, and $\mathbf{R}_1 = \mathbf{I}$, is a factored expansion of $\mathbf{H}$. It follows from the elimination theorem (with $s = 1$) that, for $i = 1$ and 2,

$$1 + \Psi_i(\mathbf{U}) = \left\{1 - \Psi_i\left(F^{-1}(\mathbf{A}) f(\mathbf{A})\right)\right\}^{-1}$$

since $\mathbf{I} - \mathbf{Q} = \mathbf{I} + \mathbf{A}f(\mathbf{A}) = F(\mathbf{A})$ and power series functions of $\mathbf{A}$ commute. Thus, by Proposition 4.3.6,

$$1 + \Psi_1(\mathbf{U}) = \left\{1 - \left(xfF^{-1} \circ \gamma(\pi_1)\right)\right\}^{-1} = \left\{F^{-1} \circ \gamma(\pi_1)\right\}^{-1}$$

and $1 + \Psi_2(\mathbf{U}) = (F^{-1} \circ \eta_q)^{-1}$. This completes the proof. □

We now turn to repeated pattern problems.

4.3.13. Definition (Sequence with Repeated Pattern)

If $\mu, \mu' \in \Omega_\Pi = \{\pi_1, \pi_2, \omega\}^*$, *then* $\sigma \in \langle \mu^* \mu' \rangle$ *is called a* **sequence with repeated pattern** μ. □

The π_1-alternating sequence problem is, of course, a sequence problem with the repeated pattern $\pi_1\pi_2$. We were able to use the maximal string decomposition theorem to solve it in **4.2.7** because such sequences may be characterized solely in terms of the lengths of maximal π_1-strings. On the other hand, sequences with repeated pattern $\pi_1^2\pi_2^2$ cannot be enumerated by the maximal string decomposition theorem because such sequences are not characterized by their unordered set of maximal π_1-string lengths. However, the elimination theorem may be applied in general for repeated pattern problems.

4.3.14. Sequences with Repeated Pattern $\pi_1^2\pi_2^2$

We consider the enumeration of $\langle(\pi_1^2\pi_2^2)^*\pi_1^k\rangle$ where $k = 0$ or 1. Following the general strategy (Remark 4.3.11) we note that the element in the set E of encodings corresponding to this problem is $\mathbf{U} = (\mathbf{I} - \mathbf{A}^2\mathbf{B}^2)^{-1}\mathbf{A}^k$ by the sum and product lemmas for incidence matrices. The required generating function is therefore $\Psi_1((\mathbf{I} - \mathbf{A}^2\mathbf{B}^2)^{-1}\mathbf{A}^k)$. Let $\mathbf{H} = \mathbf{A}^2\mathbf{B}^2$. A factored expansion of $\mathbf{I} - \mathbf{H}$ is obtained by applying a left-expansion, so

$$\begin{aligned}\mathbf{I} - \mathbf{H} &= \mathbf{I} - \mathbf{A}^2(\mathbf{W} - \mathbf{A})\mathbf{B} = \mathbf{I} - \mathbf{A}^2\mathbf{W}\mathbf{B} + \mathbf{A}^3\mathbf{B}\\ &= \mathbf{I} - \mathbf{A}^2\mathbf{W}\mathbf{B} + \mathbf{A}^3(\mathbf{W} - \mathbf{A})\\ &= \mathbf{I} - \mathbf{A}^4 - \mathbf{A}^2\mathbf{W}\mathbf{B} + \mathbf{A}^3\mathbf{W}.\end{aligned}$$

In the notation of the elimination theorem,

$$\mathbf{I} - \mathbf{H} = \mathbf{I} - \mathbf{Q} + \mathbf{L}_1\mathbf{W}\mathbf{R}_1 + \mathbf{L}_2\mathbf{W}\mathbf{R}_2,$$

where $\mathbf{I} - \mathbf{Q} = \mathbf{I} - \mathbf{A}^4$ and $\mathbf{L}_0 = \mathbf{0}$, $\mathbf{R}_0 = \mathbf{I}$; $\mathbf{L}_1 = -\mathbf{A}^2$, $\mathbf{R}_1 = \mathbf{B}$; $\mathbf{L}_2 = \mathbf{A}^3$, $\mathbf{R}_2 = \mathbf{I}$. It follows from the elimination theorem that

$$\mathbf{M}^{(l)} = \begin{bmatrix} 1 & -\Psi_l\left((\mathbf{I} - \mathbf{Q})^{-1}\mathbf{A}^2\right) & \Psi_l\left((\mathbf{I} - \mathbf{Q})^{-1}\mathbf{A}^3\right) \\ 0 & 1 - \Psi_l\left(\mathbf{B}(\mathbf{I} - \mathbf{Q})^{-1}\mathbf{A}^2\right) & \Psi_l\left(\mathbf{B}(\mathbf{I} - \mathbf{Q})^{-1}\mathbf{A}^3\right) \\ 0 & -\Psi_l\left((\mathbf{I} - \mathbf{Q})^{-1}\mathbf{A}^2\right) & 1 + \Psi_l\left((\mathbf{I} - \mathbf{Q})^{-1}\mathbf{A}^3\right) \end{bmatrix},$$

$$\mathbf{d}^{(l)} = \begin{bmatrix} \Psi_l\left((\mathbf{I} - \mathbf{Q})^{-1}\mathbf{A}^k\right) \\ \Psi_l\left(\mathbf{B}(\mathbf{I} - \mathbf{Q})^{-1}\mathbf{A}^k\right) \\ \Psi_l\left((\mathbf{I} - \mathbf{Q})^{-1}\mathbf{A}^k\right) \end{bmatrix}.$$

It remains to express $\mathbf{M}^{(l)}$ and $\mathbf{d}^{(l)}$ in terms of $\gamma(\pi_1)$ when $l = 1$. Now

$$\Psi_1\left((\mathbf{I} - \mathbf{Q})^{-1}\mathbf{A}^r\right) = x^{r+1}(1 - x^4)^{-1} \circ \gamma(\pi_1) = G_{r+1},$$

say, since $\mathbf{I} - \mathbf{Q} = \mathbf{I} - \mathbf{A}^4$. Replacing $\mathbf{B}$ by $\mathbf{W} - \mathbf{A}$ gives

$$|\mathbf{M}^{(l)}| = \begin{vmatrix} 1 & -G_3 & G_4 \\ 0 & 1 - \gamma_1 G_3 + G_4 & \gamma_1 G_4 - G_5 \\ 0 & -G_3 & 1 + G_4 \end{vmatrix}.$$

But

$$1 + G_4 = \{1 + x^4(1 - x^4)^{-1}\} \circ \boldsymbol{\gamma}(\pi_1) = (1 - x^4)^{-1} \circ \boldsymbol{\gamma}(\pi_1) = G_0,$$

$$\gamma_1 + G_5 = \{x + x^5(1 - x^4)^{-1}\} \circ \boldsymbol{\gamma}(\pi_1) = x(1 - x^4)^{-1} \circ \boldsymbol{\gamma}(\pi_1) = G_1,$$

and $\gamma_1 G_4 - G_5 = \gamma_1(G_0 - 1) - (G_1 - \gamma_1) = \gamma_1 G_0 - G_1$, so $|\mathbf{M}^{(l)}| = G_0^2 - G_1 G_3$. Also

$$[\mathbf{M}^{(l)} : \mathbf{d}^{(l)}]_0 = \begin{bmatrix} G_{k+1} & -G_3 & G_0 - 1 \\ \gamma_1 G_{k+1} - G_{k+2} & G_0 - \gamma_1 G_3 & -G_1 + \gamma_1 G_0 \\ G_{k+1} & -G_3 & G_0 \end{bmatrix}$$

$$= \begin{bmatrix} 0 & 0 & -1 \\ -G_{k+2} & G_0 & -G_1 \\ G_{k+1} & -G_3 & G_0 \end{bmatrix}$$

so $[\mathbf{M}^{(l)} : \mathbf{d}^{(l)}]_0 = G_0 G_{k+1} - G_3 G_{k+2}$.

Thus the number of sequences of type $\mathbf{i}$ and pattern $(\pi_1^2\pi_2^2)^*\pi_1^k$ is $[\mathbf{x}^{\mathbf{i}}]G(\boldsymbol{\gamma}(\pi_1))$, where $G(\boldsymbol{\gamma}(\pi_1)) = (G_0 G_{k+1} - G_3 G_{k+2})(G_0^2 - G_1 G_3)^{-1}$ and $G_r = \sum_{j \geqslant 0} \gamma_{4j+r}(\pi_1)$. □

This result is specialized to permutations as follows.

4.3.15. Permutations with Repeated Pattern $\pi_1^2\pi_2^2$, for Rises

We use the generating function $\Psi_1(\mathbf{U})$ given in **4.3.14**. The <-transformation lemma gives, for permutations with $\pi_1 = <$, $G_j = x^j(1 - x^4)^{-1} \circ \boldsymbol{\eta}$, where $\boldsymbol{\eta} = (1, x/1!, x^2/2!, \ldots)$. But $(1 - x)^{-1} \circ \boldsymbol{\eta} = e^x$ so by series quadrisection

$$G_k = \tfrac{1}{4}\{e^x + i^{-k}e^{ix} + i^{-2k}e^{i^2x} + i^{-3k}e^{i^3x}\} \qquad \text{for } 0 \leqslant k \leqslant 3$$

so $G_0 = \frac{1}{2}(\cosh x + \cos x)$, $G_1 = \frac{1}{2}(\sinh x + \sin x)$, $G_2 = \frac{1}{2}(\cosh x - \cos x)$, and $G_3 = \frac{1}{2}(\sinh x - \sin x)$. Thus, from **4.3.14**, it follows that the number of permutations on N_n with pattern in $(\pi_1^2\pi_2^2)^*\pi_1^k$, where $\pi_1 = <$, is

$$\left[\frac{x^n}{n!}\right](\tan x + \tanh x)(1 + \sec x \operatorname{sech} x)^{-1} \qquad \text{for } k = 0,$$

$$\left[\frac{x^n}{n!}\right]\tan x \tanh x(1 + \sec x \operatorname{sech} x)^{-1} \qquad \text{for } k = 1. \qquad □$$

Clearly, **4.3.14** may also be specialized to permutations with successions or c-successions with the aid of the $+$-transformation and $\oplus$-transformation lemmas.

4.3.16. Sequences with Fixed Pattern

We consider the enumeration of sequences with fixed pattern

$$\pi_1^{p_1-1}\pi_2\pi_1^{p_2-1}\pi_2\cdots\pi_2\pi_1^{p_m-1},$$

where $m, p_1,\ldots,p_m \geqslant 1$ are integers. The corresponding element of E is $\mathbf{U}=\mathbf{A}^{p_1-1}\mathbf{B}\cdots\mathbf{B}\mathbf{A}^{p_m-1}$. Let

$$a_i=\sum_{j=1}^{i}p_j \qquad \text{for } i=1,\ldots,m,\ a_0=0.$$

To obtain a system of linear equations for $\Psi_l(\mathbf{U})$, we apply the left-expansion to $\mathbf{V}_i=\mathbf{A}^{p_i-1}\mathbf{B}\cdots\mathbf{B}\mathbf{A}^{p_m-1}$. The resulting expression for $\mathbf{V}_i$ is

$$\mathbf{V}_i=(-1)^{m-i}\mathbf{A}^{a_m-a_{i-1}-1}+\sum_{j=i+1}^{m}(-1)^{j-i-1}\mathbf{A}^{a_{j-1}-a_{i-1}-1}\mathbf{W}\mathbf{V}_j \quad \text{for } 1\leqslant i\leqslant m.$$

Let $\xi_i^{(l)}=\Psi_l(\mathbf{V}_i)$ for $l=1,2$. Then applying Ψ_l to both sides gives, by the sum and product lemmas for Ψ_1 and Ψ_2,

$$\xi_i^{(l)}=(-1)^{m-i}\Psi_l(\mathbf{A}^{a_m-a_{i-1}-1})+\sum_{j=i+1}^{m}(-1)^{j-i-1}\xi_j^{(l)}\Psi_l(\mathbf{A}^{a_{j-1}-a_{i-1}-1})$$

for $1\leqslant i\leqslant m$. This is a system of linear equations for the generating functions $\xi_1^{(l)},\ldots,\xi_m^{(l)}$, where the only function required is $\xi_1^{(l)}$, since this is equal to $\Psi_l(\mathbf{V}_1)=\Psi_l(\mathbf{U})$. It follows by Cramer's rule that

$$\xi_1^{(l)}=|[\mathbf{M}^{(l)}:\mathbf{d}^{(l)}]_1|\cdot|\mathbf{M}^{(l)}|^{-1},$$

where $\mathbf{d}^{(l)}=(d_1^{(l)},\ldots,d_m^{(l)})^T$ in which $d_i^{(l)}=(-1)^{m-i}\gamma_{a_m-a_{i-1}}(\pi_1)$ for $1\leqslant i\leqslant m$, and where

$$[\mathbf{M}^{(l)}]_{ij}=\begin{cases}(-1)^{j-i}\gamma_{a_{j-1}-a_{i-1}}(\pi_1) & \text{for } 1\leqslant i\leqslant j\leqslant m\\ 0 & \text{otherwise.}\end{cases}$$

But clearly, $|\mathbf{M}^{(l)}|=1$ and it follows that the required generating function is

$$|[\mathbf{M}^{(l)}:\mathbf{d}^{(l)}]_1|. \qquad \square$$

When $\pi_1 = <$, this expression may be considerably simplified for permutations to give the following result.

4.3.17. Permutations with Fixed Pattern and Inversions

We determine the number of permutations on N_n with k inversions and with pattern

$$\pi_1^{p_1-1}\pi_2 \cdots \pi_2\pi_1^{p_m-1},$$

where $\pi_1 = <$ and $\pi_2 = \geqslant$, and $p_1 + \cdots + p_m = n$. The required number is $[q^k x^n/n!_q]\xi_1^{(2)}$, where

$$\xi_1^{(2)} = (-1)^{m-1}\left\|(-1)^{j-i+1-\delta_{jm}}\frac{x^{a_j-a_{i-1}}}{(a_j - a_{i-1})!_q}\right\|,$$

by Proposition 4.3.6, after permuting the columns of the determinant result **4.3.16**. Thus

$$\xi_1^{(2)} = x^{a_m}\left\|\frac{1}{(a_j - a_{i-1})!_q}\right\|, \qquad \text{where } a_m = n$$

in the notation of **4.3.16**. It follows that

$$\left[q^k\frac{x^n}{n!_q}\right]\xi_1^{(2)} = [q^k]n!_q\left\|\frac{1}{(a_j - a_{i-1})!_q}\right\| = [q^k]\left\|\binom{n - a_{i-1}}{a_j - a_{i-1}}_q\right\|. \qquad \square$$

Until now, we have been concerned exclusively with **bipartite problems**. These are problems that involve an arbitrary bipartition $\{\pi_1, \pi_2\}$ of N_+^2. Our approach has been to eliminate one of the blocks, π_2, and to express the desired generating function in terms of the π_1-string enumerators $\gamma_k(\pi_1)$ or $x^k/k!_q$. The idea of a pattern may be generalized to a tripartition $\{\pi_1, \pi_2, \pi_3\}$ of N_+^2. However, doing so raises difficulties in general since the theory that has been developed here permits the elimination of only one block, say π_1, of the tripartition. It is then necessary to solve a set of subproblems that consists of Ψ_1 and Ψ_2 applied to various expressions involving only $\mathsf{I}(\pi_2)$ and $\mathsf{I}(\pi_3)$. Certain tripartitions have particular properties that may be exploited usefully after interpreting these subproblems as combinatorial problems. We emphasize the important point that the theory allows us to interpret elements of E combinatorially, as well as to associate an element of E with a given problem. Accordingly, there are both combinatorial means and algebraic means for evaluating $\Psi_l(\mathbf{U})$ for $\mathbf{U} \in \mathsf{E}$. We now consider a tripartite problem for permutations.

4.3.18. A Tripartite Problem

We determine the number $c(n, t, u)$ of permutations on N_n with t successions and u rises. Let π_1 denote the set of rises, π_2 the set of nonrises, and π_3 the set of successions. But $\pi_3 \subset \pi_1$ so $\{\pi_1 - \pi_3, \pi_2, \pi_3\}$ is a tripartition of N_+^2. Thus we wish to enumerate the set

$$\langle\langle\{\pi_1-\pi_3, \pi_2, \pi_3\}^*\rangle\rangle,$$

where r and s are indeterminates marking rises and successions, respectively. Now the elements of $\pi_1-\pi_3$ are rises that are not successions, and are therefore marked with an r. The elements of π_2 are nonrises and nonsuccessions. The elements of π_3 are simultaneously rises and successions, and are therefore marked by rs. The element of E that represents the problem is therefore

$$\mathbf{U} = \{\mathsf{I} - (r(\mathbf{A} - \mathbf{C}) + \mathbf{B} + rs\mathbf{C})\}^{-1}$$

by the sum and product lemma for incidence matrices, where $\mathbf{A} = \mathsf{I}(\pi_1)$, $\mathbf{B} = \mathsf{I}(\pi_2)$, and $\mathbf{C} = \mathsf{I}(\pi_3)$.

To evaluate the sequence generating function $\Psi_1(\mathbf{U})$, let $\mathbf{H} = r(\mathbf{A} - \mathbf{C}) + \mathbf{B} + rs\mathbf{C}$. Now $\mathbf{B} = \mathbf{W} - \mathbf{A}$, so a factored expansion of $\mathbf{I} - \mathbf{H}$ is $\mathbf{I} - \mathbf{H} = \mathbf{I} - \mathbf{Q} + \mathbf{L}_1\mathbf{W}\mathbf{R}_1$, where $\mathbf{Q} = r(s-1)\mathbf{C} + (r-1)\mathbf{A}$, $\mathbf{L}_1 = -\mathbf{I}$, and $\mathbf{R}_1 = \mathbf{I}$. It follows from the elimination theorem with $s = 1$ that

$$1 + \Psi_1(\mathbf{U}) = \left(1 - \Psi_1\left(\{\mathbf{I} - r(s-1)\mathbf{C} - (r-1)\mathbf{A}\}^{-1}\right)\right)^{-1}.$$

The remaining tasks are to determine $\Psi_1(\{\mathbf{I} - r(s-1)\mathbf{C} - (r-1)\mathbf{A}\}^{-1})$ and to specialize the result to permutations. For this purpose let $r(s-1) = v$ and $r - 1 = y$. Then

$$\Psi_1\left(\{\mathbf{I} - r(s-1)\mathbf{C} - (r-1)\mathbf{A}\}^{-1}\right) = \sum_{k\geqslant 1} \Psi_1\left(\{y(\mathbf{I} - v\mathbf{C})^{-1}\mathbf{A}\}^{k-1}(\mathbf{I} - v\mathbf{C})^{-1}\right).$$

But the right-hand side of this may be reinterpreted, combinatorially, as the generating function for all $<$-strings consisting of k (not necessarily maximal) $+$-strings, for some $k \geqslant 1$, since $\mathbf{A} = \mathsf{I}(<)$ and $\mathbf{C} = \mathbf{I}(+)$. The successions internal to the $+$-paths are marked by v, and the rises between $+$-strings are marked by y. We use the following construction.

The enumerator for all $+$-strings of length at least 1 is $\sum_{j\geqslant 1} v^{j-1}\gamma_j(+)$. Now the $<$-strings with k $+$-paths consist of a set of k $+$-strings on disjoint elements of N_+, arranged in a unique order. Thus the enumerators for these appear exactly $k!$ times each as the square-free (in $x_1, \ldots$) terms in $(\sum_{j\geqslant 1} v^{j-1}\gamma_j(+))^k$. But since the remaining terms have nonlinear terms in the x_i, they will make no contribution to the linear coefficient that is extracted for permutations. Thus, summing over $k \geqslant 1$ gives

$$\sum_{k\geqslant 1} y^{k-1}\left\{\sum_{j\geqslant 1} v^{j-1}\gamma_j(+)\right\}^k / k! = \left\{-1 + \exp y \sum_{j\geqslant 1} v^{j-1}\gamma_j(+)\right\} y^{-1}$$

as the contribution of sequences with no repeated symbols to

$$\Psi_1\left(\{\mathbf{I} - v\mathbf{C} - y\mathbf{A}\}^{-1}\right),$$

so $c(n, t, u) =$

$$[r^u s^t \mathbf{x}^1]\left\{1 - (r-1)^{-1}\left\{-1 + \exp(r-1)\sum_{j\geq 1}(r(s-1))^{j-1}\gamma_j(+)\right\}\right\}^{-1}.$$

The argument is completed by applying the +-transformation lemma to obtain $c(n, t, u) =$

$$[r^u s^t x^n]\sum_{k\geq 0} k![y^k](r-1)r^{-1}\{1 - r^{-1}\exp\left((r-1)xy\{1 - r(s-1)x\}^{-1}\right)\}^{-1},$$

whence

$$c(n, t, u) = [r^u s^t x^n]\sum_{j, k\geq 1}(r-1)^{k+1}r^{-(j+1)}j^k x^k\{1 - r(s-1)x\}^{-k}.$$

The extraction of the coefficient from this generating function is a routine matter. □

We conclude this section with some instances of q-identities which may be obtained by considering permutations, with repeated pattern, enumerated with respect to inversions. Although a general result may be stated, the class of identities may be appreciated better through specific cases. The identities are obtained by using a left-expansion with elimination of **B**, on one hand, and right-expansion with elimination of **A**, on the other. The resulting generating functions are then equated, having noted that

$$\Psi_2(\mathbf{B}^k) = q^{\binom{k+1}{2}}\frac{x^{k+1}}{(k+1)!_q}.$$

The first identity concerns two q-analogues of the tangent function.

4.3.19. A q-Identity for the Tangent Function

The generating function for the number of <-alternating permutations of odd length with respect to length and number of inversions is $\Psi_2((\mathbf{I} - \mathbf{AB})^{-1})$. In **4.3.7**, using a left-expansion for $\mathbf{I} - \mathbf{AB}$, with elimination of **B**, we obtained

$$\Psi_2\left((\mathbf{I} - \mathbf{AB})^{-1}\right) = \left\{\sum_{k\geq 0}(-1)^k\frac{x^{2k}}{(2k)!_q}\right\}^{-1}\left\{\sum_{k\geq 0}(-1)^k\frac{x^{2k+1}}{(2k+1)!_q}\right\}.$$

On the other hand, a right-expansion for $\mathbf{I} - \mathbf{AB}$, with elimination of $\mathbf{A}$, is

$$\mathbf{I} - \mathbf{AB} = \mathbf{I} + \mathbf{B}^2 - \mathbf{WB} = \mathbf{I} - \mathbf{Q} + \mathbf{L}_1\mathbf{W}\mathbf{R}_1,$$

where $\mathbf{Q} = -\mathbf{B}^2$, $\mathbf{L}_1 = \mathbf{I}$, $\mathbf{R}_1 = -\mathbf{B}$, $\mathbf{C} = \mathbf{I} = \mathbf{R}_0$, $\mathbf{L}_0 = \mathbf{0}$, $\mathbf{D} = \mathbf{I}$. Thus, from the elimination theorem,

$$\Psi_2\left((\mathbf{I} - \mathbf{AB})^{-1}\right) = \left\{1 - \Psi_2\left(\mathbf{B}(\mathbf{I} + \mathbf{B}^2)^{-1}\right)\right\}^{-1}\Psi_2\left((\mathbf{I} + \mathbf{B}^2)^{-1}\right)$$

$$= \left\{\sum_{k\geqslant 0}(-1)^k q^{\binom{2k}{2}}\frac{x^{2k}}{(2k)!_q}\right\}^{-1}\left\{\sum_{k\geqslant 0}(-1)^k q^{\binom{2k+1}{2}}\frac{x^{2k+1}}{(2k+1)!_q}\right\}.$$

Equating the two expressions for $\Psi_2((\mathbf{I} - \mathbf{AB})^{-1})$, we have

$$\frac{\sum_{k\geqslant 0}(-1)^k\frac{x^{2k+1}}{(2k+1)!_q}}{\sum_{k\geqslant 0}(-1)^k\frac{x^{2k}}{(2k)!_q}} = \frac{\sum_{k\geqslant 0}(-1)^k q^{\binom{2k+1}{2}}\frac{x^{2k+1}}{(2k+1)!_q}}{\sum_{k\geqslant 0}(-1)^k q^{\binom{2k}{2}}\frac{x^{2k}}{(2k!)_q}}. \qquad \square$$

Both sides of this identity may be regarded as q-analogues of the tangent function since, when $q = 1$, they reduce to $\tan x$.

4.3.20. A q-Identity from Permutations with Repeated Pattern

We consider permutations with repeated pattern $\pi_1^3\pi_2^2$, where $\pi_1 = <$ and $\pi_2 = \geqslant$, enumerated with respect to length and inversions. The required generating function is $\Psi_2((\mathbf{I} - \mathbf{A}^3\mathbf{B}^2)^{-1})$. A left-expansion or $\mathbf{I} - \mathbf{A}^3\mathbf{B}^2$, with elimination of $\mathbf{B}$, is

$$\mathbf{I} - \mathbf{A}^3\mathbf{B}^2 = \mathbf{I} - \mathbf{A}^5 - \mathbf{A}^3\mathbf{WB} + \mathbf{A}^4\mathbf{W}.$$

It follows from the elimination theorem that

$$\Psi_2\left((\mathbf{I} - \mathbf{A}^3\mathbf{B}^2)^{-1}\right) = \begin{vmatrix} F_0 & F_2 \\ F_4 & F_1 \end{vmatrix} \cdot \begin{vmatrix} F_0 & F_1 \\ F_4 & F_0 \end{vmatrix}^{-1}, \qquad \text{where } F_j = \sum_{k\geqslant 0}\frac{x^{5k+j}}{(5k+j)!_q}.$$

Similarly, a right-expansion for $\mathbf{I} - \mathbf{A}^3\mathbf{B}^2$ with elimination of $\mathbf{A}$ is

$$\mathbf{I} - \mathbf{A}^3\mathbf{B}^2 = \mathbf{I} + \mathbf{B}^5 - \mathbf{A}^2\mathbf{WB}^2 + \mathbf{AWB}^3 - \mathbf{WB}^4.$$

It follows from the elimination theorem, after some routine manipulation, that

$$\Psi_2\left((\mathbf{I} - \mathbf{A}^3\mathbf{B}^2)^{-1}\right) = \begin{vmatrix} G_1 & -G_3 & -G_2 \\ G_3 & G_0 & -G_4 \\ G_4 & G_1 & G_0 \end{vmatrix} \cdot \begin{vmatrix} G_0 & -G_4 & -G_3 \\ G_1 & G_0 & -G_4 \\ G_2 & G_1 & G_0 \end{vmatrix}^{-1},$$

where

$$G_j = \sum_{k \geq 0} (-1)^k q^{\binom{5k+j}{2}} \frac{x^{5k+j}}{(5k+j)!_q}.$$

By equating the two expressions for $\Psi_2((\mathbf{I} - \mathbf{A}^3\mathbf{B}^2)^{-1})$, we obtain the q-identity

$$\frac{\begin{vmatrix} F_0 & F_2 \\ F_4 & F_1 \end{vmatrix}}{\begin{vmatrix} F_0 & F_1 \\ F_4 & F_0 \end{vmatrix}} = \frac{\begin{vmatrix} G_1 & -G_3 & -G_2 \\ G_3 & G_0 & -G_4 \\ G_4 & G_1 & G_0 \end{vmatrix}}{\begin{vmatrix} G_0 & -G_4 & -G_3 \\ G_1 & G_0 & -G_4 \\ G_2 & G_1 & G_0 \end{vmatrix}}.$$ □

NOTES AND REFERENCES

This section is based on Jackson and Goulden (1981b) and Jackson and Goulden (1979); other approaches have been given by Jackson and Aleliunas (1977), Gessel (1977), Stanley (1976), and Viennot (1978).

4.3.15 Carlitz and Scoville (1975); **4.3.17** MacMahon (1915), Stanley (1976); **4.3.18** Roselle (1968); [**4.3.8**] Abramson and Moser (1967), Reilly (1977); [**4.3.10**] Carlitz and Scoville (1972); [**4.3.11**] Abramson (1975); [**4.3.13**] Andrews (1975a), Jackson and Aleliunas (1977); [**4.3.14**] Carlitz and Scoville (1973); [**4.3.16–18**] Jackson, Jeffcott, and Spears (1980); [**4.3.20**] Stanley (1976); [**4.3.21**] Carlitz, Scoville, and Vaughan (1973); [**4.3.24**] Garsia and Gessel (1979); [**4.3.25**] Andrews (1971); [**4.3.26**] Gessel (1977), MacMahon (1915); [**4.3.28–30**] Jackson and Goulden (1981b); [**4.3.29**] Carlitz (1973a); [**4.3.30**] Spears, Jeffcott, and Jackson (1980).

EXERCISES

4.3.1. Prove the maximal string decomposition theorem for a distinguished final string.

4.3.2. Let $E(x) = \sum_{i \geq 1} e_i x^i$ be the length enumerator for initial maximal π_1-strings. Prove that the number of sequences of type $\mathbf{i}$, with an initial maximal π_1-string of length m, a final maximal π_1-string of length l, and k_j other maximal π_1-strings of length j, for $j \geq 1$, is

$$[e_m g_l \mathbf{f}^{\mathbf{k}} \mathbf{x}^{\mathbf{i}}]\{(EF^{-1} \circ \gamma(\pi_1))(F^{-1}G \circ \gamma(\pi_1)) - (F^{-1} \circ \gamma(\pi_1))(EF^{-1}G \circ \gamma(\pi_1))\}(F^{-1} \circ \gamma(\pi_1))^{-1},$$

where F and G are the nonfinal and final maximal π_1-string length enumerators, $\mathbf{k} = (k_1, k_2, \ldots)$.

4.3.3. (a) Show that the number of permutations on N_n with i rises and k modified maxima (see [**3.3.46**]) is

$$\left[r^i u^{k-1} \frac{x^n}{n!}\right]\left(e^{\alpha_1 x} - e^{\alpha_2 x}\right)\left(\alpha_1 e^{\alpha_2 x} - \alpha_2 e^{\alpha_1 x}\right)^{-1},$$

where $\alpha_1 + \alpha_2 = r + 1$, $\alpha_1\alpha_2 = ru$.

(b) Obtain the result of [**3.3.46(c)**] by means of [**4.3.2**].

4.3.4. (a) Prove that, under the conditions of the elimination theorem,

$$\Psi_l\left(\mathbf{R}_k(\mathbf{I} - \mathbf{H})^{-1}\mathbf{D}\right) = |[\mathbf{M}^{(l)} : \mathbf{d}^{(l)}]_{k|}\, |\mathbf{M}^{(l)}|^{-1}$$

where $[\mathbf{M}^{(l)}]_{ij} = \delta_{ij} + \Psi_l(\mathbf{R}_i(\mathbf{I} - \mathbf{Q})^{-1}\mathbf{L}_j)$ for $i, j = 1, \ldots, s$, and $d_i^{(l)} = \Psi_l(\mathbf{R}_i(\mathbf{I} - \mathbf{Q})^{-1}\mathbf{D})$ for $i = 1, \ldots, s$. Note that rows and columns are indexed from 1.

(b) Show that $\Psi_l((\mathbf{S} - \mathbf{TW})^{-1}\mathbf{U}) = \Psi_l(\mathbf{S}^{-1}\mathbf{U})\{1 - \Psi_l(\mathbf{S}^{-1}\mathbf{T})\}^{-1}$ for $l = 1, 2$, and $\mathbf{S}, \mathbf{T}, \mathbf{U}$ are arbitrary incidence matrices for which $\mathbf{S}^{-1}$ exists.

(c) Show that $1 + \Psi_l((\mathbf{S} - \mathbf{TW})^{-1}\mathbf{T}) = \{1 - \Psi_l(\mathbf{S}^{-1}\mathbf{T})\}^{-1}$ for $l = 1, 2$ and $\mathbf{S}, \mathbf{T}$ are arbitrary incidence matrices for which $\mathbf{S}^{-1}$ exists.

4.3.5. (a) By considering $<$-alternating permutations, show that

$$\left[\frac{x^{2n}}{(2n)!}\right]\sec x = \left\|\binom{2(n-i+1)}{2(j-i+1)}\right\|_{n\times n}.$$

(b) Show that

$$\left[\frac{x^{kn}}{(kn)!_q}\right]\left\{\sum_{i\geqslant 0}(-1)^i \frac{x^{ki}}{(ki)!_q}\right\}^{-1} = \left\|\binom{k(n-i+1)}{k(j-i+1)}_q\right\|_{n\times n}.$$

4.3.6. Suppose that $\pi_1^{p_1-1}\pi_2\pi_1^{p_2-1}\pi_2 \cdots \pi_1^{p_m-1} = \pi_2^{q_l-1}\pi_1\pi_2^{q_{l-1}-1}\pi_1 \cdots \pi_2^{q_2-1}\pi_1\pi_2^{q_1-1}$. Let $a_i = \sum_{j=1}^{i} p_j$, $a_0 = 0$, and $b_i = \sum_{j=1}^{i} q_j$, $b_0 = 0$. Show that

$$\left\|\binom{n - a_{i-1}}{a_j - a_{i-1}}\right\|_{m\times m} = \left\|\binom{n - b_{i-1}}{b_j - b_{i-1}}\right\|_{l\times l}.$$

4.3.7. (a) Let $c(j, k, m, \mathbf{i})$ be the number of sequences of type $\mathbf{i}$ with j rises, k falls, and m levels. Show that, if $\mathbf{A} = \mathbf{I}(<)$,

$$c(j, k, m, \mathbf{i}) = \left[r^j f^k l^m \mathbf{x}^{\mathbf{i}}\right]\Psi_1\left((\mathbf{I} - (r\mathbf{A} + l\mathbf{X} + f(\mathbf{W} - \mathbf{X} - \mathbf{A})))^{-1}\right).$$

(b) Hence derive [**2.4.17(a)**].

4.3.8. (a) Let G_m be the set of sequences on the alphabet $\{\pi_1, \ldots, \pi_m\}$. Show that $\mathsf{G}_m = \mathsf{G}_{m-1}(\pi_m \pi_m^*(\mathsf{G}_{m-1} - \mathsf{G}_0))^* \pi_m^*$, $m \geqslant 1$, where $\mathsf{G}_0 = \{\varepsilon\}$ and ε is the empty string.

(b) Show that the number of permutations on N_n with i increasing $+$-strings of length p and j decreasing $+$-strings of length s is

$$[u^i v^j \mathbf{x}^{\mathbf{1}}] \Psi_1\{\mathbf{G}(\mathbf{I} - (\mathbf{V} - \mathbf{I})(\mathbf{G} - \mathbf{I}))^{-1}\mathbf{V}\},$$

where

$$\mathbf{V} = (\mathbf{I} - \mathbf{B})^{-1}(\mathbf{I} - v\mathbf{B})^{-1}(\mathbf{I} - v\mathbf{B} + (v - 1)\mathbf{B}^{s-1}), \qquad \mathbf{G} = (\mathbf{I} - \mathbf{UC})^{-1}\mathbf{U},$$

in which

$$\mathbf{U} = (\mathbf{I} - \mathbf{A})^{-1}(\mathbf{I} - u\mathbf{A})^{-1}(\mathbf{I} - u\mathbf{A} + (u - 1)\mathbf{A}^{p-1})$$

and

$$\mathbf{A} = \mathsf{I}(+), \mathbf{B} = \mathbf{A}^T, \mathbf{C} = \mathbf{W} - \mathbf{A} - \mathbf{B}.$$

(c) Hence show that the required number is

$$[u^i v^j x^n] \sum_{k \geqslant 0} k! x^k \left\{ \frac{1 - ux + (u - 1)x^{p-1}}{1 - ux + (u - 1)x^p} + \frac{1 - vx + (v - 1)x^{s-1}}{1 - vx + (v - 1)x^s} - 1 \right\}^k.$$

(d) Replacing "$+$-strings" by "$\oplus$-strings" in (b), show that the corresponding number is

$$[u^i v^j x^n] n \sum_{k \geqslant 1} (k - 1)! x^k$$
$$\times \left\{ \frac{1 - ux + (u - 1)x^{p-1}}{1 - ux + (u - 1)x^p} + \frac{1 - vx + (v - 1)x^{s-1}}{1 - vx + (v - 1)x^s} - 1 \right\}^k.$$

4.3.9. (a) Let π_3 be the set of levels and $\Pi = \{\pi_1, \pi_2\}$ be a bipartition of N_n^2 such that $\pi_3 \subseteq \pi_2$. Show that the number of sequences of type $\mathbf{i}$ with j occurrences of π_3 and k occurrences of π_1 is

$$[l^j u^k \mathbf{z}^{\mathbf{i}}] \left\{ 1 - \sum_{i \geqslant 1} (u - 1)^{i-1} \gamma_i\left(\pi_1, \mathbf{z}(1 - (l - 1)\mathbf{z})^{-1}\right) \right\}^{-1},$$

where

$$\mathbf{z}(1 - (l - 1)\mathbf{z})^{-1} = \left(z_1(1 - (l - 1)z_1)^{-1}, z_2(1 - (l - 1)z_2)^{-1}, \ldots\right),$$

and $\gamma_i(\pi_1, \mathbf{x}) = \gamma_i(\pi_1)$.

(b) Obtain the result of [**4.3.7**(b)] by specializing part (a) to $\pi_1 = <$.

4.3.10. Show that the number of sequences $\sigma_1\sigma_2\ldots$ of type $\mathbf{i}$ such that $\sigma_1 < \sigma_2 > \sigma_3 < \cdots$ is

$$[\mathbf{x}^{\mathbf{i}}]\left\{1 + \sum_{k\geq 0} (-1)^k t_{2k+1}\right\}\left\{\sum_{k\geq 0} (-1)^k t_{2k}\right\}^{-1},$$

where $t_i = \sum_{\sigma_1<\sigma_2\leqslant\sigma_3<\sigma_4\leqslant\cdots} x_{\sigma_1} \cdots x_{\sigma_i}$, $t_0 = 1$.

4.3.11. Show that the number of permutations on N_n with k inversions and pattern $\pi_1^{p_1-1}\pi_2 \cdots \pi_1^{p_m-1}$, for $\pi_1 = <$ and $p_1 + \cdots + p_m = n$, is

$$[q^k] \sum_{i=1}^{m} (-1)^{m-i} \sum_{\alpha} \begin{bmatrix} n \\ s_1,\ldots, s_i \end{bmatrix}_q,$$

where the summation is over $\alpha = \{\alpha_1,\ldots,\alpha_i\} \subseteq \mathsf{N}_m$, with $\alpha_1 < \cdots < \alpha_i = m$, $s_j = a_{\alpha_j} - a_{\alpha_{j-1}}$, and $a_t = \sum_{l=1}^{t} p_l$, $a_{\alpha_0} = 0$.

4.3.12. (a) Show that the number of sequences of type $\mathbf{i}$ with $m_j(k_j)$ occurrences of j as the terminator (initiator) of a maximal π_1-string is $[\mathbf{y}^{\mathbf{m}}\mathbf{z}^{\mathbf{k}}\mathbf{x}^{\mathbf{i}}]\text{trace}\,\mathbf{Z}(\mathbf{I} - \mathbf{A} - \mathbf{YBZ})^{-1}\mathbf{YW}$, where $\mathbf{Y} = \text{diag}(y_1, y_2,\ldots)$, $\mathbf{Z} = \text{diag}(z_1, z_2,\ldots)$, $\mathbf{y} = (y_1, y_2,\ldots)$, $\mathbf{z} = (z_1, z_2,\ldots)$, $\mathbf{m} = (m_1, m_2,\ldots)$, and $\mathbf{k} = (k_1, k_2,\ldots)$.

(b) Show that the number of sequences of type $\mathbf{i}$ with m_j occurrences of j as the terminator of a maximal π_1-string is

$$[\mathbf{y}^{\mathbf{m}}\mathbf{x}^{\mathbf{i}}]\{1 - \text{trace}(\mathbf{I} - (\mathbf{I} - \mathbf{Y})\mathbf{A})^{-1}\mathbf{YW}\}^{-1}.$$

4.3.13. (a) Show that the number of sequences of type $\mathbf{i}$ with m_l occurrences of l as the terminator of a maximal $<$-string, for $l \geqslant 1$, is

$$[\mathbf{y}^{\mathbf{m}}\mathbf{x}^{\mathbf{i}}]\left\{1 - \sum_{j\geqslant 1} x_j y_j \prod_{k=1}^{j-1} (1 + x_k(1 - y_k))\right\}^{-1}.$$

(b) If paths are nondecreasing, show that the required number is

$$[\mathbf{y}^{\mathbf{m}}\mathbf{x}^{\mathbf{i}}]\left\{1 - \sum_{j\geqslant 1} x_j y_j \prod_{k=1}^{j} (1 - x_k(1 - y_k))^{-1}\right\}^{-1}.$$

4.3.14. Show that the number of permutations on N_n with i_2, i_1 rises beginning in positions congruent to 0, 1(mod 2) and j_2, j_1 falls beginning in positions congruent to 0, 1(mod 2) is $[r_1^{i_1}r_2^{i_2}f_1^{j_1}f_2^{j_2}x^n/n!]F$, where

$$F = \{\alpha \sinh(\alpha x) + (r_1 + f_1)(\cosh(\alpha x) - 1)\}$$
$$\times \{r_1 r_2 + f_1 f_2 - (f_1 r_2 + r_1 f_2)\cosh(\alpha x)\}^{-1}$$

and $\alpha^2 = (r_1 - f_1)(r_2 - f_2)$.

4.3.15. Show that the number of permutations on N_n with m inversions and pattern in $(\pi_1^{p_1-1}\pi_2 \cdots \pi_1^{p_s-1}\pi_2)^*\pi_1^{r-1}$, where $\pi_1 = <$, is $[q^m x^n/n!_q]|[\mathbf{M}:\mathbf{d}]_1|\,|\mathbf{M}|^{-1}$, where

$$\mathbf{d} = (d_1, \ldots, d_s)^T, \qquad d_i = \phi_{(r+a_1-a_i)}^{(a_1, s)}(x), \qquad i = 1, \ldots, s,$$

$$[\mathbf{M}]_{ij} = (-1)^{j+1+s\zeta(i,j)}\phi_{(a_j-a_i) \bmod a_1}^{(a_1, s)}(x), \qquad i, j = 1, \ldots, s,$$

in which

$$\phi_j^{(k,l)}(x) = \sum_{i \geq 0} (-1)^{li} \frac{x^{ki+j}}{(ki+j)!_q},$$

$$a_i = \sum_{j=1}^{s+1-i} p_j, \qquad i = 1, \ldots, s, \qquad \text{and } \zeta(i,j) = \begin{cases} 1 & \text{if } i < j \\ 0 & \text{if } i \geq j. \end{cases}$$

4.3.16. A (π_1, π_2)**-structure** of type (i, j), $i, j \geq 1$, is a sequence with pattern $\pi_1^i \pi_2^j$.

(a) Show that the number of sequences of type $\mathbf{i}$, with k_{ij} maximal (π_1, π_2)-structures of type (i, j), for $i, j \geq 1$, an initial maximal π_2-string of length $r \geq 1$, and a final maximal π_1-string of length $v \geq 1$ is

$$\left[t_r u_v \left(\prod_{i,j \geq 1} g_{ij}^{k_{ij}}\right)\mathbf{x}^{\mathbf{i}}\right]\Psi_1\left(\sum_{r \geq 1} t_r \mathbf{F}_r\right),$$

where

$$\mathbf{F}_r = \mathbf{B}^{r-1}\left(\mathbf{I} - \sum_{i,j \geq 1} g_{ij}\mathbf{A}^i\mathbf{B}^j\right)^{-1}\left(\sum_{v \geq 1} u_v \mathbf{A}^{v-1}\right).$$

(b) Show that $\Psi_1(\mathbf{F}_r) = |[\mathbf{M}:\mathbf{d}]_{r-1}|\,|\mathbf{M}|^{-1}$, where

$$\mathbf{d} = (d_0, d_1, \ldots), \qquad d_i = \left(x^{i+1}(1 - Q(x))^{-1}D(x)\right) \circ \gamma(\pi_1), \qquad i \geq 0,$$

$$[\mathbf{M}]_{ij} = \left((-1)^j x^{i-j}\zeta(i,j) + \sum_{l \geq j+1} (-1)^{l+j} x^{i+l-j}(1 - Q(x))^{-1} G_l(x)\right) \circ \gamma(\pi_1)$$

for $i, j \geq 0$, where $D(x) = \sum_{i \geq 1} u_i x^{i-1}$, $G_j(x) = \sum_{i \geq 1} g_{ij} x^i$,

$$Q(x) = \sum_{i \geq 1} (-1)^i G_i(x) x^i \qquad \text{and} \qquad \zeta(i,j) = \begin{cases} 1, & 0 \leq j \leq i \\ 0, & \text{otherwise.} \end{cases}$$

(c) Show that the generating function in part (b) may be expressed as a ratio of $k \times k$ determinants if $r \leq k$ and if $G_j(x) = 0$ for $j > k$.

4.3.17. (a) Show that the number of sequences of type $\mathbf{i}$, with l maximal π_1-strings of length p and m maximal π_2-strings of length s, for $p, s \geqslant 2$, is

$$[y^l z^m \mathbf{x}^{\mathbf{i}}]\Psi_1\left\{(\mathbf{I} - z\mathbf{B} + (z-1)\mathbf{B}^s)\left\{I - \sum_{i\geqslant 0} \mathbf{G}_i \mathbf{B}^i\right\}^{-1}(\mathbf{I} - y\mathbf{A} + (y-1)\mathbf{A}^p)\right\}$$

where

$$\mathbf{G}_0 = (y+1)\mathbf{A} - y\mathbf{A}^2, \qquad \mathbf{G}_2 = -z\mathbf{I} + yz\mathbf{A} - (y-1)z\mathbf{A}^p,$$

$$\mathbf{G}_1 = (z+1)\mathbf{I} - (yz+y+z)\mathbf{A} + yz\mathbf{A}^2 + (y-1)\mathbf{A}^p,$$

$$\mathbf{G}_s = (z-1)\mathbf{A} - y(z-1)\mathbf{A}^2 + (y-1)(z-1)\mathbf{A}^p, \qquad \text{all other } \mathbf{G}_i = 0.$$

(b) Thus, show that the preceding generating function may be expressed as the sum of 3 $(s+1)\times(s+1)$ determinants divided by another $(s+1)\times(s+1)$ determinant.

4.3.18. Show that the number of sequences of type $\mathbf{i}$ with no maximal (π_1, π_2)-structures of type (2, 2) is $[\mathbf{x}^{\mathbf{i}}]|\mathbf{M}_1|\,|\mathbf{M}_2|^{-1}$, where

$$\mathbf{M}_1 = \begin{bmatrix} xf & (x^3-x^5)f & (x^4-x^3)f \\ x^2 f & (x^4-x^6)f-1 & (x^5-x^4)f \\ x^3 f & (x^5-x^7)f-x & (x^6-x^5)f+1 \end{bmatrix} \circ \gamma(\pi_1)$$

$$\mathbf{M}_2 = \begin{bmatrix} (1-x)f & (x^3-x^5)f & (x^4-x^3)f \\ (x-x^2)f & (x^4-x^6)f-1 & (x^5-x^4)f \\ (x^2-x^3)f & (x^5-x^7)f-x & (x^6-x^5)f+1 \end{bmatrix} \circ \gamma(\pi_1)$$

and $f = (1 + x^4 - x^6)^{-1}$.

4.3.19. Let Φ be the generating function for sequences with f_i marking maximal π_1-strings of length i and x_j marking the occurrence of element j. If the f_j's and the x_i's are noncommutative markers that commute with each other, show that

$$1 + \Phi = \sum_{k\geqslant 0}(-1)^k\left\{\sum_{i\geqslant 1}(1-F)^i \circ \gamma(\pi_1)\right\}^k$$

where $F(x) = 1 + f_1 x + f_2 x^2 + \cdots$, $\gamma(\pi_1) = (\gamma_0(\pi_1), \gamma_1(\pi_1), \dots)$, and $\gamma_i(\pi_1) = \sum_{\sigma_1\cdots\sigma_i \in \langle \pi_1^{i-1}\rangle} x_{\sigma_1}\cdots x_{\sigma_i}$. Hence the maximal string decomposition theorem is true in the noncommutative case.

4.3.20. Consider the permutations $(\sigma_{11}, \dots, \sigma_{1n}), \dots, (\sigma_{k1}, \dots, \sigma_{kn})$ on N_n such that $\sigma_{ij} < \sigma_{i,j+1}$ for all $i = 1, \dots, k$ and all j except $j = a_1, a_2, \dots, a_{m-1}$, where

$0 = a_0 < a_1 < \cdots < a_{m-1} < a_m = n$. Show that the number of such sets of permutations with t_l inversions in permutation l, for $l = 1, \ldots, k$, is

$$[q_1^{t_1} \cdots q_k^{t_k}] \left\| \prod_{l=1}^{k} \binom{n - a_{i-1}}{a_j - a_{i-1}}_{q_l} \right\|_{m \times m}.$$

4.3.21. Show that the number of sequences $\sigma_1 \cdots \sigma_{2m}$ of type **i** on N_{2n}, with s rises, t levels, and u falls, with $\sigma_i + \sigma_{2m+1-i} = 2n + 1$ for $i = 1, \ldots, m$, is

$$[r^s l^t f^u \mathbf{z}^{\mathbf{i}}] \frac{\prod_{i=1}^{n}\{1 + (r^2 - l^2) z_i z_{2n+1-i}\} - \prod_{i=1}^{n}\{1 + (f^2 - l^2) z_i z_{2n+1-i}\}}{r\prod_{i=1}^{n}\{1 + (f^2 - l^2) z_i z_{2n+1-i}\} - f\prod_{i=1}^{n}\{1 + (r^2 - l^2) z_i z_{2n+1-i}\}}.$$

4.3.22. Let $\mu \in \{\leqslant, >\}^*$ and $\Phi(x) = \Psi_2(\mathsf{I}(\mu))$. Show that the number of compositions of m with n parts and pattern μ is $[q^m x^n]\Phi(qx(1-q)^{-1})$, and the number of permutations on N_n with m inversions and pattern μ is $[q^m x^n / n!_q]\Phi(x)$.

4.3.23. (a) Let $\mathsf{P}, \mathsf{P}_1, \mathsf{P}_2$ be sets of permutations such that

$$\mathsf{P} = \left\{ \sigma'_\alpha \sigma''_\beta \,|\, \sigma' \in \mathsf{P}_1, \quad \sigma'' \in \mathsf{P}_2, \alpha \dot{\cup} \beta = \mathsf{N}_k \text{ for some } k \geqslant 0 \right\},$$

in the notation of Section 3.2. If

$$\Phi_q(x, \mathsf{A}) = \sum_{\sigma \in \mathsf{A}} q^{I(\sigma)} \frac{x^{|\sigma|}}{|\sigma|!_q}$$

for a set of permutations A, show that

$$\Phi_q(x, \mathsf{P}) = \Phi_q(x, \mathsf{P}_1)\Phi_q(x, \mathsf{P}_2).$$

This is a **product lemma** for Eulerian generating functions.

(b) From Decomposition 2.4.19 and part (a), show that the number of permutations on N_n with k inversions and l falls is

$$\left[q^k t^l \frac{x^n}{n!_q}\right] \left\{ \sum_{i \geqslant 1} \frac{(x(1-t))^i}{i!_q} \right\} \left\{ 1 - t \sum_{i \geqslant 0} \frac{(x(1-t))^i}{i!_q} \right\}^{-1}.$$

4.3.24. (a) Let U be the set of sequences in N_+^* with no falls. In the notation of Decomposition 2.4.19, show that $(\mathsf{U} - \{\varepsilon\})0(\mathsf{U}0)^* \xrightarrow{\sim} \{\mathsf{N}_+^+ \circ (00^* \times 0^*)\}00^*$.

(b) If the falls in a sequence $\sigma = \sigma_1 \cdots \sigma_{i_1 + \cdots + i_k}$ are in positions i_1, $i_1 + i_2, \ldots, i_1 + \cdots + i_{k-1}$, for $i_1, \ldots, i_k \geqslant 1$, show that the major index of σ (see [**2.6.7**]) is $m(\sigma) = (k-1)i_1 + (k-2)i_2 + \cdots + i_{k-1}$.

(c) Show that the number of permutations on N_n with k inversions, l falls, and major index u is

$$\left[q^k t^l p^u z^n / n!_q\right]\Pi_{i=1}^n (1 - tp^i)\Sigma_{m \geq 1} t^{m-1} \Pi_{j=1}^{m-1}\left(E_q(zp^j) - \delta_{j,m-1}\right),$$

where $E_q(x) = \Sigma_{i \geq 0} x^i / i!_q$.

4.3.25. (a) Let $P(x) = \Psi_2((\mathbf{I} - \mathbf{AB})^{-1})$ and $Q(x) = 1 + \Psi_2((\mathbf{I} - \mathbf{AB})^{-1}\mathbf{A})$. Show from [**4.3.23**(a)] that $\mathsf{E}_q(P(x) - x) = P(x)P(qx)$ and $\mathsf{E}_q(Q(x) - 1) = P(x)Q(qx)$, where

$$\mathsf{E}_q\left(\sum_{i \geq 0} c_i(q)\frac{x^i}{i!_q}\right) = \sum_{i \geq 1} c_i(q)\frac{x^{i-1}}{(i-1)!_q}.$$

E_q is called the **formal Eulerian differential operator**.

(b) Show that

$$\mathsf{E}_q F(x) = (x - qx)^{-1}(F(x) - F(qx))$$

$$\mathsf{E}_q(F(x)G(x)) = F(x)\mathsf{E}_q G(x) + \{\mathsf{E}_q F(x)\}G(qx)$$

$$\mathsf{E}_q\left(F(x)^{-1}\right) = -\{\mathsf{E}_q F(x)\}(F(x)F(qx))^{-1}.$$

(c) Let $P(x) = -H(x)^{-1}\mathsf{E}_q H(x)$, where $H(0) = 1$. Show that $H(x) = \Sigma_{k \geq 0}(-1)^k x^{2k}/(2k)!_q$, and hence

$$P(x) = \left\{\sum_{k \geq 0}(-1)^k \frac{x^{2k+1}}{(2k+1)!_q}\right\}\left\{\sum_{k \geq 0}(-1)^k \frac{x^{2k}}{(2k)!_q}\right\}^{-1}.$$

(d) Show that $Q(x) = \left\{\sum_{k \geq 0}(-1)^k \frac{x^{2k}}{(2k)!_q}\right\}^{-1}$.

4.3.26. (a) Let $L_1 \geq \cdots \geq L_k \geq 1$ and $U_1 \geq \cdots \geq U_k \geq 1$. Show that the number of sequences $\sigma = \sigma_1 \cdots \sigma_{a_m}$ of type $\mathbf{i}$, with pattern $\pi_1^{p_1 - 1}\pi_2 \cdots \pi_1^{p_m - 1}$, $\pi_1 = \leq$, and $L_i \leq \sigma_i \leq U_i$ for $i = 1, \ldots, a_m$, is

$$[\mathbf{x}^{\mathbf{i}}]\|\theta_{a_j - a_{i-1}}\left(L_{a_{i-1}+1}, U_{a_j}\right)\|_{m \times m},$$

where $a_i = \Sigma_{j=1}^i p_j$, $a_0 = 0$, $\theta_k(L, U) = [z^k]\Pi_{i=L}^U (1 - zx_i)^{-1}$. (This is a special case of **5.4.5**.)

(b) Show that the number of sequences $\sigma = \sigma_1 \cdots \sigma_{a_m}$ with pattern $\pi_1^{p_1 - 1}\pi_2 \cdots \pi_1^{p_m - 1}$, $\pi_1 = \leq$, and $L_i \leq \sigma_i \leq U_i$ for $i = 1, \ldots, a_m$, is

$$\left\| \begin{pmatrix} U_{a_j} - L_{a_{i-1}+1} + a_j - a_{i-1} \\ a_j - a_{i-1} \end{pmatrix} \right\|_{m \times m}.$$

4.3.27. (a) If $\mathbf{H} = \mathbf{Q} - \Sigma_{i=1}^{t} \mathbf{L}_i \mathbf{W} \mathbf{R}_i$, show that $\Psi_2(\mathbf{C}(\mathbf{I} - \mathbf{H})\mathbf{L}_1) = |[\mathbf{N} : \mathbf{c}]_1| \, |\mathbf{N}|^{-1}$, where $[\mathbf{N}]_{ij} = \delta_{ij} + \Psi_2(\mathbf{R}_j(\mathbf{I} - \mathbf{H})^{-1}\mathbf{L}_i)$, $i, j = 1, \dots, t$, and $\mathbf{c} = (c_1, \dots, c_t)^T$, $c_i = \Psi_2(\mathbf{C}(\mathbf{I} - \mathbf{H})^{-1}\mathbf{L}_i)$, $i = 1, \dots, t$.

(b) Show that the number of permutations on N_n with m inversions and pattern in $\pi_2^{u-1}(\pi_1\pi_2^{r_t-1} \cdots \pi_1\pi_2^{r_1-1})^*$, where $\pi_1 = <$, is

$$\left[q^m \frac{x^n}{n!_q} \right] |[\mathbf{N} : \mathbf{c}]_1| \cdot |\mathbf{N}|^{-1},$$

where $\mathbf{c} = (c_1, \dots, c_t)^T$, $c_i = \psi_{(u+b_1-b_i)}^{(b_1, t)}(x)$, $i = 1, \dots, t$,

$$[\mathbf{N}]_{ij} = (-1)^{j+1+t\zeta(i,j)} \psi_{(b_j - b_i) \bmod b_1}^{(b_1, t)}(x), \qquad i, j = 1, \dots, t,$$

in which $\psi_j^{(k,l)}(x) = \Sigma_{i \geq 0} (-1)^{li} q^{\binom{ki+j}{2}} \dfrac{x^{ki+j}}{(ki+j)!_q}$, $b_i = \Sigma_{j=1}^{t+1-i} r_j$, $i = 1, \dots, t$,

and

$$\zeta(i,j) = \begin{cases} 1 & \text{if } i < j \\ 0 & \text{if } i \geq j. \end{cases}$$

(c) Suppose that $\pi_1^{p_1-1}\pi_2 \cdots \pi_1^{p_s-1}\pi_2 = \pi_1\pi_2^{r_t-1} \cdots \pi_1\pi_2^{r_1-1}$. By comparing part (b) and [**4.3.15**], show that $|[\mathbf{M} : \mathbf{d}]_1| \cdot |\mathbf{M}|^{-1} = |[\mathbf{N} : \mathbf{c}]_1| \cdot |\mathbf{N}|^{-1}$, where $\mathbf{d} = (d_1, \dots, d_s)^T$, $\mathbf{c} = (c_1, \dots, c_t)^T$, $d_i = \phi_{1+h-a_i}^{(h,s)}(x)$, $c_i = \psi_{1+h-b_i}^{(h,s)}(x)$,

$$[\mathbf{M}]_{ij} = (-1)^{s\zeta(i,j)} \phi_{(a_j - a_i) \bmod h}^{(h,s)}(x), \qquad i, j = 1, \dots, s,$$

$$[\mathbf{N}]_{ij} = (-1)^{t\zeta(i,j)} \psi_{(b_j - b_i) \bmod h}^{(h,t)}(x), \qquad i, j = 1, \dots, t,$$

in which

$$\phi_j^{(k,l)}(x) = \sum_{i \geq 0} (-1)^{li} \frac{x^{ki+j}}{(ki+j)!_q},$$

$$\psi_j^{(k,l)}(x) = \sum_{i \geq 0} (-1)^{li} q^{\binom{ki+j}{2}} \frac{x^{ki+j}}{(ki+j)!_q},$$

$$a_i = \sum_{j=1}^{s+1-i} p_j, \qquad b_i = \sum_{j=1}^{t+1-i} r_j,$$

$$h = a_1 = b_1 \qquad \text{and} \qquad \zeta(i,j) = \begin{cases} 1 & \text{if } i < j \\ 0 & \text{if } i \geq j. \end{cases}$$

4.3.28. Show that the number of permutations $\sigma_1 \cdots \sigma_{2n+1}$ on N_{2n+1} with s rises and t occurrences of $\sigma_{2i-1} < \sigma_{2i+1}$, $i = 1, \ldots, n$, is

$$\left[r^s u^t \frac{x^{2n+1}}{(2n+1)!}\right]\left\{(1-g_1)^2 - (x+g_2)g_0\right\}^{-1} g_0,$$

where $g_m(x) =$

$$(r-1)^m \sum_{k\geqslant 0} \sum_{i=0}^{k} \binom{k}{i} \{r(u-1)\}^i \{(ru-1)(r-1)\}^{k-i} \frac{x^{2k+m+1}}{(2k+m-i+1)!}.$$

4.3.29. Show that the number of $>$-alternating permutations $\sigma_1 \cdots \sigma_{2n+1}$ on N_{2n+1} with k occurrences of $\sigma_{2i-1} < \sigma_{2i+1}$, $i = 1, \ldots, n$, and m inversions is

$$\left[u^k q^m \frac{x^{2n+1}}{(2n+1)!_q}\right] F_0 (1 - F_1)^{-1}$$

where

$$F_l = \sum_{i\geqslant 0} \frac{x^{2i+l+1}}{(2i+l+1)!_q} \prod_{j=0}^{i-1} \left\{(u-1)q\binom{2j+l+1}{1}_q - 1\right\}.$$

4.3.30. Show that the number of $<$-alternating permutations $\sigma_1 \cdots \sigma_{4n+3}$ on N_{4n+3} with $\sigma_2 < \sigma_4 > \sigma_6 < \cdots < \sigma_{4n} > \sigma_{4n+2}$ and m inversions is

$$\left[q^m \frac{x^{4n+3}}{(4n+3)!_q}\right]\left\{\sum_{i\geqslant 0} (-1)^i c_i \frac{x^{4i+3}}{(4i+3)!_q}\right\}\left\{1 - \sum_{i\geqslant 0} (-1)^i c_i \frac{x^{4i+4}}{(4i+4)!_q}\right\}^{-1},$$

where

$$c_i = q^{i+1}\binom{4i+2}{1}_q \prod_{j=0}^{i-1} \binom{4j+2}{1}_q \binom{4j+5}{1}_q.$$

4.4. THE LOGARITHMIC CONNECTION FOR CIRCULAR PERMUTATIONS

So far this chapter has been concerned with linearly ordered sequences. We now consider the enumeration of circular sequences.

4.4.1. Definition (Circular Sequence)

Let $\mu = \pi_{i_1} \cdots \pi_{i_p} \in \{\pi_1, \pi_2\}^*$. *A sequence* $\sigma = \sigma_1 \cdots \sigma_p$ *over* N_n *is said to be a* **circular sequence with circular pattern** μ *if* $(\sigma_1, \sigma_2) \in \pi_{i_1}$, $(\sigma_2, \sigma_3) \in$

$\pi_{i_2}, \ldots, (\sigma_{p-1}, \sigma_p) \in \pi_{i_{p-1}}, (\sigma_p, \sigma_1) \in \pi_{i_p}$. *Two (circular) patterns or circular sequences are* **indistinguishable** *if one can be cyclically rotated into the other.* □

For example, the circular sequence $\sigma = 1\ 2\ 5\ 3\ 1\ 6\ 9$ has (circular) pattern $\sigma = \pi_1^2\pi_2^2\pi_1^2\pi_2$, where $\pi_1 = <$ and $\pi_2 = \geqslant$. The circular sequence $\sigma' = 3\ 1\ 6\ 9\ 1\ 2\ 5$ is indistinguishable from σ. Clearly, we may decompose circular sequences into maximal π_1-strings, except for circular sequences with circular pattern in π_1^+. The latter circular sequences are called π_1-**cycles**. Thus $(1\ 2\ 5), (3), (1\ 6\ 9)$ are the maximal $<$-strings of the circular sequence $3\ 1\ 6\ 9\ 1\ 2\ 5$. The circular sequence $1\ 2\ 3\ 4$ is a $\neq$-cycle. If no circular sequence over N_+ is a π_1-cycle, then π_1 is said to be **cycle-free**. For example, $<$ is cycle-free.

The purpose of this section is to examine the relationship between the enumeration of linearly ordered sequences, given in the earlier sections of this chapter, and the enumeration of circular sequences. This relationship is called the **logarithmic connection.** It can be used to extend all the general enumerative theorems in the earlier sections of this chapter.

We wish to enumerate circular permutations with pattern in μ^* where $\mu \in \{\pi_1, \pi_2\}^*$. The number of such permutations on N_n is clearly

$$[\mathbf{x}^1]\sum_{i=1}^{n}\sum_{\alpha\in\mu^*}[\mathsf{I}(\alpha)]_{ii}\,|\text{aut}\,\alpha|^{-1}$$

since each permutation begins with some $i \in \mathsf{N}_n$. Suppose now that μ is a pattern that cannot be rotated on to itself cyclically (except by the identity) so $|\text{aut}\,\mu| = 1$. Such patterns are called **prime**. It follows that $|\text{aut}(\mu^k)| = k$. Thus, the preceding expressing reduces to

$$[\mathbf{x}^1]\sum_{i=1}^{n}\sum_{k\geqslant 1}k^{-1}\left[(\mathsf{I}(\mu))^k\right]_{ii} = [\mathbf{x}^1]\,\text{trace}\log\{\mathbf{I} - \mathsf{I}(\mu)\}^{-1}.$$

If $|\text{aut}\,\mu| = g$, then $|\text{aut}\,\mu^k| = kg$, and in this case we need only divide the preceding expression, for the number of permutations with pattern in μ^*, by g. More generally, for circular sequences σ with prescribed pattern, there are additional automorphisms to be calculated, namely, aut σ. This is accounted for in the obvious way by Pólya's theorem, in addition to the preceding argument, and an example of this is given as an exercise in **[4.4.16]**. If σ is a circular permutation, then, of course, $|\text{aut}\,\sigma| = 1$, and the difficulty does not arise.

A **prime set** of patterns H is a set of primes such that each element in H^* is uniquely expressible as a concatenation of patterns in H. Clearly, if H is a prime set, then the number of permutations on N_n with pattern in H^* is

$$[\mathbf{x}^1]\,\text{trace}\log(\mathbf{I} - \mathsf{I}(\mathsf{H}))^{-1}.$$

Thus the enumerative task in this section is to evaluate $\text{trace}\log(\mathbf{I} - \mathsf{I}(\mathsf{H}))^{-1}$.

This is carried out with a factored expansion for $\mathbf{H} = \mathsf{I}(\mathsf{H})$ in the following theorem, which may be regarded as the circular counterpart of the elimination theorem. Indeed, we may consider trace $\log(\mathbf{I} - \mathbf{H})^{-1}$ to be the **circular version** of the linear problem $\Psi_1(\mathbf{C}(\mathbf{I} - \mathbf{H})^{-1}\mathbf{D})$ for permutations, when H is a prime set of patterns.

4.4.2. Theorem (Logarithmic Connection)

Let $\mathbf{H}, \mathbf{Q} \in \mathsf{M}_n(\mathsf{R}[[\mathbf{x}]]_0)$ *and* $\mathbf{L}_j, \mathbf{R}_j \in \mathsf{M}_n(\mathsf{R}[[\mathbf{x}]])$ *for* $j = 1, \ldots, s$. *If* $\mathbf{H} = \mathbf{Q} - \sum_{k=1}^{s} \mathbf{L}_k \mathbf{W} \mathbf{R}_k$, *then*

1. $|\mathbf{I} - \mathbf{H}|^{-1} = |\mathbf{I} - \mathbf{Q}|^{-1} \cdot |\mathbf{M}|^{-1}$.
2. $\operatorname{trace} \log(\mathbf{I} - \mathbf{H})^{-1} = \log|\mathbf{M}|^{-1} + \operatorname{trace} \log(\mathbf{I} - \mathbf{Q})^{-1}$,

where $[\mathbf{M}]_{ij} = \delta_{ij} + \operatorname{trace} \mathbf{R}_i(\mathbf{I} - \mathbf{Q})^{-1}\mathbf{L}_j\mathbf{W}$ *for* $1 \leqslant i, j \leqslant s$.

Proof: Now from the factored expansion for $\mathbf{H}$ we have

$$\mathbf{I} - \mathbf{H} = \mathbf{I} - \mathbf{Q} + \sum_{k=1}^{s} \mathbf{L}_k \mathbf{W} \mathbf{R}_k = (\mathbf{I} - \mathbf{Q})\left\{\mathbf{I} + (\mathbf{I} - \mathbf{Q})^{-1} \sum_{k=1}^{s} \mathbf{L}_k \mathbf{W} \mathbf{R}_k\right\}$$

whence $|\mathbf{I} - \mathbf{H}| = |\mathbf{I} - \mathbf{Q}| \cdot |\mathbf{I} + (\mathbf{I} - \mathbf{Q})^{-1}\sum_{k=1}^{s} \mathbf{L}_k \mathbf{W} \mathbf{R}_k|$. Let $\mathbf{1}$ be the column vector with n 1's. Let $\mathbf{D} = [\mathbf{D}_1 | \cdots | \mathbf{D}_s]$ where $\mathbf{D}_j = (\mathbf{I} - \mathbf{Q})^{-1}\mathbf{L}_j\mathbf{X}\mathbf{1}$ for $1 \leqslant j \leqslant s$, let $\mathbf{F} = [\mathbf{F}_1 | \cdots | \mathbf{F}_s]$ where $\mathbf{F}_i = \mathbf{R}_i^T\mathbf{1}$ for $1 \leqslant i \leqslant s$, and let $\mathbf{X} = \operatorname{diag}(x_1, \ldots, x_n)$. But

$$\sum_{k=1}^{s} (\mathbf{I} - \mathbf{Q})^{-1}\mathbf{L}_k \mathbf{W} \mathbf{R}_k = \sum_{k=1}^{s} \mathbf{D}_k \mathbf{F}_k^T = \mathbf{D}\mathbf{F}^T.$$

Moreover,

$$\begin{aligned}[\mathbf{F}^T\mathbf{D}]_{ij} &= \mathbf{F}_i^T\mathbf{D}_j = \mathbf{1}^T\mathbf{R}_i(\mathbf{I} - \mathbf{Q})^{-1}\mathbf{L}_j\mathbf{X}\mathbf{1} \\ &= \operatorname{trace} \mathbf{1}^T\mathbf{R}_i(\mathbf{I} - \mathbf{Q})^{-1}\mathbf{L}_j\mathbf{X}\mathbf{1} = \operatorname{trace} \mathbf{R}_i(\mathbf{I} - \mathbf{Q})^{-1}\mathbf{L}_j\mathbf{W}.\end{aligned}$$

Thus, by **1.1.10(5)**,

$$|\mathbf{I} + (\mathbf{I} - \mathbf{Q})^{-1} \sum_{k=1}^{s} \mathbf{L}_k \mathbf{W} \mathbf{R}_k| = |\mathbf{I} + \mathbf{F}^T\mathbf{D}| = |\mathbf{M}|,$$

so (**1**) follows. Finally, from **1.1.10(6)**,

$$\operatorname{trace} \log(\mathbf{I} - \mathbf{H})^{-1} = \log|\mathbf{I} - \mathbf{H}|^{-1} = \log|\mathbf{M}|^{-1} + \operatorname{trace} \log(\mathbf{I} - \mathbf{Q})^{-1}.$$

giving (**2**). □

Both **M**, defined in **4.4.2**, and $\mathbf{M}^{(1)}$, defined in the elimination theorem in connection with $\Psi_1(\mathbf{C}(\mathbf{I}-\mathbf{H})^{-1}\mathbf{D})$, have the same determinants. This is because the first column of $\mathbf{M}^{(1)}$ is zero, except for a 1 in the first position. The cofactor of this position in $\mathbf{M}^{(1)}$ is precisely $|\mathbf{M}|$. This fact, and the appearance of the logarithm, account for the name of Theorem 4.4.2.

Throughout this section $\mathsf{I}(\pi_1)$ is again denoted by **A** and $\mathsf{I}(\pi_2)$ by **B**. If **Q** in the factored expansion for **H** is an expression involving **A** alone, then the logarithmic connection theorem expresses the required generating function in terms of $\gamma(\pi_1)$, the π_1-string enumerators, and the π_1-cycle enumerators. This is because $\Psi_1(\mathbf{U}) = \operatorname{trace}\mathbf{UW}$ for $\mathbf{U} \in \mathsf{E}$. Since inversions do not generalize to circular permutations, no sense can be made of Ψ_2 in the present context.

As a first application of the logarithmic connection theorem, we derive a circular analogue of the maximal string decomposition theorem.

4.4.3. Theorem (Maximal String Decomposition for Circular Permutations)

Let g_i mark π_1-cycles of length $i \geqslant 1$, and f_j mark maximal π_1-strings of length $j \geqslant 1$. Let $\mathbf{f} = (f_1, \ldots)$, $\mathbf{m} = (m_1, \ldots)$ and let $F(x) = 1 + f_1x + f_2x^2 + \cdots$ be the maximal π_1-string length enumerator.

1. *The number of circular permutations on N_n with maximal π_1-string type* **m** *and t (necessarily 0 or 1) π_1-cycles of length l is*

$$[\mathbf{x}^{\mathbf{1}}\mathbf{f}^{\mathbf{m}}g_l^t]\left\{\psi + \log\left(F^{-1}\circ\gamma(\pi_1)\right)^{-1}\right\}$$

where $\psi = \operatorname{trace}\log(F(\mathbf{A}))^{-1} + \sum_{i\geqslant 1} i^{-1}g_i \operatorname{trace}\mathbf{A}^i$.

2. *If π_1 is cycle-free, then $\psi = 0$.*

Proof:

1. Each circular permutation that is not a π_1-cycle may be decomposed into at least one maximal π_1-string. Suppose that there are k maximal π_1-strings and that these are arranged in a linear order. Successive maximal π_1-strings are separated by a π_2. Now the set of patterns for maximal π_1-strings, with a terminating π_2, is $\pi_1^*\pi_2$, and this is a prime set of patterns. Thus the generating function for permutations on N_n with respect to maximal π_1-strings is

$$[\mathbf{x}^{\mathbf{1}}] \sum_{k\geqslant 1} k^{-1}\operatorname{trace}(f(\mathbf{A})\mathbf{B})^k$$

where $f(x) = f_1 + f_2x + f_3x^2 + \cdots$. The contribution from π_1-cycles of length r that are permutations on N_n is $[\mathbf{x}^{\mathbf{1}}]r^{-1}g_r\operatorname{trace}\mathbf{A}^r$, since such cycles are marked by g_r. The required number, obtained by summing over r, is $[\mathbf{x}^1\mathbf{f}^{\mathbf{m}}g_l^f]\Phi$, where

$$\Phi = \operatorname{trace}\log\{\mathbf{I} - f(\mathbf{A})\mathbf{B}\}^{-1} + \sum_{r\geqslant 1} r^{-1}g_r\operatorname{trace}\mathbf{A}^r.$$

Now $f(\mathbf{A})\mathbf{B} = -f(\mathbf{A})\mathbf{A} + f(\mathbf{A})\mathbf{W}$, a factored expansion, so, from the logarithmic connection theorem,

$$\operatorname{trace}\log\{\mathbf{I} - f(\mathbf{A})\mathbf{B}\}^{-1} = \operatorname{trace}\log F^{-1}(\mathbf{A}) + \log\{1 - \operatorname{trace} F^{-1}(\mathbf{A})f(\mathbf{A})\mathbf{W}\}^{-1}$$

since $F(x) = 1 + xf(x)$. But

$$\begin{aligned} 1 - \operatorname{trace} F^{-1}(\mathbf{A})f(\mathbf{A})\mathbf{W} &= 1 - (xfF^{-1})\circ\gamma(\pi_1) \\ &= 1 - (F-1)F^{-1}\circ\gamma(\pi_1) = F^{-1}\circ\gamma(\pi_1), \end{aligned}$$

and the result follows.

2. If π_1 is cycle-free, then $[\mathbf{x}^1]\operatorname{trace}\mathbf{A}^k = 0$, and the result follows. □

The similarity between the linear and circular forms of the maximal string decomposition theorems is striking. For linear permutations the generating function is

$$\{F^{-1}\circ\gamma(\pi_1)\}^{-1},$$

while for circular permutations the generating function is

$$\log\{F^{-1}\circ\gamma(\pi_1)\}^{-1},$$

when π_1 is cycle-free.

As an example of the use of the maximal string decomposition theorem for circular permutations, we consider the enumeration of circular permutations with respect to maxima, for $\pi_1 = <$. A **maximum** occurs at the end of a maximal π_1-path of length at least 2. For example, the maxima of the circular sequence 3169125 are 9 and 5.

4.4.4. Circular Permutations and Maxima

We determine the number, $c(n, k)$, of circular permutations on N_n with k maxima, for the bipartition $\{<, \geqslant\}$. Let z mark a maximum. Since a maximum occurs only at the end of $<$-strings of length at least 2, then the maximal $<$-string enumerator length is

$$F(x) = 1 + x + zx^2 + zx^3 + \cdots = \{1 - (1-z)x^2\}(1-x)^{-1},$$

and $F^{-1}(x) = \Sigma_{j \geq 0}(1 - z)^j\{x^{2j} - x^{2j+1}\}$. Now $<$ is cycle-free, so from the maximal string decomposition theorem for circular sequences

$$c(n, k) = [z^k \mathbf{x}^1]\log\left\{\sum_{j \geq 0} (1 - z)^j\{\gamma_{2j}(<) - \gamma_{2j+1}(<)\}\right\}^{-1}$$

$$= \left[z^k \frac{x^n}{n!}\right]\log\left(\cosh xy - y^{-1}\sinh xy\right)^{-1}$$

where $y = (1 - z)^{1/2}$, by the $<$-transformation lemma. □

We now turn our attention to a problem ascribed to Erdös concerning Hamiltonian cycles of the complete graph on n vertices, such that the cycles have a prescribed number of edges in common with a fixed Hamiltonian cycle. An analogous problem for the complete directed graph is considered first.

4.4.5. Directed Hamiltonian Cycles of the Complete Directed Graph with Distinguished Hamiltonian Cycles

Let $d(n, k, l)$ be the number of directed Hamiltonian cycles of the complete directed graph on n vertices having k edges in the Hamiltonian cycle $\overrightarrow{1\ 2\ \cdots\ n-1\ n}$ and l edges in the Hamiltonian cycle $\overrightarrow{n\ n-1\ \cdots\ 2\ 1}$.

Let u mark edges in $\{(1, 2), \ldots, (n, 1)\} = \pi_1$, let v mark edges in $\{(1, n), \ldots, (n, n-1)\} = \pi_2$, and let $\pi_3 = \mathsf{N}_n^2 - \pi_1 - \pi_2$, so $\{\pi_1, \pi_2, \pi_3\}$ is a tripartition of N_n^2. Let $\mathbf{C} = \mathbf{I}(\pi_3)$. Then

$$d(n, k, l) = [u^k v^l \mathbf{x}^1]\operatorname{trace}\log(\mathbf{I} - \mathbf{H})^{-1}$$

where $\mathbf{H} = u\mathbf{A} + v\mathbf{B} + \mathbf{C} - \mathbf{X}$ and $\mathbf{A} + \mathbf{B} + \mathbf{C} = \mathbf{W}$, so a factored expansion for $\mathbf{H}$ is $\mathbf{H} = -\mathbf{X} + (u - 1)\mathbf{A} + (v - 1)\mathbf{B} + \mathbf{W}$. Thus, by the logarithmic connection theorem,

$$d(n, k, l) = [u^k v^l \mathbf{x}^1](F_1 + F_2)$$

where $F_1 = \operatorname{trace}\log\{\mathbf{I} + \mathbf{X} - (u - 1)\mathbf{A} - (v - 1)\mathbf{B}\}^{-1}$ and $F_2 = \log\{1 - \operatorname{trace}(\mathbf{I} + \mathbf{X} - (u - 1)\mathbf{A} - (v - 1)\mathbf{B})^{-1}\mathbf{W}\}^{-1}$.

This completes the elimination of one block of the tripartition. To apply the operator $[\mathbf{x}^1]$ we note that each element in $\mathbf{ABD}$ and $\mathbf{BAD}$ consists entirely of terms that contain squared x_i's, where $\mathbf{D}$ is the incidence matrix for any nonempty pattern. This is because these matrices enumerate sequences that contain substrings of the form $i(i + 1)i$ and $i(i - 1)i$, both of which contribute x_i^2. It follows that we may replace $\mathbf{AB}$, $\mathbf{BA}$ as well as $\mathbf{X}^2$ by $\mathbf{0}$ in applying

$[\mathbf{x}^1]$, so

$$F_1 = \operatorname{trace} \sum_{j \geq 1} j^{-1}\{(u-1)\mathbf{A} + (v-1)\mathbf{B} - \mathbf{X}\}^j$$

$$= \sum_{j \geq 1} j^{-1}\{(u-1)^j \operatorname{trace} \mathbf{A}^j + (v-1)^j \operatorname{trace} \mathbf{B}^j\} - \operatorname{trace} \mathbf{X}$$

$$+ (u-1)(v-1)\operatorname{trace} \mathbf{AB}.$$

But $\operatorname{trace} \mathbf{A}^j = \gamma_n \delta_{jn} = \operatorname{trace} \mathbf{B}^j$, and $\operatorname{trace} \mathbf{AB} = \gamma_2$, where $\gamma_j = \gamma_j(\oplus)$. These quantities are nonzero because neither π_1 nor π_2 is cycle-free. Thus

$$F_1 = \frac{1}{n}\{(u-1)^n + (v-1)^n\}\gamma_n - \gamma_1 + (u-1)(v-1)\gamma_2.$$

Similarly,

$$F_2 = \log\left\{1 - \operatorname{trace} \mathbf{W} - \sum_{j \geq 1} \operatorname{trace}\{(u-1)^j \mathbf{A}^j + (v-1)\mathbf{B}^j\}\mathbf{W}\right\}^{-1}$$

$$= \log\left\{1 - \gamma_1 - \sum_{j \geq 1} \{(u-1)^j + (v-1)^j\}\gamma_{j+1}\right\}^{-1}$$

since $\operatorname{trace} \mathbf{A}^j\mathbf{W} = \operatorname{trace} \mathbf{B}^j\mathbf{W} = \gamma_{j+1}$. Thus, by the $\oplus$-transformation lemma, for $n > 2$,

$$[u^k v^l \mathbf{x}^1]F_1 = [u^k v^l x^n] x \frac{\partial}{\partial x} \frac{1}{n}\{(u-1)^n x^n + (v-1)^n x^n\}$$

$$= [u^k v^l x^n] x \frac{\partial}{\partial x}\left(\log\{1-(u-1)x\}^{-1} + \log\{1-(v-1)x\}^{-1}\right)$$

and

$$[u^k v^l \mathbf{x}^1]F_2 = [u^k v^l x^n] x \frac{\partial}{\partial x} \sum_{i \geq 1} (i-1)!$$

$$\times [y^i] \log\left\{1 - xy - \sum_{j \geq 1} \{(u-1)^j + (v-1)^j\} x^{j+1} y\right\}^{-1}$$

$$= [u^k v^l x^n] x \frac{\partial}{\partial x} \sum_{i \geq 1} (i-1)!$$

$$\times [y^i] \log\left\{1 - xy \frac{1-(u-1)(v-1)x^2}{\{1-(u-1)x\}\{1-(v-1)x\}}\right\}^{-1}$$

$$= [u^k v^l x^n] x \frac{\partial}{\partial x} \sum_{i \geq 1} i^{-1}(i-1)! x^i \left(\frac{1-(u-1)(v-1)x^2}{\{1-(u-1)x\}\{1-(v-1)x\}}\right)^i.$$

Combining the expressions gives $d(n, k, l) = n[u^k v^l x^n]\Phi(x, u, v)$, for $n > 2$, where

$$\Phi = \log\{1 - (u - 1)x\}^{-1} + \log\{1 - (v - 1)x\}^{-1}$$

$$+ \sum_{i \geqslant 1} i^{-1}(i - 1)! x^i \left\{ \frac{1 - (u - 1)(v - 1)x^2}{\{1 - (u - 1)x\}\{1 - (v - 1)x\}} \right\}^i . \qquad \square$$

This result may be specialized to the undirected case as follows.

4.4.6. Hamiltonian Cycles of the Complete Graph with a Distinguished Hamiltonian Cycle

We wish to determine the number $h(n, k)$ of Hamiltonian cycles of the complete undirected graph on n vertices that have exactly k edges in common with a distinguished Hamiltonian cycle $\overrightarrow{1\ 2 \cdots n - 1\ n}$, say.

Each undirected Hamiltonian cycle is counted twice in **4.4.5**, so $h(n, k) = \frac{1}{2}n[u^k x^n]\Phi(x, u, u)$. But

$$\Phi(x, u, u) = 2\log\{1 - (u - 1)x\}^{-1}$$

$$+ \sum_{j \geqslant 1} j^{-1}(j - 1)! x^j \{1 + (u - 1)x\}^j \{1 - (u - 1)x\}^{-j},$$

and it follows that

$$h(n, k) = (-1)^{n-k}\binom{n}{k}$$

$$+ \tfrac{1}{2}n \sum_{j=1}^{n-k} j^{-1}(j - 1)! \binom{n - j}{k} (-1)^{n-j-k} \sum_{i \geqslant 0} \binom{j}{i}\binom{n - i - 1}{j - 1}.$$

$\square$

Both the maximal string decomposition theorem for circular permutations and the Hamiltonian cycle problem (**4.4.5**) invoked the logarithmic connection theorem with $s = 1$. We conclude this section with an example in which $s = 2$.

4.4.7. The Ménage Problem

Let m_n be the number of ways of arranging n man–wife pairs of people around a circular table with a distinguished chair such that no wife sits next to her husband and such that people of the same sex are not adjacent. This number is called the **ménage number**.

Let x_i, y_i be indeterminates marking the man and woman of the ith pair, for $i = 1,\ldots, n$. Let $\mathbf{x} = (x_1,\ldots, x_n)$, $\mathbf{y} = (y_1,\ldots, y_n)$, $\mathbf{X} = \operatorname{diag}\mathbf{x}$, $\mathbf{Y} = \operatorname{diag}\mathbf{y}$, and $\mathbf{D} = \mathbf{J} - \mathbf{I}$. Since the circular arrangement may be rooted on a man or a woman we have

$$m_n = [\mathbf{x}^1\mathbf{y}^1]\{\operatorname{trace}(\mathbf{XDYD})^n + \operatorname{trace}(\mathbf{YDXD})^n\}.$$

The action of $[\mathbf{x}^1\mathbf{y}^1]$ ensures that each person is seated. The separation of $\mathbf{X}$'s and $\mathbf{Y}$'s by $\mathbf{D}$'s ensures that no husband sits next to his wife. The alternation of $\mathbf{X}$ and $\mathbf{Y}$ in the expression $(\mathbf{XDYD})^n$ ensures that no people of the same sex are adjacent. It follows that

$$m_n = 2[\mathbf{x}^1\mathbf{y}^1]\operatorname{trace}(\mathbf{XDYD})^n = 2n[\mathbf{x}^1\mathbf{y}^1]n^{-1}\operatorname{trace}(\mathbf{XDYD})^n$$

$$= 2n[\mathbf{x}^1\mathbf{y}^1]\sum_{j\geqslant 1} j^{-1}\operatorname{trace}(\mathbf{XDYD})^j = 2n[\mathbf{x}^1\mathbf{y}^1]\operatorname{trace}\log(\mathbf{I} - \mathbf{XDYD})^{-1}.$$

A factored expansion for $\mathbf{XDYD}$ is $\mathbf{XDYD} = \mathbf{XY} - \mathbf{XJY} + \mathbf{XDYJ}$, so, from the logarithmic connection theorem,

$$\operatorname{trace}\log(\mathbf{I} - \mathbf{XDYD})^{-1} = \operatorname{trace}\log(\mathbf{I} - \mathbf{XY})^{-1} + \log|\mathbf{M}|^{-1}$$

$$\text{where}\quad |\mathbf{M}| = \begin{vmatrix} 1 + \operatorname{trace}\mathbf{Y}(\mathbf{I} - \mathbf{XY})^{-1}\mathbf{XJ} & -\operatorname{trace}\mathbf{Y}(\mathbf{I} - \mathbf{XY})^{-1}\mathbf{XDYJ} \\ \operatorname{trace}(\mathbf{I} - \mathbf{XY})^{-1}\mathbf{XJ} & 1 - \operatorname{trace}(\mathbf{I} - \mathbf{XY})^{-1}\mathbf{XDYJ} \end{vmatrix}.$$

Since we are extracting the linear coefficient in x_i and y_i, set $x_i^2 = 0$ and $y_i^2 = 0$, whence $\mathbf{X}^2 = \mathbf{0}$ and $\mathbf{Y}^2 = \mathbf{0}$. Moreover, $\mathbf{X}$ and $\mathbf{Y}$ commute. Thus

$$|\mathbf{M}| = \begin{vmatrix} 1 + w & -vw \\ u & 1 + w - uv \end{vmatrix}$$

where $w = \operatorname{trace}\mathbf{XY} = x_1y_1 + \cdots + x_ny_n$, $v = \operatorname{trace}\mathbf{Y} = y_1 + \cdots + y_n$, $u = \operatorname{trace}\mathbf{X} = x_1 + \cdots + x_n$, so

$$m_n = 2n[\mathbf{x}^1\mathbf{y}^1]\left(w + \log\{(1 + w)^2 - uv\}^{-1}\right)$$

$$= 4n[\mathbf{x}^1\mathbf{y}^1]\log(1 + w)^{-1} + 2n[\mathbf{x}^1\mathbf{y}^1]\log\{1 - uv(1 + w)^{-2}\}^{-1} \quad \text{for } n > 1.$$

Now consider $[\mathbf{x}^1\mathbf{y}^1](uv)^jw^k$, for some integers j and k. This is determined combinatorially as follows. Select k distinct terms from $x_1y_1 + \cdots + x_ny_n$ in $\binom{n}{k}$ ways. The coefficient of the product of these k terms in w^k is $k!$. Now consider the complementary sets of x's and y's in u^j and v^j. The contribution of these terms is $(j!)^2\delta_{n-k,j}$, giving

$$[\mathbf{x}^1\mathbf{y}^1](uv)^jw^k = \binom{n}{k}k!(j!)^2\delta_{n-k,j} = n!(n-k)!\delta_{n-k,j},$$

to yield

$$m_n = 4n[\mathbf{x}^1\mathbf{y}^1](-1)^n n^{-1}w^n + 2n \sum_{\substack{j\geq 1\\k\geq 0}} j^{-1}\binom{-2j}{k}[\mathbf{x}^1\mathbf{y}^1](uv)^j w^k$$

$$= 4(-1)^n n! + 2n\cdot n! \sum_{k=0}^{n-1}(n-k)^{-1}\binom{-2(n-k)}{k}(n-k)!$$

$$= 2\cdot n! \sum_{k=0}^{n}(-1)^k \frac{2n}{2n-k}\binom{2n-k}{k}(n-k)! \qquad \square$$

Many of the earlier results for permutations that have been obtained by means of a factored expansion may be transformed by the logarithmic connection theorem into results for a corresponding circular permutation problem. These are left as exercises.

NOTES AND REFERENCES

This section is based on Goulden and Jackson (1983).

4.4.4 Entringer (1969); **4.4.5, 6** Nemetz (1970), Wright (1973); **4.4.7** Touchard (1934); **[4.4.4]** Tanny (1976); **[4.4.10]** Moon (1972); **[4.4.19]** Djoković (private communication).

EXERCISES

4.4.1. Show that $\Sigma_{k\geq 0}\gamma_k(\pi_1)z^k = |\mathbf{I}+z\mathbf{B}|\cdot|\mathbf{I}-z\mathbf{A}|^{-1}$, where $\mathbf{A}=\mathsf{I}(\pi_1)$, $\mathbf{B}=\mathbf{W}-\mathbf{A}$.

4.4.2. Let $\mathbf{A}=\mathsf{I}(\leqslant)$, $\mathbf{B}=\mathbf{W}-\mathbf{A}$. Show that $|\mathbf{I}-\mathbf{A}^{p-1}\mathbf{B}| = \Sigma_{i\geq 0}(-1)^i a_{ip}$, where $a_j = [t^j]\prod_{l\geq 1}(1-(tx_l)^p)(1-tx_l)^{-1}$.

4.4.3. Show that the number of π_1-alternating circular permutations on N_{2n} is

(a) $[x^{2n}/(2n)!]\log(\sec x)$ if $\pi_1 = <$.

(b) $\sum_{i\geq 1}(i-1)!\binom{n-1}{i-1}(-1)^{n-i}$ if $\pi_1 = +$.

(c) $2(-1)^n + 2n\sum_{i\geq 1}(i-1)!\binom{n-1}{i-1}(-1)^{n-i}/i$ if $\pi_1 = \oplus$.

4.4.4. Show that the number of circular permutations on N_n with k occurrences of π_1 is

(a) $[u^k x^n/n!]\log\{(u-1)(u-e^{(u-1)x})^{-1}\}$ if $\pi_1 = <$.

(b) $[u^k x^n]\Sigma_{i\geq 1}(i-1)!x^i(1-(u-1)x)^{-i}$ if $\pi_1 = +$.

(c) $[u^k x^n]\{\Sigma_{i\geq 1}(i-1)!x^i(1-(u-1)x)^{-i-1} + x^n(u-1)^n\}$ if $\pi_1 = \oplus$.

4.4.5. Show that the number of circular permutations on N_n with k maximal π_1-strings of length $\geq p$ is

(a) With $\pi_1 = <$,

$$\left[u^k \frac{x^n}{n!}\right] \log\left\{ \sum_{i\geq 0} (1-u)^i \left(\frac{x^{ip}}{(ip)!} - \frac{x^{ip+1}}{(ip+1)!} \right) \right\}^{-1}.$$

(b) With $\pi_1 = +$ and $n \equiv m (\mathrm{mod}\ p)$, $0 \leq m < p$,

$$[u^k] \sum_{j=1}^{n} (j-1)! \sum_{l\geq 0} \binom{j}{m+lp} \binom{m+lp-l-1+(n-m)/p}{j-1} \times (-1)^{j-m-lp} (1-u)^{-l+(n-m)/p}$$

(c) With $\pi_1 = \oplus$, if the circular permutation is not a $\oplus$-cycle,

$$[u^k]\left\{ -1 + \delta_{m0} p (1-u)^{n/p} + n \sum_{j=1}^{n} (j-1)! j^{-1} \sum_{l\geq 0} \binom{j}{m+lp} \times \binom{m+lp-l-1+(n-m)/p}{j-1} (-1)^{j-m-lp} (1-u)^{-l+(n-m)/p} \right\}.$$

4.4.6. Show that the number of circular permutations on N_n that are not π_1-cycles, with k maximal π_1-strings is, for $k \geq 1$,

(a) $[u^k x^n/n!] \log\{(1-u)(1-ue^{(1-u)x})^{-1}\}$ for $\pi_1 = <$.

(b) $\sum_{j=1}^{n} (j-1)! \binom{n-1}{j-1} \binom{n-j}{k-j} (-1)^{k-j}$ for $\pi_1 = +$.

(c) $\binom{n}{k}(-1)^k + n \sum_{j=1}^{n} (j-1)! j^{-1} \binom{n-1}{j-1} \binom{n-j}{k-j} (-1)^{k-j}$ for $\pi_1 = \oplus$.

4.4.7. Show that the number of circular permutations on N_{np} that are not π_1-cycles and in which all maximal π_1-strings have length divisible by p is

(a) $n^{-1}(p!)^{-n}(np)!$ for $\pi_1 = <$.

(b) $(n-1)!$ for $\pi_1 = +$.

(c) $p\{(n-1)! - 1\}$ for $\pi_1 = \oplus$.

4.4.8. Show that the number of k-tuples of circular permutations on N_n with j occurrences of

(a) Rises in the same position is

$$\left[u^j \frac{x^n}{(n!)^k}\right] \log\left\{ (u-1)\left(u - \sum_{m\geq 0} (u-1)^m \frac{x^m}{(m!)^k} \right)^{-1} \right\}.$$

(b) Successions in the same position is

$$[u^j x^n] \sum_{i\geq 1} (i!)^k i^{-1} x^i (1-(u-1)x)^{-i}.$$

(c) c-Successions in the same position is

$$[u^j x^n] n^k \sum_{i\geq 1} i^{-1}((i-1)!)^k x^i (1-(u-1)x)^{-i} + [u^j] n^{k-1} (u-1)^n.$$

4.4.9. Show that the number of circular permutations on N_n with i_1 maxima, i_2 minima, i_3 double rises, and i_4 double falls is

$$\left[u_1^{i_1}\cdots u_4^{i_4}\frac{x^n}{n!}\right]\log\{(\alpha_2-\alpha_1)(\alpha_2 e^{\alpha_1 x}-\alpha_1 e^{\alpha_2 x})^{-1}\}$$

where the single circular permutation on N_1 is a double fall and $\alpha_1+\alpha_2=u_3+u_4$, $\alpha_1\alpha_2=u_1u_2$.

4.4.10. Show that another expression for $h(n,k)$, given in **4.4.6**, is

$$(-1)^{n-k}\binom{n}{k}+\tfrac{1}{2}n\sum_{j=1}^{n}\frac{(j-1)!}{j}\binom{n-j}{k}(-1)^{n-j-k}\sum_{i=1}^{n-j}\binom{j}{i}\binom{n-j-1}{i-1}2^i.$$

4.4.11. (a) Let $\Phi_1(x)$ be the exponential generating function for circular $<$-alternating permutations and $\Phi_2(x)$ be the exponential generating function for $<$-alternating permutations of odd length. By deleting the maximum element, show that $(d/dx)\Phi_1(x)=\Phi_2(x)$.

(b) Deduce that $\Phi_1(x)=\log(\sec x)$ from $\Phi_2(x)=\tan x$. Hence we can determine circular results from linear results in special cases.

4.4.12. Show that the number of circular permutations on N_n with t successions and u rises is

$$[r^u s^t x^n]\sum_{i,k\geqslant 1}\sum_{j=0}^{i}(-1)^{i-j}i^{-1}j^k(r-1)^{k-i}x^k(1-r(s-1)x)^{-k}.$$

4.4.13. Show that the number of circular permutations on N_n, $n>2$, with i increasing $+$-strings of length p and j decreasing $+$-strings of length s is

$$[u^i v^j x^n]\sum_{k\geqslant 1}(k-1)!x^k$$

$$\times\left\{\frac{1-ux+(u-1)x^{p-1}}{1-ux+(u-1)x^p}+\frac{1-vx+(v-1)x^{s-1}}{1-vx+(v-1)x^s}-1\right\}^k.$$

4.4.14. Let $N(r_1,\ldots,r_k)$ be the number of permutations on N_n with pattern $\pi_1^{r_1-1}\pi_2\cdots\pi_1^{r_k-1}$, where $r_1+\cdots+r_k=n$. Let $M(p_1,\ldots,p_m)$ be the number of circular permutations on N_n with circular pattern $\pi_1^{p_1-1}\pi_2\cdots\pi_1^{p_m-1}\pi_2$, where $p_1+\cdots+p_m=n$, and the circular pattern has no circular automorphisms except the identity. Show that

$$M(p_1,\ldots,p_m)=\sum_{i=1}^{m}(-1)^{m-i}N\left(p_1+\sum_{j=i+1}^{m}p_j,\,p_2,\ldots,p_i\right)$$

if π_1 is cycle-free.

4.4.15. Show that the number of circular $>$-alternating permutations on N_{2n} that have k rises between adjacent maxima is

$$\left[u^k\frac{x^{2n}}{(2n)!}\right]\log\left\{1-\sum_{i\geqslant 1}\frac{x^{2i}}{(2i)!}\prod_{j=1}^{k}(2j(u-1)-1)\right\}^{-1}.$$

4.4.16. Show that the number of circular sequences of type **i**, with m rises, is

$$[r^m \mathbf{x}^{\mathbf{i}}] \sum_{k \geq 1} \frac{\phi(k)}{k} \log\left\{1 - (r^k - 1)^{-1}\left(\prod_{i \geq 1}(1 + (r^k - 1)x_i^k) - 1\right)\right\}^{-1}$$

where ϕ is the **Euler totient function.**

4.4.17. Show that the number of circular permutations on N_n with circular pattern in $(\pi_1^p \pi_2^s)^+$, $\pi_1 = <$, is $[x^n/n!]\log|\mathbf{M}|^{-1}$, where

$$[\mathbf{M}]_{ij} = \frac{x^{i-j}}{(i-j)!}(1 - \zeta(i, j)) + \sum_{k \geq 1} (-1)^{ks} \frac{x^{k(p+s)+i-j}}{(k(p+s)+i-j)!}$$

for $i, j = 1, \ldots, s$, and

$$\zeta(i, j) = \begin{cases} 1 & \text{if } i < j \\ 0 & \text{if } i \geq j. \end{cases}$$

4.4.18. Let $\alpha = \{\alpha_1, \ldots, \alpha_s\} \subseteq \mathsf{N}_k$, $\bar{\alpha} = \{\bar{\alpha}_1, \ldots, \bar{\alpha}_{k-s}\} = \mathsf{N}_k - \alpha$, $\alpha_1 < \cdots < \alpha_s$, $\bar{\alpha}_1 < \cdots < \bar{\alpha}_{k-s}$, and $\phi_j(x) = \sum_{i \geq 0} x^{ki+j}/(ki+j)!$. If $p_1 = \alpha_1$, $p_i = \alpha_i - \alpha_{i-1}$, $i = 2, \ldots, s$, and $(p_1, \ldots, p_s)$ has no cyclic automorphisms except the identity, show that

$$\|\phi_{(\alpha_i - \alpha_j) \bmod k}(-x)\|_{s \times s} = \|\phi_{(\bar{\alpha}_i - \bar{\alpha}_j) \bmod k}(x)\|_{(k-s) \times (k-s)}$$

by considering circular permutations with circular pattern in $(\pi_1^{p_1 - 1}\pi_2\pi_1^{p_2 - 1}\pi_2 \cdots \pi_1^{p_s - 1}\pi_2)^+$ for $\pi_1 = <$.

4.4.19. (a) Let $\mathbf{M}$ be a $k \times k$ matrix and $\phi_j(x) = \sum_{i \geq 0} x^{ki+j}/(ki+j)!$. A result of Jacobi states that

$$|(\operatorname{adj} \mathbf{M})[\alpha|\beta]| = (-1)^{\sigma}|\mathbf{M}|^{|\alpha|-1}|\mathbf{M}[\bar{\alpha}|\bar{\beta}]|,$$

where $\alpha = \{\alpha_1, \ldots, \alpha_s\}$, $\beta = \{\beta_1, \ldots, \beta_s\}$, $\alpha, \beta \subseteq \mathsf{N}_k$, $\bar{\alpha} = \{\bar{\alpha}_1, \ldots, \bar{\alpha}_{k-s}\} = \mathsf{N}_k - \alpha$, $\bar{\beta} = \{\bar{\beta}_1, \ldots, \bar{\beta}_{k-s}\} = \mathsf{N}_k - \beta$, and $\sigma = \sum_{i=1}^s (\alpha_i + \beta_i)$. Use this result to prove that

$$\|\phi_{(\alpha_i - \beta_j) \bmod k}(-x)\|_{s \times s} = (-1)^{\sum_{i=1}^s (\alpha_i + \beta_i)} \|\phi_{(\bar{\alpha}_i - \bar{\beta}_j) \bmod k}(x)\|_{(k-s) \times (k-s)}.$$

(b) Now let $\phi_j(x) = \sum_{i \geq 0} \dfrac{x^{ik+j}}{(ik+j)!_q}$, $\psi_j(x) = \sum_{i \geq 0} q^{\binom{ik+j}{2}} \dfrac{x^{ik+j}}{(ik+j)!_q}$. Use the result of Jacobi to prove that

$$\sum_{l \geq 0} \frac{\left\{x^k(1-q)^k(1-q^k)^{-1}\right\}^l}{l!_{q^k}} \|\psi_{(\alpha_i - \beta_j) \bmod k}(-x)\|_{s \times s}$$

$$= (-1)^{\sum_{i=1}^s (\alpha_i + \beta_i)} \|\phi_{(\bar{\alpha}_i - \bar{\beta}_j) \bmod k}(x)\|_{(k-s) \times (k-s)}.$$

4.5. PERMANENTS AND ABSOLUTE PROBLEMS

Sections 4.3 and 4.4 have shown that trace $\mathbf{C}(\mathbf{I}-\mathbf{H})^{-1}\mathbf{D}\mathbf{W}$ and trace $\log(\mathbf{I}-\mathbf{H})^{-1} = \log\det(\mathbf{I}-\mathbf{H})^{-1}$ arise as generating functions for classes of linear sequence problems and circular permutation problems. In this section combinatorial situations are examined in which $\det(\mathbf{I}-\mathbf{H})^{-1}$ itself arises as a generating function. The first such situation concerns the calculation of the permanent. The following result, which is a specialization of the MacMahon master theorem (**1.2.11**), gives the connection between the permanent and the determinant.

4.5.1. Proposition (Permanent)

Let $\mathbf{P} = [p_{ij}]_{n\times n}$, $\mathbf{X} = \operatorname{diag}(x_1,\ldots,x_n)$, *and* $\operatorname{per}\mathbf{P} = \sum_{\phi:\mathsf{N}_n \rightrightarrows \mathsf{N}_n} p_{1\phi(1)}\cdots p_{n\phi(n)}$. *Then*

$$\operatorname{per}\mathbf{P} = [\mathbf{x}^{\mathbf{1}}]\det(\mathbf{I}-\mathbf{XP})^{-1}. \qquad \square$$

In the earlier sections of this chapter each block of a partition of N_+^2 represented a relationship between pairs of adjacent symbols in sequences or permutations. We now turn our attention to the situation in which each block is to contain **positional information**.

4.5.2. Definition (Absolute Partition for Permutations)

Let $\sigma_1\cdots\sigma_n$ *be a permutation on* N_n *and let* (π_1,π_2) *be a bipartition of* N_n^2. *Then* $\sigma_1\cdots\sigma_n$ *has an* **absolute** π_i *in position* j *if* $(j,\sigma_j)\in\pi_i$. $\square$

For example, the permutation 643251 has absolute rises in positions 1 and 2, absolute levels in positions 3 and 5, and absolute falls in positions 4 and 6.

The derangement problem is an example of an **absolute problem**, since a derangement is a permutation with no absolute levels. The next result gives the connection between permanents and absolute problems.

4.5.3. Lemma (Absolute Partition)

There are $[z^k]\operatorname{per}(\mathbf{V}+z\mathbf{U})$ *permutations on* N_n *with* k *absolute* π_1*'s, where* $\mathbf{XU} = \mathsf{I}(\pi_1)$ *and* $\mathbf{XV} = \mathsf{I}(\pi_2)$. $\square$

In view of Proposition 4.5.1, the logarithmic connection theorem may now be applied to absolute problems. As before, $\mathbf{W}$ denotes the matrix $\mathbf{XJ}$.

4.5.4. Derangements

From the absolute partition lemma the number d_n of derangements on N_n is

$$d_n = \operatorname{per}(\mathbf{J} - \mathbf{I}) = [\mathbf{x}^1]\det(\mathbf{I} + \mathbf{X} - \mathbf{W})^{-1}$$

$$= [\mathbf{x}^1]\det(\mathbf{I} + \mathbf{X})^{-1}\{1 - \operatorname{trace}(\mathbf{I} + \mathbf{X})^{-1}\mathbf{W}\}^{-1}$$

by the logarithmic connection theorem. Since we are interested only in linear terms in $x_1, \ldots, x_n$, set $\mathbf{X}^2 = \mathbf{0}$. Thus, by the Jacobi identity **(1.1.10(6))**,

$$|\mathbf{I} + \mathbf{X}| = |e^{\mathbf{X}}| = e^{\operatorname{trace}\mathbf{X}} = e^{\gamma_1(<)}.$$

Also $1 - \operatorname{trace}(\mathbf{I} + \mathbf{X})^{-1}\mathbf{W} = 1 - \operatorname{trace}\mathbf{W} = 1 - \gamma_1(<)$. Applying the <-transformation lemma, we have

$$d_n = \left[\frac{x^n}{n!}\right] e^{-x}(1 - x)^{-1}. \qquad \square$$

The ménage problem has been considered in **4.4.7** as a circular permutation problem. It is now treated as an absolute problem.

4.5.5. The Ménage Problem (Absolute)

Let m_n be the number of ménage arrangements on N_n, as in **4.4.7**. We use the following construction. Suppose that the n men, labeled from 1 to n, are seated, each separated by a chair, so that the labels increase from 1 to n in a clockwise direction. Now label the gaps between two adjacent men in a clockwise direction, beginning with the man labeled 1. Wife j must not be assigned to gap j or gap $j - 1$, where gap 0 is identified with gap n. Let $\mathbf{C} = \operatorname{circ}(0, 0, 1, \ldots, 1)$ be $n \times n$, so by the absolute partition lemma $m_n = 2 \cdot n! \operatorname{per}\mathbf{C}$ since the number of assignments of the women to the gaps is per $\mathbf{C}$, the men may be permuted in $n!$ ways, and the distinguished chair may be occupied by a man or a woman.

From Proposition 4.5.1 we have $\operatorname{per}\mathbf{C} = [\mathbf{x}^1]|\mathbf{I} - \mathbf{H}|^{-1}$, where $\mathbf{H} = \mathbf{W} - \mathbf{X} - \mathbf{A}$ and $\mathbf{A} = \mathsf{I}(\oplus)$. From **4.4.5**, setting $u = 0$ and $v = 1$, we have, with $\gamma_k = \gamma_k(\oplus)$ and $\mathbf{X}^2 = \mathbf{0}$,

$$\log|\mathbf{I} - \mathbf{H}|^{-1} = -\gamma_1 + \frac{(-1)^n}{n}\gamma_n + \log\left\{1 - \sum_{k \geq 1} (-1)^{k-1}\gamma_k\right\}^{-1},$$

so

$$\operatorname{per}\mathbf{C} = (-1)^n + [\mathbf{x}^1]e^{-\gamma_1}\left\{1 - \sum_{k \geq 1} (-1)^{k-1}\gamma_k\right\}^{-1}.$$

It follows from the ⊕-transformation lemma that

$$\operatorname{per}\mathbf{C} = (-1)^n + n[x^n]\sum_{k\geq 1}(k-1)![y^k]e^{-xy}\{1-xy(1+x)^{-1}\}^{-1}$$

$$= (-1)^n + n[x^n]\sum_{k\geq 1}(k-1)!\sum_{m=0}^{k}\frac{(-x)^m}{m!}\left(\frac{x}{1+x}\right)^{k-m}$$

$$= 2(-1)^n + n[x^n]\sum_{j\geq 1}(j-1)!\left(\frac{x}{1+x}\right)^j\sum_{m\geq 0}\binom{-j}{m}x^m, \qquad \text{where } j = k-m$$

$$= 2(-1)^n + (-1)^n\sum_{j=1}^{n}(-1)^j(j-1)!\binom{n+j-1}{n-j}$$

and $m_n = 2\cdot n!\operatorname{per}\mathbf{C}$. □

We now consider some configurations that are related to derangements and ménage permutations.

4.5.6. Definition (k-Discordant Permutation)

Let $k \geqslant 1$ be a fixed integer and let A *be the set of k permutations*

$$\{(1,2,\ldots,n),(n,1,2,\ldots,n-1),\ldots,(n-k+2,\ldots,n-k+1)\}.$$

If a permutation $\sigma = \sigma_1\cdots\sigma_n$ on N_n has the property that σ_j is different from the elements in position j of each permutation in A, *for $j = 1,\ldots,n$, then σ is called* ***k-discordant***. □

A 1-discordant permutation is a derangement, and a 2-discordant permutation is a ménage permutation.

4.5.7. 3-Discordant Permutations

By the absolute partition lemma, the number of 3-discordant permutations on N_n is

$$d_3(n) = \text{per circ}(0,0,0,1,\ldots,1) = \text{per circ}(0,0,1,\ldots,1,0)$$

$$= [\mathbf{x}^1]|\mathbf{I}-\mathbf{H}|^{-1}$$

by Proposition 4.5.1, where $\mathbf{H} = \mathbf{W} - \mathbf{X} - \mathbf{A} - \mathbf{B}$, $\mathbf{A} = \mathsf{I}(\oplus)$, and $\mathbf{B} = \mathbf{A}^T$. It follows from **4.4.5**, with $u = v = 0$, that

$$\log|\mathbf{I} - \mathbf{H}|^{-1} = \gamma_2 - \gamma_1 + 2(-1)^n\gamma_n/n + \log\left\{1 - \gamma_1 - 2\sum_{k\geqslant 1}(-1)^k\gamma_{k+1}\right\}^{-1}$$

where $\gamma_k = \gamma_k(\oplus)$ and $\mathbf{X}^2 = \mathbf{0}$. Thus

$$d_3(n) = 2(-1)^n + [\mathbf{x}^1]e^{\gamma_2-\gamma_1}\left\{1 - \gamma_1 - 2\sum_{k\geqslant 1}(-1)^k\gamma_{k+1}\right\}^{-1},$$

so from the $\oplus$-transformation lemma, $d_3(n)$

$$= 2(-1)^n + n[x^n]\sum_{k\geqslant 1}(k-1)![y^k]\{1 + y(x^2 - x)(1+x)^{-1}\}^{-1}e^{(x^2-x)y}$$

$$= 2(-1)^n + n[x^n]\sum_{k\geqslant 1}(k-1)!\sum_{i=0}^{k}(-1)^i(x^2-x)^k(1+x)^{-i}/(k-i)!$$

$$= 2(-1)^n + (-1)^n n\sum_{\substack{k\geqslant 1\\ m,i\geqslant 0}}(-1)^i\binom{m+i-1}{m}\binom{k}{n-k-m}\frac{(k-1)!}{(k-i)!}. \quad \square$$

As an example of an application of the logarithmic connection theorem, with $s = 2$, to an absolute problem, consider the enumeration of $(3 \times n)$-Latin rectangles.

4.5.8. Definition (Latin Rectangle)

Let $\mathbf{M}$ *be a* $k \times n$ *matrix over* N_n *such that each row is a permutation on* N_n, *and such that* k *distinct integers appear in each column. Then* $\mathbf{M}$ *is called a* $(k \times n)$**-Latin rectangle.** □

We use the following decomposition for $(3 \times n)$-Latin rectangles.

4.5.9. Decomposition ((3 × n)-Latin Rectangles)

Let $\mathsf{L}_{3,n}$ *be the set of* $(3 \times n)$*-Latin rectangles with first row* $(1, 2, \ldots, n)$. *Let* K_n *be the set of* $n \times n$ *matrices over* $\{0, \ldots, n\}$ *with zero diagonal, whose* (i, j)*-elements are different from* i *and* j, *whose nonzero elements are distinct, and whose rows and columns each contain exactly one nonzero element. Then*

$$\mathsf{L}_{3,n} \,\tilde{\to}\, \mathsf{K}_n.$$

Proof: Let $[r_{ij}]_{n\times n} \in \mathsf{K}_n$. If r_{ij} is the nonzero element in row i, then the ith column of the corresponding Latin rectangle is $[i, j, r_{ij}]^T$, for $i = 1, \ldots, n$. □

For example,

$$\begin{bmatrix} 1 & 2 & 3 & 4 & 5 \\ 3 & 5 & 4 & 1 & 2 \\ 5 & 4 & 1 & 2 & 3 \end{bmatrix} \mapsto \begin{bmatrix} 0 & 0 & 5 & 0 & 0 \\ 0 & 0 & 0 & 0 & 4 \\ 0 & 0 & 0 & 1 & 0 \\ 2 & 0 & 0 & 0 & 0 \\ 0 & 3 & 0 & 0 & 0 \end{bmatrix}$$

under this decomposition.

4.5.10. (3 × *n*)-Latin Rectangles

Let $l_3(n)$ be the number of $(3 \times n)$-Latin rectangles, and let $b_3(n)$ be the number of these having their first row equal to $(1,\dots,n)$, so $l_3(n) = n!b_3(n)$. Let $x = x_1 + \cdots + x_n$ and

$$[\mathbf{E}]_{ij} = \begin{cases} x - (x_i + x_j) & \text{if } i \neq j \\ 0 & \text{if } i = j, \end{cases}$$

for $1 \leqslant i, j \leqslant n$, so $\mathbf{E} = x\mathbf{D} - \mathbf{XD} - \mathbf{DX}$, where $\mathbf{D} = \mathbf{J} - \mathbf{I}$. Letting $\mathbf{y} = (y_1,\dots,y_n)$ and $\mathbf{Y} = \operatorname{diag}(\mathbf{y})$ we have, by Decomposition 4.5.9 and the absolute partition lemma, $b_3(n) = [\mathbf{x}^1]\operatorname{per}\mathbf{E} = [\mathbf{x}^1\mathbf{y}^1]|\mathbf{I} - \mathbf{YE}|^{-1}$, by Proposition 4.5.1. But $\mathbf{YE} = \mathbf{Q} + \mathbf{Y}(x\mathbf{I} - \mathbf{X})\mathbf{J} - \mathbf{YJX}$, where $\mathbf{Q} = -\mathbf{Y}(x\mathbf{I} - 2\mathbf{X})$. By the logarithmic connection theorem

$$b_3(n) = [\mathbf{x}^1\mathbf{y}^1]|\mathbf{I} - \mathbf{Q}|^{-1}|\mathbf{M}|^{-1},$$

where

$$|\mathbf{M}| = \begin{vmatrix} 1 - \operatorname{trace}\mathbf{Y}(x\mathbf{I} - \mathbf{X})\mathbf{J} & \operatorname{trace}\mathbf{YJ} \\ -x\operatorname{trace}\mathbf{XYJ} & 1 + \operatorname{trace}\mathbf{XYJ} \end{vmatrix} = (1 + w)^2 - xy$$

and $y = y_1 + \cdots + y_n$, $w = x_1y_1 + \cdots + x_ny_n$. We have set $\mathbf{X}^2 = \mathbf{Y}^2 = \mathbf{0}$ since the coefficient of terms linear in $x_1,\dots,x_n, y_1,\dots,y_n$ is required. Similarly, by the Jacobi identity (**1.1.10(6)**),

$$\begin{aligned} |\mathbf{I} - \mathbf{Q}|^{-1} &= \exp\operatorname{trace}\log\{\mathbf{I} + \mathbf{Y}(x\mathbf{I} - 2\mathbf{X})\}^{-1} \\ &= \exp\operatorname{trace}\{-\mathbf{Y}(x\mathbf{I} - 2\mathbf{X})\} = e^{2w-xy}. \end{aligned}$$

Thus

$$\begin{aligned} b_3(n) &= [\mathbf{x}^1\mathbf{y}^1]e^{2w-xy}\{(1+w)^2 - xy\}^{-1} \\ &= \sum_{j,k,l,p \geqslant 0} 2^j(-1)^l\binom{-2k-2}{p}[\mathbf{x}^1\mathbf{y}^1](xy)^{k+l}w^{j+p}/l!j! \\ &= n! \sum_{\substack{j,k,l,p \geqslant 0 \\ k+l+j+p=n}} (-1)^l 2^j\binom{-2k-2}{p}(k+l)!/l!j!, \qquad \text{from } \mathbf{4.4.7} \\ &= n! \sum_{k+j \leqslant n} \sum_{p+l=n-k-j} k!2^j\binom{-(k+1)}{l}\binom{-2(k+1)}{p}\Big/ j! \end{aligned}$$

so

$$l_3(n) = (n!)^2 \sum_{k+j \leqslant n} \frac{2^j}{j!} k! \binom{-3(k+1)}{n-k-j}. \qquad \square$$

In the preceding paragraphs we have enumerated permutations with respect to absolute π_i's, where $\{\pi_1, \pi_2\}$ is a partition of N_n^2. These ideas can be extended to sequences as follows. Let $\sigma = \sigma_1 \cdots \sigma_l \in \mathsf{N}_n^*$, and let $\sigma' = \sigma_1' \cdots \sigma_l'$ be the unique sequence, of the same type as σ, such that $\sigma_1' \leqslant \cdots \leqslant \sigma_l'$. A sequence σ has an absolute π_i in position j if $(\sigma_j', \sigma_j) \in \pi_i$. For example, the sequence 2321311 has no absolute levels and is therefore called a **generalized derangement**.

4.5.11. A Correspondence for Sequences and Absolute Sequences

We determine the number, $c(s, \mathbf{k})$, of sequences of type $\mathbf{k}$ in N_n^* with s absolute π_1's. Let $\mathbf{XG} = u\mathbf{A} + \mathbf{B}$, where $\mathbf{A} = \mathbf{I}(\pi_1)$, $\mathbf{B} = \mathbf{I}(\pi_2)$, $\mathbf{G} = [g_{ij}]_{n \times n}$, and $\sigma' = \sigma_1' \cdots \sigma_l'$, where $\sigma_1' \leqslant \cdots \leqslant \sigma_l'$ and σ' has type $\mathbf{k}$. Thus

$$c(s, \mathbf{k}) = [u^s \mathbf{x}^{\mathbf{k}}] \prod_{m=1}^{l} \left(\sum_{j=1}^{n} g_{\sigma_m', j} x_j \right) = [u^s \mathbf{x}^{\mathbf{k}}] \prod_{i=1}^{n} \left(\sum_{j=1}^{n} g_{ij} x_j \right)^{k_i}$$

$$= [u^s \mathbf{x}^{\mathbf{k}}] |\mathbf{I} - \mathbf{XG}|^{-1}, \qquad \text{by the MacMahon master theorem.}$$

But from the logarithmic connection theorem

$$|\mathbf{I} - \mathbf{XG}|^{-1} = |\mathbf{I} - (u-1)\mathbf{A}|^{-1} \{1 - \operatorname{trace}(\mathbf{I} - (u-1)\mathbf{A})^{-1}\mathbf{W}\}^{-1}$$

$$= \left\{ 1 - \sum_{j \geqslant 1} (u-1)^{j-1} \gamma_j(\pi_1) \right\}^{-1}$$

if π_1 is cycle-free since, in this case,

$$|\mathbf{I} - (u-1)\mathbf{A}|^{-1} = \exp \operatorname{trace} \log(\mathbf{I} - (u-1)\mathbf{A})^{-1} = 1,$$

by the Jacobi identity (**1.1.10(6)**). Thus, comparing the preceding expression for $c(s, \mathbf{k})$ with Corollary 4.2.12, we have proved, for sequences of type $\mathbf{k}$ in N_n^*, that there exists a bijection between (i) those with s absolute π_1's and (ii) those with s π_1's, where π_1 is cycle-free. $\square$

To conclude this section we note that an interpretation of the relationship $|\mathbf{I} - \mathbf{H}|^{-1} = \exp \operatorname{trace} \log(\mathbf{I} - \mathbf{H})^{-1}$ is that $|\mathbf{I} - \mathbf{H}|^{-1}$ enumerates permutations whose cycles, considered as circular permutations, are enumerated by trace log$(\mathbf{I} - \mathbf{H})^{-1}$. As an example of this relationship consider the following permutation problem.

4.5.12. Permutations Whose Cycles Have Repeated Pattern $\pi_1^2\pi_2^2$

Let $c(n)$ be the number of permutations on N_n whose cycles have repeated circular pattern $\pi_1^2\pi_2^2$, where $\pi_1 = <$. Thus, if $\mathbf{A} = \mathsf{I}(\pi_1)$, $\mathbf{B} = \mathsf{I}(\pi_2)$, we obtain

$$c(n) = [\mathbf{x}^1]|\mathbf{I} - \mathbf{A}^2\mathbf{B}^2|^{-1} = [\mathbf{x}^1]|\mathbf{I} - \mathbf{A}^4|^{-1}\left(G_0^2 - G_1G_3\right)^{-1}$$

by **4.3.14** and the logarithmic connection theorem, where $G_i = \sum_{k\geqslant 0}\gamma_{4k+i}(<)$. But $|\mathbf{I} - \mathbf{A}^4| = 1$, since π_1 is cycle-free, so by the <-transformation lemma

$$c(n) = 2\left[\frac{x^n}{n!}\right](1 + \cos x \cosh x)^{-1}. \qquad \square$$

NOTES AND REFERENCES

4.5.7 Touchard (1953); **4.5.10** Riordan (1944); [**4.5.8–11, 13**] Riordan (1958); [**4.5.12**] Aleliunas and Anstee (private communication); [**4.5.14**] Foata and Schützenberger (1970).

EXERCISES

4.5.1. Show that the number of permutations on N_n with k absolute π_1's is
(a) $[u^kx^n/n!](u-1)(u-e^{(u-1)x})^{-1}$ if $\pi_1 = <$.
(b) $[u^kx^n]\sum_{i\geqslant 0} i!x^i\{1-(u-1)x\}^{-i}$ if $\pi_1 = +$.
(c) $[u^k](u^n + (u-1)^n - 1) + [u^kx^n]\sum_{i\geqslant 1} i!x^i\{1-(u-1)x\}^{-(i+1)}$
if $\pi_1 = \oplus$.

4.5.2. Show that the number of permutations on N_n in which all cycles are π_1-alternating circular sequences is
(a) $[x^n/n!]\sec x$ if $\pi_1 = <$.
(b) $\sum_{k=1}^{m}(-1)^{m-k}\binom{m-1}{k-1}k!$ if $\pi_1 = +$, $n = 2m$.
(c) $2(-1)^m + 2m\sum_{k=1}^{m}(-1)^{m-k}\binom{m-1}{k-1}(k-1)!$ if $\pi_1 = \oplus$, $n = 2m$.

4.5.3. Show that the number of permutations on N_n such that every cycle has circular pattern in $(\pi_1^p\pi_2^s)^+$, $\pi_1 = <$, is $[x^n/n!]|\mathbf{M}|^{-1}$ where

$$[\mathbf{M}]_{ij} = \frac{x^{i-j}}{(i-j)!}(1-\zeta(i,j)) + \sum_{k\geqslant 1}(-1)^{ks}\frac{x^{k(p+s)+i-j}}{(k(p+s)+i-j)!},$$

for $i, j = 1, \ldots, s$, and

$$\zeta(i,j) = \begin{cases} 1 & \text{if } i < j \\ 0 & \text{if } i \geqslant j. \end{cases}$$

4.5.4. (a) Show that the number of sequences on N_+ of type $\mathbf{m}$ with i absolute rises, j absolute levels, and k absolute falls is

$$\left[r^i l^j f^k \mathbf{x}^{\mathbf{m}}\right](r-f)\left\{r\prod_{t\geqslant 1}(1+(f-l)x_t)-f\prod_{t\geqslant 1}(1+(r-l)x_t)\right\}^{-1}.$$

(b) Let α be the number of $<$-alternating sequences of type $\mathbf{m}$ and even length λ. If β, γ are the number of sequences of type $\mathbf{m}$ with no absolute levels and an even (resp. odd) number of absolute falls, show that

$$\alpha = (-1)^{\lambda/2}(\beta-\gamma).$$

4.5.5. Show that the number of generalized derangements of type $\mathbf{m}$ is $[\mathbf{x}^{\mathbf{m}}]\prod_{t\geqslant 1}(1+x_t)^{-1}\{1-\sum_{i\geqslant 1}x_i(1+x_i)^{-1}\}^{-1}$.

4.5.6. By noting that the number of $(3\times n)$-Latin rectangles is $l_3(n) = [\mathbf{x}^1\mathbf{y}^1\mathbf{z}^1](\sum x_i y_j z_k)^n$, where the sum is over $1\leqslant i,j,k\leqslant n$, $i\neq j$, $j\neq k$, $k\neq i$, deduce the result in **4.5.10**.

4.5.7. (a) Let $\mathbf{C}, \mathbf{D}$ be arbitrary $n\times n$ matrices. Show that

$$\operatorname{per}(\mathbf{C}+\mathbf{D}) = \sum_{\substack{\alpha,\beta\subseteq \mathsf{N}_n\\ |\alpha|=|\beta|}} \operatorname{per}\mathbf{C}[\alpha|\beta]\operatorname{per}\mathbf{D}(\alpha|\beta).$$

(b) Use part (a) to obtain directly the number of derangements on N_n.

4.5.8. Let $\mathbf{C}$ be an $m\times n$ $\{0,1\}$-matrix. Let r_k be the number of ways of placing exactly k rooks in nontaking positions on the cells of $\mathbf{C}$ that contain 1's. Then $R_{\mathbf{C}}(x) = \sum_{k\geqslant 0} r_k x^k$ is called the **rook polynomial** of $\mathbf{C}$.

(a) Suppose that $[\mathbf{C}]_{pq} = 1$ and $\mathbf{D}$ is a $\{0,1\}$-matrix with a single 1, in position (p,q). Show that $R_{\mathbf{C}}(x) = xR_{\mathbf{C}(\{p\}|\{q\})}(x) + R_{\mathbf{C}-\mathbf{D}}(x)$.

(b) Show that $R_{\mathbf{C}_1\oplus\mathbf{C}_2}(x) = R_{\mathbf{C}_1}(x)R_{\mathbf{C}_2}(x)$, where $\mathbf{C}_1, \mathbf{C}_2$ are $\{0,1\}$-matrices.

4.5.9. (a) Let $\pi_1 \subseteq \mathsf{N}_n\times\mathsf{N}_n$ and $[\mathbf{C}]_{ij} = 1$ if and only if $(i,j)\in\pi_1$ for $i,j = 1,\dots,n$. Show that the number of permutations of N_n with k absolute π_1's is given by $[x^k]\sum_{i=0}^n (n-i)!r_i(x-1)^i$, where $R_{\mathbf{C}}(x) = \sum_{i\geqslant 0} r_i x^i$.

(b) Let $R_{\mathbf{J}_n-\mathbf{C}}(x) = \sum_{i\geqslant 0} q_i x^i$. Show that

$$q_i = \sum_{j=0}^{i}(-1)^j(i-j)!\binom{n-j}{i-j}^2 r_j.$$

4.5.10. Use the rook polynomial to find the derangement number d_n.

4.5.11. (a) Show that $R_{\mathbf{J}_n}(x) = \sum_{i=0}^n i!\binom{n}{i}^2 x^i$.

(b) Show that the number of generalized derangements of type $\mathbf{k} = (k_1,\dots,k_m)$ is given by $\sum_{i=0}^n (n-i)!(-1)^i r_i$, where $n = k_1+\cdots+k_m$, $\sum_{i\geqslant 0} r_i x^i = \prod_{j=1}^m P_{k_j}(x)$, and $P_k(x) = k!x^k L_k^{(0)}(-x^{-1})$, where $L_m^{(\alpha)}(x)$ is a Laguerre polynomial, defined in **[3.5.1]**.

4.5.12. (a) Let $[\mathbf{A}_n]_{ij} = 1$ iff $j - i = 0, 1$, and 0 otherwise and $[\mathbf{C}_n]_{ij} = 1$ iff $j - i \equiv 0, 1 (\bmod n)$ for $i, j = 1, \ldots, n$, and 0 otherwise. Let $[\mathbf{B}_n]_{ij} = 1$ iff $j - i = 0, 1$ for $i = 1, \ldots, n + 1$; $j = 1, \ldots, n$. By considering [**4.5.8**(a)], show that $U(x, y) = \sum_{n \geqslant 0} R_{\mathbf{B}_n}(x) y^{2n} + \sum_{n \geqslant 1} R_{\mathbf{A}_n}(x) y^{2n-1}$ satisfies $\{1 - y - y^2 x\} U(x, y) = 1 + xy$, and hence

$$R_{\mathbf{C}_n}(x) = \sum_{i \geqslant 0} \frac{2n}{2n - i} \binom{2n - i}{i} x^i.$$

(b) Deduce from part (a) that the ménage number is given by

$$m_n = 2 \cdot n! \sum_{i \geqslant 0} (-1)^i (n - i)! \frac{2n}{2n - i} \binom{2n - i}{i}.$$

4.5.13. (a) Let $\mathbf{T}_n$ be an $n \times n$ matrix with $[\mathbf{T}_n]_{ij} = 1$ for $i \leqslant j$ and 0 otherwise. If $T_n(x) = R_{\mathbf{T}_n}(x)$, show that $T_0(x) = 1$ and

$$T_{n+1}(x) = x^{n+1} + \sum_{k=0}^{n} \binom{n+1}{k} x^k T_{n-k}(x), \qquad n \geqslant 0.$$

(b) If $T(x, y) = \sum_{n \geqslant 0} T_n(x) y^n / n!$, show that $(\partial / \partial y) T(x, y) = (x + e^{xy}) T(x, y)$ and then that $[x^k] T_n(x) = p_{n+1, n+1-k}$, a Stirling number of the second kind, defined in **3.2.17**.

4.5.14. Let $\mathsf{V} = \{v_1, \ldots, v_k\}$ be a set of <-strings on N_n, where $v_1 = v_{i1} \cdots v_{il_i}$ and $l_i \geqslant 1$ for $i = 1, \ldots, k$, the disjoint union of the v_{ij} is N_n, and $v_{11} > v_{21} > \cdots > v_{k1}$. Let $s(\mathsf{V})$ be the number of permutations on N_n in which V is the set of maximal <-strings, and let $c(\mathsf{V})$ be the number of permutations on N_n in which V is the set of maximal <-strings in the set of circular sequences forming the disjoint cycle representation.

(a) Show that $s(\mathsf{V}) = \operatorname{per} \mathbf{C}$ and $c(\mathsf{V}) = \operatorname{per} \mathbf{C}$, where $\mathbf{C} = [c_{ij}]_{k \times k}$, $c_{ij} = 1$ if $v_{il_i} > v_{j1}$, and $c_{ij} = 0$ otherwise.

(b) Show that $s(\mathsf{V}) = c(\mathsf{V})$ by describing a bijection between the appropriate sets of permutations.

4.5.15. Derive Proposition 4.5.1 by considering the enumeration of permutations in which p_{ij} is an indeterminate marking the occurrence of j in position i for $i, j \geqslant 1$.

4.5.16. Deduce the MacMahon master theorem (**1.2.11**) from Proposition 4.5.1.

5

The Combinatorics of Paths

5.1. INTRODUCTION

In this chapter we give a detailed treatment of the enumeration of lattice paths. These paths consist of sequences of lattice points in the plane, usually beginning at the origin, in which restrictions are placed on the increments between successive points. Equivalently, we consider the sequence of **steps** associated with the path, where a step is the ordered pair giving the difference between adjacent points in the lattice path. The sequence techniques of earlier chapters are not suitable for the information recorded in this section, since we are interested in the lattice points themselves, which represent cumulative sums in the sequence of steps. In particular, we are interested in the **altitude** (y-coordinate) of the lattice points in these paths. Problems of this sort arise, for example, in ballot problems and in paths with steps restricted to $(0, 1), (1, 0)$. The latter class of paths is what is meant by the term *lattice path* in the major portion of the combinatorial and statistical literature.

Unlike the earlier chapters, this chapter does not begin with a general enumerative technique that is then applied in later sections. Rather, each section contains a treatment of particular types of lattice paths, together with decompositions that allow us to preserve the desired combinatorial information by means of an ordinary generating function. No special algebra is introduced, as it was in Chapter 4, so this chapter can be viewed as a continuation of Chapter 2.

5.1.1. Continued Fractions

The first set of paths that we enumerate, in Section 5.2, are those with steps restricted to $(1, 1), (1, 0), (1, -1)$ and in which every point has non-negative altitude. Using a simple decomposition, we show that the enumerator for these paths with respect to steps of each type at each altitude is the continued fraction of Jacobi (called a ***J*-fraction**). Thus we are able to give a combinatorial proof of the Stieltjes–Rogers theorem. Furthermore, a more elaborate construction, the **Françon–Viennot decomposition**, allows us to encode permutations on N_n as paths of the type described earlier. Thus by comparing the results of Chapter 3 and Section 5.2, we establish a number of classical identities which give, as J-fractions, the ordinary generating functions corresponding to various exponential generating functions. One of these identities involves a Jacobi elliptic function.

5.1.2. Arbitrary Steps and Nonintersecting Paths

For a set of steps in which any ordered pair of integers (except $(0, 0)$) may appear, we have a decomposition for lattice paths into three paths, a minus-path, a zero-path, and a plus-path. In Section 5.3 we identify the ordinary generating function for these three subsets of paths as the components of a unique factorization of the generating function for all lattice paths with the given set of steps. When the steps are forbidden to have an increment of less than -1 in the ordinate, we are able to use the Lagrange theorem to determine these factors.

In Section 5.4 we consider sets of n-tuples of paths with special conditions on termini and steps. The enumerator for the subset of these n-tuples in which the n paths have no pairwise intersections is obtained. This allows us to enumerate skew column-strict plane partitions with various restrictions, since a plane partition on n rows can be encoded as a set of n nonintersecting paths. The dominance theorem of Kreweras is a corollary of this enumerative result.

5.1.3. A q-Analogue of the Lagrange Theorem

In Section 5.5 we restrict attention to paths whose steps belong to the set $\{(1, i) \mid i \geqslant -1\}$. For these paths a number of decompositions are obtained, and a q-analogue of the Lagrange theorem is deduced, where q marks the **area** under a lattice path. This q-analogue is used to obtain an expansion for a special type of continued fraction.

5.2. WEIGHTED PATHS

In this section we give a combinatorial treatment of continued fractions.

5.2.1. Definition (J-Fraction, S-Fraction)

Let λ_i, κ_i, x be commutative indeterminates for $i \geqslant 0$. Let m, n be integers with $0 \leqslant m < n$. Then

1. $J_x[\kappa_k, \lambda_k : (m, n)]$

$$= \frac{1}{1 - \kappa_m x} - \frac{\lambda_m x^2}{1 - \kappa_{m+1} x} - \frac{\lambda_{m+1} x^2}{1 - \kappa_{m+2} x} - \cdots - \frac{\lambda_{n-1} x^2}{1 - \kappa_n x}$$

*is called a J-**fraction** (after Jacobi).*

2. $S_x[\lambda_k:(m,n)] = \frac{1}{1} - \frac{\lambda_m x}{1} - \frac{\lambda_{m+1}x}{1} - \cdots - \frac{\lambda_{n-1}x}{1} - \frac{\lambda_n x}{1}$ *is called an* S-**fraction** *(after Stieltjes).* □

We call $1 - \kappa_k x - \lambda_k x^2$ the kth **denominator** of $J_x[\kappa_k, \lambda_k:(0,\infty)]$, and note that

$$J_x[0, \lambda_k:(m, n+1)] = S_{x^2}[\lambda_k:(m,n)].$$

In fact, a J-fraction may be formally transformed into an S-fraction, and vice versa, by manipulating successive pairs of adjacent denominators. Transformation of an S-fraction into a J-fraction is called **contraction**.

5.2.2. Proposition (Contraction)

Let $\lambda_{-1} = 0$. *Then*

$$S_x[\lambda_k:(0,\infty)] = J_x[\lambda_{2k-1} + \lambda_{2k}, \lambda_{2k}\lambda_{2k+1}:(0,\infty)].$$ □

The next proposition gives an expansion of S-fractions as power series in x with coefficients that are polynomials in the indeterminates $\lambda_0, \lambda_1, \ldots$. The result is obtained by considering planted plane trees of given height, defined below.

5.2.3. Definition (Height of a Planted Plane Tree)

The **height** *of a planted plane tree is the maximum of the heights (see Definition 2.7.15) of any of its vertices.* □

5.2.4. Proposition (Stieltjes–Rogers Polynomials)

Let $S_x[\lambda_l:(0,\infty)] = 1 + \sum_{n\geq 1} S_n x^n$. *Then*

$$S_n = \sum_{k\geq 0} \sum_{\substack{n_0,\ldots,n_k\geq 1\\ n_0+\cdots+n_k=n}} \lambda_0^{n_0}\cdots\lambda_k^{n_k} \prod_{j=0}^{k-1}\binom{n_j+n_{j+1}-1}{n_j-1}.$$

We call S_n *a* **Stieltjes–Rogers polynomial.**

Proof: We count planted plane trees with respect to height in two different ways.

1. Let x_i be an indeterminate marking nonroot vertices at height i, and let $\Phi_k(x_1,\ldots,x_k)$ be the generating function for planted plane trees with height at most k. Then $\Phi_\infty(\mathbf{x})$, where $\mathbf{x} = (x_1, x_2, \ldots)$, is the generating function for planted plane trees with respect to (nonzero) height of vertices. Each tree of height at most $k+1$ can be obtained uniquely from a unique tree, of height at most k, by connecting each vertex at height k to some number of vertices at height $k+1$. By the composition lemma

$$\Phi_{k+1}(x_1,\ldots,x_{k+1}) = \Phi_k\left(x_1,\ldots,x_{k-1},x_k(1-x_{k+1})^{-1}\right), \qquad k>0,$$

where $\Phi_1(x_1) = (1-x_1)^{-1}$. It follows immediately that $\Phi_\infty(z\mathbf{x}) = S_z[x_l:(1,\infty)]$.

2. Suppose that there are m_j and m_{j+1} vertices at height j and $j+1$, respectively, in a planted plane tree. This tree induces a nondecreasing function from the m_{j+1}-set to the m_j-set. The number of such functions is $\binom{m_j+m_{j+1}-1}{m_j-1}$, where $j \geqslant 1$. Since the same argument may be applied independently to pairs of consecutive levels, the number of planted plane trees of height k, with $m_i \geqslant 1$ vertices at height i for $i=1,\ldots,k$ is

$$[x^n x_1^{m_1} \cdots x_k^{m_k}]S_x[x_l:(1,\infty)]$$

$$= \begin{cases} \prod_{j=1}^{k-1} \binom{m_j+m_{j+1}-1}{m_j-1}, & \text{if } m_1+\cdots+m_k = n \\ 0, & \text{otherwise.} \end{cases}$$

The result follows by setting $x_i = \lambda_{i-1}$ and $m_i = n_{i-1}$ and, finally, by replacing $k-1$ with k. □

In a similar way, the **Jacobi–Rogers polynomials** J_n are defined by

$$J_x[\kappa_k,\lambda_k:(0,\infty)] = 1 + \sum_{n\geqslant 1} J_n x^n$$

and an expression for them is given in [**5.2.2**].

We now consider the connection between continued fractions and the enumeration of paths.

5.2.5. Definition (Altitude, Path, Height)

1. *Let* $u=(i,j)\in\mathbb{Z}^2$. *Then* j *is called the* **altitude** *of* u, *and we write* $j=\operatorname{alt}(u)$.

2. *Let* $u_0\cdots u_l$ *be a sequence of points in* $\mathbb{Z}^2$ *such that*

(i) $u_0=(1,j_0)$.

(ii) $u_{k+1}=u_k+(1,\sigma_k)$, $\sigma_k\in\{-1,0,1\}$ *for* $0\leqslant k<l$.

(iii) $\operatorname{alt}(u_k)\geqslant 0$ *for* $0\leqslant k\leqslant l$.

Then $u_0\cdots u_l$ *is called a* **path** *with* **origin** u_0 *and* **terminus** u_l *and is denoted by* $(\sigma_0\cdots\sigma_{l-1})_{j_0}$.

3. *The* **height** *of* $(\sigma_0\cdots\sigma_{l-1})_{j_0}$ *is* $\max_{0\leqslant k\leqslant l}\operatorname{alt}(u_k)$. □

If $k \in \{-1, 0, 1\}$, then $(k)_j$ is called a **step** at altitude j, and is, respectively, a **fall, level,** or **rise** when $k = -1$, 0, or 1. Thus a path is a sequence of steps such that the origin of one is the terminus of the next, and such that no vertex has negative altitude.

If $\rho = (\rho_1 \cdots \rho_l)_j$, and $\sigma = (\sigma_1 \cdots \sigma_k)_i$ has terminal altitude j, then $\sigma\rho = (\sigma_1 \cdots \sigma_k \rho_1 \cdots \rho_l)_i$.

The following is a recursive decomposition for H_i, the set of all paths σ from altitude i to altitude i such that $\text{alt}(v) \geqslant i$ for each vertex $v \in \sigma$.

5.2.6. Decomposition (Path)

Let ε_i denote a single vertex at altitude i. Then

$$\mathsf{H}_i = \varepsilon_i \,\dot\cup\, \big((0)_i \,\dot\cup\, (1)_i \mathsf{H}_{i+1} (-1)_{i+1}\big) \mathsf{H}_i \qquad \textit{for } i \geqslant 0.$$

Proof: Consider an arbitrary nonempty path $\pi \in \mathsf{H}_i$. Let $\pi = \alpha\beta$, where the terminus of α is the vertex with smallest abscissa that is not the origin and that has altitude i. Clearly, $\beta \in \mathsf{H}_i$. Now α is a path from altitude i to altitude i with no internal vertices at altitude less than or equal to i. Thus

$$\alpha \in (0)_i \,\dot\cup\, (1)_i \mathsf{H}_{i+1} (-1)_{i+1},$$

and the result follows. □

Let $\boldsymbol{\alpha} = (\alpha_0, \alpha_1, \dots)$, $\boldsymbol{\beta} = (\beta_1, \beta_2, \dots)$, $\boldsymbol{\gamma} = (\gamma_0, \gamma_1, \dots)$ be sets of indeterminates where

i. α_i marks a rise at altitude $i \geqslant 0$.
ii. β_i marks a fall at altitude $i \geqslant 1$.
iii. γ_i marks a level at altitude $i \geqslant 0$.

Let $\mathbf{i} = (i_0, i_1, \dots)$, $\mathbf{j} = (j_1, j_2, \dots)$, $\mathbf{k} = (k_0, k_1, \dots)$. If a path σ has i_h rises, j_h falls, and k_h levels at altitude h, then $(\mathbf{i}, \mathbf{j}, \mathbf{k})$ is called the **step-type** of σ. We now give the main counting lemma for paths.

5.2.7. Lemma (Path)

The number of paths of step-type $(\mathbf{i}, \mathbf{j}, \mathbf{k})$, *with* n *steps, from altitude* 0 *to altitude* 0 *is*

$$\left[x^n \boldsymbol{\alpha}^{\mathbf{i}} \boldsymbol{\beta}^{\mathbf{j}} \boldsymbol{\gamma}^{\mathbf{k}}\right] J_x\left[\gamma_h, \alpha_h \beta_{h+1} : (0, \infty)\right].$$

Proof: Let $H_i(x, \boldsymbol{\alpha}, \boldsymbol{\beta}, \boldsymbol{\gamma})$ be the generating function for H_i with respect to step-type marked by $(\boldsymbol{\alpha}, \boldsymbol{\beta}, \boldsymbol{\gamma})$ and number of steps marked by x. Then from

Decomposition 5.2.6 we have

$$H_i = 1 + \left(x\gamma_i + x^2\alpha_i\beta_{i+1}H_{i+1}\right)H_i \qquad \text{for } i \geqslant 0,$$

so

$$H_i = \left(1 - x\gamma_i - x^2\alpha_i\beta_{i+1}H_{i+1}\right)^{-1} \qquad \text{for } i \geqslant 0,$$

whence

$$H_0 = J_x\left[\gamma_h, \alpha_h\beta_{h+1} : (0, \infty)\right],$$

and the result follows. □

As an example of the use of Lemma 5.2.7 we give a combinatorial proof of the Stieltjes–Rogers J-fraction theorem. The following definition is needed.

5.2.8. Definition (Addition Formula)

Let $a(x) = \sum_{i \geqslant 0} a_i x^i / i!$.

1. $\hat{a}(x)$ *is the formal power series* $\hat{a}(x) = \sum_{i \geqslant 0} a_i x^i$.
2. *If* $a(x)$ *has the property that*

$$a(x + y) = \sum_{m \geqslant 0} p_m f_m(x) f_m(y),$$

where p_m *is independent of* x *and* y *and*

$$f_m(x) = \frac{x^m}{m!} + q_{m+1}\frac{x^{m+1}}{(m+1)!} + \mathcal{O}(x^{m+2}),$$

then we say that $a(x)$ *has an* **addition formula with parameters** $\{(p_m, q_{m+1}) \mid m \geqslant 0\}$. □

The J-fraction for a formal power series $\hat{a}(x)$ is obtained from the addition formula for $a(x)$, by means of the following decomposition.

5.2.9. Decomposition (Path)

Let $\mathsf{H}_{i,j}$ *be the set of paths from altitude* i *to altitude* j *with no vertex at altitude less than* $\min(i, j)$. *Let* $\mathsf{H}_{i,j,n}$ *be the subset of paths in* $\mathsf{H}_{i,j}$ *with exactly* n *steps. Then*

1. *For any fixed non-negative integers* m, n

$$\mathsf{H}_{0,0,m+n} = \dot{\bigcup_{k \geqslant 0}} \mathsf{H}_{0,k,m}\mathsf{H}_{k,0,n}.$$

2. $\mathsf{H}_{0,k} = \mathsf{H}_0(1)_0\mathsf{H}_1(1)_1 \cdots (1)_{k-1}\mathsf{H}_k$
3. $\mathsf{H}_{k,0} = \mathsf{H}_k(-1)_k\mathsf{H}_{k-1}(-1)_{k-1} \cdots (-1)_1\mathsf{H}_0$.

Proof:

1. Consider a path $\pi \in \mathsf{H}_{0,0,m+n}$ and let $(m+1, k)$ be the unique point at which it intersects the line $x = m+1$. This point is the terminus of a subpath of π in $\mathsf{H}_{0,k,m}$, and the origin of a subpath of π in $\mathsf{H}_{k,0,n}$. The result follows.

2. Let π be a path in $\mathsf{H}_{i,k}$ for $i < k$. Let $\alpha \in \mathsf{H}_i$ be the longest subpath of π with the same origin. Then $\pi = \alpha\beta$, where β is a path in $\mathsf{H}_{i,k}$ with exactly one vertex (namely, its origin) at altitude i. Thus $\beta = (1)_i\beta'$, where $\beta' \in \mathsf{H}_{i+1,k}$, so $\mathsf{H}_{i,k} = \mathsf{H}_i(1)_i\mathsf{H}_{i+1,k}$ for $0 \leqslant i \leqslant k-1$. The result follows by repeated application of this equation.

3. Similar to (**2**). □

5.2.10. The Stieltjes–Rogers *J*-Fraction Theorem

The formal power series $f(x)$ has an addition formula with parameters $\{(p_m, q_{m+1}) \mid m \geqslant 0\}$ if and only if

$$\hat{f}(x) = J_x\left[q_{m+1} - q_m, p_{m+1}p_m^{-1} : (0, \infty)\right].$$

Proof:

1. ($\Leftarrow$): Let H_i, $H_{k,l}$, $H_{k,l,r}$ be the generating functions for $\mathsf{H}_i, \mathsf{H}_{k,l}, \mathsf{H}_{k,l,r}$ in which γ_j and α_j mark levels and rises, respectively, at altitude j, for $j \geqslant 0$. If $f_k(x) = \sum_{j \geqslant 0} H_{k,0,j} \frac{x^j}{j!}$, then

$$\hat{f}_0(x) = J_x\left[\gamma_l, \alpha_l : (0, \infty)\right]$$

by Lemma 5.2.7. From (**2**) and (**3**) of Decomposition 5.2.9

$$H_{0,k} = \alpha_0 \cdots \alpha_{k-1} H_0 \cdots H_k = p_k H_{k,0},$$

where $p_k = \alpha_0 \cdots \alpha_{k-1}$, and thus $H_{0,k,n} = p_k H_{k,0,n}$. But from (**1**) of Decomposition 5.2.9

$$H_{0,0,m+n} = \sum_{k \geqslant 0} H_{0,k,n} H_{k,0,m} = \sum_{k \geqslant 0} p_k H_{k,0,n} H_{k,0,m}.$$

Multiplying both sides by $x^n y^m / n!m!$ and summing over $m, n \geqslant 0$ gives

$$f_0(x+y) = \sum_{k \geqslant 0} p_k f_k(x) f_k(y).$$

Now $H_{k,0,j} = 0$ for $j < k$, and $H_{k,0,k} = 1$, corresponding to the path $(-1)_k(-1)_{k-1} \cdots (-1)_1$. Thus

$$f_k(x) = \frac{x^k}{k!} + q_{k+1}\frac{x^{k+1}}{(k+1)!} + \mathcal{O}(x^{k+2}),$$

where $q_{k+1} = H_{k,0,k+1}$. Since the first step in the paths counted by $H_{k,0,k+1}$ must be a rise, level, or fall at height k, then

$$H_{k,0,k+1} = \alpha_k H_{k+1,0,k} + \gamma_k H_{k,0,k} + H_{k-1,0,k} = \gamma_k + H_{k-1,0,k}.$$

Inverting $p_k = \alpha_0 \cdots \alpha_{k-1}$ and $q_{k+1} = \gamma_k + q_k$, we obtain $\alpha_k = p_{k+1} p_k^{-1}$, $\gamma_k = q_{k+1} - q_k$, and the result follows.

2. ($\Rightarrow$): Conversely, if $f(x)$ has an addition formula, then it is a straightforward matter to show that

$$f(x) = \sum_{n \geqslant 0} H_{0,0,n} \frac{x^n}{n!}$$

where $H_{0,0,n}$ is the generating function for paths of length n from altitude 0 to altitude 0, with respect to rises and levels. □

The following is a result of Euler, obtained by the use of the Stieltjes–Rogers J-fraction theorem.

5.2.11. A Continued Fraction Associated with Factorials

Let $\hat{f}(x) = \sum_{n \geqslant 0} n! x^n$, so from Definition 5.2.8 $f(x) = (1 - x)^{-1}$. To obtain an addition formula for $f(x)$ note that

$$f(x + y) = (1 - x - y)^{-1} = \{(1 - x)(1 - y) - xy\}^{-1}$$

$$= \sum_{k \geqslant 0} (k!)^2 \left\{ \frac{1}{k!} x^k (1 - x)^{-(k+1)} \right\} \left\{ \frac{1}{k!} y^k (1 - y)^{-(k+1)} \right\}.$$

Thus $p_k = (k!)^2$ and

$$f_k(x) = \frac{x^k}{k!}(1 - x)^{-(k+1)} = \frac{x^k}{k!} + (k + 1)^2 \frac{x^{k+1}}{(k + 1)!} + \mathcal{O}(x^{k+2}),$$

so $q_{k+1} = (k + 1)^2$. It follows that $q_{k+1} - q_k = 2k + 1$ and $p_{k+1} p_k^{-1} = (k + 1)^2$, and, from the Stieltjes–Rogers J-fraction theorem,

$$\sum_{n \geqslant 0} n! x^n = J_x\left[2k + 1, (k + 1)^2 : (0, \infty)\right]. \qquad □$$

Many of the permutation and partition problems of Chapter 3 have ordinary generating functions whose J-fraction representations have a simple form. A number of such J-fractions are derived in the exercises by various algebraic means, including the Stieltjes–Rogers J-fraction theorem. For example, for the set of all partitions, $f(x) = \exp(e^x - 1)$ and $\hat{f}(x) = J_x[k + 1, k + 1 : (0, \infty)]$, (from **3.2.17** and **[5.2.6]**). For the set of derangements, $f(x) = e^{-x}(1-x)^{-1}$ and $\hat{f}(x) = J_x[2k, (k+1)^2 : (0, \infty)]$, (**3.2.8** and **[5.2.5]**). Since J-fractions enumerate paths, it is reasonable to ask whether the J-fraction representation of $\hat{f}(x)$ can be obtained directly by representing permutations or partitions by paths. Such a representation for permutations is given by the Françon–Viennot decomposition (**5.2.15**) and for partitions by the Flajolet decomposition **[5.2.19]**. These require the introduction of weighted paths.

Let $w((k)_i) \in \mathsf{N}$ denote the **weight** of the step $(k)_i$. A **weighted path** is a path $(\rho_0 \cdots \rho_{l-1})_{j_0}$ with weights $m_0, \ldots, m_{l-1}$ where m_i is the weight of the step ρ_i, and is denoted by $(\rho_0 \cdots \rho_{l-1}, m_0 \cdots m_{l-1})_{j_0}$. In the representation of a weighted path in $\mathbb{Z}^2$, the weight k of a step with origin (i, j), for $j \geqslant 0$, is indicated by an asterisk at (i, k). Thus Figure 5.2.1 illustrates the weighted path

$$(1\ 1\ 0\ -1\ 1\ 1\ -1\ -1\ 0\ -1,\ 0\ 1\ 0\ 2\ 0\ 1\ 3\ 1\ 0\ 0)_0.$$

A set of weighted paths is said to have **possibility functions** $\psi_{-1}, \psi_0, \psi_1$ if the only restriction on $w((k)_i)$ is that it is an element of a fixed subset of N of size $\psi_k(i)$. The extension of Lemma 5.2.7 to weighted paths is immediate.

5.2.12. Lemma (Weighted Path)

The number of weighted paths from altitude 0 *to altitude* 0 *with n steps and with possibility functions* $\psi_{-1}, \psi_0, \psi_1$ *is*

$$[x^n] J_x[\psi_0(k), \psi_1(k)\psi_{-1}(k+1) : (0, \infty)]. \qquad \square$$

Weighted paths are now applied to the enumeration of permutations with respect to the subconfigurations defined next (see also **[3.3.46]**).

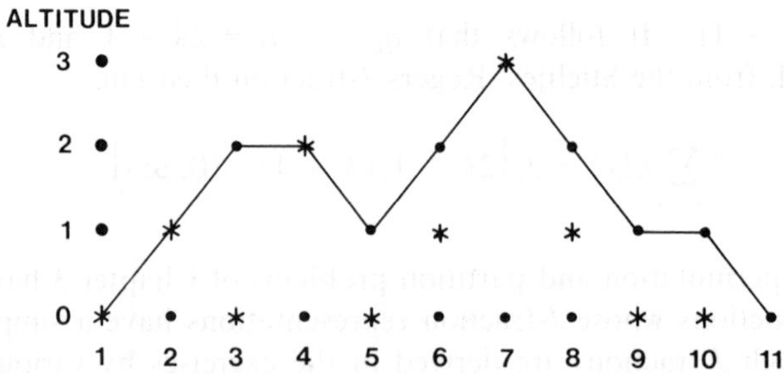

Figure 5.2.1. The weighted path $(1\ 1\ 0\ -1\ 1\ 1\ -1\ -1\ 0\ -1,\ 0\ 1\ 0\ 2\ 0\ 1\ 3\ 1\ 0\ 0)_0$.

5.2.13. Definition (Double Rise, Double Fall, Modified Maximum, Modified Minimum)

Let $\sigma = \sigma_1 \cdots \sigma_n$ be a permutation on N_n and let $\sigma_0 = \sigma_{n+1} = 0$. Let $j \in \mathsf{N}_n$. Then σ_j is called a

1. **Double rise** *if* $\sigma_{j-1} < \sigma_j < \sigma_{j+1}$.
2. **Double fall** *if* $\sigma_{j-1} > \sigma_j > \sigma_{j+1}$.
3. **Modified maximum** *if* $\sigma_{j-1} < \sigma_j > \sigma_{j+1}$.
4. **Modified minimum** *if* $\sigma_{j-1} > \sigma_j < \sigma_{j+1}$. □

A permutation σ can be represented by its **associated tree** $t(\sigma)$ defined by the following recursive algorithm.

5.2.14. Algorithm (Associated Tree)

Let $\alpha = \beta i \gamma$ be a string of distinct symbols over N_n with smallest symbol i. Then $t(\alpha)$ is given by

1. 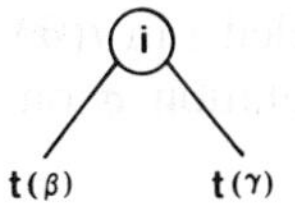

if i is a modified minimum ($\beta \neq \varepsilon, \gamma \neq \varepsilon$).

2. 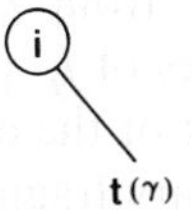

if i is a double rise ($\beta = \varepsilon, \gamma \neq \varepsilon$).

3. 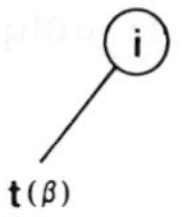

if i is a double fall ($\beta \neq \varepsilon, \gamma = \varepsilon$).

4. 

if i is a modified maximum ($\beta = \varepsilon, \gamma = \varepsilon$). □

It follows that the termini of the left-most and right-most paths of $t(\sigma)$ are, respectively, the initial symbol and final symbol of the permutation σ. Figure 5.2.2 gives the associated tree for the permutation 829613547.

The next decomposition gives a one-to-one correspondence between weighted paths and permutations.

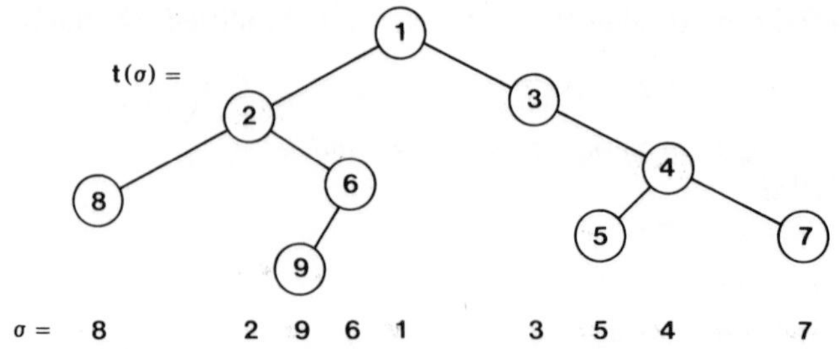

Figure 5.2.2. The associated tree for 829613547.

5.2.15. Decomposition (Françon–Viennot)

There is a one-to-one correspondence between permutations on N_n *and weighted paths, with* $n-1$ *steps, from altitude* 0 *to altitude* 0 *with possibility functions*

$$\psi_1(k)=k+1, \qquad \psi_0(k)=2(k+1), \qquad \psi_{-1}(k)=k+1.$$

Proof: Configurations **(1)**, **(2)**, **(3)**, **(4)** of **5.2.14** are called **elementary fragments** with vertex labeled i, and their incident edges are called **open edges**. Let $a_i(\sigma)$ be the elementary fragment containing the vertex labeled i in $t(\sigma)$, for $i=1,\ldots,n$, where $t(\sigma)$ is the associated tree for a permutation σ on N_n, obtained by Algorithm 5.2.14.

Weights $(W_1(\sigma),\ldots, W_n(\sigma))$ are associated with $(a_1(\sigma),\ldots, a_n(\sigma))$ as follows. Let $(f_1,\ldots,f_n)$ be configurations (called **fragments**) constructed recursively, with $f_n=t(\sigma)$. Then, for $i=n,\ldots,2$, f_{i-1} is obtained from f_i by detaching the elementary fragment $a_i(\sigma)$ from f_i. If the open edges of f_{i-1} are indexed from left to right, starting at zero, then $W_i(\sigma)$ is the index of the open edge in f_{i-1} to which $a_i(\sigma)$ was attached. Finally, f_1 is the elementary fragment $a_1(\sigma)$, and we set $W_1(\sigma)=0$. Of course, by the way in which $t(\sigma)$ is constructed, f_i is a connected structure with vertices labeled $1,\ldots, i$ for $i=1,\ldots,n$, so this procedure is well defined. Moreover, $t(\sigma)$ is uniquely reconstructible from

$$\pi=\big((a_1(\sigma),\ldots, a_n(\sigma)),(W_1(\sigma),\ldots, W_n(\sigma))\big)$$

by beginning with $f_1=a_1(\sigma)$ and successively constructing $f_2,\ldots,f_n=t(\sigma)$.

From π we uniquely construct the weighted path $(S_1\cdots S_{n-1},W_1(\sigma)\cdots W_{n-1}(\sigma))_0$, where $S_i=1,-1,0_\rho,0_\lambda$ if i is a modified minimum, modified maximum, double rise, or double fall, respectively. In the preceding, 0_ρ and 0_λ represent levels in the path that are distinguished by a marking then with indeterminates ρ and λ. This is, of course, a method for assigning a bivariate weight to levels. The two types of levels, 0_ρ and 0_λ, are called **right** and **left levels**, respectively, and their possibility functions, ψ_ρ and ψ_λ, are such that

$\psi_0 = \psi_\rho + \psi_\lambda$. In any permutation σ on N_n, the element n is a modified maximum, and f_{n-1} has a unique open edge, so $W_n(\sigma) = 0$ and π is uniquely recoverable.

The altitude of the initial vertex of the step S_i is one less than the number of open edges in f_{i-1}, and the altitudes of the initial and terminal vertices in the weighted path are zero. Thus the weighted path which we construct is from altitude 0 to altitude 0, with no vertex at negative altitude. Furthermore, if S_i is a step at altitude k, then $W_i(\sigma)$ takes on any value from 0 to k, so the possibility functions for the weighted paths in this correspondence are given by

$$\psi_{-1}(k) = \psi_1(k) = k + 1, \qquad \psi_0(k) = 2(k + 1).$$

Finally, the construction is reversible and the result follows. □

Figure 5.2.3 gives the weighted path obtained by applying the Françon–Viennot decomposition to the permutation 829613547, whose associated tree was displayed in Figure 5.2.2.

The following result of Euler is now immediate.

5.2.16. All Permutations

Counting all permutations on N_n, we have, from the Françon–Viennot decomposition and the weighted path lemma,

$$\sum_{n\geqslant 1} n!x^{n-1} = J_x[2(k + 1), (k + 1)(k + 2) : (0, \infty)]. \qquad \square$$

We have therefore obtained a combinatorial explanation for the J-fraction representation of $\Sigma_{n\geqslant 0}(n + 1)!x^n$. This is derived algebraically in [**5.2.3**].

For further applications of the decomposition, it is important to note that modified minima, double falls, double rises, and modified maxima in a permutation are represented by rises, left levels, right levels, and falls, respectively, in the corresponding path. This information is used for deriving the

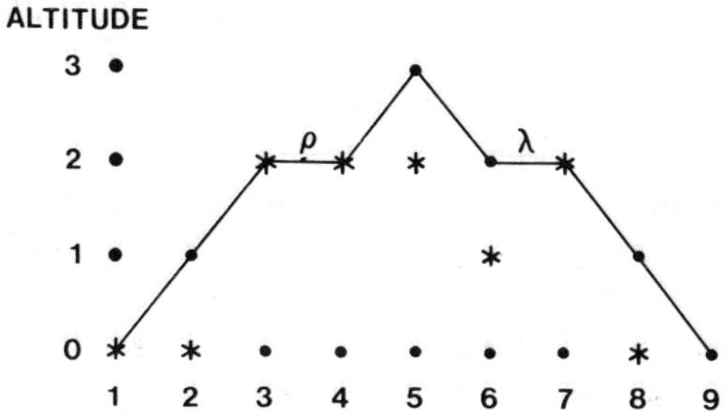

Figure 5.2.3. Weighted path corresponding to the permutation 829613547.

possibility functions for particular subsets of permutations. (Recall that the initial and final elements of a permutation are recognized as a modified minimum, double fall, double rise, or modified maximum by notionally placing a zero at both ends of the permutation.)

5.2.17. >-Alternating Permutations of Odd Length

A permutation is >-alternating (**4.2.6**) and has odd length if and only if it has no double rises or double falls so $\psi_0(k) = \psi_\rho(k) + \psi_\lambda(k) = 0$ and $\psi_1(k) = \psi_{-1}(k) = k + 1$ remain unchanged. Thus, from the weighted path lemma and the Françon–Viennot decomposition, the number of >-alternating permutations on N_{2n+1} is

$$[x^{2n}]J_x[0, (k+1)(k+2) : (0, \infty)].$$

From **3.2.22**, this number is also given by $[x^{2n+1}/(2n+1)!]\tan x$. An algebraic derivation of this J-fraction from $\tan x$ is given in [**5.2.4**]. □

A permutation σ of even length $2n$ is >-alternating if and only if $\sigma(2n+1)$ is >-alternating. This condition on the right-most element can be applied as a special case of the following corollary of the associated tree algorithm.

5.2.18. Corollary (Right-most Element in a Permutation)

Let σ be a permutation on N_n. If the application of the associated tree algorithm to σ entails attaching the elementary fragments

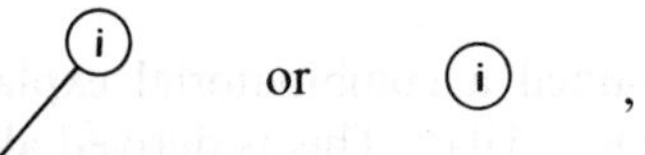

where $1 \leqslant i \leqslant n$, to a right-most open edge of a fragment, then i is the final element of σ, and conversely. □

5.2.19. >-Alternating Permutations of Even Length

Consider the set A of >-alternating permutations with odd length, terminated by their largest element. For the weighted paths corresponding to elements of A, we have $\psi_0(k) = 0$, as in **5.2.17**. Suppose that $\sigma \in \mathsf{A}$ has length $2n + 1$. From Corollary 5.2.18, we force $2n + 1$ to be the terminal element of σ by not attaching an elementary fragment corresponding to a modified maximum to the right-most open edge at any stage prior to $2n + 1$ in the construction of $t(\sigma)$. Thus the possibility function for falls in paths corresponding to elements of A is one less than in **5.2.17**, so $\psi_{-1}(k) = k$ for elements of A. Furthermore, modified minima are unaffected, so $\psi_1(k) = k + 1$. Thus the number of >-alternating permutations of length $2n$, which is equal to the number of permutations of length $2n+1$ in A, is

$$[x^{2n}]J_x\left[0,(k+1)^2:(0,\infty)\right]$$

by the weighted path lemma. From **3.2.22** this number is also given by $[x^{2n}/(2n)!]\sec x$. An algebraic derivation of this J-fraction from $\sec x$ is given in [**5.2.3**]. □

Surprisingly, a combinatorial interpretation of the Jacobi elliptic function $\mathrm{cn}(x,\alpha)$ is obtained from the preceding result, André's problem, by considering the parity of the modified minima.

5.2.20. >-Alternating Permutations of Even Length with Even-Valued Minima—the Jacobi Elliptic Function $cn(x,\alpha)$

Let $a(2n,l)$ be the number of >-alternating permutations on N_{2n} with l even-valued modified minima. For >-alternating permutations of odd length, even-valued modified minima are represented by rises at odd altitude in the corresponding path. Thus, marking rises at odd altitude in the paths of **5.2.19** by an indeterminate α, we have

$$\begin{aligned} a(2n,l) &= [\alpha^l x^{2n}]J_x\left[0,(k+1)^2\alpha^{k(\mathrm{mod}\,2)}:(0,\infty)\right] \\ &= (-1)^n\left[\alpha^l\frac{x^{2n}}{(2n)!}\right]\mathrm{cn}(x,\alpha), \qquad \text{from [5.2.8]}. \end{aligned}$$ □

An explicit expression for $a(2n,l)$ is obtained by using Stieltjes–Rogers polynomials. Thus, from Proposition 5.2.4,

$$a(2n,l)=[\alpha^l]\sum_{\substack{k\geq 0\\ n_0,\ldots,n_k\geq 1\\ n_0+\cdots+n_k=n}} 1^{2n_0}2^{2n_1}3^{2n_2}\cdots(k+1)^{2n_k}$$

$$\times\,\alpha^{n_1+n_3+n_5+\cdots}\prod_{j=0}^{k-1}\binom{n_j+n_{j+1}-1}{n_j-1}$$

so, for example, $a(4,1)=4$. These four permutations (where the single even-valued modified minimum is underlined) are $4\underline{2}31, 314\underline{2}, 3\underline{2}41, 413\underline{2}$.

So far, this section has been concerned with combinatorial interpretations of J-fractions of the form $J_x[\kappa_k,\lambda_k:(0,\infty)]$. We now turn our attention to the combinatorial use of its hth **convergent** $J_{(x)}^{\langle h\rangle}=J_x[\kappa_k,\lambda_k:(0,h)]$ and its hth **truncation** $J^{\langle h\rangle}(x)=J_x[\kappa_k,\lambda_k:(h,\infty)]$. The next proposition, which is a classical one, gives expressions for these two continued fractions in terms of the **numerator polynomial** $P_h(x)$ and **denominator polynomial** $Q_h(x)$.

5.2.21. Proposition (Recurrence Equation, Determinant Identity)

Let $J(x) = J_x[\kappa_k, \lambda_k : (0, \infty)]$, *where* $\lambda_k \neq 0$, $k \geqslant 0$. *Then*

1. $J^{\langle h\rangle}(x) = P_h(x)/Q_h(x)$ *for* $h \geqslant 0$, *where* $P_k(x)$ *and* $Q_k(x)$ *satisfy the recurrence equation*

$$u_k = (1 - \kappa_k x)u_{k-1} - \lambda_{k-1}x^2 u_{k-2} \qquad \textit{for } k \geqslant 1$$

with initial conditions $P_{-1} = 0$, $P_0 = 1$ *and* $Q_{-1} = 1$, $Q_0 = 1 - \kappa_0 x$.

2. P_h *and* Q_h *are polynomials in x of degrees at most h and* $h + 1$, *respectively.*
3. $P_k Q_{k-1} - P_{k-1}Q_k = (\lambda_0 \cdots \lambda_{k-1})x^{2k}$ *for* $k \geqslant 1$.
4. $J^{\langle h\rangle}(x) = (\lambda_{h-1}x^2)^{-1}(Q_{h-1}J - P_{h-1})/(Q_{h-2}J - P_{h-2})$ *for* $h \geqslant 1$. □

In view of the path lemma (**5.2.7**), truncations and convergents can be given combinatorial interpretations involving conditions on the height of paths enumerated with respect to step-type. These interpretations are given in the following result, which extends the path lemma (included as (**1**)).

5.2.22. Theorem (Path Enumeration)

The generating function with respect to step-type for the set of all paths from altitude a to altitude b such that

1. *Height* $\geqslant 0$, $a = b = 0$ *is* $J = J_x[\gamma_i, \alpha_i\beta_{i+1} : (0, \infty)]$.
2. *Height* $\geqslant 0$, $a = 0$, $b = h$ *is* $(\beta_1 \cdots \beta_h x^h)^{-1}(Q_{h-1}J - P_{h-1})$.
3. *Height* $\leqslant h$, $a = b = 0$ *is* P_h/Q_h.
4. *Height* $= h$, $a = b = 0$ *is* $(\alpha_0 \cdots \alpha_{h-1})(\beta_1 \cdots \beta_h)x^{2h}/Q_hQ_{h-1}$.
5. *Height* $= h$, $a = 0$, $b = h$ *is* $(\alpha_0 \cdots \alpha_{h-1})x^h/Q_h$.
6. *Height* $\geqslant h$, $a = b = h$ *is* $(\alpha_{h-1}\beta_h x^2)^{-1}(Q_{h-1}J - P_{h-1})/(Q_{h-2}J - P_{h-2})$.

P_h, Q_h *are the numerator and denominator polynomials for* J.

Proof:

1. This is the path lemma.

2. From Decomposition 5.2.9 the set of paths is $\mathsf{H}_0(1)_0\mathsf{H}_1(1)_1 \cdots (1)_{h-1}\mathsf{H}_h$. But the generating function for H_i is $J^{\langle i\rangle}(x)$, and the generating function for $(1)_i$ is $\alpha_i x$. Thus the required generating function is $J^{\langle 0\rangle}\alpha_0 x J^{\langle 1\rangle}\alpha_1 x \cdots \alpha_{h-1}xJ^{\langle h\rangle}$. The result follows from Proposition 5.2.21(4).

3. The generating function is $J^{\langle h\rangle}(x)$, by Lemma 5.2.7, since rises at altitude h are not allowed. The result follows from Proposition 5.2.21.

4. From **(3)** the required generating function is $J^{\langle h\rangle}(x) - J^{\langle h-1\rangle}(x)$. But, by Proposition 5.2.21(1, 3), this is equal to

$$P_h Q_h^{-1} - P_{h-1} Q_{h-1}^{-1} = (\alpha_0 \cdots \alpha_{h-1})(\beta_1 \cdots \beta_h) x^{2h} / Q_h Q_{h-1}.$$

5. The generating function is obtained by replacing J by $J^{\langle h\rangle}$ in **(2)**, and is therefore $(\beta_1 \cdots \beta_h x^h)^{-1}(Q_{h-1} J^{\langle h\rangle} - P_{h-1})$. The result follows from Proposition 5.2.21.

6. The generating function is $J^{\langle h\rangle}(x)$ and the result follows from Proposition 5.2.21(4). □

Let the **height** of a permutation be the height of its corresponding path. Another refinement of André's problem is obtained by restricting height and is an immediate application of the preceding theorem.

5.2.23. >-Alternating Permutations of Even Length with Respect to Height—Meixner Polynomial

From the path enumeration theorem, the number of >-alternating permutations of length $2n$ and height h is

$$c(2n, h) = [x^{2n}] x^{2h} h!^2 / Q_h Q_{h-1}$$

where, from **5.2.19**, Q_h is the denominator polynomial for the hth convergent of $J_x[0, (k+1)^2 : (0, \infty)]$. From Proposition 5.2.21, Q_k satisfies $Q_k = Q_{k-1} - k^2 x^2 Q_{k-2}$ for $k \geqslant 1$, with $Q_0(x) = Q_{-1}(x) = 1$. □

Setting $M_{k+1}(x) = x^k Q_k(x^{-1})$ in the preceding recurrence yields

$$M_{k+1}(x) = x M_k(x) - k^2 M_{k-1}(x), \qquad k \geqslant 1$$

with $M_0(x) = 1$, $M_1(x) = x$. Therefore, from **[1.1.12]**, $x^k Q_k(x^{-1})$ is a Meixner polynomial.

NOTES AND REFERENCES

This section is based on Flajolet (1980); for classical results on continued fractions see Perron (1929) and Wall (1948); for special functions that occur in connection with the three-term recurrence equations in **5.2.21**, see Chihara (1978).

5.2.4 Flajolet (1980), Rogers (1907); **5.2.7** Flajolet (1980), Good (1969), Jackson (1978), Read (1979), Touchard (1952); **5.2.10** Rogers (1907), Stieltjes

(1889); **5.2.15** Françon and Viennot (1979); **5.2.17, 19, 20** Flajolet (1980); **5.2.21** Perron (1929), Wall (1948); **5.2.22, 23** Flajolet (1980); [**5.2.1**] Rényi and Szekeres (1967); [**5.2.2**] Flajolet (1980); [**5.2.3**] Gauss; Flajolet (1980), Stieltjes (1889); [**5.2.4**] Stieltjes (1889); [**5.2.5**] Flajolet (1980); [**5.2.6**] Flajolet (1980), Stieltjes (1889); [**5.2.8**] Rogers (1907); [**5.2.9**] Heine (1846, 1847), Gauss (1813); [**5.2.11**] Jackson (1978), Rosen (1976); [**5.2.13**] Carlitz (1974); [**5.2.14**] Françon and Viennot (1979); [**5.2.15**] Flajolet (1980); [**5.2.17**] Flajolet (private communication, 1982); [**5.2.18, 19, 20**] Flajolet (1980); [**5.2.21–23**] Flajolet (1982), Gessel (1981); [**5.2.24**] Dumont (1979), Flajolet and Françon (1981), Viennot (1980).

EXERCISES

5.2.1. (a) Show that the number of labeled rooted trees on n vertices, with n_i at height i, for $i = 1, \ldots, k$, is

$$\frac{n!}{n_1! \cdots n_k!} n_1^{n_2} n_2^{n_3} \cdots n_{k-1}^{n_k}, \qquad \text{where } n_1 + \cdots + n_k = n - 1.$$

(b) Show that $\exp\{x\lambda_1 \exp\{x\lambda_2 \exp\{\cdots\}\}\} = 1 + \sum_{m \geqslant 1} A_m x^m$, where

$$A_m = \sum_{k \geqslant 1} \sum_{\substack{n_1, \ldots, n_k \geqslant 1 \\ n_1 + \cdots + n_k = m}} \frac{\lambda_1^{n_1}}{n_1!} \cdots \frac{\lambda_k^{n_k}}{n_1!} n_1^{n_2} \cdots n_{k-1}^{n_k}.$$

5.2.2. Let $J_x[\kappa_l, \lambda_l : (0, \infty)] = 1 + \sum_{n \geqslant 1} J_n x^n$. Prove that

$$J_p = \kappa_0^p + \sum_{k \geqslant 0} \sum_{\substack{m_0, \ldots, m_{k+1} \geqslant 0 \\ n_0, \ldots, n_k \geqslant 1}} (\lambda_0^{n_0} \cdots \lambda_k^{n_k})(\kappa_0^{m_0} \cdots \kappa_{k+1}^{m_{k+1}})$$

$$\times \binom{m_{k+1} + n_k - 1}{m_{k+1}} \prod_{j=0}^{k} \begin{bmatrix} m_j + n_j + n_{j-1} - 1 \\ m_j, n_j, n_{j-1} - 1 \end{bmatrix}$$

in which $n_{-1} = 1$ and the second summation is subject to the condition $2(n_0 + \cdots + n_k) + (m_0 + \cdots + m_{k+1}) = p$. (We call J_p a **Jacobi–Rogers polynomial.**)

5.2.3. Derive addition formulas to prove that

(a) $1 + \sum_{n \geqslant 1} 1.3 \cdots (2n-1) x^{2n} = J_x[0, k+1 : (0, \infty)]$.

(b) $1 + \sum_{n \geqslant 1} r(r+1) \cdots (r+n-1) x^n$

$$= J_x[r + 2k, (k+1)(k+r) : (0, \infty)],$$

(c) $\sum_{n \geqslant 0} E_{2n} x^{2n} = J_x[0, (k+1)^2 : (0, \infty)]$,

where $E_{2n} = \left[x^{2n}/(2n)!\right] \sec x$ is an **Euler number**.

5.2.4. (a) If $f(x)$ is a formal power series such that $f(0) = 0$, prove that

$$\hat{f}(x) = x\widehat{f'}(x).$$

(b) Use (a) to prove that

$$\sum_{n\geqslant 0} E_{2n+1}x^{2n+1} = xJ_x\left[0, (k+1)(k+2) : (0, \infty)\right],$$

where $E_{2n+1} = \left[x^{2n+1}/(2n+1)!\right]\tan x$ is an **Euler number**.

5.2.5. (a) Show that if $\hat{f} = J_x[\kappa_k, \lambda_k : (0, \infty)]$, then

$$\hat{g} = J_x\left[\alpha + \kappa_k, \lambda_k : (0, \infty)\right], \qquad \text{where } g(x) = e^{\alpha x}f(x).$$

(b) Use (a) to prove that $\Sigma_{n\geqslant 0} d_n x^n = J_x[2k, (k+1)^2 : (0, \infty)]$, where $d_n = [x^n/n!]e^{-x}(1-x)^{-1}$ is a derangement number.

5.2.6. By considering $\theta(x+y) = \theta(x)\theta(y) + \theta(x) + \theta(y)$, where $\theta(x) = e^x - 1$, prove that

(a) $\Sigma_{n,k\geqslant 0} S(n,k)u^k x^n = J_x[u + k, u(1+k) : (0, \infty)]$, where

$$S(n,k) = \left[u^k \frac{x^n}{n!}\right]\exp u(e^x - 1)$$

is a Stirling number of the second kind,

(b) $\Sigma_{n,k\geqslant 0} A_{n,k}u^k x^n = J_x[k + u(1+k), u(k+1)^2 : (0, \infty)]$, where

$$A_{n,k} = \left[u^k \frac{x^n}{n!}\right]\frac{1-u}{1-ue^{x(1-u)}}$$

is an Eulerian number.

5.2.7. (a) If $f(x)$ is a formal power series, prove that

$$\hat{f}(x) = x^{-1}\int_0^\infty f(t)e^{-tx^{-1}}\,dt.$$

(b) Let $t_n(x) = x^{-1}\int_0^\infty (\tan^n t)e^{-tx^{-1}}\,dt$. Show that $t_n = nx(t_{n-1} + t_{n+1})$, and hence derive [**5.2.4**(b)].

5.2.8. Let

$$x = \int_0^\phi \frac{d\theta}{(1 - m^2\sin^2\theta)^{1/2}}$$

and let

$$\mathrm{cn}(x,m) = \cos\phi, \mathrm{sn}(x,m) = \sin\phi, \mathrm{dn}(x,m) = (1 - m^2\sin^2\phi)^{1/2}.$$

These functions are called **Jacobi elliptic functions**, and for brevity we denote them by $\mathrm{cn}(x)$, $\mathrm{sn}(x)$, and $\mathrm{dn}(x)$; m is called the **amplitude**.

(a) Prove that $\mathrm{cn}(0) = \mathrm{dn}(0) = 1$, $\mathrm{sn}(0) = 0$.

(b) Prove that

(i) $\frac{d}{dx}\mathrm{cn}\, x = -\mathrm{sn}\, x\,\mathrm{dn}\, x$

(ii) $\frac{d}{dx}\mathrm{sn}\, x = \mathrm{cn}\, x\,\mathrm{dn}\, x$

(iii) $\frac{d}{dx}\mathrm{dn}\, x = -m^2\mathrm{sn}\, x\,\mathrm{cn}\, x.$

(c) Let $f(x) = \mathrm{sn}\, x$ and $g_n(x) = x^{-1}\int_0^\infty \mathrm{sn}^n t\, e^{-tx^{-1}}\,dt$. By finding a three-term recurrence equation for g_n prove that $\hat{f}(x) = xJ_{-x^2}[\kappa_j, \lambda_j : (0, \infty)]$, where

$$\kappa_j = (1 + m^2)(1 + 2j)^2, \qquad \lambda_j = m^2(1 + 2j)(2 + 2j)^2(3 + 2j).$$

(d) Let $f(x) = \mathrm{cn}\, x$ and $g_n(x) = x^{-1}\int_0^\infty \mathrm{sn}^n t\,\mathrm{cn}\, t\, e^{-tx^{-1}}\,dt$.

(i) Prove that $\hat{f}(x) = J_{-x^2}[\kappa_j, \lambda_j : (0, \infty)]$, where

$$\kappa_j = (2j)^2 m^2 + (2j + 1)^2, \qquad \lambda_j = m^2(1 + 2j)^2(2 + 2j)^2.$$

(ii) Prove that $\hat{f}(x) = S_{-x^2}[(j + 1)^2\alpha_j : (0, \infty)]$, where

$$\alpha_j = \begin{cases} 1 & \text{if } j \equiv 0 \bmod 2 \\ m^2 & \text{if } j \equiv 1 \bmod 2. \end{cases}$$

5.2.9. (a) Let

$${}_2\Phi_1\left(\begin{matrix} a, b \\ c \end{matrix} : q, x\right) = 1 + \sum_{j\geq 1} x^j \binom{a - 1 + j}{j}_q \binom{b - 1 + j}{j}_q \Big/ \binom{c - 1 + j}{j}_q.$$

Prove **Heine's theorem**, that

$${}_2\Phi_1\left(\begin{matrix} a, b + 1 \\ c + 1 \end{matrix} : q, x\right) \Big/ {}_2\Phi_1\left(\begin{matrix} a, b \\ c \end{matrix} : q, x\right) = S_x[\lambda_k : (0, \infty)],$$

where

$$\lambda_{2j} = q^{b+j}(a + j)_q(c - b + j)_q/(c + 2j)_q(c + 2j + 1)_q$$

$$\lambda_{2j-1} = q^{a+j-1}(b + j)_q(c - a + j)_q/(c + 2j - 1)_q(c + 2j)_q$$

and $(n)_q = (1 - q^n)(1 - q)^{-1}$.

(b) Deduce from Heine's theorem that

(i) $1 + \sum_{j\geq 1}(1)_q(3)_q \cdots (2j - 1)_q x^j = S_x[q^k(k + 1)_q : (0, \infty)]$,

(ii) $1 + \sum_{n\geq 1} n!_q x^n = S_x[\lambda_k : (0, \infty)]$, where $\lambda_{2j} = q^j(j + 1)_q$, $\lambda_{2j-1} = q^j(j)_q$.

5.2.10. (a) Show that

$$\sum_{n \geq 0} C_n x^n = S_x[1 : (0, \infty)],$$

where C_n is the Catalan number $\frac{1}{n+1}\binom{2n}{n}$, and hence that C_n is the number of level-free paths of length $2n + 1$ from altitude 0 to altitude 0.

(b) Show that

$$\sum_{n \geq 0} M_n x^n = \frac{1}{2x^2}\{1 - x - (1 - 2x - 3x^2)^{1/2}\} = J_x[1, 1 : (0, \infty)],$$

where M_n is the **Motzkin number**, and hence that M_n is the number of paths of length $n + 1$ from altitude 0 to altitude 0.

5.2.11. A **lead code** is a sequence of non-negative integers beginning and ending with 0 and with adjacent elements differing by 1.

(a) Show that the number of lead codes of length $2n + 1$ and type $\mathbf{i} = (i_0, i_1, \ldots)$ is

$$[x^{2n+1} x_0^{i_0} x_1^{i_1} \cdots] x x_0 S_{x^2}[x_k x_{k+1} : (0, \infty)].$$

(b) Show that there are $\frac{1}{n+1}\binom{2n}{n}$ lead codes of length $2n + 1$ (see [**2.7.13**]).

(c) For a lead code $\sigma = \sigma_1 \cdots \sigma_{2n+1}$, let $W(\sigma) = \prod_{j=1}^{2n+1}(\sigma_j + 1)$. Show that

$$\sum_{\sigma} W(\sigma) = \left[\frac{x^{2n+1}}{(2n+1)!}\right] \tan x,$$

where the sum is over all lead codes of length $2n + 1$.

5.2.12. The **area** of the path $u_0 \cdots u_l$, where $\mathrm{alt}(u_0) = \mathrm{alt}(u_l) = 0$, is defined to be $\sum_{i=0}^{l} \mathrm{alt}(u_i)$.

(a) Prove that the number of paths from altitude 0 to altitude 0 with n steps and area m is

$$[x^n q^m] J_x[q^k, q^{2k+1} : (0, \infty)].$$

(b) Prove that the number of lead codes of length $2n + 1$ whose elements sum to m is

$$[x^n q^m] S_x[q^{2k+1} : (0, \infty)].$$

(c) By considering $S_x[q^k : (0, \infty)] = N(x, q)/D(x, q)$, where $D(0, q) = 1$, show that

$$S_x[q^{2k+1} : (0, \infty)] = \frac{1 + \sum_{n \geq 1} (-x)^n q^{n(2n+1)} \prod_{i=1}^{n} (1 - q^{2i})^{-1}}{1 + \sum_{n \geq 1} (-x)^n q^{n(2n-1)} \prod_{i=1}^{n} (1 - q^{2i})^{-1}}.$$

($S_{-x}[q^k : (0, \infty)]$ is a continued fraction of Ramanujan).

5.2.13. (a) Derive a one-to-one correspondence between lead codes of length $2n+3$ and the set A_n of integer sequences $(a_1, \ldots, a_n)$ such that

$$0 \leqslant a_1 \leqslant \cdots \leqslant a_n, \qquad \text{where } a_j \leqslant j \text{ for } 1 \leqslant j \leqslant n.$$

(b) Let $b(n, m)$ be the number of elements of A_n for which $a_1 + \cdots + a_n = \binom{n+1}{2} - m$. Show that $b(n, m) = [x^{n+1}q^m]S_x[q^k : (0, \infty)]$.

(c) Prove that

$$1 - \sum_{n, m \geqslant 0} (-1)^n b(n, m) q^{n+m+1} = \prod_{i \geqslant 0} \frac{(1-q^{5i+1})(1-q^{5i+4})}{(1-q^{5i+2})(1-q^{5i+3})}.$$

5.2.14. Prove that the number of permutations on N_n in which $\pi_1, \ldots, \pi_4$ are, respectively, the sets of modified maxima, double rises, double falls, and modified minima is

$$\prod_{k=2}^{n-1}\left(1 + \sum_{j=1}^{k-1} i_j\right),$$

where $\pi_1 \,\dot\cup\, \cdots \,\dot\cup\, \pi_4 = \mathsf{N}_n$ and $i_j = \begin{cases} 1 & \text{if } j \in \pi_1 \\ 0 & \text{if } j \in \pi_2 \cup \pi_3 \\ -1 & \text{if } j \in \pi_4. \end{cases}$

5.2.15. Deduce by means of the Françon–Viennot decomposition (a) **5.2.11** and (b) [**5.2.5**(b)].

5.2.16. (a) Prove that the number of permutations on N_n in which the modified maxima appear in strictly increasing order (from the left) is

$$[x^{n-1}]J_x[2k+2, k+1 : (0, \infty)].$$

(b) Prove that there are $\frac{1}{n+1}\binom{2n}{n}$ >-alternating permutations on N_{2n+1} whose modified maxima appear in increasing order from left to right and whose modified minima appear in increasing order from left to right.

5.2.17. (a) Let $C(x)$ be the exponential generating function for permutations on N_n with no <-strings of length 3. Deduce from **4.2.15** and the <-transformation lemma that

$$C(x) = \left\{ \sum_{i \geqslant 0} \frac{x^{3i}}{(3i)!} - \frac{x^{3i+1}}{(3i+1)!} \right\}^{-1}.$$

(b) Use the Françon–Viennot decomposition to show that

$$\hat{C}(x) = J_x\left[k+1, (k+1)^2 : (0, \infty)\right].$$

(c) Show directly that $C(x)$ has an addition formula with parameters $\{((k!)^2, \frac{1}{2}(k+1)(k+2)) \mid k \geqslant 0\}$, and hence deduce the result of part (b) from the Stieltjes–Rogers J-fraction theorem.

5.2.18. A **doubled permutation** on N_n is a permutation on N_n such that $2j+1$ and $2j+2$ both appear as modified maxima, modified minima, double rises, or double falls for j such that $2j+2 \leqslant n$. Prove that the number of doubled permutations

(a) On N_{2n+1} with i rises is $(-1)^n[x^{2n+1}m^i]\widehat{\mathrm{sn}}(x,m)$.

(b) On N_{2n} with i rises is $(-1)^n[x^{2n}m^i]\widehat{\mathrm{cn}}(x,m)$, where $\mathrm{sn}(x,m)$ and $\mathrm{cn}(x,m)$ are Jacobi elliptic functions (see [**5.2.8**]).

5.2.19. Let $\Pi = \{\pi^{(1)},\ldots,\pi^{(k)}\}$ be a partition of N_n into blocks $\pi^{(i)} = \{\pi_1^{(i)},\ldots,\pi_{p_i}^{(i)}\}$, for $1 \leqslant i \leqslant k$, indexed so that $\pi_1^{(i)} < \cdots < \pi_{p_i}^{(i)}$ for $1 \leqslant i \leqslant k$ and $\pi_1^{(1)} < \cdots < \pi_1^{(k)}$. The **layer diagram** of Π consists of k horizontal line segments in the Euclidean plane. The ith of these segments, l_i, joins $(\pi_1^{(i)}, i)$ to $(\pi_{p_i}^{(i)}, i)$ and has vertices at $(\pi_1^{(i)}, i), (\pi_2^{(i)}, i), \ldots, (\pi_{p_i}^{(i)}, i)$, for $i = 1,\ldots,k$. Let y_i, the **height** of $\pi^{(i)}$, be the number of horizontal segments, including l_i itself, that are intersected by the vertical segment from $(\pi_1^{(i)}, i)$ to $(\pi_1^{(i)}, 1)$. Thus the layer diagram for $\Pi = \{\{1,6\},\{2,4\},\{3,8\},\{5,7\}\}$ is

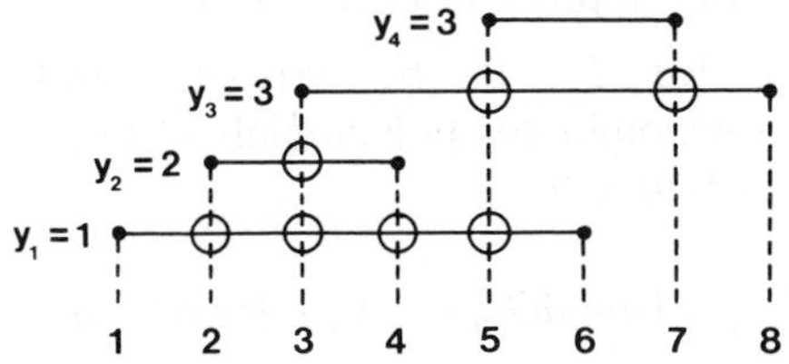

(a) By representing the initial element i of a line segment in the layer diagram by a rise and the final element j of a line segment by a fall, prove that there is a one-to-one correspondence between partitions of N_n with block sizes equal to 2, and weighted paths with n steps and possibility functions

$$\psi_{-1}(k) = k, \qquad \psi_0(k) = 0, \qquad \psi_1(k) = 1.$$

(b) By considering layer diagrams, prove that there is a one-to-one correspondence between partitions of N_n and weighted paths with n steps and possibility functions

$$\psi_{-1}(k) = k, \qquad \psi_0(k) = k+1, \qquad \psi_1(k) = 1.$$

(c) Show that the number of partitions of N_n with i subsets of size 1 and j of size at least 2 is

$$[u^i v^j x^n] J_x[k+u, (k+1)v : (0,\infty)].$$

(d) Show that the number of partitions of N_n with i subsets of size 1, j of size 2, and no others is

$$[u^i v^j x^n] J_x[u, (k+1)v : (0, \infty)].$$

5.2.20. Consider the layer diagram given in [**5.2.19**]. The dotted line in column i extends from row 0 to the line segment corresponding to the block that contains i. The number of intersections of these dotted lines with the line segments (excepting endpoints) of the layer diagram is called the **intersection number** of the partition. For Π in [**5.2.19**], the points of intersection are circled, and the intersection number is 7.

(a) Prove that the number of partitions of N_{2n}, with intersection number j and with blocks of size 2 alone is

$$[x^n q^j] S_x[q^k (k+1)_q : (0, \infty)].$$

(b) Give a combinatorial proof of [**5.2.9**(b)(ii)].

5.2.21. Let κ_l, λ_l, for $l \geqslant 0$, be integers, and let $\sum_{n \geqslant 0} J_n x^n = J_x[\kappa_l, \lambda_l : (0, \infty)]$, with denominator polynomials $Q_h(x) = \sum_{i=0}^{h+1} c(i, h) x^i$, for $h \geqslant 0$. Show that for each $m \geqslant 0$

(a) $\sum_{i=0}^{m+1} c(i, m) J_{n-i} \equiv 0 \bmod (\lambda_0 \cdots \lambda_m)$, for $n > m$.

(b) $\|J_{i+j-2}\|_{(m+1)\times(m+1)} = \prod_{i=1}^{m} \lambda_{m-i}^i$, a Hänkel determinant.

(c) The sequence $\{J_n \bmod (\lambda_0 \cdots \lambda_m) \mid n \geqslant 0\}$ is eventually periodic.

5.2.22. Let $B_n = [x^n/n!]\exp(e^x - 1)$, a Bell exponential number. Deduce from [**5.2.6**(a)] and [**5.2.21**] that

(a) $B_n - 10B_{n-1} + 5B_{n-2} + B_{n-4} \equiv 0 \bmod 24$, for $n \geqslant 4$.

(b) The sequence $\{B_n \bmod k \mid n \geqslant 0\}$ is eventually periodic for any $k > 1$.

5.2.23. (a) Deduce from **5.2.11** and [**5.2.21**] that

$$\left\| \binom{i+j-2}{i-1} \right\|_{m \times m} = 1.$$

(b) Deduce from [**5.2.3**(c)] and [**5.2.21**] that

$$E_{2n} \equiv 5^{n-1} \bmod 36, \qquad n \geqslant 1,$$

where $E_{2n} = [x^{2n}/(2n)!]\sec x$ is an Euler number.

5.2.24. Consider the edges of $t(\sigma)$, where σ is a permutation, to be directed away from the root. A left-path of σ is a subsequence of σ whose elements form a path with edges directed to the left in $t(\sigma)$. A **stratified** permutation is a nonempty permutation all of whose maximal left- and right-paths have odd (vertex) length; ***r*-stratified** and ***l*-stratified** permutations are defined similarly, with the difference that the right-most maximal right-path and left-most maximal left-path, respectively, have even length and the empty string ε belongs to both sets. Show that the number of

(a) Stratified permutations on N_{2n+1} with $2k$ rises is

$$(-1)^n\left[m^{2k}\frac{x^{2n+1}}{(2n+1)!}\right]\operatorname{sn}(x,m).$$

(b) r-Stratified permutations on N_{2n} with $2k-1$ rises is

$$(-1)^n\left[m^{2k}\frac{x^{2n}}{(2n)!}\right]\operatorname{dn}(x,m) \qquad \text{for } n>0.$$

(c) l-Stratified permutations on N_{2n} with $2k$ rises is

$$(-1)^n\left[m^{2k}\frac{x^{2n}}{(2n)!}\right]\operatorname{cn}(x,m),$$

where sn, cn, and dn are the Jacobi elliptic functions given in [**5.2.8**].

5.3. LATTICE PATHS

In Section 5.2 we considered the enumeration of weighted paths in the plane. Only three types of steps were allowed, namely, rises, levels, and falls corresponding, respectively, to the ordered pairs $(1,1)$, $(1,0)$, and $(1,-1)$ of increments in the x- and y-directions, and each point in the paths had non-negative altitude. We now turn our attention to the enumeration of lattice paths, or, more briefly, paths, with respect to a more general set of steps and with no restriction on altitude. The following definition generalizes Definition 5.2.5.

5.3.1. Definition (Altitude, Path, Step)

1. *Let $u=(i,j)\in\mathbb{Z}^2$. Then j is called the altitude of u and we write* $j=\operatorname{alt}(u)$.
2. *Let $u_0\cdots u_l$ be a sequence of lattice points in $\mathbb{Z}^2$ such that*

$$u_{k+1}=u_k+\alpha_k \qquad \text{for } 0\leqslant k<l,$$

where $\alpha_k\in\mathsf{S}$ and S is a prescribed subset of $\{\ldots,-1,0,1,\ldots\}\times\{-1,0,1,\ldots\}-(0,0)$. Then $u_0\cdots u_l$ is called a **(lattice) path** *on* **step-set** S *and is denoted by $\pi=(\alpha_0\cdots\alpha_{l-1})_{u_0}$.*

3. *The path π has* **origin** *u_0,* **terminus** *u_l, and* **length** *l. The ordered pair α_k is called a* **step**. □

In addition to removing the non-negative altitude condition of Section 5.2, we have also allowed the steps to be arbitrary pairs of integers, with the only restriction that a decrease greater than 1 in the ordinate of any step is not permissible. The latter restriction arises for algebraic reasons at a later stage (see statement of Theorem 5.3.4).

We adopt the convention that, unless otherwise stated, a path has origin $(0, 0)$. If $\pi_1 = (\alpha_0 \cdots \alpha_k)_{(0,0)}$ and $\pi_2 = (\beta_0 \cdots \beta_m)_{(0,0)}$ are two paths, then $\pi_1\pi_2$ is defined to be the path $(\alpha_0 \cdots \alpha_k \beta_0 \cdots \beta_m)_{(0,0)}$, as in Section 5.2. Thus $\pi_1\pi_2$ is obtained by translating π_2 in the plane so that its origin coincides with the terminus of π_1. If $\mathsf{P}, \mathsf{P}_1, \mathsf{P}_2$ are sets of paths such that

$$\mathsf{P} = \{\pi_1\pi_2 | \pi_1 \in \mathsf{P}_1, \pi_2 \in \mathsf{P}_2\},$$

then we write $\mathsf{P} = \mathsf{P}_1\mathsf{P}_2$.

The ordinary generating function used in this section for sets of paths is called the **path generating function**. The path generating function for a set P of paths is denoted by $\Phi_{\mathsf{P}}(x, y, z)$ and enumerates the paths in P with x and y marking, respectively, the x-coordinate and y-coordinate of the terminus of a path and z marking the length of a path. Thus if P, P_1, and P_2 are sets of paths with $\mathsf{P} = \mathsf{P}_1\mathsf{P}_2$, then $\Phi_{\mathsf{P}}(x, y, z) = \Phi_{\mathsf{P}_1}(x, y, z)\Phi_{\mathsf{P}_2}(x, y, z)$ by the product lemma for ordinary generating functions. For example, the path generating function for S^*, the set of all paths with origin $(0, 0)$ and step-set S, is

$$\Phi_{\mathsf{S}^*}(x, y, z) = \{1 - \Phi_{\mathsf{S}}(x, y, z)\}^{-1}.$$

Of course, the generating function $\Phi_{\mathsf{S}}(x, y, z)$ is straightforwardly obtained in any particular case. For instance, if $\mathsf{S} = \{(-1, 0), (1, 2), (3, -1)\}$ then $\Phi_{\mathsf{S}}(x, y, z) = x^{-1}z + xy^2z + x^3y^{-1}z$.

Our main purpose is to enumerate certain natural subsets of paths in S^* by imposing conditions on altitude. These sets are defined now.

5.3.2. Definition (Minus-, Zero-, Plus-path)

Let π be the path $u_0 \cdots u_l$ (where $u_0 = (0, 0)$ by the convention).

1. *If* $\operatorname{alt}(u_i) > \operatorname{alt}(u_l)$ *for $0 \leqslant i < l$ then π is called a* **minus-path**.
2. *If* $\operatorname{alt}(u_i) \geqslant 0$ *for $0 < i < l$ and* $\operatorname{alt}(u_l) = 0$ *then π is called a* **zero-path**.
3. *If* $\operatorname{alt}(u_i) > \operatorname{alt}(u_0) = 0$ *for $0 < i \leqslant l$, then π is called a* **plus-path**.

The sets of all minus-, zero-, and plus-paths in S^* *are denoted by* S_-, S_0, *and* S_+, *respectively.* □

Thus S_0 is the set of all paths, on the step-set S, with origin $(0, 0)$, terminus at altitude 0, and no points at negative altitude. When the step-set is

$\{(1,-1),(1,0),(1,1)\}$, these are precisely the paths considered in Section 5.2. The set S_- consists of all paths on the step-set S with origin $(0,0)$ and with terminus at a (strictly) smaller altitude than all other points in the path. In S_+, the origin $(0,0)$ has lower altitude than all other points in the path.

The single path of length 0, which is the point $(0,0)$ itself, lies in all of S_-, S_0, and S_+. Four paths P_1, P_2, P_3, P_4 in S^* for $\mathsf{S} = \{(1,1),(1,-1),(-1,1),(-1,-1)\}$ are illustrated in Figure 5.3.1. The path P_1 is in S_-, P_2 is in S_0, P_3 is in S_+, and P_4 is not in any of $\mathsf{S}_-, \mathsf{S}_0, \mathsf{S}_+$. Because negative increments in the abscissas are now allowed, these paths can be self-intersecting, and even retrace the same route.

We now state a decomposition for paths, which uniquely decomposes any path in S^* into a minus-path, a zero-path, and a plus-path.

5.3.3. Decomposition (Lattice Path)

Let S *be a set of steps. Then* $\mathsf{S}^* = \mathsf{S}_-\mathsf{S}_0\mathsf{S}_+$.

Proof: Let $\pi \in \mathsf{S}^*$ and suppose $\pi = (\alpha_0 \cdots \alpha_{l-1})_{(0,0)}$ is the path $u_0 \cdots u_l$. Let $u_{i_1}, \ldots, u_{i_k}$, where $i_1 < \cdots < i_k$ and $k \geqslant 1$, be the points of π with minimum altitude. Then $\pi = \pi_-\pi_0\pi_+$, where

$$\pi_- = \left(\alpha_0 \cdots \alpha_{i_1-1}\right)_{(0,0)} \in \mathsf{S}_-$$

$$\pi_0 = \left(\alpha_{i_1} \cdots \alpha_{i_k-1}\right)_{(0,0)} \in \mathsf{S}_0$$

$$\pi_+ = \left(\alpha_{i_k} \cdots \alpha_{l-1}\right)_{(0,0)} \in \mathsf{S}_+.$$

The paths π_-, π_0, π_+ are uniquely defined by this construction, since i_1, i_k, not necessarily distinct, are uniquely defined. The process is reversible, and the result follows. □

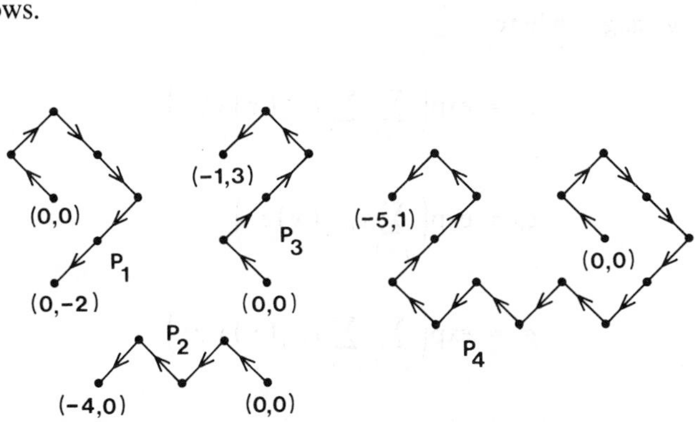

Figure 5.3.1. Four paths in S^* for $\mathsf{S} = \{(1,1),(1,-1),(-1,1),(-1,-1)\}$.

An example of this decomposition is provided by Figure 5.3.1, where $P_4 = P_1P_2P_3$ and P_1, P_2, P_3 are the unique minus-, zero-, and plus-paths determined by the preceding decomposition. The next result gives the path generating functions for $\mathsf{S}_-, \mathsf{S}_0, \mathsf{S}_+$ in terms of the path generating function for S, the set of steps.

5.3.4. Theorem (Lattice Path)

Let S *be a set of steps* (i, j), *where* $i < 0$ *for a finite number of steps and* $j = -1$ *for at least one step, so that* $\Phi_{\mathsf{S}} = y^{-1}zf(y)$ *for* $f(y) \in (\mathbb{Q}((x)))[[y]]_1$. *Then* $\Phi_{\mathsf{S}^*} = \{1 - y^{-1}zf(y)\}^{-1}$ *and*

1. $\Phi_{\mathsf{S}_-} = \{1 - y^{-1}w(z)\}^{-1}$.
2. $\Phi_{\mathsf{S}_0} = z^{-1}w(z)(f(0))^{-1}$.
3. $\Phi_{\mathsf{S}_+} = \Phi_{\mathsf{S}^*}(\Phi_{\mathsf{S}_-}\Phi_{\mathsf{S}_0})^{-1}$,

where $w = zf(w)$.

Proof: Consider the set of series of the form $1 + \sum_{j \geqslant 1}\sum_i a_{i,j}(x)y^iz^j$, where $a_{i,j}(x) \in \mathbb{Q}((x))$. Let G_-, G_0, G_+ be the sets of all such series where the second summation is over $i < 0$, $i = 0$, $i > 0$, respectively.

Step 1. We first prove that Φ_{S^*} has a unique factorization of the form $\Phi_{\mathsf{S}^*} = g_-g_0g_+$, where $g_- \in G_-$, $g_0 \in G_0$, $g_+ \in G_+$. Now $\Phi_{\mathsf{S}^*} = \{1 - y^{-1}zf(y)\}^{-1} = 1 + \sum_{i \geqslant 1} b_i(x, y)z^i$, where $b_i(x, y) \in \mathbb{Q}((x, y))$. Then $\log \Phi_{\mathsf{S}^*}$ exists in $(\mathbb{Q}((x, y)))[[z]]_0$ and

$$\log \Phi_{\mathsf{S}^*} = \sum_{j \geqslant 1} \sum_{-\infty < i < \infty} c_{i,j}(x)y^iz^j$$

$$= \sum_{j \geqslant 1}\sum_{i<0} c_{i,j}(x)y^iz^j + \sum_{j \geqslant 1} c_{0,j}(x)z^j + \sum_{j \geqslant 1}\sum_{i>0} c_{i,j}(x)y^iz^j.$$

Thus $\Phi_{\mathsf{S}^*} = g_-g_0g_+$, where

$$g_- = \exp\left(\sum_{j \geqslant 1}\sum_{i<0} c_{i,j}(x)y^iz^j\right)$$

$$g_0 = \exp\left(\sum_{j \geqslant 1} c_{0,j}(x)z^j\right)$$

$$g_+ = \exp\left(\sum_{j \geqslant 1}\sum_{i>0} c_{i,j}(x)y^iz^j\right)$$

and g_-, g_0, g_+ are in G_-, G_0, G_+, respectively.

Step 2. It follows immediately from Decomposition 5.3.3 that $\Phi_{\mathsf{S}*} = \Phi_{\mathsf{S}_-}\Phi_{\mathsf{S}_0}\Phi_{\mathsf{S}_+}$. Now by the definition of a minus-path Φ_{S_-} must contain only negative powers of y, so $\Phi_{\mathsf{S}_-} \in G_-$. Similarly, $\Phi_{\mathsf{S}_0} \in G_0$ and $\Phi_{\mathsf{S}_+} \in G_+$. But from Step 1 a factorization of this form is unique. Thus we have identified $\Phi_{\mathsf{S}_-}, \Phi_{\mathsf{S}_0}, \Phi_{\mathsf{S}_+}$ as the series g_-, g_0, g_+, respectively, determined in Step 1.

Step 3. We now use Step 1 to calculate $\Phi_{\mathsf{S}_-}, \Phi_{\mathsf{S}_0}, \Phi_{\mathsf{S}_+}$ in turn.

1. $$\Phi_{\mathsf{S}_-} = \exp \sum_{i \geqslant 1} y^{-i}[u^{-i}]\log\{1 - zu^{-1}f(u)\}^{-1}$$
$$= \exp \sum_{i \geqslant 1} y^{-i} \sum_{k \geqslant 1} k^{-1}z^k[u^{k-i}]\{f(u)\}^k.$$

By the Lagrange theorem $[z^k]w^i = ik^{-1}[u^{k-i}]\{f(u)\}^k$, where $w = zf(w)$, so

$$\Phi_{\mathsf{S}_-} = \exp \sum_{i,\, k \geqslant 1} i^{-1}z^k[z^k]\{y^{-1}w(z)\}^i = \{1 - y^{-1}w(z)\}^{-1}$$

and (**1**) is proved.

2. $\Phi_{\mathsf{S}*}^{-1} = (\Phi_{\mathsf{S}_-}\Phi_{\mathsf{S}_0}\Phi_{\mathsf{S}_+})^{-1}$,
so from (**1**),

$$1 - y^{-1}zf(y) = \left(1 - y^{-1}w(z)\right)\Phi_{\mathsf{S}_0}^{-1}\Phi_{\mathsf{S}_+}^{-1},$$

whence $y - zf(y) = (y - w(z))\Phi_{\mathsf{S}_0}^{-1}\Phi_{\mathsf{S}_+}^{-1}$. Now let $y = 0$ and note that $\Phi_{\mathsf{S}_+}^{-1}(x, 0, z) = 1$. Thus

$$\Phi_{\mathsf{S}_0}(x, y, z) = \Phi_{\mathsf{S}_0}(x, 0, z) = z^{-1}w(z)(f(0))^{-1}.$$

3. Immediate, from the lattice path decomposition. □

In the preceding result, the path generating functions for S_-, S_0, and S_+ have been determined by considering the relationship $\Phi_{\mathsf{S}*} = \Phi_{\mathsf{S}_-}\Phi_{\mathsf{S}_0}\Phi_{\mathsf{S}_+}$ and identifying the components in a unique factorization. This procedure is quite different from our usual one in which we would have tried to find two other equations to give us three equations in the three unknown path generating functions. The preceding factorization may be regarded as a discrete version of the **Wiener–Hopf factorization theorem** for random walks (Prabhu (1980)). The following example illustrates the use of the lattice path theorem.

5.3.5. Example

Let $\mathsf{S} = \{(-1, 1), (1, 1), (0, -1)\}$ be the step-set. Then

$$\Phi_{\mathsf{S}} = z\left(x^{-1}y + xy + y^{-1}\right) = zy^{-1}f(y),$$

where $f(y) = 1 + (x^{-1} + x)y^2$. From the lattice path theorem,

$$\Phi_{\mathsf{S}_-} = \left(1 - y^{-1}w(z)\right)^{-1} \quad \text{and} \quad \Phi_{\mathsf{S}_0} = z^{-1}w(z)f^{-1}(0) = z^{-1}w(z),$$

where $w = zf(w)$, so by the Lagrange theorem,

$$\Phi_{\mathsf{S}_-} = 1 + \sum_{i \geqslant 1} y^{-i} \sum_{n \geqslant i} \frac{z^n}{n}[\lambda^{n-1}]\left\{\frac{d}{d\lambda}\lambda^i\right\}(f(\lambda))^n$$

$$= 1 + \sum_{i \geqslant 1} y^{-i} \sum_{n \geqslant i} \frac{z^n}{n}[\lambda^{n-i}]i(1 + (x^{-1} + x)\lambda^2)^n$$

$$= 1 + \sum_{i \geqslant 1} \sum_{j \geqslant 0} y^{-i}z^{i+2j}\frac{i}{i+2j}\binom{i+2j}{j}(x^{-1}+x)^j, \qquad \text{where } n - i = 2j$$

$$= 1 + \sum_{i \geqslant 1} \sum_{j \geqslant 0} \sum_{k=0}^{j} y^{-i}z^{i+2j}x^{j-2k}\frac{i}{i+2j}\binom{i+2j}{j}\binom{j}{k}.$$

Also from the Lagrange theorem

$$\Phi_{\mathsf{S}_0} = \sum_{n \geqslant 1} \frac{z^{n-1}}{n}[\lambda^{n-1}]\{1 + (x^{-1} + x)\lambda^2\}^n$$

$$= \sum_{k \geqslant 0} \frac{z^{2k}}{2k+1}\binom{2k+1}{k}(x^{-1}+x)^k$$

$$= \sum_{k \geqslant 0} \sum_{i=0}^{k} \frac{1}{2k+1}\binom{2k+1}{k}\binom{k}{i}z^{2k}x^{k-2i}. \qquad \square$$

For instance, the number of paths in S_0 from $(0, 0)$ to $(1, 0)$, of length 6, with step-set $\mathsf{S} = \{(-1, 1), (1, 1), (0, -1)\}$ is $\frac{1}{7}\binom{7}{3}\binom{3}{1} = 15$ from the preceding result. In Figure 5.3.2, five of these 15 paths are displayed. They have been selected to include some self-intersecting paths, as well as some that retrace the same route.

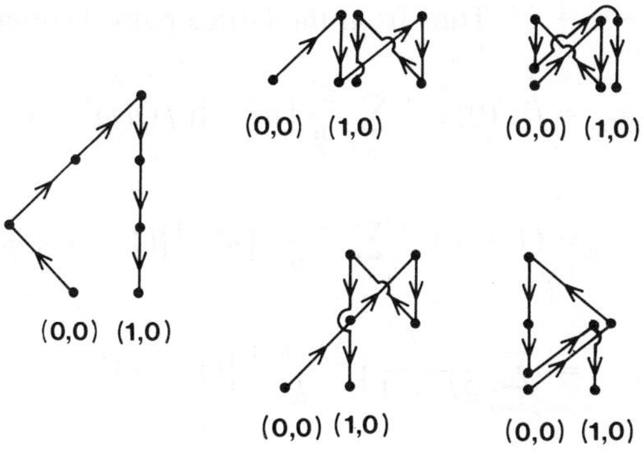

Figure 5.3.2. Some paths in S_0 for $S = \{(-1,1),(1,1),(0,-1)\}$ of length 6 from $(0,0)$ to $(1,0)$.

The main result of this section deals directly with paths for which the line $y = 0$ is a **barrier**. That is, in S_0 the paths can never descend below $y = 0$, and in S_+ the paths must stay above the line $y = 0$, after the initial point. By change of coordinates we now consider paths for which the line $y = mx$ is a barrier, for any integer m.

The change of coordinates that we use is called an ***m*-shift**, and transforms each step (i, j) in a path to the step $(i, j - im)$. Now suppose that π is a path on step-set S with origin $(0,0)$, which never descends below the line $y = mx$, and which terminates on the line $y = mx$. Then if π is m-shifted, the resulting path is a zero-path for the step-set obtained by m-shifting the steps in S. If π is a path on step-set S with origin $(0,0)$, and all other points strictly above the line $y = mx$, then by m-shifting π a plus-path for the m-shifted step-set is obtained. Finally, if π is a path on step-set S with origin $(0,0)$, which terminates on the line $y = mx - b$ for any $b \geqslant 0$, and in which all nonterminal points are above the line $y = mx - b$, then by m-shifting π we obtain a minus-path for the m-shifted step-set. Thus certain sets of paths for which $y = mx$ is a barrier can be enumerated by considering the minus-, zero-, and plus-paths for a transformed step-set, as in the following example.

5.3.6. Paths with an Oblique Barrier

We consider paths on the step-set $S = \{(0,1),(0,-1),(1,0)\}$ that terminate on $y = x$ and that remain on or above this line. By the preceding discussion, these are equivalent to zero-paths on the 1-shifted step-set $S^{(1)} = \{(0,1),(0,-1),(1,-1)\}$. Now

$$\Phi_{S^{(1)}} = z\left(y + y^{-1} + xy^{-1}\right) = zy^{-1}f(y),$$

where $f(y) = 1 + x + y^2$. Thus from the lattice path theorem

$$\Phi_{(\mathbf{S}^{(1)})_0} = f^{-1}(0)z^{-1}\sum_{n\geq 1}\frac{z^n}{n}[u^{n-1}](f(u))^n$$

$$= (1+x)^{-1}\sum_{n\geq 1}\frac{z^{n-1}}{n}[u^{n-1}](1+x+u^2)^n$$

$$= \sum_{k\geq 0}\frac{z^{2k}}{2k+1}\binom{2k+1}{k}(1+x)^k$$

$$= \sum_{k\geq 0}\sum_{i=0}^{k}\frac{1}{2k+1}\binom{2k+1}{k}\binom{k}{i}z^{2k}x^i.$$

Thus there are $\frac{1}{2k+1}\binom{2k+1}{k}\binom{k}{i}$ paths on step-set $\{(0,1),(0,-1),(1,0)\}$, of length $2k$, from $(0,0)$ to (i,i) that never go below the line $y = x$. □

The paths of length 4 in the preceding example are displayed in Figure 5.3.3. In agreement with the preceding result, 2 terminate at each of (0, 0) and (2, 2), and 4 terminate at (1, 1).

A second transformation that may be useful is the **reverse** of a path, obtained by considering the steps of the path in reverse order (though the steps themselves retain the same direction). Thus the reverse of the path $\pi = (\alpha_0 \cdots \alpha_k)_{(0,0)}$ is the path $(\alpha_k \cdots \alpha_0)_{(0,0)}$. Under this transformation zero-paths become paths that terminate on the x-axis and never go above the x-axis. The m-shift and reverse can be combined to enumerate, for example, paths that terminate on $y = x$ and never go above $y = x$. For the step-set $\{(0,1),(0,-1),(1,0)\}$ such paths of length 4 may be obtained from those in Figure 5.3.3 by reversing the order of the steps.

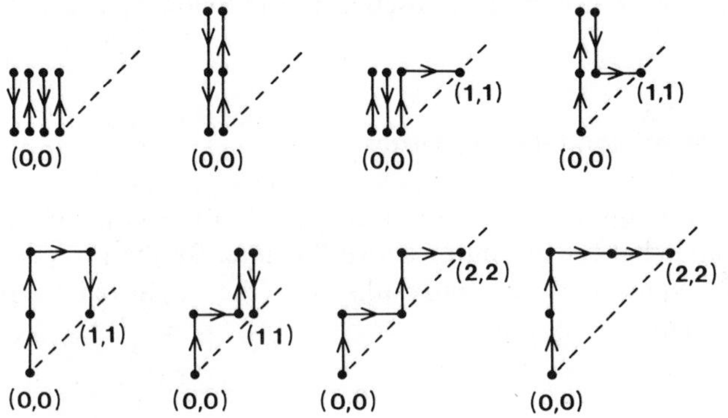

Figure 5.3.3. Paths on step-set $\{(0,1),(0,-1),(1,0)\}$ of length 4 that terminate on $y = x$ and never go below $y = x$.

NOTES AND REFERENCES

This section is based on Gessel (1980a); for further work on lattice paths see Mohanty (1979) and Narayana (1979).

[5.3.2] Aeppli (1923); **5.3.4**, **[5.3.3, 4]** Gessel (1980a); **[5.3.5]** Feller (1950), Gessel (1980a), MacMahon (1915), Takács (1967); **[5.3.7]** André (1887).

EXERCISES

5.3.1. Let $\mathsf{S} = \{(1,-1),(2,2)\}$. Show that

(a) $\Phi_{\mathsf{S}_-}(x, y, z) = 1 + \sum_{n \geqslant 1} \frac{1}{n} z^n \sum_{i \geqslant 0} (n - 3i)\binom{n}{i} x^{n+i} y^{-(n-3i)}$.

(b) $\Phi_{\mathsf{S}_0}(x, y, z) = \sum_{n \geqslant 0} \frac{1}{3n+1}\binom{3n+1}{n} x^{4n} z^{3n}$.

(c)

$$\Phi_{\mathsf{S}_+}(x, y, z) = \sum_{m, i \geqslant 0} \binom{m}{i} x^{m+i} z^m y^{3i-m}$$

$$\times \left\{1 - xzy^{-1} - \sum_{k \geqslant 0} \frac{1}{3k+2}\binom{3k+2}{k+1}(x^4 z^3)^{(k+1)}\right\}.$$

5.3.2. If $\mathsf{S} = \{(1,-1),(a+1, b-1)\}$, where $b > 1$, show that there are $\frac{1}{bk+1}\binom{bk+1}{k}$ zero-paths from $(0,0)$ to $((a+b)k, 0)$.

5.3.3. Let $\mathsf{S} = \{(0,1),(0,-1),(1,0)\}$. Show that

$$\Phi_{\mathsf{S}_0}(1, y, z) = \frac{1 - z - (1 - 2z - 3z^2)^{1/2}}{2z^2}.$$

5.3.4. Let S be the step-set $\{(i, j) \mid i, j \geqslant 0, i + j \neq 0\}$. Let c_n be the number of paths on S from $(0,0)$ to (n, n) whose points lie on or below $y = x$. Prove that

$$\sum_{n \geqslant 0} c_n x^n = \frac{1}{8x}\{1 + 2x - (1 - 12x + 4x^2)^{1/2}\}.$$

5.3.5. (a) Show that there are

$$\frac{1}{mn+1}\binom{mn+1}{n}$$

paths from $(0,0)$ to $(n,(m-1)n)$, on the step-set $\{(0,1),(1,0)\}$, whose points lie on or below the line $y = (m-1)x$.

(b) Show that there are

$$\frac{m-\mu n}{m+n}\binom{m+n}{n}$$

paths on step-set $\{(1,0),(0,1)\}$ from $(0,0)$ to (m,n), with $m > \mu n$, which never touch the line $x = \mu y$ except at $(0,0)$.

(c) Show that there are

$$\frac{m-\mu n}{m+n-r}\left[\begin{matrix} m+n-r \\ r,\, n-r,\, m-r \end{matrix}\right]$$

paths on step-set $\{(1,0),(0,1),(1,1)\}$ from $(0,0)$ to (m,n), with $m > \mu n$ that never touch the line $x = \mu y$ except at $(0,0)$ and that have r occurrences of the step $(1,1)$.

5.3.6. Show that there are $\frac{1}{n+1}\binom{2n}{n}$ lead codes of length $2n+1$ (see **[5.2.11]**).

5.3.7. Since $\frac{1}{n+1}\binom{2n}{n} = \binom{2n}{n} - \binom{2n}{n-1}$, we know from **[5.3.6]** that

$$\begin{aligned}
&(\text{number of paths from } (0,0) \text{ to } (n,n), \text{ but never above } y = x) \\
&\qquad = (\text{number of paths from } (0,0) \text{ to } (n,n)) \\
&\qquad\quad - (\text{number of paths from } (-1,1) \text{ to } (n,n)),
\end{aligned}$$

where all paths are on step-set $\{(1,0),(0,1)\}$. Find a combinatorial decomposition that yields this relationship directly.

5.4. ORDERED SETS OF PATHS

This section deals with the enumeration of ordered sets of n paths on the step-set $\mathsf{S} = \{(0,-1),(1,0)\}$ such that no pair of paths have a common vertex. Such paths are called **nonintersecting *n*-paths**, and we shall use them to enumerate plane partitions. Our immediate goal is to enumerate nonintersecting n-paths with respect to the numbers of each step at each altitude.

Accordingly, let $s_1 = (1,0)$ and $s_2 = (0,-1)$. If an n-path has m_{ij} steps s_i at altitude j, then $\mathbf{s}(w) = [m_{ij}]_{2\times\infty}$ is called the **step-type** of w. Let $\mathbf{Y} = [y_{ij}]_{2\times\infty}$ be a matrix of indeterminates and $\Phi_{\mathsf{A}}(\mathbf{Y})$ be the ordinary generating function for a set A of n-paths with respect to step-type.

Let $\mathsf{P} = (P_1,\ldots,P_n)$, $\mathsf{Q} = (Q_1,\ldots,Q_n)$ be ordered sets of points in the plane and let L_{ij} be the set of all paths on S from P_i to Q_j. Then $w \in \mathsf{L}_{1\sigma(1)} \times \cdots \times \mathsf{L}_{n\sigma(n)}$, where σ is a permutation on N_n, is called an n-path with **associated permutation** $\rho(w) = \sigma$ and **end-type** (P,Q). Let $\mathsf{L}(\mathsf{P},\mathsf{Q}) = \dot\cup_\sigma \mathsf{L}_{1\sigma(1)} \times \cdots \times \mathsf{L}_{n\sigma(n)}$, where the union is over all permutations σ on N_n, so $\mathsf{L}(\mathsf{P},\mathsf{Q})$ is

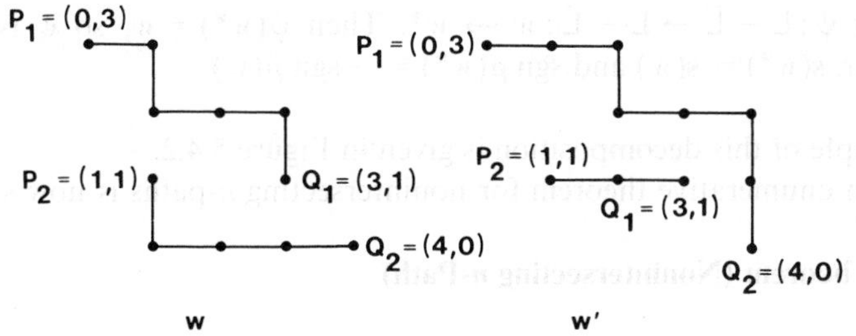

Figure 5.4.1. 2-paths with improper end-type.

the set of all n-paths with end-type (P,Q). Let $\overline{\mathsf{L}}(\mathsf{P},\mathsf{Q}) \subseteq \mathsf{L}(\mathsf{P},\mathsf{Q})$ denote the set of all nonintersecting n-paths with end-type (P,Q). If $\rho(w)$ is the identity permutation for every $w \in \overline{\mathsf{L}}(\mathsf{P},\mathsf{Q})$, then (P,Q) is said to be **proper**. For example, $(\mathsf{P}^0,\mathsf{Q}^0) = (((0,3),(1,1)),((3,1),(4,0)))$ is not proper, since $w, w' \in \overline{\mathsf{L}}(\mathsf{P}^0,\mathsf{Q}^0)$ and $\rho(w) \neq \rho(w')$, where w and w' are given in Figure 5.4.1.

By the product lemma, the n-path generating function for $\mathsf{L}(\mathsf{P},\mathsf{Q})$ is trivially $\Phi_{\mathsf{L}(\mathsf{P},\mathsf{Q})} = \operatorname{per}[\Phi_{\mathsf{L}_{ij}}]_{n\times n}$. On the other hand, we shall show in Theorem 5.4.2 that $\Phi_{\overline{\mathsf{L}}(\mathsf{P},\mathsf{Q})} = \det[\Phi_{\mathsf{L}_{ij}}]_{n\times n}$ when (P,Q) is proper. Thus the combinatorial property of nonintersection is accounted for algebraically by the alternating sign in the determinant.

5.4.1. Decomposition (Intersecting *n*-Path)

There exists a decomposition

$$\mathsf{L}(\mathsf{P},\mathsf{Q}) - \overline{\mathsf{L}}(\mathsf{P},\mathsf{Q}) \xrightarrow{\sim} \mathsf{L}(\mathsf{P},\mathsf{Q}) - \overline{\mathsf{L}}(\mathsf{P},\mathsf{Q})$$

such that $\operatorname{sgn}\rho(w) = -\operatorname{sgn}\rho(w^*)$ *and* $\mathbf{s}(w) = \mathbf{s}(w^*)$, *where* w^* *is the image of* w *under the decomposition.*

Proof: Let $w = (w_1,\ldots,w_n) \in \mathsf{L} - \overline{\mathsf{L}}$, where $\mathsf{L}, \overline{\mathsf{L}}$ denote $\mathsf{L}(\mathsf{P},\mathsf{Q})$ and $\overline{\mathsf{L}}(\mathsf{P},\mathsf{Q})$, respectively, and let $\sigma = \rho(w)$. For $r = 1,\ldots,n$ let $w_r = u_0^{(r)} \cdots u_{m_r}^{(r)}$, the sequence of points in w_r from origin $u_0^{(r)} = P_r$ to terminus $u_{m_r}^{(r)} = Q_r$. Let

$$(a,b,c,d) = \min_l \min_k \min_j \min_i \left\{(i,j,k,l) \mid u_j^{(i)} = u_l^{(k)}, i \neq k\right\}$$

and $w^* = (w_1^*,\ldots,w_n^*)$, where

$$w_r^* = \begin{cases} u_0^{(a)} \cdots u_c^{(a)} u_{d+1}^{(b)} \cdots u_{m_b}^{(b)}, & r = a \\ u_0^{(b)} \cdots u_d^{(b)} u_{c+1}^{(a)} \cdots u_{m_a}^{(a)}, & r = b \\ w_r, & r \neq a, b. \end{cases}$$

Consider $\psi: \mathsf{L} - \bar{\mathsf{L}} \to \mathsf{L} - \bar{\mathsf{L}}: w \mapsto w^*$. Then $\psi(w^*) = w$, so ψ is bijective. Moreover, $\mathbf{s}(w^*) = \mathbf{s}(w)$ and $\operatorname{sgn} \rho(w^*) = -\operatorname{sgn} \rho(w)$. □

An example of this decomposition is given in Figure 5.4.2.
The main enumerative theorem for nonintersecting n-paths is now stated.

5.4.2. Theorem (Nonintersecting *n*-Path)

If (P, Q) *is proper, then* $\Phi_{\bar{\mathsf{L}}(\mathsf{P},\mathsf{Q})}(\mathbf{Y}) = \|\Phi_{\mathsf{L}_{ij}}(\mathbf{Y})\|_{n \times n}$.

Proof: If $\mathsf{A} \subseteq \mathsf{L}(\mathsf{P}, \mathsf{Q})$, let $\Psi_{\mathsf{A}}(\mathbf{Y}) = \sum_{w \in \mathsf{A}} \operatorname{sgn} \rho(w) \mathbf{Y}^{\mathbf{s}(w)}$. Then, from the intersecting n-path decomposition, $\Psi_{\mathsf{L}-\bar{\mathsf{L}}} = -\Psi_{\mathsf{L}-\bar{\mathsf{L}}} = 0$, where L and $\bar{\mathsf{L}}$ denote $\mathsf{L}(\mathsf{P}, \mathsf{Q})$ and $\bar{\mathsf{L}}(\mathsf{P}, \mathsf{Q})$, respectively. Moreover, $\Phi_{\bar{\mathsf{L}}} = \Psi_{\bar{\mathsf{L}}}$, since (P, Q) is proper. Thus

$$\Phi_{\bar{\mathsf{L}}} = \Psi_{\bar{\mathsf{L}}} = \Psi_{\bar{\mathsf{L}}} + \Psi_{\mathsf{L}-\bar{\mathsf{L}}} = \Psi_{\mathsf{L}} = \sum_{w \in \mathsf{L}} (\operatorname{sgn} \rho(w)) \mathbf{Y}^{\mathbf{s}(w)}$$

$$= \sum_{\sigma: \mathsf{N}_n \rightleftarrows \mathsf{N}_n} (\operatorname{sgn} \sigma) \Phi_{\mathsf{L}_{1\sigma(1)}} \cdots \Phi_{\mathsf{L}_{n\sigma(n)}}, \quad \text{by the product lemma,}$$

since $\{w \in \mathsf{L} \mid \rho(w) = \sigma\} = \mathsf{L}_{1\sigma(1)} \times \cdots \times \mathsf{L}_{n\sigma(n)}$. The result follows. □

The preceding result is a combinatorial version of the **Karlin–McGregor theorem** (Karlin and McGregor (1959)) for n simultaneous stochastic processes that do not occupy the same state at the same time. We now apply the n-path theorem to the enumeration of column-strict plane partitions. These are a two-dimensional generalization of the integer partitions considered in Section 2.5. The following notation is used in connection with these configurations. Let $\boldsymbol{\alpha} = (\alpha_1, \ldots, \alpha_n)$, $\boldsymbol{\beta} = (\beta_1, \ldots, \beta_n)$, $\boldsymbol{\gamma} = (\gamma_1, \ldots, \gamma_n)$, $\boldsymbol{\delta} = (\delta_1, \ldots, \delta_n)$ be nonincreasing sequences of non-negative integers, such that $\boldsymbol{\alpha} \geqslant \boldsymbol{\beta}$, $\beta_n = 1$ and $\boldsymbol{\gamma} \geqslant \boldsymbol{\delta} \geqslant \mathbf{1}$.

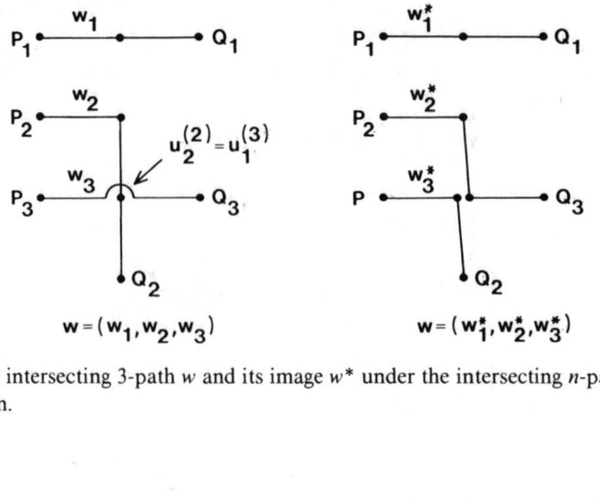

Figure 5.4.2. An intersecting 3-path w and its image w^* under the intersecting n-path decomposition.

5.4.3. Definition (Column-Strict Plane Partition, Shape, Size)

A **column-strict plane partition p** *of M with* **shape** (α, β) *and* **size** (γ, δ) *is an array of positive integers p_{ij} such that*

1. $\gamma_i \geqslant p_{i\beta_i} \geqslant p_{i\beta_i+1} \cdots \geqslant p_{i\alpha_i} \geqslant \delta_i$ *for* $i = 1, \ldots, n$.
2. $p_{ij} > p_{i+1\,j}$ *when both sides are defined in* (1).
3. $\Sigma_{i,j} p_{ij} = M$. □

The **type** of a column-strict plane partition **p** is $\tau(\mathbf{p}) = (k_1, k_2, \ldots)$, where **p** has k_i occurrences of i, for $i \geqslant 1$. If $>$ is replaced by $\geqslant$ in (2), the column condition, then **p** is called an **ordinary** plane partition. When $\boldsymbol{\beta} \neq \mathbf{1}$, **p** is often called **skew**. For example,

$$\begin{array}{ccc} & 4 & 4 & 2 \\ & 3 & & \\ 1 & & & \end{array}$$

is a (skew) column-strict plane partition of 14 with shape ((4, 2, 1), (2, 2, 1)), size ((5, 3, 1), (2, 1, 1)), and type (1, 1, 1, 2, 0, ...). Size is not uniquely defined, for, equally well, the preceding configuration has size ((4, 3, 1), (2, 2, 1). The set of all n-rowed column-strict plane partitions of shape $(\boldsymbol{\alpha}, \boldsymbol{\beta})$ and size $(\boldsymbol{\gamma}, \boldsymbol{\delta})$ is denoted by $\mathsf{C}_n(\boldsymbol{\alpha}, \boldsymbol{\beta} : \boldsymbol{\gamma}, \boldsymbol{\delta})$, and has the following decomposition.

5.4.4. Decomposition (*n*-Path: for Column-Strict Plane Partitions)

Let $P_i = (\beta_i + n - i, \gamma_i)$, $Q_i = (\alpha_i + n - i + 1, \delta_i)$ *for* $i = 1, \ldots, n$, *and let* $\mathsf{P} = (P_1, \ldots, P_n)$, $\mathsf{Q} = (Q_1, \ldots, Q_n)$. *Then*

1. (P, Q) *is proper.*
2. $\mathsf{C}_n(\boldsymbol{\alpha}, \boldsymbol{\beta} : \boldsymbol{\gamma}, \boldsymbol{\delta}) \stackrel{\sim}{\rightarrow} \overline{\mathsf{L}}(\mathsf{P}, \mathsf{Q}) : \mathbf{p} \mapsto (w_1, \ldots, w_n)$.

where w_i has $(1, 0)$-steps only at origins $(n + j - i, p_{ij})$ for $j = \beta_i, \ldots, \alpha_i$. (Clearly, the origins of the $(0, -1)$-steps in $(w_1, \ldots, w_n)$ are then uniquely determined.) □

Figure 5.4.3 gives a 4-rowed column-strict plane partition and the nonintersecting 4-path corresponding to it under the preceding decomposition.

5.4.5. Theorem (Column-Strict Plane Partition: Shape, Size, Type)

Let $\theta(l : j, i) = [x^l]\prod_{k=i}^{j}(1 - xx_k)^{-1}$. Then the number of n-rowed column-strict plane partitions of shape $(\boldsymbol{\alpha}, \boldsymbol{\beta})$, size $(\boldsymbol{\gamma}, \boldsymbol{\delta})$, and type **m** *is*

$$[\mathbf{x}^{\mathbf{m}}] \| \theta(\alpha_j - \beta_i + i - j + 1 : \gamma_i, \delta_j) \|_{n \times n}.$$

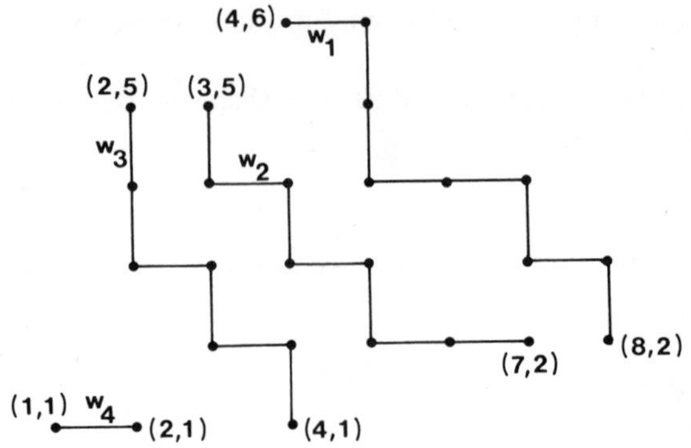

$$\begin{array}{cccc} 6 & 4 & 4 & 3 \\ 4 & 3 & 2 & 2 \\ 3 & 2 & & \\ 1 & & & \end{array}$$

Figure 5.4.3. A four-rowed column-strict plane partition of shape $((4,4,2,1),(1,1,1,1))$ and size $((6,5,5,1),(2,2,1,1))$, and its corresponding nonintersecting 4-path.

Proof: Let $y_{1j} = x_j$, $y_{2j} = 1$ for $j \geqslant 1$, $y_{1j} = y_{2j} = 0$ for $j = 0$, and let $P_i = (\beta_i + n - i, \gamma_i)$, $Q_i = (\alpha_i + n - i + 1, \delta_i)$ for $i = 1, \ldots, n$, $\mathsf{P} = (P_1, \ldots, P_n)$ and $\mathsf{Q} = (Q_1, \ldots, Q_n)$. If the n-path w corresponds to $\mathbf{p}$ under Decomposition 5.4.4, then $\tau(\mathbf{p}) = ([\mathbf{s}(w)]_{11}, [\mathbf{s}(w)]_{12}, \ldots)$. Thus, from Decomposition 5.4.4, the required number is $[\mathbf{x^m}]\Phi^{-}_{\mathsf{L}(\mathsf{P},\mathsf{Q})}(\mathbf{Y}) = [\mathbf{x^m}] \| \Phi_{\mathsf{L}_{ij}}(\mathbf{Y}) \|_{n \times n}$, from the nonintersecting n-path theorem. But $\Phi_{\mathsf{L}_{ij}}(Y) = \theta(\alpha_j - \beta_i + i - j - 1\colon\ \gamma_i, \delta_j)$ since the altitudes of the $(1,0)$-steps in an element of L_{ij} form a nonincreasing sequence of length $(\alpha_j + n - j + 1) - (\beta_i + n - i)$, bounded above by γ_i and below by δ_j. □

This result also holds for configurations with slightly less restrictive conditions on $\boldsymbol{\alpha}, \boldsymbol{\beta}, \boldsymbol{\gamma}, \boldsymbol{\delta}$. For example, the conditions $\delta_1 \geqslant \cdots \geqslant \delta_n$ and $\alpha_1 \geqslant \cdots \geqslant \alpha_n$ may be replaced by $\delta_1 > \cdots > \delta_n$ and $\alpha_i \leqslant \alpha_j + i - j$ for all $1 \leqslant j \leqslant i \leqslant n$, without affecting the properness of (P, Q) in Decomposition 5.4.4. Thus with $n = 5$, such a configuration with shape $((4,5,6,5,6),(2,2,1,1,1))$ and size $((8,8,7,5,5),(6,5,3,2,1))$ is

$$\begin{array}{cccccc} & 8 & 8 & 6 & & \\ & 8 & 7 & 5 & 5 & \\ 7 & 6 & 6 & 4 & 3 & 3. \\ 5 & 5 & 3 & 3 & 2 & \\ 5 & 3 & 2 & 2 & 1 & 1 \end{array}$$

The number of these configurations of type $\mathbf{m}$ is given by Theorem 5.4.5.

The first application of Theorem 5.4.5 is to a theorem of Kreweras for dominance systems.

5.4.6. Kreweras' Theorem for Dominance Systems

Let $\mathbf{a} = (a_1, \ldots, a_r)$, $\mathbf{b} = (b_1, \ldots, b_r)$ where $a_1 \geqslant \cdots \geqslant a_r$ and $b_1 \geqslant \cdots \geqslant b_r \geqslant 1$. Let $\mathbf{D} = [d_{ij}]_{n \times r}$, where the d_{ij} are positive integers. Then $\mathbf{D}$ is an $n \times r$ **dominance system** of size $(\mathbf{a}, \mathbf{b})$ if $d_{i1} \geqslant \cdots \geqslant d_{ir}$, $i = 1, \ldots, n$, and $a_j \geqslant d_{1j} \geqslant \cdots \geqslant d_{nj} \geqslant b_j$, $j = 1, \ldots, r$. Let $\mathsf{D}_{n,r}(\mathbf{a}, \mathbf{b})$ denote the set of all such configurations, and let $a_i' = a_i + r - i$, $b_i' = b_i + r - i$, $\mathbf{a}' = (a_1', \ldots, a_r')$, $\mathbf{b}' = (b_1', \ldots, b_r')$. Then

$$\mathsf{D}_{n,r}(\mathbf{a}, \mathbf{b}) \xrightarrow{\sim} \mathsf{C}_r(n\mathbf{1}_r, \mathbf{1}_r : \mathbf{a}', \mathbf{b}') : \left[d_{ij}\right]_{n \times r} \mapsto \left[d_{ji} + r - i\right]_{r \times n}.$$

At $\mathbf{x} = \mathbf{1}$, $\theta(n + i - j : a_i', b_j') = \binom{a_i - b_j + n}{n + i - j}$, so by Theorem 5.4.5, the number of $n \times r$ dominance systems of size $(\mathbf{a}, \mathbf{b})$ is

$$\left\| \binom{a_i - b_j + n}{i - j + n} \right\|_{r \times r}. \qquad \square$$

We conclude with two specializations of Theorem 5.4.5.

5.4.7. Column-Strict Plane Partitions: Fixed Shape

The number of n-rowed column-strict plane partitions of shape $(\boldsymbol{\alpha}, \boldsymbol{\beta})$ and type $\mathbf{m}$ is $[\mathbf{x}^{\mathbf{m}}]\|h_{\alpha_j - \beta_i + i - j + 1}\|_{n \times n}$, by Theorem 5.4.5, where $\boldsymbol{\gamma} = \infty$, $\boldsymbol{\delta} = \mathbf{1}$ and $h_r(\mathbf{x}) = [x^r]\prod_{k \geqslant 1}(1 - xx_k)^{-1}$ is called a **complete symmetric function.** $\square$

The generating function for column-strict plane partitions of shape $(\boldsymbol{\alpha}, \mathbf{1})$ is often taken to be the combinatorial definition of the **Schur function** $e_{\boldsymbol{\alpha}}$. Classically, the Schur function is defined by $e_{\boldsymbol{\alpha}} = \|x_j^{\alpha_i + n - i}\|_{n \times n} / \|x_j^{n-i}\|_{n \times n}$. The equivalence of these definitions is established by **5.4.7** with $\boldsymbol{\beta} = \mathbf{1}$ and by $\|h_{\alpha_j - j + i}\|_{n \times n} = \|x_j^{\alpha_i + n - i}\|_{n \times n} / \|x_j^{n-i}\|_{n \times n}$, which is called the **Jacobi–Trudi identity**.

5.4.8. Young Tableaux: Fixed Shape

A plane partition with N cells such that each element of N_N occurs exactly once is called a Young tableau. Let $c_n(\boldsymbol{\alpha}, \boldsymbol{\beta})$ be the number of n-rowed Young tableaux of shape $(\boldsymbol{\alpha}, \boldsymbol{\beta})$. Then from **5.4.7**, with $N = n + \sum_{i=1}^{n}(\alpha_i - \beta_i)$,

$$\begin{aligned} c_n(\boldsymbol{\alpha}, \boldsymbol{\beta}) &= [x_1 \cdots x_N]\|h_{\alpha_j - \beta_i + i - j + 1}\|_{n \times n} \\ &= N!\|(\alpha_j - \beta_i + i - j + 1)!^{-1}\|_{n \times n} \end{aligned}$$

by the $<$-transformation lemma (Lemma 4.2.5), since $h_k = \gamma_k(\leqslant)$. $\square$

NOTES AND REFERENCES

This section is based on Gessel and Viennot (1983); for further work on plane partitions and symmetric functions see MacDonald (1979) and Stanley (1971a).

5.4.2, 4, 5 Gessel and Viennot (1983); **5.4.6** Kreweras; (1965); **5.4.7** Littlewood (1950); [**5.4.1, 2**] Stanley (1971, 1972); [**5.4.3**] Frame, Robinson, and Thrall (1954); [**5.4.4, 5**] MacMahon (1915); [**5.4.6**] Glaz (1979); [**5.4.7**] Pólya (1969).

EXERCISES

5.4.1. Prove that the number of n-rowed column-strict plane partitions of N, with largest part at most m and shape $(\boldsymbol{\alpha}, \mathbf{1})$ is $[q^N]H_m(\boldsymbol{\alpha})$, where

$$H_m(\boldsymbol{\alpha}) = \left\| q^{\alpha_j - j + i} \binom{m + \alpha_j - j + i - 1}{m - 1}_q \right\|_{n \times n}.$$

5.4.2. Let $\mathbf{p}$ be a plane partition. The **hook length** of the (i, j)-cell in $\mathbf{p}$ is equal to the number of cells to the right of the (i, j)-cell, plus the number of cells directly below the (i, j)-cell, plus 1. The **content** of the (i, j)-cell is defined to be $j - i$. Let p be the number of cells in an n-rowed column-strict plane partition of shape $(\boldsymbol{\alpha}, \mathbf{1})$, where $\boldsymbol{\alpha} = (\alpha_1, \ldots, \alpha_n)$, and let c_i, d_i be the content and hook length of cell i, for $i = 1, \ldots, p$. Prove that

$$H_m(\boldsymbol{\alpha}) = q^a \frac{(m + c_1)_q \cdots (m + c_p)_q}{(d_1)_q \cdots (d_p)_q},$$

where $a = \sum_{i=1}^n i\alpha_i$ and $H_m(\boldsymbol{\alpha})$ is given in [**5.4.1**].

5.4.3. Let $\boldsymbol{\alpha} = (\alpha_1, \ldots, \alpha_n)$, where $\alpha_1 \geqslant \cdots \geqslant \alpha_n > 0$. Prove that the number of Young tableaux of shape $(\boldsymbol{\alpha}, \mathbf{1})$ is

$$f^{\boldsymbol{\alpha}} = \frac{p!}{d_1 \cdots d_p},$$

where $d_1, \ldots, d_p$ are the hook lengths associated with shape $(\boldsymbol{\alpha}, \mathbf{1})$.

5.4.4. (a) Show that the number of ordinary plane partitions of M with largest part at most m and with at most r rows and c columns is $[q^M]G_m(r, c)$, where $G_m(r, c) = q^{-c\binom{r+1}{2}} H_{m+r}(\boldsymbol{\alpha})$, $\boldsymbol{\alpha} = (\alpha_1, \ldots, \alpha_r) = (c, \ldots, c)$, and $H_{m+r}(\boldsymbol{\alpha})$ is defined in [**5.4.1**].

(b) Hence prove that

$$G_m(r, c) = \prod_{1 \leqslant j \leqslant c} \binom{m + r + j - 1}{r}_q \Big/ \binom{r + j - 1}{r}_q.$$

5.4.5. (a) Prove that the number of ordinary plane partitions of M with at most r rows is

$$[q^M]\prod_{k\geq 1}\{1-q^k\}^{-\min(k,r)}.$$

(b) Prove that the number of ordinary plane partitions of M is

$$[q^M]\prod_{k\geq 1}\{1-q^k\}^{-k}.$$

5.4.6. Prove that the number of binary sequences $\sigma_1\ldots\sigma_{mn}$ in which every substring of length m contains at least k 1's and $\sigma_{(m-1)i+1}\cdots\sigma_{mi}$ contains exactly b_i+k-1 1's for $i=1,\ldots,n$, is

$$\left\|\binom{m}{s_j+k-1-s_{i-1}}\right\|_{n\times n},$$

where $s_i=b_1+\cdots+b_i$, $i=1,\ldots,n$, and $s_0=0$.

5.4.7. Let π_1, with initial step $(0,1)$ and π_2, with initial step $(1,0)$, be paths of equal length from $(0,0)$ to the same terminus, with step-set $\{(1,0),(0,1)\}$. If π_1 and π_2 do not intersect between origin and terminus, then the resulting closed figure is called a **lattice polygon**, denoted by (π_1,π_2).

(a) Prove that the number of lattice polygons with semiperimeter $n+2$ and terminal abscissa $k+1$ is

$$\frac{1}{n+1}\binom{n+1}{k}\binom{n+1}{k+1},\qquad \text{for } n\geq k\geq 0.$$

This is called a **Runyon number**.

(b) Prove that the number of lattice polygons with semiperimeter $n+2$ is

$$\frac{1}{n+2}\binom{2n+2}{n+1},\ \text{a Catalan number}.$$

5.4.8. An n-rowed **column-bounded**, column-strict plane partition of shape $(\boldsymbol{\alpha},\boldsymbol{\beta})$ and **size** $(\boldsymbol{\gamma},\boldsymbol{\delta};\boldsymbol{\omega},\boldsymbol{\mu})$ is an n-rowed column-strict plane partition of shape $(\boldsymbol{\alpha},\boldsymbol{\beta})$ and size $(\boldsymbol{\gamma},\boldsymbol{\delta})$, with the additional restriction that $p_{1j}<\omega_j$, for $j=\beta_1,\ldots,\alpha_1$ and $p_{nj}>\mu_j$, for $j=\beta_n,\ldots,\alpha_n$ where $\boldsymbol{\omega}=(\omega_{\beta_1},\ldots,\omega_{\alpha_1})$, $\boldsymbol{\mu}=(\mu_{\beta_n},\ldots,\mu_{\alpha_n})$ are nonincreasing sequences of non-negative integers. Let

$$\theta_{ij}=\sum x_{l_{\beta_{i+n-i}}}\cdots x_{l_{\alpha_{j+n-j}}},$$

where the summation extends over all $l_{\beta_i+n-i},\ldots,l_{\alpha_j+n-j}$ such that $\gamma_i\geq l_{\beta_i+n-i}\geq\cdots\geq l_{\alpha_j+n-j}\geq\delta_j$ and $\mu_m<l_m<\omega_{m-n+1}$, where $\mu_i=0$ for $i>\alpha_n$, $\omega_i=\infty$ for $i<\beta_1$.

Prove that the number of n-rowed column-bounded, column-strict plane partitions of shape $(\boldsymbol{\alpha}, \boldsymbol{\beta})$ and size $(\boldsymbol{\gamma}, \boldsymbol{\delta}; \boldsymbol{\omega}, \boldsymbol{\mu})$, of type $\mathbf{m}$ is

$$[\mathbf{x}^{\mathbf{m}}]\|\theta_{ij}\|_{n\times n}.$$

5.5. A q-ANALOGUE OF THE LAGRANGE THEOREM

The purpose of this section is to derive a q-analogue of the Lagrange theorem by considering lattice paths with step-set $\{(1,i)|i \geqslant -1\}$. Since each step in such a path increases the x-coordinate by exactly 1, we denote the path $((1,i_1),(1,i_2),\ldots,(1,i_n))_{(0,0)}$ by the sequence $i_1 i_2 \cdots i_n$ of y-coordinate increments. Thus we consider sequences over $\mathsf{M} = \{-1,0,1,\ldots\}$ and refer to the element i_j in the sequence $w = i_1 \cdots i_n$ as the **step** $(1, i_j)$ and as the **vertex** $(j, i_1 + \cdots + i_j)$, with **altitude** $i_1 + \cdots + i_j$. The **rank** of w is the altitude of the terminal vertex and is denoted by $r(w)$. The sequence $\overline{w} = i_n \cdots i_1$ is called the **reverse** of w, and $\overline{\mathsf{A}}$ denotes the set of reverses of the elements of a set A of sequences.

The following decomposition for M^* is the main result of this section.

5.5.1. Decomposition (Additive (for Sequences))

1. *Let B be the set of all sequences in M^* of rank zero with no elements with strictly positive altitude and let $\mathsf{F} = \{-1\}\mathsf{B}$. Then*

$$\mathsf{F} = \dot{\bigcup_{n \geqslant -1}} F^{n+1}\{n\}.$$

2. *Let D be the set of all sequences in M^+ with zero rank and with no vertices at altitude 0 except the origin and terminus. Then*

$$\mathsf{M}^* = \mathsf{D}^*\mathsf{F}^* \,\dot{\cup}\, \mathsf{P} \qquad \textit{and} \quad \mathsf{D} = \dot{\bigcup_{j,\,k \geqslant 0}} \mathsf{F}^j\{j+k\}(\overline{\mathsf{F}})^k,$$

where P is the set of all sequences in M^* with strictly positive rank.

Proof: Let G_i be the set of sequences in M^* of rank $-i$, in which every vertex except the origin has negative altitude, for $i \geqslant 0$. If $w = i_1 \cdots i_l \in \mathsf{G}_n$, where $n \geqslant 1$, then w must contain falls (an occurrence of -1) from altitude $-m$ for $m = 0,\ldots, n-1$, since the altitude of successive vertices can only decrease by one at a time. Let i_{j_m} denote the right-most fall from altitude $-m$ for $m = 0,\ldots,n-1$. Then $1 = j_0 < j_1 < \cdots < j_{n-1} \leqslant l$, and $i_{j_k} \cdots i_{j_{k+1}-1} \in \mathsf{F}$ for $k = 0,\ldots,n-1$, where $j_n = l+1$. Thus $\mathsf{G}_n = \mathsf{F}^n$ for $n \geqslant 0$, since $\mathsf{G}_0 = \{\varepsilon\} = \mathsf{F}^0$, where ε is the empty string, denoting the path containing the single vertex $(0,0)$.

1. If the last step of an element of F is n, then it must occur at altitude $-n-1$, where $n \geqslant -1$, so $\mathsf{F} = \dot{\cup}_{n \geqslant -1} \mathsf{G}_{n+1}\{n\} = \dot{\cup}_{n \geqslant -1} \mathsf{F}^{n+1}\{n\}$, from the above.

2. Let $w = i_1 \cdots i_l \in \mathsf{M}^* - \mathsf{P}$ and $r(w) = s$, so $s \leqslant 0$. Suppose that $j_0, \ldots, j_n$ are the indexes of vertices in w at altitude 0, so $0 = j_0 < j_1 < \cdots < j_n \leqslant l$, and $n \geqslant 0$. Then $i_{j_k+1} \cdots i_{j_{k+1}} \in \mathsf{D}$ for $k = 0, \ldots, n-1$, and $i_{j_n} \cdots i_l \in \mathsf{G}_{-s} = \mathsf{F}^{-s}$ from the above. Thus

$$\mathsf{M}^* - \mathsf{P} = \dot{\bigcup_{n \geqslant 0}} \dot{\bigcup_{s \leqslant 0}} \mathsf{D}^n \mathsf{F}^{-s} = \mathsf{D}^* \mathsf{F}^*,$$

since this construction is reversible.

To obtain an expression for D, let $w = i_1 \cdots i_l \in \mathsf{D}$ and suppose that i_m is the index of the right-most vertex in w with nonpositive altitude, for $m < l$. Then i_m has altitude $-j$ and i_{m+1} has altitude k, for some $j, k \geqslant 0$, so $w = w_1(j+k)w_2$, where $w_1 = i_1 \cdots i_{m-1}$, $w_2 = i_{m+1} \cdots i_l$, with $r(w_1) = -j$ and $r(w_2) = -k$. Now the internal vertices of w_2 have altitude at least $-k+1$, by construction, so $w_2 \in \overline{\mathsf{G}}_k$. Moreover, all vertices in w_1 except the origin have negative altitude, since if w_1 has a vertex at positive altitude then w_1 must subsequently pass through a vertex at altitude 0 (not allowed in D) in order to reach its terminal altitude of $-j$. Thus $w_1 \in \mathsf{G}_j$, and

$$\mathsf{D} = \dot{\bigcup_{j,k \geqslant 0}} \mathsf{G}_j\{j+k\}\overline{\mathsf{G}}_k = \dot{\bigcup_{j,k \geqslant 0}} \mathsf{F}^j\{j+k\}\overline{\mathsf{F}}^k$$

from the above, since the construction is reversible. □

To demonstrate that Decomposition 5.5.1 implies the Lagrange theorem, we associate with an arbitrary subset A of M* the generating function

$$\Psi_{\mathsf{A}}(z, q, t) = \sum_{w \in \mathsf{A}} W(w),$$

where $W(w) = q^{-a(w)} z^{\lambda(w)} t^{r(w)} G(w)$, and $\lambda(w) = n$ is the **length** of $w = i_1 \cdots i_n$, $a(w) = ni_1 + (n-1)i_2 + \cdots + i_n$ is the **area** associated with w, $G(w) = g_{i_1+1} \cdots g_{i_n+1}$, where $g_0, g_1, \ldots$ are commutative indeterminates. Since $W(uv) = W(u)W(v)$ if $q = 1$, for $u, v \in \mathsf{M}^*$, it follows that

$$\Psi_{\mathsf{A}_1\mathsf{A}_2}(z, 1, t) = \Psi_{\mathsf{A}_1}(z, 1, t)\Psi_{\mathsf{A}_2}(z, 1, t),$$

where $\mathsf{A}_1, \mathsf{A}_2 \subseteq \mathsf{M}^*$. Note that $t\Psi_{\mathsf{F}}(z, q, t)$ is independent of t, since $r(u) = -1$ for $u \in \mathsf{F}$, so we may write

$$F(z, q) = t\Psi_{\mathsf{F}}(z, q, t).$$

5.5.2. A Proof of the Lagrange Theorem

From Decomposition 5.5.1(1)

$$\Psi_{\mathsf{F}}(z,1,t)=\sum_{n\geqslant -1} zt^{n+1}g_{n+1}\Psi_{\mathsf{F}}^{n+1}(z,1,t)$$

so $f(z)=F(z,1)$ satisfies the functional equation $f=zg(f)$, where $g(y)=\sum_{n\geqslant 0} g_n y^n$. From Decomposition 5.5.1(2),

$$\{1-\Psi_{\mathsf{M}}(z,1,t)\}^{-1}=\{1-\Psi_{\mathsf{D}}(z,1,t)\}^{-1}\{1-\Psi_{\mathsf{F}}(z,1,t)\}^{-1}+\Psi_{\mathsf{P}}(z,1,t),$$

and since $\Psi_{\mathsf{F}}(z,1,t)=\Psi_{\bar{\mathsf{F}}}(z,1,t)$,

$$\begin{aligned}\Psi_{\mathsf{D}}(z,1,t)&=\sum_{j,k\geqslant 0}\Psi_{\langle j+k\rangle}(z,1,t)\Psi_{\mathsf{F}}^{j+k}(z,1,t)\\&=z\sum_{j,k\geqslant 0} g_{j+k+1}f^{j+k}(z)=z\sum_{n\geqslant 0}(n+1)g_{n+1}f^n(z)=zg'(f(z)).\end{aligned}$$

But

$$\Psi_{\mathsf{M}}(z,1,t)=\sum_{n\geqslant -1} zt^n g_{n+1}=zt^{-1}g(t),$$

so

$$\{1-zt^{-1}g(t)\}^{-1}=\{1-zg'(f(z))\}^{-1}\{1-t^{-1}f(z)\}+\Psi_{\mathsf{P}}(z,1,t).$$

Now P contains only sequences with strictly positive rank, so $[z^n t^{-k}]\Psi_{\mathsf{P}}(z,1,t)=0$ for $n,k\geqslant 0$. Thus, applying $[z^n t^{-k}]$ to the preceding equation for $n,k\geqslant 0$ gives

$$[t^{n-k}]g^n(t)=[z^n]f^k(z)\{1-zg'(f(z))\}^{-1}$$

where, from the preceding, $f=zg(f)$. This is precisely the statement of the Lagrange theorem (**1.2.4(2)**). □

A q-analogue of the Lagrange theorem is obtained by carrying out **5.5.2** for arbitrary q. To do this we need the following result.

5.5.3. Proposition

Let $u, v\in \mathsf{M}^*$. Then

1. $a(uv)=a(u)+a(v)+r(u)\lambda(v)$.
2. $W(uv)=W(u)W(v)q^{-r(u)\lambda(v)}$.

3. *If* $r(u) = -i$ *and* $r(v) = -j$, *then*

$$W(u\{i+j\}\bar{v}) = W(u)q^{a(v)}z^{\lambda(v)+1}t^{r(v)+i+j}g_{i+j+1}G(v).$$

Proof:

1. The area $a(uv)$ is the sum of the altitudes of the vertices in uv. This is simply $a(u)$ for the vertices of u in uv. The altitude in uv of each of the vertices of v is increased by the terminal altitude $r(u)$ of u, since, in uv, the terminus of u is identified with the origin of v. Thus we have an additional contribution of $r(u)\lambda(v)$ to the area $a(v)$ of v. The total area is thus $a(u) + a(v) + r(u)\lambda(v)$.
2. Directly from (**1**).
3. Let $v = k_1 \cdots k_n$.

Then $a(v) + a(\bar{v})$

$$= (nk_1 + (n-1)k_2 + \cdots + k_n) + (nk_n + (n-1)k_{n-1} + \cdots + k_1)$$

$$= (n+1)r(v),$$

so $a(\bar{v}) = \{1 + \lambda(v)\}r(v) - a(v)$. Applying (**1**) twice to $u\{i+j\}\bar{v}$, we have

$$a(u\{i+j\}\bar{v}) = a(u) + (i+j) + a(\bar{v}) + r(u)\{1 + \lambda(\bar{v})\} + (i+j)\lambda(\bar{v})$$

$$= a(u) - a(v) + (1 + \lambda(v))\{r(u) + i + j + r(v)\}$$

$$= a(u) - a(v).$$

The result follows, since $G(u\{i+j\}\bar{v}) = G(u)g_{i+j+1}G(v)$, $\lambda(u\{i+j\}\bar{v}) = \lambda(u) + 1 + \lambda(v)$ and $r(u\{i+j\}\bar{v}) = r(u) + i + j + r(v) = 0$. □

For notational convenience in the q-analogue of the Lagrange theorem, denote the product $\phi(x, y)\phi(xq, y) \cdots \phi(xq^{n-1}, y)$ by $\phi^{(n)}(x, y)$, and $\phi(x, y)\phi(xq^{-1}, y) \cdots \phi(xq^{-(n-1)}, y)$ by $\phi_{(n)}(x, y)$.

5.5.4. Theorem (q-Lagrange)

The functional equation $F(z, q) = zq\sum_{n\geq 0} g_n F^{(n)}(z, q)$ *has a unique solution* $F(z, q)$. *Moreover, if* $g(x, q) = \sum_{n\geq 0} g_n x^n$, *then*

1. $[z^n]F^{(k)}(z, q)\{1 - d(z, q)\}^{-1} = q^{\binom{n+1}{2}}[t^{n-k}]g_{(n)}(q^{-1}t, q)$, *where* $d(z, q) = z\sum_{i,j\geq 0} g_{i+j+1}F^{(i)}(z, q)F_{(j)}(z, q^{-1})$.

2. $$F(z, q) = \frac{z\sum_{n\geq 0} q^{\binom{n+2}{2}} z^n [t^n] g_{(n+1)}(q^{-1}t, q)}{\sum_{n\geq 0} q^{\binom{n+1}{2}} z^n [t^n] g_{(n)}(q^{-1}t, q)}.$$

Proof:

1. We first derive the generating functions associated with the sets F^n, M^n, and $\mathsf{F}^i\{i+j\}\bar{\mathsf{F}}^j$.

(i). For F^n: From Proposition 5.5.3(2)

$$\Psi_{\mathsf{F}^n}(z,q,t) = \sum_{u\in\mathsf{F}^{n-1}} W(u) \sum_{v\in\mathsf{F}} W(v) q^{-r(u)\lambda(v)}$$

$$= \Psi_{\mathsf{F}}(q^{n-1}z,q,t)\Psi_{\mathsf{F}^{n-1}}(z,q,t)$$

since $r(u) = 1-n$ for all $u \in \mathsf{F}^{n-1}$, recalling that sequences in F have rank equal to -1. Iterating, we have

$$\Psi_{\mathsf{F}^n}(z,q,t) = t^{-n}F^{(n)}(z,q),$$

since $F(z,q) = t\Psi_{\mathsf{F}}(z,q,t)$. Similarly, from Proposition 5.5.3(2),

$$\Psi_{\mathsf{F}^{n+1}\{n\}}(z,q,t) = \sum_{u\in\mathsf{F}^{n+1}} W(u)W(n)q^{-r(u)\lambda(n)} = g_{n+1}t^{-1}qzF^{(n+1)}(z,q),$$

since $r(u) = -(n+1)$, $\lambda(n) = 1$, and $W(n) = q^{-n}t^n z g_{n+1}$.

(ii) For $\mathsf{F}^i\{i+j\}\bar{\mathsf{F}}^j$: From Proposition 5.5.3(3)

$$\Psi_{\mathsf{F}^i\{i+j\}\bar{\mathsf{F}}^j}(z,q,t) = zt^{i+j}g_{i+j+1} \sum_{w\in\mathsf{F}^i} W(w) \sum_{w'\in\mathsf{F}^j} q^{a(w')}z^{\lambda(w')}t^{r(w')}G(w')$$

$$= zt^{i+j}g_{i+j+1}\Psi_{\mathsf{F}^i}(z,q,t)\Psi_{\mathsf{F}^j}(z,q^{-1},t)$$

$$= zg_{i+j+1}F^{(i)}(z,q)F_{(j)}(z,q^{-1}), \qquad \text{from (i)}.$$

(iii) For M^n: From Proposition 5.5.3(2), since $\lambda(v) = 1$ for $v \in \mathsf{M}$,

$$\Psi_{\mathsf{M}^n}(z,q,t) = \sum_{u\in\mathsf{M}^{n-1}} W(u)q^{-r(u)} \sum_{v\in\mathsf{M}} W(v) = zqt^{-1}g(q^{-1}t)\Psi_{\mathsf{M}^{n-1}}(z,q,tq^{-1}).$$

It follows by iterating this equation that

$$\Psi_{\mathsf{M}^n}(z,q,t) = (zt^{-1})^n q^{\binom{n+1}{2}} g_{(n)}(q^{-1}t,q).$$

We may now apply the main decomposition to obtain an expression relating the generating functions of F, M, D, and P, as follows. From Decomposition 5.5.1(1)

$$\Psi_{\mathsf{F}}(z,q,t) = \sum_{n\geq -1} \Psi_{\mathsf{F}^{n+1}\{n\}}(z,q,t),$$

whence $F(z,q) = zq\Sigma_{i\geqslant 0} g_i F^{(i)}(z,q)$, from **(i)**. From Decomposition 5.5.1(2),

$$\begin{aligned}\Psi_{\mathsf{M}^*}(z,q,t) &= \Psi_{\mathsf{D}^*\mathsf{F}^*}(z,q,t) + \Psi_{\mathsf{P}}(z,q,t)\\ &= \{1 - \Psi_{\mathsf{D}}(z,q,t)\}^{-1}\Psi_{\mathsf{F}^*}(z,q,t) + \Psi_{\mathsf{P}}(z,q,t),\end{aligned}$$

from Proposition 5.5.3(2), since $r(u) = 0$ for $u \in \mathsf{D}$. But

$$\Psi_{\mathsf{D}}(z,q,t) = \sum_{i,j\geqslant 0} \Psi_{\mathsf{F}^i(i+j)\bar{\mathsf{F}}^j}(z,q,t) = z\sum_{i,j\geqslant 0} g_{i+j+1}F^{(i)}(z,q)F_{(j)}(z,q^{-1}),$$

so, from **(iii)** and **(i)**,

$$\sum_{n\geqslant 0} q^{\binom{n+1}{2}}(zt^{-1})^n g_{(n)}(q^{-1}t,q) = \{1 - d(z,q)\}^{-1}\sum_{k\geqslant 0} t^{-k}F^{(k)}(z,q) + \Psi_{\mathsf{P}}(z,q,t).$$

Now apply $[z^n t^{-k}]$ to this equation, for $n, k \geqslant 0$, noting that $[t^{-k}]\Psi_{\mathsf{P}}(z,q,t) = 0$, since all sequences in P have strictly positive rank, so

$$[t^{n-k}]q^{\binom{n+1}{2}} g_{(n)}(q^{-1}t,q) = [z^n]\{1 - d(z,q)\}^{-1}F^{(k)}(z,q).$$

The result follows, since the equation that $F(z,q)$ satisfies has a unique solution.

2. To complete the proof, note that $F^{(0)}(z,q) = 1$ and $F^{(1)}(z,q) = F(z,q)$. Thus

$$F(z,q) = \frac{F^{(1)}(z,q)}{1 - d(z,q)}\left\{\frac{F^{(0)}(z,q)}{1 - d(z,q)}\right\}^{-1}$$

and the result follows from **(1)**. □

The following corollary indicates how continued fractions may occur in a special case and gives a special form of continued fraction as a rational function of power series.

5.5.5. Corollary

Let $F(z,q) = zqJ_z[g_1q^k, g_2q^{2k+1} : (1,\infty)]$, $g(y,q) = 1 + g_1y + g_2y^2$ *and*

$$G_k(z,q) = \sum_{n\geqslant 0} q^{\binom{n+k+1}{2}} z^{n+k}[t^n]g_{(n+k)}(q^{-1}t,q).$$

Then

1. $F(z,q) = G_1(z,q)/G_0(z,q)$.
2. $g_1z + g_2z\{F(z,q) + F(z,q^{-1})\} = 1 - G_0^{-1}(z,q)$.

Proof:

1. It is immediate that

$$F(z,q) = zq\{1 - g_1 zq - g_2 zqF(zq,q)\}^{-1},$$

whence $F(z,q) = zq\{F^{(0)}(z,q) + g_1 F^{(1)}(z,q) + g_2 F^{(2)}(z,q)\}$. The result is a restatement of (**2**) of the q-Lagrange theorem for $g(y,q) = 1 + g_1 y + g_2 y^2$.

2. Setting $k = 0$ in (**1**) of the q-Lagrange theorem yields $\{1 - d(z,q)\}^{-1} = G_0(z,q)$, whence $d(z,q) = 1 - G_0^{-1}(z,q)$. The result follows by directly evaluating $d(z,q)$. □

Although (**1**) of Corollary 5.5.5 gives an expression for the generating function $F(z,q)$, (**2**) gives a simpler expression that is equally useful, since $F(z,q)$ admits only non-negative powers of q. Thus $[q^k]F(z,q^{-1}) = 0$ for $k > 0$, so the coefficients of $F(z,q)$ can be recovered from $g_1 z + g_2 z\{F(z,q) + F(z,q)^{-1}\}$. As an application of Corollary 5.5.5, we consider the enumeration of lead codes that are compositions of a given integer. This gives a simpler form of the generating function than the one given in [**5.2.12**] and is included since it exhibits certain technical details about Gaussian coefficients.

5.5.6. Lead Codes

Let $b(l,k)$ be the number of lead codes of length l whose elements sum to $k - l$. From [**5.2.12**(b)], the required generating function is

$$B(z,q) = \sum_{l,k\geqslant 0} b(l,k) z^l q^k = qzJ_z\left[0, q^{2j+1} : (1,\infty)\right],$$

so Corollary 5.5.5 can be applied with $g(y,q) = 1 + y^2$. Thus

$$G_k(z,q) = z^k \sum_{n\geqslant 0} q^{\binom{n+k+1}{2}} z^n [t^n] \prod_{i=1}^{n+k} (1 + q^{-2i}t^2)$$

$$= z^k \sum_{n\geqslant 0} q^{\binom{n+k+1}{2}} z^n [t^n] \sum_{i=0}^{n+k} q^{-i(i+1)} \binom{n+k}{i}_{q^{-2}} t^{2i}, \quad \text{from } \mathbf{2.6.12(1)}$$

$$= z^k \sum_{m\geqslant 0} q^{\binom{2m+k+1}{2}} z^{2m} q^{-m(m+1)} \binom{2m+k}{m}_{q^{-2}}, \quad \text{where } n = 2m,\ i = m$$

$$= z^k q^{\binom{k+1}{2}} \sum_{m\geqslant 0} q^{-m^2} \binom{2m+k}{m}_{q^2} z^{2m}.$$

From Corollary 5.5.5(1), the desired generating function is therefore

$$B(z,q) = zq\left(\sum_{m\geqslant 0} q^{-m^2}\binom{2m+1}{m}_{q^2} z^{2m}\right)\Big/ \sum_{m\geqslant 0} q^{-m^2}\binom{2m}{m}_{q^2} z^{2m}.$$

We may, however, obtain a simpler expression. From Corollary 5.5.5(2)

$$zB(z,q) + zB(z,q^{-1}) = 1 - \left\{\sum_{m\geqslant 0} q^{-m^2}\binom{2m}{m}_{q^2} z^{2m}\right\}^{-1}$$

so

$$b(l,k) = -[z^{l-1}q^k]\left\{\sum_{m\geqslant 0} q^{-m^2}\binom{2m}{m}_{q^2} z^{2m}\right\}^{-1}. \qquad \square$$

NOTES AND REFERENCES

This section is based on Gessel (1980b); other q-analogues of the Lagrange theorem have been given by Andrews (1975b) and Garsia (1983).

5.5.1, 2 Gessel (1980b); **5.5.4** Garsia (1983), Gessel (1980b); **5.5.5, 6** Gessel (1980b); [**5.5.1**] Gessel (1980b), Raney (1960); [**5.5.2**] Gessel (1980b), Pólya (1969); [**5.5.3**] Levine (1959), Narayana (1959); [**5.5.4**] Gessel (1980b).

EXERCISES

5.5.1. (a) Let P be the set of all sequences in M* with strictly positive rank. Prove that M* = (ε ∪ S)BF*, where B is the set of all sequences in M* of rank zero with no elements at strictly positive altitude, and where F = {−1}B. This decomposition is called the **multiplicative decomposition** (for sequences).

(b) Use (a) to prove the Lagrange theorem, that if $f(z)$ is the unique formal power series solution of $f = zg(f)$, then $[t^{n-k}]g^n(t)/n = [z^n]f^k(z)/k$.

5.5.2. The **area** of a lattice polygon (π_1, π_2) is the number of grid squares that it contains (see [**5.4.7**]) and is denoted by area(π_1, π_2). Let $p(i,j,k)$ be the number of lattice polygons with semiperimeter i, area j, and terminal abscissa k, and let $P(z,q,s) = \sum_{i,j,k\geqslant 0} p(i,j,k) z^i q^j s^k$. Prove that

(a) $P(z,q,s) = z^2 qsJ_z[(1+s)q^k, sq^{2k+1} : (1,\infty)]$,

(b) $$P(z,q,s) = z^2qs\frac{\sum_{n\geqslant 0}\sum_{l=0}^{n} q^{-l(n-l)}\binom{n+1}{l}_q\binom{n+1}{l+1}_q s^l z^n}{\sum_{n\geqslant 0}\sum_{l=0}^{n} q^{-l(n-l)}\binom{n}{l}_q^2 s^l z^n},$$

(c) $$p(i,j,k) = -[z^i q^j s^k]\left\{\sum_{n\geqslant 0}\sum_{l=0}^{n} q^{-l(n-l)}\binom{n}{l}_q^2 s^l z^n\right\}^{-1}.$$

5.5.3. Prove that the number of lattice polygons with semiperimeter $n + 2$ and terminal abscissa $k + 1$ is

$$\frac{1}{n+1}\binom{n+1}{k}\binom{n+1}{k+1}.$$

5.5.4. Let $J(x, q) = x(\partial/\partial x)A(x, q)$, where $A(x, q)$ is given in [**3.3.48**].

(a) Show that $J(x, q) = x \sum_{i \geqslant 0} \frac{1}{i!} J^{(i)}(x, q)$.

(b) Deduce from part (a) that

$$J(x, q) = \frac{x\sum_{n\geqslant 0}\frac{x^n}{n!}q^{-\binom{n+1}{2}}(1 + q + \cdots + q^n)^n}{\sum_{n\geqslant 0}\frac{x^n}{n!}q^{-\binom{n+1}{2}}(1 + q + \cdots + q^{n-1})^n}.$$

Solutions

SECTION 1.1

1.1.1. Let $f(x) = \Sigma_{i \geqslant 1} f_i x^i$ and $g(x) = \Sigma_{i \geqslant 1} g_i x^i$. Then

$$f(x)g^{-1}(x) = \left(\sum_{i \geqslant 1} f_i x^i\right) x^{-1} \left(\sum_{i \geqslant 1} g_i x^{i-1}\right)^{-1} = \left(\sum_{i \geqslant 1} f_i x^{i-1}\right)\left(\sum_{i \geqslant 1} g_i x^{i-1}\right)^{-1},$$

so

$$f(x)g^{-1}(x)|_{x=0} = f_1 g_1^{-1}$$

But

$$f'(x)\{g'(x)\}^{-1} = \left(\sum_{i \geqslant 1} i f_i x^{i-1}\right)\left(\sum_{i \geqslant 1} i g_i x^{i-1}\right)^{-1}, \text{ so } f'(x)\{g'(x)\}^{-1}|_{x=0} = f_1 g_1^{-1}$$

also.

1.1.2. (a)

$$\binom{-n}{k} = \frac{1}{k!}(-n)(-n-1)\cdots(-n-k+1) = \binom{n+k-1}{k}(-1)^k,$$

as required.

(b)

$$\binom{-\frac{1}{2}}{k} = \frac{1}{k!}\left(-\frac{1}{2}\right)\left(-\frac{3}{2}\right)\cdots\left(-\frac{(2k-1)}{2}\right) = \frac{(-2)^{-k}}{k!} 1.3 \cdots (2k-1)$$

$$= \frac{(-4)^{-k}}{k!}\frac{(2k)!}{k!} = (-4)^{-k}\binom{2k}{k},$$

as required.

1.1.3. $$\sum_{i=a}^{b-1} i x^i = x \sum_{i=a}^{b-1} \frac{d}{dx} x^i = x \frac{d}{dx}\{(x^a - x^b)(1-x)^{-1}\}$$

$$= x(1-x)^{-2}\{(a x^{a-1} - b x^{b-1})(1-x) + (x^a - x^b)\},$$

by the product rule, and the result follows.

1.1.4. Multiplying the given recurrence equation by $t^n/n!$ and summing for $n \geqslant 0$ yields

$$\sum_{n\geqslant 0} b_n \frac{t^n}{n!} = \sum_{n\geqslant 0} \sum_{i=0}^{n} \frac{a_i t^n}{i!(n-i)!}$$

or $B(t) = \exp(t)A(t)$. The result is obtained by multiplying on both sides of this equation by $\exp(-t)$ and comparing coefficients of $t^n/n!$.

1.1.5. (a) Let $A(t) = \sum_{n=0}^{p}(p-n)!a_n t^n$ and $B(t) = \sum_{n=0}^{p}(p-n)!b_n t^n$ be such that $A(t) = e^t B(t)$. Then $B(t) = e^{-t}A(t)$ and the result follows by applying $[(p-n)!t^n]$ to these expressions.

(b) Let $A(t) = \sum_{n\geqslant 0} a_n t^n$ and $B(t) = \sum_{n\geqslant 0} b_n t^n$ be such that $A(t) = (1-4t)^{-1/2}B(t)$. Then $B(t) = (1-4t)^{1/2}A(t)$ and the result follows by applying $[t^n]$ to these expressions.

1.1.6. The result follows by applying $[x^0]$ to both sides of the identity

$$\{1 + (1+x)(1+yx^{-1})(1+y^{-1})\}^n = \{1 + y^{-1}(1+x)\}^n\{1 + y(1+x^{-1})\}^n.$$

1.1.7. Let $\operatorname{Re}(\omega)$ denote the real part of the complex number ω. Then

$$\sum_{k=0}^{n}\binom{n}{k}\cos(k\theta) = \operatorname{Re}\sum_{k=0}^{n}\binom{n}{k}\exp(ik\theta), \qquad \text{where } i^2 = -1,$$

$$= \operatorname{Re}(1+e^{i\theta})^n = \operatorname{Re}\left\{\exp\left(in\frac{\theta}{2}\right)\left(\exp\left(-i\frac{\theta}{2}\right) + \exp\left(i\frac{\theta}{2}\right)\right)^n\right\}$$

$$= \left(2\cos\left(\frac{\theta}{2}\right)\right)^n \operatorname{Re}\left\{\exp\left(in\frac{\theta}{2}\right)\right\}, \qquad \text{and the result follows.}$$

1.1.8.

$$\sum_{n\geqslant 0} c_n t^n = \sum_{i\geqslant 0}\sum_{n\geqslant 0}\binom{n+il}{m+ik}t^n = \sum_{i\geqslant 0}\sum_{n\geqslant 0}\binom{n+il}{n-m+i(l-k)}t^n$$

$$= \sum_{i\geqslant 0} t^{m+i(k-l)} \sum_{n\geqslant 0}\binom{-(m+ik+1)}{n-m+i(l-k)}(-t)^{n-m+i(l-k)}$$

$$= \sum_{i\geqslant 0} t^{m+i(k-l)}(1-t)^{-(m+ik+1)}$$

$$= t^m(1-t)^{-(m+1)}\sum_{i\geqslant 0}\{t^{k-l}(1-t)^{-k}\}^i$$

$$= t^m(1-t)^{-(m+1)}\{1 - t^{k-l}(1-t)^{-k}\}^{-1}, \qquad \text{and the result follows.}$$

1.1.9. (a) We have immediately $\omega^{mn} = e^{2\pi i n} = 1$, so $\omega^{mn} - 1 = 0$. Factoring this gives

$$(\omega^n - 1)\sum_{k=0}^{m-1} \omega^{kn} = 0.$$

But $\omega^n - 1 \neq 0$ unless $m|n$, so $\sum_{k=0}^{m-1} \omega^{kn} = 0$ if $m \nmid n$. If $m|n$, then $\omega^n = 1$, and $\sum_{k=0}^{m-1} \omega^{kn} = \sum_{k=0}^{m-1} 1 = m$.

(b)
$$\frac{1}{m}\sum_{j=0}^{m-1} f(x\omega^j)\omega^{-lj} = \frac{1}{m}\sum_{k\geq 0} f_k x^k \sum_{j=0}^{m-1} \omega^{j(k-l)}$$
$$= \frac{1}{m}\sum_{m|(k-l)} f_k x^k m, \qquad \text{from part (a)}$$
$$= \sum_{k\equiv l(\text{mod } m)} f_k x^k.$$

1.1.10. (a) Let $\omega = e^{2i\pi/3}$. Then

$$\sum_{k=0}^{n}\binom{3n}{3k} = \frac{1}{3}\sum_{j=0}^{2}(1+\omega^j)^{3n}, \qquad \text{from } [\mathbf{1.1.9}],$$
$$= \tfrac{1}{3}\{2^{3n} + (-\omega^2)^{3n} + (-\omega)^{3n}\}, \qquad \text{since } 1+\omega+\omega^2 = 0$$
$$= \tfrac{1}{3}\{2^{3n} + 2(-1)^n\}, \qquad \text{since } \omega^3 = 1.$$

(b) By [**1.1.9**] the required number is

$$\frac{1}{m}\sum_{j=0}^{m-1}(1+\omega^j)^n\omega^{-lj}$$
$$= \frac{1}{m}\sum_{j=0}^{m-1}(\omega^{-j/2}+\omega^{j/2})^n\omega^{(1/2)nj-lj}$$
$$= \frac{1}{m}\sum_{j=0}^{m-1}\left\{\exp\left(\frac{-i\pi j}{m}\right)+\exp\left(\frac{i\pi j}{m}\right)\right\}^n \exp\left(\frac{i\pi}{m}(n-2l)j\right)$$
$$= \frac{1}{m}\sum_{j=0}^{m-1}\left(2\left(\cos\left(\frac{\pi j}{m}\right)\right)\right)^n\left\{\cos\left(\frac{(n-2l)j\pi}{m}\right)+i\sin\left(\frac{(n-2l)j\pi}{m}\right)\right\}$$

where $\omega = e^{2i\pi/m}$. The result follows, since this is a real number, so the imaginary part is zero.

1.1.11. (a) Integrating by parts gives $J_{m,k} = (k/(m+1))J_{m+1,k-1}$ for $k \geq 1$. Iterating this yields $J_{m,k} = \binom{m+k}{k}^{-1} J_{m+k,0}$ and the result follows, since $J_{m+k,0} = (m+k+1)^{-1}(1+x)^{m+k+1}|_{-1}^{1} = (m+k+1)^{-1}2^{m+k+1}$.

(b) From part (a)

$$\sum_{k=0}^{n}\binom{n}{k}^{-1} = (n+1)2^{-(n+1)}\int_{-1}^{1}\sum_{k=0}^{n}(1+x)^{n-k}(1-x)^{k}\,dx$$

$$= (n+1)2^{-(n+1)}\int_{-1}^{1}\frac{1}{2x}\{(1+x)^{n+1}-(1-x)^{n+1}\}\,dx$$

$$= (n+1)2^{-(n+1)}\sum_{j\geqslant 0}\binom{n+1}{2j+1}\int_{-1}^{1}x^{2j}\,dx, \qquad \text{by bisection}$$

$$= (n+1)2^{-n}\sum_{j\geqslant 0}\frac{1}{2j+1}\binom{n+1}{2j+1}.$$

1.1.12. Multiplying both sides of the recurrence by $u^k/k!$ and summing over $k \geqslant 0$ yields

$$\sum_{k\geqslant 0} M_{k+1}(x)\frac{u^k}{k!} = x\sum_{k\geqslant 0} M_k(x)\frac{u^k}{k!} - \sum_{k\geqslant 1} k^2 M_{k-1}(x)\frac{u^k}{k!}.$$

If $M(u, x) = \sum_{k\geqslant 0} M_k(x)\frac{u^k}{k!}$ then the preceding equation becomes

$$\frac{\partial}{\partial u}M(x,u) = xM(x,u) - u\sum_{k\geqslant 1}(k-1+1)M_{k-1}(x)\frac{u^{k-1}}{(k-1)!}$$

$$\frac{\partial}{\partial u}M(x,u) = xM(x,u) - u^2\frac{\partial}{\partial u}M(x,u) - uM(x,u)$$

$$\frac{\partial}{\partial u}\log M(x,u) = \frac{x-u}{1+u^2}.$$

Integrating,

$$\log M(x,u) = x\int_0^u (1+t^2)^{-1}\,dt - \int_0^u t(1+t^2)^{-1}\,dt$$

$$= x\tan^{[-1]}(u) - \tfrac{1}{2}\log(1+u^2),$$

and the result follows by applying *exp* to both sides.

1.1.13. $$\sum_{\mathbf{n}\geqslant \mathbf{1}} \min(\mathbf{n})\mathbf{t}^{\mathbf{n}} = \sum_{m\geqslant 1}\sum_{\min(\mathbf{n})\geqslant m}\mathbf{t}^{\mathbf{n}}$$

$$= \sum_{m\geqslant 1}\prod_{i=1}^{k} t_i^m(1-t_i)^{-1}$$

$$= (1-t_1\cdots t_k)^{-1}\prod_{i=1}^{k}t_i(1-t_i)^{-1}, \quad \text{as required.}$$

1.1.14. Let $\mathbf{C} = \begin{pmatrix} \mathbf{I}_n & \mathbf{A} \\ -\mathbf{B} & \mathbf{I}_k \end{pmatrix}$ and $\mathbf{D} = \begin{pmatrix} \mathbf{I}_n & \mathbf{0} \\ \mathbf{B} & \mathbf{I}_k \end{pmatrix}$.

Then

$$\mathbf{CD} = \begin{pmatrix} \mathbf{I}_n + \mathbf{AB} & \mathbf{A} \\ \mathbf{0} & \mathbf{I}_k \end{pmatrix} \quad \text{and} \quad \mathbf{DC} = \begin{pmatrix} \mathbf{I}_n & \mathbf{A} \\ \mathbf{0} & \mathbf{I}_k + \mathbf{BA} \end{pmatrix},$$

so

$$|\mathbf{I}_n + \mathbf{AB}| = |\mathbf{CD}| = |\mathbf{DC}| = |\mathbf{I}_k + \mathbf{BA}|.$$

1.1.15. Suppose that $(\mathbf{I} + \mathbf{M})^{-1} = \mathbf{I} + \alpha\mathbf{M}$, where α is a scalar. Then $(\mathbf{I} + \mathbf{M})(\mathbf{I} + \alpha\mathbf{M}) = \mathbf{I}$ so $(1 + \alpha)\mathbf{M} = -\alpha\mathbf{M}^2$. But $\mathbf{M} = \mathbf{u}\mathbf{v}^T$ for some column vectors $\mathbf{u}, \mathbf{v}$, since $\mathbf{M}$ has rank 1. Thus $\mathbf{M}^2 = \mathbf{u}\mathbf{v}^T\mathbf{u}\mathbf{v}^T = (\mathbf{v}^T\mathbf{u})\mathbf{u}\mathbf{v}^T$, since $\mathbf{v}^T\mathbf{u}$ is a scalar. But $\operatorname{trace}\mathbf{M} = \operatorname{trace}(\mathbf{u}\mathbf{v}^T) = \operatorname{trace}(\mathbf{v}^T\mathbf{u}) = \mathbf{v}^T\mathbf{u}$, and thus $\mathbf{M}^2 = (\operatorname{trace}\mathbf{M})\mathbf{M}$. Then if $(\mathbf{I} + \mathbf{M})^{-1}$ exists, α must satisfy $1 + \alpha = -\alpha \operatorname{trace}\mathbf{M}$, so

$$\alpha = -(1 + \operatorname{trace}\mathbf{M})^{-1}, \qquad \operatorname{trace}\mathbf{M} \neq -1.$$

But if $\operatorname{trace}\mathbf{M} = -1$, then $|\mathbf{I} + \mathbf{M}| = 1 + \operatorname{trace}\mathbf{M} = 0$, and $(\mathbf{I} + \mathbf{M})^{-1}$ does not exist.

1.1.16. Let $[\mathbf{C}]_{ij} = \delta_{i1}\delta_{1j}$ for $i, j = 1, \ldots, n + 1$. Then $\mathbf{M}^k = [z^k]f(z)$, where

$$\begin{aligned} f(z) &= \sum_{k \geqslant 0} z^k\mathbf{M}^k = (\mathbf{I} - z\mathbf{M})^{-1} \\ &= (\mathbf{D} - z\mathbf{J})^{-1}, \qquad \text{where } \mathbf{D} = \mathbf{I} + z\mathbf{C} \\ &= (\mathbf{I} - z\mathbf{D}^{-1}\mathbf{J})^{-1}\mathbf{D}^{-1} \\ &= \mathbf{D}^{-1} + z\{1 - z \operatorname{trace}\mathbf{D}^{-1}\mathbf{J}\}^{-1}\mathbf{D}^{-1}\mathbf{J}\mathbf{D}^{-1}, \qquad \text{from [1.1.15]}. \end{aligned}$$

But $\mathbf{D}^{-1} = \Sigma_{i \geqslant 0}(-1)^i z^i\mathbf{C}^i = \mathbf{I} - z(1 + z)^{-1}\mathbf{C}$, since $\mathbf{C}^i = \mathbf{C}$ for all $i \geqslant 1$, so

$$\begin{aligned} 1 - z \operatorname{trace}\mathbf{D}^{-1}\mathbf{J} &= 1 - z\{(n + 1) - z(1 + z)^{-1}\} \\ &= \{1 - nz(1 + z)\}(1 + z)^{-1}. \end{aligned}$$

Also $\mathbf{CJC} = \mathbf{C}$, so making the preceding substitutions for $\mathbf{D}^{-1}$ and $1 - z \operatorname{trace}\mathbf{D}^{-1}\mathbf{J}$ in $f(z)$, gives

$$f(z) = \mathbf{I} - z(1 + z)^{-1}\mathbf{C} + z(1 + z)\{1 - nz(1 + z)\}^{-1} \times \{\mathbf{J} - z(1 + z)^{-1}(\mathbf{CJ} + \mathbf{JC}) + z^2(1 + z)^{-2}\mathbf{C}\}.$$

Thus $[\mathbf{M}^k]_{11}$

$$= [z^k]\{1 - z(1+z)^{-1} + z(1+z)\{1 - nz(1+z)\}^{-1}(1 - z(1+z)^{-1})^2\}$$

$$= [z^k]\{(1+z)^{-1} + z(1+z)^{-1}\{1 - nz(1+z)\}^{-1}\}$$

$$= (-1)^k + [z^{k-1}] \sum_{l=0}^{k-1} n^l z^l (1+z)^{l-1}$$

$$= \sum_{l=1}^{k-1} n^l \binom{l-1}{k-l-1}, \qquad \text{for } k > 1,$$

$$[\mathbf{M}^k]_{i1} = [\mathbf{M}^k]_{1j} = [z^k] z(1+z)\{1 - nz(1+z)\}^{-1}(1 - z(1+z)^{-1}),$$

for $i, j > 1$

$$= [z^{k-1}]\{1 - nz(1+z)\}^{-1}$$

$$= \sum_{l=1}^{k-1} n^l \binom{l}{k-l-1}, \qquad \text{for } k > 1,$$

$$[\mathbf{M}^k]_{ij} = [z^k] z(1+z)\{1 - nz(1+z)\}^{-1}, \qquad \text{for } i, j > 1$$

$$= \sum_{l=0}^{k-1} n^l \binom{l+1}{k-l-1}, \qquad \text{for } k \geqslant 1.$$

1.1.17. Let $\mathbf{C} = \operatorname{circ}(x_1, \ldots, x_n)$, $\mathbf{z}_i = (1, \omega^i, \ldots, \omega^{(n-1)i})^T$ and $\lambda_i = \sum_{j=0}^{n-1} \omega^{ij} x_j$, for $i = 0, \ldots, n-1$. Now $\mathbf{C}\mathbf{z}_i = \lambda_i \mathbf{z}_i$ for $i = 0, \ldots, n-1$, so $\lambda_0, \ldots, \lambda_{n-1}$ are the distinct eigenvalues of $\mathbf{C}$, and $|\mathbf{C}| = \prod_{i=0}^{n-1} \lambda_i$, as required.

1.1.18. (a) Let $a(x) = \sum_{i \geqslant 0} a_i x^i$, $b(x) = \sum_{i \geqslant 0} b_i x^i$, where $a_0 = b_0 = 1$, and $\mathbf{A} = [a_{j-i}]_{(p+q)\times(p+q)}$, $\mathbf{B} = [b_{j-i}]_{(p+q)\times(p+q)}$ and assume without loss of generality that $q \geqslant p$. Suppose that $\mathbf{A}$ and $\mathbf{B}$ are partitioned into blocks $\mathbf{A}_{ij} = \mathbf{A}[l(i)|l(j)]$, $\mathbf{B}_{ij} = \mathbf{B}[l(i)|l(j)]$, for $i, j = 1, 2, 3$, where $l(1) = \mathsf{N}_p$, $l(2) = \mathsf{N}_q - \mathsf{N}_p$, $l(3) = \mathsf{N}_{p+q} - \mathsf{N}_q$. Thus $\mathbf{A}_{ij}$ and $\mathbf{B}_{ij}$ are $\mathbf{0}$ matrices when $i > j$, and $\mathbf{A}_{ii}$ and $\mathbf{B}_{ii}$ are upper triangular matrices with 1's on the diagonal. Let

$$\mathbf{P} = \begin{pmatrix} \mathbf{A}_{12} & \mathbf{A}_{13} \\ \mathbf{A}_{22} & \mathbf{A}_{23} \end{pmatrix} \quad \text{and} \quad \mathbf{Q} = \begin{pmatrix} \mathbf{B}_{22} & \mathbf{B}_{23} \\ \mathbf{0} & \mathbf{B}_{33} \end{pmatrix}.$$

Then $|\mathbf{Q}| = 1$, so

$$|\mathbf{P}| = |\mathbf{P}| \cdot |\mathbf{Q}| = |\mathbf{PQ}| = \begin{vmatrix} \mathbf{A}_{12}\mathbf{B}_{22} & \mathbf{A}_{12}\mathbf{B}_{23} + \mathbf{A}_{13}\mathbf{B}_{33} \\ \mathbf{A}_{22}\mathbf{B}_{22} & \mathbf{A}_{22}\mathbf{B}_{23} + \mathbf{A}_{23}\mathbf{B}_{33} \end{vmatrix}.$$

But $\mathbf{AB} = \mathbf{I}_{p+q}$, since $a(x)a^{-1}(x) = 1$, yielding $\mathbf{A}_{22}\mathbf{B}_{22} = \mathbf{I}_{q-p}$, $\mathbf{A}_{11}\mathbf{B}_{13} + \mathbf{A}_{12}\mathbf{B}_{23} + \mathbf{A}_{13}\mathbf{B}_{33} = \mathbf{0}$, $\mathbf{A}_{22}\mathbf{B}_{23} + \mathbf{A}_{23}\mathbf{B}_{33} = 0$, so by the Laplace expansion

$$|\mathbf{P}| = \begin{vmatrix} \mathbf{A}_{12}\mathbf{B}_{22} & -\mathbf{A}_{11}\mathbf{B}_{13} \\ \mathbf{I} & \mathbf{0} \end{vmatrix} = (-1)^{(q-p)p}|-\mathbf{A}_{11}\mathbf{B}_{13}|,$$

$$= (-1)^{pq}|\mathbf{B}_{13}|, \qquad \text{since } p^2 - p \equiv 0 \pmod 2.$$

The result follows, since $[\mathbf{P}]_{ij} = [x^{p+j-i}]a(x)$ and $[\mathbf{B}_{13}]_{ij} = [x^{q+i-j}]a^{-1}(x)$.

(b) Put $p = 1$ in (a).

SECTION 1.2

1.2.1. We want $c(t) = \sum_{n\geq 0} c_n t^n$, where $c_n = \binom{2n}{n-k} = [\lambda^n]F(\lambda)\{\phi(\lambda)\}^n$ for $F(\lambda) = \lambda^k$, $\phi(\lambda) = (1+\lambda)^2$. From **(2)** of the Lagrange theorem, $c(t) = w^k\{1 - 2t(1+w)\}^{-1}$, where $w = t(1+w)^2$. The unique $w(t) \in \mathsf{R}[[t]]_0$ that satisfies this functional equation is $w = \{1 - 2t - (1-4t)^{1/2}\}/2t$. Thus $1 - 2t(1+w) = (1-4t)^{1/2}$ so $c(t) = (1-4t)^{-1/2}w^k$, and the result follows.

1.2.2. Let $F(\lambda) = e^{a\lambda}$ and $x = y\phi(x)$, where $\phi(x) = e^x$. Then from **(2)** of the Lagrange theorem

$$\frac{F(x)}{1 - y\phi'(x)} = \sum_{m\geq 0} y^m[\lambda^m]e^{(a+m)\lambda} = \sum_{m\geq 0} \frac{(a+m)^m}{m!}y^m.$$

But $\phi'(x) = \phi(x)$ so $F(x)\{1 - y\phi'(x)\}^{-1} = e^{\alpha x}(1-x)^{-1}$ and the result follows, since $y = xe^{-x}$.

1.2.3. Let $xt^{-1} = y$. Then the functional equation becomes

$$w = t\{1 + yw^2(1-w)^{-1}\},$$

so by **(1)** of the Lagrange theorem

$$w = \sum_{n\geq 1} \frac{t^n}{n}[\lambda^{n-1}]\{1 + y\lambda^2(1-\lambda)^{-1}\}^n$$

$$= \sum_{n\geq 1}\sum_{k=0}^{n} \frac{t^n}{n}y^k\binom{n}{k}[\lambda^{n-2k-1}](1-\lambda)^{-k}$$

$$= \sum_{i>k\geq 0} \frac{t^{i+k}y^k}{k+i}\binom{k+i}{k}\binom{i-2}{i-k-1}, \qquad \text{where } n-k=i.$$

1.2.4. If $v = w - a$, then v satisfies $v = t\phi(v + a)$, so from **(1)** of the Lagrange theorem

$$v = \sum_{n \geqslant 1} \frac{t^n}{n}[\lambda^{n-1}]\phi^n(\lambda + a)$$

and the result follows, since $v = w - a$.

1.2.5. Let $\Psi(\lambda) = \Phi^{[-1]}(\lambda)$ and $[\mathbf{B}]_{ij} = [\lambda^i]\Psi^j(\lambda)$, for $i, j \geqslant 1$. Then

$$[\mathbf{AB}]_{ij} = [\lambda^i] \sum_{k \geqslant 1} \Phi^k(\lambda)\{[\lambda^k]\Psi^j(\lambda)\} = [\lambda^i]\Psi^j(\Phi(\lambda))$$

$$= [\lambda^i]\lambda^j = \delta_{ij}$$

so $\mathbf{B} = \mathbf{A}^{-1}$. But if $\Phi(w) = t$ then $w = \Psi(t)$ and by **(1)** of the Lagrange theorem, with $\Phi(w) = w\phi^{-1}(w)$,

$$[\mathbf{A}^{-1}]_{ij} = [t^i]w^j = \frac{1}{i}[\lambda^{i-1}]j\lambda^{j-1}\phi^i(\lambda) = \frac{j}{i}[\lambda^{-j}]\Phi^{-i}(\lambda).$$

1.2.6. (a) Let $G(t) = \sum_{n \geqslant 1} \binom{an}{n-1} t^n$ and $F(\lambda) = \lambda$, $\phi(\lambda) = (1 + \lambda)^a$. Then

$$G(t) = \sum_{n \geqslant 1} t^n[\lambda^n]F(\lambda)\phi^n(\lambda) = w\{1 - t\phi'(w)\}^{-1}$$

by **(2)** of the Lagrange theorem. But

$$w = \sum_{i \geqslant 1} \frac{1}{i} t^i[\lambda^{i-1}]\phi^i(\lambda) = \sum_{i \geqslant 1} \frac{1}{i}\binom{ai}{i-1} t^i,$$

and

$$\{1 - t\phi'(w)\}^{-1} = \sum_{k \geqslant 0} t^k[\lambda^k]\phi^k(\lambda) = \sum_{k \geqslant 0} \binom{ak}{k} t^k$$

by **(1)** and **(2)** of the Lagrange theorem, respectively. The identity results from equating coefficients of t^n on both sides of $G(t) = w\{1 - t\phi'(w)\}^{-1}$.

(b) Let

$$H(t) = \sum_{n \geqslant 1} \frac{n^{n-1}}{(n-1)!} t^n \quad \text{and} \quad F(\lambda) = \lambda, \quad \phi(\lambda) = e^\lambda.$$

Then

$$H(t) = \sum_{n \geqslant 1} t^n[\lambda^n]F(\lambda)\phi^n(\lambda) = w\{1 - t\phi'(w)\}^{-1}$$

by (**2**) of the Lagrange theorem. But

$$w = \sum_{i\geq 1} \frac{1}{i} t^i [\lambda^{i-1}]\phi^i(\lambda) = \sum_{i\geq 1} \frac{i^{i-1}}{i!} t^i,$$

and $\{1 - t\phi'(w)\}^{-1} = \sum_{k\geq 0} t^k [\lambda^k]\phi^k(\lambda) = \sum_{k\geq 0} t^k k^k/k! = 1 + H(t)$, by (**1**) and (**2**) of the Lagrange theorem, respectively. The identity follows by equating coefficients of $t^n/(n-1)!$ on both sides of $H(t) - w = wH(t)$.

1.2.7. Let $A(t) = \sum_{n\geq 0} a_n t^n$, $B(t) = \sum_{n\geq 0} b_n t^n$. Then $a_n = \sum_{k\geq 0} \binom{n}{k} b_{n-ck}$ can be written as $a_n = [\lambda^n]\phi^n(\lambda)B(\lambda)$, where $\phi(\lambda) = 1 + \lambda^c$. Then, from (**2**) of the Lagrange theorem, with $w = t\phi(w)$,

$$A(t) = B(w)\{1 - ctw^{c-1}\}^{-1},$$

so

$$B(w) = \{1 - ctw^{c-1}\}A(t),$$

whence

$$\begin{aligned} b_n &= [w^n]\{1 - ctw^{c-1}\}A(t) \\ &= \sum_{i\geq 0} a_i\{[w^n]t^i - c[w^{n-c+1}]t^{i+1}\} \\ &= \sum_{i\geq 0} a_i\{[w^{n-i}](1 + w^c)^{-i} - c[w^{n-c-i}](1 + w^c)^{-i-1}\}, \end{aligned}$$

since $t = w(1 + w^c)^{-1}$. Now c must divide $n - i$, so let $i = n - ck$, and the result follows.

1.2.8. Let $A(y) = \sum_{n\geq 0} a_n y^n/n!$ and $B(y) = \sum_{n\geq 0} b_n y^n/n!$ be such that $A(y) = e^{\alpha x}B(y)$, where $x = ye^x$. Then $B(y) = e^{-\alpha x}A(y)$ and the result follows from **1.2.7**, by equating coefficients of $y^n/n!$ in the preceding equations.

1.2.9. Let $A(x) = \sum_{n\geq 0} a_n x^n$ and $f_n = [x^n]f(x)$. If $f(x) = \sum_{n\geq 0} a_n J_n(x)$, then equating coefficients of x^n gives

$$f_n = \frac{2^{-n}}{n!} \sum_{i\geq 0} a_{n-i}(-1)^i \binom{n}{i} = \frac{2^{-n}}{n!}[x^n]A(x)(1 - x^2)^n,$$

so $n!2^n f_n = [t^n]A(w)\{1 + 2tw\}^{-1}$ by (**2**) of the Lagrange theorem, where $w = t(1 - w^2)$. Now let $g(x) = \sum_{n\geq 0} n! f_n (2x)^n$. Then

$$g(t) = A(w)\{1 + 2tw\}^{-1}, \qquad \text{so } A(w) = g(t)\{1 + 2tw\}$$

and

$$
\begin{aligned}
a_n &= [w^n](1+w^2)(1-w^2)^{-1}g\big(w(1-w^2)^{-1}\big)\\
&= \sum_{i\geqslant 0} i!2^i f_i [w^{n-i}](1+w^2)(1-w^2)^{-i-1}\\
&= \sum_{j\geqslant 0} (n-2j)!2^{n-2j} f_{n-2j}\left\{\binom{n-j}{j} + \binom{n-j-1}{j-1}\right\},
\end{aligned}
$$

where $n - 2j = i$, and the result follows.

1.2.10. Let B denote the expression on the left side of the given equation. Then by the chain rule, with $\lambda = x - 1$,

$$
\begin{aligned}
B &= \frac{d^{n-1}}{d\lambda^{n-1}}\left(\frac{\lambda}{(1+\lambda)^a - 1}\right)^n\Bigg|_{\lambda=0}\\
&= (n-1)![\lambda^{n-1}]\phi^n(\lambda), \qquad \text{where } \phi(\lambda) = \frac{\lambda}{(1+\lambda)^a - 1}\\
&= n![t^n]w,
\end{aligned}
$$

where $w = t\phi(w)$, by (**1**) of the Lagrange theorem. Solving this functional equation directly gives $w = (1+t)^{1/a} - 1$, so $B = n!\binom{1/a}{n}$, and the result follows.

1.2.11. From the definition

$$
P_n(h) = \frac{1}{2^n n!}\frac{d^n}{d\lambda^n}\{(h+\lambda)^2 - 1\}^n\big|_{\lambda=0} = [\lambda^n]F(\lambda)\phi^n(\lambda),
$$

where $F(\lambda) = 1$, $\phi(\lambda) = \frac{1}{2}\{(h+\lambda)^2 - 1\}$. Then from (**2**) of the Lagrange theorem, $\sum_{n\geqslant 0} P_n(h)t^n = \{1 - t(h+w)\}^{-1}$, where $w = t\phi(w)$. Solving this quadratic functional equation for w yields $h + w = t^{-1}\{1 - (1 - 2ht + t^2)^{1/2}\}$, where the negative root has been selected since $w \in \mathsf{R}[[t]]_0$. Thus

$$
\{1 - t(h+w)\}^{-1} = (1 - 2ht + t^2)^{-1/2},
$$

and the result follows.

1.2.12. Let A denote the expression on the left side of the equation. Then

$$
\begin{aligned}
A &= \frac{d^n}{d\lambda^n}\{(x+\lambda)^{n-1}f\big((x+\lambda)^{-1}\big)\}\big|_{\lambda=0}\\
&= -n![\lambda^n]\phi^{n+1}(\lambda)\frac{d}{d\lambda}\{g\big((x+\lambda)^{-1}\big)\},
\end{aligned}
$$

where $\phi(\lambda) = x + \lambda$ and $g'(x) = f(x)$. If $w = t\phi(w)$, then $(x+w)^{-1} = (1-t)x^{-1}$, and from (**1**) of the Lagrange theorem,

$$\begin{aligned} A &= -(n+1)![t^{n+1}]g\big((x+w)^{-1}\big) \\ &= -\frac{d^{n+1}}{dt^{n+1}}\{g((1-t)x^{-1})\}|_{t=0} \\ &= (-1)^n x^{-n-1} g^{(n+1)}(x^{-1}) \\ &= (-1)^n x^{-n-1} f^{(n)}(x^{-1}), \qquad \text{since } g'(x) = f(x). \end{aligned}$$

1.2.13. (a) Let S denote the given summation. Then, as in **1.2.12**,

$$\begin{aligned} S &= (-1)^{m+n+k}\left[x^{m+n}y^{n+k}z^{k+m}\right](y-z)^{m+n}(z-x)^{n+k}(x-y)^{k+m} \\ &= (-1)^{m+n+k}\left[x^{m+n}y^{n+k}z^{k+m}\right](1+xy+yz+zx)^{-1} \\ &= \frac{(m+n+k)!}{m!n!k!}. \end{aligned}$$

(b) Let $T = \sum_i \binom{m+n}{m+i}\binom{n+k}{n+i}\binom{k+m}{k+i}u^{m+n+k+3i}$. Then

$$\begin{aligned} T &= (-1)^{m+n+k}\left[x^{m+n}y^{n+k}z^{k+m}\right](y-zu)^{m+n}(z-xu)^{n+k}(x-yu)^{k+m} \\ &= (-1)^{m+n+k}\left[x^{m+n}y^{n+k}z^{k+m}\right]|\mathbf{I}-\mathbf{XA}|^{-1} \end{aligned}$$

by the MacMahon master theorem, where $\mathbf{X} = \operatorname{diag}(x, y, z)$ and

$$\mathbf{A} = \begin{pmatrix} 0 & 1 & -u \\ -u & 0 & 1 \\ 1 & -u & 0 \end{pmatrix}.$$

Thus

$$\begin{aligned} T &= (-1)^{m+n+k}\left[x^{m+n}y^{n+k}z^{k+m}\right]\{1+(xy+yz+zx)u+xyz(u^3-1)\}^{-1} \\ &= (-1)^{m+n+k}\sum_{j\geqslant 0}(u^3-1)^{2j}\left[x^{m+n-2j}y^{n+k-2j}z^{k+m-2j}\right] \\ &\qquad\qquad \times\{1+(xy+yz+zx)u\}^{-2j-1} \\ &= \sum_{j\geqslant 0}(-1)^j(u^3-1)^{2j}\binom{m+n+k-j}{2j}\left[\begin{matrix} m+n+k-3j \\ m-j,\, n-j,\, k-j \end{matrix}\right]u^{m+n+k-3j}. \end{aligned}$$

The result follows by dividing both sides of the resulting identity by u^{m+n+k}, and then setting $t = u^3$.

1.2.14. Let $x_i = t_i\phi_i(\mathbf{x})$, where $\phi_i(\mathbf{x}) = \exp(\Sigma_{j=1}^n a_{ij}x_j)$ for $i = 1,\ldots, n$. If $F(\boldsymbol{\lambda}) = 1$ then, from **(2)** of the multivariate Lagrange theorem,

$$\|\delta_{ij} - t_i a_{ij}\phi_i(\mathbf{x})\|^{-1} = \sum_{\mathbf{m}\geqslant\mathbf{0}} \mathbf{t}^{\mathbf{m}}[\boldsymbol{\lambda}^{\mathbf{m}}]\boldsymbol{\phi}^{\mathbf{m}}(\boldsymbol{\lambda}).$$

But

$$[\boldsymbol{\lambda}^{\mathbf{m}}]\boldsymbol{\phi}^{\mathbf{m}}(\boldsymbol{\lambda}) = \frac{\overline{m}_1^{m_1}}{m_1!}\cdots\frac{\overline{m}_n^{m_n}}{m_n!}, \qquad \mathbf{t}^{\mathbf{m}} = u_1^{m_1}\cdots u_n^{m_n}$$

and $t_i\phi_i(\mathbf{x}) = x_i$, so the result follows, by expressing both sides in terms of the x_i's.

SECTION 2.3

2.3.1. (a) The required generating function is, from Decomposition 2.3.7,

$$\sum_{n\geqslant 0} c_n x^n = \sum_{k\geqslant 0}\left(x^l + x^{m+l} + x^{2m+l} + \cdots\right)^k = \sum_{k\geqslant 0} x^{lk}(1 - x^m)^{-k}$$

$$= \left\{1 - x^l(1 - x^m)^{-1}\right\}^{-1} = (1 - x^m)(1 - x^l - x^m)^{-1}$$

Thus

$$(1 - x^l - x^m)\sum_{n\geqslant 0} c_n x^n = 1 - x^m,$$

so

$$c_n - c_{n-l} - c_{n-m} = \begin{cases} 1 & \text{if } n = 0 \\ -1 & \text{if } n = m \\ 0 & \text{otherwise,} \end{cases}$$

and $c_n = 0$, for $n < 0$.

(b) From part (a) the required number is

$$[x^n](1 - x^2)(1 - x - x^2)^{-1} = [x^{n-1}](1 - x - x^2)^{-1} = F_n$$

from **2.2.23.**

2.3.2. Clearly, the required number is the number of compositions of n into k parts, each $\leqslant 6$. This is simply

$$[x^n](x + x^2 + \cdots + x^6)^k = [x^{n-k}](1 - x^6)^k(1 - x)^{-k}$$

$$= \sum_{i \geqslant 0} \binom{k}{i}(-1)^i[x^{n-k-6i}](1 - x)^{-k}$$

$$= \sum_{i \geqslant 0} \binom{k}{i}(-1)^i\binom{n - 6i - 1}{k - 1}.$$

2.3.3. (a) The set of solutions for all n is given by N^k, and we wish to keep track of the sum of the k components. Thus the required number is

$$[x^n](1 + x + x^2 + \cdots)^k = [x^n](1 - x)^{-k} = \binom{n + k - 1}{n}.$$

(b) The set of solutions for all n is $\times_{i=1}^{k}(\mathsf{N}_{b_i-1} - \mathsf{N}_{a_i-1})$, so the required number is

$$[x^n]\prod_{i=1}^{k}(x^{a_i} + x^{a_i+1} + \cdots + x^{b_i-1}) = [x^n]\prod_{i=1}^{k}(x^{a_i} - x^{b_i})(1 - x)^{-1}$$

$$= [x^n]x^{a_1+\cdots+a_k}(1 - x)^{-k}\prod_{i=1}^{k}(1 - x^{b_i-a_i}).$$

2.3.4. The set of solutions for all n is given by $\mathsf{N}^k \times \mathsf{N}^k$, where we keep track of the sum of the first k components, and of twice the other k components. Thus the required number is

$$[x^n](1 + x + x^2 + \cdots)^k(1 + x^2 + x^4 + \cdots)^k$$

$$= [x^n](1 - x)^{-k}(1 - x^2)^{-k} = \sum_{i \geqslant 0}\binom{k + i - 1}{k - 1}\binom{n + k - 2i - 1}{k - 1}.$$

A different expansion yields

$$[x^n](1 - x)^{-k}(1 - x^2)^{-k} = [x^n](1 - x)^{-2k}(1 + x)^{-k}$$

$$= \sum_{i \geqslant 0}\binom{2k + i - 1}{i}\binom{n + k - i - 1}{k - 1}(-1)^{n-i}.$$

2.3.5. (a) Each such representation $(x_0, \ldots, x_{k-1})$ is an element of $\times_{i=1}^{k}\{0, 1, \ldots, a_i - 1\}$, where j in the ith position contributes $j\alpha_i$ where $\alpha_i = \prod_{l=1}^{i-1} a_l$ to the sum, for $i > 1$, and j to the sum for $i = 1$. Thus the number of such representations of m is

$$[x^m]\{1 + x + \cdots + x^{a_1 - 1}\}\prod_{i=2}^{k}\{1 + x^{\alpha_i} + \cdots + x^{(a_i - 1)\alpha_i}\}$$

$$= [x^m]\frac{1 - x^{a_1}}{1 - x}\prod_{i=2}^{k}\left(\frac{1 - x^{\alpha_{i+1}}}{1 - x^{\alpha_i}}\right) = [x^m](1 - x^{\alpha_{k+1}})(1 - x)^{-1}$$

$$= 1 \qquad \text{for} \qquad 0 \leqslant m < \prod_{l=1}^{k} a_l.$$

(b) and (c) These follow from part (a) by setting $k = \infty$, and $a_j = l$ and $j + 1$, respectively.

(d) Set $a_j = n - l + j$ and $k = l$ in part (a).

2.3.6. The proof by induction proceeds as follows.

$$S_{1,m} = \sum_{m > x_1 \geqslant 0} z^{\binom{x_1}{1}} = \left\{1 - z^{\binom{m}{1}}\right\}(1 - z)^{-1}.$$

Assume that $S_{l,m} = \{1 - z^{\binom{m}{l}}\}(1 - z)^{-1}$ for all $l < k$, so

$$S_{k,m} = \sum_{m > x_k \geqslant k-1} z^{\binom{x_k}{k}} \sum_{x_k > x_{k-1} > \cdots \geqslant 0} z^{\binom{x_1}{1} + \cdots + \binom{x_{k-1}}{k-1}}$$

$$= \sum_{m > x_k \geqslant k-1} z^{\binom{x_k}{k}}\left(1 - z^{\binom{x_k}{k-1}}\right)(1 - z)^{-1} \qquad \text{by the induction hypothesis}$$

$$= (1 - z)^{-1} \sum_{m > x_k \geqslant k-1} \left\{z^{\binom{x_k}{k}} - z^{\binom{x_k + 1}{k}}\right\} = \left(1 - z^{\binom{m}{k}}\right)(1 - z)^{-1},$$

as required, for $k \geqslant 2$.

Now the number of representations of n as $\binom{x_1}{1} + \binom{x_2}{2} + \cdots + \binom{x_k}{k}$, $0 \leqslant x_1 < x_2 < \cdots < x_k$, for fixed $k \geqslant 1$, is clearly

$$[z^n] \sum_{x_k > x_{k-1} > \cdots > x_1 \geqslant 0} z^{\binom{x_1}{1} + \binom{x_2}{2} + \cdots + \binom{x_k}{k}}$$

$$= [z^n] \lim_{m \to \infty} \sum_{m > x_k > \cdots > x_1 \geqslant 0} z^{\binom{x_1}{1} + \cdots + \binom{x_k}{k}}$$

$$= [z^n] \lim_{m \to \infty} \left(1 - z^{\binom{m}{k}}\right)(1 - z)^{-1} = 1 \qquad \text{for } n \geqslant 0.$$

2.3.7. (a) Let each person, independently, be either an alpha or a beta. Then in each riding of $2k+1$ people there are $\binom{2k+1}{i}$ ways of choosing i of them to be alphas. Marking alphas by x and a riding with a majority ($\geqslant k+1$) of alphas by u gives the enumerator for each riding as

$$\sum_{i=0}^{k}\binom{2k+1}{i}x^i+\sum_{i=k+1}^{2k+1}\binom{2k+1}{i}x^iu.$$

Thus the required number is $[x^pu^q]F(x,u)$, where

$$F(x,u)=\left\{\sum_{i=0}^{k}\binom{2k+1}{i}x^i+u\sum_{i=k+1}^{2k+1}\binom{2k+1}{i}x^i\right\}^n.$$

(b) If we consider each of the arrangements of the p alphas in the population to be equally likely, then the expected number of ridings in which alphas have the majority is

$$\left\{[x^p]\frac{\partial}{\partial u}F(x,u)|_{u=1}\right\}\{[x^p]F(x,1)\}^{-1}.$$

But

$$[x^p]F(x,1)=[x^p](1+x)^{(2k+1)n}=\binom{(2k+1)n}{p}$$

and

$$[x^p]\frac{\partial}{\partial u}F(x,u)|_{u=1}=[x^p]n\left\{\sum_{i=k+1}^{2k+1}\binom{2k+1}{i}x^i\right\}(1+x)^{(2k+1)(n-1)},$$

so the result follows.

2.3.8. Record the successions as in **2.3.15**, and then mark the occurrence of $\{1,n\}$ by y. But $\{1,n\}$ occurs only when the first element of N_+ in $\mathsf{N}_+^k\times\mathsf{N}$ is 1 and when the element of N is 0. Thus the required number is

$$[x^ny^m](yx+x^2+x^3+\cdots)^{k-1}\{(x+x^2+\cdots)\times(1+x+\cdots)+(y-1)x\}$$

$$=[x^ny^m]\left(yx+\frac{x^2}{1-x}\right)^{k-1}\left\{\frac{x}{(1-x)^2}-x+yx\right\}$$

$$=\binom{k-1}{m}\left\{\binom{n-k+1}{k-m}-\binom{n-k-1}{k-m-2}\right\}+\binom{k-1}{m-1}\binom{n-k-1}{k-m-1}$$

$$=\frac{n}{k}\binom{k}{m}\binom{n-k-1}{k-m-1}.$$

2.3.9. (a) From Decomposition 2.3.14, such subsets may be decomposed as $\mathsf{N}_+ \times (\mathsf{N}_{b-1} - \mathsf{N}_{a-1})^{k-1} \times \mathsf{N}$, where the size of the subsets is recorded by the sum of the components. Thus the required number is

$$[x^n](x + x^2 + \cdots)(x^a + x^{a+1} + \cdots + x^{b-1})^{k-1}(1 + x + x^2 + \cdots)$$

$$= [x^n]x^{a(k-1)+1}(1-x)^{-(k+1)}(1-x^{b-a})^{k-1}$$

$$= \sum_{i \geqslant 0} (-1)^i \binom{k-1}{i} \binom{n-(a-1)(k-1)+i(a-b)}{k}.$$

(b) Following **2.3.22**, the required number is

$$\frac{n}{k}[x^n](x^a + x^{a+1} + \cdots + x^{b-1})^k = \frac{n}{k}[x^n]x^{ak}(1-x)^{-k}(1-x^{b-a})^k$$

$$= \frac{n}{k} \sum_{i \geqslant 0} (-1)^i \binom{k}{i} \binom{n-(a-1)k-i(b-a)-1}{k-1}.$$

2.3.10. The number of such subsets with $\sigma_k - \sigma_1 = m$ is $(n-m)[x^m](x^a + x^{a+1} + \cdots + x^{b-1})^{k-1}$ from the decomposition in [**2.3.9(a)**], since there are $n - m$ choices for σ_1. Thus the number required is

$$[x^n] \sum_{m=n-c+1}^{n-d} (n-m)x^{n-m}(x^a - x^b)^{k-1}(1-x)^{-(k-1)}$$

$$= [x^n]x^{a(k-1)}(1-x^{b-a})^{k-1}(1-x)^{-(k-1)} \sum_{i=d}^{c-1} ix^i$$

and the result follows by evaluating the sum as in [**1.1.3**].

2.3.11. (a) Let $\mathsf{S}(\mathbf{l}, \mathbf{q})$ be the set of $(\mathbf{l}, \mathbf{q})$-subsets of N_n for all n. Then, from Decomposition 2.3.14,

$$\mathsf{S}(\mathbf{l}, \mathbf{q}) \,\tilde{\rightarrow}\, \left\{ \mathop{\times}_{i=1}^{k} \{ j \mid j \equiv (1 + l_i) \bmod q_i \} \right\} \times \mathsf{N},$$

where n is recorded by the sum of the $k + 1$ components in the decomposition. Then the required number is

$$[x^n]\left\{ \prod_{i=1}^{k} (x^{1+l_i} + x^{1+l_i+q_i} + x^{1+l_i+2q_i} + \cdots) \right\}(1 + x + x^2 + \cdots)$$

$$= [x^n]\left\{ \prod_{i=1}^{k} x^{1+l_i}(1-x^{q_i})^{-1} \right\}(1-x)^{-1}, \qquad \text{and the result follows.}$$

(b) From part (a), the required number is

$$[x^n]x^{k+l_1+\cdots+l_k}(1-x)^{-1}(1-x^p)^{-k}$$
$$=[x^{n-(l_1+\cdots+l_k)}]x^k(1-x)^{-1}(1-x^p)^{-k}.$$

But this is simply the number of Skolem subsets with index p (Definition 2.3.16) of $\mathsf{N}_{n-(l_1+\cdots+l_k)}$, and the result follows from **2.3.18.**

2.3.12. In [**2.3.11**(b)] set $p=2$, and if the first block is of odd parity, $l_i=0$ for $i\equiv 1$ or $(\alpha+1)\bmod(\alpha+\beta)$, and $l_i=1$ otherwise. If the first block is even, then $l_1=1$ instead of 0.

2.3.13. If $\sigma_1\equiv t(\bmod p)$, $0\leqslant t\leqslant p-1$, then the required number of subsets is $\binom{\lfloor\{n-k-(t+l_2+\cdots+l_k)\}/p\rfloor+k}{k}$ from [**2.3.11**(b)]. Thus the total number of subsets is

$$\sum_{t=0}^{p-1}\binom{\lfloor\{n-k-(t+l_2+\cdots+l_k)\}/p\rfloor+k}{k}.$$

But

$$\lfloor\{n-k-(t+l_2+\cdots+l_k)\}/p\rfloor=\begin{cases}u & \text{if } t\leqslant v\\ u-1 & \text{if } t>v,\end{cases}$$

so the number is

$$\sum_{t=0}^{v}\binom{u+k}{k}+\sum_{t=v+1}^{p-1}\binom{u-1+k}{k}$$
$$=(v+1)\binom{u+k}{k}+(p-v-1)\frac{u}{u+k}\binom{u+k}{k}$$
$$=\frac{kv+pu+k}{u+k}\binom{u+k}{k}.$$

2.3.14. Suppose that the $2\times n$ board is covered by placing the tiles sequentially, in a canonical order. The order is determined by always placing a domino or square on a partially completed board to cover the left-most uncovered square on the board. If there are two left-most uncovered squares (in the same column), cover the square in the top row. Following this procedure leads to coverings, at intermediate stages, of four shapes, given in the diagram.

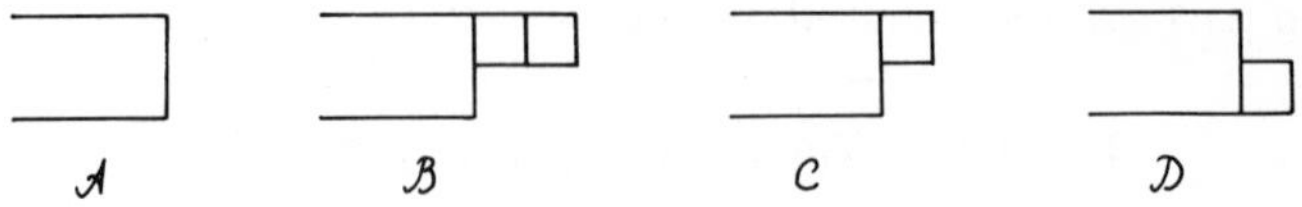

Let $A(x, y)$, $B(x, y)$, $C(x, y)$, and $D(x, y)$ be the generating functions for the sets A, B, C, and D of coverings, with x marking dominos and y marking squares. The single covering ε of the empty 2×0 board belongs to A, so $A(0,0) = 1$. If d and s denote a domino and square, respectively, then considering how boards of each shape arise from boards at one stage earlier leads to four simultaneous decompositions. These are

$$\mathsf{A} - \{\varepsilon\} \overset{\sim}{\rightarrow} \{d\} \times \mathsf{A} \,\dot{\cup}\, \{d\} \times \mathsf{B} \,\dot{\cup}\, \{s\} \times \mathsf{C} \,\dot{\cup}\, \{s\} \times \mathsf{D},$$

$$\mathsf{B} \overset{\sim}{\rightarrow} \{d\} \times \mathsf{A}, \quad \mathsf{C} \overset{\sim}{\rightarrow} \{s\} \times \mathsf{A} \,\dot{\cup}\, \{s\} \times \mathsf{B} \,\dot{\cup}\, \{d\} \times \mathsf{D}, \quad \mathsf{D} \overset{\sim}{\rightarrow} \{d\} \times \mathsf{C}$$

and yield immediately the four linear equations

$$A - 1 = xA + xB + yC + yD, \qquad B = xA$$

$$C = yA + yB \qquad + xD, \qquad D = xC.$$

The required number is $[x^n y^k]A(x, y)$, and the result follows by solving this system for A.

2.3.15. A polyomino consists of columns of unit squares, in which the sides of adjacent columns overlap in at least one edge. Let $F(x, y)$ be the generating function for polyominos in which x marks area and y marks the area of the right-most column of squares. Now consider attaching a column of m squares to the right-most end of a polyomino whose right-most column has k squares. This can be done in $m + k - 1$ ways to form a new polyomino, and all polyominos can be constructed in this way except those with a single column of squares. Thus

$$F(x, y) - \sum_{m \geq 1} (xy)^m = \sum_{m \geq 1} (xy)^m \left\{ y \frac{\partial}{\partial y} y^{m-1} F(x, y)|_{y=1} \right\}.$$

But $y(\partial/\partial y) y^{m-1} F(x, y)|_{y=1} = (m - 1)F(x, 1) + G(x)$, where $G(x) = (\partial/\partial y)F(x, y)|_{y=1}$, so

$$F(x, y) = x^2y^2(1 - xy)^{-2}F(x, 1) + xy(1 - xy)^{-1}\{1 + G(x)\}. \quad (1)$$

Setting $y = 1$ in (1) gives

$$F(x, 1) = x^2(1 - x)^{-2}F(x, 1) + x(1 - x)^{-1}\{1 + G(x)\},$$

and applying $\partial/\partial y$ to (1) and then setting $y = 1$ gives $G(x) = 2x^2(1-x)^{-3}F(x, 1) + x(1-x)^{-2}\{1 + G(x)\}$. Eliminating $G(x)$ from this system of linear equations for $G(x)$ and $F(x, 1)$ yields

$$F(x, 1) = \tfrac{1}{16}\{-5 + 4x + (5 - 13x + 7x^2)(1 - 5x + 7x^2 - 4x^3)^{-1}\},$$

and the result follows.

SECTION 2.4

2.4.1. (a) The required number, from Decomposition 2.4.5, is

$$\begin{aligned}
&[x^ny^m](1 + yx + yx^2 + \cdots)\{1 - (yx + yx^2 + \cdots)^2\}^{-1} \\
&\qquad \times(1 + yx + yx^2 + \cdots) = [x^ny^m]\left(1 + \frac{yx}{1-x}\right)^2\left\{1 - \left(\frac{yx}{1-x}\right)^2\right\}^{-1} \\
&\qquad = [x^ny^m]\left(1 + \frac{yx}{1-x}\right)\left(1 - \frac{yx}{1-x}\right)^{-1} \\
&\qquad = 2[x^n]x^m(1-x)^{-m} = 2\binom{n-1}{m-1}.
\end{aligned}$$

(b) Let $F(x, y)$ be the generating function in part (a). Then the required number, assuming all such sequences equally likely, is

$$\begin{aligned}
&\left([x^n]\left\{\frac{\partial}{\partial y}F(x, y)\right\}\bigg|_{y=1}\right) \bigg/ ([x^n]F(x, 1)) \\
&= 2^{-n}[x^n]\left\{\frac{x}{1-x}\left(1 - \frac{x}{1-x}\right)^{-1} + \frac{x}{1-x}\left(1 + \frac{x}{1-x}\right)\left(1 - \frac{x}{1-x}\right)^{-2}\right\} \\
&= 2^{-n}[x^n]\{x(1-2x)^{-1} + x(1-2x)^{-2}\} = \frac{n+1}{2}.
\end{aligned}$$

2.4.2. The set of $\{0, 1\}$-sequences with no adjacent 0's is given by $1^*(011^*)^*(\varepsilon \cup 0)$. Since a maximal block of k 1's contributes $k - 1$ pairs of adjacent 1's, the required number is

$$[x^ny^r](1 + x + yx^2 + y^2x^3 + \cdots)\{1 - x(x + yx^2 + y^2x^3 + \cdots)\}^{-1}(1 + x)$$

and the result follows.

2.4.3. The set of sequences with only even maximal blocks of 1's and odd maximal blocks of 0's is given by $(1^2)^*(0(0^2)^*1^2(1^2)^*)^*(\varepsilon \cup 0(0^2)^*)$. Thus, if length is marked by x, the required number is

$$[x^n](1-x^2)^{-1}\{1-x^3(1-x^2)^{-2}\}^{-1}\left(1+x(1-x^2)^{-1}\right)$$

$$=[x^n]\sum_{i\geqslant 0}x^{3i}(1-x^2)^{-(2i+1)}\left(1+x(1-x^2)^{-1}\right)$$

$$=\sum_{i\geqslant 0}\{[x^{n-3i}](1-x^2)^{-(2i+1)}+[x^{n-3i-1}](1-x^2)^{-(2i+2)}\}.$$

Now n and i must have the same parity, so let $n-i=2j$ in the first term and $n-i=2j+1$ in the second. Thus the number is

$$\sum_{j\geqslant 0}\{[x^{6j-2n}](1-x^2)^{-(2n-4j+1)}+[x^{6j-2n+2}](1-x^2)^{-(2n-4j)}\}$$

$$=\sum_{j\geqslant 0}\left\{\binom{n-j}{3j-n}+\binom{n-j}{3j-n+1}\right\}=\sum_{j\geqslant 0}\binom{n-j+1}{3j-n+1}.$$

2.4.4. The sequence of heads and tails is a $\{0,1\}$-sequence, where heads are 0's and tails 1's. The possible sequences are those with no maximal blocks of 0's of length at least r before the last maximal block of 0's, which has length r. These sequences are given by $1^*(\{0,\ldots,0^{r-1}\}1^*1)^*0^r$, so the number of such sequences of length k is given by

$$[x^k](1-x)^{-1}\{1-(x-x^r)(1-x)^{-1}x(1-x)^{-1}\}^{-1}x^r$$

$$=[x^{k-r}](1-x)(1-2x+x^{r+1})^{-1}.$$

Since each such sequence has probability 2^{-k}, the required probability is

$$2^{-k}[x^{k-r}](1-x)(1-2x+x^{r+1})^{-1}$$

$$=2^r[x^{k-r}]\left(1-\frac{x}{2}\right)\left(1-x+\left(\frac{x}{2}\right)^{r+1}\right)^{-1}$$

$$=[x^{k-r}](2-x)(2^{r+1}-2^{r+1}x+x^{r+1})^{-1}.$$

2.4.5. By Decomposition 2.3.1, a k-subset of N_n may be represented as a $\{0,1\}$-sequence of length n, with k ones. Moreover, a succession occurs for each pair of adjacent 1's. Thus a maximal block of k 1's contains exactly $k-1$ successions. If x marks length, y marks the number of 1's, and z marks successions, then the required number, from Decomposition 2.4.5, is

$$[x^n y^k z^m]\Big(1+\sum_{i\geq 1} z^{i-1}y^i x^i\Big)\Big\{1-x(1-x)^{-1}\sum_{i\geq 1} z^{i-1}y^i x^i\Big\}^{-1}(1-x)^{-1}$$

$$= [x^n y^k z^m]\big(1+yx(1-zyx)^{-1}\big)$$

$$\times\{1-yx^2(1-x)^{-1}(1-xyz)^{-1}\}^{-1}(1-x)^{-1}$$

$$= [x^n y^k z^m]\{1+xy(1-xyz)^{-1}\}^{-1}\{1-x\{1+xy(1-xyz)^{-1}\}\}^{-1}$$

$$= [x^n y^k z^m]\sum_{l=0}^{\infty} x^l\{1+xy(1-xyz)^{-1}\}^{l+1}$$

$$= [x^n y^k z^m]\sum_{l,p,q}\binom{l+1}{p}\binom{p+q-1}{q}x^{p+q+l}y^{q+p}z^q$$

$$= \binom{k-1}{m}\binom{n-k+1}{k-m} \qquad \text{since } l=n-k, p=k-m, q=m.$$

2.4.6. The set of such sequences is given by

$$(\varepsilon \cup 11^* \cup 22^*)\{\{3,4\}\{3,4\}^*(11^* \cup 22^*)\}^*\{3,4\}^*.$$

If the length is marked by x, then the required number is

$$[x^n]\Big(1+\frac{2x}{1-x}\Big)\Big\{1-\frac{2x}{1-2x}\frac{2x}{1-x}\Big\}^{-1}(1-2x)^{-1}$$

$$= [x^n](1+x)(1-3x-2x^2)^{-1}$$

$$= [x^n]\big(A_1(1-\alpha_1 x)^{-1}+A_2(1-\alpha_2 x)^{-1}\big),$$

where $\alpha_1, \alpha_2 = (3 \pm \sqrt{17})/2$; $A_1, A_2 = (5 \pm \sqrt{17})/2\sqrt{17}$ and the result follows.

2.4.7. Let $F(x_1, x_2, \ldots, x_{k+2}) = \{1-(x_1+\cdots+x_{k+2})\}^{-1}$ be the generating function for sequences on N_{k+2}. By series bisection the sequences with an even number of 1's are enumerated by

$$G(x_1, x_2, \ldots, x_{k+2}) = \tfrac{1}{2}\{F(x_1, x_2, \ldots, x_{k+2}) + F(-x_1, x_2, \ldots, x_{k+2})\},$$

and those that also have an odd number of 2's are enumerated by $\frac{1}{2}\{G(x_1, x_2, \ldots, x_{k+2}) - G(x_1, -x_2, \ldots, x_{k+2})\}$. Thus, if length is marked by

x, the required number is

$$\tfrac{1}{4}[x^n]\{(1-(k+2)x)^{-1}+(1-kx)^{-1}-(1-kx)^{-1}-(1-(k-2)x)^{-1}\}$$

$$=\tfrac{1}{4}\{(k+2)^n-(k-2)^n\}.$$

2.4.8. By **2.4.16**, the generating function for sequences on N_{2n} with distinct adjacent elements is $\{1-\sum_{i=1}^{2n} x_i(1+x_i)^{-1}\}^{-1}$. Sequences in which odd numbers can be adjacent to themselves are obtained by replacing each isolated odd number by a block of that odd number. The generating function with respect to type for the required set of sequences is thus $\{1-\sum_{i=1}^{n} x_{2i}(1+x_{2i})^{-1}-\sum_{i=1}^{n} x_{2i-1}\}^{-1}$. The required number is obtained by recording only length, and is

$$[x^l]\{1-nx(1+x)^{-1}-nx\}^{-1}=[x^l](1+x)\{1-(2n-1)x-nx^2\}^{-1}.$$

2.4.9. By Definition 2.4.14, we want the number of Smirnov sequences of type $(2,\ldots,2)$. The required number is thus, by **2.4.16**,

$$[x_1^2\cdots x_n^2]\left\{1-\sum_{i=1}^{n} x_i(1+x_i)^{-1}\right\}^{-1}$$

$$=[x_1^2\cdots x_n^2]\left\{1-\sum_{i=1}^{n} x_i+\sum_{i=1}^{n} x_i^2\right\}^{-1}$$

$$=[x_1^2\cdots x_n^2]\sum_{k=0}^{n}(s_1-s_2)^{2n-k}, \qquad \text{where } s_j=\sum_{i=1}^{n} x_i^j, j=1,2,$$

$$=[x_1^2\cdots x_n^2]\sum_{k=0}^{n} s_1^{2n-2k}s_2^k\binom{2n-k}{k}(-1)^k.$$

But

$$[x_1^2\cdots x_n^2]s_1^a s_2^b=\begin{cases}\binom{n}{b}b!a!(2!)^{-a/2}, & a+2b=2n\\ 0, & \text{otherwise,}\end{cases}$$

and the result follows.

2.4.10. We forbid maximal blocks of size 1 and thus obtain the required number from **2.4.18** by setting $f_1=0$, $f_i=1$ for $i \geqslant 2$. The number is accordingly

$$[x_1^5 \cdots x_n^5]\left((1 + x^2 + x^3 + \cdots)^{-1} \circ \mathbf{s}\right)^{-1}$$

$$= [x_1^5 \cdots x_n^5]\left(\left(\frac{1 - x + x^2}{1 - x}\right)^{-1} \circ \mathbf{s}\right)^{-1}$$

$$= [x_1^5 \cdots x_n^5]\left((1 - x^2 - x^3 + x^5) \circ \mathbf{s}\right)^{-1}$$

$$= [x_1^5 \cdots x_n^5](1 - s_2 - s_3 + s_5)^{-1}, \qquad \text{where } s_j = \sum_{i=1}^{n} x_i^j$$

$$= [x_1^5 \cdots x_n^5] \sum_{i=0}^{n} (-1)^i s_5^i s_2^{n-i} s_3^{n-i} (2n - i)!/\{(n - i)!\}^2 i!.$$

But

$$[x_1^5 \cdots x_n^5] s_2^a s_3^b s_5^c = \begin{cases} \binom{n}{c} c!a!b!, & \text{if } a = b, 2a + 3b + 5c = 5n \\ 0, & \text{otherwise,} \end{cases}$$

and the result follows.

2.4.11. (a) Sequences in which maximal blocks have length less than k are enumerated by **2.4.18** with $f_i = 0$ for $i < k$ and $f_i = 1$ for $i \geqslant k$. If we mark length by x, then the required number is

$$[x^l]\left\{1 - nx^k(1 - x)^{-1}\left(1 + x^k(1 - x)^{-1}\right)^{-1}\right\}^{-1}$$

$$= [x^l](1 - x + x^k)(1 - x - (n - 1)x^k)^{-1}.$$

(b) Sequences in which maximal blocks have length less than k are enumerated by **2.4.18** with $f_i = 1$ for $i < k$ and $f_i = 0$ for $i \geqslant k$. Marking the length by x, we obtain the required number as

$$[x^l]\left(1 - n(x - x^k)(1 - x^k)^{-1}\right)^{-1} = [x^l](1 - x^k)(1 - nx + (n - 1)x^k)^{-1}.$$

2.4.12. We follow **2.4.18**, but note that a maximal block of l's of length $t \geqslant k_l$ contains $t - k_l + 1$ blocks of l's, for $l = 1, \ldots, n$. If we mark such a block by y_l, then the required number is $[\mathbf{x}^{\mathbf{i}}\mathbf{y}^{\mathbf{j}}]D(g_1, \ldots, g_n)$, where

$$g_l = x_l + x_l^2 + \cdots + x_l^{k_l - 1} + y_l x_l^{k_l} + y_l^2 x_l^{k_l + 1} + \cdots$$

$$= \{x_l(1 - y_l x_l) - x_l^{k_l}(1 - y_l)\}(1 - x_l)^{-1}(1 - y_l x_l)^{-1}$$

for $l = 1, \ldots, n$. The result follows.

2.4.13. The sequences in which adjacent elements are distinct mod p are obtained by considering Smirnov sequences on N_p and replacing each isolated element i by an arbitrary element in $\{i, i+p, i+2p, \dots\}$. Thus, from **2.4.16**, sequences over N_+ with distinct adjacent elements mod p are enumerated by

$$\left\{1 - \sum_{i=1}^{p} (x_i + x_{i+p} + \cdots)(1 + x_i + x_{i+p} + \cdots)^{-1}\right\}^{-1}.$$

Accordingly, from Remark 2.4.11, the number of compositions of n with adjacent parts that are distinct mod p is

$$[x^n]\left\{1 - \sum_{i=1}^{p} (x^i + x^{i+p} + \cdots)(1 + x^i + x^{i+p} + \cdots)^{-1}\right\}^{-1},$$

and the result follows.

2.4.14. We employ Decomposition 2.4.19, replacing fall by nonsuccession and nonfall by succession. A sequence of length l with k successions has $l-k-1$ nonsuccessions, and thus, following **2.4.20**, the required number is $[x^l t^{l-k-1}]G(x,t)$, where $t(1+t)^{-1}G(x(1+t), t(1+t)^{-1}) = \{1 - tD(x)\}^{-1} - 1$, and $D(x)$ is the generating function for sequences on N_n that have no nonsuccessions. These sequences are of the form $(i, i+1, \dots, i+k)$ and there are thus $n-k+1$ of them of length k. Thus $D(x) = \sum_{k=1}^{n} (n-k+1)x^k = \{nx - (n+1)x^2 + x^{n+2}\}(1-x)^{-2}$. Setting $z = x(1+t)$, $u = t(1+t)^{-1}$ gives $t = u(1-u)^{-1}$, $x = (1-u)z$, so

$$uG(z,u) = \left\{1 - u\frac{nz - (n+1)(1-u)z^2 - (1-u)^{n+1}z^{n+2}}{(1-(1-u)z)^2}\right\}^{-1} - 1$$

and the result follows.

2.4.15. (a) Follow the solution to [**2.4.14**], the only difference being that there are n sequences of length k that consist entirely of circular successions, for $k \geqslant 1$. Thus $D(x) = \sum_{k\geqslant 1} nx^k = nx(1-x)^{-1}$, so the required number is

$$[z^l u^{l-k}]\left\{1 - \frac{u}{1-u}\,\frac{n(1-u)z}{1-(1-u)z}\right\}^{-1}$$

$$= [z^l u^{l-k}]\{1 - (1-u)z\}\{1 - z(1+(n-1)u)\}^{-1}$$

$$= [u^{l-k}]\{(1+(n-1)u)^l - (1-u)(1+(n-1)u)^{l-1}\}$$

$$= [u^{l-k}]nu\{1+(n-1)u\}^{l-1} = n\binom{l-1}{k}(n-1)^{l-k-1}.$$

(b) The sequence contains $l-1$ pairs of adjacent elements. Choose one of the $\binom{l-1}{k}$ k-subsets of these to be circular successions. The second element of each of these adjacent pairs is then determined by the first element of the pair. There are $l-k$ remaining elements to be chosen, with n choices for the first and $n-1$ for each of the remaining $l-k-1$, since they cannot form a circular succession with the preceding element. The result follows.

2.4.16. Consider the set of all sequences in N_n^*, with generating function $S(\mathbf{x}) = \{1 - \sum_{i=1}^n x_i\}^{-1}$. If we replace each element in such a sequence by a block of that element, and mark the levels internal to that block by y, then we have the generating function $S(x_1(1-yx_1)^{-1}, \ldots, x_n(1-yx_n)^{-1})$. By construction, this is an "at-least" generating function for levels, and, from the Principle of Inclusion and Exclusion, the number of sequences of type $\mathbf{i}$ with k levels is

$$[\mathbf{x}^{\mathbf{i}}y^k]S\big(x_1(1-(y-1)x_1)^{-1}, \ldots, x_n(1-(y-1)x_n)^{-1}\big)$$

$$= \left\{1 - \sum_{i=1}^{n} \frac{x_i}{1-(y-1)x_i}\right\}^{-1}.$$

2.4.17. (a) Use Decomposition 2.4.15 and follow **2.4.17**. If $\Gamma(\mathbf{x}, f)$ enumerates sequences with respect to type and falls, then $\Gamma(x_1(1+x_1)^{-1}, \ldots, x_n(1+x_n)^{-1}, f)$ enumerates these without levels, and if l marks levels, then $G(\mathbf{x}, l, f) = \Gamma(x_1(1-(l-1)x_1)^{-1}, \ldots, x_n(1-(l-1)x_n)^{-1}, f)$ enumerates sequences with respect to type, levels, and falls. If $e(\sigma)$ and $d(\sigma)$ are the number of levels and falls in σ, then from **2.4.20**

$$G(\mathbf{x}, l, f) = \sum_{\sigma \in \mathsf{N}_n^+} \mathbf{x}^{\tau(\sigma)} l^{e(\sigma)} f^{d(\sigma)}$$

$$= \frac{\prod_{i=1}^n \{1-(l-f)x_i\}^{-1} - \prod_{i=1}^n \{1-(l-1)x_i\}^{-1}}{\prod_{i=1}^n \{1-(l-1)x_i\}^{-1} - f\prod_{i=1}^n \{1-(l-f)x_i\}^{-1}}.$$

But if $c(\sigma)$ is the number of rises, then

$$c(\sigma) = i_1 + \cdots + i_n - e(\sigma) - d(\sigma) - 1,$$

so that the required number is $[\mathbf{x}^{\mathbf{m}} r^i l^j f^k] r^{-1} G(r\mathbf{x}, r^{-1}l, r^{-1}f)$ and the result follows.

(b) Set $l = 0$ and $f = 1$ in part (a).

(c) Since we are not interested in rises, substitute $r = 1$ in part (b). This is admissible since the generating function is bounded in r, but gives the form

0/0. Thus use L'Hôpital's rule [**1.1.1**] to obtain

$$\lim_{r\to 1}\left\{\frac{\Sigma_{j=1}^{n} x_j \Pi_{i=1,i\neq j}^{n}(1+rx_i)}{\Pi_{i=1}^{n}(1+x_i)-\Sigma_{j=1}^{n} x_j \Pi_{i=1,i\neq j}^{n}(1+rx_i)}\right\}$$

$$=\left\{1-\sum_{j=1}^{n} x_j(1+x_j)^{-1}\right\}^{-1}-1.$$

2.4.18. (a) Apply Remark 2.4.11 to the result of [**2.4.17**(a)].

(b) Apply Remark 2.4.29 to the result of [**2.4.17**(a)].

2.4.19. Let P be the set of permutations on N_n, for all $n \geqslant 0$, and I be the set of indecomposable permutations on N_n, for all $n \geqslant 1$. Define a **component** of a permutation $\sigma_1 \cdots \sigma_n$ to be a substring $\sigma_i \cdots \sigma_j$, $i \leqslant j$, which is a permutation on $\{i,\dots,j\}$. Then

$$\mathsf{P} \mathrel{\tilde{\to}} \mathsf{I}^* : \sigma_1 \cdots \sigma_n \mapsto \left(\rho_{11} \cdots \rho_{1i_1}, \dots, \rho_{k1} \cdots \rho_{ki_k}\right)$$

is a length-preserving decomposition, where

$$\sigma_1 \cdots \sigma_{i_1}, \sigma_{i_1+1} \cdots \sigma_{i_1+i_2}, \dots, \sigma_{i_1+\cdots+i_{k-1}+1} \cdots \sigma_{i_1+\cdots+i_k}$$

are the minimal components of $\sigma_1 \cdots \sigma_n$, and

$$\rho_{lj} = \sigma_{i_1+\cdots+i_{l-1}+j} - (i_1 + \cdots + i_{l-1}) \qquad \text{for } l = 1,\dots,k, j = 1,\dots,i_l.$$

Let $P(x)$, $I(x)$ be the ordinary generating functions for P, I, respectively, with x marking length. From the preceding decomposition, $P(x) = \{1 - I(x)\}^{-1}$ so $I(x) = 1 - P^{-1}(x)$. But $P(x) = \Sigma_{i\geqslant 0}\, i!x^i$, and the result follows.

2.4.20. (a) Let M be the set of monic polynomials over GF(p), and let $r_1, r_2, \dots$ be the irreducible polynomials in M. Since each monic polynomial can be factored uniquely into a collection of irreducible monic polynomials, we have immediately $\mathsf{M} \mathrel{\tilde{\to}} r_1^* r_2^* r_3^* \dots$, where the empty string on the right-hand side corresponds to the single monic polynomial of degree 0 on the left-hand side. Thus if $M(z)$ is the generating function for M with respect to degree, and d_i is the degree of r_i, this decomposition yields

$$M(z) = \prod_{n\geqslant 1}(1 - z^{d_n})^{-1} = \prod_{i\geqslant 1}(1 - z^i)^{-m_i}.$$

But there are p choices for each of the coefficients of $x^0, \dots, x^{n-1}$ in a monic polynomial of degree n, so $M(z) = (1 - pz)^{-1}$ and the result follows.

(b) Applying *log* to both sides of part (a) gives

$$\sum_{j\geqslant 1}\sum_{i\geqslant 1} im_i z^{ij}/ij = \sum_{n\geqslant 1} p^n z^n/n.$$

Comparing coefficients of z^n/n yields $\Sigma_{i|n} im_i = p^n$, so that $nm_n \leqslant p^n$, $n \geqslant 1$. Therefore

$$nm_n \geqslant p^n - \sum_{i=1}^{n-1} im_i \geqslant p^n - \sum_{i=1}^{n-1} p^i$$

$$= p^n - (p^n - p)(p-1)^{-1} = p\{p^{n-2}(p-2)+1\}(p-1)^{-1} > 0,$$

since $p \geqslant 2$. Thus $m_n > 0$ and $m_n \geqslant 1$ since m_n is an integer. Since a $\mathrm{GF}(p^i)$ can be constructed from an irreducible monic polynomial of degree i over $\mathrm{GF}(p)$, this gives a combinatorial proof for the existence of Galois fields of every order.

2.4.21. (a) Let $\mathsf{S}_{\alpha,l}$ be the set of sequences on N_n with first element α and last element l. Let $f_{\alpha,l}(\mathbf{x},\mathbf{A})$ be the generating function for sequences in $\mathsf{S}_{\alpha,l}$ with x_i marking the occurrence of i as an element of the sequence and a_{ij} marking the occurrence of ij as a substring of the sequence, for $i, j = 1,\ldots, n$. Then by considering the first element of such a sequence

$$\mathsf{S}_{i,l} = \dot{\bigcup_{j=1}^{n}} i\mathsf{S}_{j,l} \qquad \text{for } i = 1,\ldots,n,\ i \neq l,$$

$$\mathsf{S}_{l,l} = \{l\} \,\dot{\cup}\, \left\{\dot{\bigcup_{j=1}^{n}} l\mathsf{S}_{j,l}\right\}.$$

This immediately gives the system of functional equations

$$f_{i,l} = x_i\left\{\delta_{il} + \sum_{j=1}^{n} a_{ij} f_{j,l}\right\} \qquad \text{for } i = 1,\ldots,n,$$

where $c_{\alpha,l}(\mathbf{k},\mathbf{M}) = [\mathbf{x}^{\mathbf{k}}\mathbf{A}^{\mathbf{M}}] f_{\alpha,l}(\mathbf{x},\mathbf{A})$.

(b) The system of functional equations in part (a) can be solved by applying the multivariate Lagrange theorem for monomials (Corollary 1.2.13). To carry this out consider the unknown functions $f_{1,l},\ldots, f_{n,l}$ as power series in $x_1,\ldots, x_n$ with polynomial coefficients in the a_{ij}. Thus

$$c_{\alpha,l}(\mathbf{k},\mathbf{M}) = [\mathbf{A}^{\mathbf{M}}]\big([\mathbf{x}^{\mathbf{k}}] f_{\alpha,l}(\mathbf{x},\mathbf{A})\big)$$

$$= [\mathbf{A}^{\mathbf{M}}](k_1 \cdots k_n)^{-1} \sum_{\boldsymbol{\mu}} \|\delta_{ij}k_i - \mu_{ij}\| \prod_{i=1}^{n}\big([f_{1,l}^{\mu_{i1}} \cdots f_{n,l}^{\mu_{in}}]\phi_i^{k_i}\big),$$

where $\phi_i = \delta_{il} + \sum_{j=1}^n a_{ij} f_{j,l}$ for $i = 1,\ldots, n$, and where the summation is over $\mu = [\mu_{ij}]_{n\times n}$ such that

$$\sum_{i=1}^{n} \mu_{ij} = k_j - \delta_{j\alpha} \qquad \text{for } j = 1,\ldots, n.$$

This yields $c_{\alpha,l}(\mathbf{k}, \mathbf{M})$

$$= [\mathbf{A}^{\mathbf{M}}](k_1 \cdots k_n)^{-1} \sum_{\mu} \|\delta_{ij} k_i - \mu_{ij}\| \prod_{i=1}^{n} k_i! \frac{a_{i1}^{\mu_{i1}}}{\mu_{i1}!} \cdots \frac{a_{in}^{\mu_{in}}}{\mu_{in}!} \frac{\delta_{il}^{k_i - \sum_{j=1}^n \mu_{ij}}}{\left(k_i - \sum_{j=1}^n \mu_{ij}\right)!},$$

where, for a nonzero contribution, $\sum_{j=1}^n \mu_{ij} = k_i$ for $i \neq l$. The column restrictions on μ mean that $\sum_{j=1}^n \sum_{i=1}^n \mu_{ij} = k_1 + \cdots + k_n - 1$, so $\sum_{j=1}^n \mu_{lj} = k_l - 1$ and μ satisfies the row restrictions $\sum_{j=1}^n \mu_{ij} = k_i - \delta_{il}$. But $\mathbf{M}$ satisfies the same row and column restrictions as μ, by counting occurrences of each element as an initiator or terminator of substrings of length 2, and the result follows.

2.4.22. (a) A directed Hamiltonian cycle in the digraph can be represented as a sequence by listing the labels of the vertices that are traversed in succession, starting and finishing at an arbitrary vertex l. Such a sequence starts and finishes with l, contains no other occurrence of l, and has exactly one occurrence of each of the remaining elements of N_n. The result follows since in the Hamiltonian cycles corresponding to such a sequence there are d_{ij} choices of edges in the graph for each substring ij in the sequence.

(b) From the multivariate Lagrange theorem (**1.2.9**) and part (a),

$$h(\mathbf{D}) = \left[f_{1,l} \cdots f_{l,l}^2 \cdots f_{n,l}\right] f_{l,l} \phi_l \|\delta_{ij}\phi_i - d_{ij} f_{j,l}\|,$$

where $\phi_i = \delta_{il} + \psi_i$, $\psi_i = \sum_{j=1}^n d_{ij} f_j$, for $i = 1,\ldots, n$. Thus

$$h(\mathbf{D}) = [f_{1,l} \cdots f_{n,l}] \phi_l \left\{\|\delta_{ij}\psi_i - d_{ij} f_j\| + \operatorname{cof}_{ll}\left[\delta_{ij}\psi_i - d_{ij} f_j\right]_{n\times n}\right\}$$

by expanding the determinant using row l. But all the row sums in $[\delta_{ij}\psi_i - d_{ij} f_j]_{n\times n}$ are zero, so

$$h(\mathbf{D}) = [f_{1,l} \cdots f_{n,l}](\psi_l + 1)\left(0 + \operatorname{cof}_{ll}\left[\delta_{ij}\psi_i - d_{ij} f_j\right]_{n\times n}\right)$$

$$= [f_{1,l} \cdots f_{n,l}] \psi_l \operatorname{cof}_{ll}\left[\delta_{ij}\psi_i - d_{ij} f_j\right]_{n\times n},$$

since $\operatorname{cof}_{cc}[\delta_{ij}\psi_i - d_{ij} f_j]_{n\times n}$ is homogeneous of degree $(n-1)$ in $f_{1,l}, \ldots, f_{n,l}$ and $f_{1,l} \cdots f_{n,l}$ is of degree n. Finally, expanding the cofactor as the determinant of a sum (**1.1.10(4)**),

$$h(\mathbf{D}) = [f_{1,l} \cdots f_{n,l}] \sum_{\beta \subseteq \mathrm{N}_n - \{l\}} (-1)^{|\beta|} (\det \mathbf{D}[\beta|\beta]) \Big(\prod_{j \in \beta} f_{j,n} \Big) \Big(\prod_{i \in \mathrm{N}_n - \beta} \psi_i \Big),$$

and the result follows.

2.4.23. A closed directed Euler trail in the digraph can be represented as a sequence by listing the labels of the vertices that are traversed in succession, starting at an arbitrary vertex l, and terminating there as well. Such a sequence begins and ends with l, has d_{ij} occurrences of the substring ij, and $b_i + \delta_{il}$ occurrences of element i, for $i, j = 1, \ldots, n$. Thus, in the notation of [**2.4.21**], there are $c_{l,l}(\mathbf{b} + \boldsymbol{\delta}_l, \mathbf{D})$ such sequences, where $\boldsymbol{\delta}_l = (\delta_{1l}, \ldots, \delta_{nl})$. However, since edges in the trail are distinct, and the sequence representation of the trail can be started at any of the b_l occurrences of vertex l, then

$$e(\mathbf{D}) = b_l^{-1} \mathbf{D}! c_{l,l}(\mathbf{b} + \boldsymbol{\delta}_l, \mathbf{D}) = (\mathbf{b} - \mathbf{1})! \| \delta_{ij}(b_i + \delta_{il}) - d_{ij} \|$$

from [**2.4.21**(b)]. Expanding this determinant on row l gives

$$e(\mathbf{D}) = (\mathbf{b} - \mathbf{1})! \big\{ \| \delta_{ij} b_i - d_{ij} \| + \operatorname{cof}_{ll} \big[\delta_{ij} b_i - d_{ij} \big]_{n \times n} \big\}.$$

But $[\delta_{ij} b_i - d_{ij}]_{n \times n}$ has row and column sums equal to zero, so $\| \delta_{ij} b_i - d_{ij} \| = 0$ and the result follows.

SECTION 2.5

2.5.1. (a) From **2.5.5** the number of partitions of n with distinct parts is given by

$$\begin{aligned}
[q^n] \prod_{i \geq 1} (1 + q^i) &= [q^n] \prod_{i \geq 1} (1 - q^{2i})(1 - q^i)^{-1} \\
&= [q^n] \prod_{i \geq 1} (1 - q^{2i}) \prod_{j \geq 1} (1 - q^j)^{-1} \\
&= [q^n] \prod_{k \geq 1} (1 - q^{2k-1})^{-1}.
\end{aligned}$$

But this is also the number of partitions of n with only odd parts, from Decomposition 2.5.2.

(b) Let $\alpha = (\alpha_1, \ldots, \alpha_k)$ be a partition of n, with distinct parts, where $\alpha_j = o_j 2^{i_j}$, $i_j \geq 0$, o_j is odd, for $j = 1, \ldots, k$. If β is a partition containing 2^{i_j} parts equal to o_j, for $j = 1, \ldots, k$, then β is a partition of n with odd parts.

Though the o_j are not necessarily distinct, this construction is reversible since the number of times which each different part of β is repeated has a unique representation as a sum of different powers of 2. Thus this is an appropriate bijection.

(c) The number of partitions of n into parts not divisible by m is given by

$$[q^n]\sum_{i\nmid m}(1-q^i)^{-1} = [q^n]\prod_{j\geq 1}(1-q^{mj})\prod_{i\geq 1}(1-q^i)^{-1}$$
$$= [q^n]\prod_{i\geq 1}(1-q^{mi})(1-q^i)^{-1}$$
$$= [q^n]\prod_{i\geq 1}(1+q^i+q^{2i}+\cdots+q^{(m-1)i}).$$

But this is the number of partitions of n with no part repeated m times or more.

2.5.2. The set of partitions in which only the odd parts can be repeated is decomposed as $(1^*)\langle\varepsilon, 2\rangle(3^*)\langle\varepsilon, 4\rangle\cdots$ so the number of partitions of n in which only the odd parts can be repeated is

$$[q^n]\prod_{i\geq 1}(1+q^{2i})(1-q^{2i-1})^{-1}$$
$$= [q^n]\prod_{i\geq 1}(1-q^{4i})(1-q^{2i})^{-1}(1-q^{2i-1})^{-1}$$
$$= [q^n]\prod_{j\geq 1}(1-q^{4j})(1-q^j)^{-1}$$
$$= [q^n]\prod_{j\geq 1}(1+q^j+q^{2j}+q^{3j}).$$

But this is also the number of partitions of n with no part repeated more than three times.

2.5.3. The number of partitions of n into an even number of parts minus the number with an odd number of parts is given by

$$\sum_{i\geq 1}([q^n t^{2i}]-[q^n t^{2i-1}])\prod_{k\geq 1}(1-tq^k)^{-1}$$
$$= [q^n]\prod_{k\geq 1}(1-(-1)q^k)^{-1} = [q^n]\prod_{k\geq 1}(1-q^k)(1-q^{2k})^{-1}$$
$$= [q^n]\prod_{j\geq 1}(1-q^{2j-1}).$$

But the number of partitions of n into m odd, distinct parts is $[q^n t^m]\prod_{j\geq 1}(1+tq^{2j-1})$. Now if n is even, then m must be even, and if n is odd, then m must be odd, because each part is odd. Therefore the number of partitions of n into odd, distinct parts is also given by $(-1)^n[q^n]\prod_{j\geq 1}(1-q^{2j-1})$. Thus the two quantities agree up to sign and have equal absolute values.

2.5.4. Let α be a partition of n in which consecutive integers do not both appear as parts. Then the conjugate partition $\tilde{\alpha}$ is a partition of n with no part appearing exactly once. This is reversible, so the result follows.

2.5.5. (a) The number of partitions of n in which each part appears 2, 3, or 5 times is given by

$$\begin{aligned}[q^n]\prod_{i\geqslant 1}(1+q^{2i}+q^{3i}+q^{5i}) &= [q^n]\prod_{i\geqslant 1}(1+q^{2i})(1+q^{3i})\\ &= [q^n]\prod_{i\geqslant 1}(1-q^{4i})(1-q^{2i})^{-1}(1-q^{6i})(1-q^{3i})^{-1}\\ &= [q^n]\prod_{i\geqslant 1}(1-q^{4i})\prod_{j\geqslant 1}(1-q^{2j})^{-1}\prod_{k\geqslant 1}(1-q^{6k})\prod_{l\geqslant 1}(1-q^{3l})^{-1}\\ &= [q^n]\prod_{j\geqslant 0}(1-q^{4j+2})^{-1}\prod_{k\geqslant 0}(1-q^{6k+3})^{-1}.\end{aligned}$$

But this is the number of partitions of n each of whose parts is congruent to 2, 3, 6, 9, or 10 mod 12.

(b) The number of partitions of n in which each part appears a, b, or $a+b$ times is given by

$$\begin{aligned}[q^n]\prod_{i\geqslant 1}(1+q^{ai}+q^{bi}+q^{(a+b)i}) &= [q^n]\prod_{i\geqslant 1}(1+q^{ai})(1+q^{bi})\\ &= [q^n]\prod_{i\geqslant 1}(1-q^{2ai})(1-q^{ai})^{-1}(1-q^{2bi})(1-q^{bi})^{-1}\\ &= [q^n]\prod_{k\geqslant 0}(1-q^{a(2k+1)})^{-1}\prod_{j\geqslant 0}(1-q^{b(2j+1)})^{-1}\end{aligned}$$

But we can never have $a(2k+1)=b(2j+1)$ for any j, k under the given conditions on a, b. The result is now immediate.

2.5.6. The number of partitions of n in which no part appears exactly once is given by

$$\begin{aligned}[q^n]\prod_{i\geqslant 1}(1+q^{2i}+q^{3i}+\cdots) &= [q^n]\prod_{i\geqslant 1}\left(1+q^{2i}(1-q^i)^{-1}\right)\\ &= [q^n]\prod_{i\geqslant 1}(1-q^i+q^{2i})(1-q^i)^{-1}\\ &= [q^n]\prod_{i\geqslant 1}(1+q^{3i})(1+q^i)^{-1}(1-q^i)^{-1}\\ &= [q^n]\prod_{i\geqslant 1}(1-q^{2i})^{-1}(1-q^{3i})^{-1}(1-q^{6i})\\ &= [q^n]\prod_{j\geqslant 1}(1-q^{6j-4})^{-1}(1-q^{6j-3})^{-1}(1-q^{6j-2})^{-1}(1-q^{6j})^{-1}.\end{aligned}$$

But this is the number of partitions of n in which no parts are congruent to 1 or 5 mod 6.

2.5.7. The number of partitions of n with unique smallest part and largest part at most twice the smallest part is given by

$$[q^n]\sum_{m\geq 1} q^m \prod_{i=m+1}^{2m} (1-q^i)^{-1}$$

$$= [q^n]\sum_{m\geq 1} \{(1-q^{2m})(1-q^m)^{-1}-1\} \prod_{i=m+1}^{2m} (1-q^i)^{-1}$$

$$= [q^n]\left\{\sum_{m\geq 1}\prod_{i=m}^{2m-1}(1-q^i)^{-1} - \sum_{m\geq 1}\prod_{i=m+1}^{2m}(1-q^i)^{-1}\right\}$$

$$= [q^n]\left\{\sum_{m\geq 1}\prod_{i=m}^{2m-1}(1-q^i)^{-1} - \sum_{l\geq 2}\prod_{i=l}^{2l-2}(1-q^i)^{-1}\right\}$$

$$= [q^n]\left\{(1-q)^{-1} + \sum_{m\geq 2}\{(1-q^{2m-1})^{-1}-1\}\prod_{i=m}^{2m-2}(1-q^i)^{-1}\right\}$$

$$= [q^n]\left\{1 + \sum_{m\geq 1} q^{2m-1}\prod_{i=m}^{2m-1}(1-q^i)^{-1}\right\}.$$

But this is the number of partitions of n in which the largest part is odd and the smallest part is larger than one-half of the largest part.

2.5.8. Let $F(q, y) = \sum_{i,n} f(i, n)q^n y^i = (1-yq)^{-1}\prod_{j\geq 2}(1-q^j)^{-1}$, where $f(i, n)$ is the number of partitions of n with i 1's. Thus the sum of the number of 1's in all partitions of n is

$$\sum_i if(i, n) = [q^n]\frac{\partial}{\partial y}F(q, y)|_{y=1} = [q^n]q(1-q)^{-1}\prod_{i\geq 1}(1-q^i)^{-1}.$$

Similarly, let $G(q, y) = \sum_{i,n\geq 0} g(i, n)y^i q^n = \prod_{j\geq 1}(1 + yq^j(1-q^j)^{-1})$, where $g(i, n)$ is the number of partitions of n with i different parts. Then the sum of the number of different parts in all partitions of n is

$$\sum_i ig(i, n) = [q^n]\frac{\partial}{\partial y}G(q, y)|_{y=1}$$

$$= [q^n]\sum_{i\geq 1} q^i \prod_{j\geq 1}(1-q^j)^{-1} = [q^n]q(1-q)^{-1}\prod_{i\geq 1}(1-q^i)^{-1}.$$

The two quantities are thus equal.

2.5.9. Use Decomposition 2.5.12, marking number of parts by z and size of largest part by w. The Durfee square D_m contributes m parts, m to the largest part, and m^2 dots. Thus the generating function for D_m is $q^{(m^2)}z^m w^m$. The element of Q_m contributes nothing to the largest part and is thus enumerated with respect to number of parts so the generating function for Q_m is $\prod_{i=1}^{m}(1-zq^i)^{-1}$. The element of R_m contributes nothing to the number of parts and is enumerated with respect to size of largest part. Of course, the largest part of the element of R_m is added to m to give the largest part of the resulting partition, so the generating function for R_m is $\prod_{i=1}^{m}(1-wq^i)^{-1}$. Accordingly, the number of partitions of n with largest part k and l parts is

$$[q^n w^k z^l] \sum_{m\geqslant 1} \frac{q^{(m^2)}z^m w^m}{(1-zq)\cdots(1-zq^m)(1-wq)\cdots(1-wq^m)}.$$

We determine another expression for this generating function by considering those partitions with largest part m and Decomposition 2.5.6. The largest part is enumerated by $zq^m w^m$. The remaining parts contribute nothing to the largest part, and are thus enumerated by $\prod_{i=1}^{m}(1-zq^i)^{-1}$, so the number of partitions of n with largest part k and l parts is

$$[q^m w^k z^l] \sum_{m\geqslant 1} \frac{q^m z w^m}{(1-zq)\cdots(1-zq^m)}.$$

The result follows.

2.5.10. Use Decomposition 2.5.16, marking number of parts by z and size of largest part by w. The triangle T_m contributes m to number of parts, m to the largest part, and contains $\binom{m+1}{2}$ dots, so the generating function for T_m is $q^{\binom{m+1}{2}}z^m w^m$. The element of R_m contributes nothing to the number of parts and is thus enumerated with respect to largest part so the generating function for R_m is $\prod_{i=1}^{m}(1-wq^i)^{-1}$. Thus the number of partitions of n into distinct parts, with largest part k and l parts, is

$$[q^n w^k z^l] \sum_{m\geqslant 1} \frac{q^{\binom{m+1}{2}}z^m w^m}{(1-wq)\cdots(1-wq^m)}.$$

We determine another expression for this generating function by considering partitions into distinct parts, with largest part m. The generating function for the largest part is clearly $q^m z w^m$, while the other parts of the partition are enumerated with respect to number of parts alone. This generating function is $\prod_{i=1}^{m-1}(1+zq^i)$. Thus the number of partitions of n into distinct parts, with largest part k and l parts, is

$$[q^n w^k z^l] \sum_{m\geqslant 1} q^m z w^m (1+zq)\cdots(1+zq^{m-1}).$$

The result follows immediately.

2.5.11. Let $F(x) = \prod_{k\geqslant 0}(1 - xq^k)^{-1}$. Then

$$F(x) = 1 + \sum_{j\geqslant 1} x^j \prod_{l=1}^{j}(1 - q^l)^{-1}$$

from **2.5.9** and $F^{-1}(-x) = 1 + \sum_{j\geqslant 1} x^j q^{\binom{j}{2}} \prod_{l=1}^{j}(1 - q^l)^{-1}$ from **2.5.17**. By series bisection, with $i^2 = -1$, the left side of the given identity is $i^{-1}\{F(ix) - F(-ix)\}\{F(ix) + F(-ix)\}^{-1}$ and the right side is $i^{-1}\{F^{-1}(-ix) - F^{-1}(ix)\}\{F^{-1}(-ix) + F^{-1}(ix)\}^{-1}$. But the latter expression is obtained from the former by multiplying the numerator and denominator by $F^{-1}(ix)F^{-1}(-ix)$, so the result follows.

2.5.12. If there are a_1 1's in the partition, then there must be no other number less than $a_1 + 1$, and there must be at least one copy of $a_1 + 1$. If there are a_2 copies of $a_1 + 1$, then there can be no other number less than $(a_2 + 1)(a_1 + 1)$. Carrying on, if there are a_1 1's, a_2 $(a_1 + 1)$'s,..., a_m $(a_1 + 1)\cdots(a_{m-1} + 1)$'s, then the next largest part must be $(a_1 + 1)\cdots(a_m + 1)$, and we choose an arbitrary number a_{m+1} of these. Thus a perfect partition of n must consist of $n_i - 1$ copies of $\prod_{j=1}^{i-1} n_j$, for $i = 1,\ldots, k$, where $\prod_{j=1}^{k} n_j - 1 = n$, and $n_i = a_i + 1 \geqslant 2$, $i = 1,\ldots, k$.

2.5.13. Let D be the set of partitions with distinct parts, and let U be the set of partitions $(\alpha_1,\ldots, \alpha_r)$ in which $s(\alpha) = d(\alpha) = r$ or $s(\alpha) - 1 = d(\alpha) = r$, for $r \geqslant 1$. With $\alpha \in \mathsf{D} - \mathsf{U}$ we associate a partition $\alpha' \in \mathsf{D} - \mathsf{U}$ as follows.

Case 1. $s(\alpha) \leqslant d(\alpha)$: We obtain α' by removing the smallest part, $s(\alpha)$, from α and adding 1 to each of the largest $s(\alpha)$ parts of α.

Case 2. $s(\alpha) > d(\alpha)$: We obtain α' by subtracting 1 from each of the largest $d(\alpha)$ parts of α and adding a new (smallest) part of size $d(\alpha)$.

Thus we have obtained **Franklin's decomposition** $\mathsf{D} - \mathsf{U} \stackrel{\sim}{\to} \mathsf{D} - \mathsf{U} : \alpha \mapsto \alpha'$. Let $F(\mathsf{S})$ be the generating function for the set S of partitions, with the sum of the parts marked by q and number of parts by -1. The partitions α and α' have the same sum and their numbers of parts have opposite (odd–even) parity. Thus the preceding decomposition yields $F(\mathsf{D} - \mathsf{U}) = -F(\mathsf{D} - \mathsf{U})$, so $F(\mathsf{D} - \mathsf{U}) = 0$, and $F(\mathsf{D}) = F(\mathsf{U})$. But

$$F(\mathsf{D}) = \prod_{i\geqslant 1}(1 + zq^i)|_{z=-1} = \prod_{i\geqslant 1}(1 - q^i)$$

and

$$F(\mathsf{U}) = 1 + \sum_{r\geqslant 1}(-1)^r\{q^{r(3r-1)/2} + q^{r(3r+1)/2}\}$$

$$= \sum_{m=-\infty}^{\infty}(-1)^m q^{m(3m-1)/2},$$

since a partition α with r parts has sum

(i) $\sum_{i=r}^{2r-1} i = \frac{1}{2}r(3r-1)$ if $s(\alpha) = d(\alpha) = r$.

(ii) $\sum_{i=r+1}^{2r} i = \frac{1}{2}r(3r+1)$ if $s(\alpha) - 1 = d(\alpha) = r$.

2.5.14. (a) From the Jacobi triple product identity

$$\prod_{m\geqslant 1}(1-q^{2m+1}y)(1-q^{2m-1}y^{-1})(1-q^{2m}) = (1-qy)^{-1}\sum_{i=-\infty}^{+\infty}(-y)^i q^{(i^2)}$$

$$= (1-qy)^{-1}\left\{1+\sum_{i=1}^{\infty} q^{(i^2)}\left((-y)^i+(-y)^{-i}\right)\right\}$$

$$= 1+\sum_{m\geqslant 1} q^m\left\{\sum_{i=1}^{M}\left((-y)^i+(-y)^{-i}\right)y^{m-i^2}+y^m\right\}, \qquad \text{where } M = \lfloor\sqrt{m}\rfloor.$$

Let $y = q^{-1}$; then

$$\prod_{m\geqslant 1}(1-q^{2m})^3 = 1+\sum_{m\geqslant 1}\left(1+\sum_{i=1}^{M}(-1)^i(q^i+q^{-i})q^{i^2}\right)$$

$$= 1+\sum_{m\geqslant 1}\left\{1+\sum_{i=1}^{M}(-1)^i\left(q^{i(i+1)}+q^{i(i-1)}\right)\right\}$$

$$= 1+\sum_{m\geqslant 1}\left\{\sum_{i=1}^{M}(-1)^i q^{i(i+1)}+\sum_{k=1}^{M-1}(-1)^{k+1}q^{k(k+1)}\right\}$$

$$= 1+\sum_{m\geqslant 1}(-1)^M q^{M(M+1)} = \sum_{k\geqslant 0}(-1)^k q^{k(k+1)}(2k+1).$$

The result follows by replacing q^2 with q.

(b) From part (a),

$$F(q) = q\prod_{i\geqslant 1}(1-q^i)\prod_{k\geqslant 1}(1-q^i)^3$$

$$= q\sum_{m=-\infty}^{\infty}(-1)^m q^{m(3m-1)/2}\sum_{k\geqslant 0}(-1)^k(2k+1)q^{\binom{k+1}{2}}.$$

Now we have a contribution to $[q^n]F(q)$ for $n \equiv 0 \pmod 5$, whenever $1 + \frac{1}{2}m(3m-1) + \frac{1}{2}k(k+1) \equiv 2(m-1)^2 + (2k+1)^2 \equiv 0 \pmod 5$. But this requires that $m - 1 \equiv 0 \pmod 5$ and $2k + 1 \equiv 0 \pmod 5$, so $(-1)^{k+m}(2k+1)$, the contribution to $[q^n]F(q)$, is a multiple of 5. Thus $[q^n]F(q)$ is divisible by 5 for all $n \equiv 0 \pmod 5$. Now $q\Pi_{i\geq 1}(1-q^i)^{-1} = F(q)\Pi_{k\geq 1}(1-q^k)^{-5}$. But by the multinomial theorem, all coefficients in $\{\Pi_{k\geq 1}(1-q^k)^{-1}\}^5$ are divisible by 5 except the coefficients of q^{5j}, $j \geq 0$. Thus 5 must divide $[q^{5j}]q\Pi_{i\geq 1}(1-q^i)^{-1} = p(5j-1)$, so $p(5n+4) \equiv 0 \pmod 5$.

(c) $$G(q) = q^2\left\{\sum_{k\geq 0}(-1)^k(2k+1)q^{\binom{k+1}{2}}\right\}^2, \quad \text{by part (a)}$$
$$= q^2\sum_{k\geq 0}\sum_{j\geq 0}(-1)^{k+j}(2k+1)(2j+1)q^{\binom{k+1}{2}+\binom{j+1}{2}}.$$
We have contributions to $[q^{7i}]G(q)$ whenever $2 + \frac{1}{2}k(k+1) + \frac{1}{2}j(j+1) \equiv (2k+1)^2 + (2j+1)^2 \equiv 0 \pmod 7$, so $2k+1 \equiv 0 \pmod 7$ and $2j+1 \equiv 0 \pmod 7$. But the contribution of such a term is $(-1)^{k+j}(2k+1)(2j+1)$, which must be divisible by 7 by the above. Therefore 7 divides $[q^{7i}]G(q)$ for all $i \geq 0$. But $q^2\Pi_{i\geq 1}(1-q^i)^{-1} = G(q)\,\{\Pi_{i\geq 1}(1-q^i)^{-1}\}^7$, and, by the same reasoning as in part (b), 7 divides $[q^{7i}]q^2\Pi_{i\geq 1}(1-q^i)^{-1} = p(7i-2)$. Hence $p(7n+5) \equiv 0 \pmod 7$.

2.5.15. (a) $$F(tq) = \prod_{i\geq 0}(1-tq^{i+1})^{-1} = \prod_{i\geq 1}(1-tq^i)^{-1}$$
$$= (1-t)\prod_{i\geq 0}(1-tq^i)^{-1} = (1-t)F(t).$$

(b) From (a), $\Sigma_{i\geq 0}c(i)t^iq^i = (1-t)\Sigma_{i\geq 0}c(i)t^i$. Equating coefficients of t^n gives $c(n)q^n = c(n) - c(n-1)$, $n \geq 1$, so $c(n) = (1-q^n)^{-1}c(n-1)$, $n \geq 1$, and $c(0) = 1$. Thus

$$c(n) = c(0)\prod_{i=1}^{n}(1-q^i)^{-1} = \prod_{i=1}^{n}(1-q^i)^{-1}, \qquad n \geq 1.$$

2.5.16. Let $F(t) = \Pi_{i\geq 0}(1-atq^i)(1-tq^i)^{-1} = \Sigma_{i\geq 0}c(i)t^i$. Then

$$F(tq) = \prod_{i\geq 0}(1-atq^{i+1})(1-tq^{i+1})^{-1} = (1-t)(1-at)^{-1}F(t),$$

so $(1-at)F(tq) = (1-t)F(t)$. Then by Euler's device

$$c(n) = (1-aq^{n-1})(1-q^n)^{-1}c(n-1), \qquad n \geq 1,$$

and $c(0) = 1$, so

$$c(n) = c(0)\prod\nolimits_{i=1}^{n}(1-aq^{i-1})(1-q^i)^{-1}$$

and the result follows.

2.5.17. Let $F_n(t) = \Pi_{i=0}^{n}(1-tq^i)^{-1} = \Sigma_{k\geq 0}c_n(k)t^k$. Then

$$F_n(tq) = \prod_{i=0}^{n}(1-tq^{i+1})^{-1} = (1-t)(1-tq^{n+1})^{-1}F_n(t),$$

so $(1 - tq^{n+1})F_n(tq) = (1 - t)F_n(t)$. Then by Euler's device

$$c_n(m) = (1 - q^{n+m})(1 - q^m)^{-1}c_n(m - 1), \qquad m \geq 1, c_n(0) = 1,$$

so $$c_n(k) = \prod_{i=1}^{k}(1 - q^{n+i})(1 - q^i)^{-1}.$$

But $c_n(k) = [t^k]F_n(t) = \sum_{j=0}^{k}[t^j]\prod_{i=1}^{n}(1 - tq^i)^{-1}$ and is thus the generating function for partitions with at most k parts and largest part at most n. The polynomial $c_n(k)$ is called a **Gaussian coefficient** and is considered in detail in Section 2.6.

2.5.18. Let $F(t) = \prod_{m\geq 1}(1 + q^{2m-1}t)(1 + q^{2m-1}t^{-1})(1 - q^{2m}) = \sum_{i=-\infty}^{+\infty}c(i)t^i$. Then

$$\begin{aligned} F(tq^2) &= \prod_{m\geq 1}(1 + q^{2m+1}t)(1 + q^{2m-3}t^{-1})(1 - q^{2m}) \\ &= (1 + qt)^{-1}(1 + q^{-1}t^{-1})F(t) = q^{-1}t^{-1}F(t). \end{aligned}$$

By Euler's device $c(n + 1) = c(n)q^{2n+1}$, $n \geq 0$, and $c(-n) = c(n)$ by symmetry, so

$$c(n) = c(0)\prod_{i=0}^{n-1}q^{2i+1} = c(0)q^{(n^2)}, \qquad n \geq 0.$$

But

$$\begin{aligned} c(0) = [t^0]F(t) &= \prod_{m\geq 1}(1 - q^{2m})\sum_{i\geq 0}\left\{[t^i]\prod_{l\geq 1}(1 + q^{2l-1}t)\right\}^2 \\ &= \prod_{m\geq 1}(1 - q^{2m})\left\{1 + \sum_{i\geq 1}q^{2i^2}\prod_{j=1}^{i}(1 - q^{2j})^{-2}\right\}, \qquad \text{from } \mathbf{2.5.17} \\ &= \prod_{m\geq 1}(1 - q^{2m})\prod_{l\geq 1}(1 - q^{2l})^{-1}, \qquad \text{from } \mathbf{2.5.14} \\ &= 1. \end{aligned}$$

Thus $c(n) = q^{(n^2)}$ for $n \geq 0$, and $c(-n) = q^{(n^2)} = q^{((-n)^2)}$, so that $c(i) = q^{(i^2)}$ for all i, as required.

2.5.19. (a) The results follow immediately from **2.5.24** by replacing q by q^α and setting $y = -q^\beta$, where

$$\text{(i)}\ \ \alpha = \tfrac{5}{2},\ \ \beta = \tfrac{1}{2}, \qquad \text{(ii)}\ \ \alpha = \tfrac{5}{2},\ \ \beta = \tfrac{3}{2}.$$

(b) Let $P(x) = \prod_{i \geq 1}(1 - xq^i)$ and $G(x)$ represent the left side of the given identity, so

$$P(x)G(x) = 1 + \sum_{n \geq 1} (-1)^n x^{2n} q^{n(5n-1)/2} \times (1 - xq^{2n}) \frac{(1 - xq) \cdots (1 - xq^{n-1})}{(1 - q) \cdots (1 - q^n)}. \tag{1}$$

Now writing $1 - xq^{2n}$ as $(1 - q^n) + q^n(1 - xq^n)$ gives

$$\begin{aligned} P(x)G(x) &= 1 + \sum_{m \geq 1} (-1)^m x^{2m} q^{m(5m-1)/2} \frac{(1 - xq) \cdots (1 - xq^{m-1})}{(1 - q) \cdots (1 - q^{m-1})} \\ &\quad + \sum_{n \geq 1} (-1)^n x^{2n} q^{n(5n+1)/2} \frac{(1 - xq) \cdots (1 - xq^n)}{(1 - q) \cdots (1 - q^n)} \\ &= \sum_{n \geq 0} (-1)^n x^{2n} q^{n(5n+1)/2} (1 - x^2 q^{4n+2}) \\ &\quad \times \frac{(1 - xq) \cdots (1 - xq^n)}{(1 - q) \cdots (1 - q^n)}, \qquad \text{setting } m = n + 1. \end{aligned} \tag{2}$$

Let $D = P(x)\{G(x) - G(xq)\}$ and use representation (2) for $P(x)G(x)$ and (1) with x replaced by xq for $P(xq)G(xq)$ to get

$$D = \sum_{n \geq 0} (-1)^n x^{2n} q^{n(5n+1)/2} \frac{(1 - xq) \cdots (1 - xq^n)}{(1 - q) \cdots (1 - q^n)} \times \{1 - x^2 q^{4n+2} - q^n(1 - xq^{2n+1})\}.$$

But $1 - x^2 q^{4n+2} - q^n(1 - xq^{2n+1}) = (1 - q^n) + xq^{3n+1}(1 - xq^{n+1})$, so

$$\begin{aligned} D &= \sum_{m \geq 1} (-1)^m x^{2m} q^{m(5m+1)/2} \frac{(1 - xq) \cdots (1 - xq^m)}{(1 - q) \cdots (1 - q^{m-1})} \\ &\quad + qx \sum_{n \geq 0} (-1)^n x^{2n} q^{n(5n+7)/2} \frac{(1 - xq) \cdots (1 - xq^{n+1})}{(1 - q) \cdots (1 - q^n)} \\ &= qx \sum_{n \geq 0} (-1)^n x^{2n} q^{n(5n+7)/2} (1 - xq^{2n+2}) \\ &\quad \times \frac{(1 - xq) \cdots (1 - xq^{n+1})}{(1 - q) \cdots (1 - q^n)}, \qquad \text{setting } m = n + 1 \\ &= xqP(x)G(xq^2), \qquad \text{from (1).} \end{aligned}$$

Thus $G(x) = G(xq) + xqG(xq^2)$, and $G(0) = 1$. Let $G(x) = \sum_{n\geqslant 0} c_n x^n$ so by Euler's device

$$c_n = q^{2n-1}(1-q^n)^{-1}c_{n-1}, \qquad n \geqslant 1, c_0 = 1,$$

whence $c_n = q^{(n^2)}\prod_{i=1}^{n}(1-q^i)^{-1}$, and the result follows.

(c) (i) Set $x = 1$ in (b), yielding

$$1 + \sum_{k\geqslant 1} q^{(k^2)}\prod_{i=1}^{k}(1-q^i)^{-1}$$

$$= \prod_{i\geqslant 1}(1-q^i)^{-1}\left\{1 + \sum_{n\geqslant 1}(-1)^n q^{n(5n-1)/2}(1-q^{2n})(1-q^n)^{-1}\right\}$$

$$= \prod_{i\geqslant 1}(1-q^i)^{-1}\left\{\sum_{j=-\infty}^{\infty}(-1)^j q^{j(5j+1)/2}\right\}$$

since $(1-q^{2n})(1-q^n)^{-1} = 1 + q^n$, and the result follows from (i) of (a).

(ii) Set $x = q$ in (b), yielding

$$1 + \sum_{k\geqslant 1} q^{k^2+k}\prod_{i=1}^{k}(1-q^i)^{-1}$$

$$= \prod_{j\geqslant 2}(1-q^j)^{-1}\left\{1 + \sum_{n\geqslant 1}(-1)^n q^{n(5n+3)/2}(1-q^{2n+1})(1-q)^{-1}\right\}$$

$$= \prod_{j\geqslant 1}(1-q^j)^{-1}\left\{1 - q + \sum_{n\geqslant 1}(-1)^n q^{n(5n+3)/2}(1-q^{2n+1})\right\}$$

$$= \prod_{j\geqslant 1}(1-q^j)^{-1}\left\{\sum_{i=-\infty}^{\infty}(-1)^i q^{i(5i+3)/2}\right\}$$

and the result follows from (ii) of (a).

SECTION 2.6

2.6.1. In the inversion algorithm, the element $n + 1$ is outstanding only if it is inserted in front of a permutation of N_n. If $f_n(q)$ is the generating function for permutations on N_n, with q marking outstanding elements, then

$$f_{n+1}(q) = (n+q)f_n(q).$$

However, $f_1(q) = q$, so $f_n(q) = \prod_{i=0}^{n-1}(i+q)$, and the result follows.

2.6.2. (a) Consider between-set inversions in ordered partitions of N_n of type $(m, k-m, n-k)$. We count these in two ways. The between-set inversions involving the $(n-k)$-set are enumerated by $\binom{n}{k}_q$, and those between the m-set and $(k-m)$-set are enumerated by $\binom{k}{m}_q$. Thus the between-set inversions are enumerated by $\binom{n}{k}_q\binom{k}{m}_q$. On the other hand, the inversions involving the m-set are enumerated by $\binom{n}{m}_q$, and those between the $(k-m)$-set and $(n-k)$-set are enumerated by $\binom{n-m}{k-m}_q$. Thus

$$\binom{n}{k}_q\binom{k}{m}_q=\binom{n}{m}_q\binom{n-m}{k-m}_q.$$

(b) From part (a)

$$\sum_{k=m}^{n}\binom{n}{k}_q\binom{k}{m}_q(-1)^{k-m}q^{\binom{k-m}{2}}=\binom{n}{m}_q\sum_{k=m}^{n}\binom{n-m}{k-m}_q(-1)^{k-m}q^{\binom{k-m}{2}}$$

$$=\binom{n}{m}_q\prod_{i=0}^{n-m-1}(1-q^i) \qquad \text{by Lemma 2.6.9}$$

$$=\delta_{n,m}.$$

2.6.3. (a) Count ordered partitions of N_n of type $(n-k, k)$ with respect to between-set inversions. The required generating function is $\binom{n}{n-k}_q=\binom{n}{k}_q$. We determine another expression for this generating function by considering the location of n. If n is in the $(n-k)$-set, then it contributes k inversions, since it is larger than all the objects in the k-set. The remaining elements contribute $\binom{n-1}{n-k-1}_q=\binom{n-1}{k}_q$ so that this case contributes $q^k\binom{n-1}{k}_q$ to the generating function. If n is in the k-set, then it is involved in 0 inversions. The remaining elements contribute $\binom{n-1}{n-k}_q=\binom{n-1}{k-1}_q$, and this case contributes $\binom{n-1}{k-1}_q$ to the generating functions. The result follows.

(b) Count ordered partitions of N_{n+m+1} of type $(m+1, n)$ with respect to between-set inversions. The required generating function is $\binom{n+m+1}{m+1}_q$. We determine another expression for this function by considering the maximum element in the $(m+1)$-set. This element is $m+k+1$ for some $k=0,\ldots,n$. If $m+k+1$ is the maximum element in the $(m+1)$-set, then the $n-k$ objects $m+k+2,\ldots,m+n+1$ are in the n-set. These objects contribute no inversions, and $m+k+1$ contributes k inversions, since it is larger than the other k objects in the n-set. The remaining $m+k$ objects form an ordered (m, k) partition of N_{m+k} and thus contribute $\binom{m+k}{m}_q$ to the generating function. The result follows by summing over k.

(c) Count ordered partitions of N_{a+b} of type $(n, a+b-n)$ with respect to between-set inversions. Suppose that the n-set contains k elements of N_a and $n-k$ elements of $\mathsf{N}_{a+b} - \mathsf{N}_a$, so the $(a+b-n)$-set contains $a-k$ elements of N_a and $b+k-n$ elements of $\mathsf{N}_{a+b} - \mathsf{N}_a$. There are no inversions between the k-set and the $(b+k-n)$-set, and $(n-k)(a-k)$ between the $(n-k)$-set and the $(a-k)$-set. This is because all elements of $\mathsf{N}_{a+b} - \mathsf{N}_a$ are larger than all elements of N_a. The k-set and $(a-k)$-set form an ordered partition of N_a of type $(k, a-k)$ and thus contribute $\binom{a}{k}_q$ to the between-set inversion generating function. The $(n-k)$-set and $(b+k-n)$-set form an ordered partition of $\mathsf{N}_{a+b} - \mathsf{N}_a$ of type $(n-k, b+k-n)$, and thus contribute $\binom{b}{n-k}_q$ to the between-set generating function. Thus the ordered partitions with k elements of N_a in the n-set contribute $q^{(a-k)(n-k)}\binom{a}{k}_q\binom{b}{n-k}_q$ to the between-set generating function. The result follows by summing over all values of k.

2.6.4. We have, from **2.6.11** with $z = 1$, $y = 0$, and $x = -u$,

$$\sum_{i=0}^{2n} \binom{2n}{i}_q u^i q^{\binom{i}{2}} = \prod_{j=0}^{2n-1} (1 + uq^j).$$

Thus by bisection on u,

$$\sum_{k=0}^{n} \binom{2n}{2k}_q u^{2k} q^{\binom{2k}{2}} = \frac{1}{2}\left\{ \prod_{i=0}^{2n-1} (1 + uq^i) + \prod_{i=0}^{2n-1} (1 - uq^i) \right\},$$

and the result follows by setting $u = 1$.

2.6.5. (a) From **2.6.12(1)**

$$\prod_{i=1}^{n} (1 + q^i x) \prod_{j=0}^{n-1} (1 + q^j x^{-1})$$

$$= \left\{ \sum_{m=0}^{n} (qx)^m q^{\binom{m}{2}} \binom{n}{m}_q \right\} \left\{ \sum_{l=0}^{n} x^{-l} q^{\binom{l}{2}} \binom{n}{l}_q \right\}$$

$$= \sum_{k=-n}^{n} x^k q^{\binom{k+1}{2}} \sum_{j=0}^{n} q^{(n-j)(n+k-j)} \binom{n}{n+k-j}_q \binom{n}{j}_q,$$

where $k = m - l, j = n - l$

$$= \sum_{k=-n}^{n} x^k q^{\binom{k+1}{2}} \binom{2n}{n+k}_q, \qquad \text{by } [\mathbf{2.6.3(c)}].$$

The result follows from replacing q by q^2, and then x by xq^{-1}.

(b) Taking the limit as $n \to \infty$ in part (a) gives

$$\prod_{j\geqslant 1}(1+xq^{2j-1})(1+x^{-1}q^{2j-1}) = \sum_{k=-\infty}^{\infty} q^{(k^2)}x^k \prod_{i\geqslant 1}(1-q^{2i})^{-1}$$

from **2.6.13**, and the result follows.

2.6.6. (a) Let $G(t) = \prod_{i\geqslant 0}\{1 - t(1-q)q^i\}$. Then, from **2.6.14(1)** and **(2)**

$$G(t) = \sum_{k\geqslant 0}(-t)^k \frac{q^{\binom{k}{2}}}{k!_q} \quad \text{and} \quad G^{-1}(t) = \sum_{k\geqslant 0}\frac{t^k}{k!_q},$$

and the result follows.

(b) Let $A(t) = \sum_{k\geqslant 0} a_k t^k/k!_q$ and $B(t) = \sum_{k\geqslant 0} b_k t^k/k!_q$. From part (a), if $A(t) = B(t)\sum_{i\geqslant 0} t^i/i!_q$, then $B(t) = \{\sum_{i\geqslant 0}(-1)^i q^{\binom{i}{2}} t^i/i!_q\}A(t)$, and the result follows by applying $[t^n/n!_q]$ to these equations. $A(t)$ is called the **Eulerian generating function** for $\{a_0, a_1, \ldots\}$, whose combinatorial properties are considered in Chapter 4.

(c)
$$\sum_{n\geqslant 0}\frac{F_n(x)t^n}{(1-q)\cdots(1-q^n)} = \sum_{n\geqslant 0}\sum_{j=0}^{n}\binom{n}{j}_q x^j \frac{\left(t(1-q)^{-1}\right)^n}{n!_q}$$

$$= \left\{\sum_{i\geqslant 0}\frac{\left(t(1-q)^{-1}\right)^i}{i!_q}\right\}\left\{\sum_{j\geqslant 0}\frac{\left(xt(1-q)^{-1}\right)^j}{j!_q}\right\}$$

$$= \prod_{m\geqslant 0}(1-tq^i)^{-1}(1-xtq^i)^{-1}, \qquad \text{from part (a).}$$

(d) $\sum_{j=0}^{n}\binom{n}{j}_q(-1)^j = F_n(-1)$, in the notation of part (c). But

$$\sum_{n\geqslant 0}\frac{F_n(-1)t^n}{(1-q)\cdots(1-q^n)} = \prod_{i\geqslant 0}(1-q^it)^{-1}(1+q^it)^{-1}, \qquad \text{from part (c)}$$

$$= \prod_{i\geqslant 0}(1-q^{2i}t^2)^{-1} = \sum_{m\geqslant 0} t^{2m}\prod_{i=1}^{m}(1-q^{2i})^{-1}, \qquad \text{from part (a).}$$

Comparing coefficients of t^n gives

$$F_{2m}(-1) = \prod_{j=1}^{2m}(1-q^j)\prod_{i=1}^{m}(1-q^{2i})^{-1} = \prod_{i=1}^{m}(1-q^{2i-1})$$

and $F_{2m+1}(-1) = 0$, for $m \geqslant 0$.

2.6.7. Let $(j_1, j_2, \ldots, j_{f(\sigma)})$, with $1 \leqslant j_1 < j_2 < \cdots < j_{f(\sigma)} \leqslant n$, be the set of values of i for which $\sigma_i > \sigma_{i+1}$. Call this the fall set of σ, and note that $j_1 + j_2 + \cdots + j_{f(\sigma)} = m(\sigma)$.

(a) If $n+1$ is inserted in front of σ, then the fall set of $\hat{\sigma}$ is $(1, j_1+1, \ldots, j_{f(\sigma)}+1)$, so $m(\hat{\sigma}) = 1 + (j_1+1) + \cdots + (j_{f(\sigma)}+1) = m(\sigma) + f(\sigma) + 1$.

(b) If $n+1$ is inserted in the ith fall from the right of σ, then the fall set of $\hat{\sigma}$ is $(j_1, j_2, \ldots, j_{f(\sigma)-i}, j_{f(\sigma)-i+1}+1, \ldots, j_{f(\sigma)}+1)$, so $m(\hat{\sigma}) = j_1 + \cdots + j_{f(\sigma)-i} + (j_{f(\sigma)-i+1}+1) + \cdots + (j_{f(\sigma)}+1) = m(\sigma) + i$, for $i = 1, 2, \ldots, f(\sigma)$.

(c) If $n+1$ is inserted in the ith rise from the left of σ, then the fall set of $\hat{\sigma}$ is $(j_1, \ldots, j_l, i+l+1, j_{l+1}+1, \ldots, j_{f(\sigma)}+1)$, where l is the unique value such that $j_l < i+l+1 < j_{l+1}+1$, for $0 \leqslant l \leqslant f(\sigma)$. Thus

$$m(\hat{\sigma}) = j_1 + \cdots + j_l + (i+l+1) + (j_{l+1}+1) + \cdots + (j_{f(\sigma)}+1)$$

$$= m(\sigma) + (i+l+1) + (f(\sigma) - l) = m(\sigma) + f(\sigma) + i + 1$$

for $i = 1, \ldots, n - f(\sigma) - 1$.

(d) If $n+1$ is inserted at the right end of σ, then the fall set of $\hat{\sigma}$ is the same as the fall set of σ, so $m(\hat{\sigma}) = m(\sigma)$.

(e) Let $g_n(q)$ be the ordinary generating function for permutations of N_n with respect to major index. Then from parts (a)–(d) the values of $m(\hat{\sigma})$ for all $\hat{\sigma}$ constructed by inserting $n+1$ in σ are given by $m(\sigma)+i$ for $i = 0, \ldots, n$. Thus

$$g_{n+1}(q) = \sum_{\hat{\sigma} \in \mathsf{P}_{n+1}} q^{m(\hat{\sigma})} = \sum_{i=0}^{n} \sum_{\sigma \in \mathsf{P}_n} q^{m(\sigma)+i}$$

$$= (1 - q^{n+1})(1-q)^{-1} g_n(q).$$

But $g_1(q) = 1$, so $g_n(q) = (1-q)^{-n} \prod_{i=1}^{n} (1-q^i) = n!_q$.

2.6.8. (a) Suppose that σ is such a sequence, and that $\alpha \subseteq \mathsf{N}_n$ is the set of positions in σ that contain the element 1. Then the number of inversions in σ is equal to the number of between-set inversions in the ordered partition $(\alpha, \mathsf{N}_n - \alpha)$ of type $(k, n-k)$. Thus the number of sequences with m inversions is given by $[q^m]\binom{n}{k}_q$, the number of ordered partitions of N_n of type $(k, n-k)$, with m between-set inversions, from Proposition 2.6.6(2).

(b) The generating function for $\{0,1\}$-sequences with k 1's and $(n-k)$ 0's, with respect to inversions, is $\binom{n}{k}_q$, by part (a). If the last element in such a sequence σ is 0, then it contributes k inversions, one for each 1 in σ. The remaining sequence contains k 1's and $(n-k-1)$ 0's, so inversions are enumerated by $\binom{n-1}{k}_q$. If the last element in σ is 1, then it contributes no inversions, and the inversions in the remaining $n-1$ elements are enumerated by $\binom{n-1}{k-1}_q$. Thus

$$\binom{n}{k}_q = q^k\binom{n-1}{k}_q + \binom{n-1}{k-1}_q.$$

(c) The generating function for $\{0, 1\}$-sequences of length $(a+b)$ with n 1's, with respect to inversions, is $\binom{a+b}{n}_q$. Suppose that the final a positions contain k 1's and $(a-k)$ 0's, so that the remaining b positions contain $(n-k)$ 1's and $(b+k-n)$ 0's. The inversions internal to the final a positions are enumerated by $\binom{a}{k}_q$, and those internal to the first b positions are enumerated by $\binom{b}{n-k}_q$. There are exactly $(a-k)(n-k)$ other inversions since the $(n-k)$ 1's in the first b positions precede the $(a-k)$ 0's in the final a positions. Sum over k to obtain the result.

2.6.9. (a) From the proof of Proposition 2.6.13, the number of ordered partitions of N_{m+i} of type (m, i) with n between-set inversions is

$$[q^n]\binom{m+i}{m}_q = [q^n] \sum_{\substack{\beta_1,\ldots,\beta_m \geqslant 0 \\ \beta_1 + \cdots + \beta_m \leqslant i}} q^{\beta_m + 2\beta_{m-1} + \cdots + m\beta_1},$$

where $\beta_j = \alpha_j - 1$ for $j = 1,\ldots, m$. But if $\beta_m + 2\beta_{m-1} + \cdots + m\beta_1 = n$, then $(\beta_1,\ldots, \beta_m)$ represents a partition of n with β_i copies of $m+1-i$, for $i = 1,\ldots, m$. Thus this partition has largest part at most m. Moreover, since $\beta_1 + \cdots + \beta_m \leqslant i$, the partition has at most i parts. The result follows, and agrees with [**2.5.17**].

(b) Let α be a partition with $i+1$ parts, with largest part $m+1$. If the first row and column of the Ferrers graph $F(\alpha)$ are deleted, then the remaining dots form the Ferrers graph of a partition with at most i parts, each at most m. But there is a total of $m+i+1$ dots in the first row and column, and the result follows from part (a).

2.6.10. (a) The partitions with at most $n-k$ parts, each at most k, are enumerated by $\binom{n}{k}_q$, from [**2.6.9**(a)]. If the largest part is exactly k, then there are at most $n-k-1$ other parts, each at most k. Thus, in this case the partitions are enumerated by $q^k\binom{n-1}{k}_q$. The remaining partitions have at most $n-k$ parts, each at most $k-1$. These are enumerated by $\binom{n-1}{k-1}_q$, and thus

$$\binom{n}{k}_q = q^k\binom{n-1}{k}_q + \binom{n-1}{k-1}_q.$$

(b) Consider partitions with distinct parts each at most n. Then the generating function for these partitions is $\prod_{i=1}^{n}(1+q^i x)$, where x marks the number of parts. If such a partition has exactly k parts, then it is formed by the row abutment of the maximal triangle T_k and a partition with at most k parts, each at most $n-k$. The enumerator for these is $q^{\binom{k+1}{2}}\binom{n}{k}_q x^k$, since T_k is a partition of $\binom{k+1}{2}$. Summing over k gives the identity.

(c) Again consider partitions with distinct parts, each at most n. If x marks number of parts and w marks the size of the largest part, then the generating function for these partitions with largest part i is $w^i q^i x \prod_{j=1}^{i-1}(1+q^j x)$. Summing over i gives $\sum_{i=1}^{n} w^i q^i x \prod_{j=1}^{i-1}(1+q^j x)$ as the generating function for all such partitions. If such a partition has exactly k parts, then it is formed by the row abutment of T_k and a partition with at most k parts, with largest part l, for some $l = 0, \ldots, n-k$. The result follows by summing over l and k.

2.6.11. Let α_j be the number of 1's following the jth occurrence of 0 in such a sequence, for $j = 1, \ldots, n-k$. Then $k \geqslant \alpha_1 \geqslant \alpha_2 \geqslant \cdots \geqslant \alpha_{n-k} \geqslant 0$, so $(\alpha_1, \alpha_2, \ldots, \alpha_{n-k})$ is a partition of $\alpha_1 + \alpha_2 + \cdots + \alpha_{n-k}$ with at most $n-k$ parts, and some 0's appended. Each part is at most k. However, $\alpha_1 + \alpha_2 + \cdots + \alpha_{n-k}$ is also the area under the path, so the paths with area m correspond to partitions of m with at most $n-k$ parts, each at most k. The number of such paths is accordingly $[q^i]\binom{n}{k}_q$ from [**2.6.9**(a)].

2.6.12. The elements of the vector space $\mathbf{V}_n(q)$ of dimension n over GF(q) have the form $(x_1, x_2, \ldots, x_n)$, where $x_1, x_2, \ldots, x_n \in \mathrm{GF}(q)$.

(a) Let $f_{m,n}(q)$ be the number of m-tuples of linearly independent vectors in $\mathbf{V}_n(q)$. Given a set of m linearly independent vectors, there are q^m vectors in $\mathbf{V}_n(q)$ that can be obtained as linear combinations of these vectors, and hence $q^n - q^m$ vectors that are not in the span of the set of m vectors. Thus $f_{m+1,n}(q) = (q^n - q^m) f_{m,n}(q)$, and $f_{1,n}(q) = q^n - 1$, so that the number of linearly independent m-tuples is $f_{m,n}(q) = (q^n - 1)(q^n - q) \cdots (q^n - q^{m-1})$.

(b) From part (a), each m-dimensional subspace is spanned by $f_{m,m}(q)$ of the $f_{m,n}(q)$ linearly independent m-tuples of vectors in $\mathbf{V}_n(q)$. Thus the number of m-dimensional subspaces of $\mathbf{V}_n(q)$ is $f_{m,n}(q)/f_{m,m}(q) = \binom{n}{m}_q$.

2.6.13. We count the $\binom{n}{k}_q$ subspaces of $\mathbf{V}_n(q)$ of dimension k by considering a fixed nonzero vector v. Let $(v_1, \ldots, v_{n-1}, v)$ be a basis (ordered) for $\mathbf{V}_n$ and suppose that $(v_1, \ldots, v_{n-1})$ is a basis for $\mathbf{V}_{n-1}$, an $(n-1)$-dimensional subspace of $\mathbf{V}_n$. The k-dimensional subspaces $\mathbf{V}_k$ that contain v are such that $\mathbf{V}_k \cap \mathbf{V}_{n-1}$ is a subspace of $\mathbf{V}_{n-1}$ of dimension $k-1$. Thus there are $\binom{n-1}{k-1}_q$ such subspaces, from [**2.6.12**(b)]. Now consider those subspaces $\mathbf{V}_k$ that do not contain v. Construct a unique basis for such a subspace by considering one of the $\binom{n-1}{k}_q$ k-dimensional subspaces of $\mathbf{V}_{n-1}$, and adding some multiple of v to each of the k vectors in the basis. Thus there are $q^k\binom{n-1}{k}_q$ subspaces that do not contain v. Therefore

$$\binom{n}{k}_q = q^k\binom{n-1}{k}_q + \binom{n-1}{k-1}_q$$

is a polynomial identity for all prime powers, and is thus true for an indeterminate q.

2.6.14. Suppose that $\mathbf{V}_n$ is an n-dimensional subspace of $\mathbf{V}_{a+b}$. Then there are $\binom{a+b}{n}_q$ such subspaces. Suppose that $\mathbf{V}_a$ is a fixed a-dimensional subspace of $\mathbf{V}_{a+b}$, with (ordered) basis $(v_1,\ldots, v_a)$, and that $\mathbf{V}_n \cap \mathbf{V}_a$ is a k-dimensional subspace with basis $(v_1,\ldots, v_k)$. If $\mathbf{V}_a \oplus \mathbf{V}_b = \mathbf{V}_{a+b}$, then $\mathbf{V}_n \cap \mathbf{V}_b$ is an $(n-k)$-dimensional subspace. Suppose that $\mathbf{V}_n \cap \mathbf{V}_b$ has basis $(u_1,\ldots, u_{n-k})$ and that $(u_1,\ldots, u_b)$ is a basis for $\mathbf{V}_b$. Then there are $\binom{a}{k}_q$ choices for $\mathbf{V}_n \cap \mathbf{V}_a$ and $\binom{b}{n-k}_q$ choices for $\mathbf{V}_n \cap \mathbf{V}_b$. We can construct a unique basis for $\mathbf{V}_n$ by considering $(v_1,\ldots, v_k)$ and one of the q^{a-k} linear combinations of $(v_{k+1},\ldots, v_a)$ added to each of the $n-k$ elements of $(u_1,\ldots, u_{n-k})$. There are accordingly $q^{(a-k)(n-k)}\binom{a}{k}_q\binom{b}{n-k}_q$ subspaces $\mathbf{V}_n$ that intersect $\mathbf{V}_a$ in a k-dimensional subspace. The identity follows by summing over k.

2.6.15. Let $(v_1,\ldots, v_n)$ be a basis (ordered) for $\mathbf{V}_n(q)$. A linear transformation from $\mathbf{V}_n$ to $\mathbf{Z}$ is determined uniquely by its action on the basis vectors. There are z choices for the image of each element of $(v_1,\ldots, v_n)$ and thus z^n linear transformations. Suppose that $\mathbf{V}_{n-k}$ is an $(n-k)$-dimensional subspace of $\mathbf{V}_n$, and that $(v_1,\ldots, v_n)$ is chosen so that $(v_{k+1},\ldots, v_n)$ is a basis for $\mathbf{V}_{n-k}$. If $\mathbf{V}_{n-k}$ is the null space of a linear transformation, then $v_{k+1},\ldots, v_n$ are all mapped to the 0 vector in $\mathbf{Z}$. The basis elements $(v_1,\ldots, v_k)$ must then be mapped to a k-tuple of linearly independent vectors in $\mathbf{Z}$. There are $(z-1)\cdots(z-q^{k-1})$ such k-tuples, from [**2.6.12**(a)], and thus the same number of linear transformations whose null space is $\mathbf{V}_{n-k}$. The identity follows by summing over k, since there are $\binom{n}{n-k}_q = \binom{n}{k}_q$ choices for $\mathbf{V}_{n-k}$.

2.6.16. Let $(v_1,\ldots, v_n)$ be a basis (ordered) for $\mathbf{V}_n$. Then a 1–1 linear transformation $f: \mathbf{V}_n \to \mathbf{Z}$, in which $f(\mathbf{V}_n) \cap \mathbf{X}$ has dimension 0, is determined by the n-tuple of linearly independent vectors in $\mathbf{Z} - \mathbf{X}$, which contains the images of $(v_1,\ldots, v_n)$. In the notation of **2.6.11** there are $Q_n(-x, z)$ such n-tuples, from [**2.6.12**(a)], and thus $Q_n(-x, z)$ such linear transformations. Suppose that $\mathbf{U}$ is a k-dimensional subspace of $\mathbf{V}_n$ such that $f(\mathbf{U}) = (\mathbf{Y} - \mathbf{X}) \cup \{0\}$ and $f(\mathbf{V}_n - \mathbf{U}) = \mathbf{Z} - \mathbf{Y}$ and that $(v_1,\ldots, v_n)$ is chosen so that $(v_1,\ldots, v_k)$ is a basis for $\mathbf{U}$. The number of linear transformations for each such $\mathbf{U}$ is determined by the linearly independent images of $(v_1,\ldots, v_k)$ in $\mathbf{Y} - \mathbf{X}$, and $(v_{k+1},\ldots, v_n)$ in $\mathbf{Z} - \mathbf{Y}$. There are $Q_k(-x, y)$ k-tuples of linearly independent images for $(v_1,\ldots, v_k)$, and $Q_{n-k}(-y, z)$ $(n-k)$-tuples of images for $(v_{k+1},\ldots, v_n)$. The identity follows by summing over k, since there are $\binom{n}{k}_q$ choices of $\mathbf{U}$.

SECTION 2.7

2.7.1. (a) The required number is $[x^{mk+1}]A(x)$, where $A = x(1 - A^k)^{-1}$, from the branch decomposition. But, by the Lagrange theorem,

$$[x^{mk+1}]A(x) = \frac{1}{mk+1}[\lambda^{mk}](1-\lambda^k)^{-(mk+1)} = \frac{1}{mk+1}\binom{m(k+1)}{m}.$$

(b) Let $F(x, y)$ be the generating function for planted plane trees with x marking nonroot vertices and y marking nonroot vertices of odd degree. Then, by the branch decomposition, $F = x(y + F + yF^2 + F^3 + \cdots) = x(y + F)(1 - F^2)^{-1}$. By the Lagrange theorem, the required number is

$$\begin{aligned}[x^{2n}y^{2m+1}]F(x,y) &= \frac{1}{2n}[\lambda^{2n-1}y^{2m+1}]\{(y+\lambda)(1-\lambda^2)^{-1}\}^{2n}\\ &= \frac{1}{2n}\binom{2n}{2m+1}[\lambda^{2m}](1-\lambda^2)^{-2n}\\ &= \frac{1}{2n+m}\binom{2n}{2m+1}\binom{2n+m}{m}.\end{aligned}$$

2.7.2. (a) Let $F(x, u)$ be the generating function for planted plane trees in which the nonroot vertices are marked by x and the nonroot vertices of degree $s + 1$ are marked by u. Then, by the branch decomposition,

$$F = x\{(1-F)^{-1} + (u-1)F^s\}$$

Thus the required number, by the Lagrange theorem, is

$$\begin{aligned}[u^k x^n]F(x,u) &= \frac{1}{n}[\lambda^{n-1}u^k]\{(1-\lambda)^{-1} + (u-1)\lambda^s\}^n\\ &= \frac{1}{n}[u^k]\sum_{i=0}^{n}\binom{n}{i}(u-1)^i\binom{-(n-i)}{n-1-is}(-1)^{n-1-is}\\ &= \frac{1}{n}\sum_{i=k}^{n-1}\binom{n}{i}\binom{i}{k}(-1)^{i-k}\binom{2n-2-i(s+1)}{n-i-1}.\end{aligned}$$

(b) Let $F(x, u, v)$ be the generating function for planted plane trees with x marking nonroot vertices, u marking nonroot vertices of degree $s + 1$, and v marking nonroot vertices of degree $t + 1$. Then, by the branch decomposition, $F = x\{(1 - F)^{-1} + (u - 1)F^s + (v - 1)F^t\}$. Thus the required number, by the Lagrange theorem, is

$$[u^k v^m x^n]F(x,u,v) = \frac{1}{n}[u^k v^m \lambda^{n-1}]\{(1-\lambda)^{-1} + (u-1)\lambda^s + (v-1)\lambda^t\}^n$$

$$= \frac{1}{n}[u^k v^m \lambda^{n-1}] \sum_{i+j+l=n} \begin{bmatrix} n \\ i, j, l \end{bmatrix} (u-1)^j (v-1)^l (1-\lambda)^{-i} \lambda^{sj+lt},$$

and the result follows.

2.7.3. (a) Let $P(x)$ be the generating function for planted plane trees with respect to nonroot vertices. Then, by the branch decomposition, $P = x(1-P)^{-1}$ and the required number is

$$[x^n]xP^{m-1}(x) = [x^{n-1}]P^{m-1}(x)$$

$$= \frac{1}{n-1}[\lambda^{n-2}](m-1)\lambda^{m-2}(1-\lambda)^{-(n-1)}$$

$$= \frac{m-1}{n-1}\binom{2n-m-2}{n-2}.$$

(b) By the branch decomposition the required number is

$$[x^n]\sum_{i\geq 0} xP^{2i}(x) = [x^{n-1}](1-P^2(x))^{-1}$$

$$= \frac{1}{n-1}[\lambda^{n-2}]2\lambda(1-\lambda^2)^{-2}(1-\lambda)^{-(n-1)}$$

$$= \frac{2}{n-1}[\lambda^{n-3}](1+\lambda)^{-2}(1-\lambda)^{-(n+1)},$$

from the Lagrange theorem, and the result follows.

2.7.4. From **2.7.10**, following **2.7.11**, the required number is

$$[x^n]H(x+x^2) = [x^n]P\left((x+x^2)(1+x+x^2)^{-1}\right)$$

$$= [x^n]\sum_{m\geq 0}\frac{1}{m+1}\binom{2m}{m}(x+x^2)^{m+1}(1+x+x^2)^{-(m+1)}$$

$$= [x^n]\sum_{m\geq 0}\sum_{i\geq 0}\frac{1}{m+1}\binom{2m}{m}\binom{-(m+1)}{i}(x+x^2)^{m+i+1},$$

and the result follows.

2.7.5. From **2.7.18** the required number is

$$\sum_{d\geqslant 1} c_h(n,d) = [x^{n+h-1}] \sum_{d\geqslant 1} P^{2h+d-1}(x)$$

$$= [x^{n+h}]P^{2h+1}(x), \qquad \text{since } P = x(1-P)^{-1}$$

$$= \frac{1}{n+h}[\lambda^{n+h-1}](2h+1)\lambda^{2h}(1-\lambda)^{-(n+h)}$$

$$= \frac{2h+1}{n+h}\binom{2n-2}{n-h-1},$$

from the Lagrange theorem.

2.7.6. Let $b(m, n)$ be the number of planted plane cubic trees with n trivalent vertices, and in which the sum of the heights of the trivalent vertices equals m. If $B(z, x) = \sum_{m,\, n\geqslant 0} b(m, n)z^m x^n$, then $B(z, x) = 1 + xB(z, zx)^2$ from Decomposition 2.7.2. Let $A(x) = B(1, x)$ and $D(x) = \{(\partial/\partial z)B(z, x)\}|_{z=1}$, so $h_n = \{[x^n]D(x)\}/\{[x^n]A(x)\}$. Setting $z = 1$ in the functional equation for B and solving the resulting quadratic gives $A(x) = (1 - \sqrt{1-4x})/2x$, so $[x^n]A(x) = \binom{2n}{n}/(n+1)$.

Differentiating the equation for B with respect to z and setting $z = 1$ yields $D(x) = xA(x)\{xA'(x) + D(x)\}$, whence

$$D(x) = 2x^2A(x)A'(x)\{1 - 2xA(x)\}^{-1}.$$

But $A'(x) = A^2(x)\{1 - 2xA(x)\}^{-1}$ from the equation for $B(1, x)$, and $\{1 - 2xA(x)\}^2 = 1 - 4x$ from above, so $D(x) = 2(1-4x)^{-1}x^2A^3(x)$. Finally

$$x^2A^3(x) = xA(x)(A(x) - 1) = (1 - x)A(x) - 1$$

from the equation for B, and

$$[x^n]D(x) = [x^n]\{x^{-1} + (1-4x)^{-1} - x^{-1}(1-4x)^{-(1/2)} + (1-4x)^{-(1/2)}\}$$

$$= 4^n - \binom{2n+2}{n+1} + \binom{2n}{n}.$$

The result follows by dividing this by $\binom{2n}{n}/(n+1)$.

2.7.7. The required number is $[t^n\mathbf{f}^{\mathbf{i}}]h(t, \mathbf{f})$, where h is the generating function for planted plane trees in which a left-most path of length j is marked by tf_j, for $j \geqslant 2$, and in which $\mathbf{f} = (f_2, f_3, \ldots)$. Thus, from Decomposition 2.7.21, h

satisfies the functional equation

$$h = t \sum_{k \geqslant 2} f_k \left(\sum_{i \geqslant 0} h^i \right)^{k-2} = t \sum_{k \geqslant 2} f_k (1-h)^{-(k-2)}.$$

Then, by the Lagrange theorem

$$[t^n \mathbf{f}^{\mathbf{i}}] h(t, \mathbf{f}) = \frac{1}{n} [\mathbf{f}^{\mathbf{i}} \lambda^{n-1}] \left\{ \sum_{k \geqslant 2} f_k (1-h)^{-(k-2)} \right\}^n$$

$$= \frac{(n-1)!}{\mathbf{i}!} \binom{m+n-2}{n-1}.$$

2.7.8. Apply the edge subdivision of Decomposition 2.7.9 to the set of planted plane trees. Let $P(x)$ be the generating function for planted plane trees, with x marking nonroot vertices. Marking the bivalent vertices introduced in edge subdivision by xy gives the generating function $P(x(1 - xy)^{-1})$. This is an "at-least" generating function for bivalent vertices since planted plane trees are counted a number of times by this construction, once for each subset of bivalent vertices that are introduced by edge subdivision. By the Principle of Inclusion and Exclusion the number of planted plane trees on n nonroot vertices, m of which are bivalent, is $[x^n y^m] P(x\{1 - (y-1)x\}^{-1})$.

2.7.9. (a) This is the left-most path decomposition (**2.7.21**) restricted to C since, when a left-most path of length $k + 2$ is deleted, there remains an ordered collection of k elements of C.

In fact, we may continue, as follows, to obtain a functional equation. The only monovalent vertex introduced in the right side of the decomposition is the terminal vertex of the left-most path. Thus, if $C(x)$ is the generating function for planted plane cubic trees, then $C(x) - x = x\sum_{k \geqslant 1} C(x)^k$ so $C(x) = x(1 - C(x))^{-1}$.

(b) Any planted plane tree t, except for ε, has a nonempty branch list, $\Lambda(t) = (t_1, \ldots, t_{p-1}, t_p)$. Decompose such a tree into the pair of planted plane trees $(\Lambda^{-1}(t_1, \ldots, t_{p-1}), t_p)$. This is clearly a reversible procedure. Moreover, no nonroot vertices in addition to those of the two trees are introduced on the right-hand side of this decomposition. Thus, if $P(x)$ is the generating function for planted plane trees with respect to nonroot vertices, then $P(x) - x = P^2(x)$.

(c) Comparing the decomposition for C in part (a), and the branch decomposition for P, we deduce the following correspondence. Let $c \in \mathsf{C}$ and $p \in \mathsf{P}$. Then the vertex adjacent to the root of p is the monovalent vertex on the left-most path of c. The ordered branches $p_1, \ldots, p_k$ in p are constructed by repeating this procedure with the ordered collection $c_1, \ldots, c_k$ of elements of C that are obtained by deleting the left-most path of c. This procedure is

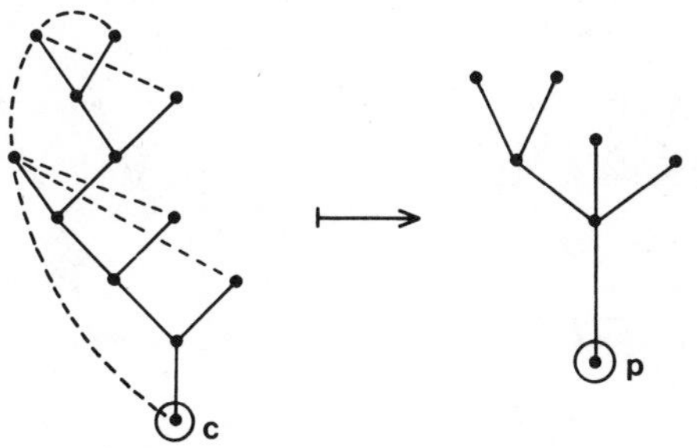

repeated until all monovalent vertices in c are used. The correspondence is illustrated.

2.7.10. (a) Any 2-chromatic planted plane tree t with root color 1, except for ε_1, has a nonempty branch list $\Lambda(t) = (t_1, \ldots, t_{p-1}, t_p)$. We can decompose such a tree into the pair of trees $(\Lambda^{-1}(t_1, \ldots, t_{p-1}), t_p)$, where $\Lambda^{-1}(t_1, \ldots, t_{p-1}) \in \mathsf{P}_1$, and t_p has root color 2. Then $\Lambda(t_p)$, the branch list of t_p, is an ordered collection of elements of P_1. Thus $t \mapsto (\Lambda^{-1}(t_1, \ldots, t_{p-1}), \varepsilon_2, \Lambda(t_p))$ gives an appropriate decomposition. If $h_1(x_1, x_2)$ is the generating function for P_1, with x_i marking nonroot vertices of color i, for $i = 1, 2$, then we have directly from this decomposition that

$$h_1(x_1, x_2) - x_2 = \sum_{k \geq 1} x_1 h_1^k(x_1, x_2), \quad \text{so } h_1 = x_2 + x_1 h_1 (1 - h_1)^{-1},$$

since ε_1, ε_2 have single nonroot vertices of colors 2, 1 respectively.

From the branch decomposition, if $M(x, y)$ is the generating function for planted plane trees with x marking monovalent nonroot vertices and y marking all other nonroot vertices, then M satisfies the functional equation $M = x + yM(1 - M)^{-1}$. Thus $M(x, y) = h_1(y, x)$, and the 1–1 correspondence is established.

(b) The branch decompositions for P_1 and P_2 (**2.7.13**) yield

$$\mathsf{P}_1 \xrightarrow{\sim} \dot{\bigcup_{k \geq 0}} \{\varepsilon_1\} \times \mathsf{P}_2^k \xrightarrow{\sim} \dot{\bigcup_{k \geq 0}} \{\varepsilon_1\} \times \left(\dot{\bigcup_{l \geq 0}} \{\varepsilon_2\} \times \mathsf{P}_1^l \right)^k$$

and the left-most path decomposition for the set P of planted plane trees (**2.7.21**) is

$$\mathsf{P} \to \dot{\bigcup_{k \geq 0}} \left\{ \dot{\bigcup_{l \geq 0}} P^l \right\}^k \times \{\pi_{k+2}\}.$$

Inspection of these provides the needed identification between monovalent vertices in P and vertices of color 2 in P_1, and other nonroot vertices (interior

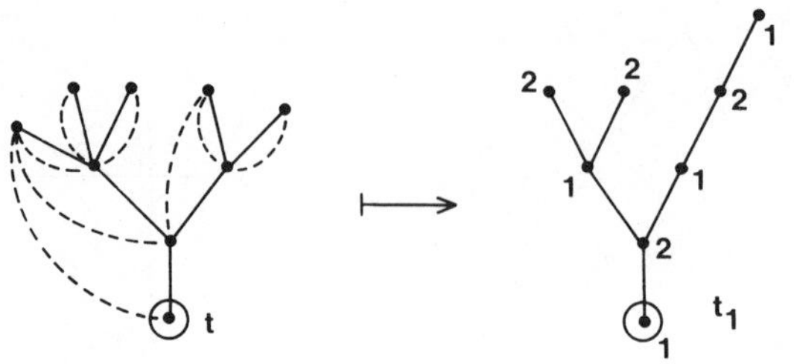

to left-most paths) in P and vertices of color 1 in P_1. This bijection is illustrated here for a particular pair of trees, $t \in \mathsf{P}$ and $t_1 \in \mathsf{P}_1$.

2.7.11. Let the s-objects of trees in P be the nonroot vertices. The elements of P' fall into two classes. In the first class the distinguished vertex is the unique vertex adjacent to the root, so this class is trivially decomposed as P itself. In the second class the distinguished vertex is a nonroot vertex in a tree in the branch list. If this tree and the trees to the right of it in the branch list are deleted, then a tree in P remains. Similarly, if this tree and those to the left are deleted, a tree in P remains. Thus the second class can be decomposed as $\mathsf{P}^2 \times \mathsf{P}'$. Marking nonroot vertices with x gives $xP' = P + x^{-1}P^2(xP')$.

2.7.12. (a) Let B be the set of bracketings of any number n of objects, for $n \geqslant 1$. Let $\{\varepsilon\}$ be the empty bracketing of a single element. Then any nonempty bracketing b can be written as $b = (b_1 b_2)$, where b_1, b_2 are arbitrary bracketings. Therefore we have the decomposition $\mathsf{B} - \{\varepsilon\} \stackrel{\sim}{\to} \mathsf{B}^2$. If x records the number of objects, then $\{\varepsilon\}$ has generating function x, and since no additional objects are introduced on the right-hand side, $B(x) - x = B^2(x)$, so $B(x) = x + B^2(x)$. Of course, by **2.7.3**, we obtain

$$b(n) = \frac{1}{n}\binom{2n-2}{n-1}.$$

(b) If A is again the set of bracketings, then any nonempty bracketing b may be written as $b = (b_1 \dots b_k)$, for some $k \geqslant 2$, and where $b_1, \dots, b_k$ are arbitrary bracketings. Therefore we have the decomposition $\mathsf{A} - \{\varepsilon\} \to \dot\cup_{k \geqslant 2} \mathsf{A}^k$. If x records the number of objects, then the generating function $A(x)$ satisfies $A(x) - x = \sum_{k \geqslant 2} A(x)^k$, so $A(x) = x(1 - A(x))(1 - 2A(x))^{-1}$. Therefore, by the Lagrange theorem, the required number is

$$\begin{aligned}[x^n]A(x) &= \frac{1}{n}[\lambda^{n-1}](1-\lambda)^n(1-2\lambda)^{-n} \\ &= \frac{1}{n}\sum_{i=0}^{n-1}\binom{n}{i}(-1)^i[\lambda^{n-i-1}](1-2\lambda)^{-n} \\ &= \frac{1}{n}\sum_{i=0}^{n-1}(-1)^i 2^{n-1-i}\binom{n}{i}\binom{2n-2-i}{n-1}.\end{aligned}$$

By the branch decomposition for planted plane trees, this is also the number of homeomorphically irreducible planted plane trees with n nonroot monovalent vertices.

(c) In this case, if C is the set of bracketings, then any nonempty bracketing b can be written as $(b_1 b_2)$, where b_1, b_2 is an unordered pair of bracketings. The generating function for pairs of identical bracketings is $C(x^2)$, and for ordered pairs of distinct bracketings is $C(x)^2 - C(x^2)$. But each ordered pair of distinct bracketings is counted twice, so that $\frac{1}{2}\{C(x)^2 - C(x^2)\}$ is the generating function for unordered pairs of distinct bracketings. Thus

$$C(x) - x = \tfrac{1}{2}\{C(x)^2 - C(x^2)\} + C(x^2) = \tfrac{1}{2}\left(C(x)^2 + C(x^2)\right).$$

Alternatively, $\mathsf{B} - \{\varepsilon\} \xrightarrow{\sim} \mathsf{B}^2/S_2$, where S_2 is the symmetric group of order 2. The result follows immediately from Pólya's theorem since the cycle index polynomial for S_2 is $\frac{1}{2}(x_1^2 + x_2)$.

2.7.13. Let the set of such sequences, for all $n \geqslant 1$, be S. An element σ in S must begin with a 1 and end with (-1), so that $\sigma = 1\hat{\sigma}(-1)$, where $\hat{\sigma}$ is a sequence of an equal number of 1's and (-1)'s, in which partial sums are non-negative. Thus $\hat{\sigma}$ either is empty or can be uniquely expressed as $\alpha_1 \ldots \alpha_k$, where $\alpha_1, \ldots, \alpha_k$ are elements of S for some $k \geqslant 1$. Thus $\mathsf{S} \xrightarrow{\sim} 1(\mathsf{S})^*(-1)$, so that $C(x) = x\{1 - C(x)\}^{-1}$. But, from **2.7.3**, $c(n) = \dfrac{1}{n}\dbinom{2n-2}{n-1}$.

2.7.14. Let D be the set of dissected n-gons, and let $d \in \mathsf{D}$. Let e be the edge in the exterior face, called the root edge of d, which is distinguished. Now either e is the only edge in d or e lies on an interior polygon p of d with $(j+1)$ edges, for some $j \geqslant 2$. If the edges of p, in clockwise order, are $e, e_1, \ldots, e_j$, then $e_1, \ldots, e_j$ are the root edges for disjoint arbitrary diagonalized polygons $d_1, \ldots, d_j$. Thus

$$\mathsf{D} \xrightarrow{\sim} \dot{\bigcup_{\substack{j \geqslant 0 \\ j \neq 1}}} \mathsf{D}^j.$$

Now let $D(x_1, x_3, x_4, \ldots)$ be the generating function for D in which x_1 marks exterior nonroot edges and x_j marks an interior j-gon. Now in the decomposition d contains the interior j-gon p, which is not recorded in $d_1, \ldots, d_j$. Moreover, if d_i consists of the single edge e_i, then e_i is an exterior edge in both d_i and d, but a root edge only in d_i. Otherwise, e_i is an exterior edge in d_i, but not in d. Thus $D = 1 + \sum_{j \geqslant 2} x_{j+1}\{x_1 + (D-1)\}^j$. Now let $E(x_1, x_3, \ldots) = D(x_1, x_3, \ldots) + x_1 - 1$, so E satisfies $E = x_1 + \sum_{j \geqslant 2} x_{j+1} E^j$. But by the branch decomposition, E must enumerate homeomorphically irreducible planted plane trees with nonroot vertices of degree j marked by x_j, for $j = 1, 3, 4, \ldots$. The result follows from **2.7.7**.

2.7.15. In each half-edge structure distinguish one incident half-edge as a "bottom" edge, and any others as "top" edges. Since $i_1 + i_2 + \cdots = n$ and

$i_1 + 2i_2 + \cdots = 2n - 1$, there are n bottom edges and $n - 1$ top edges in this collection of half-edge structures. Label the bottom edges 1 to n and the top edges 1 to $n - 1$ in arbitrary order. We construct a planted plane tree as follows. First attach the bottom edge of any of the $n - 1$ half-edge structures that do not contain the top edge labeled 1 to the top edge labeled 1, forming a single half-edge structure with two vertices and a single bottom edge. This leaves us with $n - 1$ half-edge structures, $n - 1$ unattached bottom edges, and $n - 2$ unattached top edges. For $i = 2, \ldots, n - 1$ we continue this process, where at stage i the bottom edge of any of the $n - i$ half-edge structures that do not contain the top edge labeled i is attached to the top edge labeled i. After the $(n - 1)$-st stage there remains a single half-edge structure, with one unattached bottom edge and no unattached top edges. A planted plane tree with degree sequence $\mathbf{i}$ is obtained by attaching this bottom edge to a root vertex.

By the preceding process $\prod_{i=1}^{n-1}(n - i) = (n - 1)!$ planted plane trees are constructed. These consist of $\prod_{j \geqslant 1} i_j!$ copies of each planted plane tree with degree sequence $\mathbf{i}$ since in the construction all half-edge structures are regarded as distinct. Thus there are $(n - 1)!(\mathbf{i}!)^{-1}$ planted plane trees with degree sequence $\mathbf{i}$, as required.

2.7.16. (a) Let P_i be the set of planted plane k-chromatic trees with root color i, for $i = 1, \ldots, k$. By extending Decomposition 2.7.13 to k colors, we have the simultaneous decompositions

$$\mathsf{P}_i - \{\varepsilon_i\} \xrightarrow{\sim} \dot{\bigcup_{\substack{j=1 \\ j \neq i}}^{k}} \bigcup_{m \geqslant 1} \{\varepsilon_i\} \times \mathsf{P}_j^m, \qquad \text{for } i = 1, \ldots, k,$$

where ε_i is the trivial tree in P_i. Let $f_i(\mathbf{W}, \mathbf{x})$ be the generating function for P_i in which a vertex of color j and degree m is marked by $x_j w_{jm}$. Then $c_r(\mathbf{L}, \mathbf{n}) = [\mathbf{W}^{\mathbf{L}}\mathbf{x}^{\mathbf{n}}]f_r$ and, from the preceding decomposition, $f_1, \ldots, f_k$ satisfies the given system of equations.

(b) Let $y_j = x_j g_j(f_j)$, for $j = 1, \ldots, k$. Then from part (a) we have $f_i = y - y_i$, so $y_1, \ldots, y_k$ satisfies $y_i = x_i g_i(y - y_i)$, for $i = 1, \ldots, k$.

(c) Applying the Lagrange theorem for monomials (**1.2.13**) to the system of equations in part (b), we obtain

$$c_r(\mathbf{L}, \mathbf{n}) = (n_1 \cdots n_k)^{-1} \sum_{\substack{\alpha=1 \\ \alpha \neq r}}^{k} \sum_{\mathbf{M}} \|\delta_{ij} n_i - m_{ij}\|$$

$$\times \prod_{i=1}^{k} [w_{i1}^{l_{i1}} \cdots][y_1^{m_{i1}} \cdots y_k^{m_{ik}}] g_i^{n_i}(y - y_i)$$

where the summation is over $\mathbf{M} = [m_{ij}]_{k \times k}$ such that $\sum_{i=1}^{k} m_{ij} = n_j - \delta_{\alpha j}$ and

$m_{jj} = 0$, for $j = 1, \ldots, k$. Thus

$$c_r(\mathbf{L}, \mathbf{n}) = (\mathbf{n} - \mathbf{1})!(\mathbf{L}!)^{-1} \sum_{\substack{\alpha=1 \\ \alpha \neq r}}^{k} \sum_{\mathbf{M}} \|\delta_{ij} n_i - m_{ij}\| \prod_{i=1}^{k} [y_1^{m_{i1}} \cdots y_k^{m_{ik}}](y - y_i)^{q_i},$$

where $q_i = \sum_{j \geq 1} (j - 1) l_{ij}$, for $i = 1, \ldots, k$. If $\mathbf{q} = (q_1, \ldots, q_k)$, then

$$c_r(\mathbf{L}, \mathbf{n}) = \mathbf{q}!(\mathbf{n} - \mathbf{1})!(\mathbf{L}!)^{-1} \sum_{\substack{\alpha=1 \\ \alpha \neq r}}^{k} \sum_{\mathbf{M}} \|\delta_{ij} n_i - m_{ij}\| (\mathbf{M}!)^{-1},$$

where the summation is over $\mathbf{M}$ such that $\sum_{i=1}^{k} m_{ij} = n_j - \delta_{\alpha j}$, $\sum_{i=1}^{k} m_{ji} = q_j$, and $m_{jj} = 0$, for $j = 1, \ldots, k$. Let $\beta = \mathbf{q}!(\mathbf{n} - \mathbf{1})!(\mathbf{L}!)^{-1}$ and $\mathbf{u} = (u_1, \ldots, u_k)$, $\mathbf{v} = (v_1, \ldots, v_k)$. Then

$$c_r(\mathbf{L}, \mathbf{n}) = \beta[\mathbf{u}^{\mathbf{q}} \mathbf{v}^{\mathbf{n}}] \sum_{\substack{\alpha=1 \\ \alpha \neq r}}^{k} v_\alpha \sum_{\substack{\mathbf{M} \geq \mathbf{0} \\ \text{trace}(\mathbf{M}) = 0}} \|\delta_{ij}(m_{1j} + \cdots + m_{kj} + \delta_{\alpha j}) - m_{ij}\|$$

$$\times \prod_{i, j=1}^{k} \frac{(u_i v_j)^{m_{ij}}}{m_{ij}!}.$$

But $\sum_{\mathbf{i} \geq \mathbf{0}} h(\mathbf{i}) \mathbf{x}^{\mathbf{i}} / \mathbf{i}! = h(\mathbf{x}) \exp(x_1 + \cdots + x_n)$ if h is multilinear, where $\mathbf{i} = (i_1, \ldots, i_n)$, $\mathbf{x} = (x_1, \ldots, x_n)$, so

$$c_r(\mathbf{L}, \mathbf{n}) = \beta[\mathbf{u}^{\mathbf{q}} \mathbf{v}^{\mathbf{n}}] \exp\left\{ \sum_{\substack{i, j=1 \\ i \neq j}}^{k} u_i v_j \right\} \sum_{\substack{\alpha=1 \\ \alpha \neq r}}^{k} \|\delta_{ij}(v_j u + \delta_{\alpha j}) - u_i v_j\| v_\alpha,$$

where $u = u_1 + \cdots + u_k$, $v = v_1 + \cdots + v_k$. Expanding this determinant on row α, we have

$$\begin{aligned}
\|\delta_{ij}(v_j u + \delta_{\alpha j}) - u_i v_j\| &= \|\delta_{ij} v_j u - u_i v_j\| + \text{cof}_{\alpha\alpha}[\delta_{ij} v_j u - u_i v_j]_{k \times k} \\
&= 0 + \frac{v_1 \cdots v_k}{v_\alpha} \text{cof}_{\alpha\alpha}[\delta_{ij} u - u_i]_{k \times k} \\
&= \frac{v_1 \cdots v_k}{v_\alpha} u^{k-1} \text{cof}_{\alpha\alpha}\left[\delta_{ij} - \frac{u_i}{u}\right]_{k \times k} \\
&= \frac{v_1 \cdots v_k}{v_\alpha} u^{k-1} \left\{1 - \frac{u - u_\alpha}{u}\right\}, \qquad \text{from } \mathbf{1.1.10(5)}.
\end{aligned}$$

Substituting this into the expression for $c_r(\mathbf{L}, \mathbf{n})$ gives

$$c_r(\mathbf{L},\mathbf{n}) = \beta[\mathbf{u}^{\mathbf{q}}\mathbf{v}^{\mathbf{n}-\mathbf{1}}](u-u_r)u^{k-2}\exp\left\{\sum_{\substack{i,j=1\\ i\neq j}}^{n} u_i v_j\right\}$$

$$= \beta[\mathbf{u}^{\mathbf{q}}\mathbf{v}^{\mathbf{n}-\mathbf{1}}]\frac{\partial}{\partial v_r}u^{k-2}\exp\left\{\sum_{\substack{i,j=1\\ i\neq j}}^{n} u_i v_j\right\}$$

$$= (\mathbf{L}!)^{-1}\mathbf{q}!(\mathbf{n}-\mathbf{1}+\boldsymbol{\delta}_r)![\mathbf{u}^{\mathbf{q}}\mathbf{v}^{\mathbf{n}-\mathbf{1}+\boldsymbol{\delta}_r}]u^{k-2}\exp\left(uv-\sum_{j=1}^{k}u_jv_j\right),$$

where $\boldsymbol{\delta}_r = (\delta_{r1}, \ldots, \delta_{rk})$. Let $\mathbf{m} = (m_1, \ldots, m_k)$ and expand

$$\exp\left(uv-\sum_{j=1}^{k}u_jv_j\right) \text{ as } \exp(uv)\prod_{j=1}^{k}\exp(-u_jv_j), \qquad \text{giving}$$

$$c_r(\mathbf{L},\mathbf{n}) = (\mathbf{L}!)^{-1}\mathbf{q}!(\mathbf{n}-\mathbf{1}+\boldsymbol{\delta}_r)!\sum_{\mathbf{m}\geqslant\mathbf{0}}[\mathbf{u}^{\mathbf{q}-\mathbf{m}}\mathbf{v}^{\mathbf{n}-\mathbf{1}+\boldsymbol{\delta}_r-\mathbf{m}}]$$

$$\times(-1)^{m_1+\cdots+m_k}\frac{u^{k-2}}{\mathbf{m}!}\exp(uv)$$

$$= (\mathbf{L}!)^{-1}\sum_{\mathbf{m}\geqslant\mathbf{0}}\frac{(\mathbf{n}-\mathbf{1}+\boldsymbol{\delta}_r)!}{\mathbf{m}!(\mathbf{n}-\mathbf{1}+\boldsymbol{\delta}_r-\mathbf{m})!}\frac{\mathbf{q}!}{(\mathbf{q}-\mathbf{m})!}(N-i)!(-1)^i,$$

where $N = n_1 + \cdots + n_k - 1$, $i = m_1 + \cdots + m_k$. Thus

$$c_r(\mathbf{L},\mathbf{n}) = (\mathbf{L}!)^{-1}\sum_{i\geqslant 0}(N-i)![x^k]P(x),$$

where

$$P(x) = \prod_{j=1}^{k}\sum_{m\geqslant 0}\binom{n_j-1+\delta_{rj}}{m}\binom{q_j}{m}m!(-x)^m,$$

as required.

SECTION 2.8

2.8.1. We have the case $p = 1$ and record the sequence length only, so $C(x, y) = (1 - y(x^3 + x^6))^{-1}yx^7$. Therefore, by Theorem 2.8.6, the required number is

$$[x^n]\{1 - 2x + x^7(1 + x^3 + x^6)^{-1}\}^{-1}$$

$$= [x^n](1 + x^3 + x^6)\{1 - 2x + x^3 - 2x^4 + x^6 - x^7\}^{-1}.$$

2.8.2. The cluster generating function is $C(x, y) = (1 - y(x^3 + x^4))^{-1}yx^5$, so the required number is

$$[x^l]\{1 - nx + x^5(1 + x^3(1 + x))^{-1}\}^{-1}$$

$$= \sum_{k \geqslant 0} \sum_{i=0}^{k} (-1)^i \binom{k}{i} n^{k-i} [x^{l-k-4i}](1 + x^3(1 + x))^{-i},$$

and the result follows.

2.8.3. By Decomposition 2.3.1, a k-subset of N_n can be encoded as a $\{0, 1\}$-sequence of length n, which contains k 1's. Moreover, a succession is recognized by the occurrence of the substring 11. If x marks 0's, z marks 1's, and y marks occurrences of the distinguished substring 11, then the cluster generating function is $(1 - yz)^{-1}yz^2$. Then, by Theorem 2.8.6, the required number is

$$[x^{n-k}z^k y^m]\{1 - (x + z) - (y - 1)z^2(1 - (y - 1)z)^{-1}\}^{-1}$$

$$= [z^k y^m]\{1 - z - (y - 1)z^2(1 - (y - 1)z)^{-1}\}^{-(n-k+1)}$$

$$= [z^{k-m}(zy)^m]\{1 + z(1 - yz)^{-1}\}^{n-k+1},$$

and the result follows.

2.8.4. If $A_1 = 11101011$, $A_2 = 101111$, then

$$\mathbf{V} = \begin{pmatrix} x^6 + x^7 & x^4 + x^7 \\ x^3 + x^4 + x^5 & x^5 \end{pmatrix} \quad \text{and} \quad \mathbf{L} = \begin{pmatrix} x^8 & 0 \\ 0 & x^6 \end{pmatrix},$$

so

$$(\mathbf{I} + \mathbf{V})^{-1} = \{(1 + x^5)(1 + x^6 + x^7) - (x^4 + x^7)(x^3 + x^4 + x^5)\}^{-1}\mathbf{W},$$

where

$$\mathbf{W} = \begin{pmatrix} 1 + x^5 & -(x^4 + x^7) \\ -(x^3 + x^4 + x^5) & 1 + x^6 + x^7 \end{pmatrix}.$$

Then, by Theorem 2.8.6 and Lemma 2.8.10, the required number is

$$[x^l]\left\{1 - 2x + \frac{x^8(1 - x^3 - x^4) + x^6(1 + x^6 - x^4)}{1 + x^5 + x^6 - x^8 - x^9 - x^{10}}\right\}^{-1},$$

and the result follows.

2.8.5. The distinguished substrings are $A_i = i^k$, for $i = 1, \ldots, n$. Thus the elements of the connector matrix are $v_{ii} = \sum_{j=1}^{k-1} x_i^j$ for $i = 1, \ldots, n$ and $v_{ij} = 0$ for $i \neq j$. Therefore, if $\mathbf{X} = \text{diag}(x_1, \ldots, x_n)$,

$$\mathbf{V} = \sum_{j=1}^{k-1} \mathbf{X}^j = (\mathbf{X} - \mathbf{X}^k)(\mathbf{I} - \mathbf{X})^{-1}, \qquad \mathbf{L} = \mathbf{X}^k.$$

Thus from Lemma 2.8.10 the cluster generating function is

$$C(\mathbf{x}, \mathbf{y}) = \text{trace}\left(\mathbf{I} - \mathbf{Y}(\mathbf{X} - \mathbf{X}^k)(\mathbf{I} - \mathbf{X})^{-1}\right)^{-1}\mathbf{Y}\mathbf{X}^k\mathbf{J}.$$

From Theorem 2.8.6, the required number is

$$[\mathbf{x}^{\mathbf{i}}\mathbf{y}^{\mathbf{m}}]\left\{1 - \sum_{j=1}^{n}\left(x_j + (y_j - 1)x_j^k\left\{1 - (y_j - 1)(x_j - x_j^k)(1 - x_j)^{-1}\right\}^{-1}\right)\right\}^{-1},$$

and the result follows.

2.8.6. (a) The set of distinguished substrings is $\{12, 23, \ldots, (n-1)n\}$. Thus a k-cluster is a sequence of the form $j(j+1)\cdots(j+k)$ for $j = 1, \ldots, n-k$. Accordingly, there are $(n-k)$ k-clusters, each on a sequence of length $k+1$, for $k = 1, \ldots, n-1$, so

$$\begin{aligned} C(x, y) &= \sum_{k=1}^{n-1} y^k(n-k)x^{k+1} \\ &= \{(n-1)x^2y - nx^3y^2 + x^{n+2}y^{n+1}\}(1 - xy)^{-2}. \end{aligned}$$

From Theorem 2.8.6, the required number is $[x^l]\{1 - nx - C(x, y-1)\}^{-1}$, and the result follows by replacing x by xt and y by t^{-1}.

(b) The set of distinguished substrings is $\{12, \ldots, (n-1)n, n1\}$, so that there are n k-clusters, each of length $k+1$, given by $12\cdots(k+1), \ldots, n1 \cdots k$. Thus, since k is any positive integer,

$$C(x, y) = \sum_{k \geq 1} ny^kx^{k+1} = nyx^2(1 - yx)^{-1}.$$

From Theorem 2.8.6, the required number is

$$[x^l y^k]\left(1 - nx - n(y-1)x^2(1-(y-1)x)^{-1}\right)^{-1}$$

$$= [x^l y^k](1-(y-1)x)\{1 - (y-1)x - nx\}^{-1}$$

$$= [y^k]\{(y-1+n)^l - (y-1)(y-1+n)^{l-1}\}$$

$$= n\binom{l-1}{k}(n-1)^{l-k-1}.$$

2.8.7. Let $\mathsf{A}_1 = \{ij \mid 1 \leqslant i < j \leqslant n\}$ be the set of rises and $\mathsf{A}_2 = \{ii \mid 1 \leqslant i \leqslant n\}$ be the set of levels. Then $\mathsf{A}_1 \cup \mathsf{A}_2$ is a reduced set of distinguished substrings. The k-clusters are nondecreasing sequences of length $(k+1)$, for $k \geqslant 1$, so the clusters are elements of $(1^*)(2^*)\cdots(n^*) - (\varepsilon \cup 1 \cup \cdots \cup n)$. If elements of A_1 are marked by r and A_2 by l, then

$$C(\mathbf{x}, l, r) = r^{-1}\left\{\prod_{i=1}^{n}(1 + rx_i + rlx_i^2 + \cdots) - (1 + rx_1 + \cdots + rx_n)\right\}$$

$$= r^{-1}\left\{\prod_{i=1}^{n}(1+(r-l)x_i)(1-lx_i)^{-1} - (1 + rx_1 + \cdots + rx_n)\right\}.$$

Thus, from Theorem 2.8.6, the number of sequences of type $\mathbf{i}$ with k rises and m levels is

$$[\mathbf{x}^{\mathbf{i}} r^k l^m]\{1 - (x_1 + \cdots + x_n) - C(\mathbf{x}, l-1, r-1)\}^{-1} = [\mathbf{x}^{\mathbf{i}} r^k l^m]\Phi(\mathbf{x}, r, l),$$

where $\Phi(\mathbf{x}, r, l) = (r-1)\{r - \prod_{i=1}^{n}(1+(r-l)x_i)(1+(1-l)x_i)^{-1}\}^{-1} - 1$. But the number of rises, levels, and falls sum to one less than the length of the sequence. Thus the number of sequences of type $\mathbf{i}$ with k rises, m levels, and p falls is $[\mathbf{x}^{\mathbf{i}} r^k l^m f^p] f^{-1}\Phi(f\mathbf{x}, f^{-1}r, f^{-1}l)$, giving the result.

2.8.8. If x_i is replaced by x, for $i = 1, \ldots, n$, then $\mathbf{L}$ becomes $\Lambda(x) = \operatorname{diag}(x^{|A_1|}, \ldots, x^{|A_p|})$, and $\mathbf{I} + \mathbf{V}$ becomes $\Lambda(x)\chi(x^{-1})$. Thus, from Theorem 2.8.6 and Lemma 2.8.10, the required number is

$$[x^l]\{1 - (x_1 + \cdots + x_n) + \operatorname{trace}(\mathbf{I} + \mathbf{V})^{-1}\mathbf{L}\mathbf{J}\}^{-1}\Big|_{\mathbf{x} = x\mathbf{1}}$$

$$= [x^l]\{1 - nx + \operatorname{trace}\chi(x^{-1})^{-1}\Lambda(x)^{-1}\Lambda(x)\mathbf{J}\}^{-1}$$

$$= [x^l]\{1 - nx + \operatorname{trace}(\chi(x^{-1}))^{-1}\mathbf{J}\}^{-1}.$$

2.8.9. We follow the proof of Theorem 2.8.6, constructing an "at-least" generating function by means of the cluster generating function. Those circular permutations consisting of elements of N_n^* separated by elements of $\mathsf{D(A)}$ are enumerated by $\sum_{k \geq 1} k^{-1}(x_1 + \cdots + x_n + C(\mathbf{x}, \mathbf{y}))^k$. We divide by k because rotations are considered equivalent, and because we are extracting the coefficient of $x_1 \cdots x_n$, all elements are distinct. The other possible permutations are **circular clusters**, which are elements of $\mathsf{D(A)}$ in which the end of the terminal distinguished substring overlaps with the initial segment of the first distinguished substring in the cluster. The contribution of these to the "at-least" generating function is $\sum_{k \geq 1} k^{-1} \sum_{i=1}^{p} [(\mathbf{YV})^k]_{ii} = \operatorname{trace} \log(\mathbf{I} - \mathbf{YV})^{-1}$. The result follows by the Principle of Inclusion and Exclusion.

SECTION 2.9

2.9.1. **1.** We first prove this for maps whose defining graphs are trees. Such maps have only a single face, so $n_3 = 1$. If the tree has n_1 vertices, then it has $n_2 = n_1 - 1$ edges, which is proved by induction on n_1 as follows. It is clearly true for $n_1 = 1$. All trees on $n_1 + 1$ vertices may be constructed by joining a vertex in a tree on n_1 vertices to a new monovalent vertex, by a new edge. Thus if the trees on n_1 vertices have $n_1 - 1$ edges, then the trees on $n_1 + 1$ vertices have n_1 edges, and the result is true by induction. Therefore in the case of trees $n_1 - n_2 + n_3 = n_1 - (n_1 - 1) + 1 = 2$.

We now prove that $n_2 - n_3 = n_1 - 2$ for all planar maps on n_1 vertices, by induction on n_2. This is true for $n_2 = n_1 - 1$ from the preceding. Now every planar map on $n_2 + 1 \geq n_1$ edges can be constructed by joining two vertices (not necessarily distinct) in a planar map on n_2 edges by a new edge. This new edge splits one face into two faces and so the planar map on $n_2 + 1$ edges has $n_3 + 1$ faces if the map on n_2 edges has n_3 faces. Thus $n_2 - n_3$ is constant for $n_2 \geq n_1 - 1$, so the result is proved.

2. In both sums each of the n_2 edges is counted twice. The edges are counted once for each end in the sum over the vertices, and once for each "side" in the sum over the faces.

2.9.2. (a) Consider the removal of the nonroot edge incident with the root face from a rooted near-triangulation M with root face of degree 2, for $M \neq \delta$. Then the map M' obtained in this way is a planar triangulation. Furthermore, the number of faces of M' is equal to the number of inner faces of M, and the construction is reversible. Thus $S(y) - 1$ is the required generating function, since 1 is the generating function with respect to inner faces for δ.

(b) Consider an arbitrary strict triangulation M, and its edges $e_1, \ldots, e_{3n}$. Suppose that, for each of these edges, we uniquely specify a "tail" end and a "head" end. Now we construct a unique planar triangulation M' by performing the following operation for each edge e_i of M. We remove e_i and identify the vertex at the tail end of e_i with the root vertex of an arbitrary rooted near-triangulation with outer-face degree 2, say M_i. The other vertex on the outer face of M_i is identified with the head end of e_i. Of course, we can have $M_i = \delta$, in which case the preceding procedure has no effect on e_i. If e_i is the root edge of M, then the root edge of M' is defined to be the edge in the outer face of M_i, which becomes incident with the outer face of M in this operation.

Now if M has $3n$ edges, as above, then it must have $2n$ faces, so we have, immediately from the preceding construction, the composition lemma and part (a),

$$S(y) - 1 = T(yS^{3/2}(y)).$$

(c) The parametric equations for $S(y)$ given in **2.9.6** are

$$\left.\begin{aligned} y &= \alpha(1-2\alpha^2) \\ S(y) &= (1-3\alpha^2)(1-2\alpha^2)^{-2}. \end{aligned}\right\}$$

Thus, changing variables to $t = yS^{3/2}(y)$ gives the parametric equations

$$\left.\begin{aligned} t^2 &= \alpha^2(1-3\alpha^2)^3(1-2\alpha^2)^{-4} \\ T(t) &= S(y) - 1 = (1-3\alpha^2)(1-2\alpha^2)^{-2} - 1. \end{aligned}\right\}$$

Now change parameters to $\theta = \alpha^2(1-2\alpha^2)^{-1}$, so that $\alpha^2 = \theta(1+2\theta)^{-1}$, $(1-2\alpha^2)^{-1} = 1+2\theta$ and $1-3\alpha^2 = (1-\theta)(1+2\theta)^{-1}$. The parametric equations become

$$\left.\begin{aligned} t^2 &= \theta(1-\theta)^3 \\ T &= \theta(1-2\theta), \end{aligned}\right\}$$

so, by the Lagrange theorem,

$$\begin{aligned} [t^{2n}]T(t) &= \frac{1}{n}[\lambda^{n-1}]\left\{\frac{d}{d\lambda}\lambda(1-2\lambda)\right\}(1-\lambda)^{-3n} \\ &= \frac{1}{n}[\lambda^{n-1}](1-4\lambda)(1-\lambda)^{-3n} = \frac{2(4n-3)!}{n!(3n-1)!}. \end{aligned}$$

This is the number of strict triangulations with $2n$ faces, for $n \geqslant 1$. There are no strict triangulations with an odd number of faces.

(d) Consider an arbitrary simple triangulation M with more than one nonroot face, and its nonroot faces $f_1, \ldots, f_{2n-1}$. Suppose that, for each of these triangular faces, we uniquely specify an incident edge as "distinguished." We now construct a unique strict triangulation M' by performing the following operation for each nonroot face f_i of M. We take an arbitrary strict triangulation, say M_i, and place it inside f_i. The vertices and edges in the outer face of M_i are identified with those in f_i, oriented so that the root edge of M_i is identified with the distinguished edge of f_i. If M_i is the strict triangulation with one inner face, then f_i is unaffected by this procedure. The root edge of M' is defined to be the root edge of M. By this construction we obtain all strict triangulations with more than one nonroot face uniquely, and the construction is reversible.

Let $U(y)$ be the generating function for simple triangulations in which y marks faces. Then the preceding construction yields

$$T(y) - y^2 = y\left(y^{-1}T(y)\right)^{-1}\left\{U\left(y^{-1}T(y)\right) - \left(y^{-1}T(y)\right)^2\right\},$$

where $T(y)$ satisfies the parametric equations $y^2 = \theta(1-\theta)^3$, $T = \theta(1-2\theta)$, from the solution to part (c). Now we change variables to $z = y^{-1}T(y)$, and the preceding equation becomes

$$U(z) = 2z^2 - T,$$

so $[z^{2n}]U(z) = -[z^{2n}]T(z)$ for $n > 1$. The parametric equations are transformed by this change of variables to

$$\left.\begin{aligned} T &= \theta(1-2\theta) \\ \theta &= z^2(1-2\theta)^{-2}(1-\theta)^3. \end{aligned}\right\}$$

We solve this by the Lagrange theorem, giving, for $n > 1$,

$$\begin{aligned} [z^{2n}]U(z) &= -\frac{1}{n}[\lambda^{n-1}]\left\{\frac{d}{d\lambda}\lambda(1-2\lambda)\right\}(1-\lambda)^{3n}(1-2\lambda)^{-2n} \\ &= -\frac{1}{n}[\lambda^{n-1}](1-4\lambda)(1-\lambda)^n\left\{1-\lambda(1-\lambda)^{-1}\right\}^{-2n} \\ &= -\frac{1}{n}[\lambda^{n-1}](1-4\lambda)\sum_{i\geqslant 0}\binom{2n+i-1}{i}\lambda^i(1-\lambda)^{n-i} \\ &= \frac{1}{n}\sum_{i=0}^{n-1}(-1)^{n-i}\binom{2n+i-1}{i}\binom{2n-2i}{2}. \end{aligned}$$

This is the number of simple triangulations on $2n$ faces, for $n > 1$. There is one simple triangulation on two faces, and none with an odd number of faces.

2.9.3. (a) The decomposition we use is that given in the solution to [**2.9.2**(b)]. In this case, the number of edges in the outer face is preserved. Furthermore, the number of edges in the strict near-triangulation is calculated by

$$2 \times \text{no. of edges} = (3 \times \text{no. of inner faces}) + (\text{no. of edges in outer face}),$$

from Proposition 2.9.3(2). Thus $T(x, y) = B(xS^{1/2}(y), yS^{3/2}(y))$.

(b) From **2.9.6** and **2.9.7**

$$[x^n]T(x, y) = \frac{(2n-4)!}{(n-2)!(n-1)!}\frac{y^{n-2}}{(1-2\alpha^2)^{2n-3}} - \frac{(2n-2)!}{(n-1)!n!}\frac{y^n}{(1-2\alpha^2)^{2n}},$$

where $y = \alpha(1 - 2\alpha^2)$ and $S(y) = (1 - 3\alpha^2)(1 - 2\alpha^2)^{-2}$. The required number is $[z^n w^{n+2j}]B(z, w)$, so changing variables to $z = xS^{1/2}(y)$, $w = yS^{3/2}(y)$ gives $B(z, w) = T(zS^{-1/2}(y), wS^{-3/2}(y))$, and from the preceding,

$$[z^n]B(z, w) = c_{n-2}w^{n-2}S^{-(2n-3)}(y)(1-2\alpha^2)^{-(2n-3)}$$
$$-c_{n-1}w^n S^{-2n}(y)(1-2\alpha^2)^{-2n}$$

where $c_n = \frac{1}{n+1}\binom{2n}{n}$. From the parametric equations for y, $S(y)$ and the change of variables $w = yS^{3/2}(y)$ we obtain, after making the substitution $\theta = \alpha^2(1 - 2\alpha^2)^{-1}$ (as in the solution to [**2.9.2**(c)]), $w^2 = \theta(1 - \theta)^3$, $(1 - 2\alpha^2)S(y) = 1 - \theta$. Thus

$$[z^n]B(z, w) = c_{n-2}w^{n-2}(1-\theta)^{-(2n-3)} - c_{n-1}w^n(1-\theta)^{-2n},$$

and the required number is

$$[z^n w^{n+2j}]B(z, w) = c_{n-2}[w^{2(j+1)}](1-\theta)^{-(2n-3)} - c_{n-1}[w^{2j}](1-\theta)^{-2n},$$

where θ satisfies the functional equation $\theta = w^2(1 - \theta)^{-3}$. We can thus apply the Lagrange theorem, to obtain

$$[z^n w^{n+2j}]B(z, w) = c_{n-2}(j+1)^{-1}[\lambda^j](2n-3)(1-\lambda)^{-(2n+3j+1)}$$
$$-c_n j^{-1}[\lambda^{j-1}]2n(1-\lambda)^{-(2n+3j+1)}$$
$$= \frac{2(2n-3)!(2n+4j-1)!}{(n-1)!(n-3)!(j+1)!(2n+3j)!}, \qquad n \geqslant 3, j \geqslant 1.$$

If $j = 0$, then the required number is

$$c_{n-2}(2n-3) - c_{n-1} = \frac{(2n-3)!}{(n-3)!n!}, \qquad n \geqslant 3,$$

which agrees with the preceding formula. If $j = -1$, then the required number is c_{n-2}, $n \geqslant 2$.

2.9.4. (a) The decomposition that we use is given in the solution to [**2.9.2**(d)]. The number of edges in the outer face is unaffected by our construction.

(b) We want to find $[x^n z^{n+2j}]U(x, z)$, so we change the variable y in part (a) by $z = y^{-1}T(y)$. This gives $U(x, z) = B(x, y) - x^3y + x^3z$. From the solution to [**2.9.3**(b)], we have

$$[x^n]B(x, y) = c_{n-2}y^{n-2}(1-\theta)^{-(2n-3)} - c_{n-1}y^n(1-\theta)^{-2n}.$$

Thus, for $n > 3$, $[x^n]U(x, z)$

$$= c_{n-2}T^{n-2}(y)z^{-(n-2)}(1-\theta)^{-(2n-3)} - c_{n-1}T^n(y)z^{-n}(1-\theta)^{-2n}$$

where, from [**2.9.2**(c)], $T(y)$ satisfies the parametric equations $T = \theta(1-2\theta)$, $y^2 = \theta(1-\theta)^3$. We can thus express z in terms of θ as

$$z^2 = \theta(1-2\theta)^2(1-\theta)^{-3}.$$

The substitution $\alpha = \theta(1-\theta)^{-1}$ then yields $T = \alpha(1-\alpha)(1+\alpha)^{-2}$, $z^2 = \alpha(1-\alpha)^2$, $1 - \theta = (1+\alpha)^{-1}$, so

$$[x^n]U(x, z) = c_{n-2}z^{n-2}(1+\alpha)(1-\alpha)^{-(n-2)} - c_{n-1}z^n(1-\alpha)^{-n}.$$

The required number is thus

$$[x^n z^{n+2j}]U(x, z) = c_{n-2}[z^{2(j+1)}](1+\alpha)(1-\alpha)^{-(n-2)} - c_{n-1}[z^{2j}](1-\alpha)^{-n}$$

where α satisfies the functional equation $\alpha = z^2(1-\alpha)^{-2}$. We can now apply the Lagrange theorem to get

$$[x^n z^{n+2j}]U(x, z) = c_{n-2}\frac{1}{j+1}[\lambda^j]\left\{\frac{d}{d\lambda}(1+\lambda)(1-\lambda)^{-(n-2)}\right\}(1-\lambda)^{-2(j+1)}$$

$$-c_{n-1}\frac{1}{j}[\lambda^{j-1}]\left\{\frac{d}{d\lambda}(1-\lambda)^{-n}\right\}(1-\lambda)^{-2j}$$

$$= c_{n-2}\frac{1}{j+1}\left\{\binom{n+3j-1}{j} + (n-2)\left(\binom{n+3j}{j}\right.\right.$$

$$\left.\left. + \binom{n+3j-1}{j-1}\right)\right\} - c_{n-1}\frac{n}{j}\binom{n+3j-1}{j-1}$$

$$= \frac{(2n-4)!(n+3j-1)!}{(j+1)!(n-1)!(n-4)!(n+2j)!}, \qquad n \geqslant 4, j \geqslant 1.$$

If $j = 0$, then we obtain

$$[x^n z^n]U(x, z) = \frac{(2n-4)!}{(n-1)!(n-2)!}(n-1) - \frac{(2n-2)!}{(n-1)!n!}$$

$$= \frac{(2n-4)!}{n!(n-4)!}, \qquad n \geqslant 4.$$

in agreement with the preceding expression. The simple near-triangulations with $j = -1$ are the same as strict near-triangulations with $j = -1$ and are counted in [**2.9.3**(b)]. If $n = 3$, then these are simple triangulations, counted in [**2.9.2**(d)]. Of course, δ is the only simple near-triangulation with $n = 2$.

2.9.5. (a) Consider an arbitrary strong near-triangulation M, and suppose that the outer face contains k edges. Let these edges be directed counterclockwise around the outer face. Now consider the following procedure for each of these directed edges except the root edge, denoting them by $e_1, \ldots, e_{k-1}$. We take an arbitrary strict near-triangulation M_i and identify its root edge with edge e_i, drawing M_i in the outer face of M, for $i = 1, \ldots, k-1$. Of course, edge e_i is unchanged if $M_i = \delta$, and the root edge of the resulting map is chosen to be the root edge of M. The rooted map M' constructed in this way is any strict near-triangulation except δ. The construction is reversible and M' has the same number of inner faces as the sum of the number of inner faces in $M, M_1, \ldots, M_{k-1}$. M' also has $k-2$ fewer edges in the outer face than the sums of the corresponding numbers for $M_1, \ldots, M_{k-1}$. The given equation follows directly.

(b) To obtain the required generating function $A(t, w)$, we change variables in (a) by $z^{-1}B(z, w) = t$, giving

$$A(t, w) = t^2 - tz. \tag{1}$$

From **2.9.6**, $yx^{-1}T^2(x, y) + (yx^{-1} - 1)T(x, y) + x^2 - xyS(y) = 0$ and from [**2.9.3**(a)],

$$T(x, y) = B(z, w)$$

where $z = xS^{1/2}(y)$, $w = yS^{3/2}(y)$. Eliminating $T(x, y)$ yields

$$wz^{-1}S^{-1}(y)B^2(z, w) + \left(wz^{-1}S^{-1}(y) - 1\right)B(z, w)$$

$$+z^2S^{-1}(y) - zwS^{-1}(y) = 0.$$

But $B(z, w) = zt$, and multiplying by $w^2S(y)$ gives

$$(wz)^2 + (wz)\left((wt)^2 - (wt)S(y) - w^2\right) + (tw)w^2 = 0.$$

Now we know that $S(y)$ is given parametrically in terms of w by $S(y) = (1-\theta)(1+2\theta)$, $w^2 = \theta(1-\theta)^3$, from [2.9.3(b)], so

$$(wz)^2 + (wz)\{(wt)^2 - wt(1-\theta)(1+2\theta) - \theta(1-\theta)^3\} + wt\theta(1-\theta)^3 = 0. \tag{2}$$

The discriminant D for this quadratic is given by

$$D = \{(wt)^2 - wt(1-\theta)(1+2\theta) - \theta(1-\theta)^3\}^2 - 4wt\theta(1-\theta)^3,$$

which can be simplified by letting $v = wt(1-\theta)^{-1}$. This gives

$$\begin{aligned} D &= (1-\theta)^4\{(v^2 - v(1+2\theta) - \theta(1-\theta))^2 - 4v\theta\} \\ &= (1-\theta)^4(v-\theta)^2(1-\theta+v)^2\{1 - 4v(1-\theta+v)^{-2}\} \end{aligned}$$

so

$$\begin{aligned} D^{1/2} &= (1-\theta)^3\big(wt(1-\theta)^{-1} - \theta\big)\big(1 + wt(1-\theta)^{-2}\big) \\ &\qquad \times\Big\{1 - 4wt(1-\theta)^{-3}\big(1 + wt(1-\theta)^{-2}\big)^{-2}\Big\}^{1/2} \\ &= \big(wt(1-\theta)^{-1} - \theta\big)\Big\{(1-\theta)^3\big(1 + wt(1-\theta)^{-2}\big) \\ &\qquad - 2\sum_{i\geqslant 1} c_{i-1} w^i t^i (1-\theta)^{-3(i-1)}\big(1 + wt(1-\theta)^{-2}\big)^{-(2i-1)}\Big\} \\ &= \big(wt(1-\theta)^{-1} - \theta\big)\Big\{(1-\theta)^3\big(1 + wt(1-\theta)^{-2}\big) \\ &\qquad - 2\sum_{i\geqslant 1} c_{i-1} \sum_{m\geqslant 0} (-1)^m \binom{m+2i-2}{m} w^{i+m} t^{i+m} (1-\theta)^{-(2m+3i-3)}\Big\} \end{aligned}$$

where $c_n = \dfrac{1}{n+1}\dbinom{2n}{n}$. But the solution of (2) is

$$2wz = -\{(wt)^2 - wt(1-\theta)(1+2\theta) - \theta(1-\theta)^3\} + D^{1/2}, \tag{3}$$

where the positive root has been selected to give positive solutions. The required number is $[t^n w^{n+2j}]A(t, w)$, which is given, from (1) and (3), by

$$-[t^n w^{n+2j}]tz = -[t^{n-1}w^{n+2j+1}]wz$$

$$= [w^{2(j+1)}]\Bigg\{ \sum_{i+m=n-2} (-1)^m c_{i-1}\binom{m+2i-2}{m}(1-\theta)^{-(2m+3i-2)}$$

$$- \sum_{i+m=n-1} (-1)^m c_{i-1}\binom{m+2i-2}{m}\theta(1-\theta)^{-(2m+3i-3)}\Bigg\},$$

for $n > 2$, where θ satisfies the functional equation $\theta = w^2(1-\theta)^{-3}$. Thus, by the Lagrange theorem,

$$[w^{2k}](1-\theta)^{-l} = \frac{1}{k}[\lambda^{k-1}]\left\{\frac{d}{d\lambda}(1-\lambda)^{-l}\right\}(1-\lambda)^{-3k} = \frac{l}{k}\binom{4k+l-1}{k-1},$$

so the solution is

$$\sum_{i+m=n-2} (-1)^m c_{i-1}\binom{m+2i-2}{m}\frac{2m+3i-2}{j+1}\binom{4j+3i+2m+1}{j}$$

$$- \sum_{i+m=n-1} (-1)^m c_{i-1}\binom{m+2i-2}{m}\frac{2m+3i}{j}\binom{4j+3i+2m-1}{j-1},$$

and the result follows.

(c) Consider an arbitrary strong near-triangulation M, where M is not the rooted triangle. Suppose that v is the vertex, not incident with the root edge of M, which lies on the nonroot face containing the root edge. Let $e_1, \ldots, e_k$, listed in the order in which they are encountered in the counterclockwise circulation of v, be the edges that are incident both with v and with vertices $v_1, \ldots, v_k$, respectively, which are in the outer face of M, where v_1 is the root vertex of M. Then the root edge of M is directed from v_1 to v_2, and $k \geqslant 2$. Suppose that we obtain the rooted map M' by deleting the root edge from M and letting the root edge of the resulting map be the edge from v to v_2. Then the generating function for all M' obtained in this way is $xy^{-1}(A(x, y) - x^3y)$, since the triangle, enumerated by x^3y, is disallowed and since M' has one fewer face than M but one more edge incident with its outer face.

We can also decompose M' uniquely into $k-1$ arbitrary strong triangulations $M_2, \ldots, M_k$, whose root edges are identified with $\overrightarrow{vv_2}, \ldots, \overrightarrow{vv_k}$, respectively, as illustrated here.

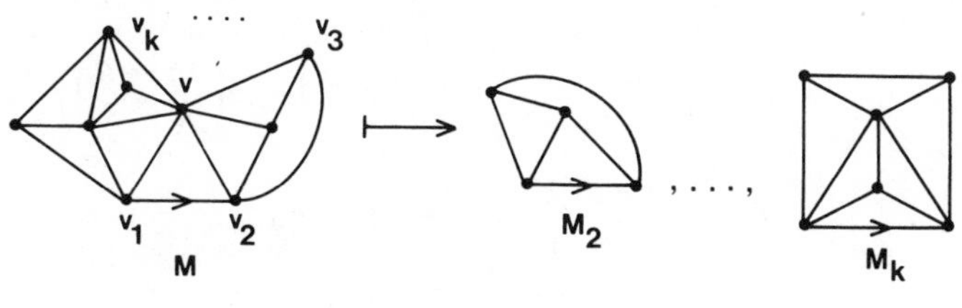

Now the sum of the number of edges in the outer faces of $M_2, \ldots, M_k$ is $2k - 4$ fewer than the number of edges in the outer face of M'. The sum of the number of inner faces in $M_2, \ldots, M_k$ is the same as the number of inner faces in M'. Finally, if $k = 2$, then $M_2 = M'$ cannot have three edges in the outer face. But $x^3y^{-1}T(y)$ is the generating function for such maps, so we obtain the alternative expression for the preceding generating function

$$\sum_{k \geqslant 2} x^{-(2k-4)} A^{k-1}(x, y) - x^3y^{-1}T(y)$$
$$= A(x, y)\{1 - x^{-2}A(x, y)\}^{-1} - x^3y^{-1}T(y).$$

This leads to the functional equation

$$xy^{-1}(A(x, y) - x^3y) + x^3y^{-1}T(y) = A(x, y)\{1 - x^{-2}A(x, y)\}^{-1},$$

and the result follows.

2.9.6. (a) We simply restrict Decomposition 2.9.12 to 2-edge-connected maps. The effect is to replace the set M of all rooted planar maps by the set C of 2-edge-connected rooted planar maps, yielding the decomposition

$$\mathsf{C} \stackrel{\sim}{\to} \mathsf{P} \circ (\mathsf{C}^2),$$

where P is the set of nonseparable rooted planar maps. This immediately leads to the functional equation $b(y) = P(yb^2(y))$.

(b) From **2.9.11** we have the parametric equations

$$\left.\begin{aligned} y &= \zeta(1 + \zeta)^{-4} \\ b(y) &= (1 - \zeta)(1 + \zeta)^2. \end{aligned}\right\}$$

Now let $t = yb^2(y)$ and express t in terms of the parameter ζ, yielding the functional equations

$$\left.\begin{aligned} \zeta &= t(1 - \zeta)^{-2} \\ P(t) = b(y) &= (1 - \zeta)(1 + \zeta)^2. \end{aligned}\right\}$$

Applying the Lagrange theorem to these equations, we obtain, for $n \geqslant 1$,

$$\begin{aligned} [t^n]P(t) &= \frac{1}{n}[\lambda^{n-1}]\left\{\frac{d}{d\lambda}(1 - \lambda)(1 + \lambda)^2\right\}(1 - \lambda)^{-2n} \\ &= \frac{1}{n}\left\{\binom{3n-2}{n-1} - 2\binom{3n-3}{n-2} - 3\binom{3n-4}{n-3}\right\} \\ &= \frac{2(3n-3)!}{n!(2n-1)!}, \qquad \text{as required.} \end{aligned}$$

2.9.7. (a) Let P be the set of nonseparable rooted planar maps. Let Q consist of those maps in P that yield a nonseparable planar map when the root edge is deleted. Thus Q contains l, but not v or δ. The only map in Q for which the outer face has degree 1 is l. Let Q_2 be the set of maps in Q with outer-face degree 2, so that the maps in Q with root-face degree at least 3 belong to $Q - \{l\} - Q_2$. If we apply the root-shift operator Q to these maps, we obtain uniquely the maps in Q for which the inner-face incident with the root edge has degree at least 3. The single map in Q for which this inner face has degree 1 is l. Thus $Q - \{l\} - Q_2 \tilde{\to} Q - \{l\} - F_2 : M \mapsto \Delta(M)$, where F_2 is the set of maps in Q for which this inner face has degree 2.

Suppose that M' is obtained by deleting the root edge of the rooted planar map M and choosing the root edge of M' to be the edge that follows the deleted edge in the clockwise circulation of the root vertex. Then

$$Q_2 \tilde{\to} \{\delta\} \times (P - \{v\} - \{l\}) : M \mapsto M'$$

and

$$F_2 \tilde{\to} \{\delta\} \times (P - \{v\} - \{l\}) : M \mapsto M'.$$

Now mark edges by y and edges in the root face by x, and the given equation follows from the three preceding decompositions, by eliminating the generating functions for Q_2 and F_2.

(b) If the root edge r of a map M in $P - Q - \{\delta\} - \{v\}$ is deleted, then the resulting map contains k cut-vertices, for some $k \geqslant 1$. These vertices must be incident with the outer face of M. Let $v_1, \ldots, v_k$ denote these vertices, listed in the order in which they are encountered as the outer face is traversed in a counterclockwise direction, beginning at the root edge r. Suppose further that r is directed from vertex v_{k+1} to v_0. Now in the inner face of M that is incident with r, add directed edges $r_1, \ldots, r_{k+1}$, with r_i directed from v_i to v_{i-1} for $i = 1, \ldots, k+1$, and delete the root edge r. Then the resulting object consists of the rooted planar maps $M_1, \ldots, M_{k+1}$, where M_i has root edge r_i, for $i = 1, \ldots, k+1$. These maps are joined by identifying the root vertex of M_i with the nonroot vertex incident to the root edge of M_{i+1}, resulting in vertex v_i, for $i = 1, \ldots, k$. This procedure is illustrated here.

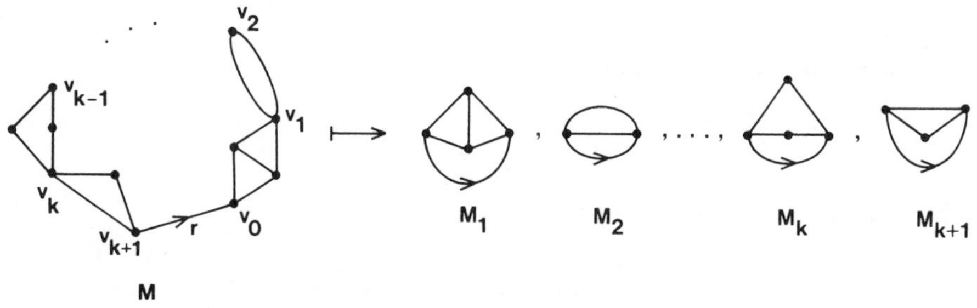

Now M_i is a unique element of $\mathsf{Q} - \{l\}$, and this procedure is reversible, so

$$\mathsf{P} - \mathsf{Q} - \{\delta\} - \{v\} \tilde{\to} (\mathsf{Q} - \{l\})^2(\mathsf{Q} - \{l\})^*.$$

In this decomposition the number of edges in M is k less than the sum of the numbers of edges in $M_1, \ldots, M_{k+1}$. Similarly, the number of edges in the outer face of M is k less than the sum of the numbers of edges in the outer faces of $M_1, \ldots, M_{k+1}$. Thus, if edges are marked by y and edges in the outer face by x, this decomposition leads to the functional equation

$$P(x, y) - Q(x, y) - x^2y - 1 = \sum_{k \geqslant 1} (xy)^{-k}(Q(x, y) - xy)^{k+1},$$

and the result follows.

(c) Eliminating $Q(x, y)$ between the equations in (a) and (b) gives

$$(P(x, y) - x^2y - 1)(P(x, y) - xP(1, y)) = xy(1 - x).$$

We now proceed with the quadratic method (**2.9.1(1)**) to obtain $P(1, y)$. First the square is completed in the preceding quadratic, giving

$$(2P(x, y) - xh - x^2y - 1)^2 = 4xy(1 - x) + (xh - x^2y - 1)^2 = D(x, y),$$

where $h = P(1, y)$. The substitution $x = \alpha(y)$ is now made such that

$$(D/x^2)|_{x=\alpha} = (\partial/\partial x)(D/x^2)|_{x=\alpha} = 0,$$

giving the equations

$$4y(\alpha^{-1} - 1) + (h - \alpha y - \alpha^{-1})^2 = 0 \tag{1}$$

$$-4y\alpha^{-2} + 2(h - \alpha y - \alpha^{-1})(\alpha^{-2} - y) = 0. \tag{2}$$

Now from (2) we have

$$h - \alpha y - \alpha^{-1} = 2y(1 - \alpha^2 y)^{-1}, \tag{3}$$

and substituting this to eliminate h in (1) yields

$$y = (1 - \alpha^{-1})(1 - \alpha^2 y)^2.$$

This quadratic factorizes to give

$$\alpha^4(1 - \alpha)\{y + \alpha^{-1}(1 - \alpha)^{-1}\}\{y + \alpha^{-3}(1 - \alpha)\} = 0$$

and the root $y = \alpha^{-3}(\alpha - 1)$ is chosen, since $\alpha(0) = 1$ from (3). Finally, eliminate y from (3), yielding $h = \alpha^{-2}(4\alpha - 3)$, and let $\theta = 1 - \alpha^{-1}$, to give the simultaneous equations

$$\left.\begin{aligned} y &= \theta(1-\theta)^2 \\ h &= (1-\theta)(1+3\theta). \end{aligned}\right\}$$

But this is precisely the system of parametric equations that was solved in **2.9.13** to give the number of nonseparable rooted planar maps on n edges as

$$[y^n]h = \frac{2(3n-3)!}{n!(2n-1)!}, \qquad n \geqslant 1.$$

(d) We now apply **2.9.1(2)** to obtain $P(x, y)$. First express D in terms of x and θ, giving

$$\begin{aligned} D &= 4x(1-x)\theta(1-\theta)^2 + \big(x(1-\theta)(1+3\theta) - x^2\theta(1-\theta)^2 - 1\big)^2 \\ &= (1-(1-\theta)x)^2\{(1-\theta(1-\theta)x)^2 - 4\theta^2(1-\theta)x\}, \end{aligned}$$

so

$$\begin{aligned} D^{1/2} &= (1-(1-\theta)x)(1-\theta(1-\theta)x) \\ &\quad \times \left\{1 - 2\sum_{i\geqslant 1} \frac{(2i-2)!}{i!(i-1)!} x^i(1-\theta)^i\theta^{2i}(1-\theta(1-\theta)x)^{-2i}\right\} \\ &= (1-(1-\theta)x) \\ &\quad \times \left\{1 - \theta(1-\theta)x - 2\sum_{i\geqslant 1}\sum_{l\geqslant 0} \frac{(2i+l-2)!}{i!(i-1)!l!} x^{i+l}(1-\theta)^{i+l}\theta^{2i+l}\right\}. \end{aligned}$$

The quadratic for $P(x, y)$ in part (c) can now be solved, giving

$$\begin{aligned} P(x, y) &= \tfrac{1}{2}\{1 + x(1+3\theta)(1-\theta) + x^2y \pm D^{1/2}\} \\ &= 1 + xy + x^2y - \sum_{n\geqslant 2}\sum_{j=n+1}^{2n} \frac{(j-2)!x^n(1-\theta)^n\theta^j}{(j-n)!(j-n-1)!(2n-j)!} \\ &\quad + \sum_{n\geqslant 2}\sum_{j=n}^{2n-2} \frac{(j-2)!x^n(1-\theta)^n\theta^j}{(j-n+1)!(j-n)!(2n-j-2)!}, \end{aligned}$$

where the positive root has been chosen, and the transformations $n = i + l$, $j = 2i + l$ and $n = i + l + 1$, $j = 2i + l$ have been made in the summations.

Thus the required number is, for $n, m \geqslant 2$,

$$[x^n y^m]P(x, y) = [y^m] \sum_{j=n}^{2n} \frac{(j-2)!n(3n-2j-1)}{(j-n+1)!(j-n)!(2n-j)!} y^j (1-\theta)^{n-2j},$$

where θ satisfies the functional equation $\theta = y(1-\theta)^{-2}$. But by the Lagrange theorem we obtain

$$[y^m] y^j (1-\theta)^{n-2j} = [y^{m-j}](1-\theta)^{n-2j} = \frac{(2j-n)(3m-n-j-1)!}{(m-j)!(2m-n)!}$$

for $m \geqslant j$, and the solution follows.

2.9.8. (a) Let $yzQ(x, y, z)$ be the ordinary generating function with respect to edges in the outer face, faces, and vertices for those nonseparable rooted planar maps that, when the root edge is deleted, yield a nonseparable planar map. Then from the decomposition given in the solutions to [**2.9.7**(a), (b)],

$$\begin{aligned} &yzQ(x, y, z) - xy^2z - x^2(yzP(1, y, z) - yz - y^2z) \\ &\quad = x\{yzQ(x, y, z) - xy^2z - y(yzP(x, y, z) - yz - xy^2z)\} \end{aligned}$$

and

$$yzP(x, y, z) - x^2yz^2 - yz = xy^2z\{1 - (xy^2z)^{-1}(yzQ(x, y, z) - xy^2z)\}^{-1}.$$

The result follows by eliminating $Q(x, y, z)$.

(b) We now use the quadratic method (**2.9.1(1)**) to obtain $P(1, y, z)$. First the square is completed to give

$$\begin{aligned} (2P(x, y, z) - xh - x^2z - 1)^2 &= 4xy(1-x) + (xh - x^2z - 1)^2 \\ &= D(x, y, z), \end{aligned}$$

where $h = P(1, y, z)$. As in the solution to [**2.9.7**(c)], we make the substitution $x = \alpha(y, z)$ such that

$$\left(\frac{D}{x^2}\right)\bigg|_{x=\alpha} = \frac{\partial}{\partial x}\left(\frac{D}{x^2}\right)\bigg|_{x=\alpha} = 0,$$

giving the equations

$$4y(\alpha^{-1} - 1) + (h - \alpha z - \alpha^{-1})^2 = 0 \tag{1}$$

$$-4y\alpha^{-2} + 2(h - \alpha z - \alpha^{-1})(\alpha^{-2} - z) = 0. \tag{2}$$

Eliminating h yields

$$y = (1 - \alpha^{-1})(1 - \alpha^2 z)^2, \tag{3}$$

and eliminating y from (2) and (3) gives

$$h = \alpha z + \alpha^{-1} + 2(1 - \alpha^{-1})(1 - \alpha^2 z). \tag{4}$$

Now let $\theta = 1 - \alpha^{-1}$ in (3) and (4), yielding the following simultaneous functional equations for $P(1, y, z) = h$,

$$\left.\begin{aligned} \theta &= y\left(1 - z(1-\theta)^{-2}\right)^{-2} \\ P(1, y, z) &= 1 - \theta + z(1-\theta)^{-1} + 2y\left(1 - z(1-\theta)^{-2}\right)^{-1}. \end{aligned}\right\}$$

By considering $P(1, y, z)$ as a power series in y, with power series in z as coefficients, we use the Lagrange theorem to obtain the required number, which is $[y^{n+1}z^{m+1}]yzP(1, y, z)$. Thus $[y^n z^m]P(1, y, z)$

$$\begin{aligned} &= \frac{1}{n}[\lambda^{n-1}z^m]\left\{\frac{\partial}{\partial\lambda}\left(1 - \lambda + z(1-\lambda)^{-1}\right)\right\}\left(1 - z(1-\lambda)^{-2}\right)^{-2n} \\ &\qquad + \frac{2}{n-1}[\lambda^{n-2}z^m]\left\{\frac{\partial}{\partial\lambda}\left(1 - z(1-\lambda)^{-2}\right)^{-1}\right\}\left(1 - z(1-\lambda)^{-2}\right)^{-2(n-1)} \\ &= -\frac{1}{n}[\lambda^{n-1}z^m]\left(1 - z(1-\lambda)^{-2}\right)^{-(2n-1)} \\ &\qquad + \frac{4}{n-1}[\lambda^{n-2}z^{m-1}](1-\lambda)^{-3}\left(1 - z(1-\lambda)^{-2}\right)^{-2n} \\ &= \frac{(2n+m-2)!(2m+n-2)!}{n!m!(2n-1)!(2m-1)!}, \qquad \text{for } n \geqslant 2. \end{aligned}$$

Also $[yz^m]P(1, y, z) = -[z^m](1 - z)^{-1} + 2 = 1$, which agrees with the preceding formula.

2.9.9. (a) If we let $x = 1$ in [**2.9.7**(b)], then

$$P(y) - y - 1 = y\left(1 - y^{-1}(Q(y) - y)\right)^{-1}$$

and the result follows immediately.

(b) We define the **planar dual** $\tilde{M}$ of a rooted planar map M in the following way. The vertices of $\tilde{M}$ are identified with the faces of M such that an edge is incident with a pair of vertices in $\tilde{M}$ when an edge is incident with the corresponding faces in M. The root edge of $\tilde{M}$ is directed from the vertex corresponding to the face on the left of the root edge of M to the vertex corresponding to the face on the right. The dual of the vertex map is itself. Thus M and $\tilde{M}$ have the same number of edges, and the dual of $\tilde{M}$ is simply M with the root edge directed in the reverse direction. An example of such a pair of maps is illustrated here.

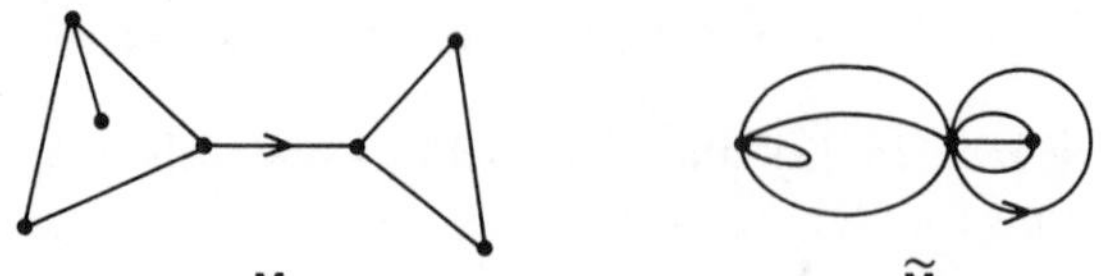

M $\tilde{M}$

Now the maps that are enumerated by $Q(y)$ are simply the planar duals of those enumerated by $D(y)$, and since M and $\tilde{M}$ have the same number of edges, we immediately obtain $D(y) = Q(y)$.

An alternative derivation is given by the following. We construct all maps enumerated by $P(y)$ except v and l uniquely. Consider the rooted map M with two vertices and $k+1$ edges between the vertices, for some $k \geqslant 0$. Let $e_1, \ldots, e_k$ denote the nonroot edges of M, directed from the root vertex of M to the other vertex. Now take any map M_i except δ in the set enumerated by $D(y)$. Identify the root edge of M_i with e_i and remove e_i. Repeat this procedure for $i = 1, \ldots, k$ to obtain the nonseparable map M', which has $k-1$ fewer edges than the total number of edges in $M_1, \ldots, M_k$. This procedure is reversible and is illustrated for $k = 2$.

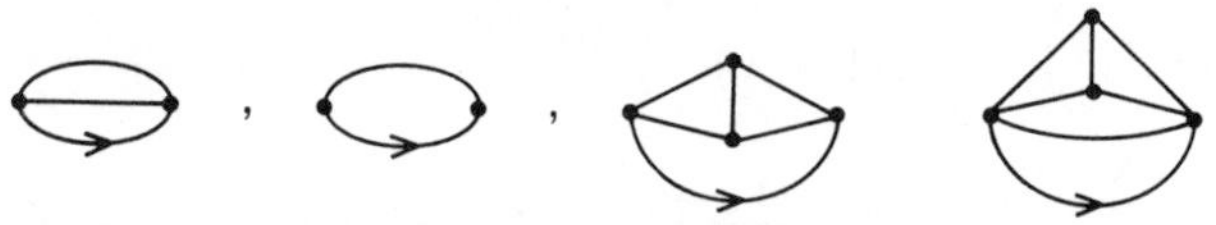

M M$_1$ M$_2$ M′

Thus

$$P(y) - y - 1 = \sum_{k \geqslant 0} y^{1-k}(D(y) - y)^k = y\{1 - y^{-1}(D(y) - y)\}^{-1}.$$

But from part (a) we have

$$P(y) - y - 1 = y\{1 - y^{-1}(Q(y) - y)\}^{-1}$$

and comparison yields $D(y) = Q(y)$.

(c) Consider an arbitrary rooted c-net M. For each nonroot edge in M we identify the edge, directed in a unique way, with the root edge of any nonseparable rooted planar map except v, l, and δ, and then we delete the edge. The generating function for maps constructed in this way is

$y(y^{-1}F(y))^{-1}E(y^{-1}F(y))$, where $F(y) = P(y) - 2y - 1$. The maps that we construct uniquely in this way are the nonseparable rooted planar maps except those in three classes.

Case 1. The first class consists of the nonseparable maps with less than three edges. These are v, l, δ and the map consisting of a pair of edges between two vertices, and thus this class is enumerated by $1 + 2y + y^2$.

Case 2. The second class consists of those such that, when the root edge is deleted, the resulting map contains at least one cut-vertex. This class is enumerated by $P(y) - Q(y) - y - 1$, from part (a).

Case 3. The third class consists of those that, when the root edge is deleted, yield a map for which the pair of vertices incident with the root edge is a 2-separator. This class is enumerated by $P(y) - D(y) - y - 1$, from part (b).

Thus we have the functional equation $y(y^{-1}F(y))^{-1}E(y^{-1}F(y))$

$$= P(y) - \{(1 + 2y + y^2) + (P(y) - Q(y) - y - 1)$$
$$+ (P(y) - D(y) - y - 1)\}$$
$$= Q(y) + D(y) - F(y) - y^2 - 2y.$$

Now we eliminate $Q(y)$, $D(y)$ by using the results of parts (a) and (b). This gives

$$y(y^{-1}F(y))^{-1}E(y^{-1}F(y)) = 2y - y^2 - 2y^2(y + F(y))^{-1} - F(y),$$

which yields the required equation.

(d) The required number is $[z^n]E(z)$. Making the change of variable $z = y^{-1}F(y)$ in the equation of part (c) gives

$$E(z) = z^2 - 2z^3(1 + z)^{-1} - zy,$$

so

$$[z^n]E(z) = 2(-1)^n - [z^{n-1}]y, \qquad \text{for } n \geqslant 4.$$

But from **2.9.13** we have the parametric equations $P(y) = (1 - \theta)(1 + 3\theta)$, $y = \theta(1 - \theta)^2$, so

$$z = y^{-1}(P(y) - 2y - 1) = \theta(1 - 2\theta)(1 - \theta)^{-2}.$$

Setting $\alpha = (1 - \theta)^{-1}$ gives

$$\left.\begin{aligned} \alpha &= z(1 - \alpha)^{-1} \\ y &= \alpha(1 + \alpha)^{-3}. \end{aligned}\right\}$$

so, by the Lagrange theorem, the required number is $[z^n]E(z)$

$$= 2(-1)^n - \frac{1}{n-1}[\lambda^{n-2}]\left\{\frac{d}{d\lambda}\lambda(1+\lambda)^{-3}\right\}(1-\lambda)^{-(n-1)}$$

$$= 2(-1)^n - \frac{1}{n-1}[\lambda^{n-2}](1+\lambda)^{-4}\{(1-\lambda)^{-(n-2)} - \lambda(1-\lambda)^{-(n-1)}\}$$

$$= 2(-1)^n - \frac{1}{n-1}\sum_{i\geq 0}\binom{-4}{i}\left\{\binom{2n-i-5}{n-3} - \binom{2n-i-5}{n-2}\right\},$$

and the result follows by letting $j = i - 1$.

SECTION 3.3

3.3.1. Each way of writing the summation induces an ordered partition on N_n into k nonempty blocks, and thus lies in the set $\mathsf{P} \circledast \mathsf{U}_+$, where the tagged objects are the indices $1, \ldots, n$. The required number is thus

$$\left[\frac{x^n}{n!}\right](1-x)^{-1} \circ (e^x - 1) = \left[\frac{x^n}{n!}\right](2 - e^x)^{-1}.$$

3.3.2. (a) An appropriate decomposition is for $\alpha_i \subseteq \mathsf{N}_m$ to be the preimage of $i \in \mathsf{N}_n$ for $i = 1, \ldots, n$.

(b) From part (a), if x marks tagged objects, then the number of functions from N_m to N_n is

$$\left[\frac{x^m}{m!}\right](e^x)^n = n^m.$$

(c) A bijection is a function in which the preimage of each element of the range has cardinality 1. Thus, from part (a), the required number is

$$\left[\frac{x^m}{m!}\right](x)^n = n!\delta_{n,m}.$$

(d) An injection is a function in which the preimage of each element of the range has cardinality at most 1. Thus, from part (a), the required number is

$$\left[\frac{x^m}{m!}\right](1+x)^n = m!\binom{n}{m}.$$

(e) A surjection is a function in which the preimage of each element of the range has cardinality at least 1. Thus, from part (a), the required number is

$$\left[\frac{x^m}{m!}\right]\left(\sum_{i\geq 1}\frac{x^i}{i!}\right)^n = \left[\frac{x^m}{m!}\right](e^x - 1)^n = \sum_{k=0}^{n}(-1)^{n-k}\binom{n}{k}k^m.$$

(f) The preimages of $1, \ldots, r$ must have cardinality at least 1, and those of $r+1, \ldots, n$ can be of any size. Thus, by part (a), the required number is

$$\left[\frac{x^m}{m!}\right](e^x - 1)^r (e^x)^{n-r} = \sum_{k=0}^{r} (-1)^k \binom{r}{k} (n-k)^m.$$

(g) If k is the highest value such that $f(i) = k$, for some i in the domain, then such functions are enumerated by $[x^n/n!](e^x - 1)^k$ from part (e). But k can take on any value, so the required number is

$$\left[\frac{x^n}{n!}\right] \sum_{k \geqslant 0} (e^x - 1)^k = \left[\frac{x^n}{n!}\right](2 - e^x)^{-1}.$$

3.3.3. (a) If S is the set of surjections, then

$$\mathsf{S} \tilde{\rightarrow} \mathsf{U} \circledast \left(\dot{\bigcup_{k \geqslant 1}} (\mathsf{N}_1, \mathsf{N}_k) \right) : f \mapsto \{(1, \alpha_1), \ldots, (n, \alpha_n)\},$$

where α_i is the preimage of i for $i = 1, \ldots, n$, and $\alpha_1 \dot{\cup} \cdots \dot{\cup} \alpha_n = \mathsf{N}_m$ for some $m \geqslant n$. Further, the number of tagged s-objects in f is m, and the number of tagged t-objects in f is n, which are additively preserved in the decomposition. The exponential generating function for $(\mathsf{N}_1, \mathsf{N}_k)$ is $yx^k/k!$, and for $\dot{\bigcup}_{k \geqslant 1} (\mathsf{N}_1, \mathsf{N}_k)$ is thus $\sum_{k \geqslant 1} yx^k/k! = y(e^x - 1)$. Thus the required generating function is $S(x, y) = \exp\{y(e^x - 1)\}$, and the number of surjections from N_m to N_n is $[x^m y^n / m! n!] S(x, y)$.

(b) Distinguish each tagged s-object in turn, for each element of S. The resulting set, $(sd/ds)\mathsf{S}$ is enumerated by $x(\partial S/\partial x)$ (Remark 3.3.18) and may be constructed in another way as follows. If the distinguished s-object, say k, is the only element in the preimage of some t-object l, then the rest of the function is a surjection, and this set is accordingly $(\mathsf{N}_1, \mathsf{N}_1) * \mathsf{S}$, enumerated by xyS. Otherwise, we can construct the function by adding an element, which becomes the distinguished element, to the preimage of a distinguished t-object. This set is $(\varnothing, \mathsf{N}_1) * (td/dt)\mathsf{S}$ and is enumerated by $xy\partial S/\partial y$. Thus we obtain $x\partial S/\partial x = xyS + xy\partial S/\partial y$, whence $\partial S/\partial x = yS + y\partial S/\partial y$.

3.3.4. A graph in G consists uniquely of an unordered collection of connected components, so that $\mathsf{G} \tilde{\rightarrow} \mathsf{U} \circledast \mathsf{B}$ is tag-weight preserving. Thus we immediately have $G(x) = \exp B(x)$.

3.3.5. (a) A covering of the complete graph with complete graphs is a graph whose components are complete graphs. There is exactly one complete graph on k labeled vertices, for $k \geqslant 1$, so by **[3.3.4]** the required number is

$$\left[\frac{x^n}{n!}\right] \exp \sum_{k \geqslant 1} \frac{x^k}{k!} = \left[\frac{x^n}{n!}\right] \exp(e^x - 1).$$

(b) A tag-weight preserving decomposition for complete bipartite graphs whose labeled vertices are tagged objects is $\mathsf{N}_2 \circledast \mathsf{U}_+$, so that the exponential generating function is $\frac{1}{2}(e^x - 1)^2$. Thus the required number is

$$\left[\frac{x^n}{n!}\right]\exp\{\tfrac{1}{2}(e^x - 1)^2\}.$$

3.3.6. (a) The number of paths that can be constructed on k labeled vertices is $\frac{1}{2}k!$ for $k \geqslant 2$, and 1 for $k = 1$, since the paths are undirected. Thus the exponential generating function for paths is

$$x + \tfrac{1}{2}\sum_{k \geqslant 2} \frac{k!x^k}{k!} = \tfrac{1}{2}(1 - x)^{-1}(2x - x^2),$$

and the required number is

$$\left[\frac{x^n}{n!}\right]\exp\left\{\frac{x}{2}\frac{(2 - x)}{(1 - x)}\right\}.$$

(b) The number of undirected cycles that can be constructed on k labeled vertices is $\frac{1}{2}(k - 1)!$ for $k \geqslant 3$, and zero for $k < 3$, since the complete graph is simple. Thus the exponential generating function for cycles is

$$\sum_{k \geqslant 3} \tfrac{1}{2}(k - 1)!\frac{x^k}{k!} = \tfrac{1}{2}\log(1 - x)^{-1} - \frac{x}{2} - \frac{x^2}{4},$$

and the required number is

$$\left[\frac{x^n}{n!}\right]\exp\left\{\tfrac{1}{2}\log(1 - x)^{-1} - \frac{x}{2} - \frac{x^2}{4}\right\} = \left[\frac{x^n}{n!}\right](1 - x)^{-1/2}\exp\left(-\frac{x}{2} - \frac{x^2}{4}\right).$$

3.3.7. There are k star graphs that can be formed on k labeled vertices, for $k \neq 2$, and a single star graph on two labeled vertices. Thus the exponential generating function for star graphs is

$$x + \frac{x^2}{2!} + \sum_{k \geqslant 3} k\frac{x^k}{k!} = x\left(e^x - \frac{x}{2}\right),$$

so that the required number is

$$\left[\frac{x^n}{n!}\right]\exp\left\{x\left(e^x - \frac{x}{2}\right)\right\}.$$

3.3.8. Let tagged s-objects be the labeled vertices in the n-set and tagged t-objects be the labeled vertices in the m-set. Then the generating function for complete graphs between tagged s-objects, marked by x and t-objects, marked by y, is

$$\sum_{i\geqslant 1}\sum_{j\geqslant 1}\frac{x^i}{i!}\frac{y^j}{j!} = (e^x - 1)(e^y - 1),$$

since there is a single complete bipartite graph on a set of i vertices and a set of j vertices. Thus the required number is

$$\left[\frac{x^n}{n!}\frac{y^m}{m!}\right]\exp\{(e^x - 1)(e^y - 1)\}$$

since a covering of $K_{n,m}$ with complete bipartite graphs is a graph whose connected components consist of complete bipartite graphs.

3.3.9. (a) A generalized weak order must correspond to an ordered partition $(b_1,\ldots, b_k)$ of N_n for some $k \geqslant 1$, in which either both (i, j) and (j, i) are in R or neither are in R for $i, j \in b_l$ and $l = 1,\ldots, k$, and, moreover, $(i, j) \in R$ and $(j, i) \notin R$, for all $i \in b_l, j \in b_m$ with $1 \leqslant l < m \leqslant k$. Transitivity in *caR* is easily seen to be violated by any nonsymmetry internal to a block and any missing relation from a lower block to a higher block or symmetry between blocks. Thus the set is decomposed as $\mathsf{P}\circledast\mathsf{S}$, where S is a symmetric relation on a set of (tagged) objects. There are $2^{\binom{i+1}{2}}$ symmetric relations on a set of i objects, since, for each of the $\binom{i}{2}$ distinct pairs (a, b) of elements either both $(a, b) \in R$ and $(b, a) \in R$, or both $(a, b) \notin R$ and $(b, a) \notin R$. Moreover, for each of the i elements a, either $(a, a) \in R$ or $(a, a) \notin R$. Thus in each of $\binom{i}{2} + i = \binom{i+1}{2}$ cases, we have two choices, so the required number is

$$\left[\frac{x^n}{n!}\right](1 - x)^{-1}\circ\left\{\sum_{i\geqslant 1} 2^{\binom{i+1}{2}}\frac{x^i}{i!}\right\},$$

and the result follows.

(b) If the generalized weak order is transitive, then each of the blocks must have a transitive, symmetric relation. Thus these relations decompose as $\mathsf{P}\circledast\mathsf{A}$, where A is the set of arbitrary transitive, symmetric relations on a set of i (tagged) objects, for all $i \geqslant 1$. But A must consist of a (possibly empty) set of elements that are related to no element in A, and a set of blocks, each nonempty, which form equivalence relations, so

$$\mathsf{A} \mathrel{\tilde{\to}} \mathsf{U}*\{\mathsf{U}\circledast\mathsf{U}_+\} - \{\varepsilon\}$$

and is enumerated by $e^x\exp(e^x - 1) - 1$. Thus the required number is

$$\left[\frac{x^n}{n!}\right](1 - x)^{-1}\circ\{e^x\exp(e^x - 1) - 1\},$$

and the result follows.

(c) If the generalized weak order is asymmetric, then each of the blocks in the ordered partition must have the empty relation. Accordingly, asymmetric generalized weak orders correspond to ordered partitions and are enumerated by

$$\left[\frac{x^n}{n!}\right](1-x)^{-1}\circ(e^x-1)=\left[\frac{x^n}{n!}\right](2-e^x)^{-1}.$$

3.3.10. (a) For a regular chain $c \in \mathsf{A}$ on N_n, let the tagged objects be the elements of N_n. Then if $n > 1$, c is a set of cardinality a, whose elements are themselves regular chains on disjoint subsets of N_n. There is a single chain c_1 on N_1, so we have the recursive tag-weight preserving decomposition

$$\mathsf{A} - \{c_1\} \stackrel{\sim}{\rightarrow} \mathsf{N}_a \circledast \mathsf{A},$$

so

$$C(t) - t = \frac{1}{a!}(C(t))^a.$$

(b) From part (a),

$$C(t) = t\left\{1 - \frac{C(t)^{a-1}}{a!}\right\}^{-1},$$

so by the Lagrange theorem,

$$c_n = \left[\frac{t^n}{n!}\right]C(t) = n!\frac{1}{n}[\lambda^{n-1}]\left(1 - \frac{\lambda^{a-1}}{a!}\right)^{-n},$$

and the result follows.

3.3.11. Let $P(x, t_1, t_2, \ldots)$ be the generating function for P in which the tagged objects are marked by x, and the (untagged) cycles of length j are marked by t_j, for $j \geqslant 1$. Then from the cycle decomposition **(3.3.5)** for permutations

$$P(x, t_1, t_2, \ldots) = \exp\left\{\sum_{j\geqslant 1}(j-1)!t_j\frac{x^j}{j!}\right\} = \exp\left\{t_1 x + t_2\frac{x^2}{2} + \cdots\right\}$$

since there are $(j-1)!$ circular permutations on N_j, for $j \geqslant 1$, and each such circular permutation represents a cycle of length j in the element of P. Thus the number of permutations on N_n with i_j cycles of length j, for $j \geqslant 1$, is

$$\left[\frac{x^n}{n!}t_1^{i_1}t_2^{i_2}\cdots\right]P(x, t_1, t_2, \ldots) = n!\left\{\prod_{j\geqslant 1} i_j!j^{i_j}\right\}^{-1},$$

in which, of course, $n = i_1 + 2i_2 + \cdots$.

P is the generating function for the **cycle index polynomials** of the symmetric groups.

3.3.12. (a) From [**3.3.11**], the required generating function is $P(x, t_1, t_2, \dots)$ in which $t_j = 1$ if $d|j$, and $t_j = 0$ otherwise. Thus the required number is

$$\left[\frac{x^n}{n!}\right]\exp\sum_{j\geq 1}\frac{x^{jd}}{jd} = \left[\frac{x^n}{n!}\right]\exp\left\{\frac{1}{d}\log(1-x^d)^{-1}\right\}$$

$$= \left[\frac{x^n}{n!}\right](1-x^d)^{-1/d}$$

$$= \begin{cases} (md)!\binom{-1/d}{m}(-1)^m, \text{ if } n = md, \\ 0 \quad \text{otherwise,} \end{cases}$$

and the result follows.

(b) From [**3.3.11**], the required generating function is $P(x, t_1, t_2, \dots)$ in which $t_j = 0$ if $d|j$, and $t_j = 1$ otherwise. Thus the required number is

$$\left[\frac{x^n}{n!}\right]\exp\left\{\log(1-x)^{-1} - \frac{1}{d}\log(1-x^d)^{-1}\right\}$$

$$= \left[\frac{x^n}{n!}\right](1-x)^{-1}(1-x^d)^{1/d} = n!\sum_{i=0}^{\lfloor n/d\rfloor}\binom{1/d}{i}(-1)^i.$$

3.3.13. If u marks odd cycles and v marks even cycles, then the required number, from [**3.3.11**], is $[u^k v^j \frac{x^n}{n!}]P(x, u, v, u, v, \dots)$. But

$$P(x, u, v, u, v, \dots) = \exp\left\{u\sum_{i \text{ odd}}\frac{x^i}{i} + v\sum_{i \text{ even}}\frac{x^i}{i}\right\}$$

$$= \exp\left\{\frac{u}{2}\left(\log(1-x)^{-1} - \log(1+x)^{-1}\right)\right.$$

$$\left. + \frac{v}{2}\left(\log(1-x)^{-1} + \log(1+x)^{-1}\right)\right\},$$

by series bisection, and the result follows.

3.3.14. The permutations σ such that σ^m is the identity are those in which the length of each cycle divides m. Thus, by the cycle decomposition, the required number is obtained by setting $t_d = 1$ if $d|m$, and $t_d = 0$ otherwise. This gives

$$\left[\frac{x^n}{n!}\right]\exp\sum_{d|m}\frac{x^d}{d}$$

as required.

3.3.15. If $\sigma^2 = \rho$, then the cycles of ρ are obtained by considering every second element of cycles of σ. Thus a cycle of length $2k+1$ in σ gives rise to a cycle of length $2k+1$ in ρ, and a cycle of length $2k$ in σ gives rise to two disjoint cycles, each of length k, in ρ. Accordingly, ρ must have an even number of cycles of length $2m$, for each $m \geqslant 1$. Furthermore, this condition is also sufficient for at least one σ to exist such that $\sigma^2 = \rho$. Thus the number of square permutations on N_n is given by

$$\sum_{\substack{m_i \geqslant 0 \\ i \geqslant 1}} \left[\frac{x^n}{n!} t_2^{2m_1} t_4^{2m_2} \cdots \right] P(x, 1, t_2, 1, t_4, \ldots)$$

$$= \left[\frac{x^n}{n!}\right] \exp\left(\sum_{k \geqslant 0} \frac{x^{2k+1}}{2k+1}\right) \prod_{i \geqslant 1} \sum_{l \geqslant 0} \left(\frac{x^{2i}}{2i}\right)^{2l} \frac{1}{(2l)!},$$

and the result follows.

3.3.16. (a) A typical term in the expansion of $|\mathbf{A}|$ contains the monomial $a_{1\sigma_1} a_{2\sigma_2} \cdots a_{n\sigma_n}$, where $\sigma \in \mathsf{P}$. If

$$\{i_{11} i_{12} \cdots i_{1j_1}, i_{21} i_{22} \cdots i_{2j_2}, \ldots, i_{k1} i_{k2} \cdots i_{kj_k}\} = \{c_1, \ldots, c_k\}$$

are the cycles of σ, then the monomial can be written as $\prod_{l=1}^{k} a_{i_{l1} i_{l2}} \cdots a_{i_{lj_l} i_{l1}}$. But this monomial is produced by any permutation $\rho \in \mathsf{P}$, each of whose cycles is either in $\{c_1, \ldots, c_k\}$ or in $\{\bar{c}_1, \ldots, \bar{c}_k\}$, where $\bar{c}_i$ is c_i traversed in the reverse direction. It is easy to see that this monomial will be produced only by such permutations ρ, and thus

$$b_n = \left[\frac{x^n}{n!}\right] \exp\left\{x + \frac{x^2}{2} + \frac{1}{2} \sum_{i \geqslant 3} \frac{x^i}{i}\right\} = \left[\frac{x^n}{n!}\right] (1-x)^{-1/2} \exp\left(\frac{x}{2} + \frac{x^2}{4}\right),$$

since there are only $\frac{1}{2}(i-1)!$ different circular permutations on N_i when reverse order is indistinguishable, for $i \geqslant 3$. There are still single circular permutations on N_1 and N_2 when order can be reversed.

(b) If $B(x) = \sum_{n \geqslant 0} b_n \dfrac{x^n}{n!}$, then from part (a)

$$(1-x) B'(x) = \left(1 - \tfrac{1}{2}x^2\right) B(x),$$

so comparing coefficients of $x^n/n!$, for $n \geqslant 0$,

$$b_{n+1} - n b_n = b_n - \tfrac{1}{2} n(n-1) b_{n-2}.$$

Rearranging, we obtain $b_{n+1} = (n+1) b_n - \binom{n}{2} b_{n-2}$, $n \geqslant 0$, $b_0 = 1$, $b_i = 0$ for $i < 0$.

3.3.17. (a) If u marks cycles, then the required number, from [**3.3.11**], is

$$\left[u^k\frac{x^n}{n!}\right]P(x,u,u,\ldots) = \left[u^k\frac{x^n}{n!}\right]\exp u\log(1-x)^{-1},$$

and the result follows.

(b) Suppose that in a permutation $\sigma = \sigma_1 \cdots \sigma_n$ on N_n, there are k upper records, namely, $1 = i_1 < i_2 < \cdots < i_k \leqslant n$. Then the sequences $\sigma_1 \cdots \sigma_{i_2-1}, \sigma_{i_2} \cdots \sigma_{i_3-1}, \ldots, \sigma_{i_k} \cdots \sigma_n$ are each in $\hat{\mathsf{C}}$, where C is the set of circular permutations, and the leading element in each sequence is the largest. Moreover, the leading elements $\sigma_1 < \sigma_{i_2} < \cdots < \sigma_{i_k}$ are in strictly increasing order. This procedure is reversible, and so

$$\mathsf{P} \xrightarrow{\sim} \mathsf{U} \circledast \mathsf{C} : \sigma \mapsto \{c_1, c_2, \ldots, c_k\}$$

is tag-weight preserving. Furthermore, in this decomposition σ is the permutation $c_1c_2 \cdots c_k$, where the largest elements of $c_1,\ldots,c_k$ are in increasing order. Moreover, they appear as the leading elements in $c_1,\ldots,c_k$ and are the k upper records of σ. Thus, if x marks the tagged objects $\sigma_1,\ldots,\sigma_n$ and u marks upper records, the number of permutations on N_n with k upper records is, by the $*$-composition lemma,

$$\left[u^k\frac{x^n}{n!}\right]\exp\left\{u\sum_{i\geqslant 1}\frac{x^i}{i}\right\},$$

as in part (a). This decomposition is the **Foata–Schützenberger decomposition.**

(c) An appropriate correspondence is obtained by comparing the solutions to parts (a) and (b). Under the cycle decomposition, $\sigma = 539614287$ corresponds to the collection $\{15, 2397, 46, 8\}$ of four disjoint cycles. This collection becomes $\{51, 9723, 64, 8\}$ when largest elements are written first. Under the decomposition of part (b), the unique permutation with four upper records that corresponds to this unordered collection of circular permutations is $\sigma' = 516489723$.

3.3.18. (a) From [**3.3.17**], $G = (1-x)^{-u}$, so $\partial G/\partial x = u(1-x)^{-u-1}$, whence $(1-x)\,\partial G/\partial x = uG$.

(b) Decomposition 3.3.16 is appropriate, since $d\mathsf{C}/ds$ contributes $u(1 + x + x^2 + \cdots) = u(1-x)^{-1}$, so $\partial G/\partial x = u(1-x)^{-1}G$. More directly,

$$\frac{d}{ds}(\mathsf{P} - \{\varepsilon\}) \xrightarrow{\sim} \frac{sd}{ds}(\mathsf{P} - \{\varepsilon\}) \mathbin{\dot\cup} \mathsf{P}.$$

In this decomposition $(d/ds)(\mathsf{P} - \{\varepsilon\})$ represents the deletion of the largest tag in each $\sigma \in \mathsf{P} - \{\varepsilon\}$. Thus $\partial G/\partial x$ is the generating function for the left-hand side. If the largest tag appears on a cycle of length 1, then the remaining elements are in P and have one less cycle than σ. Thus P on the right-hand side contributes uG. Otherwise the largest tag appears on a cycle containing at least one other element. When the largest tag is deleted, we can form a new permutation on the same number of cycles by joining the object on the cycle that precedes the largest tag to the object on the cycle that follows the largest tag. So that the position of the largest tag will be recoverable, we distinguish the element that precedes the largest tag. The elements that remain when the largest tag is deleted are in $(sd/ds)(\mathsf{P} - \{\varepsilon\})$, which contributes $x\,\partial G/\partial x$ to the right-hand side. Thus

$$\frac{\partial}{\partial x}G = x\frac{\partial}{\partial x}G + uG.$$

3.3.19. From Decomposition 3.3.16, we have immediately that if the required number is a_n, then

$$\frac{d}{dx}\left\{\sum_{n\geqslant 1} a_n\frac{x^n}{n!}\right\} = \left\{\frac{d}{dx}\frac{x^m}{m}\right\}(1-x)^{-1}.$$

Comparing coefficients of $x^{n-1}/(n-1)!$, we obtain

$$a_n = \left[\frac{x^{n-1}}{(n-1)!}\right]x^{m-1}(1-x)^{-1} = (n-1)!, \qquad \text{for } n \geqslant m.$$

3.3.20. (a) Consider permutations of the set $\{a_1,\ldots, a_k, b_1,\ldots, b_m\}$ for $k, m \geqslant 0$. Let $a_1,\ldots, a_k$ be tagged s-objects, marked by x, and $b_1,\ldots, b_m$ be tagged t-objects, marked by y. Then from the cycle decomposition for permutations, the generating function for permutations on $\{a_1,\ldots, a_k, b_1,\ldots, b_m\}$, with cycles containing i a's and j b's marked by u_{ij}, is

$$\left[\frac{x^k}{k!}\frac{y^m}{m!}\right]\exp\left\{\sum_{\substack{i\geqslant 0\\(i,j)\neq(0,0)}}\sum_{j\geqslant 0} u_{ij}(i+j-1)!\frac{x^i}{i!}\frac{y^j}{j!}\right\},$$

since there are $(i+j-1)!$ circular permutations on $i+j$ elements. Thus the number of such permutations in which each cycle contains at least one a_l is

$$\left[\frac{x^k}{k!}\frac{y^m}{m!}\right]\exp\left\{\sum_{i\geqslant 1}\sum_{j\geqslant 0}(i+j-1)!\frac{x^i}{i!}\frac{y^j}{j!}\right\}$$

$$= \left[\frac{x^k}{k!}\frac{y^m}{m!}\right]\exp\left\{\sum_{i\geqslant 1}\frac{x^i}{i}\sum_{j\geqslant 0}\binom{i+j-1}{j}y^j\right\}$$

$$= \left[\frac{x^k}{k!}\frac{y^m}{m!}\right]\exp\left\{\sum_{i\geqslant 1}\frac{x^i}{i}(1-y)^{-i}\right\}$$

$$= k!\left[\frac{y^m}{m!}\right](1-y)^{-k} = k(k+m-1)!$$

The result follows by identifying a_i with i, for $i = 1, \ldots, k$ and b_j with $k + j$, for $j = 1, \ldots, m$.

(b) Following part (a), the required number is

$$\left[\frac{x^k}{k!}\frac{y^m}{m!}\right]\exp\left\{\sum_{i\geqslant 1}\sum_{j\geqslant 1}(i+j-1)!\frac{x^i}{i!}\frac{y^j}{j!}\right\}$$

$$= \left[\frac{x^k}{k!}\frac{y^m}{m!}\right]\exp\left\{\sum_{i\geqslant 1}\frac{x^i}{i}\left((1-y)^{-i}-1\right)\right\}$$

$$= \left[\frac{x^k}{k!}\frac{y^m}{m!}\right](1-x)\left(1-x(1-y)^{-1}\right)^{-1}$$

$$= \left[\frac{x^k}{k!}\frac{y^m}{m!}\right]\left\{1+xy(1-x-y)^{-1}\right\} = km(k+m-2)!.$$

(c) Following part (a), the required number is

$$\left[\frac{x^k}{k!}\frac{y^m}{m!}\right]\exp\left\{\sum_{i\text{ odd}}\sum_{j\text{ even}}(i+j-1)!\frac{x^i}{i!}\frac{y^j}{j!}\right\}$$

$$= \left[\frac{x^k}{k!}\frac{y^m}{m!}\right]\exp\left\{\sum_{i\text{ odd}}\frac{x^i}{2i}\left((1-y)^{-i}+(1+y)^{-i}\right)\right\}, \qquad \text{by series bisection,}$$

$$= \left[\frac{x^k}{k!}\frac{y^m}{m!}\right]\exp\left\{\tfrac{1}{4}\left(\log\left(1-x(1-y)^{-1}\right)^{-1}-\log\left(1+x(1-y)^{-1}\right)^{-1}\right.\right.$$

$$\left.\left.+\log\left(1-x(1+y)^{-1}\right)^{-1}-\log\left(1+x(1+y)^{-1}\right)^{-1}\right)\right\},$$

by series bisection, and the result follows.

3.3.21. (a) From the labeled branch decomposition, the required number is $[x^n/n!]x(T^m/m!)$, where $T = xe^T$. Then by the Lagrange theorem, the answer is

$$\frac{n!}{m!}[x^{n-1}]T^m = \frac{n!}{m!}\frac{1}{n-1}[\lambda^{n-2}]m\lambda^{m-1}(e^\lambda)^{n-1}$$

$$= n\binom{n-2}{m-1}(n-1)^{n-m-1}.$$

(b) From the labeled branch decomposition, the required number is

$$\left[\frac{x^n}{n!}\right]x\sum_{i\geqslant 0}\frac{F^{2i+1}}{(2i+1)!},$$

where $F = x\sum_{i\geqslant 0}\dfrac{F^{2i}}{(2i)!} = \dfrac{x}{2}(e^F + e^{-F})$. Thus, by the Lagrange theorem, the answer is

$$n![x^{n-1}]\tfrac{1}{2}\{e^F - e^{-F}\} = n(n-2)![\lambda^{n-2}]\left(\frac{e^\lambda + e^{-\lambda}}{2}\right)^n$$

$$= n(n-2)!2^{-n}[\lambda^{n-2}]e^{n\lambda}(1+e^{-2\lambda})^n$$

$$= n2^{-n}\sum_{k=0}^{n}\binom{n}{k}(n-2k)^{n-2}.$$

3.3.22. From the labeled branch decomposition the required number is

$$\left[\frac{x^n}{n!}\right]x\left(e^H - \frac{H^d}{d!}\right),$$

where $H = x(e^H - H^{d-1}/(d-1)!)$. By the Lagrange theorem, this is

$$n![x^{n-1}](e^H - H^d/d!) = n(n-2)![\lambda^{n-2}](e^\lambda - \lambda^{d-1}/(d-1)!)^n,$$

and the result follows.

3.3.23. From the labeled branch decomposition, the required number is

$$\left[t_1^{i_1}t_2^{i_2}\cdots\frac{x^n}{n!}\right]x\left(1 + t_1F + t_2\frac{F^2}{2!} + \cdots\right),$$

where $F = x(t_1 + t_2F + t_3\frac{F^2}{2!} + \cdots)$ is the generating function in which x marks the labeled vertices, which are tagged objects, and t_j marks those of degree j, for $j \geqslant 1$. By the Lagrange theorem, the number is

$$n!\left[t_1^{i_1}t_2^{i_2}\cdots x^{n-1}\right]\left(1 + t_1F + t_2\frac{F^2}{2!} + \cdots\right)$$

$$= n(n-2)!\left[t_1^{i_1}t_2^{i_2}\cdots\lambda^{n-2}\right]\left(t_1 + t_2\lambda + t_3\frac{\lambda^2}{2!} + \cdots\right)^n,$$

and the result follows. Of course, when $n = 1$, there is only one labeled rooted tree, with a single vertex of degree 0.

3.3.24. (a) Differentiating $T = xe^T$ with respect to x, we obtain $dT/dx = e^T + xe^T dT/dx$. Multiplying through by x and applying $T = xe^T$ gives the required equation.

(b) The labeled vertices of $t \in \mathsf{T}$ are tagged s-objects, marked by x in $T(x)$. Consider the set $(sd/ds)\mathsf{T}$ obtained by distinguishing a single vertex in all ways in all elements of $t \in \mathsf{T}$. This set may be obtained in another way by considering two cases. If the distinguished vertex is the root vertex of t, then we have the set T itself. If the distinguished vertex is not the root vertex, then it must lie in some rooted tree $t_0 \in \hat{\mathsf{T}}$ which is in the labeled branch list of t. In this case, t_0 is a uniquely specified labeled rooted tree with a single distinguished vertex, and the tree that remains when t_0 and incident edge are deleted is itself a rooted tree in $\hat{\mathsf{T}}$. Thus in this case we have the set $\mathsf{T} * (sd\mathsf{T}/ds)$, so that

$$\frac{sd}{ds}\mathsf{T} \tilde{\to} \mathsf{T} \cup \left\{T * \left(\frac{sd}{ds}\mathsf{T}\right)\right\}$$

is tag-weight preserving. The differential equation follows directly.

3.3.25. (a) The equations $A_{h,d}(x, z) = x \exp A_{h-1,d}(x, z)$, $h \geqslant 2$, $A_{0,d-1}(x, z) = x\{e^{T(x)} + (z-1)T(x)^{d-1}/(d-1)!\}$, $T(x) = xe^{T(x)}$, where $T(x)$ is the generating function for labeled rooted trees, are immediate from the labeled branch decomposition. The equation

$$A_{1,d}(x, z) = x \exp\{A_{0,d-1}(x, z)\}$$

follows from the labeled branch decomposition, and by noting that vertices of degree d at height 1 in t correspond to root vertices in the trees of the labeled branch list of t, and hence to vertices of degree $d-1$ at height 0. Following **2.7.16**, the required number is

$$t_n(h, d) = \left[\frac{x^n}{n!}\right]\left\{\frac{\partial}{\partial z}A_{h,d}(x, z)\right\}\Big|_{z=1}.$$

Let $\phi(t) = xe^t$ and $\psi(t) = x(e^t + (z-1)t^{d-1}/(d-1)!)$. Then

$$A_{h,d}(x, z) = \phi^{[h]}(\psi(T(x))$$

Applying the chain rule iteratively as in **2.7.16** gives

$$\left\{\frac{\partial}{\partial z}A_{h,d}(x, z)\right\}\Big|_{z=1} = \{\phi'(T(x))\}^h x \frac{T^{d-1}(x)}{(d-1)!}.$$

However, $\phi'(t) = \phi(t)$, so $\phi'(T(x)) = \phi(T(x)) = T(x)$, whence

$$\left\{\frac{\partial}{\partial z}A_{h,d}(x, z)\right\}\Big|_{z=1} = x\frac{T^{h+d-1}(x)}{(d-1)!},$$

so

$$t_n(h, d) = \frac{n!}{(d-1)!}[x^{n-1}]T^{h+d-1}(x).$$

Thus from the Lagrange theorem

$$t_n(h,d) = \frac{n!}{(d-1)!}\frac{1}{n-1}[\lambda^{n-2}](h+d-1)\lambda^{h+d-2}e^{(n-1)\lambda},$$

and the result follows.

(b) In a labeled rooted tree, if we distinguish a vertex of degree d at height h, then there is a unique path from the root vertex to the distinguished vertex. When the path is deleted (Remark 2.7.19), each of the h vertices in it, other than the distinguished vertex, is the root vertex of an arbitrary element of $\hat{\mathsf{T}}$. The distinguished vertex is the root of a tree with root degree $d-1$ in $\hat{\mathsf{T}}$, so that its branch list contains $d-1$ labeled rooted trees. Thus

$$\frac{d}{ds}\mathsf{T} \xrightarrow{\sim} \underbrace{\mathsf{T}*\cdots*\mathsf{T}}_{h\text{ times}}*(\mathsf{N}_{d-1}\circledast\mathsf{T}),$$

so

$$t_n(h,d) = \left[\frac{x^n}{n!}\right]T(x)^h x\frac{T(x)^{d-1}}{(d-1)!} = \frac{n!}{(d-1)!}[x^{n-1}]T^{h+d-1},$$

where $T = xe^T$. This is in agreement with part (a).

3.3.26. (a) We obtain this equation by using the labeled branch decomposition. In this case the labeled branch list consists of planted labeled trees, whose untagged monovalent vertices are identified as tagged vertices adjacent to the root in the decomposition. Thus, if H is the set of planted labeled homeomorphically irreducible trees, $\mathsf{H} \xrightarrow{\sim} \{\nu\}*(\mathsf{U}-\mathsf{N}_1)\circledast\mathsf{H}$, since a single tree in the labeled branch list would induce a bivalent vertex. Here ν represents the planted labeled vertex tree. Accordingly, $H = x(e^H - H)$, since the decomposition is tag-weight preserving. The result follows by applying $x(d/dx)$ to this equation.

(b) Consider the set obtained by distinguishing, in turn, each tagged vertex of each tree $t \in \mathsf{H}$. This set is $(sd/ds)\mathsf{H}$, where the labeled vertices are tagged s-objects. This set may be obtained in another way by considering two cases. If the distinguished vertex is adjacent to the untagged monovalent vertex, then this set is in one-to-one correspondence with H. If the distinguished vertex is not adjacent to the untagged monovalent vertex, then it lies in some tree t_1 in the branch list of t. Thus t_1 is an arbitrary element of $\overline{(sd/ds)\mathsf{H}}$. The tree that remains when t_1 and incident edge are deleted from t is any element of $\hat{\mathsf{H}}$ except $\{\nu\}$, the planted labeled tree with a single tagged vertex, or is a labeled tree in which there is a single bivalent vertex, adjacent to the untagged monovalent vertex. The latter trees are decomposed as $\{\nu\}*\mathsf{H}$, so

$$\frac{sd}{ds}\mathsf{H} \xrightarrow{\sim} \mathsf{H} \cup \left(\frac{sd}{ds}\mathsf{H}\right)*\{\mathsf{H} - \{\nu\} \cup (\{\nu\}*\mathsf{H})\}$$

is tag-weight preserving, leading directly to the differential equation of part (a).

3.3.27. From the cycle decomposition for functions, the required number is $[x^n/n!]\exp\{T^k/k\}$, where $T = xe^T$. By the Lagrange theorem, this is

$$n!\frac{1}{n}[\lambda^{n-1}]\lambda^{k-1}\exp\left(\frac{\lambda^k}{k}\right)e^{n\lambda} = (n-1)![\lambda^{n-k}]\sum_{i\geq 0}\frac{\lambda^{ki}}{k^i i!}e^{n\lambda},$$

and the result follows.

3.3.28. From the cycle decomposition for functions, the required number is

$$\left[z^k\frac{x^n}{n!}\right]\exp\left(zT + z\frac{T^2}{2} + \cdots\right), \qquad \text{where } T = xe^T$$

$$= \left[z^k\frac{x^n}{n!}\right](1-T)^{-z} = n!\frac{1}{n}[z^k\lambda^{n-1}]z(1-\lambda)^{-z-1}e^{n\lambda},$$

by the Lagrange theorem, and the result follows.

3.3.29. From the cycle decomposition for functions the required number is

$$\left[\frac{x^n}{n!}\right]\exp\left(\frac{T^2}{2} + \frac{T^4}{4} + \cdots\right) = \left[\frac{x^n}{n!}\right](1-T^2)^{-1/2}, \qquad \text{where } T = xe^T$$

$$= n!\frac{1}{n}[\lambda^{n-1}]\lambda(1-\lambda^2)^{-3/2}e^{n\lambda}, \qquad \text{from the Lagrange theorem}$$

$$= (n-1)!\sum_{i=0}^{\left\lfloor\frac{n-2}{2}\right\rfloor}\binom{-\frac{3}{2}}{i}(-1)^i\frac{n^{n-2-2i}}{(n-2-2i)!},$$

and the result follows.

3.3.30. From **3.3.14** the required number is

$$c_m(r,n) = \frac{1}{n^n}\sum_{k\geq m}k(k-1)\cdots(k-m+1)d_k(r,n)$$

$$= \frac{1}{n^n}\left[\frac{x^n}{n!}\right]\left\{\frac{\partial^m}{\partial z^m}\sum_{n,k\geq 0}d_k(r,n)z^k\frac{x^n}{n!}\right\}\Bigg|_{z=1}$$

$$= \frac{1}{n^n}\left[\frac{x^n}{n!}\right]\left(\frac{1}{r}T^r\right)^m(1-T)^{-1},$$

since

$$\sum_{n,k\geq 0}d_k(r,n)z^k\frac{x^n}{n!} = \exp\left(T + \frac{1}{2}T^2 + \cdots + \frac{z}{r}T^r + \cdots\right).$$

But we again have $xT' = T(1-T)^{-1}$, so that $T^{rm}(1-T)^{-1} = x(T^{rm})'/rm$.

It follows that

$$c_m(r,n) = \frac{n!}{n^n}\frac{r^{-m}}{rm}[x^{n-1}](T^{rm})' = \frac{n!}{n^{n-1}}\frac{r^{-m}}{rm}[x^n]T^{rm},$$

$$= \frac{n!}{n^{n-1}}\frac{r^{-m}}{rm}\frac{1}{n}[\lambda^{n-1}]rm\lambda^{rm-1}e^{n\lambda}, \qquad \text{by the Lagrange theorem,}$$

and the result follows.

3.3.31. (a) If $f^{[m+k]}(i) = f^{[k]}(i)$ for all i, then $f^{[k]}(i)$ is a recurrent element for all i. Thus each transient vertex is a distance at most k from its recurrent root. The lengths of all cycles divide m, so $f^{[m]}(f^{[k]}(i)) = f^{[k]}(i)$. The result follows from the cycle decomposition for functions.

(b) From the labeled branch decomposition for labeled rooted trees, we have immediately that $T_k = x\exp(T_{k-1})$, $k \geqslant 1$. Since $T_0(x)$ counts the labeled single vertex only, we obtain $T_0(x) = x$.

3.3.32. From the cycle decomposition for functions the required number is

$$\left[z^k\frac{x^n}{n!}\right]\exp\left\{\sum_{i\geqslant 1}\frac{(zT)^i}{i}\right\} = \left[z^k\frac{x^n}{n!}\right](1-zT)^{-1} = \left[\frac{x^n}{n!}\right]T^k, \quad \text{where } T = xe^T$$

$$= (n-1)![\lambda^{n-1}]k\lambda^{k-1}e^{n\lambda}, \qquad \text{from the Lagrange theorem,}$$

and the result follows.

3.3.33. From the cycle decomposition for functions, the required number is, with $T = xe^T$,

$$\left[\frac{x^n}{n!}\right]\exp\{T(x)\} = \frac{n!}{n}[\lambda^{n-1}]e^{\lambda}(e^{\lambda})^n = (n+1)^{n-1},$$

from the Lagrange theorem. This is also, of course, the number of **forests** of rooted labeled trees with a total of n vertices.

3.3.34. We construct an undirected graph on vertex set $\{1,\ldots,n\}$ corresponding to a frame as follows. The graph contains the edge between vertex i and vertex j iff the point of intersection of lines l_i and l_j is contained in the frame. Thus the graph is simple and loopless and contains n edges; in addition, each vertex has degree $\leqslant 2$. But since the sum of the degrees equals twice the number of edges, every vertex has degree exactly equal to 2. Hence the graphs corresponding to frames are the cycle covers of the complete graph K_n. But the latter are enumerated in the solution to [**3.3.6**(b)], giving the required number.

3.3.35. The exponential generating function for permutations, in which u is an ordinary marker for cycles of length 1, is $e^{(u-1)x}(1-x)^{-1}$ by the cycle decomposition for permutations. Thus the expected value of the number of cycles of length 1 is given by

$$\mu = \frac{1}{n!}\left[\frac{x^n}{n!}\right]\left\{\frac{\partial}{\partial u}e^{(u-1)x}(1-x)^{-1}\right\}\Bigg|_{u=1}$$

$$= [x^n]x(1-x)^{-1} = 1, \qquad \text{for } n \geqslant 1.$$

The variance is given by

$$\left(\frac{1}{n!}\left[\frac{x^n}{n!}\right]\left\{\frac{\partial^2}{\partial u^2}e^{(u-1)x}(1-x)^{-1}\right\}\bigg|_{u=1}\right)+\mu-\mu^2$$

$$=[x^n]x^2(1-x)^{-1}=1, \qquad \text{for } n \geqslant 2.$$

3.3.36. (a) The required number, from **3.2.13** is

$$\left[z^k\frac{x^m}{m!}\right](1+z(e^x-1))^n=\left[\frac{x^m}{m!}\right]\binom{n}{k}(e^x-1)^k.$$

(b) The required number, from part (a), is

$$\frac{1}{n^m}\left[\frac{x^m}{m!}\right]\left\{\frac{\partial}{\partial z}(1+z(e^x-1))^n\right\}\bigg|_{z=1}=\frac{1}{n^m}\left[\frac{x^m}{m!}\right]ne^{x(n-1)}(e^x-1),$$

and the result follows.

3.3.37. From the matrix-tree theorem the generating function for the spanning trees of K_n, with u marking edges in K_m, is

$$\operatorname{cof}_{nn}(\{(m(u-1)+n)\mathbf{I}_m\oplus n\mathbf{I}_{n-m}\}-\{(u-1)\mathbf{J}_m\oplus\mathbf{0}_{n-m}\}-\mathbf{J}_n)$$

$$=\alpha\left|\mathbf{I}_{n-1}-\left\{\frac{1}{m(u-1)+n}\mathbf{I}_m\oplus\frac{1}{n}\mathbf{I}_{n-m-1}\right\}\mathbf{A}^T\mathbf{B}\right|,$$

where

$$\mathbf{A}=\begin{pmatrix}\overbrace{u-1\ \cdots\ u-1}^{m} & 0\ \cdots\ 0\\ 1\ \cdots\ 1 & 1\ \cdots\ 1\end{pmatrix}_{2\times(n-1)},\qquad \mathbf{B}=\begin{pmatrix}\overbrace{1\ \cdots\ 1}^{m} & 0\ \cdots\ 0\\ 1\ \cdots\ 1 & 1\ \cdots\ 1\end{pmatrix}_{2\times(n-1)}$$

and $\alpha=(m(u-1)+n)^m n^{n-m-1}$. But from **1.1.10(5)** this equals

$$\alpha\left|\mathbf{I}_2-\mathbf{B}\left\{\frac{1}{m(u-1)+n}\mathbf{I}_m\oplus\frac{1}{n}\mathbf{I}_{n-m-1}\right\}\mathbf{A}^T\right|$$

$$=\alpha\begin{vmatrix}\dfrac{n}{m(u-1)+n} & \dfrac{-m}{m(u-1)+n}\\[2ex] \dfrac{n}{m(u-1)+n}-1 & \dfrac{m+1}{n}-\dfrac{m}{m(u-1)+n}\end{vmatrix}$$

$$=(m(u-1)+n)^{m-1}n^{n-m-1}.$$

The result follows by extracting the coefficient of u^i.

3.3.38. From the matrix-tree theorem, the required number is

$$\operatorname{cof}_{n+m,n+m}\left(\begin{array}{c|c} m\mathbf{I}_n & -\mathbf{J}_{n,m} \\ \hline -\mathbf{J}_{m,n} & n\mathbf{I}_m \end{array}\right) = |(m\mathbf{I}_n \oplus n\mathbf{I}_{m-1}) - \mathbf{A}^T\mathbf{B}|,$$

where

$$\mathbf{A} = \begin{pmatrix} \overbrace{0 \cdots 0}^{n} & 1 \cdots 1 \\ 1 \cdots 1 & 0 \cdots 0 \end{pmatrix}_{2\times(n+m-1)}, \qquad \mathbf{B} = \begin{pmatrix} \overbrace{1 \cdots 1}^{n} & 0 \cdots 0 \\ 0 \cdots 0 & 1 \cdots 1 \end{pmatrix}_{2\times(n+m-1)}$$

Then from **1.1.10(5)** the number of spanning trees is

$$m^n n^{m-1}\begin{vmatrix} 1 & -1 \\ \dfrac{1-m}{m} & 1 \end{vmatrix} = m^{n-1}n^{m-1}.$$

3.3.39. From the matrix-tree theorem the required number is

$$\operatorname{cof}_{nn}\left[\begin{array}{c|c} (n\mathbf{I}_{n-m-1} \oplus (n-1)\mathbf{I}_m) - \mathbf{J}_{n-1} & \begin{matrix} -1 \\ \vdots \\ -1 \\ 0 \\ \vdots \\ 0 \end{matrix} \\ \hline \underbrace{-1 \ \cdots \ -1}_{n-m-1} \ \underbrace{0 \ \cdots \ 0}_{m} & n-m-1 \end{array}\right]$$

$$= |(n\mathbf{I}_{n-m-1} \oplus (n-1)\mathbf{I}_m) - \mathbf{J}_{n-1}|$$

$$= n^{n-m-1}(n-1)^m\left(1 - \frac{n-m-1}{n} - \frac{m}{n-1}\right)$$

$$= n^{n-m-2}(n-1)^{m-1}(n-m-1),$$

from **1.1.10(5)**.

3.3.40. (a) From the matrix-tree theorem we have

$$t_n = \operatorname{cof}_{n+1,n+1}\left(\begin{array}{c|c} \operatorname{circ}(3,-1,0,\ldots,0,-1) & \begin{matrix} -1 \\ \vdots \\ -1 \end{matrix} \\ \hline -1 \ \cdots \ -1 & n \end{array}\right)$$

$$= |\operatorname{circ}(3,-1,0,\ldots,0,-1)|$$

$$= \prod_{k=0}^{n-1}(3 - \omega^k - \omega^{k(n-1)})$$

from [**1.1.17**], where $\omega = e^{2\pi i/n}$. Thus

$$t_n = \prod_{k=0}^{n-1} (3 - \omega^k - \omega^{-k}) = \alpha^{-n} \prod_{k=0}^{n-1} (\alpha - \omega^k)(\alpha - \omega^{-k}),$$

where $\alpha = (3 - \sqrt{5})/2 = 2/(3 + \sqrt{5})$. But

$$\prod_{k=0}^{n-1} (\alpha - \omega^k) = \prod_{k=0}^{n-1} (\alpha - \omega^{-k}) = \alpha^n - 1,$$

since $\{\omega^0, \omega^1, \ldots, \omega^{n-1}\} = \{\omega^0, \omega^{-1}, \ldots, \omega^{-(n-1)}\}$ is the set of roots of the polynomial $x^n - 1$. Thus $t_n = \alpha^{-n}(1 - \alpha^n)^2 = \alpha^n + \alpha^{-n} - 2$, and the result follows.

(b) From part (a)

$$t_n = \prod_{k=0}^{n-1} (3 - \omega^k - \omega^{-k}) = \prod_{k=0}^{n-1} (3 - e^{2k\pi i/n} - e^{-2k\pi i/n})$$

$$= \prod_{k=0}^{n-1} \left(3 - 2\cos\frac{2k\pi}{n}\right).$$

But $\cos(2k\pi/n) = 1 - 2\sin^2(k\pi/n)$, and the result follows.

3.3.41. (a) The spanning trees of g that do not contain edge e are the spanning trees of $g - e$. Those which do contain e may be obtained uniquely, in the obvious way, from the spanning trees of $g \cdot e$. The result follows.

(b) The number of spanning trees of g_e^λ is equal to $t(g - e)$, those with no copy of edge e, plus λ times $t(g) - t(g - e)$, those that contain a copy of edge e. Thus

$$t(g_e^\lambda) = \lambda(t(g) - t(g - e)) + t(g - e)$$

$$= \lambda t(g) - (\lambda - 1)t(g - e).$$

(c) Let V_n be the graph obtained by joining a single vertex to each vertex in a path with n vertices, as shown here.

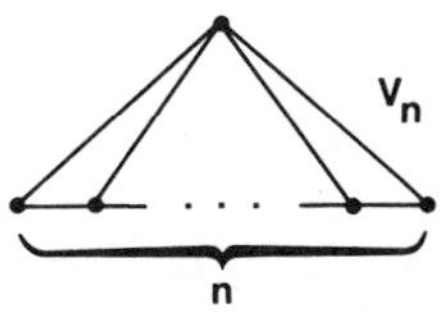

Similarly, define the following graphs.

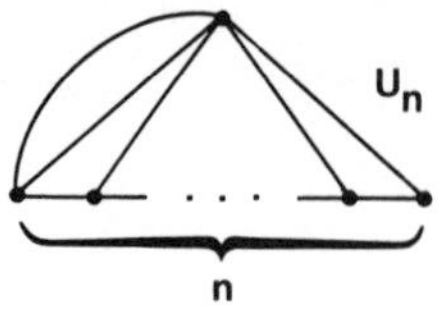

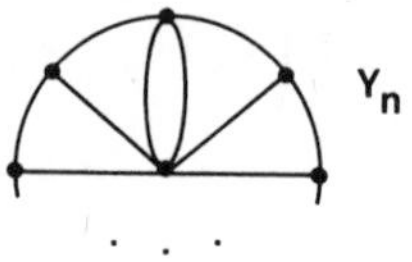

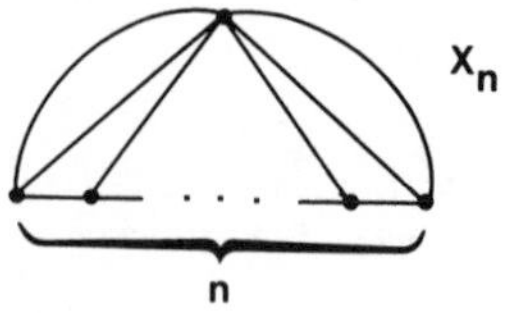

Let the lower-case letters w_n, v_n, u_n, x_n, y_n denote the numbers of spanning trees in the graphs defined by the corresponding upper-case letters.

First we consider an edge in the outer cycle of W_n and apply part (a) to get

$$w_n = v_n + y_{n-1}.$$

Using part (b) with $\lambda = 2$ for the doubled edge yields

$$y_{n-1} = 2w_{n-1} - (w_{n-2} + v_{n-2}).$$

Now eliminate y_{n-1} to get

$$w_n - 2w_{n-1} + w_{n-2} = v_n - v_{n-2}, \qquad \text{for } n \geqslant 3. \tag{1}$$

Next we consider a "spoke" of the wheel, and apply part (a), giving

$$w_n = (w_{n-1} + v_{n-1}) + x_{n-1}.$$

Again using part (b) twice with $\lambda = 2$, we have

$$x_{n-1} = 2u_{n-1} - u_{n-2} = 2(2v_{n-1} - v_{n-2}) - (2v_{n-2} - v_{n-3})$$
$$= 4v_{n-1} - 4_{n-2} + v_{n-3}.$$

Eliminating x_{n-1} from w_n gives the equation

$$w_n - w_{n-1} = 5v_{n-1} - 4v_{n-2} + v_{n-3}, \qquad \text{for } n \geqslant 4. \tag{2}$$

Finally we use parts (a) and (b) to get

$$v_n = v_{n-1} + u_{n-1}, \qquad u_{n-1} = 2v_{n-1} - v_{n-2},$$

and eliminate u_{n-1} to give

$$0 = v_n - 3v_{n-1} + v_{n-2}, \qquad \text{for } n \geqslant 3. \tag{3}$$

Eliminate v_n from (1) by applying (3) and v_{n-3} from (2) by applying (3) with n replaced by $n - 1$. This gives

$$w_n - 2w_{n-1} + w_{n-2} = 3v_{n-1} - 2v_{n-2}$$

$$w_n - w_{n-1} = 4v_{n-1} - v_{n-2}, \tag{4}$$

and v_{n-2} is eliminated between these to get

$$w_n - w_{n-2} = 5v_{n-1}. \tag{5}$$

Finally we eliminate v_{n-1} from (4) by applying (5), and v_{n-2} from (4) by applying (5) with n replaced by $n - 1$. This gives

$$5w_n - 5w_{n-1} = 4(w_n - w_{n-2}) - (w_{n-1} - w_{n-3}), \qquad \text{for } n \geqslant 4,$$

and the result follows. This linear recurrence can be routinely solved to give the solution to [**3.3.40**(a)]. Similarly, cofactor expansions of the determinant of the matrix-tree theorem also yield the recurrence equation of this problem.

(d) Define the graph H_n as shown here.

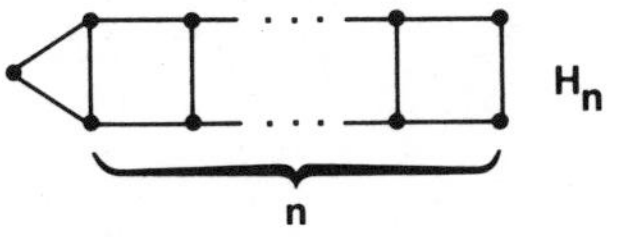

Applying part (a) to L_n, we obtain

$$t(L_n) = t(L_{n-1}) + t(H_{n-1})$$

and applying part (a) and part (b) with $\lambda = 2$ gives

$$t(H_{n-1}) = t(L_{n-1}) + (2t(L_{n-1}) - t(L_{n-2}))$$

$$= 3t(L_{n-1}) - t(L_{n-2}).$$

We eliminate $t(H_{n-1})$ to get the result.

(e) The solution is $t(L_n) = a\alpha_1^n + b\alpha_2^n$, where α_1, α_2 are the roots of $x^2 - 4x + 1 = 0$, and a, b are chosen so $t(L_1) = 1$, $t(L_2) = 4$. Thus

$$a\alpha_1 + b\alpha_2 = 1, \qquad a\alpha_1^2 + b\alpha_2^2 = 4,$$

which has the solution

$$a = \frac{4 - \alpha_2}{\alpha_1(\alpha_1 - \alpha_2)}, \qquad b = \frac{\alpha_1 - 4}{\alpha_2(\alpha_1 - \alpha_2)},$$

where $\alpha_1 = 2 + \sqrt{3}$, $\alpha_2 = 2 - \sqrt{3}$. The result follows.

3.3.42. (a) From Lemma 3.3.22 the required number is

$$\theta_c(\mathbf{D}) = \left[\mathbf{A}^{\mathbf{D}}\mathbf{x}^{\mathbf{n}}\frac{z^N}{N!}\right] f_c(\mathbf{A}, \mathbf{x}, z),$$

where $f_i = zx_i\exp(a_{i1}f_1 + \cdots + a_{ik}f_k)$, for $i = 1,\ldots, k$. But from the multivariate Lagrange theorem for monomials (**1.2.13**) we obtain

$$\theta_c(\mathbf{D}) = N!(n_1 \cdots n_k)^{-1}\sum_{\boldsymbol{\mu}}\|\delta_{ij}n_i - \mu_{ij}\|$$

$$\times \prod_{i=1}^{k}\left\{[a_{i1}^{d_{i1}} \cdots a_{ik}^{d_{ik}} f_1^{\mu_{i1}} \cdots f_k^{\mu_{ik}}]\exp\left(n_i \sum_{j=1}^{k} a_{ij}f_j\right)\right\},$$

where $\sum_{i=1}^{k}\mu_{ij} = n_j - \delta_{jc}$. Thus the only nonzero contribution to this sum comes from the term with $\boldsymbol{\mu} = \mathbf{D}$, and the result follows.

(b) From part (a) the required number is $\lambda_c(\mathbf{n}, \mathbf{T}) = \sum_{\mathbf{D}} \theta_c(\mathbf{D})$, where the sum is over all $\mathbf{D}$ such that $\sum_{i=1}^{k} d_{ij} = n_j - \delta_{jc}$ and $d_{ij} = 0$ when $T_{ij} = 0$. If $\boldsymbol{\delta}_c = (\delta_{1c},\ldots, \delta_{kc})$, then

$$\lambda_c(\mathbf{n}, \mathbf{T}) = N!(n_1 \cdots n_k)^{-1}[\mathbf{x}^{\mathbf{n}-\boldsymbol{\delta}_c}] \sum_{\mathbf{D}\geqslant\mathbf{0}} \left\{\prod_{i=1}^{k}\prod_{j=1}^{k}(n_ix_iT_{ij})^{d_{ij}}(d_{ij}!)^{-1}\right\}$$

$$\times \operatorname{cof}_{cc}\left[\delta_{ij}\left(\sum_{l=1}^{k} d_{lj}T_{lj}\right) - d_{ij}T_{ij}\right]_{k\times k}$$

$$= N!(n_1 \cdots n_k)^{-1}[\mathbf{x}^{\mathbf{n}-\boldsymbol{\delta}_c}]\exp\left\{\sum_{i=1}^{k}\sum_{j=1}^{k} n_ix_jT_{ij}\right\}$$

$$\times \operatorname{cof}_{cc}\left[\delta_{ij}\left(\sum_{l=1}^{k} n_lx_jT_{lj}\right) - n_ix_jT_{ij}\right]_{k\times k},$$

since $\sum_{\mathbf{i}\geqslant\mathbf{0}} \mathbf{x}^{\mathbf{i}}(\mathbf{i}!)^{-1}g(\mathbf{i}) = g(\mathbf{x})\exp(\sum_j x_j)$ if $g(\mathbf{x})$ is multilinear. Thus

$$\lambda_c(\mathbf{n}, \mathbf{T}) = N!(n_1 \cdots n_k)^{-1}\operatorname{cof}_{cc}\left[\delta_{ij}\left(\sum_{l=1}^{k} n_lT_{lj}\right) - n_iT_{ij}\right]_{k\times k}$$

$$\times [\mathbf{x}^{\mathbf{n}-\mathbf{1}}]\exp\left\{\sum_{j=1}^{k} x_j \sum_{i=1}^{k} n_iT_{ij}\right\},$$

and the result follows.

3.3.43. This is simply the statement of **[2.4.21]** with the determinant identified as the number of in-directed spanning arborescences by the matrix-tree theorem.

3.3.44. Following **3.3.25**, the required generating function is obtained by replacing a_{ij} by $x_i x_j$ in the matrix-tree theorem, since the edge a_{ij} contributes 1 to each of the degrees of vertices i and j. Thus if $x = x_1 + \cdots + x_n$,

$$\begin{aligned} T(x_1, \ldots, x_n) &= \operatorname{cof}_{11}\left[x_i x \delta_{ij} - x_i x_j\right]_{n \times n} \\ &= x_2^2 \cdots x_n^2 \det\{\operatorname{diag}(x_2^{-1}x, \ldots, x_n^{-1}x) - \mathbf{J}_{n-1}\} \\ &= x_2 \cdots x_n x^{n-1}\{1 - x^{-1}(x_2 + \cdots + x_n)\}, \qquad \text{from } \mathbf{1.1.10(5)} \\ &= x_1 \cdots x_n x^{n-2}. \end{aligned}$$

3.3.45. (a) For $\sigma \in \mathsf{B} - \{\varepsilon\}$ we define σ_1 to be the sequence preceding the first occurrence of 1 in σ, σ_2 to be the sequence between the occurrences of 1, and σ_3 to be the sequence following the second occurrence of 1. Then $\sigma_1, \sigma_2, \sigma_3 \in \hat{\mathsf{B}}$, since no element $i > 1$ can occur in two of $\sigma_1, \sigma_2, \sigma_3$. Thus if an s-object is an element of the alphabet, we have the decomposition

$$\frac{d}{ds}(\mathsf{B} - \{\varepsilon\}) \tilde{\to} \mathsf{B}^3 : \sigma \mapsto (\sigma_1, \sigma_2, \sigma_3).$$

Now the first occurrence of 1 introduces a fall if σ_1 is nonempty, and the second introduces a fall if σ_2 is nonempty, so

$$\frac{\partial}{\partial x}(B - 1) = (f(B-1) + 1)^2 B, \qquad \text{where } B(0, f) = 1.$$

(b) Expanding the left-hand side of $B'(f(B-1)+1)^{-2}B^{-1} = 1$ by partial fractions, we obtain

$$B'\{B^{-1} - f(1 - f + fB)^{-1} - f(1-f)(1 - f + fB)^{-2}\} = (1-f)^2.$$

Integrating and applying the initial condition $B(0, f) = 1$ gives

$$\log B - \log(1 - f + fB) + (1-f)(1 - f + fB)^{-1} = (1-f)^2 x + 1 - f,$$

or

$$\log\{B(1 - f + fB)^{-1}\} - f\{B(1 - f + fB)^{-1} - 1\} = (1-f)^2 x.$$

Now substitute $H(x, f) = \log\{B(1 - f + fB)^{-1}\}$ in the preceding equation, so H satisfies $H = a + f\phi(H)$, where $a = (1-f)^2 x$ and $\phi(H) = e^H - 1$. Thus by [**1.2.4**], we obtain

$$1 + f(1-f)^{-1}B = \{1 - f\phi'(H)\}^{-1} = \sum_{n\geq 0} f^n[\lambda^n](e^{\lambda+a} - 1)^n$$

$$= \sum_{n\geq 0} f^n \sum_{k\geq 0} S(n+k, n)(1-f)^{2k}\frac{x^k}{k!},$$

and the result follows.

3.3.46. (a) Let P_+ denote the set of all permutations on the s-tag-set N_n for $n \geq 1$, and let ε denote the empty string. Then we have

$$\frac{d}{ds}(\mathsf{P}_+ - \{1\}) \tilde{\to} (\{\varepsilon\} * \mathsf{P}_+) \cup (\mathsf{P}_+ * \mathsf{P}_+) \cup (\mathsf{P}_+ * \{\varepsilon\}) : \sigma_1 1 \sigma_2 \mapsto (\sigma_1, \sigma_2).$$

The terms $\{\varepsilon\} * \mathsf{P}_+$ and $\mathsf{P}_+ * \{\varepsilon\}$ arise when 1 occurs as the left-most and right-most element, respectively. But 1 terminates a fall in $\sigma_1 1 \sigma_2$ unless $\sigma_1 = \varepsilon$, so we have immediately

$$\frac{\partial}{\partial x}(g - x) = g + fg^2 + fg, \qquad \text{where } g(0, f) = 0.$$

(b) Let $h = 1 + g$, so h satisfies $\partial h/\partial x = fg^2 + (1-f)g$, where $h(0, f) = 1$. This is the special case of equation **1.1.8(1)** with $a = f$, $b = 1 - f$, $\alpha = 1$, and thus has the solution

$$h(x, z) = e^{(1-f)x}\{1 - f(1-f)^{-1}(e^{(1-f)x} - 1)\}^{-1},$$

from **1.1.8(2)**. The result follows since $g = h - 1$.

(c) Let $F = F(x, u_1, u_2, u_3, u_4)$ be the generating function for nonempty permutations, where u_1 marks modified minima, u_2 modified maxima, u_3 double rises, and u_4 double falls, and where x is an exponential marker for the length of a permutation. Then from the decomposition in part (a)

$$\frac{\partial}{\partial x}(F - u_2 x) = u_3 F + u_1 F^2 + F u_4,$$

so

$$\frac{\partial}{\partial x}F = u_2 + (u_3 + u_4)F + u_1 F^2.$$

This is a Riccati equation, and we solve it by letting $F = (-1/u_1 G)(\partial G/\partial x)$, where $G(0, u_1, u_2, u_3, u_4) = 1$, so

$$\frac{\partial^2 G}{\partial x^2} - (u_3 + u_4)\frac{\partial G}{\partial x} + u_1 u_2 G = 0.$$

The solution to the preceding equation is $G = Ae^{\alpha_1 x} + Be^{\alpha_2 x}$, where $\alpha_1\alpha_2 = u_1u_2$ and $\alpha_1 + \alpha_2 = u_3 + u_4$. But $G(0, u_1, u_2, u_3, u_4) = 1$, so $A + B = 1$, and $F(0, u_1, u_2, u_3, u_4) = 0$, so

$$-\frac{1}{u_1 G}\frac{\partial G}{\partial x}\bigg|_{x=0} = 0, \qquad \text{and} \qquad \frac{\partial G}{\partial x}\bigg|_{x=0} = A\alpha_1 + B\alpha_2 = 0.$$

Thus

$$A = \frac{\alpha_2}{\alpha_2 - \alpha_1} \qquad \text{and} \qquad B = 1 - A = \frac{-\alpha_1}{\alpha_2 - \alpha_1},$$

so

$$G = \frac{\alpha_2 e^{\alpha_1 x} - \alpha_1 e^{\alpha_2 x}}{\alpha_2 - \alpha_1}, \qquad \frac{\partial G}{\partial x} = \alpha_1\alpha_2\left(\frac{e^{\alpha_1 x} - e^{\alpha_2 x}}{\alpha_2 - \alpha_1}\right),$$

and the result follows.

3.3.47. (a) The sets of sequences in $\mathsf{A}_n, \mathsf{B}_n$ that contain at least one occurrence of n are $\mathsf{A}_n - \mathsf{A}_{n-1}, \mathsf{B}_n - \mathsf{B}_{n-1}$. We obtain the following decompositions by considering the first occurrence of n from the left in such a sequence:

$$\mathsf{A}_n - \mathsf{A}_{n-1} - \{n\} \tilde{\rightarrow} \mathsf{A}_{n-1} \times \{n\} \times \mathsf{A}_n$$

$$\mathsf{B}_n - \mathsf{B}_{n-1} \tilde{\rightarrow} \mathsf{B}_{n-1} \times \{n\} \times \mathsf{B}_n.$$

The first of these follows because the first occurrence of n must be preceded by an alternating sequence of odd length, containing no n's, and followed by an alternating sequence of odd length that might contain some n's. This fails only if the sequence is of length 1, and the only such sequence in $\mathsf{A}_n - \mathsf{A}_{n-1}$ is $\{n\}$. The second decomposition is similar. Now the generating function for $\{n\}$ is x_n, so

$$A_n - A_{n-1} = x_n(A_{n-1}A_n + 1), \qquad n \geqslant 1$$

$$B_n - B_{n-1} = x_n A_{n-1} B_n, \qquad n \geqslant 1.$$

In the decomposition, we have allowed B_n to contain the empty sequence, so that $B_n(\mathbf{0}) = 1$, $n \geqslant 0$. A_n cannot be empty because it has odd length, so $A_n(\mathbf{0}) = 0$, $n \geqslant 0$.

(b) By definition, we have

$$\nabla C_n D_n = C_n D_n - C_{n-1}D_{n-1} = C_n(D_n - D_{n-1}) + D_{n-1}(C_n - C_{n-1})$$
$$= C_n \nabla D_n + D_{n-1}\nabla C_n.$$

$$\nabla\left(\frac{1}{C_n}\right) = \frac{1}{C_n} - \frac{1}{C_{n-1}} = \frac{C_{n-1} - C_n}{C_n C_{n-1}} = \frac{-\nabla C_n}{C_n C_{n-1}}.$$

(c) From parts (a) and (b)

$$\nabla\left(\frac{\nabla H_{n+1}}{-x_{n+1}H_n}\right) = x_n\left(\frac{\nabla H_{n+1}}{x_{n+1}H_n}\frac{\nabla H_n}{x_nH_{n-1}} + 1\right)$$

But

$$\nabla\left(\frac{\nabla H_{n+1}}{-x_{n+1}H_n}\right) = (\nabla^2 H_{n+1})\left(\frac{-1}{x_nH_{n-1}}\right) - (\nabla H_{n+1})\nabla\left(\frac{1}{x_{n+1}H_n}\right)$$

$$= -\frac{\nabla^2 H_{n+1}}{x_nH_{n-1}} + \frac{(\nabla H_{n+1})\nabla(x_{n+1}H_n)}{x_{n+1}H_nx_nH_{n-1}}$$

$$= \frac{-\nabla^2 H_{n+1}}{x_nH_{n-1}} + \frac{(\nabla H_{n+1})(x_n(\nabla H_n) + H_n(x_{n+1} - x_n))}{x_nx_{n+1}H_nH_{n-1}}$$

so

$$\frac{-\nabla^2 H_{n+1}}{x_nH_{n-1}} + \frac{(\nabla H_{n+1})(x_{n+1} - x_n)}{x_nx_{n+1}H_{n-1}} = x_n$$

whence $x_{n+1}\nabla^2 H_{n+1} - \nabla H_{n+1}(x_{n+1} - x_n) + x_n^2 x_{n+1}H_{n-1} = 0.$

But

$$\nabla\gamma_{n+1}(k) = x_{n+1}\gamma_n(k-1),$$

so

$$x_{n+1}\nabla^2\gamma_{n+1}(k) = \nabla\gamma_{n+1}(k)(x_{n+1} - x_n) + x_n^2 x_{n+1}\gamma_{n-1}(k-2),$$

and comparing these, we have

$$H_n = \sum_{k\geqslant 0}(-1)^k\{c_1\gamma_n(2k) + c_2\gamma_n(2k+1)\}.$$

But $H_n(\mathbf{0}) = 1$, so $c_1 = 1$, and $A_n(\mathbf{0}) = 0$, so $(\nabla H_{n+1})/x_{n+1}|_{\mathbf{x}=\mathbf{0}} = 0$. Therefore, $c_2 = 0$, and $H_n = \sum_{k\geqslant 0}(-1)^k\gamma_n(2k)$. Finally, $A_n = (\nabla H_{n+1})/(-x_{n+1}H_n)$, and the result follows.

(d) In the preceding solution

$$\nabla\left(\frac{1}{H_n}\right) = \frac{-\nabla H_n}{H_nH_{n-1}} = x_n\left(\frac{1}{H_n}\right)\left(\frac{\nabla H_n}{-x_nH_{n-1}}\right), \text{ so } \nabla\left(\frac{1}{H_n}\right) = x_n\left(\frac{1}{H_n}\right)A_{n-1}.$$

But $\nabla B_n = x_nB_nA_{n-1}$, so $B_n = c/H_n$, where c is determined by $B_n(\mathbf{0}) = 1$. Since $H_n(\mathbf{0}) = 1$, we obtain

$$B_n(\mathbf{x}) = \frac{1}{H_n(\mathbf{x})} = \left\{\sum_{k\geqslant 0}(-1)^k\gamma_n(2k)\right\}^{-1}.$$

3.3.48. (a) The labeled branch decomposition (**3.3.9**) gives (i) immediately.

For (ii), let B_2 and B_M be the subsets of B in which the vertex adjacent to the root has label 2 and the maximum label, respectively. For $t \in \mathsf{B} - \mathsf{B}_M$, suppose that the vertex adjacent to the root has label i, and define $t_1 \in \mathsf{B} - \mathsf{B}_2$ to be the tree obtained from t by interchanging labels i and $i + 1$. Then

$$\mathsf{B} - \mathsf{B}_M \mathrel{\tilde{\to}} \mathsf{B} - \mathsf{B}_2 : t \mapsto t_1$$

and t_1 has one more tree inversion than t. Thus

$$B(x, q) - B_M(x, q) = q^{-1}\{B(x, q) - B_2(x, q)\}, \tag{1}$$

where $B_2(x, q)$ and $B_M(x, q)$ are the generating functions for B_2 and B_M, respectively.

For $t \in \mathsf{B}_2$, let $t_2 \in \mathsf{A}$ be obtained by deleting the root vertex of t and subtracting 1 from all remaining labels. For $t \in \mathsf{B}_M$, let $t_3 \in \mathsf{A}$ be obtained by deleting the root vertex and replacing the maximum label by 1. Then

$$\frac{d}{ds}\mathsf{B}_2 \mathrel{\tilde{\to}} \mathsf{A} : t \mapsto t_2$$

$$\frac{d}{ds}\mathsf{B}_M \mathrel{\tilde{\to}} \mathsf{A} : t \mapsto t_3,$$

where the differentiations denote deleting the root vertex. Moreover, t_2 has the same number of tree inversions as t, and t_3 has $n - 2$ less tree inversions than t (where the maximum label in t is n), so

$$\frac{\partial}{\partial x} B_2(x, q) = A(x, q), \qquad \frac{\partial}{\partial x} B_M(x, q) = q^{-1}A(xq, q).$$

Differentiating (1) with respect to x and substituting in the preceding equations, we obtain (ii).

(b) Solving for $(\partial/\partial x)B(x, q)$ in (ii) of part (a) gives

$$\frac{\partial}{\partial x} B(x, q) = (1 - q)^{-1}\{T(x, q) - T(xq, q)\},$$

so the elimination of $(\partial/\partial x)B(x, q)$ between this equation and (i) of part (a) yields

$$\frac{\partial}{\partial x} A(x, q) = \exp\{(1 - q)^{-1}(A(x, q) - A(xq, q))\}$$

or

$$(q - 1)\frac{\partial}{\partial x}\exp\left\{\frac{-A(x, q)}{1 - q}\right\} = \exp\left\{\frac{-A(xq, q)}{1 - q}\right\}. \tag{2}$$

Now let $\exp\{-A(x,q)/(1-q)\} = \Sigma_{n\geq 0} a_n(q)x^n/n!$. Then, applying $[x^n/n!]$ to (2) for $n \geq 0$, we have $(q-1)a_{n+1}(q) = q^n a_n(q)$, so $a_n(q) = (q-1)^{-n}q^{\binom{n}{2}}$, since $a_0(q) = 1$, and the result follows.

3.3.49. (a) Consider the tree consisting of the path T whose ordered set of labels is $(2, 3, \ldots, n-1, n, 1)$. Now each of the $\binom{n-1}{2}$ pairs i, j, for $2 \leq i < j \leq n$ contributes exactly one tree inversion. Thus T is a graph on n vertices with the greatest number of tree inversions, so $A_n(q)$ is a polynomial of degree $\binom{n-1}{2}$ in q, since there are a finite number of labeled trees on n vertices.

(b) From [**3.3.48**(b)] we have

$$A_n(0) = \left[\frac{x^n}{n!}\right](-1)\log(1-x) = (n-1)!.$$

(c) $A_n(1)$ is the number of labeled trees on n vertices. By Cayley's result (**3.3.10**) this is equal to n^{n-2}.

(d) From [**3.3.48**(c)],

$$\sum_{n\geq 0} A_{n+1}(-1)\frac{x^n}{n!} = \frac{\partial}{\partial x}A(x,-1) = \frac{\Sigma_{i\geq 0}\frac{1}{i!}(-1)^{\binom{i}{2}}\left(\frac{x}{2}\right)^i}{\Sigma_{i\geq 0}\frac{1}{i!}(-1)^{\binom{i}{2}}\left(\frac{-x}{2}\right)^i}$$

$$= \frac{\cos\left(\frac{x}{2}\right) + \sin\left(\frac{x}{2}\right)}{\cos\left(\frac{x}{2}\right) - \sin\left(\frac{x}{2}\right)},$$

and the result follows by multiplying in the numerator and denominator by $\cos(x/2) + \sin(x/2)$.

SECTION 3.4

3.4.1. Consider Decomposition 3.4.1, and let x and y mark tagged rows and columns, z mark 1's, and u and v mark empty rows and columns. Now B contains no empty row or columns, and this is enumerated by $e^{-x-y}\Sigma_{i,j\geq 0} x^i y^j (1+z)^{ij}/i!j!$, from **3.4.3**. Every row and column in elements of Z is empty, and thus Z is enumerated by e^{ux+vy}. Accordingly, from Decomposition 3.4.1, the required number is

$$\left[z^p u^k v^l \frac{x^m}{m!}\frac{y^n}{n!}\right] e^{x(u-1)}e^{y(v-1)} \sum_{i,j\geq 0} \frac{x^i}{i!}\frac{y^j}{j!}(1+z)^{ij}$$

$$= \sum_{i,j\geq 0} \binom{m}{i}\binom{n}{j}\binom{ij}{p}[u^k v^l](u-1)^{m-i}(v-1)^{n-j},$$

and the result follows.

3.4.2. From **3.4.9** the generating function for A_2 is

$$A_2(x, y) = \exp\left\{-x - \frac{x^2}{2}y\right\} \sum_{m \geqslant 0} \frac{x^m}{m!}(1+y)^{\binom{m}{2}}.$$

Then the generating function for C_2 with x and y marking tagged rows and columns, and u marking distinct columns, is $C_2(x, y, u) = A_2(x, u(e^y - 1))$, from the decomposition in **3.4.9**. Thus the required number is

$$\left[u^k \frac{x^m}{m!} \frac{y^n}{n!}\right] \exp\left\{-x - \frac{x^2}{2}u(e^y - 1)\right\} \sum_{i \geqslant 0} \frac{x^i}{i!}(1 + u(e^y - 1))^{\binom{i}{2}}$$

$$= \left[\frac{x^m}{m!} \frac{y^n}{n!}\right](e^y - 1)^k e^{-x} \sum_{j \geqslant 0} \frac{(-x^2)^j}{2^j j!} \sum_{i \geqslant 0} \frac{x^i}{i!} \binom{\binom{i}{2}}{k-j},$$

and the result follows.

3.4.3. We use Decomposition 3.4.7, marking rows and columns by x and y, distinct columns by u, and distinct rows by v. From [**3.4.2**], the generating function for C_2 is

$$\exp\left\{-vx - u\frac{v^2x^2}{2}(e^y - 1)\right\} \sum_{i \geqslant 0} \frac{v^i x^i}{i!}(1 + u(e^y - 1))^{\binom{i}{2}},$$

since all rows are distinct. The generating function for $\mathsf{U} \circledast \mathsf{E}_2$ is

$$\exp\{uv\tfrac{1}{2}x^2(e^y - 1)\},$$

and we ignore W, since rows are nonempty. Thus the required number is

$$\left[u^k v^l \frac{x^m}{m!} \frac{y^n}{n!}\right] \exp\left\{-vx - uv\frac{x^2}{2}(e^y - 1)(v - 1)\right\} \sum_{i \geqslant 0} v^i \frac{x^i}{i!}(1 + u(e^y - 1))^{\binom{i}{2}}$$

$$= \left[v^l \frac{x^m}{m!} \frac{y^n}{n!}\right](e^y - 1)^k e^{-vx} \sum_{j \geqslant 0} \frac{(v(1-v)x^2)^j}{2^j j!} \sum_{i \geqslant 0} \frac{v^i x^i}{i!} \binom{\binom{i}{2}}{k-j}$$

$$= \left\{\sum_{t=0}^{k} \binom{k}{t} t^n (-1)^{k-t}\right\} m!$$

$$\times \sum_{i, j \geqslant 0} \binom{\binom{i}{2}}{k-j} \frac{(-1)^{m-i-2j} 2^{-j}}{i! j! (m-i-2j)!} [v^l] v^{m-j} (1-v)^j,$$

and the result follows.

3.4.4. The generating function for choosing distinct subsets of N_k, with x_j marking subsets of size j, for $j \geqslant 1$, is $\prod_{j \geqslant 1}(1 + x_j)^{\binom{k}{j}}$. This is because each of the $\binom{k}{j}$ subsets of size j is either chosen or not chosen, and thus has generating function $1 + x_j$. Considering all choices of k, we have generating functions

$$F(z, x_1, x_2, \ldots) = \sum_{k \geqslant 0} \frac{z^k}{k!} \prod_{j=1}^{k} (1 + x_j)^{\binom{k}{j}},$$

where z marks the tagged elements of N_k. But the set for which F is the generating function consists uniquely of a cover of some subset α of N_k, and the elements of $\mathsf{N}_k - \alpha$, none of whose elements are contained in any of the subsets. Thus $\mathsf{F} \tilde{\to} \mathsf{E} * \mathsf{C}$, where F is the set constructed earlier, with generating function F, C is the set of covers, whose generating function $C(z, x_1, x_2, \ldots)$ we wish to know, and E is the set containing elements not covered by elements of F. Thus E has generating function e^z, so that

$$C(z, x_1, x_2, \ldots) = e^{-z} F(z, x_1, x_2, \ldots).$$

The required number is

$$\left[x_1^{i_1} x_2^{i_2} \cdots \frac{z^n}{n!}\right] e^{-z} \sum_{k \geqslant 0} \frac{z^k}{k!} \prod_{j=1}^{k} (1 + x_j)^{\binom{k}{j}} = \sum_{k=0}^{n} (-1)^{n-k} \binom{n}{k} \prod_{j=1}^{k} \binom{\binom{k}{j}}{i_j}.$$

3.4.5. Let $M(x, y)$ be the generating function for matrices over N_n with x an exponential marker for rows and y an exponential marker for columns. Let $D(x, y)$ be the generating function for those of the preceding matrices that have distinct rows and distinct columns. Then we have

$$M(x, y) = \sum_{i \geqslant 0} \sum_{j \geqslant 0} n^{ij} \frac{x^i}{i!} \frac{y^j}{j!} = \sum_{i \geqslant 0} \frac{x^i}{i!} \exp(y n^i).$$

By repeating rows and columns (in either order) of the matrices with distinct rows and columns, we obtain uniquely all matrices over N_n. Thus

$$D(e^x - 1, e^y - 1) = M(x, y),$$

so that $D(z, w) = M(\log(1 + z), \log(1 + w))$. The required number is given by

$$\left[\frac{z^m}{m!} \frac{w^k}{k!}\right] D(z, w) = \left[\frac{z^m}{m!} \frac{w^k}{k!}\right] \sum_{i \geqslant 0} \frac{\log^i(1 + z)}{i!} (1 + w)^{n^i}$$

$$= \left[\frac{z^m}{m!}\right] \sum_{i \geqslant 0} k! \binom{n^i}{k} [u^i](1 + z)^u.$$

3.4.6. Let $M(x, y)$ be the generating function for matrices over N_n with x and y exponential markers for rows and columns, respectively. We have immediately that

$$M(x, y) = \sum_{i \geqslant 0} \sum_{j \geqslant 0} \frac{x^i}{i!} \frac{y^j}{j!} n^{ij}.$$

Let $D(u, x, y)$ be the generating function for these matrices in which u is an "at-least" marker for rows or columns consisting entirely of i's for some $i = 1, \ldots, n$. By adjoining rows and columns that consist entirely of i's to a matrix, we get

$$D = \{1 + n(e^{ux} - 1) + n(e^{uy} - 1) + (e^{nux} - 1)(e^{nuy} - 1)\} M(x, y),$$

so the required number is

$$c(m, k) = \left[\frac{x^m}{m!} \frac{y^k}{k!} \right] D(-1, x, y)$$

from the Principle of Inclusion and Exclusion. The result follows.

3.4.7. (a) Let the labeled vertices of a graph be tagged objects marked by x and the edges be untagged objects marked by y. Then the generating function for all simple graphs is

$$G(x, y) = \sum_{m \geqslant 0} \frac{x^m}{m!} (1 + y)^{\binom{m}{2}}$$

since between each of the $\binom{m}{2}$ distinct pairs of vertices we can either choose to have no edge or one edge, with generating function $1 + y$. Let $B(x, y)$ be the generating function for the subset of these graphs that are connected. Then since a graph can be expressed uniquely as an unordered set of connected components, we have $G(x, y) = \exp B(x, y)$, and thus

$$B(x, y) = \log G(x, y) = \log\left(\sum_{m \geqslant 0} \frac{x^m}{m!} (1 + y)^{\binom{m}{2}} \right).$$

(b) From part (a), the graphs with k components are represented uniquely as an unordered collection of k components, each enumerated by $B(x, y)$. Thus the generating function for graphs with k components is

$$\frac{1}{k!} (B(x, y))^k = \frac{1}{k!} \left(\log\left\{ \sum_{m \geqslant 0} \frac{x^m}{m!} (1 + y)^{\binom{m}{2}} \right\} \right)^k.$$

3.4.8. (a) From [3.4.7(b)], if z is an ordinary marker for components, then the required generating function is

$$\sum_{k\geqslant 0} z^k \frac{1}{k!}\left(\log\left\{\sum_{m\geqslant 0}\frac{x^m}{m!}(1+y)^{\binom{m}{2}}\right\}\right)^k$$

$$= \exp\left(z\log\left\{\sum_{m\geqslant 0}\frac{x^m}{m!}(1+y)^{\binom{m}{2}}\right\}\right) = \left\{\sum_{m\geqslant 0}\frac{x^m}{m!}(1+y)^{\binom{m}{2}}\right\}^z.$$

(b) If $m_1 = m_2 = \cdots = m_k = 0$, then $m_1x_1 + \cdots + m_kx_k$ is identically 0, so each of $x_1,\ldots, x_k$ can take on any of the p values in GF(p). Thus there are p^k solutions. Otherwise, suppose that $m_i \neq 0$ for some i. Then if each of $x_1,\ldots, x_{i-1}, x_{i+1},\ldots, x_k$ takes on any value in GF(p) we can determine x_i uniquely by

$$x_i = m_i^{-1}\sum_{\substack{j=1\\ j\neq i}}^{k} m_jx_j,$$

since m_i^{-1} exists when $m_i \neq 0$. Thus there are p^{k-1} solutions in this case.

(c) Consider the $\binom{n}{2}$ properties $x_i = x_j$, $1 \leqslant i < j \leqslant n$. We are interested in solutions having none of these properties. The number of solutions having at least a given set of k properties $x_{i_1} = x_{j_1},\ldots, x_{i_k} = x_{j_k}$ is determined by constructing the graph on n labeled vertices with edge set $(i_1, j_1),\ldots,(i_k, j_k)$. If this graph has i components, then solutions having at least this set of properties must have exactly i disjoint groups of x's that are forced to be equal. Solutions with this constraint are solutions to a new equation, with i x's, each of whose coefficients is the sum of the m's corresponding to the vertices in the appropriate component. But these sums of m's cannot be 0, unless there is $i = 1$ component in the graph, and so there are $p^{i-1+\delta_{i1}}$ solutions for each such set of properties, from part (b). Accordingly, there are

$$N(\mathbf{m}) = \sum_{k=0}^{\binom{n}{2}}(-1)^k\sum_{i=1}^{n} p^{i-1+\delta_{i1}}g_n(i,k)$$

solutions with distinct x_i's, by the Principle of Inclusion and Exclusion, where $\mathbf{m} = (m_1,\ldots, m_n)$.

(d) From part (a), we know that

$$\sum_{n\geqslant 0}\frac{x^n}{n!}\sum_{k=0}^{\binom{n}{2}} y^k\sum_{i=1}^{n} z^i g_n(i,k) = \left\{\sum_{m\geqslant 0}\frac{x^m}{m!}(1+y)^{\binom{m}{2}}\right\}^z.$$

Now we want $N(\mathbf{m})$

$$= p^{-1} \sum_{k=0}^{\binom{n}{2}} (-1)^k \sum_{i=1}^{n} p^i g_n(i,k) + (p-1) \sum_{k=0}^{\binom{n}{2}} (-1)^k g_n(1,k)$$

$$= p^{-1}\left[\frac{x^n}{n!}\right]\left\{\sum_{m\geqslant 0} \frac{x^m}{m!}(1-1)^{\binom{m}{2}}\right\}^p + (p-1)\left[\frac{x^n}{n!}\right]\log\left\{\sum_{m\geqslant 0}\frac{x^m}{m!}(1-1)^{\binom{m}{2}}\right\}$$

$$= \left[\frac{x^n}{n!}\right]\{p^{-1}(1+x)^p + (p-1)\log(1+x)\},$$

and the result follows.

3.4.9. (a) We construct such graphs by distinguishing a vertex in each block and by identifying these vertices to form a distinguished vertex incident with k blocks. The construction of all such labeled graphs is completed by identifying each of the nonroot vertices in the resulting graph with the root vertex of an arbitrary connected simple rooted graph. The first stage of this process is enumerated by $x(B'(x))^k/k!$, where the single x represents the root vertex. The second stage is completed by replacing the x's marking nonroot vertices by $xC'(x)$, the generating function for connected simple graphs rooted at a distinguished vertex. This yields the required generating function by the $*$-composition lemma.

(b) A rooted connected simple graph either is a single vertex or has the root vertex connected to k blocks, for some unique $k \geqslant 1$. Thus we have two expressions for the generating function for rooted connected simple labeled graphs. The first, $xC'(x)$, is immediate, and the second, $x + \sum_{k\geqslant 1} x(B'(xC'(x)))^k/k! = x\exp B'(xC'(x))$, follows from part (a). Thus $xC'(x) = x\exp B'(xC'(x))$, so $C'(x) = \exp B'(xC'(x))$, and taking logarithms of both sides, $B'(xC'(x)) = \log C'(x)$.

3.4.10. We associate a unique tree $T \in \mathsf{A}$ with a graph $g \in \mathsf{C}$, the set of all connected graphs in G, by the following depth first search algorithm.

Suppose T has n vertices. Let $i = 0$, $v_0 = 1$, and $\mathbf{E}_0 = \varnothing$. Let $\mathbf{N}(v_i)$ denote the set of vertices adjacent to v_i. If $\mathbf{N}(v_i) \subset \{v_0, \ldots, v_i\}$, then backtrack to the previous vertex. Otherwise let v_{i+1} be the vertex in $\mathbf{N}(v_i) - \{v_0, \ldots, v_i\}$ with smallest label. Let $\mathbf{E}_{i+1} = \mathbf{E}_i \cup \{v_i v_{i+1}\}$, where $v_i v_{i+1}$ denotes the edge joining v_i and v_{i+1}. Then $\mathbf{E}_n$ is the edge set for a unique spanning tree T of g.

Let (k, l) be a tree inversion in T, where $k < l$. Let l' be the unique vertex adjacent to l on the path from l to 1 in T. This associates a unique edge kl' with k. Let $\mathbf{E}_T$ be the set of all such edges associated with T. Let $\mathbf{E}(g), \mathbf{V}(g)$ denote the edge sets and vertex sets of g, respectively. If $\mathbf{V}(g') = \mathsf{N}_n$ and $\mathbf{E}(g') \subset \mathbf{E}(T) \cup \mathbf{E}_T$, then T is the spanning tree associated with g' by the recursive algorithm. Thus

$$\sum_{\substack{g\in\mathsf{C}\\ |\mathbf{V}(g)|=n}} q^{|\mathbf{E}(g)|} = \sum_{\substack{T\in\mathsf{A}\\ |\mathbf{V}(T)|=n}} q^{n-1}(1+q)^{I(T)} = q^{n-1}A_n(1+q),$$

where $I(T)$ denotes the number of tree inversions in T. Then

$$\sum_{n\geqslant 1}\frac{x^n}{n!}q^{n-1}A_n(1+q)=\sum_{n\geqslant 1}\frac{x^n}{n!}\sum_{\substack{g\in \mathsf{C}\\ |\mathbf{V}(g)|=n}} q^{|\mathbf{E}(g)|}$$

$$=\log\left\{\sum_{i\geqslant 0}\frac{x^i}{i!}(q+1)^{i(i-1)/2}\right\},$$

as required.

As an example of the recursive algorithm note that the unique spanning tree associated with

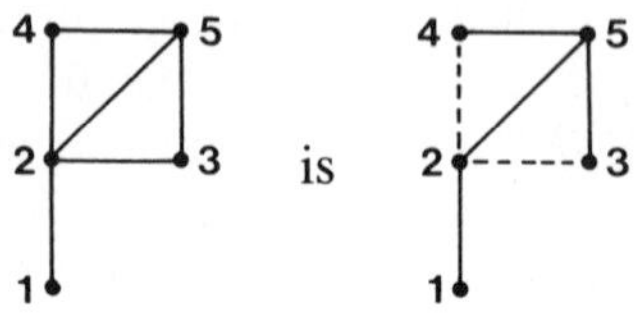

where the dotted lines are the edges in $\mathbf{E}_T$. Note that the tree inversions of T are $(3,5),(4,5)$ and the vertex adjacent to 5 on the path from 5 to 1 is 2. The dotted edges are 42 and 32.

3.4.11. We modify Decomposition 3.4.12 and follow **3.4.14** in the following way.

(i) For C: The cycles can now have one edge or more. There are $\frac{1}{2}(i-1)!$ cycles on i vertices for $i\geqslant 3$, and 1 for $i=1,2$. Thus

$$\Psi_{\mathsf{C}}(x,y,z)=\tfrac{1}{2}\{(xyz)+\tfrac{1}{2}(xyz)^2+\cdots\}+\tfrac{1}{2}xyz+\tfrac{1}{4}(xyz)^2$$

$$=\tfrac{1}{2}\log(1-xyz)^{-1}+\tfrac{1}{2}xyz+\tfrac{1}{4}(xyz)^2.$$

(ii) For L: The subdivided loops can now contain one initial vertex. There are $\frac{1}{2}i!$ elements of L on i noninitial vertices for $i\geqslant 2$, and one element with $i=1$. We also allow any number of loops to be placed at each initial vertex. Thus

$$\Psi_{\mathsf{U}\circledast\mathsf{L}}(x,y,z)=(1-x)^{-1}\exp\{\tfrac{1}{2}(x^2yz+x^3y^2z^2+\cdots)+\tfrac{1}{2}x^2yz\}$$

$$=(1-x)^{-1}\exp\{\tfrac{1}{2}x^2yz(1+(1-xyz)^{-1})\}.$$

(iii) For M: We can now have any number of unsubdivided edges between any pair of vertices, enumerated by $(1-x)^{-1}$. M_1 remains the same, so

$$\Psi_{\mathsf{M}}(x,y,z)=(1-x)^{-1}\exp\{x^2yz(1-xyz)\}^{-1}.$$

Thus, in the notation of **3.4.14**,

$$A(x,y,z)=(\exp\Psi_{\mathsf{C}})\sum_{j\geq 0}\frac{y^j}{j!}\{\Phi_{\mathsf{U}\circledast\mathsf{L}}\}^j\Psi_{\mathsf{M}}^{\binom{j}{2}}$$

and the required generating function is

$$H(x,y,0)=A(x,y,-1)$$

$$=(1+xy)^{-1/2}\exp(-\tfrac{1}{2}xy+\tfrac{1}{4}x^2y^2)$$

$$\times\sum_{j\geq 0}\frac{y^j}{j!}\left\{(1-x)^{-1}\exp\left(\frac{-x^2y}{1+xy}\right)\right\}^{\binom{j}{2}}\left\{(1-x)^{-1}\exp\left(\frac{-\frac{1}{2}x^2y}{1+xy}-\tfrac{1}{2}x^2y\right)\right\}^j,$$

and the result follows.

3.4.12. We follow the solution to [**3.4.11**], with two modifications. Cycles cannot have a single vertex, so that Ψ_{C} becomes

$$\tfrac{1}{2}(xyz)^2+\tfrac{1}{2}\{\tfrac{1}{3}(xyz)^3+\cdots\}=\tfrac{1}{2}\log(1-xyz)^{-1}-\tfrac{1}{2}xyz+\tfrac{1}{4}(xyz)^2.$$

We cannot place any loops at initial vertices, so that $\Psi_{\mathsf{U}\circledast\mathsf{L}}$ becomes $\exp\{\frac{1}{2}x^2yz(1+(1-xyz)^{-1})\}$. Thus the required generating function is

$$H(x,y,0)=A(x,y,-1)$$

$$=(1+xy)^{-1/2}\exp(\tfrac{1}{2}xy+\tfrac{1}{4}x^2y^2)$$

$$\times\sum_{j\geq 0}\frac{y^j}{j!}\left\{(1-x)^{-1}\exp\left(\frac{-x^2y}{1+xy}\right)\right\}^{\binom{j}{2}}\left\{\exp\left(\frac{-\frac{1}{2}x^2y}{1+xy}-\tfrac{1}{2}x^2y\right)\right\}^j,$$

and the result follows.

3.4.13. We follow the solution to [**3.4.11**] with slight modifications. We can no longer allow more than one edge that is not subdivided in M, so Ψ_{M} is $(1+x)\exp\{x^2yz(1-xyz)^{-1}\}$, as in **3.4.14**. The subdivided loops placed at each vertex must have at least two noninitial vertices, and at most one loop can be placed at each vertex, so $\Psi_{\mathsf{U}\circledast\mathsf{L}}$ becomes $(1+x)\exp\{\frac{1}{2}x^3y^2z^2(1+xyz)^{-1}\}$.

Finally, cycles with two edges are not allowed in C, so Ψ_{C} becomes

$$xyz + \tfrac{1}{2}\{\tfrac{1}{3}(xyz)^3 + \tfrac{1}{4}(xyz)^4 + \cdots\} = \tfrac{1}{2}\log(1 - xyz)^{-1} + \tfrac{1}{2}xyz - \tfrac{1}{4}(xyz)^2.$$

Thus the required function is

$$\begin{aligned} H(x, y, 0) = A(x, y, -1) &= (1 + xy)^{-1/2}\exp(-\tfrac{1}{2}xy - \tfrac{1}{4}x^2y^2) \\ &\times \sum_{j\geqslant 0} \frac{y^j}{j!}\{(1 + x)\exp(-x^2y(1 + xy)^{-1})\}^{\binom{j}{2}} \\ &\times \{(1 + x)\exp\{\tfrac{1}{2}x^2y^2(1 + xy)^{-1}\}^j\}, \end{aligned}$$

and the result follows.

3.4.14. From [**3.4.13**],

$$f(n) = \left[\frac{y^n}{n!}\right] \sum_{j\geqslant 0} \frac{y^j}{j!} 2^{\binom{j+1}{2}} g_j(y),$$

where

$$\begin{aligned} \log g_j(y) &= -\tfrac{1}{2}\log(1 + y) - \tfrac{1}{2}y - \tfrac{1}{4}y^2 - \frac{y}{2(1 + y)}j^2 + \frac{y}{2}j \\ &= -\tfrac{1}{2}\log(1 + y) + \tfrac{1}{2}y(j - 1) - \tfrac{1}{4}y^2 - \tfrac{1}{2}j^2 + \frac{j^2}{2(1 + y)}. \end{aligned}$$

Differentiating with respect to y gives

$$\frac{g_j'(y)}{g_j(y)} = -\tfrac{1}{2}(1 + y)^{-1} + \tfrac{1}{2}(j - 1) - \tfrac{1}{2}y - \tfrac{1}{2}j^2(1 + y)^{-2}.$$

Multiplying both sides by $2(1 + y)^2 g_j(y)$ yields

$$\begin{aligned} &2(1 + 2y + y^2)g_j'(y) \\ &\qquad + \{2 - j + j^2 + (4 - 2j)y + (3 - j)y^2 + y^3\}g_j(y) = 0. \end{aligned}$$

Now let $g_j(y) = \sum_{k\geqslant 0} b_j(k)y^k$ and compare coefficients of y^m in the preceding equation to get

$$\begin{aligned} &2(m + 1)b_j(m + 1) + 4m\, b_j(m) + 2(m - 1)b_j(m - 1) + (2 - j + j^2)b_j(m) \\ &\qquad + (4 - 2j)b_j(m - 1) + (3 - j)b_j(m - 2) + b_j(m - 3) = 0. \end{aligned}$$

The result follows.

3.4.15. Let $\mathsf{M}, \mathsf{M}_{00}, \mathsf{M}_{02}, \mathsf{M}_{20}, \mathsf{M}_{11}, \mathsf{T}$ be sets of matrices with tagged s- and t-objects that are, respectively, rows and columns. Consider the $\{0, 1\}$-matrices with i row sums and j column sums equal to 1 and all other row and column sums equal to 2. Then M_{ij} is the set of all such matrices that have no proper submatrix induced by the nonzero elements in a subset of the rows and columns. The set T contains only the 1×1 matrix with single element 2, and M is the set of matrices over $\{0, 1, 2\}$ with all row and column sums equal to 1 or 2. Then, if $\mathsf{U} = \{\varnothing, \mathsf{N}_1, \mathsf{N}_2, \dots\}$, we have the decomposition

$$\mathsf{M} \xrightarrow{\sim} \mathsf{U} \circledast (\mathsf{T} \,\dot\cup\, \mathsf{M}_{00} \,\dot\cup\, \mathsf{M}_{02} \,\dot\cup\, \mathsf{M}_{20} \,\dot\cup\, \mathsf{M}_{11}),$$

by finding the permutation of rows and columns of an element of M that yields a direct sum of matrices in $\mathsf{T}, \mathsf{M}_{00}, \mathsf{M}_{02}, \mathsf{M}_{20}, \mathsf{M}_{11}$ alone. The $*$-composition is with respect to s- and t-objects. Let x and y be exponential markers for rows and columns, and x_i, y_i be ordinary markers for rows and columns with sum equal to i, for $i = 1, 2$. The required number is thus

$$\left[x_1^{m_1} x_2^{m_2} \frac{x^{m_1+m_2}}{(m_1 + m_2)!} y_1^{n_1} y_2^{n_2} \frac{y^{n_1+n_2}}{(n_1 + n_2)!} \right] \exp(T + M_{00} + M_{02} + M_{20} + M_{22}),$$

where T and M_{ij} are the generating functions for T and M_{ij}, respectively, and are determined below.

(i) We immediately obtain $T = x x_2 y y_2$.

(ii) From permutations $\rho_1 \cdots \rho_n, \sigma_1 \cdots \sigma_n$ on N_n we construct the $n \times n$ element of M_{00} with 1's in positions (ρ_j, σ_j) and (ρ_j, σ_{j+1}) for $j = 1, \dots, n$, where $\sigma_{n+1} = \sigma_1$. But if we carry this out for all $(n!)^2$ pairs of such permutations, we construct every element of M_{00} exactly $2n$ times for $n \geqslant 2$. Thus

$$\begin{aligned} M_{00} &= \sum_{n \geqslant 2} \frac{(n!)^2}{2n} \frac{(x x_2)^n}{n!} \frac{(y y_2)^n}{n!} \\ &= \tfrac{1}{2} \log(1 - x x_2 y y_2)^{-1} - \tfrac{1}{2} x x_2 y y_2. \end{aligned}$$

(iii) From permutations $\rho_1 \cdots \rho_n$ on N_n and $\sigma_1 \cdots \sigma_{n+1}$ on N_{n+1} we construct the $n \times (n+1)$ element of M_{02} with 1's in positions (ρ_j, σ_j) and (ρ_j, σ_{j+1}) for $j = 1, \dots, n$. If we carry this out for all $n!(n+1)!$ pairs of such permutations, we construct every element of M_{02} exactly twice for $n \geqslant 1$. Thus

$$\begin{aligned} M_{02} &= \tfrac{1}{2} \sum_{n \geqslant 1} n!(n+1)! x_2^n \frac{x^n}{n!} y_1^2 y_2^{n-1} \frac{y^{n+1}}{(n+1)!} \\ &= \tfrac{1}{2} x x_2 (y y_1)^2 (1 - x x_2 y y_2)^{-1}. \end{aligned}$$

(iv) The matrices in M_{20} are the transposes of those in M_{02}, so

$$M_{20} = \tfrac{1}{2} yy_2 (xx_1)^2 (1 - xx_2 yy_2)^{-1}.$$

(v) A similar construction yields

$$M_{11} = xx_1 yy_1 (1 - xx_2 yy_2)^{-1}.$$

The result follows by setting $x = y = 1$.

SECTION 3.5

3.5.1. We have $T(\mathbf{x}) = G(\mathbf{s}(\mathbf{x}))$, where $G(\mathbf{s}) = \{1 - \Sigma_{i \geqslant 1} (-1)^{i-1} s_i\}^{-1}$, so

$$\frac{\partial G}{\partial s_i} = (-1)^{i+1} G^{-2}, \qquad \text{for } i \geqslant 1.$$

Thus

$$\frac{\partial G}{\partial s_i} = (-1)^{i+1} \frac{\partial G}{\partial s_1}, \qquad \text{for } i \geqslant 1,$$

so from the Γ-series theorem,

$$E_i\left(\frac{\partial}{\partial \mathbf{y}}\right)\Gamma(G) = (-1)^{i+1} E_1\left(\frac{\partial}{\partial \mathbf{y}}\right)\Gamma(G) = (-1)^{i+1}\frac{\partial}{\partial y_1}\Gamma(G).$$

It follows that

$$\sum_{i \geqslant 1} z^i E_i\left(\frac{\partial}{\partial \mathbf{y}}\right)\Gamma(G) = \sum_{i \geqslant 1} (-1)^{i+1} z^i \frac{\partial}{\partial y_1}\Gamma(G),$$

whence, from Proposition 3.5.6,

$$\left(\log\left\{1 + \sum_{i \geqslant 1} z^i \frac{\partial}{\partial y_i}\right\}\right)\Gamma(G) = z(1+z)^{-1}\frac{\partial}{\partial y_1}\Gamma(G).$$

Thus

$$\left\{1 + \sum_{i \geqslant 1} z^i \frac{\partial}{\partial y_i}\right\}\Gamma(G) = \left\{\exp\left(z(1+z)^{-1}\frac{\partial}{\partial y_1}\right)\right\}\Gamma(G).$$

Extracting the coefficient of z^i, we have

$$\frac{\partial}{\partial y_i}\Gamma(G) = P_i\left(\frac{\partial}{\partial y_1}\right)\Gamma(G), \qquad \text{where } P_i(x) = (-1)^i L_i^{(-1)}(x)$$

so

$$c(\mathbf{i}) = \left[\frac{\mathbf{y}^{\mathbf{f}}}{\mathbf{f}!}\right]\Gamma(G) = \left\{\prod_{i\geqslant 1}\frac{\partial^{f_i}}{\partial y_i^{f_i}}\right\}\Gamma(G)\Bigg|_{\mathbf{y}=\mathbf{0}}$$

$$= \left\{\prod_{i\geqslant 1} P_i^{f_i}\left(\frac{\partial}{\partial y_1}\right)\right\}\Gamma(G)\Bigg|_{\mathbf{y}=\mathbf{0}}$$

$$= \sum_{m\geqslant 0}\left\{[x^m]\prod_{i\geqslant 1}P_i^{f_i}(x)\right\}\frac{\partial^m}{\partial y_1^m}\Gamma(G)\Bigg|_{\mathbf{y}=\mathbf{0}}$$

$$= \sum_{m\geqslant 0}\left\{[x^m]\prod_{i\geqslant 1}P_i^{f_i}(x)\right\}\left[\frac{y_1^m}{m!}\right]\Gamma(G).$$

But $[y_1^m/m!]\Gamma(G)$ is equal to the number of Smirnov sequences that contain one of each of m distinct objects. There are $m!$ of these, since every permutation of the m objects has distinct adjacent elements. The result follows.

3.5.2. (a) From **2.4.18**, the required number is $[\mathbf{x}^{\mathbf{i}}]\{1 + \sum_{k\geqslant 1} f_k s_k\}^{-1}$, where $1 + \sum_{k\geqslant 1} f_k x^k = (1 + x^2 + x^3 + \cdots)^{-1} = (1 - x^2)(1 + x^3)^{-1}$, since maximal blocks cannot have length 1. The result follows.

(b) We have the generating function

$$G(\mathbf{s}) = \left\{1 + \sum_{k\geqslant 1}(-1)^k(s_{3k-1} + s_{3k})\right\}^{-1}$$

so

$$\frac{\partial}{\partial s_{3k}}G(\mathbf{s}) = \frac{\partial}{\partial s_{3k-1}}G(\mathbf{s}) = (-1)^k G(\mathbf{s})^{-2}$$

and

$$\frac{\partial}{\partial s_{3k-2}}G(\mathbf{s}) = 0, \qquad \text{for } k \geqslant 1.$$

All of these derivatives can be related to $(\partial/\partial s_2)G(s)$ as follows.

$$\frac{\partial}{\partial s_{3k}}G(\mathbf{s}) = (-1)^{k-1}\frac{\partial}{\partial s_2}G$$

$$\frac{\partial}{\partial s_{3k-1}}G(\mathbf{s}) = (-1)^{k-1}\frac{\partial}{\partial s_2}G$$

$$\frac{\partial}{\partial s_{3k-2}}G(\mathbf{s}) = 0, \qquad k \geqslant 1.$$

Multiplying by the appropriate power of z and summing gives

$$\sum_{n\geq 1} z^n \frac{\partial}{\partial s_n} G(\mathbf{s}) = \frac{z^2(1+z)}{1+z^3} \frac{\partial}{\partial s_2} G(\mathbf{s}).$$

Now, if $\mathbf{f} = (f_1, f_2, \ldots) = \tau(\mathbf{i})$ is the type of $\mathbf{i}$, then $[\mathbf{x}^{\mathbf{i}}]G(\mathbf{s}) = [\mathbf{y}^{\mathbf{f}}/\mathbf{f}!]\Gamma(G)$, and applying the Γ-series theorem to the preceding differential equation, we have

$$\left\{ \sum_{n\geq 1} z^n E_n\left(\frac{\partial}{\partial \mathbf{y}}\right)\right\} \Gamma(G) = \frac{z^2(1+z)}{1+z^3}\left(\frac{\partial}{\partial y_2} - \frac{1}{2}\frac{\partial^2}{\partial y_1^2}\right)\Gamma(G).$$

But $\sum_{n\geq 1} z^n E_n(\mathbf{x}) = \log(1 + zx_1 + z^2x_2 + \cdots)$, and $\partial G/\partial s_1 = 0$, so $(\partial^2/\partial y_1^2)\Gamma(G) = 0$, and

$$\log\left(1 + z\frac{\partial}{\partial y_1} + z^2\frac{\partial}{\partial y_2} + \cdots\right)\Gamma(G) = \frac{z^2(1+z)}{1+z^3}\frac{\partial}{\partial y_2}\Gamma(G).$$

Now the operators applied to $\Gamma(G)$ on both sides commute, so taking the exponential of both,

$$\left(1 + z\frac{\partial}{\partial y_1} + z^2\frac{\partial}{\partial y_2} + \cdots\right)\Gamma(G) = \exp\left(\frac{z^2(1+z)}{1+z^3}\frac{\partial}{\partial y_2}\right)\Gamma(G).$$

Taking coefficients of z^i on both sides, we have

$$\begin{aligned}
\frac{\partial}{\partial y_i}\Gamma(G) &= [z^i]\left(\sum_{k\geq 0} \frac{z^{2k}(1+z)^k}{k!(1+z^3)^k}\frac{\partial^k}{\partial y_2^k}\right)\Gamma(G) \\
&= [z^i]\left(\sum_{k\geq 0}\sum_{l\geq 0}\binom{-k}{l}\frac{z^{2k+3l}}{k!}(1+z)^k\frac{\partial^k}{\partial y_2^k}\right)\Gamma(G) \\
&= \sum_{k\geq 0}\sum_{l\geq 0}\binom{-k}{l}\binom{k}{i-2k-3l}\frac{1}{k!}\frac{\partial^k}{\partial y_2^k}\Gamma(G).
\end{aligned}$$

Now let

$$P_i(x) = \sum_{k\geq 0}\sum_{l\geq 0}\binom{-k}{l}\binom{k}{i-2k-3l}\frac{x^k}{k!},$$

so that

$$\begin{aligned}
\left[\frac{\mathbf{y}^{\mathbf{f}}}{\mathbf{f}!}\right]\Gamma(\mathbf{G}) &= \left\{\frac{\partial^{f_1+f_2+\cdots}}{\partial y_1^{f_1}\partial y_2^{f_2}\cdots}\Gamma(G)\right\}\Bigg|_{\mathbf{y}=\mathbf{0}} \\
&= \sum_{m\geq 0}\left\{\frac{\partial^m}{\partial y_2^m}\Gamma(G)\right\}\Bigg|_{\mathbf{y}=\mathbf{0}}[x^m]P_1^{f_1}(x)P_2^{f_2}(x)\cdots.
\end{aligned}$$

However,

$$\left.\frac{\partial^m}{\partial y_2^m}\Gamma(G)\right|_{\mathbf{y}=\mathbf{0}} = \left[\frac{y_2^m}{m!}\right]\Gamma(G) = m!,$$

since each of m types of objects appears in adjacent pairs, with all $m!$ permutations of these pairs giving the set of allowable sequences. The result follows.

3.5.3. The generating function for labeled graphs (loops and multiple edges allowed) with t_i marking the degree of vertex i is

$$\prod_{1 \leqslant i \leqslant j} (1 - t_i t_j)^{-1} = G(\mathbf{s}(\mathbf{t})),$$

where $G(\mathbf{s}) = \exp\{\Sigma_{k \geqslant 1}(s_k^2 + s_{2k})/2k\}$. Thus G satisfies the system

$$\left.\begin{aligned} \frac{\partial G}{\partial s_{2k+1}} &= \frac{1}{2k+1} s_{2k+1} G \qquad && \text{for } k \geqslant 0 \\ \frac{\partial G}{\partial s_{2k}} &= \frac{1}{2k}(1 + s_{2k})G \qquad && \text{for } k \geqslant 1. \end{aligned}\right\}$$

Let $V(y_1, y_2, y_3) = \Gamma(G)(y_1, y_2, y_3, 0, \dots)$ and apply the Γ-series theorem to the preceding system, giving

$$\left.\begin{aligned} V_1 &= y_1 V + y_2 V_1 + y_3 V_2 \\ 2V_2 - V_{11} &= (1 + y_2)V + y_3 V_1 \\ 3V_3 - 3V_{12} + V_{111} &= y_3 V, \end{aligned}\right\}$$

where V_i denotes $\partial V / \partial y_i$. The result follows by following the elimination scheme used in **3.5.9**.

3.5.4. From the Γ-series theorem we obtain

$$V_1 = y_1 V + y_2 V_1 + y_3 V_2 + y_4 V_3 \tag{1}$$

$$2V_2 - V_{11} = -(1 + y_2)V - y_3 V_1 - y_4 V_2 \tag{2}$$

$$3V_3 - 3V_{12} + V_{111} = y_3 V + y_4 V_1 \tag{3}$$

$$4V_4 - 4V_{13} - 2V_{22} + 4V_{112} - V_{1111} = (1 - y_4)V, \tag{4}$$

where V_i denotes $\partial V / \partial y_i$. Equations (1), (2), (3), and $(\partial / \partial y_1)$(2) form a system of linear equations, from which we eliminate V_{12}, V_2, and V_3 to give a linear equation, (5), for V, V_1, V_{11}, V_{111}. We thus express each of V_{111}, V_2, and V_3 in terms of V, V_1, V_{11} by means of equations (5), (2), and (1), respectively. Using these equations we then express $V_{13}, V_{22}, V_{112}, V_{1111}$ in terms of V, V_1, V_{11}, and thus eliminate $V_{13}, V_{22}, V_{112}, V_{1111}$ from (4) to give an expression for V_4 in terms of V_{11}, V_1, V. Continuing, we express V_{44} in terms of V_{11}, V_1, V. When $y_1 = y_2 = y_3 = 0$, V_1 has coefficient 0 in both these expressions, so we eliminate V_{11} between these expressions to give an equation involving V, V_4, and V_{44}, where $R_4(x) = V$ at $y_1 = y_2 = y_3 = 0$, $y_4 = x$. This equation is

$$\sum_{i=0}^{2} \phi_i(x) \frac{d^i}{dx^i} R_4(x) = 0$$

where

$$\phi_0(x) = -x^4(x^5 + 2x^4 + 2x^2 + 8x - 4)^2,$$

$$\phi_2(x) = 16x^2(x-1)^2(x+2)^2(x^5 + 2x^4 + 2x^2 + 8x - 4),$$

$$\phi_1(x) = -4(x^{13} + 4x^{12} - 16x^{10} - 10x^9 - 36x^8 - 220x^7 - 348x^6$$
$$-48x^5 + 200x^4 - 336x^3 - 240x^2 + 416x - 96),$$

and was obtained by use of the computer and an algebraic manipulation system.

3.5.5. From the Γ-series theorem we obtain

$$V_{1,} = y_1V + y_2V_{,1} + y_3V_{,2} \quad (1); \qquad V_{,1} = x_1V + x_2V_{1,} + x_3V_{2,} \quad (1)'$$

$$2V_{2,} - V_{11,} = y_2V + y_3V_{,1} \quad (2); \qquad 2V_{,2} - V_{,11} = x_2V + x_3V_{1,} \quad (2)'$$

$$3V_{3,} - 3V_{12,} + V_{111,} = y_3V \quad (3); \qquad 3V_{,3} - 3V_{,12} + V_{,111} = x_3V \quad (3)'$$

where $V_{i,j}$ denotes $\partial^2 V/\partial x_i \partial y_j$. Eliminating $V_{2,}$ between (1)′ and (2) we get a linear equation, (4), for $V_{11,}, V_{,1}, V_{1,}, V$. Similarly, we eliminate $V_{,2}$ between (1) and (2)′ to get a linear equation, (4)′, for $V_{,11}, V_{,1}, V_{1,}, V$. Thus we express $V_{,2}, V_{2,}, V_{,11}, V_{11,}$ in terms of $\mathsf{S} = \{V, V_{1,}, V_{,1}, V_{1,1}\}$ by means of (1), (1)′, (4)′, (4). Using these equations, we express $V_{12,}, V_{,12}, V_{111,}, V_{,111}$ in terms of S and substitute into (3) and (3)′ to obtain expressions for $V_{,3}$ and $V_{3,}$ in terms of S. Continuing, we express $V_{3,3}$ and $V_{33,33}$ linearly in terms of the elements of S. When $x_1 = x_2 = y_1 = y_2 = 0$, $V_{,1}$ and $V_{1,}$ have coefficients equal to 0 in both these expressions, so we eliminate $V_{1,1}$ to give an equation involving $V_{33,33}$, $V_{3,3}$, and V. But $V(0,0,x_3;0,0,y_3) = F(x_3y_3)$, and this equation yields the differential equation

$$\sum_{i=0}^{4} \phi_i(x) \frac{d^i}{dx^i} F(x) = 0,$$

where $\phi_4(x) = 81x^5(x^4 - x^2 + x + 4)$, $\phi_3(x) = 324x^4(x^4 - x^2 + x + 4)$,

$$\phi_2(x) = -9x(x^{10} - 4x^9 + 22x^8 - 8x^7 - 22x^6 + 8x^5 + 106x^4 + 234x^3 + 48x^2 - 320x + 64),$$

$$\phi_1(x) = -9(x^{10} - 4x^9 + 22x^8 - 8x^7 - 4x^6 + 8x^5 + 88x^4 + 252x^3 + 120x^2 - 320x + 64),$$

$$\phi_0(x) = x^{11} - 7x^{10} + 30x^9 - 16x^8 - 43x^7 + 51x^6 + 238x^5 + 630x^4 + 36x^3 - 1944x^2 - 1152x + 576.$$

This equation was obtained by use of the computer and an algebraic manipulation system.

3.5.6. (a) A proper k-cover of N_n is a set of distinct subsets that contain, together, k copies of each element of N_n. Each subset $\{\alpha_1,\ldots,\alpha_i\}$ is either chosen or not chosen to appear in the k-cover. If the occurrences of each object $j \in \mathsf{N}_n$ are marked by x_j, then the generating function for $\{\alpha_1,\ldots,\alpha_i\}$ is $1 + x_{\alpha_1}\cdots x_{\alpha_i}$, and for all subsets is $\prod_{i\geqslant 1}\prod_\alpha(1 + x_{\alpha_1}\cdots x_{\alpha_i})$. We extract the coefficient of $x_1^k \cdots x_n^k$, since we want k copies of each element of N_n.

(b) Applying *exp log* to the generating function in part (a), we obtain

$$\exp \sum_{i\geqslant 1} \sum_{\substack{\{\alpha_1,\ldots,\alpha_i\}\subset \mathsf{N}_+\\ \alpha_1<\cdots<\alpha_i}} \sum_{m\geqslant 1} (-1)^{m-1} x_{\alpha_1}^m \cdots x_{\alpha_i}^m/m$$

$$= \exp \sum_{m\geqslant 1} \frac{(-1)^{m-1}}{m}\left(\left\{\sum_{j\geqslant 1}(1 + x_j^m)\right\} - 1\right)$$

$$= \exp\left\{\sum_{m\geqslant 1} \frac{(-1)^{m-1}}{m}\left\{\exp\left(\sum_{i\geqslant 1}\frac{(-1)^{i-1}}{i}\sum_{j\geqslant 1} x_j^{mi}\right) - 1\right\}\right\},$$

and the result follows.

(c) From part (b),

$$\frac{\partial G}{\partial s_1} = \exp\left(\sum_{i\geqslant 1}(-1)^{i-1}s_i/i\right)G$$

$$\frac{\partial G}{\partial s_2} = \left\{-\tfrac{1}{2}\exp\left(\sum_{i\geqslant 1}(-1)^{i-1}s_i/i\right) - \tfrac{1}{2}\exp\left(\sum_{i\geqslant 1}(-1)^{i-1}s_{2i}/i\right)\right\}G$$

$$= -\frac{1}{2}\frac{\partial G}{\partial s_1} - \tfrac{1}{2}\exp\left(\sum_{i\geqslant 1}(-1)^{i-1}s_{2i}/i\right)G.$$

Applying the Γ-series theorem to both sides, and setting $y_3 = y_4 = \cdots = 0$, we obtain

$$\left(\frac{\partial}{\partial y_2} - \frac{1}{2}\frac{\partial^2}{\partial y_1^2}\right)V = -\frac{1}{2}\frac{\partial V}{\partial y_1} - \frac{1}{2}\exp(y_2)V,$$

as required, since

$$\Gamma\left(\exp(s_2 - \tfrac{1}{2}s_4 + \cdots)G\right)\big|_{y_3=y_4=\cdots=0} = \Gamma\left(\exp(s_2)G\right)\big|_{y_3=y_4=\cdots=0} = \exp(y_2)V.$$

Thus

$$\left(\frac{\partial}{\partial y_2} + \frac{1}{2}e^{y_2}\right)V = -\frac{1}{2}\left(\frac{\partial}{\partial y_1} - \frac{\partial^2}{\partial y_1^2}\right)V.$$

Of course, from part (a), $[x_1^2 \cdots x_n^2]G(\mathbf{s}) = [y_2^n/n!]V$ is the number of proper 2-covers of N_n.

3.5.7. From [**3.5.6**(a)], the required number is

$$[x^t x_1^k \cdots x_n^k] \prod_{i \geqslant 1} \prod_{\substack{\{\alpha_1, \ldots, \alpha_i\} \subset \mathsf{N}_+ \\ \alpha_1 < \cdots < \alpha_i}} (1 + x x_{\alpha_1} \cdots x_{\alpha_i}).$$

Following [**3.5.6**(b)], this becomes

$$[x^t x_1^k \cdots x_n^k] \exp \sum_{m \geqslant 1} \frac{(-1)^{m-1}}{m} x^m \left\{ \prod_{j \geqslant 1} (1 + x_j^m) - 1 \right\}$$

$$= [x^t x_1^k \cdots x_n^k] \exp \sum_{m=1}^{k} (-1)^{m-1} x^m \left\{ \prod_{j \geqslant 1} (1 + x_j^m) - 1 \right\} / m$$

$$= [x^t x_1^k \cdots x_n^k] \exp\left(\sum_{m=1}^{k} (-x)^m / m \right) \exp \sum_{m=1}^{k} \left\{ \frac{1}{m} (-1)^{m-1} x^m \prod_{j=1}^{n} (1 + x_j^m) \right\}$$

$$= [x^t x_1^k \cdots x_n^k] \exp\left(\sum_{m=1}^{k} (-x)^m / m \right)$$

$$\times \prod_{m=1}^{k} \sum_{u_m \geqslant 0} \frac{(-1)^{u_m} (-x)^{m u_m}}{u_m! m^{u_m}} \prod_{j=1}^{n} (1 + x_j^m)^{u_m}$$

$$= [x^t]\exp\left(\sum_{m=1}^{k}(-x)^m/m\right)\sum_{u_1,\ldots,u_k\geq 0}\frac{(-1)^{u_1+\cdots+u_k}(-x)^{u_1+2u_2+\cdots+ku_k}}{u_1!\cdots u_k!2^{u_2}\cdots k^{u_k}}$$

$$\times\prod_{j=1}^{n}[x_j^k](1+x_j)^{u_1}\cdots(1+x_j^k)^{u_k}$$

$$= [x^t]\exp\left(\sum_{m=1}^{k}(-x)^m/m\right)\sum_{u_1,\ldots,u_k\geq 0}\frac{(-1)^{u_1+\cdots+u_k}(-x)^{u_1+2u_2+\cdots+ku_k}}{u_1!\cdots u_k!2^{u_2}\cdots k^{u_k}}$$

$$\times\{[z^k](1+z)^{u_1}\cdots(1+z^k)^{u_k}\}^n,$$

and the result follows.

3.5.8. From [**3.5.7**], the number of proper 3-covers of N_n of order t is

$$\left[x^t\frac{y^n}{n!}\right]\exp\left(-x+\frac{x^2}{2}-\frac{x^3}{3}\right)\sum_{u_1,u_2,u_3\geq 0}\frac{(-1)^{u_1+u_2+u_3}(-x)^{u_1+2u_2+3u_3}}{u_1!u_2!u_3!2^{u_2}3^{u_3}}$$

$$\times\exp(y[z^3](1+z)^{u_1}(1+z^2)^{u_2}(1+z^3)^{u_3}.$$

But $[z^3](1+z)^{u_1}(1+z^2)^{u_2}(1+z^3)^{u_3} = u_3 + u_1u_2 + \binom{u_1}{3}$, so the number is

$$\left[x^t\frac{y^n}{n!}\right]\exp\left(-x+\frac{x^2}{2}-\frac{x^3}{3}\right)\sum_{u_1\geq 0}\frac{x^{u_1}}{u_1!}\exp\left\{\binom{u_1}{3}y\right\}\exp\left(\frac{x^3}{3}e^y\right)\exp\left(-\frac{x^2}{2}e^{u_1y}\right).$$

Replace u_1 by r and rearrange, to obtain

$$\left[x^t\frac{y^n}{n!}\right]\exp\left(-x+\frac{x^3}{3}(e^y-1)\right)\sum_{r\geq 0}\frac{x^r}{r!}\exp\left\{\binom{r}{3}y+\frac{x^2}{2}(e^{ry}-1)\right\}.$$

But by the construction of **3.4.9**, we obtain the generating function for restricted proper 3-covers by substituting $\log(1+y)$ for y in the preceding expression. The result follows.

SECTION 4.2

4.2.1. (a) Let γ_j denote $\gamma_j(<)$. If $\Phi(\gamma) = \sum_{\mathbf{i}} a_{\mathbf{i}}\gamma^{\mathbf{i}}$, where $\mathbf{i} = (i_1, i_2, \ldots)$, then $[x_1\cdots x_n]\Phi(\gamma) = \sum_{\mathbf{i}} a_{\mathbf{i}}[x_1\cdots x_n]\gamma^{\mathbf{i}}$. Consider $[x_1\cdots x_n]\gamma^{\mathbf{i}} = [x_1\cdots x_n]\gamma_1^{i_1}\cdots\gamma_n^{i_n}$. This coefficient is clearly 0 if $n \neq i_1 + 2i_2 + \cdots + ni_n$. Now γ_i is the ordinary generating function for subsets of size i. Thus, if $n = i_1 + 2i_2 + \cdots + ni_n$, we have $[x_1\cdots x_n]\gamma_1^{i_1}\cdots\gamma_n^{i_n} = n!(1!^{i_1}\cdots n!^{i_n})^{-1}$, the multinomial coefficient. Substituting above, we get

$$[x_1 \cdots x_n]\Phi(\gamma) = \sum_{\substack{\mathbf{i} \\ i_1+2i_2+\cdots+ni_n=n}} a_{\mathbf{i}} n!(1!^{i_1} \cdots n!^{i_n})^{-1}$$

$$= \left[\frac{x^n}{n!}\right]\sum_{\mathbf{i}} a_{\mathbf{i}}\left(\frac{x^1}{1!}\right)^{i_1} \cdots \left(\frac{x^n}{n!}\right)^{i_n} = \left[\frac{x^n}{n!}\right]\Phi\left(1, \frac{x^1}{1!}, \frac{x^2}{2!}, \ldots\right).$$

(b) Let γ_j denote $\gamma_j(+)$, and $\Phi(\gamma) = \Sigma_{\mathbf{i}}\, a_{\mathbf{i}}\gamma^{\mathbf{i}}$. Now γ_j is the enumerator for subsets of N_n consisting of j consecutive elements. Thus $[x_1 \cdots x_n]\gamma_1^{i_1} \cdots \gamma_n^{i_n} = 0$ if $n \neq i_1 + 2i_2 + \cdots + ni_n$. Otherwise, $[x_1 \cdots x_n]\gamma_1^{i_1} \cdots \gamma_n^{i_n}/i_1! \cdots i_n!$ is the number of ways of partitioning N_n into $i_1 + \cdots + i_n = t$ subsets of consecutive integers such that there are i_j subsets of size j for $j = 1,\ldots, n$. This partitioning can be carried out in $t!(i_1!\ldots i_n!)^{-1}$ ways, once for each sequence of i_1 1's,..., $i_n n$'s. The partition that corresponds to the sequence $\sigma_1\sigma_2 \cdots \sigma_t$ in this way is $\{\{1,\ldots, \sigma_1\},\{\sigma_1 + 1,\ldots, \sigma_1 + \sigma_2\},\ldots,\{\sigma_1 + \cdots + \sigma_{t-1} + 1,\ldots, \sigma_1 + \cdots + \sigma_t\}\}$, where $\sigma_1 + \cdots + \sigma_t = n$. Thus $[x_1 \cdots x_n]\gamma_1^{i_1} \cdots \gamma_n^{i_n} = t!$ if $n = i_1 + 2i_2 + \cdots + ni_n$, so that

$$[x_1 \cdots x_n]\Phi(\gamma) = \sum_{\substack{\mathbf{i} \\ i_1+2i_2+\cdots+ni_n=n}} a_{\mathbf{i}}(i_1 + \cdots + i_n)!$$

$$= [x^n]\sum_{\mathbf{i}} a_{\mathbf{i}}(x)^{i_1}(x^2)^{i_2} \cdots (x^n)^{i_n}(i_1 + \cdots + i_n)!$$

$$= \sum_{\mathbf{i}} a_{\mathbf{i}} \sum_{k \geqslant 0} k![x^n y^k](xy)^{i_1}(x^2y)^{i_2} \cdots (x^ny)^{i_n}$$

$$= [x^n] \sum_{k \geqslant 0} k![y^k]\Phi(1, xy, x^2y,\ldots),$$

where, in applying $[y^k]$, we consider $\Phi(1, xy, x^2y,\ldots)$ as a power series in y with coefficients that are power series in x.

(c) Let γ_j denote $\gamma_j(\oplus)$ and $\Phi(\gamma) = \Sigma_{\mathbf{i}}\, a_{\mathbf{i}}\gamma^{\mathbf{i}}$. Now γ_j is the enumerator for subsets of N_n consisting of j elements that are circularly consecutive. Again we have $[x_1 \cdots x_n]\gamma_1^{i_1} \cdots \gamma_n^{i_n} = 0$ if $n \neq i_1 + 2i_2 + \cdots + ni_n$. Otherwise, $[x_1 \cdots x_n]\gamma_1^{i_1} \cdots \gamma_n^{i_n}/i_1! \cdots i_n!$ is the number of ways of partitioning N_n into $i_1 + \cdots + i_n = t$ subsets of circularly consecutive integers such that there are i_j subsets of size j for $j = 1,\ldots, n$. There are n such partitions corresponding to each of the $(t-1)!(i_1! \cdots i_n!)^{-1}$ distinct circular sequences of i_1 1's,..., i_n n's (Two circular sequences are distinct if one cannot be cyclically rotated to give the other.) The partitions that correspond to the circular sequence

$\sigma_1 \cdots \sigma_t$ are

$$\{\{(i \bmod n)+1,\ldots,((\sigma_1+\mathrm{i}) \bmod n)+1\},$$

$$\{((\sigma_1+1+i) \bmod n)+1,\ldots,((\sigma_1+\sigma_2+i) \bmod n)+1\},\ldots$$

$$\{((\sigma_1+\cdots+\sigma_{t-1}+1+i) \bmod n)+1,\ldots,((\sigma_1+\cdots+\sigma_t+i) \bmod n)+1\}\},$$

for $i=0,\ldots,n-1$. Thus $[x_1 \cdots x_n]\gamma_1^{i_1} \cdots \gamma_n^{i_n} = n(t-1)!$ if $n = i_1 + 2i_2 + \cdots + ni_n$, where $t = i_1 + \cdots + i_n$, so

$$\begin{aligned}[x_1 \cdots x_n]\Phi(\gamma) &= [x^n]\sum_{\mathbf{i}} a_{\mathbf{i}} n(i_1+\cdots+i_n-1)!(x^1)^{i_1}\cdots(x^n)^{i_n}\\ &= \sum_{\mathbf{i}} a_{\mathbf{i}} \sum_{k\geq 1}(k-1)!n[x^n y^k](xy)^{i_1}\cdots(x^n y)^{i_n}\\ &= [x^n]\sum_{k\geq 1}(k-1)![y^k]x\frac{\partial}{\partial x}\Phi(1,xy,x^2y,\ldots),\end{aligned}$$

where, in applying $[y^k]$, we regard $\Phi(1, xy, x^2y, \ldots)$ as a power series in y with coefficients that are power series in x.

4.2.2. (a) Now $[x_1^2 \cdots x_n^2]\Phi(\gamma(<)) = 2^{-n}[v_1w_1 \cdots v_nw_n]\Phi(t_0, t_1, \ldots)$, where

$$t_k = [u^k]\prod_{i=1}^{n}(1+(v_i+w_i)u),$$

since $[v_iw_i]x_i^2 = 2$ if $x_i = v_i + w_i$. Thus

$$\begin{aligned}t_k &= [u^k]\prod_{i=1}^{n}\{(1+v_iu)(1+w_iu)-v_iw_iu^2\}\\ &= \sum_{j\geq 0}(-1)^j\beta_j[u^{k-2j}]\prod_{i=1}^{n}(1+v_iu)(1+w_iu)+d_0\end{aligned}$$

in which $\beta_j = [u^j]\prod_{i=1}^{n}(1+v_iw_iu)$, and each term in the expansion of d_0 contains squares in v_i or w_i. We then have

$$t_k = \sum_{j\geq 0}(-1)^j\frac{x^j}{j!}\frac{z^{k-2j}}{(k-2j)!}+d_1,$$

where $x = v_1w_1 + \cdots + v_nw_n$, $z = v_1 + \cdots + v_n + w_1 + \cdots + w_n$, and each term in the expansion of d_1 contains a square in v_i or w_i. Now

$$[v_1w_1 \cdots v_nw_n]x^jz^m = \binom{n}{j} j![v_1w_1 \cdots v_{n-j}w_{n-j}]z^m = \binom{n}{j} j!m!\delta_{m,2n-2j}.$$

Moreover, d_1 makes no contribution to $[v_1w_1 \cdots v_nw_n]t_k$. So setting $d_1 = 0$,

$$[x_1^2 \cdots x_n^2]\Phi(\gamma(<)) = 2^{-n}\sum_{j=0}^{n}\binom{n}{j} j!(2n-2j)![x^jz^{2n-2j}]\Phi(t_0, t_1, \ldots)$$

$$= 2^{-n}\sum_{j=0}^{n}\binom{n}{j}\left[\frac{x^j}{j!}\frac{z^{2n-2j}}{(2n-2j)!}\right]\Phi(t_0, t_1, \ldots).$$

(b) From part (a) and Corollary 4.2.7 the required number is

$$[x_1^2 \cdots x_n^2]\left\{\sum_{k\geqslant 0}(-1)^k\gamma_{2k}(<)\right\}^{-1}$$

$$= 2^{-n}\sum_{j=0}^{n}\binom{n}{j}\left[\frac{x^j}{j!}\frac{z^{2n-2j}}{(2n-2j)!}\right]\left\{\sum_{k\geqslant 0}(-1)^k t_{2k}\right\}^{-1}.$$

By series bisection,

$$\sum_{k\geqslant 0} t_{2k}u^{2k} = \tfrac{1}{2}\{\exp(zu - xu^2) + \exp(-zu - xu^2)\},$$

so that $\sum_{k\geqslant 0}(-1)^k t_{2k} = e^x\cos z$, by choosing $u^2 = -1$. Thus the number is

$$2^{-n}\sum_{j=0}^{n}\binom{n}{j}\left[\frac{x^j}{j!}\frac{z^{2n-2j}}{(2n-2j)!}\right]e^{-x}\sec z = 2^{-n}\sum_{k=0}^{n}\binom{n}{k}(-1)^{n-k}\left[\frac{z^{2k}}{(2k)!}\right]\sec z,$$

where $k = n - j$.

4.2.3. In the maximal string decomposition theorem we have $F(x) = 1 + x^k$, so that $F(x)^{-1} = \sum_{i\geqslant 0}(-1)^i x^{ki}$, and $\{F^{-1}\circ\gamma\}^{-1} = \{\sum_{i\geqslant 0}(-1)^k\gamma_{ki}(\pi_1)\}^{-1}$. The number of sequences of type $\mathbf{i}$ in which all maximal π_1-strings have length k is thus

$$[\mathbf{x}^{\mathbf{i}}]\{F^{-1}\circ\gamma\}^{-1} = [\mathbf{x}^{\mathbf{i}}]\left\{\sum_{i\geqslant 0}(-1)^k\gamma_{ki}(\pi_1)\right\}^{-1}.$$

4.2.4. (a) In the maximal string decomposition theorem we have $F(x) = 1 + \sum_{j\geqslant 1}x^{jk} = (1-x^k)^{-1}$, so that $F(x)^{-1} = 1 - x^k$ and $F^{-1}\circ\eta_q =$

$1 - x^k/k!_q$. The required number is thus

$$\left[q^m \frac{x^{nk}}{(nk)!_q}\right]\left\{1 - \frac{x^k}{k!_q}\right\}^{-1} = \left[q^m \frac{x^{nk}}{(nk)!_q}\right] \sum_{i \geqslant 0} x^{ik}(k!_q)^{-i}$$

$$= [q^m](nk)!_q(k!_q)^{-n}.$$

(b) The permutations in part (a) consist of an ordered arrangement of increasing strings of length k. These do not need to be maximal. The result follows directly by the interpretation of the q-multinomial coefficient in Lemma 2.6.6.

4.2.5. The maximal π_1-string length enumerator, with u marking maximal π_1-strings is $F(x) = 1 + u\Sigma_{i \geqslant 1} x^i = \{1 - (1-u)x\}(1-x)^{-1}$. Thus

$$F(x)^{-1} = (1-x)\sum_{i \geqslant 0}(1-u)^i x^i = 1 + \sum_{j \geqslant 1} x^j\{(1-u)^j - (1-u)^{j-1}\}$$

$$= 1 - u\sum_{j \geqslant 1} x^j(1-u)^{j-1}$$

$$= (1-u)^{-1}\left\{1 - u\sum_{j \geqslant 0} x^j(1-u)^j\right\},$$

so that $(F^{-1} \circ \gamma)^{-1} = (1-u)\{1 - u\Sigma_{j \geqslant 0}(1-u)^j\gamma_j(\pi_1)\}^{-1}$.

(a) By the <-transformation lemma, the required number is

$$\left[u^k \frac{x^n}{n!}\right](1-u)\left\{1 - u\sum_{j \geqslant 0}(1-u)^j \frac{x^j}{j!}\right\}^{-1} = \left[u^k \frac{x^n}{n!}\right](1-u)\{1 - ue^{(1-u)x}\}^{-1}.$$

(b) By the +-transformation lemma, the number is

$$[u^k x^n] \sum_{i \geqslant 0} i![y^i]\left\{1 - u\sum_{j \geqslant 1}(1-u)^{j-1} y x^j\right\}^{-1}$$

$$= [u^k x^n] \sum_{i \geqslant 0} i!\left(u\sum_{j \geqslant 1}(1-u)^{j-1}x^j\right)^i$$

$$= [u^k x^n] \sum_{i \geqslant 0} i!\{ux(1 - (1-u)x)^{-1}\}^i$$

$$= [u^k] \sum_{i \geqslant 1} i! u^i (1-u)^{n-i}\binom{n-1}{i-1}.$$

(c) By the ⊕-transformation lemma, the number is

$$[u^k x^n] \sum_{i\geq 1} (i-1)![y^i]x\frac{\partial}{\partial x}\{1 - uyx(1-(1-u)x)^{-1}\}^{-1}$$

$$= [u^k x^n] n \sum_{i\geq 1} (i-1)! u^i x^i (1-(1-u)x)^{-i}$$

$$= [u^k] n \sum_{i\geq 1} (i-1)! u^i (1-u)^{n-i} \binom{n-1}{i-1}.$$

4.2.6. (a) From [**4.2.5**], the required number is

$$[u^k \mathbf{x}^{\mathbf{i}}](1-u)\left\{1 - u\sum_{l\geq 0}(1-u)^l \gamma_l(<)\right\}^{-1}$$

$$= [u^k \mathbf{x}^{\mathbf{i}}](1-u)\left\{1 - u\prod_{j\geq 1}(1+(1-u)x_j)\right\}^{-1},$$

since $\prod_{j\geq 1}(1+zx_j) = \sum_{l\geq 0}\gamma_l(<)z^l$.

(b) The result follows directly from part (a) and Remark 2.4.11.

(c) From [**4.2.5**] and the maximal string decomposition theorem, the required number is

$$\left[u^k q^m \frac{x^n}{n!_q}\right](1-u)\left\{1 - u\sum_{j\geq 0}(1-u)^j \frac{x^j}{j!_q}\right\}^{-1}.$$

But from **2.6.14(2)** $\sum_{j\geq 0}((1-u)x)^j/j!_q = \prod_{i\geq 0}(1-(1-u)x(1-q)q^i)^{-1}$, and the result follows.

4.2.7. The maximal π_1-string length enumerator is

$$F(x) = 1 + \sum_{i>p} x^i = 1 + x^{p+1}(1-x)^{-1} = (1 - x + x^{p+1})(1-x)^{-1},$$

so that

$$F(x)^{-1} = (1-x)(1-x(1-x^p))^{-1} = (1-x)\sum_{m\geq 0} x^m(1-x^p)^m$$

$$= (1-x)\sum_{m\geq 0}\sum_{i=0}^{m}\binom{m}{i}(-1)^i x^{pi+m}.$$

The required number, from the maximal string decomposition theorem, is

$$[\mathbf{x^i}](F^{-1}\circ\gamma)^{-1} = [\mathbf{x^i}]\left\{\sum_{m\geqslant 0}\sum_{i=0}^{m}\binom{m}{i}(-1)^i(\gamma_{pi+m}-\gamma_{pi+m+1})\right\}^{-1},$$

where $\gamma_i = \gamma_i(\pi_1)$, $i \geqslant 0$.

4.2.8. (a) The maximal π_1-string length enumerator is

$$F(x) = \sum_{i=0}^{p-1} x^i + u\sum_{i\geqslant p} x^i = \{1-(1-u)x^p\}(1-x)^{-1},$$

where u marks maximal π_1-strings of length $\geqslant p$. Thus

$$F(x)^{-1} = (1-x)\sum_{i\geqslant 0}(1-u)^i x^{pi},$$

so that

$$F^{-1}\circ\eta_q = \sum_{i\geqslant 0}(1-u)^i\left(\frac{x^{pi}}{(pi)!_q}-\frac{x^{pi+1}}{(pi+1)!_q}\right).$$

By the maximal string decomposition theorem, the required number is $[u^k q^m x^n/n!_q](F^{-1}\circ\eta_q)^{-1}$, and the result follows.

(b) From part (a), with $p = 2$, and the <-transformation lemma, the required number is

$$\left[u^k\frac{x^n}{n!}\right]\left\{\sum_{i\geqslant 0}(1-u)^i\left(\frac{x^{2i}}{(2i)!}-\frac{x^{2i+1}}{(2i+1)!}\right)\right\}^{-1}$$

$$= \left[u^k\frac{x^n}{n!}\right](\cosh xz - z^{-1}\sinh xz)^{-1},$$

where $z = (1-u)^{1/2}$.

4.2.9. From Corollaries 4.2.7 and 4.2.20, the required number is

$$[\mathbf{x^i}]\left\{1+\sum_{k\geqslant 0}(-1)^k\gamma_{2k+1}(<)\right\}\left\{\sum_{k\geqslant 0}(-1)^k\gamma_{2k}(<)\right\}^{-1}.$$

But

$$\sum_{k\geqslant 0}(-1)^k\gamma_{2k}(<) = \sum_{k\geqslant 0}\rho^{2k}\gamma_{2k}(<) = \frac{1}{2}\left\{\prod_{j\geqslant 1}(1+\rho x_j)+\prod_{j\geqslant 1}(1-\rho x_j)\right\}$$

by series bisection. Similarly,

$$\sum_{k\geqslant 0}(-1)^k\gamma_{2k+1}(<)=\frac{1}{2\rho}\Big\{\prod_{j\geqslant 1}(1+\rho x_j)-\prod_{j\geqslant 1}(1-\rho x_j)\Big\}.$$

Thus the required number is

$$[\mathbf{x^i}]\Big\{2-\rho\prod_{j\geqslant 1}(1+\rho x_j)+\rho\prod_{j\geqslant 1}(1-\rho x_j)\Big\}\Big\{\prod_{j\geqslant 1}(1+\rho x_j)+\prod_{j\geqslant 1}(1-\rho x_j)\Big\}^{-1}.$$

4.2.10. In the maximal string decomposition theorem for a distinguished final string

$$F(x)=1+u\sum_{i\geqslant 1}x^{it}=(1-(1-u)x^t)(1-x^t)^{-1},$$

$$G(x)=u\sum_{i\geqslant 1}x^{it-j}=ux^{t-j}(1-x^t)^{-1}$$

where $j<t$, and u marks maximal strings. Now

$$F(x)^{-1}G(x)=u\sum_{i\geqslant 0}(1-u)^i x^{ti+t-j}$$

and

$$F(x)^{-1}=1-u\sum_{i\geqslant 1}x^{ti}(1-u)^{i-1}.$$

Thus by the <-transformation lemma, the required number is

$$\left[u^k\frac{x^n}{n!}\right]u\sum_{i\geqslant 0}(1-u)^{i+1}\frac{x^{t(i+1)-j}}{(t(i+1)-j)!}\left\{1-u\sum_{i\geqslant 0}(1-u)^i\frac{x^{ti}}{(ti)!}\right\}^{-1}$$

$$=\left[u^{k-1}\frac{x^n}{n!}\right](1-u)^{j/t}\phi_t^{(t-j)}\big(x(1-u)^{1/t}\big)\big\{1-u\phi_t^{(0)}\big(x(1-u)^{1/t}\big)\big\}^{-1}.$$

4.2.11. By the maximal string decomposition theorem for a distinguished final string, the required number is $[u^k x_1\cdots x_n](F^{-1}G\circ\gamma(<))(F^{-1}\circ\gamma(<))^{-1}$, where $F(x)=1+u\Sigma_{i\geqslant t}x^i=(1-x+ux^t)(1-x)^{-1}$ and $G(x)=u\Sigma_{i\geqslant s}x^i=ux^s(1-x)^{-1}$. Thus $F(x)^{-1}G(x)=ux^s(1-x+ux^t)^{-1}$ and $F(x)^{-1}=(1-x)(1-x+ux^t)^{-1}$. Let $\beta_1,\dots,\beta_t$ be the roots of $P(x)=1-x+ux^t=0$. These are distinct. To see this, suppose that $\beta_1=\beta_2$. Thus β_1 is a root of $P'(x)=0$, so that $tu\beta_1^{t-1}=1$. But $u\beta_1^t-\beta_1+1=0$ so $tu\beta_1^t-t\beta_1+t=0$, and substituting $tu\beta_1^t=\beta_1$, we have $\beta_1=t/(t-1)$. But $P(t/(t-1))$ is clearly not zero, so the roots must be distinct. Let $\alpha_1=\beta_1^{-1},\dots,\alpha_t=\beta_t^{-1}$ be the distinct roots of $z^t-z^{t-1}+u=0$. Then $P(x)=(1-\alpha_1x)\cdots(1-\alpha_tx)$, and $F(x)^{-1}G(x) = ux^s(1-\alpha_1 x)^{-1}\cdots(1-\alpha_t x)^{-1}$, $F(x)^{-1} = (1-x)(1-\alpha_1 x)^{-1}\cdots(1-\alpha_t x)^{-1}$. Now expand these by partial fractions, to get

$$F(x)^{-1}G(x) = \sum_{i=1}^{t} \frac{A_i}{1-\alpha_i x}, \qquad F(x)^{-1} = \sum_{i=1}^{t} \frac{B_i}{1-\alpha_i x}.$$

By the <-transformation lemma, since A_i, B_i are independent of x, the required number is

$$\left[u^k \frac{x^n}{n!}\right]\left\{\sum_{i=1}^{t} A_i e^{\alpha_i x}\right\}\left\{\sum_{i=1}^{t} B_i e^{\alpha_i x}\right\}^{-1}.$$

To determine A_i, B_i for $i = 1,\ldots, t$, multiply by $P(x)$ on both sides of the expression for $F^{-1}G$, giving

$$ux^s = \sum_{j=1}^{t} A_j P(x)(1-\alpha_j x)^{-1}.$$

Substituting $x = \alpha_i^{-1}$, we get

$$A_i = u\alpha_i^{-s}\left\{\lim_{x\to\alpha_i^{-1}}\left(\frac{P(x)}{1-\alpha_i x}\right)\right\}^{-1} = u\alpha_i^{-s}\frac{P'(\alpha_i^{-1})}{-\alpha_i}$$

$$= -u\alpha_i^{-s+1}\left(tu\alpha_i^{-(t-1)}-1\right)^{-1} = \alpha_i^{t-s}\left(\alpha_i^{t-1}u^{-1}-t\right)^{-1}.$$

Similarly,

$$B_i = (\alpha_i - 1)\alpha_i^{t-1}\left(\alpha_i^{t-1} - ut\right)^{-1} = \left(t - u^{-1}\alpha_i^{t-1}\right)^{-1},$$

because $(\alpha_i - 1)\alpha_i^{t-1} = -u$, since α_i is a root of $z^t - z^{t-1} + u = 0$. Thus the required number is

$$\left[u^k \frac{x^n}{n!}\right]\left\{\sum_{i=1}^{t} \frac{\alpha_i^{t-s}e^{\alpha_i x}}{\alpha_i^{t-1}u^{-1}-t}\right\}\left\{\sum_{i=1}^{t} \frac{e^{\alpha_i x}}{t-u^{-1}\alpha_i^{t-1}}\right\}^{-1}.$$

4.2.12. Maxima occur as the final elements in nonfinal maximal <-strings of length at least 2. Thus the required number is

$$\left[r^i u^k x_1 \cdots x_n\right]\left(F^{-1}G\circ\gamma(<)\right)\left(F^{-1}\circ\gamma(<)\right)^{-1}$$

where u marks maxima, r marks rises, and

$$F(x) = 1 + x + u\sum_{i\geq 2} r^{i-1}x^i = \left(1-(r-1)x+(u-1)rx^2\right)(1-rx)^{-1}$$

$$G(x) = \sum_{i\geq 1} r^{i-1}x^i = x(1-rx)^{-1}.$$

We now have

$$F(x)^{-1}G(x) = x(1-(r-1)x+(u-1)rx^2)^{-1}$$

$$F(x)^{-1} = (1-rx)(1-(r-1)x+(u-1)rx^2)^{-1}.$$

If β_1, β_2 are the roots of $z^2-(r-1)z+(u-1)r=0$, then β_1, β_2 are distinct, and $1-(r-1)x+(u-1)rx^2 = (1-\beta_1 x)(1-\beta_2 x)$. Again we consider partial fraction expansions,

$$F(x)^{-1}G(x) = \frac{A_1}{1-\beta_1 x}+\frac{A_2}{1-\beta_2 x}, \qquad F(x)^{-1} = \frac{C_1}{1-\beta_1 x}+\frac{C_2}{1-\beta_2 x}.$$

Thus $A_1\beta_2 + A_2\beta_1 = -1$, $A_1 + A_2 = 0$, $C_1\beta_2 + C_2\beta_1 = r$, $C_1 + C_2 = 1$. Solving these, we obtain $A_1 = -(\beta_2-\beta_1)^{-1}$, $A_2 = (\beta_2-\beta_1)^{-1}$, $C_1 = (r-\beta_1)(\beta_2-\beta_1)^{-1}$, $C_2 = (\beta_2-r)(\beta_2-\beta_1)^{-1}$, so, by the <-transformation lemma, the required number is

$$\left[r^i u^k \frac{x^n}{n!}\right]\left(A_1 e^{\beta_1 x} + A_2 e^{\beta_2 x}\right)\left(C_1 e^{\beta_1 x} + C_2 e^{\beta_2 x}\right)^{-1}$$

$$= \left[r^i u^k \frac{x^n}{n!}\right]\left(e^{\beta_2 x} - e^{\beta_1 x}\right)\left((\beta_2 - r)e^{\beta_2 x} + (r-\beta_1)e^{\beta_1 x}\right)^{-1},$$

where $\beta_1\beta_2 = r(u-1)$, $\beta_1+\beta_2 = r-1$. Now let $\alpha_1 = r-\beta_1$, $\alpha_2 = r-\beta_2$, so that the solution becomes

$$\left[r^i u^k \frac{x^n}{n!}\right]\left(e^{(r-\alpha_2)x} - e^{(r-\alpha_1)x}\right)\left(-\alpha_2 e^{(r-\alpha_2)x} + \alpha_1 e^{(r-\alpha_1)x}\right)^{-1}$$

$$= \left[r^i u^k \frac{x^n}{n!}\right]\left(e^{\alpha_1 x} - e^{\alpha_2 x}\right)\left(\alpha_1 e^{\alpha_2 x} - \alpha_2 e^{\alpha_1 x}\right)^{-1},$$

where $\alpha_1+\alpha_2 = (r-\beta_1)+(r-\beta_2) = 2r-(\beta_1+\beta_2) = r+1$, $\alpha_1\alpha_2 = (r-\beta_1)(r-\beta_2) = r^2 - r(\beta_1+\beta_2)+\beta_1\beta_2 = ur$.

4.2.13. The maximal <-string enumerators are

$$G(x) = x^2 + x^4 + \cdots = x^2(1-x^2)^{-1} \qquad \text{and} \qquad F(x) = 1+x^2.$$

Thus from Theorem 4.2.19 the required number is

$$\left[\frac{x^n}{n!}\right]\left\{\sum_{i\geqslant 0}\frac{x^{2+4i}}{(2+4i)!}\right\}\left\{\sum_{i\geqslant 0}(-1)^i\frac{x^{2i}}{(2i)!}\right\}^{-1} = \tfrac{1}{2}\left[\frac{x^n}{n!}\right]\left\{-1+\frac{\cosh x}{\cos x}\right\},$$

and the result follows.

4.2.14. (a) The $<$-transformation lemma and Corollary 4.2.12 give the required number as

$$\left[u^j\frac{x^n}{n!}\right]\left\{1-\sum_{i\geqslant 1}(u-1)^{i-1}\frac{x^i}{i!}\right\}^{-1}=\left[u^j\frac{x^n}{n!}\right](u-1)(u-e^{(u-1)x})^{-1}.$$

(b) The $+$-transformation lemma and Corollary 4.2.12 give the required number as

$$[u^jx^n]\sum_{k\geqslant 0}k![y^k]\left\{1-\sum_{i\geqslant 1}(u-1)^{i-1}yx^i\right\}^{-1}$$

$$=[u^jx^n]\sum_{k\geqslant 0}k!x^k(1-(u-1)x)^{-k}.$$

(c) The $\oplus$-transformation lemma and Corollary 4.2.12 give the required number as

$$[u^jx^n]\sum_{k\geqslant 1}(k-1)![y^k]x\frac{\partial}{\partial x}\left\{1-\sum_{i\geqslant 1}(u-1)^{i-1}yx^i\right\}^{-1}$$

$$=[u^jx^n]\sum_{k\geqslant 1}(k-1)!x\frac{\partial}{\partial x}\{x(1-(u-1)x)^{-1}\}^k$$

$$=[u^jx^n]\sum_{k\geqslant 1}k!x^k(1-(u-1)x)^{-k}\{1+(u-1)x(1-(u-1)x)^{-1}\}$$

$$=[u^jx^n]\sum_{k\geqslant 1}k!x^k(1-(u-1)x)^{-(k+1)}.$$

4.2.15. (a) Consider the bipartition (α_1, α_2) of $\mathsf{K}\times\mathsf{K}$, $\mathsf{K}=\mathsf{N}_+^k$ in which $((i_1,\ldots,i_k)^T,(j_1,\ldots,j_k)^T)\in\alpha_1$ if $(i_l, j_l)\in\pi_1$ for $l=1,\ldots,k$. Suppose further that the sequence element $\mathbf{i}=(i_1,\ldots,i_k)^T$ is marked by $z_{\mathbf{i}}$. Then in all our previous work, the alphabet N_+ is replaced by K, π_1 is replaced by α_1, and the x_i's are replaced by $z_{\mathbf{i}}$'s. The α_1-string enumerators on K are $\gamma_i(\alpha_1)=\sum_{\sigma\in\langle\alpha_1^{i-1}\rangle}\mathbf{z}^{\tau(\sigma)}$, where $\mathbf{z}$ is a vector of the $z_{\mathbf{i}}$'s in a canonical order, and $\tau(\sigma)$ contains the frequency of occurrence of $\mathbf{i}$ in σ, in the same order. Now if we mark the occurrence of $\mathbf{i}$ by $z_{\mathbf{i}}=x_{1i_1}x_{2i_2}\cdots x_{ki_k}$, then $\gamma_i(\alpha_1)=\gamma_i(\pi_1,\mathbf{x}_1)\cdots\gamma_i(\pi_1,\mathbf{x}_k)$, where x_{ij} marks the appearance of $j\in\mathsf{N}_+$ in sequence i of the k-tuple. Thus the required number is

$$[u^j\mathbf{x}_1^{\mathbf{i}_1}\cdots\mathbf{x}_k^{\mathbf{i}_k}]\left\{1-\sum_{m\geqslant 1}(u-1)^{m-1}\gamma_m(\pi_1,\mathbf{x}_1)\cdots\gamma_m(\pi_1,\mathbf{x}_k)\right\}^{-1},$$

directly from Corollary 4.2.12.

(b) In each case we want

$$[u^j x_{11} \cdots x_{1n} x_{21} \cdots x_{2n} \cdots x_{k1} \cdots x_{kn}]$$

$$\times \left\{1 - \sum_{m \geqslant 1} (u-1)^{m-1} \gamma_m(\pi_1, \mathbf{x}_1) \cdots \gamma_m(\pi_1, \mathbf{x}_k)\right\}^{-1}.$$

(i) If the <-transformation lemma is applied simultaneously to the vectors $\mathbf{x}_1, \ldots, \mathbf{x}_k$, the required number is

$$\left[u^j \frac{z_1^n}{n!} \cdots \frac{z_k^n}{n!}\right](u-1)\left\{u - \sum_{m \geqslant 0} (u-1)^m \frac{z_1^m}{m!} \cdots \frac{z_k^m}{m!}\right\}^{-1}.$$

If we let $z_1 \cdots z_k = x$, then this becomes

$$\left[u^j \frac{x^n}{(n!)^k}\right](u-1)\left\{u - \sum_{m \geqslant 0} (u-1)^m \frac{x^m}{(m!)^k}\right\}^{-1}.$$

(ii) Applying the +-transformation lemma, we get

$$[u^j z_1^n \cdots z_k^n] \sum_{i_1, \ldots, i_k \geqslant 0} i_1! \cdots i_k! [y_1^{i_1} \cdots y_k^{i_k}]$$

$$\times \left\{1 - \sum_{m \geqslant 1} (u-1)^{m-1} y_1 z_1^m \cdots y_k z_k^m\right\}^{-1}$$

$$= [u^j x^n] \sum_{i \geqslant 0} (i!)^k [y^i] \left\{1 - \sum_{m \geqslant 1} (u-1)^{m-1} y x^m\right\}^{-1}$$

$$= [u^j x^n] \sum_{i \geqslant 0} (i!)^k x^i \{1 - (u-1)x\}^{-i}.$$

(iii) Applying the ⊕-transformation lemma, we get

$$[u^j z_1^n \cdots z_k^n] \sum_{i_1, \ldots, i_k \geqslant 1} (i_1 - 1)! \cdots (i_k - 1)! [y_1^{i_1} \cdots y_k^{i_k}] z_1 \frac{\partial}{\partial z_1} \cdots z_k \frac{\partial}{\partial z_k}$$

$$\times \left\{1 - \sum_{m \geqslant 1} (u-1)^{m-1} y_1 z_1^m \cdots y_k z_k^m\right\}^{-1}$$

$$= n^k [u^j (z_1 \cdots z_k)^n] \sum_{i \geqslant 1} ((i-1)!)^k (z_1 \cdots z_k)^i (1 - (u-1) z_1 \cdots z_k)^{-i}$$

$$= n^k [u^j x^n] \sum_{i \geqslant 1} ((i-1)!)^k x^i (1 - (u-1)x)^{-i}.$$

4.2.16. Following [**4.2.15**] and from Corollary 4.2.12, the required number is

$$\left[u^t \mathbf{x}_1^{\mathbf{i}_1} \mathbf{x}_2^{\mathbf{i}_2}\right]\left\{1 - \sum_{m \geqslant 1} (u-1)^{m-1} \gamma_m(<, \mathbf{x}_1) \gamma_m(=, \mathbf{x}_2)\right\}^{-1}$$

$$= \left[u^t \mathbf{x}_1^{\mathbf{i}_1} \mathbf{x}_2^{\mathbf{i}_2}\right]\left\{1 + (1-u)^{-1} \sum_{m \geqslant 1} (u-1)^m \gamma_m(<, \mathbf{x}_1) \sum_{i \geqslant 1} x_{2i}^m\right\}^{-1},$$

since $\gamma_m(=, \mathbf{x}_2) = x_{21}^m + x_{22}^m + \cdots$. Now

$$\sum_{m \geqslant 1} \gamma_m(<, \mathbf{x}_1) z^m = \prod_{l \geqslant 1} (1 + z x_{1l}) - 1,$$

so reversing the order of summation and letting $z = (u-1)x_{2i}$ for $i = 1, 2, \ldots$, in turn, we obtain the solution

$$\left[u^t \mathbf{x}_1^{\mathbf{i}_1} \mathbf{x}_2^{\mathbf{i}_2}\right]\left(1 + (1-u)^{-1} \sum_{i \geqslant 1}\left\{\prod_{l \geqslant 1}(1 + (u-1)x_{2i}x_{1l}) - 1\right\}\right)^{-1}.$$

4.2.17. Let $\mathbf{x}_l = (x_{a_l+1}, \ldots, x_{a_l+i_l})$ for $l = 1, \ldots, m$, and $\mathbf{x} = (\mathbf{x}_1, \ldots, \mathbf{x}_m)$. Then the required number is

$$[u^j \mathbf{x}^{\mathbf{i}}]\left\{1 - \sum_{l \geqslant 1} (u-1)^{l-1} \gamma_l(+, \mathbf{x})\right\}^{-1}$$

from Corollary 4.2.12. But $\gamma_l(+, \mathbf{x}) = \gamma_l(+, \mathbf{x}_1) + \cdots + \gamma_l(+, \mathbf{x}_m)$ in the notation of [**4.2.15**], since $a_{k+1} > a_k + i_k$ for $k = 1, \ldots, m-1$. Applying the $+$-transformation lemma simultaneously on the m alphabets of consecutive integers yields

$$[u^j \mathbf{z}^{\mathbf{i}}] \sum_{\mathbf{k} \geqslant \mathbf{0}} \mathbf{k}! [\mathbf{y}^{\mathbf{k}}]\left\{1 - \sum_{l \geqslant 1} (u-1)^{l-1}\left(y_1 z_1^l + \cdots + y_m z_m^l\right)\right\}^{-1},$$

where $\mathbf{z} = (z_1, \ldots, z_m)$. This coefficient becomes

$$[u^j \mathbf{z}^{\mathbf{i}}] \sum_{\mathbf{k} \geqslant \mathbf{0}} s(\mathbf{k})! \mathbf{z}^{\mathbf{k}} (1 - (u-1)z_1)^{-k_1} \cdots (1 - (u-1)z_m)^{-k_m},$$

and the result follows.

SECTION 4.3

4.3.1. Let g_j and f_j be indeterminates marking, respectively, final and nonfinal maximal π_1-strings of length j for $j \geqslant 1$. Then, following **4.3.12**, the required generating functions are

$$\Psi_i\left(\left(\mathbf{I}-\sum_{j\geq 0} f_{j+1}\mathbf{A}^j\mathbf{B}\right)^{-1}\sum_{l\geq 0} g_{l+1}\mathbf{A}^l\right)$$

for $i = 1, 2$. Thus we want $\Psi_i(\mathbf{C}(\mathbf{I}-\mathbf{H})^{-1}\mathbf{D})$, where $\mathbf{C}=\mathbf{I}$, $\mathbf{H}=f(\mathbf{A})\mathbf{B}$, $\mathbf{D}=g(\mathbf{A})$, $f(x)=f_1+f_2x+\cdots$, $g(x)=g_1+g_2x+\cdots$. A factored expansion is $\mathbf{H}=\mathbf{Q}-\mathbf{L}_1\mathbf{W}\mathbf{R}_1$, where $\mathbf{Q}=-f(\mathbf{A})\mathbf{A}$, $\mathbf{L}_1=-f(\mathbf{A})$, $\mathbf{R}_1=\mathbf{I}$. Thus, from the elimination theorem (with $s=1$), the required generating functions are

$$\Psi_1\left(F^{-1}(\mathbf{A})g(\mathbf{A})\right)\left\{1-\Psi_1\left(F^{-1}(\mathbf{A})f(\mathbf{A})\right)\right\}^{-1}$$

$$=\left(\left(xF^{-1}g\right)\circ\gamma(\pi_1)\right)\left\{1-\left(xF^{-1}f\right)\circ\gamma(\pi_1)\right\}^{-1}$$

$$=\left(F^{-1}G\circ\gamma(\pi_1)\right)\left(F^{-1}\circ\gamma(\pi_1)\right)^{-1},$$

and, similarly,

$$\Psi_2\left(F^{-1}(\mathbf{A})g(\mathbf{A})\right)\left\{1-\Psi_2\left(F^{-1}(\mathbf{A})f(\mathbf{A})\right)\right\}^{-1}=\left(F^{-1}G\circ\eta_q\right)\left(F^{-1}\circ\eta_q\right)^{-1}.$$

4.3.2. Let e_j, g_j, and f_j be indeterminates marking, respectively, initial, final, and other maximal π_1-strings of length j for $j\geq 1$. Then, following **4.3.12**, the required generating function is

$$\Psi_1\left(\sum_{i\geq 0} e_{i+1}\mathbf{A}^i\mathbf{B}\left(\mathbf{I}-\sum_{j\geq 0} f_{j+1}\mathbf{A}^j\mathbf{B}\right)^{-1}\sum_{k\geq 0} g_{k+1}\mathbf{A}^k\right).$$

This is of the form $\Psi_1(\mathbf{C}(\mathbf{I}-\mathbf{H})^{-1}\mathbf{D})$, where $\mathbf{C}=e(\mathbf{A})\mathbf{B}$, $\mathbf{H}=f(\mathbf{A})\mathbf{B}$, $\mathbf{D}=g(\mathbf{A})$, $e(x)=e_1+e_2x+\cdots$, $f(x)=f_1+f_2x+\cdots$, $g(x)=g_1+g_2x+\cdots$. A factored expansion was obtained for $\mathbf{H}$ in **4.3.12** and is $\mathbf{H}=\mathbf{Q}-\mathbf{L}_1\mathbf{W}\mathbf{R}_1$, where $\mathbf{Q}=-f(\mathbf{A})\mathbf{A}$, $\mathbf{L}_1=-f(\mathbf{A})$, $\mathbf{R}_1=\mathbf{I}$. Thus $\mathbf{I}-\mathbf{Q}=F(\mathbf{A})$, and from the elimination theorem,

$$\Psi_1\left(\mathbf{C}(\mathbf{I}-\mathbf{H})^{-1}\mathbf{D}\right)$$

$$=\left\{\Psi_1\left(F(\mathbf{A})^{-1}\right)\right\}^{-1}\begin{vmatrix}\Psi_1\left(e(\mathbf{A})\mathbf{B}F(\mathbf{A})^{-1}g(\mathbf{A})\right) & -\Psi_1\left(e(\mathbf{A})\mathbf{B}F(\mathbf{A})^{-1}f(\mathbf{A})\right)\\ \Psi_1\left(F(\mathbf{A})^{-1}g(\mathbf{A})\right) & 1-\Psi_1\left(F(\mathbf{A})^{-1}f(\mathbf{A})\right)\end{vmatrix}.$$

We evaluate the entries in the determinant by letting $\mathbf{B} = \mathbf{W} - \mathbf{A}$, so that

$$\Psi_1\big(e(\mathbf{A})\mathbf{B}F(\mathbf{A})^{-1}g(\mathbf{A})\big) = (E \circ \gamma)(F^{-1}G \circ \gamma) - (EF^{-1}G \circ \gamma)$$

$$\Psi_1\big(e(\mathbf{A})\mathbf{B}F(\mathbf{A})^{-1}f(\mathbf{A})\big) = (E \circ \gamma)\{1 - (F^{-1} \circ \gamma)\} - (E \circ \gamma) + (EF^{-1} \circ \gamma)$$

$$= (EF^{-1} \circ \gamma) - (E \circ \gamma)(F^{-1} \circ \gamma)$$

$$\Psi_1\big(F(\mathbf{A})^{-1}g(\mathbf{A})\big) = F^{-1}G \circ \gamma$$

$$1 - \Psi_1\big(F(\mathbf{A})^{-1}f(\mathbf{A})\big) = 1 - \{1 - (F^{-1} \circ \gamma)\} = F^{-1} \circ \gamma.$$

The generating function now becomes

$$(F^{-1} \circ \gamma)^{-1}\{(F^{-1} \circ \gamma)\{(E \circ \gamma)(F^{-1}G \circ \gamma) - (EF^{-1}G \circ \gamma)\}$$
$$+ (F^{-1}G \circ \gamma)\{(EF^{-1} \circ \gamma) - (E \circ \gamma)(F^{-1} \circ \gamma)\}\}$$
$$= (F^{-1} \circ \gamma)^{-1}\{(EF^{-1} \circ \gamma)(F^{-1}G \circ \gamma) - (F^{-1} \circ \gamma)(EF^{-1}G \circ \gamma)\},$$

as required. Of course, for permutations and inversions we replace $\gamma(\pi_1)$ by η_q.

4.3.3. (a) Let r mark rises and u mark modified maxima. An initial string of length i contains a single modified maximum and $i - 1$ rises. Any other maximal $<$-string of length i contains no rises and no modified maxima if $i = 1$, and a single modified maximum and $i - 1$ rises if $i \geqslant 2$. Thus, using the notation of [**4.3.2**], we have $E(x) = u\Sigma_{i \geqslant 1} r^{i-1}x^i = ux(1 - rx)^{-1}$ and $F(x) = 1 + G(x) = 1 + x + u\Sigma_{i \geqslant 2} r^{i-1}x^i = (1 - (r-1)x + r(u-1)x^2)(1 - rx)^{-1}$. Simplifying the result of [**4.3.2**] by letting $F = 1 + G$ and adding E, we get the required generating function as

$$(EF^{-1} \circ \gamma(<))(F^{-1} \circ \gamma(<))^{-1},$$

where

$$E(x)F(x)^{-1} = ux(1 - (r-1)x + r(u-1)x^2)^{-1}$$

$$F(x)^{-1} = (1 - rx)(1 - (r-1)x + r(u-1)x^2)^{-1}.$$

Comparing these expressions with those in [**4.2.12**], we have the solution

$$\left[r^i u^{k-1}\frac{x^n}{n!}\right](e^{\alpha_1 x} - e^{\alpha_2 x})(\alpha_1 e^{\alpha_2 x} - \alpha_2 e^{\alpha_1 x})^{-1},$$

where $\alpha_1 + \alpha_2 = r + 1$, $\alpha_1\alpha_2 = ru$.

(b) As in part (a), $E(x)=\sum_{k\geqslant 1}u_2u_3^{k-1}x^k=u_2x(1-u_3x)^{-1}$ and

$$F(x)=1+G(x)=1+u_4x+\sum_{k\geqslant 2}u_1u_2u_3^{k-2}x^k$$

$$=\{1+(u_4-u_3)x+(u_1u_2-u_3u_4)x^2\}(1-u_3x)^{-1}.$$

Let β_1,β_2 be the roots of $z^2+(u_4-u_3)z+(u_1u_2-u_3u_4)=0$, so that $EF^{-1}=u_2x(1-\beta_1x)^{-1}(1-\beta_2x)^{-1}$ and $F^{-1}=(1-u_3x)(1-\beta_1x)^{-1}(1-\beta_2x)^{-1}$. The result follows by considering partial fraction expansions of EF^{-1} and F^{-1}, as in [**4.2.12**], and letting $\alpha_1=u_3-\beta_1$, $\alpha_2=u_3-\beta_2$.

4.3.4. (a) By letting $\mathbf{T}=\mathbf{R}_1,\ldots,\mathbf{R}_s$ in the proof of Theorem 4.3.8, we obtain the linear system

$$\xi_i^{(l)}+\sum_{m=1}^{s}\Psi_l\left(\mathbf{R}_i(\mathbf{I}-\mathbf{Q})^{-1}\mathbf{L}_m\right)\xi_m^{(l)}=d_i^{(l)}$$

for $i=1,\ldots,s$, where $\xi_m^{(l)}=\Psi_l(\mathbf{R}_m(\mathbf{I}-\mathbf{H})^{-1}\mathbf{D})$. We want $\xi_k^{(l)}$ and the result follows by Cramer's rule.

(b) The result follows directly from part (a) with $s=1$, since $\mathbf{R}_1=\mathbf{I}$, $\mathbf{L}_1=-\mathbf{T}$, $\mathbf{D}=\mathbf{U}$, $\mathbf{I}-\mathbf{Q}=\mathbf{S}$.

(c) From part (b) we obtain

$$\Psi_l\left((\mathbf{S}-\mathbf{TW})^{-1}\mathbf{T}\right)=\Psi_l(\mathbf{S}^{-1}\mathbf{T})\{1-\Psi_l(\mathbf{S}^{-1}\mathbf{T})\}^{-1}=\{1-\Psi_l(\mathbf{S}^{-1}\mathbf{T})\}^{-1}-1.$$

4.3.5. (a) From **4.2.8**, the number of <-alternating permutations on N_{2n} is $[x^{2n}/(2n)!]\sec x$. But these are permutations with fixed pattern $(\pi_1\pi_2)^{n-1}\pi_1$, so we can apply **4.3.17** with $m=n$, $p_j=2$ for $j=1,\ldots,n$, and $a_i=\sum_{j=1}^{i}p_j=2i$, for $i=0,\ldots,n$. Thus the number of permutations with fixed pattern $(\pi_1\pi_2)^{n-1}\pi_1$, from **4.3.17** with $q=1$, is $\left\|\binom{2n-2(i-1)}{2j-2(i-1)}\right\|_{n\times n}$. These two expressions are equal, and the result follows.

(b) From [**4.2.3**], the generating function with respect to inversions for permutations on N_{nk} in which every maximal <-string has length k is

$$\left[\frac{x^{nk}}{(nk)!_q}\right]\left\{\sum_{i\geqslant 0}(-1)^i\frac{x^{ik}}{(ik)!_q}\right\}^{-1}.$$

But these permutations have pattern $(\pi_1^{k-1}\pi_2)^{n-1}\pi_1^{k-1}$, so we can apply **4.3.17** with $m=n$, $p_j=k$ for $j=1,\ldots,n$, and $a_i=\sum_{j=1}^{i}p_j=ki$, for $i=0,\ldots,n$. Thus the generating function with respect to inversions for permutations with fixed pattern $(\pi_1^{k-1}\pi_2)^{n-1}\pi_1^{k-1}$, from **4.3.17** is $\left\|\binom{nk-(i-1)k}{jk-(i-1)k}_q\right\|_{n\times n}$. These expressions are equal, and the result follows.

4.3.6. The number of permutations with pattern $\pi_1^{p_1-1}\pi_2\pi_1^{p_2-1}\pi_2 \cdots \pi_1^{p_m-1}$ for $\pi_1 = <$, is $\left\|\binom{n-a_{i-1}}{a_j - a_{i-1}}\right\|_{m\times m}$, from **4.3.17** with $q = 1$. The number of permutations with pattern $\pi_2^{q_l-1}\pi_1 \cdots \pi_2^{q_2-1}\pi_1\pi_2^{q_1-1}$ is the same as the number of permutations with pattern $\pi_1^{q_1-1}\pi_2\pi_1^{q_2-1}\pi_2 \cdots \pi_1^{q_l-1}$, by reading the permutations backward. But the latter number is $\left\|\binom{n-b_{i-1}}{b_j - b_{i-1}}\right\|_{l\times l}$, from **4.3.17** with $q = 1$. The result follows.

4.3.7. (a) Each sequence in N_+^* has a unique pattern in $\{\pi_1, \pi_2, \pi_3\}^*$, where π_1 is the set of rises, π_2 the set of levels, and π_3 the set of falls. If r marks rises, f marks falls, and l marks levels, then the required number is

$$[r^j f^k l^m \mathbf{x}^{\mathbf{i}}]\Psi_1\left((\mathbf{I} - (r\mathbf{I}(\pi_1) + l\mathbf{I}(\pi_2) + f\mathbf{I}(\pi_3)))^{-1}\right).$$

But if $\mathbf{I}(\pi_1) = \mathbf{A}$ and $\mathbf{I}(\pi_2) = \mathbf{X}$, then $\mathbf{I}(\pi_3) = \mathbf{W} - \mathbf{I}(\pi_1) - \mathbf{I}(\pi_2) = \mathbf{W} - \mathbf{A} - \mathbf{X}$, and the result follows.

(b) We want $\Psi_1((\mathbf{I} - \mathbf{H})^{-1})$, where $\mathbf{H} = \mathbf{Q} - \mathbf{L}_1\mathbf{W}\mathbf{R}_1$, and $\mathbf{Q} = (r-f)\mathbf{A} + (l-f)\mathbf{X}$, $\mathbf{L}_1 = -f\mathbf{I}$, $\mathbf{R}_1 = \mathbf{I}$. Thus by [**4.3.4**(c)] we have

$$1 + f\Psi_1\left((\mathbf{I} - \mathbf{H})^{-1}\right) = \left\{1 - f\Psi_1\left((\mathbf{I} - (r-f)\mathbf{A} - (l-f)\mathbf{X})^{-1}\right)\right\}^{-1}.$$

But $\Psi_1((\mathbf{I} - (r-f)\mathbf{A} - (l-f)\mathbf{X})^{-1})$ is the ordinary generating function for nondecreasing sequences with $r-f$ marking rises and $l-f$ marking levels. The generating function for strictly increasing sequences of this type is clearly

$$(r-f)^{-1}\left\{\prod_{i\geqslant 1}(1 + (r-f)x_i) - 1\right\}.$$

Now to incorporate levels, we replace each occurrence of i by a nonempty block of i's, and the generating function for this substitution is $\sum_{k\geqslant 1} x_i^k (l - f)^{k-1} = x_i(1-(l-f)x_i)^{-1}$, since a block of k i's contains $k-1$ levels. Thus the generating function for nondecreasing sequences is

$$(r-f)^{-1}\left\{\prod_{i\geqslant 1}\left(1 + (r-f)x_i(1-(l-f)x_i)^{-1}\right) - 1\right\}$$
$$= (r-f)^{-1}\left\{\prod_{i\geqslant 1}(1 + (r-l)x_i)(1 + (f-l)x_i)^{-1} - 1\right\}.$$

Substituting this into the expression for $\Psi_1((\mathbf{I} - \mathbf{H})^{-1})$, we obtain

$$1 + f\Psi_1\left((\mathbf{I} - \mathbf{H})^{-1}\right) = (r-f)\left\{r - f\prod_{i\geqslant 1}(1 + (r-l)x_i)(1 + (f-l)x_i)^{-1}\right\}^{-1}.$$

Simplifying this yields the required result.

4.3.8. (a) In any sequence over $\{\pi_1, \ldots, \pi_m\}$, the maximal blocks of π_m's, in $\pi_m\pi_m^*$, are separated by nonempty sequences on $\{\pi_1, \ldots, \pi_{m-1}\}$, in $\mathsf{G}_{m-1} - \mathsf{G}_0$. The result follows immediately from this observation.

(b) Let π_2 be the set of increasing successions, π_3 be the set of decreasing successions, and $\pi_1 = \omega - \pi_2 - \pi_3$. Then each sequence in N_n^* has a unique pattern in $\{\pi_1, \pi_2, \pi_3\}^*$, and from part (a) we have

$$\{\pi_1, \pi_2, \pi_3\}^* = \mathsf{G}_2\big(\pi_3\pi_3^*(\mathsf{G}_2 - \mathsf{G}_0)\big)^*\pi_3^*,$$

where $\mathsf{G}_2 = \pi_1^*(\pi_2\pi_2^*\pi_1\pi_1^*)^*\pi_2^* = (\pi_2^*\pi_1)^*\pi_2^*$. This decomposition has exposed the maximal π_2-strings and π_3-strings. Now a maximal π_2-string of length l contains $l - p + 1$ π_2-strings of length p if $l \geqslant p$, and a maximal π_3-string of length l contains $l - s + 1$ π_3-strings of length s for $l \geqslant s$. Thus, if u marks maximal π_2-strings of length p and v marks maximal π_3-strings of length s, the required number is

$$\big[u^i v^j x_1 \cdots x_n\big]\Psi_1\big(\mathbf{G}(\mathbf{I} - (\mathbf{V} - \mathbf{I})(\mathbf{G} - \mathbf{I}))^{-1}\mathbf{V}\big),$$

where $\mathbf{V} = \sum_{m=0}^{s-2}\mathbf{B}^m + \sum_{m \geqslant s-1} v^{m-s+2}\mathbf{B}^m = (\mathbf{I} - \mathbf{B})^{-1}(\mathbf{I} - v\mathbf{B})^{-1}\{\mathbf{I} - v\mathbf{B} + (v - 1)\mathbf{B}^{s-1}\}$ and $\mathbf{G} = (\mathbf{I} - \mathbf{UC})^{-1}\mathbf{U}$, in which

$$\mathbf{U} = \sum_{m=0}^{p-2}\mathbf{A}^m + \sum_{m \geqslant p-1} u^{m-p+2}\mathbf{A}^m$$

$$= (\mathbf{I} - \mathbf{A})^{-1}(\mathbf{I} - u\mathbf{A})^{-1}\{\mathbf{I} - u\mathbf{A} + (u - 1)\mathbf{A}^{p-1}\}.$$

(c) If we let the generating function in part (b) be Φ, then

$$\Phi = \Psi_1\big(\mathbf{G}(\mathbf{I} - (\mathbf{V} - \mathbf{I})(\mathbf{G} - \mathbf{I}))^{-1}\mathbf{V}\big) = \Psi_1\big(\mathbf{G}(\mathbf{G} - \mathbf{V}(\mathbf{G} - \mathbf{I}))^{-1}\mathbf{V}\big)$$

$$= \Psi_1\big((\mathbf{V}^{-1} - (\mathbf{I} - \mathbf{G}^{-1}))^{-1}\big) = \Psi_1\big((\mathbf{V}^{-1} - \mathbf{I} + \mathbf{U}^{-1}(\mathbf{I} - \mathbf{UC}))^{-1}\big)$$

$$= \Psi_1\big((\mathbf{V}^{-1} + \mathbf{U}^{-1} - \mathbf{I} - \mathbf{C})^{-1}\big).$$

This is of the form $\Psi_1(\mathbf{I} - \mathbf{H})^{-1}$, where $\mathbf{H} = (\mathbf{I} - \mathbf{V}^{-1}) + (\mathbf{I} - \mathbf{U}^{-1}) + \mathbf{C}$. By letting $\mathbf{C} = \mathbf{W} - \mathbf{B} - \mathbf{A}$, we have the factored expansion $\mathbf{H} = \mathbf{Q} - \mathbf{L}_1\mathbf{W}\mathbf{R}_1$, where $\mathbf{Q} = (\mathbf{I} - \mathbf{V}^{-1}) + (\mathbf{I} - \mathbf{U}^{-1}) - (\mathbf{A} + \mathbf{B})$, $\mathbf{L}_1 = -\mathbf{I}$, $\mathbf{R}_1 = \mathbf{I}$. Thus, by the elimination theorem with $s = 1$,

$$1 + \Phi = \big\{1 - \Psi_1\big((\mathbf{V}^{-1} + \mathbf{U}^{-1} - \mathbf{I} + \mathbf{A} + \mathbf{B})^{-1}\big)\big\}^{-1}.$$

Note that since $[x_1, \ldots, x_n]\Phi$ is the solution to this problem, only square-free terms in x_i need to be considered. Now each element in the incidence matrices $\mathbf{ABT}$ and $\mathbf{BAT}$ contains only terms with squared x_i's, where $\mathbf{T}$ is the incidence matrix for any nonempty pattern. This is because **AB** and **BA** are incidence matrices for sequences of the form $i, i+1, i$ and $i, i-1, i$, respectively. Thus we may disregard occurrences of **AB** and **BA** in evaluating $\Psi_1((\mathbf{V}^{-1}+\mathbf{U}^{-1}-\mathbf{I}+\mathbf{A}+\mathbf{B})^{-1})$. Hence

$$\begin{aligned}
&\Psi_1\big((\mathbf{V}^{-1}+\mathbf{U}^{-1}-\mathbf{I}+\mathbf{A}+\mathbf{B})^{-1}\big)\\
&\quad=\Psi_1\big((\mathbf{I}-(\mathbf{I}-\mathbf{A}-\mathbf{U}^{-1})-(\mathbf{I}-\mathbf{B}-\mathbf{V}^{-1}))^{-1}\big)\\
&\quad=\Psi_1\big((\mathbf{I}-(\mathbf{I}-\mathbf{A}-\mathbf{U}^{-1}))^{-1}\big)+\Psi_1\big((\mathbf{I}-(\mathbf{I}-\mathbf{B}-\mathbf{V}^{-1}))^{-1}\big)-\Psi_1(\mathbf{I})\\
&\quad=\Psi_1\big((\mathbf{I}+\mathbf{UA})^{-1}\mathbf{U}\big)+\Psi_1\big((\mathbf{I}+\mathbf{VB})^{-1}\mathbf{V}\big)-\Psi_1(\mathbf{I}).
\end{aligned}$$

Let $\mathbf{U}=f(\mathbf{A})$, $\mathbf{V}=g(\mathbf{B})$. Then by the +-transformation lemma, since increasing +-strings have the same enumerators as decreasing +-strings, the required number is

$$\begin{aligned}
&[u^iv^jx^n]\sum_{k\geqslant 0}k![y^k]\{1-yxf(x)(1+xf(x))^{-1}\\
&\qquad\qquad -yxg(x)(1+xg(x))^{-1}+yx\}^{-1}\\
&\quad=[u^iv^jx^n]\sum_{k\geqslant 0}k!x^k\{f(x)(1+xf(x))^{-1}+g(x)(1+xg(x))^{-1}-1\}^k.
\end{aligned}$$

But

$$\begin{aligned}
1+xf(x)&=1+x(1-ux+(u-1)x^{p-1})(1-x)^{-1}(1-ux)^{-1}\\
&=(1-ux+(u-1)x^p)(1-x)^{-1}(1-ux)^{-1},
\end{aligned}$$

so that

$$f(x)(1+xf(x))^{-1}=(1-ux+(u-1)x^{p-1})(1-ux+(u-1)x^p)^{-1}.$$

Similarly, we obtain

$$g(x)(1+xg(x))^{-1}=(1-vx+(v-1)x^{s-1})(1-vx+(v-1)x^s)^{-1},$$

and the result follows.

(d) In this case we simply let $\mathbf{A}=\mathbf{I}(\oplus)$, $\mathbf{B}=\mathbf{A}^T$, $\mathbf{C}=\mathbf{W}-\mathbf{A}-\mathbf{B}$ and proceed as in parts (b) and (c). We use the $\oplus$-transformation lemma in this case so that the answer becomes

$$[u^i v^j x^n] \sum_{k \geqslant 1} (k-1)! [y^k] x \frac{\partial}{\partial x} \{1 - yxf(x)(1 + xf(x))^{-1} - yxg(x)(1 + xg(x))^{-1} + yx\}^{-1}$$

$$= [u^i v^j x^n] n \sum_{k \geqslant 1} (k-1)! x^k \{f(x)(1 + xf(x))^{-1} + g(x)(1 + xg(x))^{-1} - 1\}^k,$$

and the result follows as in part (c).

4.3.9. (a) Suppose that $G(u, l, x_1, x_2, \dots)$ is the generating function for sequences, in which l marks levels, u marks occurrences of π_1, and x_i marks occurrences of i, for $i \geqslant 1$. Let $F(u, x_1, x_2, \dots)$ be the corresponding generating function for sequences with no levels. Then the sequences enumerated by G may be constructed from the sequences enumerated by F by means of Decomposition 2.4.14, in which each sequence element is replaced by a nonempty block of that element. If such a block of i's has length k, then $k-1$ levels are introduced, $k-1$ more i's are introduced, and the number of π_1's is not affected, for $i, k \geqslant 1$. Thus we may replace x_i by $z_i \sum_{k \geqslant 1} (lz_i)^{k-1} = z_i(1 - lz_i)^{-1}$ in F to obtain G, so that

$$G(u, l, z_1, z_2, \dots) = F\left(u, z_1(1 - lz_1)^{-1}, z_2(1 - lz_2)^{-1}, \dots\right).$$

If we let $l = 1$ and $y_i = z_i(1 - z_i)^{-1}$, so that $z_i = y_i(1 + y_i)^{-1}$, then

$$F(u, y_1, y_2, \dots) = G\left(u, 1, y_1(1 + y_1)^{-1}, y_2(1 + y_2)^{-1}, \dots\right).$$

Now if $y_i = z_i(1 - lz_i)^{-1}$, then $y_i(1 + y_i)^{-1} = z_i(1 - (l-1)z_i)^{-1}$. The preceding relationships then give

$$G(u, l, z_1, z_2, \dots) = G\left(u, 1, z_1(1 - (l-1)z_1)^{-1}, z_2(1 - (l-1)z_2)^{-1}, \dots\right).$$

But $G(u, 1, x_1, x_2, \dots)$ is the generating function for sequences with respect to occurrences of π_1, so from Corollary 4.2.12

$$G(u, 1, x_1, x_2, \dots) = \left\{1 - \sum_{j \geqslant 1} (u-1)^{j-1} \gamma_j(\pi_1)\right\}^{-1}.$$

The result is obtained by replacing x_i in $\gamma_j(\pi_1)$ by $z_i(1 - (l-1)z_i)^{-1}$.

(b) If $F(r, l, \mathbf{x})$ is the generating function for sequences with r marking rises, l marking levels, and x_i marking i's, then, from part (a) and **4.2.13**, we have

$$F(r,l,\mathbf{x}) = (r-1)\Big\{r - \prod_{i\geq 1}\big(1+(r-1)x_i(1-(l-1)x_i)^{-1}\big)\Big\}^{-1}$$

$$= (r-1)\prod_{i\geq 1}(1+(1-l)x_i)\Big\{r\prod_{i\geq 1}(1+(1-l)x_i) - \prod_{i\geq 1}(1+(r-l)x_i)\Big\}^{-1}.$$

Let $G(r, l, f, \mathbf{x})$ be the generating function as above, but where f marks falls. Then

$$G(r,l,f,\mathbf{x}) = f^{-1}\{F(rf^{-1}, lf^{-1}, f\mathbf{x}) - 1\},$$

since the sum of the number of rises, levels, and falls is one less than the length of the sequence. The result follows by making the preceding substitutions.

4.3.10. The required number of sequences is $[\mathbf{x}^{\mathbf{i}}]\Psi_1((\mathbf{I}-\mathbf{AB})^{-1}(\mathbf{A}+\mathbf{I}))$, where $\mathbf{A} = \mathbf{I}(<)$ and $\mathbf{B} = \mathbf{I}(>)$. But $\mathbf{H} = \mathbf{AB} = \mathbf{A}(\mathbf{W}-\mathbf{X}-\mathbf{A}) = \mathbf{Q} - \mathbf{L}_1\mathbf{W}\mathbf{R}_1$, where $\mathbf{Q} = -\mathbf{A}(\mathbf{X}+\mathbf{A})$, $\mathbf{L}_1 = -\mathbf{A}$, $\mathbf{R}_1 = \mathbf{I}$, $\mathbf{D} = \mathbf{A}+\mathbf{I}$. Thus the desired generating function is

$$\Psi_1\big((\mathbf{I}+\mathbf{A}(\mathbf{X}+\mathbf{A}))^{-1}(\mathbf{A}+\mathbf{I})\big)\big\{1 - \Psi_1\big((\mathbf{I}+\mathbf{A}(\mathbf{X}+\mathbf{A}))^{-1}\mathbf{A}\big)\big\}^{-1}.$$

Now $\Psi_1((\mathbf{I}+\mathbf{A}(\mathbf{X}+\mathbf{A}))^{-1}\mathbf{A}) = \sum_{k\geq 1}(-1)^{k-1}t_{2k}$ and $\Psi_1((\mathbf{I}+\mathbf{A}(\mathbf{X}+\mathbf{A}))^{-1}) = \sum_{k\geq 1}(-1)^{k-1}t_{2k-1}$, where $t_i = \sum_\sigma \mathbf{x}^{\tau(\sigma)}$, and the summation is over sequences $\sigma = \sigma_1\sigma_2\cdots\sigma_i$ for which $\sigma_1 < \sigma_2 \leqslant \sigma_3 < \sigma_4 \leqslant \cdots$. The answer is thus

$$[\mathbf{x}^{\mathbf{i}}]\Big\{\sum_{k\geq 1}(-1)^{k-1}(t_{2k-1}+t_{2k})\Big\}\Big\{\sum_{k\geq 0}(-1)^k t_{2k}\Big\}^{-1}$$

$$= [\mathbf{x}^{\mathbf{i}}]\left(\Big\{1+\sum_{k\geq 0}(-1)^k t_{2k+1}\Big\}\Big\{\sum_{k\geq 0}(-1)^k t_{2k}\Big\}^{-1} - 1\right).$$

4.3.11. The number that we require is

$$\left[q^k\frac{x^n}{n!_q}\right]\Psi_2\big(\mathbf{A}^{p_1-1}\mathbf{B}\mathbf{A}^{p_2-1}\mathbf{B}\cdots\mathbf{A}^{p_m-1}\big),$$

where $\mathbf{A} = \mathbf{I}(<)$. Now replace each occurrence of $\mathbf{B}$ by $\mathbf{W}-\mathbf{A}$, so that

$$\mathbf{A}^{p_1-1}(\mathbf{W}-\mathbf{A})\mathbf{A}^{p_2-1}(\mathbf{W}-\mathbf{A})\cdots(\mathbf{W}-\mathbf{A})\mathbf{A}^{p_m-1}$$

$$= \sum_{i=1}^{m}(-1)^{m-i}\sum_{\alpha}\mathbf{A}^{s_1-1}\mathbf{W}\mathbf{A}^{s_2-1}\mathbf{W}\cdots\mathbf{W}\mathbf{A}^{s_i-1},$$

where $\alpha = \{\alpha_1, \ldots, \alpha_i\} \subseteq \mathsf{N}_m$, $\alpha_1 < \cdots < \alpha_i = m$, $s_j = a_{\alpha_j} - a_{\alpha_{j-1}}$, and $a_k = \sum_{l=1}^{k} p_l$, $a_{\alpha_0} = 0$. Applying Ψ_2 to this expression, we obtain

$$\Psi_2(\mathbf{A}^{p_1-1}\mathbf{B} \cdots \mathbf{A}^{p_m-1}) = \sum_{i=1}^{m} (-1)^{m-i} \sum_{\alpha} \prod_{j=1}^{i} \frac{x^{s_j}}{s_j!_q}.$$

But $s_1 + \cdots + s_i = p_1 + \cdots + p_m = n$, and the answer is

$$[q^k]n!_q \sum_{i=1}^{m} (-1)^{m-i} \sum_{\alpha} \prod_{j=1}^{i} (s_j!_q)^{-1}.$$

4.3.12. (a) Each sequence in N_+^* has a unique pattern in $\{\pi_1, \pi_2\}^*$. An element that initiates a π_2 terminates a maximal π_1-string, and an element that follows a π_2 initiates a maximal π_1-string. Thus if y_i and z_i mark the occurrence of i as terminator and initiator, respectively, of a maximal π_1-path, then the incidence matrix for π_1 is $\mathbf{A}$ and for π_2 is $\mathbf{YBZ}$. Furthermore, the initial element of the sequence initiates a maximal π_1-string and the final element terminates a maximal π_1-string. Thus by Lemma 4.3.5, the required number is

$$[\mathbf{y}^{\mathbf{m}}\mathbf{z}^{\mathbf{k}}\mathbf{x}^{\mathbf{i}}]\operatorname{trace} \mathbf{Z}(\mathbf{I} - \mathbf{A} - \mathbf{YBZ})^{-1}\mathbf{YW}.$$

(b) From part (a), with $\mathbf{Z} = \mathbf{I}$, the required number is

$$[\mathbf{y}^{\mathbf{m}}\mathbf{x}^{\mathbf{i}}]\operatorname{trace}(\mathbf{I} - \mathbf{A} - \mathbf{YB})^{-1}\mathbf{YW}.$$

Substituting $\mathbf{B} = \mathbf{W} - \mathbf{A}$ gives the generating function as

$$\operatorname{trace}(\mathbf{I} - (\mathbf{I} - \mathbf{Y})\mathbf{A} - \mathbf{YW})^{-1}\mathbf{YW}.$$

But from [**4.3.4**(c)], or the elimination theorem, this equals

$$\{1 - \operatorname{trace}(\mathbf{I} - (\mathbf{I} - \mathbf{Y})\mathbf{A})^{-1}\mathbf{YW}\}^{-1} - 1.$$

Thus the required number is $[\mathbf{y}^{\mathbf{m}}\mathbf{x}^{\mathbf{i}}]\{1 - \operatorname{trace}(\mathbf{I} - (\mathbf{I} - \mathbf{Y})\mathbf{A})^{-1}\mathbf{YW}\}^{-1}$.

4.3.13. (a) From [**4.3.12**], we must evaluate $\operatorname{trace}((\mathbf{I} - (\mathbf{I} - \mathbf{Y})\mathbf{A})^{-1}\mathbf{YW})$ when $\mathbf{A} = \mathsf{I}(<)$. This is the generating function for $<$-strings in which i, as the final element, is marked by $y_i x_i$ and, as any other element, by $(1 - y_i)x_i$. Such sequences that end in the element j are enumerated by

$$x_j y_j \prod_{k=1}^{j-1} (1 + (1 - y_k)x_k),$$

and the answer is

$$[\mathbf{y}^{\mathbf{m}}\mathbf{x}^{\mathbf{i}}]\left\{1-\sum_{j\geq 1} x_j y_j \prod_{k=1}^{j-1}(1+(1-y_k)x_k)\right\}^{-1},$$

since j can be any positive integer.

(b) The enumerator for nondecreasing strings in which i, as a terminal element, is marked by $x_i y_i$ and, as any other element, by $(1-y_i)x_i$ is $\Sigma_{j\geq 1} x_j y_j \Pi_{k=1}^{j}(1-(1-y_k)x_k)^{-1}$. Thus the answer is

$$[\mathbf{y}^{\mathbf{m}}\mathbf{x}^{\mathbf{i}}]\left\{1-\sum_{j\geq 1} x_j y_j \prod_{k=1}^{j}(1-(1-y_k)x_k)^{-1}\right\}^{-1}.$$

4.3.14. The required number is clearly $[r_1^{i_1} r_2^{i_2} f_1^{j_1} f_2^{j_2} x_1 \cdots x_n]\Phi$ where

$$\Phi = \Psi_1\big((\mathbf{I}-(r_1\mathbf{A}+f_1\mathbf{B})(r_2\mathbf{A}+f_2\mathbf{B}))^{-1}(\mathbf{I}+r_1\mathbf{A}+f_1\mathbf{B})\big)$$

$$= \Psi_1\big((\mathbf{I}-\mathbf{H})^{-1}\mathbf{D}\big),$$

and $\mathbf{A} = \mathsf{I}(<)$. Now, by a right expansion with $\alpha_1 = r_1 - f_1$, $\alpha_2 = r_2 - f_2$,

$$\mathbf{H} = (r_1\mathbf{A}+f_1\mathbf{B})(\alpha_2\mathbf{A}+f_2\mathbf{W})$$

$$= \alpha^2\mathbf{A}^2 + f_2(r_1\mathbf{A}+f_1\mathbf{B})\mathbf{W} + f_1\alpha_2\mathbf{W}\mathbf{A} = \mathbf{Q} - \mathbf{L}_1\mathbf{W}\mathbf{R}_1 - \mathbf{L}_2\mathbf{W}\mathbf{R}_2,$$

where $\mathbf{Q} = \alpha^2\mathbf{A}^2$, $\mathbf{L}_1 = -f_2(r_1\mathbf{A}+f_1\mathbf{B})$, $\mathbf{R}_1 = \mathbf{I}$, $\mathbf{L}_2 = -f_1\alpha_2\mathbf{I}$, $\mathbf{R}_2 = \mathbf{A}$. Now $\mathbf{D} = (\mathbf{I}+r_1\mathbf{A}+f_1\mathbf{B})$, and $\mathbf{C} = \mathbf{R}_1 = \mathbf{I}$, so by [**4.3.4**(a)]

$$\Phi = \begin{vmatrix} \Psi_1\big((\mathbf{I}-\mathbf{Q})^{-1}\mathbf{D}\big) & -f_1\alpha_2\Psi_1\big((\mathbf{I}-\mathbf{Q})^{-1}\big) \\ \Psi_1\big(\mathbf{A}(\mathbf{I}-\mathbf{Q})^{-1}\mathbf{D}\big) & 1-f_1\alpha_2\Psi_1\big(\mathbf{A}(\mathbf{I}-\mathbf{Q})^{-1}\big) \end{vmatrix} \cdot |\mathbf{M}|^{-1},$$

where

$$|\mathbf{M}| = \begin{vmatrix} 1-f_2\Psi_1\big((\mathbf{I}-\mathbf{Q})^{-1}(r_1\mathbf{A}+f_1\mathbf{B})\big) & -f_1\alpha_2\Psi_1\big((\mathbf{I}-\mathbf{Q})^{-1}\big) \\ -f_2\Psi_1\big(\mathbf{A}(\mathbf{I}-\mathbf{Q})^{-1}(r_1\mathbf{A}+f_1\mathbf{B})\big) & 1-f_1\alpha_2\Psi_1\big(\mathbf{A}(\mathbf{I}-\mathbf{Q})^{-1}\big) \end{vmatrix}.$$

Let $\mathbf{B} = \mathbf{W} - \mathbf{A}$ in both numerator and denominator. Add $\alpha_2^{-1}(\gamma_1(<)+f_1^{-1})$ times column 2 to column 1 in the numerator, giving

$$\begin{vmatrix} \alpha_1\Psi_1\big((\mathbf{I}-\mathbf{Q})^{-1}\mathbf{A}\big) & -f_1\alpha_2\Psi_1\big((\mathbf{I}-\mathbf{Q})^{-1}\big) \\ (\gamma_1+f_1^{-1})\alpha_2^{-1}+\alpha_1\Psi_1\big(\mathbf{A}(\mathbf{I}-\mathbf{Q})^{-1}\mathbf{A}\big) & 1-f_1\alpha_2\Psi_1\big(\mathbf{A}(\mathbf{I}-\mathbf{Q})^{-1}\big) \end{vmatrix}.$$

Subtract $f_2\alpha_2^{-1}\gamma_1(<)$ times column 2 from column 1 in the denominator, giving

$$|\mathbf{M}| = \begin{vmatrix} 1 - f_2\alpha_1\Psi_1\left((\mathbf{I}-\mathbf{Q})^{-1}\mathbf{A}\right) & -f_1\alpha_2\Psi_1\left((\mathbf{I}-\mathbf{Q})^{-1}\right) \\ -f_2\gamma_1\alpha_2^{-1} - f_2\alpha_1\Psi_1\left(\mathbf{A}(\mathbf{I}-\mathbf{Q})^{-1}\mathbf{A}\right) & 1 - f_1\alpha_2\Psi_1\left(\mathbf{A}(\mathbf{I}-\mathbf{Q})^{-1}\right) \end{vmatrix}.$$

Under the $<$-transformation, $\gamma_j(<)$ becomes $x^j/j!$, so $\Psi_1((\mathbf{I}-\mathbf{Q})^{-1})$, $\Psi_1((\mathbf{I}-\mathbf{Q})^{-1}\mathbf{A}) = \Psi_1(\mathbf{A}(\mathbf{I}-\mathbf{Q})^{-1})$ and $\Psi_1(\mathbf{A}(\mathbf{I}-\mathbf{Q})^{-1}\mathbf{A})$ are replaced by, respectively, $\alpha^{-1}\sinh(\alpha x)$, $\alpha^{-2}(\cosh(\alpha x) - 1)$, and $\alpha^{-3}(\sinh(\alpha x) - \alpha x)$. Thus the numerator becomes $\alpha^{-2}(\sinh(\alpha x) + (r_1 + f_1)(\cosh(\alpha x) - 1))$, and $|\mathbf{M}|$ becomes $\alpha^{-2}(r_1r_2 + f_1f_2 - (r_1f_2 + f_1r_2)\cosh(\alpha x))$, so the result follows.

4.3.15. The required generating function is $\Psi_2((\mathbf{I}-\mathbf{H})^{-1}\mathbf{D})$, where $\mathbf{H} = \mathbf{A}^{p_1-1}\mathbf{B}\cdots\mathbf{A}^{p_s-1}\mathbf{B}$, $\mathbf{D} = \mathbf{A}^{r-1}$, $\mathbf{A} = \mathbf{I}(<)$, $\mathbf{B} = \mathbf{W} - \mathbf{A}$. Now $\mathbf{H}$ has the left-expansion

$$\mathbf{H} = (-1)^s\mathbf{A}^{a_1} - \sum_{i=1}^{s}(-1)^{s+1-i}\mathbf{A}^{a_i-1}\mathbf{W}\mathbf{A}^{p_{s-i+2}-1}\mathbf{B}\cdots\mathbf{A}^{p_s-1}\mathbf{B},$$

so $\mathbf{Q} = (-1)^s\mathbf{A}^{a_1}$, $\mathbf{L}_i = (-1)^{s+1-i}\mathbf{A}^{a_i-1}$, $\mathbf{R}_i = \mathbf{A}^{p_{s-i+2}-1}\mathbf{B}\cdots\mathbf{A}^{p_s-1}\mathbf{B}$, for $i = 1,\dots,s$, where $\mathbf{R}_1 = \mathbf{I}$. Thus, by [**4.3.4**(a)], the generating function is $|[\mathbf{M}:\mathbf{d}]_1|\,|\mathbf{M}|^{-1}$, where $[\mathbf{M}]_{ij} = \delta_{ij} + \Psi_2(\mathbf{R}_i(\mathbf{I}-\mathbf{Q})^{-1}\mathbf{L}_j)$, and $\mathbf{d} = (d_1,\dots,d_s)^T$, $d_i = \Psi_2(\mathbf{R}_i(\mathbf{I}-\mathbf{Q})^{-1}\mathbf{D})$. We now replace row i in the numerator and denominator by (row i) $+ \sum_{j=1}^{i-1}(-1)^{i-j}\Psi_2(\mathbf{A}^{a_j-a_i-1}) \times$ (row j), beginning at $i = s$ and continuing until $i = 1$. By this procedure, d_i becomes $(-1)^{i-1}\Psi_2(\mathbf{A}^{a_1-a_i}(\mathbf{I}-\mathbf{Q})^{-1}\mathbf{A}^{r-1})$ and $[\mathbf{M}]_{ij}$ becomes

$$(-1)^{s+i-j}\Psi_2\left(\mathbf{A}^{a_1-a_i}(\mathbf{I}-\mathbf{Q})^{-1}\mathbf{A}^{a_j-1}\right) + \left(1 - \zeta(i,j) - \delta_{ij}\right)(-1)^{i-j}\Psi_2(\mathbf{A}^{a_j-a_i-1}) + \delta_{ij}.$$

Multiplying row i by $(-1)^{i-1}$ for $i = 1,\dots,s$ in both the numerator and denominator leaves the result unchanged, so d_i becomes

$$\Psi_2\left(\mathbf{A}^{a_1-a_i}(\mathbf{I}-\mathbf{Q})^{-1}\mathbf{A}^{r-1}\right) = \phi^{(a_1,s)}_{r+a_1-a_i}(x), \qquad i = 1,\dots,s,$$

and $[\mathbf{M}]_{ij}$ becomes

$$(-1)^{s+1+j}\Psi_2\left(\mathbf{A}^{a_1-a_i+a_j-1}(\mathbf{I}-\mathbf{Q})^{-1}\right) + \left(1 - \zeta(i,j) - \delta_{ij}\right)(-1)^{1+j}\Psi_2(\mathbf{A}^{a_j-a_i-1}) + \delta_{ij}(-1)^{i+1}.$$

But if $i > j$, then

$$\mathbf{A}^{a_j-a_i-1} + (-1)^s\mathbf{A}^{a_1+a_j-a_i-1}(\mathbf{I}-\mathbf{Q})^{-1} = \mathbf{A}^{a_j-a_i-1}(\mathbf{I}-\mathbf{Q})^{-1}.$$

Thus $[\mathbf{M}]_{ij}$ becomes

$$(-1)^{1+j+s\zeta(i,j)}\phi^{(a_1,s)}_{(a_j-a_i)\bmod a_1}(x), \qquad i,j = 1,\dots,s.$$

4.3.16. (a) Let t_i mark an initial maximal π_2-string of length i, u_i mark a final maximal π_1-string of length i, and g_{ij} mark a maximal (π_1, π_2)-structure of type (i, j). Each sequence has a unique pattern in $\{\pi_1, \pi_2\}^* = \pi_2^*\{\pi_1\pi_1^*\pi_2\pi_2^*\}^*\pi_1^*$ and thus the required generating function is

$$\Psi_1\left(\sum_{i\geq 1} t_i \mathbf{B}^{i-1}\left(\mathbf{I} - \sum_{j\geq 1}\sum_{k\geq 1} g_{jk}\mathbf{A}^j\mathbf{B}^k\right)^{-1}\sum_{l\geq 1} u_l\mathbf{A}^{l-1}\right) = \Psi_1\left(\sum_{i\geq 1} t_i\mathbf{F}_i\right),$$

where $\mathbf{F}_i = \mathbf{B}^{i-1}(\mathbf{I} - \mathbf{H})^{-1}\mathbf{D}$ for $i \geq 1$ and

$$\mathbf{H} = \sum_{i\geq 1}\sum_{j\geq 1} g_{ij}\mathbf{A}^i\mathbf{B}^j, \qquad \mathbf{D} = \sum_{i\geq 1} u_i\mathbf{A}^{i-1}.$$

(b) Let $\mathbf{G}_j = \sum_{i\geq 1} g_{ij}\mathbf{A}^i$. Now, using a left-expansion,

$$\mathbf{B}^j = (-1)^j\mathbf{A}^j + \sum_{k=0}^{j-1}(-1)^{j-k-1}\mathbf{A}^{j-k-1}\mathbf{W}\mathbf{B}^k,$$

so

$$\mathbf{H} = \sum_{j\geq 1}\mathbf{G}_j\mathbf{B}^j = \mathbf{Q} + \sum_{k\geq 0}\sum_{j\geq k+1}(-1)^{j-k-1}\mathbf{G}_j\mathbf{A}^{j-k-1}\mathbf{W}\mathbf{B}^k.$$

where $\mathbf{Q} = \sum_{j\geq 1}(-1)^j\mathbf{G}_j\mathbf{A}^j$. This is a factored expansion, and from the proof of Theorem 4.3.8 we obtain, with $\mathbf{T} = \mathbf{A}^m$,

$$\Psi_1\left(\mathbf{A}^m(\mathbf{I} - \mathbf{Q})^{-1}\mathbf{D}\right) = \Psi_1\left(\mathbf{A}^m(\mathbf{I} - \mathbf{H})^{-1}\mathbf{D}\right) - \sum_{k\geq 0} S(m, k)\xi_k,$$

where

$$\xi_k = \Psi_1\left(\mathbf{B}^k(\mathbf{I} - \mathbf{H})^{-1}\mathbf{D}\right), \qquad \text{for } k \geq 0,$$

and

$$S(m, k) = \sum_{j\geq k+1}(-1)^{j-k-1}\Psi_1\left(\mathbf{A}^m(\mathbf{I} - \mathbf{Q})^{-1}\mathbf{G}_j\mathbf{A}^{j-k-1}\right), \qquad \text{for } m, k \geq 0.$$

But

$$\mathbf{A}^m = (-1)^m\mathbf{B}^m + \sum_{k=0}^{m-1}(-1)^k\mathbf{A}^{m-k-1}\mathbf{W}\mathbf{B}^k,$$

so

$$\Psi_1\left(\mathbf{A}^m(\mathbf{I} - \mathbf{H})^{-1}\mathbf{D}\right) = (-1)^m\xi_m + \sum_{k=0}^{m-1}(-1)^k\Psi_1(\mathbf{A}^{m-k-1})\xi_k.$$

Substituting this in the preceding gives

$$\Psi_1\big(\mathbf{A}^m(\mathbf{I}-\mathbf{Q})^{-1}\mathbf{D}\big) = \sum_{k=0}^{m-1}\{(-1)^k\Psi_1(\mathbf{A}^{m-k-1}) - S(m,k)\}\xi_k$$

$$+\{(-1)^m - S(m,m)\}\xi_m - \sum_{k\geqslant m+1} S(m,k)\xi_k, \qquad \text{for } m \geqslant 0.$$

We solve this system of equations by Cramer's rule for $\Psi_1(\mathbf{F}_r) = \xi_{r-1}$.

(c) Under the given conditions $[\mathbf{M}]_{ij} = 0$ and $[[\mathbf{M}:\mathbf{d}]_{r-1}]_{ij} = 0$ for $i<j$ and $j \geqslant k$, and the diagonal elements are 1 for $j \geqslant k$. Thus the denominators are equal to the determinants of the submatrices consisting of rows and columns $0, 1, \ldots, k-1$.

4.3.17. (a) A maximal π_1-string of length l contains $l-p+1$ π_1-strings of length p for $l \geqslant p$, and a maximal π_2-string of length l contains $l-s+1$ π_2-strings of length s for $l \geqslant s$. Let y mark π_1-strings of length p and z mark π_2-strings of length s. Then in the notation of **[4.3.16]** we have

$$g_{ij} = \begin{cases} y^{i-p+2}z^{j-s+2} & \text{for } i \geqslant p-1,\ j \geqslant s-1 \\ y^{i-p+2} & \text{for } i \geqslant p-1,\ j < s-1 \\ z^{j-s+2} & \text{for } i < p-1,\ j \geqslant s-1 \\ 1 & \text{for } i < p-1,\ j < s-1 \end{cases}$$

$$t_j = \begin{cases} z^{j-s+1} & \text{for } j \geqslant s \\ 1 & \text{for } j < s \end{cases}, \qquad u_i = \begin{cases} y^{i-p+1} & \text{for } i \geqslant p \\ 1 & \text{for } i < p. \end{cases}$$

The required generating function is

$$\Phi = \Psi_1\big(f(s,z,\mathbf{B})\{\mathbf{I} - \mathbf{A}f(p-1,y,\mathbf{A})\mathbf{B}f(s-1,z,\mathbf{B})\}^{-1}f(p,y,\mathbf{A})\big),$$

where $f(r,\alpha,\mathbf{C}) = (\mathbf{I}-\mathbf{C})^{-1}(\mathbf{I}-\alpha\mathbf{C})^{-1}(\mathbf{I}-\alpha\mathbf{C}+(\alpha-1)\mathbf{C}^r)$. Thus

$$\Phi = \Psi_1\big((\mathbf{I} - z\mathbf{B} + (z-1)\mathbf{B}^s)$$

$$\times\{(\mathbf{I}-y\mathbf{A})(\mathbf{I}-\mathbf{A})(\mathbf{I}-z\mathbf{B})(\mathbf{I}-\mathbf{B}) - \mathbf{A}(\mathbf{I}-y\mathbf{A}+(y-1)\mathbf{A}^{p-1})$$

$$\times\mathbf{B}(1 - z\mathbf{B} + (z-1)\mathbf{B}^{s-1})\}^{-1}(\mathbf{I}-y\mathbf{A}+(y-1)\mathbf{A}^p)\big)$$

$$= \Psi_1\left((\mathbf{I} - z\mathbf{B} + (z-1)\mathbf{B}^s)\Big\{\mathbf{I} - \sum_{j\geqslant 0}\mathbf{G}_j\mathbf{B}^j\Big\}^{-1}\mathbf{D}\right),$$

where $\mathbf{D} = \mathbf{I} - y\mathbf{A} + (y-1)\mathbf{A}^p$.

(b) Now **[4.3.16]** is very easily modified to incorporate the nonzero values of g_{0j} and g_{i0} that are required by this modified problem. The value of $Q(x)$ becomes $\Sigma_{i\geqslant 0}(-1)^i G_i(x)x^i$ and $G_j(x) = \Sigma_{i\geqslant 0} g_{ij}x^i$. We may write

$$\Phi = \Psi_1(\mathbf{U}) - z\Psi_1(\mathbf{BU}) + (z-1)\Psi_1(\mathbf{B}^s\mathbf{U}),$$

where $\mathbf{U} = (\mathbf{I} - \Sigma_{i\geqslant 0}\mathbf{G}_j\mathbf{B}^j)^{-1}\mathbf{D}$. Since $G_j(x) = 0$ for $j \geqslant s+1$, each of the three generating functions may be expressed as the ratio of $(s+1)\times(s+1)$ determinants. The denominator is the same for all three of them.

4.3.18. Instead of applying **[4.3.16]** with $g_{ij} = 1 - \delta_{i2}\delta_{2j}$, to give a ratio of infinite-dimensional determinants, we write the generating function as

$$\begin{aligned}\Phi &= \Psi_1\Big((\mathbf{I}-\mathbf{B})^{-1}\big\{\mathbf{I} - \mathbf{A}(\mathbf{I}-\mathbf{A})^{-1}\mathbf{B}(\mathbf{I}-\mathbf{B})^{-1} + \mathbf{A}^2\mathbf{B}^2\big\}^{-1}(\mathbf{I}-\mathbf{A})^{-1}\Big)\\ &= \Psi_1\Big(\big\{(\mathbf{I}-\mathbf{A})(\mathbf{I}-\mathbf{B}) - \mathbf{AB} + \mathbf{A}^2(\mathbf{I}-\mathbf{A})\mathbf{B}^2(\mathbf{I}-\mathbf{B})\big\}^{-1}\Big)\\ &= \Psi_1\Big(\big\{\mathbf{I} - \mathbf{A} - \mathbf{B} + \mathbf{A}^2\mathbf{B}^2 - \mathbf{A}^3\mathbf{B}^2 - \mathbf{A}^2\mathbf{B}^3 + \mathbf{A}^3\mathbf{B}^3\big\}^{-1}\Big).\end{aligned}$$

We modify **[4.3.16]** as in **[4.3.17(b)]** and have

$$G_0(x) = x, \qquad G_1(x) = 1, \qquad G_2(x) = x^3 - x^2, \qquad G_3(x) = x^2 - x^3,$$

$$D(x) = 1, \qquad Q(x) = x - x + (x^3 - x^2)x^2 - (x^2 - x^3)x^3 = x^6 - x^4.$$

Since $G_i(x) = 0$ for $i > 3$ we may express the solution as the ratio of 4×4 determinants by ignoring rows and columns after 4. By considering the first column, we transform these immediately into 3×3 determinants.

4.3.19. The product lemmas for incidence matrices and Ψ_1, Ψ_2 are clearly true in the noncommutative case, if $\mathbf{x}^{\tau(\sigma)}$ in Ψ_1 is replaced by $x_{\sigma_1}\cdots x_{\sigma_l}$. Thus, from **4.3.12**, $\Phi = \Psi_1((I - f(\mathbf{A})\mathbf{B})^{-1}f(\mathbf{A}))$, so

$$\Big\{1 - \Psi_1\Big(\sum_{l\geqslant 0}(-1)^l(f(\mathbf{A})\mathbf{A})^l f(\mathbf{A})\Big)\Big\}\Phi = \Psi_1\Big(\sum_{l\geqslant 0}(-1)^l(f(\mathbf{A})\mathbf{A})^l f(\mathbf{A})\Big).$$

Finally,

$$1 + \Phi = \sum_{k\geqslant 0}\Big\{\Psi_1\Big(\sum_{l\geqslant 0}(-1)^l(f(\mathbf{A})\mathbf{A})^l f(\mathbf{A})\Big)\Big\}^k.$$

But

$$\Psi_1\big((f(\mathbf{A})\mathbf{A})^l f(\mathbf{A})\big) = (xf)^{l+1}\circ\gamma(\pi_1),$$

so

$$\Psi_1\left(\sum_{l\geqslant 0}(-1)^l(f(\mathbf{A})\mathbf{A})^l f(\mathbf{A})\right) = -(1-F)^{l+1}\circ\gamma(\pi_1)$$

and

$$1+\Phi = \sum_{k\geqslant 0}(-1)^k\left\{\sum_{i\geqslant 1}(1-F)^i\circ\gamma(\pi_1)\right\}^k.$$

4.3.20. We follow [**4.2.15**] and **4.3.17**. If we modify Ψ_2 to be the generating function for k-tuples of permutations, with q_l marking inversions in permutation l, and replace $x^{|\sigma|}/|\sigma|!_q$ by $x^{|\sigma|}/\prod_{l=1}^k|\sigma|!_{q_l}$ then the product and sum lemmas hold for the modified Ψ_2, where $\mathbf{W}$ is now the incidence matrix for $\mathsf{N}_+^k\times\mathsf{N}_+^k$. Thus, following **4.3.17**, the required number is

$$\left[q_1^{t_1}\cdots q_k^{t_k}\right]n!_{q_1}\cdots n!_{q_k}\left\|\left\{\prod_{l=1}^k(a_j-a_{i-1})!_{q_l}\right\}^{-1}\right\|_{m\times m}$$

$$=\left[q_1^{t_1}\cdots q_k^{t_k}\right]\left\|\prod_{l=1}^k\binom{n-a_{i-1}}{a_j-a_{i-1}}_{q_l}\right\|_{m\times m}.$$

4.3.21. Consider such a sequence $\sigma=\sigma_1\cdots\sigma_{2m}$. This sequence may be uniquely recovered by considering the initial segment $\sigma'=\sigma_1\cdots\sigma_m$. The occurrence of i in σ' is marked by $x_i=z_iz_{2n+1-i}$, where z_i marks occurrences of i in σ. A rise, level, or fall (σ_i,σ_{i+1}) in σ' leads to another rise, level, or fall, respectively, in σ. Thus rises, levels, and falls in σ' are marked by r^2, l^2, and f^2. Finally, (σ_m,σ_{m+1}) is a rise in σ if $\sigma_m\leqslant n$, and a fall if $\sigma_m>n$. Let incidence matrices be $2n\times 2n$, and $\mathbf{M}=\operatorname{diag}(r,\ldots,r,f,\ldots,f)$, where there are n r's and n f's. Then the required generating function is

$$\Phi=\operatorname{trace}\{\mathbf{I}-(r^2\mathbf{A}+l^2\mathbf{B}+f^2\mathbf{C})\}^{-1}\mathbf{MW},$$

where $\mathbf{A}=\mathsf{I}(<)$, $\mathbf{B}=\mathsf{I}(=)$, $\mathbf{C}=\mathbf{W}-\mathbf{A}-\mathbf{B}$. Thus

$$\Phi=\operatorname{trace}\{\mathbf{I}-(r^2-f^2)\mathbf{A}-(l^2-f^2)\mathbf{B}-f^2\mathbf{W}\}^{-1}\mathbf{MW},$$

$$=\operatorname{trace}\{\mathbf{I}-v\mathbf{A}-y\mathbf{B}\}^{-1}\mathbf{MW}\{1-f^2\operatorname{trace}(\mathbf{I}-v\mathbf{A}-y\mathbf{B})^{-1}\mathbf{W}\}^{-1},$$

from [**4.3.4**(b)], where $v=r^2-f^2, y=l^2-f^2$. Now trace $(\mathbf{I}-v\mathbf{A}-y\mathbf{B})^{-1}\mathbf{W}$ is the generating function for nondecreasing sequences, with v marking rises and y marking levels. This was evaluated in the solution to [**4.3.7**(b)] and is

$F(2n)$, where

$$F(k) = v^{-1}\left(\prod_{i=1}^{k}\{1 + vx_i(1 - yx_i)^{-1}\} - 1\right)$$

$$= (r^2 - f^2)^{-1}(P(r,k) - P(f,k))P(f,k)^{-1}$$

and $P(g,k) = \prod_{i=1}^{k}\{1 + (g^2 - l^2)z_i z_{2n+1-i}\}$. To evaluate the numerator of Φ, note that it is the enumerator for nondecreasing sequences on N_{2n}, with rises marked by v, levels by y, and the last element by r or f, depending on whether it is $\leqslant n$ or $> n$, respectively. Thus the numerator is

$$rF(n) + f(F(2n) - F(n)) = (r - f)F(n) + fF(2n).$$

But $P(g, 2n) = P(g,n)^2$, so

$$\Phi = \{P(r,n) - P(f,n)\}\{rP(f,n) - fP(r,n)\}^{-1},$$

as required.

4.3.22. If $\Psi_1(\mu) = F(\gamma_1(\leqslant), \gamma_2(\leqslant), \ldots)$, then the number of permutations on N_n with m inversions and pattern in μ is

$$\left[q^m \frac{x^n}{n!_q}\right]\Phi(x), \qquad \text{where } \Phi(x) = F\left(\frac{x^1}{1!_q}, \frac{x^2}{2!_q}, \ldots\right).$$

Now $\gamma_k(\leqslant) = [z^k]\prod_{i\geqslant 1}(1 - zx_i)^{-1}$, and for compositions, with $x_i = xq^i$, by Remark 2.4.11,

$$\gamma_k(\leqslant) = [z^k]\prod_{i\geqslant 1}(1 - zxq^i)^{-1}$$

$$= (xq)^k \prod_{i=1}^{k}(1 - q^i)^{-1}, \qquad \text{from } \mathbf{2.5.9}$$

$$= \left(xq(1-q)^{-1}\right)^k (k!_q)^{-1}.$$

Thus the number of compositions of m with n parts and pattern in μ is $[q^m x^n]\Phi(qx(1-q)^{-1})$.

4.3.23. (a) From the definition of Φ_q,

$$\Phi_q(x, \mathsf{P}) = \sum_{\sigma\in\mathsf{P}} q^{I(\sigma)} \frac{x^{|\sigma|}}{|\sigma|!_q}$$

$$= \sum_{\sigma'\in\mathsf{P}_1} \sum_{\sigma''\in\mathsf{P}_2} \sum q^{I(\sigma') + I(\sigma'') + I(\alpha,\beta)} \frac{x^{|\sigma'| + |\sigma''|}}{|\sigma|!_q}$$

(where the right-most summation is over $\alpha \cup \beta = \mathsf{N}_n$, $|\alpha| = |\sigma'|$, $|\beta| = |\sigma''|$)

$$= \sum_{\sigma' \in \mathsf{P}_1} \sum_{\sigma'' \in \mathsf{P}_2} q^{I(\sigma')+I(\sigma'')} \frac{x^{|\sigma'|+|\sigma''|}}{|\sigma|!_q} \frac{|\sigma|!_q}{|\sigma'|!_q |\sigma''|!_q}, \qquad \text{from Lemma 2.6.6}$$

$$= \Phi_q(x, \mathsf{P}_1)\Phi_q(x, \mathsf{P}_2).$$

Of course, when $\mathsf{P}, \mathsf{P}_1, \mathsf{P}_2$ are specified as permutations having patterns $\mu, \mu_1, \mu_2 \in \{<, \geqslant\}^*$, then this result is simply the product lemma for Ψ_2, since $\mu = \mu_1 \omega \mu_2$.

(b) Let $F(x, q, t)$ be the generating function for permutations on N_k, for $k > 0$, with x marking length, q marking inversions, and t marking falls. If sequences on N_n are replaced by sequences on N_+ in Decomposition 2.4.19, and we specialize to permutations, then from part (a)

$$\sum_{k \geqslant 1} \left(\sum_{i \geqslant 1} \frac{z^i}{i!_q} \right)^k f^k = f(1+f)^{-1} F\big(z(1+f), q, f(1+f)^{-1}\big),$$

where z marks length, q marks inversions, and f marks 0's, following **2.4.20**. Let $t = f(1+f)^{-1}$ and $x = z(1+f)$, so that $f = t(1-t)^{-1}$, $z = x(1-t)$. Substituting in the preceding equation,

$$F(x, q, t) = \left\{ \sum_{i \geqslant 1} \frac{(x(1-t))^i}{i!_q} \right\} \left\{ 1 - t \sum_{i \geqslant 0} \frac{(x(1-t))^i}{i!_q} \right\}^{-1}.$$

4.3.24. (a) The objects on both sides are sequences in N_+^+ with a nonempty string of 0's in every fall, and a (possibly empty) string of 0's in nonfalls and at the end of the sequence.

(b) By definition $m(\sigma) = \sum_{j=1}^{k-1} \sum_{l=1}^{j} i_l = \sum_{l=1}^{k-1} (k-l) i_l$.

(c) Let P_n be the set of permutations on N_n. In part (a), let z mark the length of a permutation, let q mark inversions, and let tp^i mark each 0 that lies between σ_i and σ_{i+1}, and tp^n mark each 0 that occurs at the end of an element of P_n. Then from parts (a) and (b)

$$\sum_{k \geqslant 1} t^k \{E_q(zp^{k-1}) - 1\} \prod_{j=1}^{k-2} E_q(zp^j)$$

$$= \sum_{n \geqslant 1} \sum_{\sigma \in \mathsf{P}_n} q^{I(\sigma)} \frac{z^n}{n!_q} \frac{t^{f(\sigma)+1} p^{m(\sigma)}}{(1-tp) \cdots (1-tp^n)},$$

where $f(\sigma)$ is the number of falls in σ. Thus

$$\sum_{\sigma \in \mathsf{P}_n} q^{I(\sigma)} t^{f(\sigma)} p^{m(\sigma)} = \left[\frac{z^n}{n!_q} \right] \prod_{i=1}^{n} (1 - tp^i) \sum_{l \geqslant 1} t^{l-1} \prod_{j=1}^{l-1} \big(E_q(zp^j) - \delta_{j,l-1}\big).$$

4.3.25. (a) We have $I(\sigma_1 n \sigma_2) = I(\sigma_1) + I(\sigma_2) + I(\alpha, \beta) + |\sigma_2|$, where $\sigma_1 = \sigma_\alpha$, $\sigma_2 = \sigma'_\beta$, σ, σ' are permutations, and $\alpha \mathbin{\dot\cup} \beta = \mathsf{N}_n - \{1\}$. We follow **3.2.22** using [**4.3.23**(a)] and recording the contribution of n to the number of inversions, $|\sigma_2|$, by the argument qx.

(b) If $F(x) = \Sigma_{i\geq 0} f_i x^i / i!_q$, then

$$F(x) - F(qx) = \sum_{i\geq 1} f_i x^i \frac{(1-q^i)}{i!_q} = (1-q)x \sum_{i\geq 1} f_i \frac{x^{i-1}}{(i-1)!_q}.$$

Thus

$$\sum_{i\geq 1} f_i \frac{x^{i-1}}{(i-1)!_q} = (x-qx)^{-1}(F(x) - F(qx)).$$

Now

$$\begin{aligned}\mathsf{E}_q(F(x)G(x)) &= (x-qx)^{-1}(F(x)G(x) - F(qx)G(qx)) \\ &= (x-qx)^{-1}\{F(x)(G(x)-G(qx)) + (F(x)-F(qx))G(qx)\} \\ &= F(x)\mathsf{E}_q G(x) + \{\mathsf{E}_q F(x)\}G(qx),\end{aligned}$$

and

$$\begin{aligned}\mathsf{E}_q\left(F(x)^{-1}\right) &= (x-qx)^{-1}\left(F(x)^{-1} - F(qx)^{-1}\right) \\ &= (x-qx)^{-1}(F(qx) - F(x))F(x)^{-1}F(qx)^{-1} \\ &= -\{\mathsf{E}_q F(x)\}F(x)^{-1}F(qx)^{-1}.\end{aligned}$$

(c) From part (b)

$$\begin{aligned}\mathsf{E}_q P(x) &= -\mathsf{E}_q\left(H(x)^{-1}\mathsf{E}_q H(x)\right) \\ &= -\left\{\mathsf{E}_q\left(H(x)^{-1}\right)\right\}\mathsf{E}_q H(qx) - \{\mathsf{E}_q^2 H(x)\}H(x)^{-1} \\ &= H(x)^{-1}H(qx)^{-1}\{\mathsf{E}_q H(x)\}\mathsf{E}_q H(qx) - H(x)^{-1}\mathsf{E}_q^2 H(x).\end{aligned}$$

Substituting this in part (a) gives

$$\begin{aligned}&H(x)^{-1}H(qx)^{-1}\{\mathsf{E}_q H(x)\}\mathsf{E}_q H(qx) - H(x)^{-1}\mathsf{E}_q^2 H(x) \\ &\qquad = H(x)^{-1}H(qx)^{-1}\{\mathsf{E}_q H(x)\}\mathsf{E}_q H(qx) + 1.\end{aligned}$$

This reduces to $\mathsf{E}_q^2 H(x) = -H(x)$. But $\mathsf{E}_q^2(x^i/i!_q) = x^{i-2}/(i-2)!_q$, $i \geqslant 2$, so

$$H(x) = a \sum_{k\geqslant 0} (-1)^k \frac{x^{2k}}{(2k)!_q} + b \sum_{k\geqslant 0} (-1)^k \frac{x^{2k+1}}{(2k+1)!_q}.$$

Now $H(0) = 1$, so $a = 1$, and $P(0) = 0$, giving $\{\mathsf{E}_q H(x)\}|_{x=0} = 0$, and $b = 0$. Thus

$$H(x) = \sum_{k\geqslant 0} (-1)^k \frac{x^{2k}}{(2k)!_q},$$

so

$$\begin{aligned} P(x) &= -\{\mathsf{E}_q H(x)\} H(x)^{-1} \\ &= \left\{ \sum_{k\geqslant 0} (-1)^k \frac{x^{2k+1}}{(2k+1)!_q} \right\} \left\{ \sum_{k\geqslant 0} (-1)^k \frac{x^{2k}}{(2k)!_q} \right\}^{-1}. \end{aligned}$$

(d) In the preceding solution

$$\mathsf{E}_q\left(H(x)^{-1}\right) = -H(x)^{-1} H(qx)^{-1} \mathsf{E}_q(H(x)) = P(x) H(qx)^{-1}.$$

Comparing this with part (a) gives

$$Q(x) = H(x)^{-1} = \left\{ \sum_{k\geqslant 0} (-1)^k \frac{x^{2k}}{(2k)!_q} \right\}^{-1},$$

since $Q(0) = H(0)^{-1} = 1$.

4.3.26. (a) Define the generating function

$$F\left(\mathsf{I}(\mu); s_1, \ldots, s_{|\mu|}; t_1, \ldots, t_{|\mu|}\right) = \sum_{\sigma} \mathbf{x}^{\tau(\sigma)},$$

where the summation is over $\sigma \in \langle \mu \rangle$ with $s_i \leqslant \sigma_i \leqslant t_i$ for $i = 1, \ldots, |\mu|$. The sum lemma for Ψ_1 clearly holds for F, but the product lemma must be modified, giving

$$\begin{aligned} &F\left(\mathsf{I}(\mu_1 \omega \mu_2); s_1 \cdots s_{|\mu_1 \omega \mu_2|}; t_1 \cdots t_{|\mu_1 \omega \mu_2|}\right) \\ &\quad = F\left(\mathsf{I}(\mu_1); s_1 \cdots s_{|\mu_1|}; t_1 \cdots t_{|\mu_1|}\right) \\ &\quad\quad \times F\left(\mathsf{I}(\mu_2); s_{|\mu_1|+1} \cdots s_{|\mu_1|+|\mu_2|}; t_{|\mu_1|+1} \cdots t_{|\mu_1|+|\mu_2|}\right). \end{aligned}$$

Now in the notation of **4.3.16**, let

$$\zeta_i = F\left(\mathbf{V}_i;\, L_{a_{i-1}+1} \cdots L_{a_m};\, U_{a_{i-1}+1} \cdots U_{a_m}\right),$$

for $i = 1, \ldots, m$. Thus, following **4.3.16** and using the modified product and sum lemmas for F

$$\zeta_i = (-1)^{m-i} F\left(\mathbf{A}^{a_m - a_{i-1} - 1};\, L_{a_{i-1}+1} \cdots L_{a_m};\, U_{a_{i-1}+1} \cdots U_{a_m}\right)$$

$$+ \sum_{j=i+1}^{m} (-1)^{j-i-1} \zeta_j F\left(\mathbf{A}^{a_{j-1} - a_{i-1} - 1};\, L_{a_{i-1}+1} \cdots L_{a_{j-1}};\, U_{a_{i-1}+1} \cdots U_{a_{j-1}}\right).$$

But, since $L_{a_{i-1}+1} \geqslant \cdots \geqslant L_{a_{j-1}}$, $U_{a_{i-1}+1} \geqslant \cdots \geqslant U_{a_{j-1}}$, the coefficients in this linear system become

$$F\left(\mathbf{A}^{a_{j-1} - a_{i-1} - 1};\, L_{a_{i-1}+1} \cdots L_{a_{j-1}};\, U_{a_{i-1}+1} \cdots U_{a_{j-1}}\right) = \sum_{\sigma} \mathbf{x}^{\tau(\sigma)},$$

where the summation is over $\sigma = \sigma_1 \cdots \sigma_{a_{j-1} - a_{i-1}} \in \langle \mathbf{A}^{a_{j-1} - a_{i-1} - 1} \rangle$, $\pi_1 = \leqslant$, with $L_{a_{i-1}+1} \leqslant \sigma_1 \leqslant \cdots \leqslant \sigma_{a_{j-1} - a_{i-1}} \leqslant U_{a_{j-1}}$. Thus the preceding generating function is $\theta_{a_{j-1} - a_{i-1}}(L_{a_{i-1}+1}, U_{a_{j-1}})$, so the required number is, following **4.3.16** and **4.3.17**,

$$[\mathbf{x}^{\mathbf{i}}] \| \theta_{a_j - a_{i-1}}\left(L_{a_{i-1}+1}, U_{a_j}\right) \|_{m \times m}.$$

(b) If $x_i = x$ for $i \geqslant 1$, then

$$\theta_k(L, U) = [z^k](1 - zx)^{-(U-L+1)} = \binom{U - L + k}{k} x^k,$$

and the result follows.

4.3.27. (a) Premultiplying both sides of $\mathbf{I} - \mathbf{H} = \mathbf{I} - \mathbf{Q} + \Sigma_{j=1}^{t} \mathbf{L}_j \mathbf{W} \mathbf{R}_j$ by $\mathbf{C}(\mathbf{I} - \mathbf{H})^{-1}$ and postmultiplying by $(\mathbf{I} - \mathbf{Q})^{-1} \mathbf{L}_i$, we obtain

$$\mathbf{C}(\mathbf{I} - \mathbf{Q})^{-1} \mathbf{L}_i = \mathbf{C}(\mathbf{I} - \mathbf{H})^{-1} \mathbf{L}_i + \sum_{j=1}^{t} \mathbf{C}(\mathbf{I} - \mathbf{H})^{-1} \mathbf{L}_j \mathbf{W} \mathbf{R}_j (\mathbf{I} - \mathbf{Q})^{-1} \mathbf{L}_i$$

for $i = 1, \ldots, t$. Apply Ψ_2 to both sides of this equation, and let $\Psi_2(\mathbf{C}(\mathbf{I} - \mathbf{H})^{-1} \mathbf{L}_i) = \xi_i$, $i = 1, \ldots, t$. This yields the equations

$$\Psi_2\left(\mathbf{C}(\mathbf{I} - \mathbf{Q})^{-1} \mathbf{L}_i\right) = \sum_{j=1}^{t} \xi_j \left\{ \delta_{ij} + \Psi_2\left(\mathbf{R}_j (\mathbf{I} - \mathbf{Q})^{-1} \mathbf{L}_i\right) \right\}$$

for $i = 1, \ldots, t$. The result follows by Cramer's rule.

(b) The required generating function is $\Psi_2(\mathbf{C}(\mathbf{I}-\mathbf{H})^{-1})$, where $\mathbf{H} = \mathbf{A}\mathbf{B}^{r_t-1}\cdots\mathbf{A}\mathbf{B}^{r_1-1}$, $\mathbf{C}=\mathbf{B}^{u-1}$, $\mathbf{A} = \mathbb{I}(<)$, $\mathbf{B}=\mathbf{W}-\mathbf{A}$. Now $\mathbf{H}$ has the right-expansion

$$\mathbf{H} = (-1)^t\mathbf{B}^{b_1} - \sum_{i=1}^{t} \mathbf{A}\mathbf{B}^{r_t-1}\cdots\mathbf{A}\mathbf{B}^{r_{t-i+2}-1}\mathbf{W}\mathbf{B}^{b_i-1}(-1)^{t+1-i},$$

so $\mathbf{Q}=(-1)^t\mathbf{B}^{b_1}$, $\mathbf{L}_i = \mathbf{A}\mathbf{B}^{r_t-1}\cdots\mathbf{A}\mathbf{B}^{r_{t-i+2}-1}$, $\mathbf{R}_i = (-1)^{t+1-i}\mathbf{B}^{b_i-1}$, for $i=1,\ldots,t$, where $\mathbf{L}_1 = \mathbf{I}$. Thus by part (a), the generating function required is $|[\mathbf{N}:\mathbf{c}]_1|\cdot|\mathbf{N}|^{-1}$, where $[\mathbf{N}]_{ij} = \delta_{ij} + \Psi_2(\mathbf{R}_j(\mathbf{I}-\mathbf{Q})^{-1}\mathbf{L}_i)$, and $\mathbf{c} = (c_1,\ldots,c_t)^T$, $c_i = \Psi_2(\mathbf{C}(\mathbf{I}-\mathbf{Q})^{-1}\mathbf{L}_i)$. We now replace row i in the numerator and denominator by (row i) $+ \sum_{j=1}^{i-1}(-1)^{i-j}\Psi_2(\mathbf{B}^{b_j-b_i-1})\times$ (row j), beginning at $i=t$ and continuing until $i=1$. By this procedure c_i becomes $(-1)^{i-1}\Psi_2(\mathbf{B}^{u-1}(\mathbf{I}-\mathbf{Q})^{-1}\mathbf{B}^{b_1-b_i})$ and $[\mathbf{N}]_{ij}$ becomes

$$(-1)^{t+i-j}\Psi_2\left(\mathbf{B}^{b_j-1}(\mathbf{I}-\mathbf{Q})^{-1}\mathbf{B}^{b_1-b_i}\right) + \left(1-\zeta(i,j)-\delta_{ij}\right)(-1)^{i-j}\Psi_2(\mathbf{B}^{b_j-b_i-1}) + \delta_{ij}.$$

Multiplying row i by $(-1)^{i-1}$ for $i=1,\ldots,t$ in both the numerator and denominator leaves the result unchanged, so c_i becomes

$$\Psi_2\left(\mathbf{B}^{u-1}(\mathbf{I}-\mathbf{Q})^{-1}\mathbf{B}^{b_1-b_i}\right) = \Psi^{(b_1,t)}_{u+b_1-b_i}(x), \qquad i=1,\ldots,t,$$

and $[\mathbf{N}]_{ij}$ becomes

$$(-1)^{t+1+j}\Psi_2\left(\mathbf{B}^{b_1+b_j-b_i-1}(\mathbf{I}-\mathbf{Q})^{-1}\right) + \left(1-\zeta(i,j)-\delta_{ij}\right)(-1)^{j+1}\Psi_2(\mathbf{B}^{b_j-b_i-1}) + (-1)^{i-1}\delta_{ij}.$$

But if $i>j$, then

$$\mathbf{B}^{b_j-b_i-1} + (-1)^t\mathbf{B}^{b_1+b_j-b_i-1}(\mathbf{I}-\mathbf{Q})^{-1} = \mathbf{B}^{b_j-b_i-1}(\mathbf{I}-\mathbf{Q})^{-1},$$

so $[\mathbf{N}]_{ij}$ becomes

$$(-1)^{1+j+t\zeta(i,j)}\Psi^{(b_1,t)}_{(b_j-b_i)\bmod b_1}(x), \qquad i,j=1,\ldots,t.$$

(c) Under the condition given, we have

$$\Psi_2\left((\mathbf{I}-\mathbf{A}^{p_1-1}\mathbf{B}\cdots\mathbf{A}^{p_s-1}\mathbf{B})^{-1}\right) = \Psi_2\left((\mathbf{I}-\mathbf{A}\mathbf{B}^{r_t-1}\cdots\mathbf{A}\mathbf{B}^{r_1-1})^{-1}\right).$$

These generating functions are evaluated in **[4.3.15]**, with $r=1$, and in part (b), with $u = 1$. The factor of $(-1)^{j+1}$ in $[\mathbf{M}]_{ij}$ and $[\mathbf{N}]_{ij}$ can be removed, since it appears in both the numerator and denominator of each side.

4.3.28. Let $\mathbf{M}_i$, $i = 1,\ldots, 6$, be the incidence matrices for sequences $\sigma_1\sigma_2\sigma_3$ on N_n with restrictions as given in the following list.

$$\mathbf{M}_1: \sigma_1 \geqslant \sigma_2, \sigma_2 < \sigma_3, \sigma_1 < \sigma_3$$

$$\mathbf{M}_2: \sigma_1 < \sigma_2, \sigma_2 < \sigma_3$$

$$\mathbf{M}_3: \sigma_1 < \sigma_2, \sigma_2 \geqslant \sigma_3, \sigma_1 < \sigma_3$$

$$\mathbf{M}_4: \sigma_1 \geqslant \sigma_2, \sigma_2 < \sigma_3, \sigma_1 \geqslant \sigma_3$$

$$\mathbf{M}_5: \sigma_1 \geqslant \sigma_2, \sigma_2 \geqslant \sigma_3$$

$$\mathbf{M}_6: \sigma_1 < \sigma_2, \sigma_2 \geqslant \sigma_3, \sigma_1 \geqslant \sigma_3.$$

Thus, for example $[\mathbf{M}_1]_{ij} = \sum_k x_i x_k$, where the summation extends over $i \geqslant k$, $k < j$, for $i < j$, and $[\mathbf{M}_1]_{ij} = 0$ for $i \geqslant j$. The preceding six sets of restrictions represent the only patterns that three consecutive elements in a sequence may have. Thus, if r marks rises between adjacent elements and u marks rises between elements in adjacent odd positions, the required number is $[r^s u^t x_1 \cdots x_{2n+1}]\Psi_1((\mathbf{I}-\mathbf{H})^{-1})$, where

$$\mathbf{H} = ru\mathbf{M}_1 + r^2u\mathbf{M}_2 + ru\mathbf{M}_3 + r\mathbf{M}_4 + \mathbf{M}_5 + r\mathbf{M}_6.$$

Furthermore, the following relationships hold between the $\mathbf{M}_i$, where $\mathbf{A} = \mathsf{I}(<)$, $\mathbf{B} = \mathsf{I}(\geqslant)$,

$$\mathbf{M}_1 + \mathbf{M}_4 = \mathbf{BA}, \qquad \mathbf{M}_3 + \mathbf{M}_6 = \mathbf{AB}, \qquad \mathbf{M}_2 = \mathbf{A}^2,$$

$$\mathbf{M}_5 = \mathbf{B}^2, \qquad \mathbf{M}_1 + \mathbf{M}_2 + \mathbf{M}_3 = \mathbf{A}\gamma_1(<).$$

Eliminating $\mathbf{M}_1,\ldots, \mathbf{M}_6$ we have

$$\mathbf{H} = r(u-1)\{\mathbf{A}\gamma_1 + (r-1)\mathbf{A}^2\} + r^2\mathbf{A}^2 + r\mathbf{AB} + r\mathbf{BA} + \mathbf{B}^2.$$

A right-expansion of $\mathbf{H}$ is therefore, by eliminating $\mathbf{B}$, $\mathbf{H} = \mathbf{Q} - \mathbf{L}_1\mathbf{W}\mathbf{R}_1 - \mathbf{L}_2\mathbf{W}\mathbf{R}_2$, where $\mathbf{L}_1 = \mathbf{I}$, $\mathbf{R}_1 = -(r-1)\mathbf{A}$, $\mathbf{L}_2 = r\mathbf{A} + \mathbf{B}$, $\mathbf{R}_2 = -\mathbf{I}$, and $\mathbf{Q} = r(u-1)\mathbf{A}\gamma_1 + (r-1)(ru-1)\mathbf{A}^2$. We may now apply the elimination theorem. Let $\mathbf{F}_k = (r-1)^k\Psi_1(\mathbf{A}^k(\mathbf{I}-\mathbf{Q})^{-1})$. Then $\Psi_1((\mathbf{I}-\mathbf{H})^{-1})$

$$= -\begin{vmatrix} 1-F_1 & -F_1 \\ -F_0 & -F_0 \end{vmatrix} \cdot \begin{vmatrix} 1-F_1 & -(r-1)\Psi_1\big(\mathbf{A}(\mathbf{I}-\mathbf{Q})^{-1}(r\mathbf{A}+\mathbf{B})\big) \\ -F_0 & 1-\Psi_1\big((\mathbf{I}-\mathbf{Q})^{-1}(r\mathbf{A}+\mathbf{B})\big) \end{vmatrix}^{-1}$$

$$= F_0 \cdot \begin{vmatrix} 1-F_1 & -\gamma_1 - F_2 \\ -F_0 & 1-F_1 \end{vmatrix}^{-1}.$$

But F_m

$$= (r-1)^m \sum_{k\geqslant 0}\sum_{i=0}^{k}\binom{k}{i}\{r(u-1)\}^i\{(ru-1)(r-1)\}^{k-i}\gamma_{2k+m-i+1}(<)\gamma_1^i(<).$$

Specializing to permutations, by the <-transformation lemma, we have the required result.

4.3.29. The required number is

$$\left[u^k q^m \frac{x^{2n+1}}{(2n+1)!_q}\right]\Phi,$$

where $\Phi = \Psi_2((\mathbf{I} - u\mathbf{M}_1 - \mathbf{M}_4)^{-1})$ and $\mathbf{M}_1, \mathbf{M}_4$ are given in the solution to [**4.3.28**]. But $\mathbf{M}_1 + \mathbf{M}_4 = \mathbf{BA} = \mathbf{WA} - \mathbf{A}^2$, where $\mathbf{A} = \mathsf{I}(<)$. Thus

$$\begin{aligned}\Phi &= \Psi_2\left((\mathbf{I} - (u-1)\mathbf{M}_1 + \mathbf{A}^2 - \mathbf{WA})^{-1}\right)\\ &= \Psi_2\left((\mathbf{I} - \mathbf{Q})^{-1}\right)\left\{1 - \Psi_2\left(\mathbf{A}(\mathbf{I} - \mathbf{Q})^{-1}\right)\right\}^{-1}\\ &= F_0(1 - F_1)^{-1},\end{aligned}$$

where $\mathbf{Q} = (u-1)\mathbf{M}_1 - \mathbf{A}^2$, from the elimination theorem, and

$$F_l = \Psi_2\left(\mathbf{A}^l(\mathbf{I} - \mathbf{Q})^{-1}\right) = \sum_{i\geqslant 0}\frac{x^{2i+l+1}}{(2i+l+1)!_q}c_i.$$

By considering the location of successive maxima from the right,

$$c_i = \left\{(u-1)q\binom{2i+l-1}{1}_q - 1\right\}c_{i-1},$$

for $i \geqslant 1$, and $c_0 = 1$. The result follows.

4.3.30. The required number is

$$\left[q^m \frac{x^{4n+3}}{(4n+3)!_q}\right]\Psi_2\left(\mathbf{A}(\mathbf{I} - \mathbf{H})^{-1}\mathbf{B}\right),$$

where $\mathbf{H} = \mathbf{M}_1\mathbf{M}_4$, $\mathbf{A} = \mathsf{I}(<)$, and $\mathbf{M}_1, \mathbf{M}_4$ are given in the solution to [**4.3.28**]. But $\mathbf{H} = -\mathbf{M}_1^2 - \mathbf{M}_1\mathbf{A}^2 + \mathbf{M}_1\mathbf{WA}$, so

$$\Psi_2\left(\mathbf{A}(\mathbf{I} - \mathbf{H})^{-1}\mathbf{B}\right) = \Psi_2\left(\mathbf{A}(\mathbf{I} - \mathbf{Q})^{-1}\mathbf{B}\right)\left\{1 - \Psi_2\left(\mathbf{A}(\mathbf{I} - \mathbf{Q})^{-1}\mathbf{M}_1\right)\right\}^{-1}$$

by the elimination theorem, where $\mathbf{Q} = -\mathbf{M}_1(\mathbf{M}_1 + \mathbf{A}^2)$. Now

$$\Psi_2\left(\mathbf{A}(\mathbf{I}-\mathbf{Q})^{-1}\mathbf{B}\right) = \sum_{i\geq 0} (-1)^i c_i \frac{x^{4i+3}}{(4i+3)!_q},$$

where

$$c_i = \left\{\prod_{j=0}^{i-1} q\binom{4j+2}{1}_q \left\{q\binom{4j+4}{1}_q + 1\right\}\right\} q\binom{4i+2}{1}_q,$$

again by considering successive maxima from the right. We can simplify this by noting that $q\binom{4j+4}{1}_q + 1 = \binom{4j+5}{1}_q$, giving

$$c_i = q^{i+1}\binom{4i+2}{1}_q \prod_{j=0}^{i-1}\binom{4j+2}{1}_q\binom{4j+5}{1}_q.$$

But

$$\Psi_2\left(\mathbf{A}(\mathbf{I}-\mathbf{Q})^{-1}\mathbf{M}_1\right) = \sum_{i\geq 0}(-1)^i c_i \frac{x^{4i+4}}{(4i+4)!_q},$$

and the result follows.

SECTION 4.4

4.4.1. From Lemma 4.3.5 and Proposition 4.3.6, we have

$$\sum_{k\geq 0} \gamma_k(\pi_1) z^k = 1 + z\,\mathrm{trace}(\mathbf{I} - z\mathbf{A})^{-1}\mathbf{W}$$

$$= |\mathbf{I} + z(\mathbf{I} - z\mathbf{A})^{-1}\mathbf{W}| \qquad \text{from } \mathbf{1.1.10(5)}$$

$$= |\mathbf{I} - z\mathbf{A} + z\mathbf{W}| \cdot |\mathbf{I} - z\mathbf{A}|^{-1} = |\mathbf{I} + z\mathbf{B}| \cdot |\mathbf{I} - z\mathbf{A}|^{-1}.$$

4.4.2. From **1.1.10(5)**, $|\mathbf{I} - \mathbf{A}^{p-1}\mathbf{B}|$

$$= |\mathbf{I} + \mathbf{A}^p - \mathbf{A}^{p-1}\mathbf{W}|$$

$$= |\mathbf{I} + \mathbf{A}^p| \cdot \left\{1 - \mathrm{trace}(\mathbf{I} + \mathbf{A}^p)^{-1}\mathbf{A}^{p-1}\mathbf{W}\right\}$$

$$= \prod_{l\geq 1}(1 + x_l^p)\left\{\sum_{j\geq 0}(-1)^j [t^{jp}] \prod_{l\geq 1}(1 - tx_l)^{-1}\right\}$$

$$= \left\{ \sum_{m\geqslant 0} (-1)^m [t^{mp}] \prod_{l\geqslant 1} (1 - (tx_l)^p) \right\} \left\{ \sum_{j\geqslant 0} (-1)^j [t^{jp}] \prod_{l\geqslant 1} (1 - tx_l)^{-1} \right\}$$

$$= \sum_{i\geqslant 0} (-1)^i [t^{ip}] \prod_{l\geqslant 1} (1 - (tx_l)^p)(1 - tx_l)^{-1}$$

where $i = m + j$, since $[t^k]\prod_{l\geqslant 1}(1 - (tx_l)^p) = 0$ unless $p|k$.

4.4.3. (a) From the maximal string decomposition theorem for circular permutations, and Corollary 4.2.7, the required number is

$$[x_1 \cdots x_{2n}]\left\{ \log\left(\sum_{k\geqslant 0} (-1)^k \gamma_{2k}(\pi_1) \right)^{-1} + \psi \right\},$$

where $\psi = \operatorname{trace}\log(\mathbf{I} + \mathbf{A}^2)^{-1}$, $\mathbf{A} = \mathbf{I}(\pi_1)$. If $\pi_1 = <$ then $\psi = 0$, and by the $<$-transformation lemma, the solution is

$$\left[\frac{x^{2n}}{(2n)!}\right] \log\left(\sum_{k\geqslant 0} (-1)^k \frac{x^{2k}}{(2k)!} \right)^{-1} = \left[\frac{x^{2n}}{(2n)!}\right] \log(\sec x).$$

(b) If $\pi_1 = +$ then $\psi = 0$, and by the $+$-transformation lemma, the solution is

$$[x^{2n}] \sum_{i\geqslant 0} i! [y^i] \log\left(1 - y \sum_{k\geqslant 1} (-1)^{k-1} x^{2k} \right)^{-1}$$

$$= [x^{2n}] \sum_{i\geqslant 1} (i-1)! x^{2i} (1 + x^2)^{-i} = \sum_{i\geqslant 1} (-1)^{n-i} (i-1)! \binom{n-1}{i-1}.$$

(c) If $\pi_1 = \oplus$, then π_1 is not cycle-free, and

$$[x_1 \cdots x_{2n}]\psi = [x_1 \cdots x_{2n}] \sum_{i\geqslant 1} i^{-1} \operatorname{trace}(-\mathbf{A}^2)^i = 2(-1)^n.$$

From the $\oplus$-transformation lemma, the solution is

$$2(-1)^n + [x^{2n}] 2n \sum_{i\geqslant 1} (i-1)! [y^i] \log\left(1 - y \sum_{k\geqslant 1} (-1)^{k-1} x^{2k} \right)^{-1}$$

$$= 2(-1)^n + [x^{2n}] 2n \sum_{i\geqslant 1} (i-1)! i^{-1} x^{2i} (1 + x^2)^{-i}$$

$$= 2(-1)^n + 2n \sum_{i\geqslant 1} (-1)^{n-i} (i-1)! \binom{n-1}{i-1} / i.$$

4.4.4. (a) Let u mark occurrences of π_1. Then in the maximal string decomposition for circular permutations we have $f_i = u^{i-1}$, $g_i = u^i$, $i \geqslant 1$. Thus $F(x) = 1 + \Sigma_{i \geqslant 1} u^{i-1} x^i = (1-(u-1)x)(1-ux)^{-1}$ and the required number is, from Corollary 4.2.12,

$$[u^k x_1 \cdots x_n]\left(\log\left\{1 - \sum_{i \geqslant 1} (u-1)^{i-1} \gamma_i(\pi_1)\right\}^{-1} + \psi\right),$$

where

$$\psi = \operatorname{trace}\log(\mathbf{I} - u\mathbf{A}) + \operatorname{trace}\log(\mathbf{I} - (u-1)\mathbf{A})^{-1} + \sum_{i \geqslant 1} i^{-1} u^i \operatorname{trace}\mathbf{A}^i$$

$$= \sum_{i \geqslant 1} i^{-1}\left(u^i + (u-1)^i - u^i\right) \operatorname{trace}\mathbf{A}^i, \qquad \mathbf{A} = \mathbf{I}(\pi_1)$$

If $\pi_1 = <$, π_1 is cycle-free, and $\psi = 0$. Thus by the $<$-transformation lemma the required number is

$$\left[u^k \frac{x^n}{n!}\right] \log\{(u-1)(u - e^{(u-1)x})^{-1}\}.$$

(b) If $\pi_1 = +$, π_1 is cycle-free, and by the $+$-transformation lemma the required number is

$$[u^k x^n] \sum_{i \geqslant 0} i![y^i] \log\left\{1 - y \sum_{i \geqslant 1} (u-1)^{i-1} x^i\right\}^{-1}$$

$$= [u^k x^n] \sum_{i \geqslant 1} (i-1)! x^i (1 - (u-1)x)^{-i}.$$

(c) If $\pi_1 = \oplus$, π_1 is not cycle-free, and $[x_1 \cdots x_n]\operatorname{trace}\mathbf{A}^j = n\delta_{n,j}$. Thus $[u^k x_1 \cdots x_n]\psi = [u^k](u^n + (u-1)^n - u^n)$, and by the $\oplus$-transformation lemma, the required number is

$$[u^k x^n]\left\{\sum_{i \geqslant 1} (i-1)! x \frac{\partial}{\partial x} [y^i] \log\left(1 - yx(1-(u-1)x)^{-1}\right)^{-1} + x^n (u-1)^n\right\}$$

$$= [u^k x^n]\left\{\sum_{i \geqslant 1} (i-1)! i^{-1} x \frac{\partial}{\partial x} x^i (1-(u-1)x)^{-i} + x^n (u-1)^n\right\}$$

$$= [u^k x^n]\left\{\sum_{i \geqslant 1} (i-1)! x^i (1-(u-1)x)^{-i-1} + x^n (u-1)^n\right\}.$$

4.4.5. (a) Let u mark the occurrence of maximal π_1-strings of length $\geqslant p$. Then in the maximal string decomposition theorem for circular permutations we have $\pi_1 = <$ and $f_i = u$ for $i \geqslant p$, $f_i = 1$ for $i < p$. Thus

$$F(x) = \sum_{i=0}^{p-1} x^i + u \sum_{i \geqslant p} x^i = (1-x)^{-1}\{1 - (1-u)x^p\}.$$

Since π_1 is cycle-free, the required number is

$$[u^k x_1 \cdots x_n] \log\left\{ \sum_{i \geqslant 0} (1-u)^i \left(\gamma_{ip}(\pi_1) - \gamma_{ip+1}(\pi_1)\right)\right\}^{-1}$$

$$= \left[u^k \frac{x^n}{n!}\right] \log\left\{ \sum_{i \geqslant 0} (1-u)^i \left(\frac{x^{ip}}{(ip)!} - \frac{x^{ip+1}}{(ip+1)!}\right)\right\}^{-1},$$

from the <-transformation lemma.

(b) From part (a), with $\pi_1 = +$, the required number is

$$[u^k x_1 \cdots x_n] \log\left\{ \sum_{i \geqslant 0} (1-u)^i \left(\gamma_{ip}(+) - \gamma_{ip+1}(+)\right)\right\}^{-1},$$

since π_1 is cycle-free. Applying the +-transformation lemma, we obtain

$$[u^k x^n] \sum_{j \geqslant 0} j! [y^j] \log\left\{ 1 - y \sum_{i \geqslant 1} \left((1-u)^{i-1} x^{(i-1)p+1} - (1-u)^i x^{ip}\right)\right\}^{-1}$$

$$= [u^k x^n] \sum_{j \geqslant 1} (j-1)! \{(x - (1-u)x^p)(1 - (1-u)x^p)^{-1}\}^j$$

$$= [u^k x^n] \sum_{j \geqslant 1} (j-1)! \sum_{l \geqslant 0} \binom{j}{m+lp}$$

$$\times (u-1)^{j-m-lp} x^{m+lp+p(j-m-lp)} (1-(1-u)x^p)^{-j},$$

and the result follows.

(c) From part (a), with $\pi_1 = \oplus$, the required number is

$$[u^k x_1 \cdots x_n] \left(\log\left\{ \sum_{i \geqslant 0} (1-u)^i \left(\gamma_{ip}(\oplus) - \gamma_{ip+1}(\oplus)\right)\right\}^{-1} + \psi \right),$$

where

$$\psi = \operatorname{trace} \log F(\mathbf{A})^{-1}, \quad \mathbf{A} = \mathbf{I}(\oplus)$$

$$= \operatorname{trace} \log(\mathbf{I} - \mathbf{A}) + \operatorname{trace} \log(\mathbf{I} - (1-u)\mathbf{A}^p)^{-1}$$

so

$$[u^k x_1 \cdots x_n]\psi = \begin{cases} [u^k]\left(-1 + p(1-u)^{n/p}\right), & \text{if } p \mid n \\ [u^k](-1), & \text{otherwise.} \end{cases}$$

From the $\oplus$-transformation lemma

$$[u^k x_1 \cdots x_n]\log\left\{1 - \sum_{i\geqslant 1}\left((1-u)^{i-1}\gamma_{(i-1)p+1}(\oplus) - (1-u)^i\gamma_{ip}(\oplus)\right)\right\}^{-1}$$

$$= [u^k x^n] n \sum_{j\geqslant 1}(j-1)!$$

$$\times [y^j]\log\left\{1 - y\sum_{i\geqslant 1}\left((1-u)^{i-1}x^{(i-1)p+1} - (1-u)^i x^{ip}\right)\right\}^{-1}$$

from part (b), and the result follows.

4.4.6. Put $p = 1$, $m = 0$ in [**4.4.5**].

4.4.7. (a) Since all maximal π_1-strings have lengths divisible by p, then the maximal π_1-string length enumerator is $F(x) = 1 + \sum_{i\geqslant 1} x^{pi} = (1-x^p)^{-1}$. Then by the maximal string decomposition theorem for circular permutations the required number of circular permutations is

$$[x_1 \cdots x_{np}]\left\{\psi + \log\left(1 - \gamma_p(\pi_1)\right)^{-1}\right\},$$

where $\psi = \text{trace } \log(\mathbf{I} - \mathbf{A}^p) = -\sum_{i\geqslant 1} i^{-1}\text{trace } \mathbf{A}^{pi}$, $\mathbf{A} = \mathbf{I}(\pi_1)$. If $\pi_1 = <$, then $\psi = 0$ and by the $<$-transformation lemma the answer is

$$\left[\frac{x^{np}}{(np)!}\right]\log\left(1 - \frac{x^p}{p!}\right)^{-1} = n^{-1}(p!)^{-n}(np)!.$$

(b) If $\pi_1 = +$, then $\psi = 0$ and by the $+$-transformation lemma the answer is

$$[x^{np}]\sum_{k\geqslant 0} k![y^k]\log(1 - yx^p)^{-1} = n!n^{-1} = (n-1)!.$$

(c) If $\pi_1 = \oplus$, then π_1 is not cycle-free, and

$$[x_1 \cdots x_{np}]\psi = -n^{-1}(np) = -p.$$

Thus, by the $\oplus$-transformation lemma, the required number is

$$-p + [x^{np}]np\sum_{k\geqslant 1}(k-1)![y^k]\log(1 - yx^p)^{-1}$$

$$= -p + np(n-1)!n^{-1} = p\{(n-1)! - 1\}.$$

4.4.8. (a) By applying the maximal string decomposition theorem for circular permutations with $F(x) = 1 + x(1 - ux)^{-1}$, and the solution to [**4.2.15**], the solution is

$$\left[u^j x_{11} \cdots x_{1n} \cdots x_{k1} \cdots x_{kn}\right]$$

$$\log\left\{1 - \sum_{m \geq 1} (u-1)^{m-1} \gamma_m(\pi_1, \mathbf{x}_1) \cdots \gamma_m(\pi_1, \mathbf{x}_k)\right\}^{-1},$$

if π_1 is cycle-free. Applying the $<$-transformation lemma simultaneously in the vectors $\mathbf{x}_1, \ldots, \mathbf{x}_k$, the required number is

$$\left[u^j \frac{z_1^n}{n!} \cdots \frac{z_k^n}{n!}\right] \log\left\{1 - \sum_{m \geq 1} (u-1)^{m-1} \frac{z_1^m}{m!} \cdots \frac{z_k^m}{m!}\right\}^{-1}$$

$$= \left[u^j \frac{x^n}{(n!)^k}\right] \log\left\{1 - \sum_{m \geq 1} (u-1)^{m-1} \frac{x^m}{(m!)^k}\right\}^{-1}.$$

(b) Applying the $+$-transformation lemma, we get

$$\left[u^j z_1^n \cdots z_k^n\right] \sum_{i_1, \ldots, i_k \geq 0} i_1! \cdots i_k!$$

$$\times \left[y_1^{i_1} \cdots y_k^{i_k}\right] \log\left\{1 - \sum_{m \geq 1} (u-1)^{m-1} y_1 z_1^m \cdots y_k z_k^m\right\}^{-1}$$

$$= [u^j x^n] \sum_{i \geq 0} (i!)^k i^{-1} x^i (1 - (u-1)x)^{-i},$$

since $+$ is cycle-free.

(c) Applying the $\oplus$-transformation lemma, we obtain

$$\left[u^j z_1^n \cdots z_k^n\right] n^k \sum_{i_1, \ldots, i_k \geq 1} (i_1 - 1)! \cdots (i_k - 1)!$$

$$\times \left[y_1^{i_1} \cdots y_k^{i_k}\right] \log\left\{1 - \sum_{m \geq 1} (u-1)^{m-1} y_1 z_1^m \cdots y_k z_k^m\right\}^{-1}$$

$$= [u^j x^n] n^k \sum_{i \geq 1} ((i-1)!)^k i^{-1} x^i (1 - (u-1)x)^{-i}.$$

There is a further contribution since $\oplus$ is not cycle-free. This is from π_1-cycles and is

$$\left[u^j x_{11} \cdots x_{1n} \cdots x_{k1} \cdots x_{kn}\right] \frac{1}{n} (u-1)^n \operatorname{trace} \mathbf{A}^n,$$

where $\mathbf{A}$ is the incidence matrix for π_1-cycles on the alphabet of k-tuples. Thus $[x_{11} \cdots x_{1n} \cdots x_{k1} \cdots x_{kn}]$ trace $\mathbf{A}^n = n^k$, since there are n starting points for the single $\oplus$-cycle in each entry of the k-tuple. The result follows.

4.4.9. Let maxima, minima, double rises, and double falls be marked by u_1, u_2, u_3, u_4, respectively. Then a maximal $<$-string of length 1 is a double fall, and a maximal $<$-string of length $i \geqslant 2$ contains a maximum, a minimum, and $i-2$ double rises. Thus we can use the maximal string decomposition theorem for circular permutations with $F(x) = 1 + u_4 x + u_1 u_2 \sum_{i \geqslant 2} u_3^{i-2} x^i = (1-u_3x)^{-1}\{1 + (u_4 - u_3)x + (u_1u_2 - u_3u_4)x^2\}$. But from the solution of [**4.3.3**(b)],

$$F(x)^{-1} = (\alpha_2 - \alpha_1)^{-1}\{\alpha_2(1-\alpha_2 x)^{-1} - \alpha_1(1-\alpha_1 x)^{-1}\},$$

where $\alpha_1 + \alpha_2 = u_3 + u_4$ and $\alpha_1\alpha_2 = u_1u_2$. The solution follows by the $<$-transformation lemma.

4.4.10.

$$\begin{aligned} h(n,k) &= (-1)^{n-k}\binom{n}{k} + \frac{n}{2}[u^k x^n]\sum_{j\geqslant 1} j^{-1}(j-1)!x^j \\ &\qquad \times\{1 + 2(u-1)x(1-(u-1)x)^{-1}\}^j \\ &= (-1)^{n-k}\binom{n}{k} + \frac{n}{2}\sum_{j\geqslant 1} j^{-1}(j-1)![u^k x^{n-j}] \\ &\qquad \times \sum_{i\geqslant 0}\binom{j}{i}(2(u-1)x)^i(1-(u-1)x)^{-i} \\ &= (-1)^{n-k}\binom{n}{k} + \frac{n}{2}\sum_{j=1}^{n} j^{-1}(j-1)!\binom{n-j}{k}(-1)^{n-j-k} \\ &\qquad \times \sum_{i=0}^{n-j}\binom{j}{i}2^i\binom{-i}{n-j-i}(-1)^{n-j-i}, \end{aligned}$$

and the result follows.

4.4.11. (a) If we delete the maximum element from a circular $<$-alternating permutation, we obtain a unique $<$-alternating permutation by starting with the element following the maximum and moving clockwise to the element preceding the maximum. This is reversible, so $(d/dx)\Phi_1(x) = \Phi_2(x)$.

(b) From **4.2.22** we obtain $\Phi_2(x) = \tan x$. Thus $(d/dx)\Phi_1(x) = \tan x$, so $\Phi_1(x) = \log(\sec x)$, since $\Phi_1(0) = 0$.

4.4.12. The solution, following **4.3.18**, is

$$[s^t r^u x_1 \cdots x_n] \text{trace} \log(\mathbf{I} - \mathbf{H})^{-1},$$

where $\mathbf{H} = r(\mathbf{A} - \mathbf{C}) + \mathbf{B} + rs\mathbf{C}$, $\mathbf{A} = \mathbf{I}(<)$, $\mathbf{C} = \mathbf{I}(+)$, $\mathbf{B} = \mathbf{W} - \mathbf{A}$. Since $<$ and $+$ are cycle-free, the answer is, from the logarithmic connection theorem and **4.3.18**,

$$[s^t r^u x_1 \cdots x_n] \log\left(1 - \text{trace}\left(\{\mathbf{I} - r(s-1)\mathbf{C} - (r-1)\mathbf{A}\}^{-1}\mathbf{W}\right)\right)^{-1}.$$

But from **4.3.18**, since trace $\mathbf{UW} = \Psi_1(\mathbf{U})$, and we are only interested in linear terms in $x_1, \ldots, x_n$, we replace trace $(\{\mathbf{I} - r(s-1)\mathbf{C} - (r-1)\mathbf{A}\}^{-1}\mathbf{W})$ by

$$(r-1)^{-1}\left\{-1 + \exp\left\{(r-1)\sum_{j \geqslant 1}(r(s-1))^{j-1}\gamma_j(+)\right\}\right\}.$$

Thus we obtain, by the $+$-transformation lemma,

$$\begin{aligned}
&[r^u s^t x^n] \sum_{k \geqslant 0} k! [y^k] \log\left(1 - (r-1)^{-1}\right. \\
&\qquad\qquad \left. \times\left(-1 + \exp\{(r-1)xy(1 - r(s-1)x)^{-1}\}\right)\right)^{-1} \\
&= [r^u s^t x^n] \sum_{k \geqslant 1} k! [y^k] \sum_{i \geqslant 1} i^{-1}(r-1)^{-i} \\
&\qquad\qquad \times \left(\exp\{(r-1)xy(1 - r(s-1)x)^{-1}\} - 1\right)^i \\
&= [r^u s^t x^n] \sum_{k \geqslant 1} k! \sum_{i \geqslant 1} i^{-1}(r-1)^{-i} \sum_{j=0}^{i} (-1)^{i-j} \frac{((r-1)x)^k}{k!} \\
&\qquad\qquad \times (1 - r(s-1)x)^{-k} j^k,
\end{aligned}$$

and the result follows.

4.4.13. Using the notation of **[4.3.8]**, the required number is $[u^i v^j x_1 \cdots x_n]\Phi$, where

$$\begin{aligned}
\Phi &= \text{trace} \log(\mathbf{I} - (\mathbf{G} - \mathbf{I})(\mathbf{V} - \mathbf{I}))^{-1} \\
&\quad + \text{trace} \log(\mathbf{I} - \mathbf{UC})^{-1} + \text{trace} \log(\mathbf{I} - \mathbf{A})^{-1} \\
&= \text{trace} \log(\mathbf{V}^{-1} + \mathbf{U}^{-1} - \mathbf{I} - \mathbf{C})^{-1} \\
&\quad + \text{trace} \log \mathbf{U}^{-1} + \text{trace} \log \mathbf{V}^{-1} + \text{trace} \log(\mathbf{I} - \mathbf{A})^{-1}.
\end{aligned}$$

We now follow [4.3.8], since we are interested in permutations, and obtain

$$[u^i v^j x_1 \cdots x_n]\,\text{trace}\log(\mathbf{V}^{-1}+\mathbf{U}^{-1}-\mathbf{I}-\mathbf{C})^{-1}$$

$$= [u^i v^j x^n]\sum_{k\geq 0} k![y^k]\log\{1-yxf(x)(1+xf(x))^{-1}$$

$$-yxg(x)(1+xg(x))^{-1}+yx\}^{-1}$$

$$= [u^i v^j x^n]\sum_{k\geq 1}(k-1)!x^k\left\{\frac{1-ux+(u-1)x^{p-1}}{1-ux+(u-1)x^p}\right.$$

$$\left.+\frac{1-vx+(v-1)x^{s-1}}{1-vx+(v-1)x^s}-1\right\}^k.$$

To evaluate the remaining expressions, we know that + is cycle-free so,

$$\text{trace}\log\mathbf{U}^{-1}=\text{trace}\log\mathbf{V}^{-1}=\text{trace}\log(\mathbf{I}-\mathbf{A})^{-1}=0,$$

and the result follows.

4.4.14. We have $M(p_1,\ldots,p_m)=\text{trace}(\mathbf{A}^{p_1-1}\mathbf{B}\cdots\mathbf{A}^{p_m-1}\mathbf{B})$, where $\mathbf{A}=\mathbf{I}(\pi_1)$, $\mathbf{B}=\mathbf{W}-\mathbf{A}$. Applying a right-expansion to $\mathbf{A}^{p_1-1}\mathbf{B}\cdots\mathbf{A}^{p_m-1}\mathbf{B}$, we obtain $M(p_1,\ldots,p_m)$

$$=(-1)^m\text{trace}\,\mathbf{A}^{p_1+\cdots+p_m}+\sum_{i=1}^{m}(-1)^{m-i}\text{trace}\left(\mathbf{A}^{p_1-1}\mathbf{B}\cdots\mathbf{A}^{p_i-1}\mathbf{W}\mathbf{A}^{\Sigma_{j=i+1}^m p_j}\right)$$

$$=\sum_{i=1}^{m}(-1)^{m-i}\text{trace}\left(\mathbf{A}^{p_1+\Sigma_{j=i+1}^m p_j-1}\mathbf{B}\mathbf{A}^{p_2-1}\mathbf{B}\cdots\mathbf{A}^{p_i-1}\mathbf{W}\right),$$

since π_1 is cycle-free and trace $\mathbf{CD}$ = trace $\mathbf{DC}$. But $N(r_1,\ldots,r_k)=\text{trace}(\mathbf{A}^{r_1-1}\mathbf{B}\cdots\mathbf{A}^{r_k-1}\mathbf{W})$, so that

$$M(p_1,\ldots,p_m)=\sum_{i=1}^{m}(-1)^{m-i}N\left(p_1+\sum_{j=i+1}^{m}p_j,p_2,\ldots,p_i\right).$$

4.4.15. This is a circular version of [4.3.29], so the solution is, in the notation of [4.3.29],

$$[u^k x_1\cdots x_{2n}]\,\text{trace}\log(\mathbf{I}-u\mathbf{M}_1-\mathbf{M}_4)^{-1}$$

$$=[u^k x_1\cdots x_{2n}]\left\{\text{trace}\log(\mathbf{I}-\mathbf{Q})^{-1}+\log\left\{1-\text{trace}\left(\mathbf{A}(\mathbf{I}-\mathbf{Q})^{-1}\mathbf{W}\right)\right\}^{-1}\right\}.$$

But trace $\log(\mathbf{I}-\mathbf{Q})^{-1}=0$. Also, by the <-transformation lemma we may replace $\operatorname{trace}(\mathbf{A}(\mathbf{I}-\mathbf{Q})^{-1}\mathbf{W})$ by

$$F_1|_{q=1}=\sum_{i\geqslant 1}\frac{x^{2i}}{(2i)!}\prod_{j=1}^{i}(2j(u-1)-1),$$

and the result follows.

4.4.16. From **[4.4.4]**, the number of circular permutations on N_n with m rises is $[r^m z^n x_1\cdots x_n]\Omega(r,\{x_i\},z)$, where

$$\Omega(r,\{x_i\},z)=\log\left\{1-\sum_{j\geqslant 1}(r-1)^{j-1}\gamma_j(<)\right\}^{-1}$$

$$=\log\left\{1-(r-1)^{-1}\left(\prod_{i\geqslant 1}(1+(r-1)x_i)-1\right)\right\}^{-1}.$$

Now let $\chi(r,\{x_i\},z)=z(\partial/\partial z)\Omega(r,\{x_i\},z)$. Then by Pólya's theorem,

$$z\frac{\partial}{\partial z}\Phi(r,\{x_i\},z)=\sum_{k\geqslant 1}\phi(k)\chi\left(r^k,\{x_i^k\},z^k\right),$$

where $\Phi(r,\{x_i\},1)$ is the desired generating function, since automorphisms are in the cyclic group, which has cycle index polynomial $\sum_{k\geqslant 1}\phi(k)z^k$. Thus, integrating, we obtain

$$\Phi(r,\{x_i\},1)=\sum_{k\geqslant 1}k^{-1}\phi(k)\Omega\left(r^k,\{x_i^k\},1\right)$$

$$=\sum_{k\geqslant 1}k^{-1}\phi(k)\log\left\{1-(r^k-1)^{-1}\left(\prod_{i\geqslant 1}\left(1+(r^k-1)x_i^k\right)-1\right)\right\}^{-1}.$$

4.4.17. We want trace $\log(\mathbf{I}-\mathbf{H})^{-1}$, where $\mathbf{H}=\mathbf{A}^p\mathbf{B}^s$, $\mathbf{A}=\mathsf{I}(<)$. Thus $\mathbf{H}=\mathbf{A}^{p_1-1}\mathbf{B}\mathbf{A}^{p_2-1}\mathbf{B}\cdots\mathbf{A}^{p_s-1}\mathbf{B}$, where $p_1=p+1$, $p_2=\cdots=p_s=1$. In the notation of **[4.3.15]** we have $a_i=\sum_{j=1}^{q+1-i}p_j=p+q+1-i$. Since $\pi_1=<$ is cycle-free, the answer, by the logarithmic connection theorem and **[4.3.15]** with $q=1$, is

$$[\mathbf{M}]_{ij}=\frac{x^{a_j-a_i}}{(a_j-a_i)!}(1-\zeta(i,j))+\sum_{k\geqslant 1}(-1)^{k(p+s)}\frac{x^{k(p+s)+a_j-a_i}}{\left(k(p+s)+a_j-a_i\right)!},$$

for $i,j=0,\ldots,s$. But $a_j-a_i=i-j$, and the result follows.

4.4.18. From **[4.3.15]**, with $q=1$, and the logarithmic connection theorem, the number of circular permutations with circular pattern in

$$\left(\pi_1^{p_1-1}\pi_2\cdots\pi_1^{p_s-1}\pi_2\right)^+$$

is $[x^n/n!]\log|\mathbf{M}|^{-1}$, where

$$[\mathbf{M}]_{ij} = (-1)^{j+1+s\zeta(i,j)} \sum_{m\geq 0} (-1)^{sm} \frac{x^{km+(a_j-a_i)\bmod k}}{(km+(a_j-a_i)\bmod k)!},$$

and

$$k = p_1 + \cdots + p_s, \quad \zeta(i,j) = \begin{cases} 1, & i<j \\ 0, & i \geq j \end{cases}; \quad a_i = \sum_{l=1}^{s+i-1} p_l.$$

But also from **[4.3.15]**, with $q = 1$, the number of circular permutations with circular pattern in $(\pi_1^{p_1-1}\pi_2 \cdots \pi_1^{p_s-1}\pi_2)^+ = (\pi_1\pi_2^{r_t-1} \cdots \pi_1\pi_2^{r_1-1})^+$ is, by reading the permutations in reverse order, $[x^n/n!]\log|\mathbf{N}|^{-1}$, where

$$[\mathbf{N}]_{ij} = (-1)^{j+1+t\zeta(i,j)} \sum_{m\geq 0} (-1)^{tm} \frac{x^{km+(b_j-b_i)\bmod k}}{(km+(b_j-b_i)\bmod k)!},$$

$b_i = \sum_{l=1}^{t+1-i} r_l$. Now $s + t = k$, and

$$\{k+1-b_1,\ldots,k+1-b_t\} \,\dot\cup\, \{a_1,\ldots,a_s\} = \mathsf{N}_k,$$

so we let $\alpha_i = a_{s+1-i}$ and $\bar{\alpha}_i = k+1-b_i$. Now

$$[\mathbf{N}]_{ij} = (-1)^{j+1+t\zeta(i,j)+(\bar{\alpha}_i-\bar{\alpha}_j)(\bmod k)t/k} \sum_{m\geq 0} \frac{\{(-1)^{t/k}x\}^{km+(\bar{\alpha}_i-\bar{\alpha}_j)\bmod k}}{(km+(\bar{\alpha}_i-\bar{\alpha}_j)\bmod k)!}$$

and

$$[\mathbf{M}]_{ij} = (-1)^{j+1+(k-t)\zeta(i,j)+(\alpha_{s+1-j}-\alpha_{s+1-i})(\bmod k)(1-t/k)}$$

$$\times \sum_{m\geq 0} \frac{\{(-1)^{t/k}(-x)\}^{km+(\alpha_{s+1-j}-\alpha_{s+1-i})\bmod k}}{(km+(\alpha_{s+i-j}-\alpha_{s+1-i})\bmod k)!}.$$

But $|\mathbf{M}| = |\mathbf{N}|$, and transposing $\mathbf{M}$, reversing the orders of rows and columns, and canceling powers of (-1), we obtain the required identity.

4.4.19. (a) Let $\mathbf{C} = [c_{ij}]_{k\times k}$, where $c_{21} = c_{32} = \cdots = c_{1k} = 1$, and $c_{ij} = 0$ otherwise. Now $\exp(x\mathbf{C}) = \mathbf{M}(x)$, where $[\mathbf{M}(x)]_{ij} = \phi_{(i-j)\bmod k}(x)$, so

$$|\mathbf{M}(x)| = |\exp(x\mathbf{C})| = \exp \operatorname{trace}(x\mathbf{C}) = 1,$$

since trace $\mathbf{C} = 0$. Therefore

$$\operatorname{adj}\mathbf{M}(x) = \mathbf{M}(x)^{-1}|\mathbf{M}(x)| = \mathbf{M}(x)^{-1}$$

$$= (\exp x\mathbf{C})^{-1} = \exp(-x\mathbf{C}) = \mathbf{M}(-x),$$

and by the result of Jacobi $|\mathbf{M}(-x)[\alpha|\beta]| = (-1)^{\sigma}|\mathbf{M}(x)[\bar{\alpha}|\bar{\beta}]|$. But $[\mathbf{M}(-x)[\alpha|\beta]]_{ij} = \phi_{(\alpha_i-\beta_j)\bmod k}(-x)$ and $[\mathbf{M}(x)[\bar{\alpha}|\bar{\beta}]]_{ij} = \phi_{(\bar{\alpha}_i-\bar{\beta}_j)\bmod k}(x)$ so the result follows. This is a generalization of [**4.4.18**].

(b) Define $\mathbf{C}$ as in part (a), and let $\mathbf{M}(x) = \sum_{i\geq 0}(x\mathbf{C})^i/i!_q$, so that $[\mathbf{M}(x)]_{ij} = \phi_{(i-j)\bmod k}(x)$. Now, since $\mathbf{M}(x)$ is a power series in the single matrix $\mathbf{C}$, and commutative indeterminates x, q, we have

$$\mathbf{M}(x)^{-1} = \sum_{i\geq 0}(-1)^i q^{\binom{i}{2}}(x\mathbf{C})^i/i!_q$$

from [**2.6.7**(a)], so $[\mathbf{M}(x)^{-1}]_{ij} = \psi_{(i-j)\bmod k}(-x)$. But

$$\mathbf{M}(x) = \prod_{j\geq 0}(\mathbf{I} - x(1-q)q^j\mathbf{C})^{-1}$$

from **2.5.9**, so that

$$|\mathbf{M}(x)| = \prod_{j\geq 0}|\mathbf{I} - x(1-q)q^j\mathbf{C}|^{-1}.$$

Now

$$|\mathbf{I} - z\mathbf{C}|^{-1} = \exp\operatorname{trace}\log(\mathbf{I} - z\mathbf{C})^{-1}, \qquad \text{from } \mathbf{1.1.10(6)}$$

$$= \exp\sum_{i\geq 1}\frac{z^i}{i}\operatorname{trace}\mathbf{C}^i = \exp\sum_{l\geq 1}\frac{z^{kl}}{l} = (1-z^k)^{-1}.$$

Thus

$$|\mathbf{M}(x)| = \prod_{j\geq 0}\left(1 - x^k(1-q)^k q^{jk}\right)^{-1} = \sum_{i\geq 0}\frac{\left\{x^k(1-q)^k\right\}^i}{(1-q^k)^i\, i!_{q^k}}$$

from **2.5.9** again. Now Jacobi's identity may be rewritten as

$$|\mathbf{M}|\cdot|\mathbf{M}^{-1}[\alpha|\beta]| = (-1)^{\sigma}|\mathbf{M}[\bar{\alpha}|\bar{\beta}]|$$

and we obtain

$$\sum_{l\geq 0}\frac{\left\{x^k(1-q)^k(1-q^k)^{-1}\right\}^l}{l!_{q^k}}\|\psi_{(\alpha_i-\beta_j)\bmod k}(-x)\|_{s\times s}$$

$$= (-1)^{\sum_{l=1}^{s}(\alpha_l+\beta_l)}\|\phi_{(\bar{\alpha}_i-\bar{\beta}_j)\bmod k}(x)\|_{(k-s)\times(k-s)}.$$

SECTION 4.5

4.5.1. (a) Apply the <-transformation lemma to **4.5.11**.

(b) Apply the +-transformation lemma to **4.5.11**, since + is cycle-free.

(c) Applying the ⊕-transformation lemma to **4.5.11**, following [**4.4.7**(c)], we obtain the solution as

$$[u^k x^n] \sum_{i \geq 1} (i-1)! x \frac{\partial}{\partial x} [y^i] \{1 - yx(1-(u-1)x)^{-1}\}^{-1}$$

$$\times \exp\left\{1 + \frac{yx^n}{n}(u^n + (u-1)^n - 1)\right\}$$

$$= [u^k](u^n + (u-1)^n - 1) + [u^k x^n] \sum_{i \geq 1} (i-1)!$$

$$\times x \frac{\partial}{\partial x} [y^i] \{1 - yx(1-(u-1)x)^{-1}\}^{-1}$$

and the results follows, as in [**4.2.14**(c)].

4.5.2. (a) and (b) follow immediately from the remark preceding **4.5.12**, and **4.2.8**, **4.2.9**.

(c) Following [**4.4.3**(c)], the solution is

$$[x_1 \cdots x_{2m}] \exp(2(-1)^m x_1 \cdots x_{2m}) \left\{ \sum_{k \geq 0} (-1)^k \gamma_{2k}(\oplus) \right\}^{-1}$$

$$= 2(-1)^m + [x_1 \cdots x_{2m}] \left\{ \sum_{k \geq 0} (-1)^k \gamma_{2k}(\oplus) \right\}^{-1}$$

$$= 2(-1)^m + 2m \sum_{k=1}^{m} (-1)^{m-k} \binom{m-1}{k-1} (k-1)!, \qquad \text{from } \mathbf{4.2.10},$$

since $F^{-1} \circ \gamma = \sum_{k \geq 0} (-1)^k \gamma_{2k}(\oplus)$.

4.5.3. The result follows immediately from the remark preceding **4.5.12** and [**4.4.17**], since $\pi_1 = <$ is cycle-free.

4.5.4. (a) From the absolute partition lemma and Proposition 4.5.1, the required number is $[r^i l^j f^k \mathbf{x}^{\mathbf{m}}]\Phi(r, l, f; \mathbf{x})$, where

$$\Phi = |\mathbf{I} - (r\mathbf{A} + l\mathbf{X} + f(\mathbf{W} - \mathbf{X} - \mathbf{A}))|^{-1},$$

and $\mathbf{A} = \mathsf{I}(<)$, $\mathbf{X} = \mathsf{I}(=)$. But by the logarithmic connection theorem

$$\Phi = |\mathbf{I} - (r-f)\mathbf{A} - (l-f)\mathbf{X}|^{-1}$$

$$\times\left\{1 - f\,\mathrm{trace}(\mathbf{I} - (r-f)\mathbf{A} - (l-f)\mathbf{X})^{-1}\mathbf{W}\right\}^{-1}$$

$$= \left\{\prod_{t\geq 1}(1-(l-f)x_t)^{-1}\right\}(r-f)$$

$$\times\left\{r - f\prod_{t\geq 1}(1+(r-l)x_t)(1+(f-l)x_t)^{-1}\right\}^{-1},$$

from **[4.3.7(b)]**, and the result follows.

(b) From **[4.2.9]**

$$\alpha = [\mathbf{x}^{\mathbf{m}}]2\left\{\prod_{t\geq 1}(1+\rho x_t) + \prod_{t\geq 1}(1-\rho x_t)\right\}^{-1},$$

where $\rho^2 = -1$. Thus, in the notation of part (a)

$$\alpha = [\mathbf{x}^{\mathbf{m}}]\Phi(\rho, 0, -\rho; \mathbf{x}).$$

Now let $c(i, j, k; \mathbf{m})$ be the number of sequences with i absolute rises, j absolute levels, k absolute falls, and type $\mathbf{m}$. Then

$$[\mathbf{x}^{\mathbf{m}}]\Phi(\rho, 0, -\rho; \mathbf{x}) = \sum_{i+k=\lambda} c(i, 0, k; \mathbf{m})\rho^i(-\rho)^k$$

$$= (-1)^{\lambda/2}\sum_{i+k=\lambda} c(i, 0, k; \mathbf{m})(-1)^k$$

$$= (-1)^{\lambda/2}(\beta - \gamma)$$

4.5.5. We can either set $l = 0$ and let r tend to f in **[4.5.4]**, or we can use the absolute partition lemma and **4.5.1** to give the solution as

$$[\mathbf{x}^{\mathbf{m}}]|\mathbf{I} + \mathbf{X} - \mathbf{W}|^{-1} = [\mathbf{x}^{\mathbf{m}}]|\mathbf{I} + \mathbf{X}|^{-1}\left\{1 - \Psi_1(\mathbf{I} + \mathbf{X})^{-1}\right\}^{-1}$$

$$= [\mathbf{x}^{\mathbf{m}}]\prod_{t\geq 1}(1+x_t)^{-1}\left\{1 - \sum_{t\geq 1} x_t(1+x_t)^{-1}\right\}^{-1}.$$

4.5.6. If elements in rows 1, 2, and 3 are marked by x's, y's, and z's, then the enumerator for each of n columns is $\sum x_i y_j z_k$, where $i \neq j$, $j \neq k$, $k \neq i$. We extract the coefficient of $x_1 y_1 z_1 \cdots x_n y_n z_n$ to ensure that all elements of N_n

appear in each row. But

$$\sum_{\substack{i\neq j\\ j\neq k\\ k\neq i}} x_i y_j z_k = \alpha_1\alpha_2\alpha_3 - \alpha_{12}\alpha_3 - \alpha_{13}\alpha_2 - \alpha_{23}\alpha_1 + 2\alpha_{123},$$

where $\alpha_1 = x_1 + \cdots + x_n$, $\alpha_2 = y_1 + \cdots + y_n$, $\alpha_3 = z_1 + \cdots + z_n$, $\alpha_{12} = x_1y_1 + \cdots + x_ny_n$, $\alpha_{23} = y_1z_1 + \cdots + y_nz_n$, $\alpha_{13} = x_1z_1 + \cdots + x_nz_n$, $\alpha_{123} = x_1y_1z_1 + \cdots + x_ny_nz_n$. Thus we use a multinomial expansion to obtain

$$\begin{aligned} l_3(n) &= [\mathbf{x}^1\mathbf{y}^1\mathbf{z}^1](\alpha_1\alpha_2\alpha_3 - \alpha_{12}\alpha_3 - \alpha_{23}\alpha_1 - \alpha_{13}\alpha_2 + 2\alpha_{123})^n \\ &= \sum_{i_1+\cdots+i_5=n} (-1)^{i_2+i_3+i_4}\begin{bmatrix} n \\ i_1,\ldots,i_5\end{bmatrix}2^{i_5} \\ &\qquad \times [\mathbf{x}^1\mathbf{y}^1\mathbf{z}^1]\,\alpha_1^{i_1+i_3}\alpha_2^{i_1+i_4}\alpha_3^{i_1+i_2}\alpha_{12}^{i_2}\alpha_{23}^{i_3}\alpha_{13}^{i_4}\alpha_{123}^{i_5}. \end{aligned}$$

But this coefficient may be extracted easily, giving

$$\begin{aligned} &\begin{bmatrix} n \\ i_1,\ldots,i_5\end{bmatrix} i_5!i_4!i_3!i_2!(i_1+i_2)!(i_1+i_3)!(i_1+i_4)! \\ &\qquad = \frac{n!}{i_1!}(i_1+i_2)!(i_1+i_3)!(i_1+i_4)! \end{aligned}$$

so the required number is $l_3(n)$

$$\begin{aligned} &= n! \sum_{i_1+\cdots+i_5=n} \begin{bmatrix} n \\ i_1,\ldots,i_5\end{bmatrix} 2^{i_5}(-1)^{i_2+i_3+i_4}(i_1+i_4)!(i_1+i_3)!(i_1+i_2)!/i_1! \\ &= (n!)^2 \sum_{i_1+m+i_5=n} 2^{i_5}(-1)^m(i_1!/i_5!) \sum_{i_2+i_3+i_4=m} \binom{i_1+i_4}{i_4}\binom{i_1+i_3}{i_3}\binom{i_1+i_2}{i_2}. \end{aligned}$$

But the second sum is

$$[x^m]\left(\sum_{i\geq 0}\binom{i_1+1}{i}x^i\right)^3 = [x^m](1-x)^{-3(i_1+1)},$$

so

$$\begin{aligned} l_3(n) &= (n!)^2[x^n]\sum_{i_5\geq 0}\frac{(2x)^{i_5}}{i_5!}\sum_{i_1\geq 0} i_1!x^{i_1}(1+x)^{-3(i_1+1)} \\ &= (n!)^2[x^n]e^{2x}\sum_{i\geq 0} i!x^i(1+x)^{-3(i+1)}, \end{aligned}$$

in agreement with **4.5.10**.

4.5.7. (a) Each term in the permanent contains either an element of **C** or an element of **D** for each row and column. If α and β are the subsets of rows and columns in which elements of **C** are chosen, we must have $|\alpha| = |\beta|$, and elements of **D** must be chosen from rows in $\mathsf{N}_n - \alpha$ and columns in $\mathsf{N}_n - \beta$. The Laplace expansion and the cofactor expansion are immediate corollaries.

(b) From **4.5.3**, the number of permutations on N_n with no absolute levels is $d_n = \operatorname{per}(\mathbf{J}_n - \mathbf{I}_n)$. But, from part (a), we have

$$d_n = \sum_{\substack{\alpha,\beta\subseteq \mathsf{N}_n\\ |\alpha|=|\beta|}} \operatorname{per}(-\mathbf{I}_n)[\alpha|\beta]\operatorname{per}\mathbf{J}_n(\alpha|\beta).$$

Now $\operatorname{per}\mathbf{J}_n(\alpha|\beta) = (n - |\alpha|)!$ for all $\alpha, \beta \subseteq \mathsf{N}_n$ with $|\alpha| = |\beta|$. Also $\operatorname{per}(-\mathbf{I}_n)[\alpha|\beta] = (-1)^{|\alpha|}$ if $\alpha = \beta$, and is zero otherwise. Thus

$$d_n = \sum_{\alpha\subseteq \mathsf{N}_n} (n-|\alpha|)!(-1)^{|\alpha|}$$

$$= \sum_{k=0}^{n} (n-k)!(-1)^k \sum_{\substack{\alpha\subseteq \mathsf{N}_n\\ |\alpha|=k}} 1 = n!\sum_{k=0}^{n}\frac{(-1)^k}{k!}.$$

4.5.8. (a) Consider the (p, q)-cell of **C**. If a rook is placed in this position we mark it by x, and must place the remaining rooks in $\mathbf{C}(\{p\}|\{q\})$, the submatrix of **C** obtained by deleting row p and column q. If a rook is not placed in this position, then we can replace the 1 by a 0.

(b) Rooks may be distributed on $\mathbf{C}_1$ and $\mathbf{C}_2$ independently, since $\mathbf{C}_1$ and $\mathbf{C}_2$ occupy disjoint sets of rows and columns in $\mathbf{C}_1 \oplus \mathbf{C}_2$, and $\mathbf{C}_1 \oplus \mathbf{C}_2$ contains no 1's other than those in $\mathbf{C}_1$ and $\mathbf{C}_2$.

4.5.9. (a) From the absolute partition lemma, the required number is $[x^k]\operatorname{per}(\mathbf{J}_n + (x-1)\mathbf{C})$. But from [**4.5.7**(a)], this is

$$[x^k]\sum_{i=0}^{n} (n-i)!(x-1)^i \sum_{\substack{\alpha,\beta\subseteq \mathsf{N}_n\\ |\alpha|=|\beta|=i}} \operatorname{per}\mathbf{C}[\alpha|\beta].$$

But $\operatorname{per}\mathbf{C}[\alpha|\beta]$ is the number of ways of placing i rooks in nontaking positions in the subsets α and β of the rows and columns of **C**. Thus

$$\sum_{\substack{\alpha,\beta\subseteq \mathsf{N}_n\\ |\alpha|=|\beta|=i}} \operatorname{per}\mathbf{C}[\alpha|\beta] = r_i$$

by definition. The result follows immediately.

(b) If a permutation on N_n contains k absolute π_1's, then it must also contain $n-k$ absolute π_2's, where $\pi_1 \mathbin{\dot\cup} \pi_2 = \omega$. Thus, from part (a),

$$[x^{n-k}]\sum_{i=0}^{n}(n-i)!q_i(x-1)^i = [x^k]\sum_{j=0}^{n}(n-j)!r_j(x-1)^j,$$

for $k = 0,\dots,n$, so

$$\sum_{i=0}^{n}(n-i)!q_i(x-1)^i = \sum_{j=0}^{n}(n-j)!r_j x^n(x^{-1}-1)^j$$

$$= \sum_{j=0}^{n}(n-j)!r_j(-1)^j x^{n-j}(x-1)^j.$$

Now let $y = x-1$, giving

$$\sum_{i=0}^{n}(n-i)!q_i y^i = \sum_{j=0}^{n}(n-j)!r_j(-1)^j(1+y)^{n-j}y^j.$$

Finally, compare coefficients of y^i, yielding

$$(n-i)!q_i = \sum_{j=0}^{i}(n-j)!r_j(-1)^j\binom{n-j}{i-j},$$

and the result follows.

4.5.10. From [**4.5.9**(a)], the required number is

$$d_n = \sum_{i=0}^{n}(n-i)!(-1)^i r_i,$$

where $\sum_{i\geqslant 0} r_i x^i = R_{\mathbf{I}_n}(x)$. But from [**4.5.8**(b)], we have $R_{\mathbf{I}_n}(x) = (R_{\mathbf{I}_1}(x))^n = (1+x)^n$, so that

$$r_i = [x^i](1+x)^n = \binom{n}{i}.$$

Thus $d_n = n!\sum_{i=0}^{n}(-1)^i/i!$.

4.5.11. (a) Combinatorially, we may place i rooks in nontaking positions by choosing a subset of i rows, a subset of i columns, and then placing them in the resulting submatrix $\mathbf{J}_i$ in $i!$ ways. Thus $r_i = i!\binom{n}{i}^2$ and

$$\sum_{i\geqslant 0} r_i x^i = \sum_{i=0}^{n} i!\binom{n}{i}^2 x^i.$$

(b) From [**4.5.9**(a)], the required number is $c(\mathbf{k}) = \sum_{i=0}^{n}(n-i)!(-1)^i r_i$, where

$$\sum_{i\geqslant 0} r_i x^i = R_{\mathbf{J}_{k_1}\oplus\cdots\oplus\mathbf{J}_{k_m}}(x).$$

But from [**4.5.8**(b)],

$$R_{\mathbf{J}_{k_1}\oplus\cdots\oplus\mathbf{J}_{k_m}}(x) = \prod_{j=1}^{m} R_{\mathbf{J}_{k_j}}(x) = \prod_{j=1}^{m}\sum_{l=0}^{k_j} l!\binom{k_j}{l}^2 x^l$$

from part (a).

4.5.12. (a) Let $A_n(x) = R_{\mathbf{A}_n}(x)$, $B_n(x) = R_{\mathbf{B}_n}(x)$, $C_n(x) = R_{\mathbf{C}_n}(x)$. Then by considering position $(n, 1)$ and [**4.5.8**(a)]

$$C_n(x) = xA_{n-1}(x) + A_n(x).$$

Also, by considering position $(1, 1)$,

$$\left.\begin{aligned} A_n(x) &= xA_{n-1}(x) + B_{n-1}(x), \qquad & n\geqslant 2 \\ B_n(x) &= xB_{n-1}(x) + A_n(x), \qquad & n\geqslant 1. \end{aligned}\right\}$$

Now let $U_{2n}(x) = B_n(x)$, $U_{2n-1}(x) = A_n(x)$, and $U(x, y) = \sum_{n\geqslant 0} U_n(x)y^n$. But $U_0(x) = B_0(x) = 1$, $U_1(x) = A_1(x) = x + 1$, so

$$\left.\begin{aligned} U_{2n-1}(x) &= xU_{2n-3}(x) + U_{2n-2}(x), \qquad & n\geqslant 2 \\ U_{2n}(x) &= xU_{2n-2}(x) + U_{2n-1}(x), \qquad & n\geqslant 1. \end{aligned}\right\}$$

Multiply the first of these equations by y^{2n-1} and sum over $n \geqslant 2$; multiply the second by y^{2n} and sum over $n \geqslant 1$. Adding the resulting pair of equations gives

$$U(x, y) - U_1(x)y - U_0(x) = xy^2U(x, y) + y\{U(x, y) - U_0(x)\}$$

so

$$U(x, y) = (1 + xy)\{1 - y(1 + xy)\}^{-1} = \sum_{i\geqslant 0} y^i(1 + xy)^{i+1}.$$

Thus

$$U_n(x) = [y^n]U(x, y) = \sum_{i=0}^{n}\binom{i+1}{n-i}x^{n-i} = \sum_{j=0}^{n}\binom{n-j+1}{j}x^j.$$

Finally,

$$C_n(x) = xA_{n-1}(x) + A_n(x) = xU_{2n-3}(x) + U_{2n-1}(x)$$

$$= \sum_{j=0}^{2n-3} \binom{2n-2-j}{j} x^{j+1} + \sum_{i=0}^{2n-1} \binom{2n-i}{i} x^i,$$

and the result follows.

(b) From **4.4.7** and [**4.5.9**(a)], the required number is

$$m_n = 2 \cdot n! \sum_{i=0}^{n} (n-i)!(-1)^i r_i,$$

where $\sum_{i \geqslant 0} r_i x^i = R_{\mathbf{C}_n}(x)$ is obtained from part (a).

4.5.13. (a) If rooks are placed on every entry in the diagonal, then the contribution is x^{n+1}, and no other rooks can be placed. If k rooks are placed in any of the $\binom{n+1}{k}$ subsets of the diagonal entries then the matrix that remains, after deleting the corresponding rows and columns, is $\mathbf{T}_{n-k}$, for $k = 0, \ldots, n$.

(b) If we multiply both sides of the equation in part (a) by $y^{n+1}/(n+1)!$, and sum over $n \geqslant 0$, we obtain

$$T(x, y) - 1 = e^{xy} - 1 + e^{xy} \sum_{i \geqslant 0} T_i(x) \frac{y^{i+1}}{(i+1)!}$$

so

$$\sum_{i \geqslant 0} T_i(x) \frac{y^{i+1}}{(i+1)!} = e^{-xy} T(x, y) - 1. \tag{1}$$

Differentiating with respect to y yields

$$T(x, y) = e^{-xy} \left\{ \frac{\partial}{\partial y} T(x, y) - xT(x, y) \right\},$$

so

$$\frac{\partial}{\partial y} \{\log T(x, y)\} = x + e^{xy}$$

whence $T(x, y) = \exp\{xy + x^{-1}(e^{xy} - 1)\}$ since $T(x, 0) = T_0(x) = 1$. Thus $e^{-xy} T(x, y) = \exp\{x^{-1}(e^{xy} - 1)\}$ and from equation (1),

$$e^{-xy} T(x, y) = 1 + \sum_{i \geqslant 0} T_i(x) \frac{y^{i+1}}{(i+1)!},$$

so

$$[x^k]T_n(x) = \left[x^k \frac{y^{n+1}}{(n+1)!}\right] \exp\{x^{-1}(e^{xy}-1)\}$$

$$= \left[x^{k-(n+1)} \frac{w^{n+1}}{(n+1)!}\right] \exp\{x^{-1}(e^{w}-1)\}$$

$$= \left[z^{n+1-k} \frac{w^{n+1}}{(n+1)!}\right] \exp\{z(e^{w}-1)\}.$$

4.5.14. (a) Let $\mathbf{x} = (x_1, \ldots, x_k)$ and $\mathbf{X} = \text{diag}(\mathbf{x})$. If $v_i v_j$ appears as a substring in a permutation σ, then we must have $v_{il_i} > v_{j1}$ for v_i, v_j to be maximal $<$-strings in σ. Thus $s(\mathsf{V}) = [\mathbf{x}^1]\text{trace}(\mathbf{I} - \mathbf{XC})^{-1}\mathbf{XJ}$, and

$$1 + \text{trace}(\mathbf{I} - \mathbf{XC})^{-1}\mathbf{XJ} = |\mathbf{I} + (\mathbf{I} - \mathbf{XC})^{-1}\mathbf{XJ}|, \qquad \text{by } \mathbf{1.1.10(5)}$$

$$= |\mathbf{I} - \mathbf{XC}|^{-1} \cdot |\mathbf{I} + \mathbf{X}(\mathbf{J} - \mathbf{C})|.$$

But $\mathbf{J} - \mathbf{C}$ is strictly upper triangular, so $|\mathbf{I} + \mathbf{X}(\mathbf{J} - \mathbf{C})| = 1$, and

$$s(\mathsf{V}) = [\mathbf{x}^1]|\mathbf{I} - \mathbf{XC}|^{-1} = \text{per}(\mathbf{C}).$$

Similarly,

$$c(\mathsf{V}) = [\mathbf{x}^1]\exp \text{trace} \log(\mathbf{I} - \mathbf{XC})^{-1} = [\mathbf{x}^1]|\mathbf{I} - \mathbf{XC}|^{-1} = \text{per}(\mathbf{C}).$$

(b) If σ is a permutation on N_n, then let σ' be the permutation on N_n whose ith entry is formed by subtracting the ith entry of σ from $n+1$. Let $\sigma' \mapsto \{c_1, \ldots, c_k\}$ under the Foata–Schützenberger decomposition, given in [**3.3.17**(b)], and let $\{c_1, \ldots, c_k\}$ be the set of disjoint cycles of the permutation τ on N_n. Then the maximal $<$-strings of σ are precisely the maximal $<$-strings of the disjoint cycles of τ', and the construction is reversible, so $c(\mathsf{V}) = s(\mathsf{V})$.

4.5.15. Let Φ be the generating function for permutations on N_n with p_{ij} marking the occurrence of j in position i, for $i, j = 1, \ldots, n$. Then, by definition, $\Phi = \text{per}\,\mathbf{P}$. On the other hand, the generating function for the cycles containing $\{\alpha_1, \ldots, \alpha_r\} \subseteq \mathsf{N}_n$ is $[x_{\alpha_1} \cdots x_{\alpha_r}]r^{-1}\text{trace}(\mathbf{XP})^r = [x_{\alpha_1} \cdots x_{\alpha_r}]\text{trace} \log(\mathbf{I} - \mathbf{XP})^{-1}$, and since a permutation consists of an unordered collection of disjoint cycles, we obtain

$$\Phi = [\mathbf{x}^1]\exp \text{trace} \log(\mathbf{I} - \mathbf{XP})^{-1} = [\mathbf{x}^1]|\mathbf{I} - \mathbf{XP}|^{-1},$$

as required.

4.5.16. Let $z_i = z_{i1} + \cdots + z_{ik_i}$ for $1 \leqslant i \leqslant n$. Then

$$[\mathbf{x}^{\mathbf{k}}](\mathbf{P}\mathbf{x}^T)^{\mathbf{k}} = [z_1^{k_1} \cdots z_n^{k_n}] \prod_{i=1}^{n} (p_{i1}z_1 + \cdots + p_{in}z_n)^{k_i}$$

$$= (\mathbf{k}!)^{-1}[z_{11}z_{12} \cdots z_{nk_n}] \prod_{i=1}^{n} (p_{i1}z_{11} + \cdots + p_{i1}z_{1k_1} + \cdots + p_{in}z_{nk_n})^{k_i}$$

$$= (\mathbf{k}!)^{-1}[z_{11} \cdots z_{nk_n}]|\mathbf{I} - \mathbf{ZB}|^{-1},$$

by Proposition 4.5.1, where $\mathbf{Z} = \operatorname{diag}(z_{11}, \ldots, z_{1k_1}, \ldots, z_{n1}, \ldots, z_{nk_n})$ and

$$\mathbf{B} = \begin{bmatrix} p_{11}\mathbf{J}_{k_1, k_1} & \cdots & p_{1n}\mathbf{J}_{k_1, k_n} \\ \vdots & & \vdots \\ p_{n1}\mathbf{J}_{k_n, k_1} & \cdots & p_{nn}\mathbf{J}_{k_n, k_n} \end{bmatrix}.$$

But $|\mathbf{I} - \mathbf{ZB}|$ is a polynomial in $\{z_{ij}\}$ that is symmetric in the elements of $\{z_{i1}, \ldots, z_{ik_i}\}$ for each i and has the property that $[z_{il}z_{im}]|\mathbf{I} - \mathbf{ZB}| = 0$ for $1 \leqslant l, m \leqslant k_i$. Replacing each z_{i1} by z_i and z_{ij} by 0 for $j \geqslant 2$ we have $|\mathbf{I} - \mathbf{ZB}| = |\mathbf{I} - \mathbf{\Lambda P}|$, where $\mathbf{\Lambda} = \operatorname{diag}(z_1, \ldots, z_n)$. Thus

$$\mathbf{k}!^{-1}[z_{11} \cdots z_{nk_n}]|\mathbf{I} - \mathbf{ZB}|^{-1} = \mathbf{k}!^{-1}[z_{11} \cdots z_{nk_n}]|\mathbf{I} - \mathbf{\Lambda P}|^{-1}$$

$$= [z_1^{k_1} \cdots z_n^{k_n}]|\mathbf{I} - \mathbf{\Lambda P}|^{-1},$$

and the result follows.

SECTION 5.2

5.2.1. (a) There are $\left[\begin{matrix} n \\ 1, n_1, \ldots, n_k \end{matrix}\right]$ ways of choosing the label for the root vertex and the n_i labels for the vertices at height i, for $i = 1, \ldots, k$. The edges joining the vertices at height i and $i + 1$ are uniquely determined by a function from a set of size n_{i+1} to a set of size n_i, for $i = 1, \ldots, k - 1$. Since there are $n_i^{n_{i+1}}$ such functions, the number of trees is

$$\left[\begin{matrix} n \\ 1, n_1, \ldots, n_k \end{matrix}\right] \prod_{i=1}^{k-1} n_i^{n_{i+1}}.$$

(b) We determine the generating function for labeled rooted trees with x an exponential marker for vertices and λ_i an ordinary marker for vertices at height i, for $i \geqslant 0$. First, by the labeled branch decomposition **(3.3.9)** this generating function is $x\lambda_0\exp\{x\lambda_1\exp\{x\lambda_2\exp\{\dots\}\}\}$. But by part (a), this generating function is also given by $x\lambda_0\{1+\sum_{m\geq 1}A_m x^m\}$, where $m=n-1$. The result follows. The polynomials A_m can be regarded as labeled versions of the Stieltjes–Rogers polynomials (**5.2.4**) since the latter count planted plane trees with respect to vertices at given height. Furthermore the S-fraction in **5.2.4** is an iterated geometric series, and is replaced in the preceding result by an iterated exponential series.

5.2.2. Count paths from altitude 0 to altitude 0 in two different ways. First, from the path lemma the generating function for paths from altitude 0 to altitude 0 with respect to rises and levels is $J_x[\gamma_l, \alpha_l:(0,\infty)]$. To evaluate this generating function in another way construct all paths from altitude 0 to altitude 0 from level-free paths by inserting levels in all permissible ways at each altitude. Accordingly, we first consider level-free paths. Replacing x by x^2 and λ_i by α_i in Proposition 5.2.4 gives

$$J_x[0,\alpha_l:(0,\infty)]$$

$$=1+\sum_{n\geq 0}x^{2n}\sum_{k\geq 0}\sum_{\substack{n_0,\dots,n_k\geq 1\\ n_0+\cdots+n_k=n}}\alpha_0^{n_0}\cdots\alpha_k^{n_k}\prod_{j=0}^{k-1}\binom{n_j+n_{j+1}-1}{n_j-1}.$$

Thus, there are $\prod_{j=0}^{k-1}\binom{n_j+n_{j+1}-1}{n_j-1}$ level-free paths of length $2(n_0+\cdots+n_k)$ and height $k+1$, from altitude 0 to altitude 0, with n_j rises at altitude j, where $0\leq j\leq k$. Such a path contains n_j+n_{j-1} vertices at altitude j for $j=0,\dots,k+1$, where $n_{-1}=1$ and $n_{k+1}=0$. This follows since a vertex at altitude j is the origin of a rise n_j times and the origin of a fall n_{j-1} times.

We now insert levels in all possible ways by replacing each vertex at altitude j by a path consisting entirely of levels at altitude j. The number of ways to insert m_j levels at altitude j in a level-free path containing n_j+n_{j-1} vertices at altitude j is thus

$$[x^{m_j}](1-x)^{-(n_j+n_{j-1})}=\binom{m_j+n_j+n_{j-1}-1}{m_j}.$$

It follows that there are

$$\left\{\prod_{j=0}^{k-1}\binom{n_j+n_{j+1}-1}{n_j-1}\right\}\left\{\prod_{j=0}^{k+1}\binom{m_j+n_j+n_{j-1}-1}{m_j}\right\}$$

paths of height $k+1$, from altitude 0 to altitude 0, of length $2(n_0+\cdots+n_k)+(m_0+\cdots+m_{k+1})$ with m_j levels at altitude j, for $0\leq j\leq k+1$, and n_j

rises at altitude j for $0 \leqslant j \leqslant k$. This product simplifies to

$$\binom{m_{k+1}+n_k-1}{m_{k+1}} \prod_{j=0}^{k} \begin{bmatrix} m_j+n_j+n_{j-1}-1 \\ m_j, n_{j-1}-1, n_j \end{bmatrix}$$

and the monomial associated with this is $(\alpha_0^{n_0} \cdots \alpha_k^{n_k})(\gamma_0^{m_0} \cdots \gamma_{k+1}^{m_{k+1}})$.

The excluded case consists of paths of height 0. For any length greater than or equal to one, there is exactly one of these from altitude 0 to altitude 0. The result follows by replacing α_i by λ_i and γ_i by κ_i.

5.2.3. (a) We want $\hat{f}(x)$ where $f(x) = e^{x^2/2}$, so

$$f(x+y) = e^{xy}e^{x^2/2}e^{y^2/2} = \sum_{k \geqslant 0} k!\left(\frac{1}{k!}x^k e^{x^2/2}\right)\left(\frac{1}{k!}y^k e^{y^2/2}\right).$$

But $x^k e^{x^2/2}/k! = x^k/k! + \mathcal{O}(x^{k+2})$ and we have obtained an addition formula with parameters $\{(k!, 0) | k \geqslant 0\}$, so the result follows from the Stieltjes–Rogers J-fraction theorem.

(b) We want $\hat{f}(x)$ where $f(x) = (1-x)^{-r}$. Now following **5.2.11**,

$$f(x+y) = (1-x)^{-r}(1-y)^{-r}\{1-(1-x)^{-1}(1-y)^{-1}xy\}^{-r}$$

$$= \sum_{k \geqslant 0} k!^2 \binom{k+r-1}{k} \left\{\frac{1}{k!}x^k(1-x)^{-(k+r)}\right\}\left\{\frac{1}{k!}y^k(1-y)^{-(k+r)}\right\}.$$

But

$$\frac{1}{k!}x^k(1-x)^{-(k+r)} = \frac{x^k}{k!} + (k+1)(k+r)\frac{x^{k+1}}{(k+1)!} + \mathcal{O}(x^{k+2}),$$

and the result follows from the Stieltjes–Rogers J-fraction theorem.

(c) We want $\hat{f}(x)$ where $f(x) = \sec x$. But

$$f(x+y) = \frac{\sec x \sec y}{1 - \tan x \tan y}$$

$$= \sum_{k \geqslant 0} k!^2 \left(\frac{1}{k!} \sec x \tan^k x\right)\left(\frac{1}{k!} \sec y \tan^k y\right).$$

Now $\sec x \tan^k x/k! = x^k/k! + \mathcal{O}(x^{k+2})$, and the result follows from the Stieltjes–Rogers J-fraction theorem.

5.2.4. (a) Let $f(x) = a_1 x/1! + a_2 x^2/2! + \cdots$. Thus $f'(x) = a_1 + a_2 x/1! + \cdots$, so $\widehat{f'}(x) = a_1 + a_2 x + \cdots$, and the result follows.

(b) If $f(x) = \tan x$, then $f'(x) = \sec^2 x$, so

$$f'(x+y) = \sec^2 x \sec^2 y (1 - \tan x \tan y)^{-2}$$

$$= \sum_{k \geqslant 0} (k+1)(k!)^2 \left(\frac{1}{k!} \sec^2 x \tan^k x\right)\left(\frac{1}{k!} \sec^2 y \tan^k y\right).$$

Now $\sec^2 x \tan^k x / k! = x^k/k! + \mathcal{O}(x^{k+2})$, and the result follows from (a) and the Stieltjes–Rogers J-fraction theorem.

5.2.5. (a) Since $\hat{f} = J_x[\kappa_k, \lambda_k : (0, \infty)]$ we know from the Stieltjes–Rogers J-fraction theorem that $f(x)$ has an addition formula with parameters $\{(p_k, q_{k+1}) | k \geqslant 0\}$, where $f(x+y) = \sum_{k \geqslant 0} p_k f_k(x) f_k(y)$ and

$$\kappa_k = q_{k+1} - q_k, \qquad \lambda_k = p_{k+1} p_k^{-1}, \qquad \text{for } k \geqslant 0.$$

Thus $g(x+y) = \sum_{k \geqslant 0} p_k (e^{\alpha x} f_k(x))(e^{\alpha y} f_k(y))$, where

$$e^{\alpha x} f_k = \frac{x^k}{k!} + \{q_{k+1} + \alpha(k+1)\} \frac{x^{k+1}}{(k+1)!} + \mathcal{O}(x^{k+2}),$$

so $g(x)$ has an addition formula with parameters

$$\{(p_k, q_{k+1} + \alpha(k+1)) | k \geqslant 0\}$$

and the result follows from the Stieltjes–Rogers J-fraction theorem.

(b) From **5.2.11**, $\hat{f}(x) = J_x[2k+1, (k+1)^2 : (0, \infty)]$, where $f(x) = (1-x)^{-1}$. The result follows from (a) with $\alpha = -1$.

5.2.6. (a) Let $\theta(x) = e^x - 1$ and $f(x) = \exp u(e^x - 1)$. Then

$$f(x+y) = \sum_{k \geqslant 0} u^k k! \left\{\frac{1}{k!} \theta^k(x) e^{u\theta(x)}\right\}\left\{\frac{1}{k!} \theta^k(y) e^{u\theta(y)}\right\},$$

where

$$\frac{1}{k!} \theta^k(x) e^{u\theta(x)} = \frac{x^k}{k!} + \tfrac{1}{2}(k+1)(2u+k) \frac{x^{k+1}}{(k+1)!} + \mathcal{O}(x^{k+2}),$$

and the result follows from the Stieltjes–Rogers J-fraction theorem.

(b) Let

$$f(x) = \frac{1-u}{1-ue^{x(1-u)}} = \left(1 - \frac{u}{1-u}\{e^{x(1-u)} - 1\}\right)^{-1},$$

and $\theta(x) = e^{x(1-u)} - 1$, so

$$f(x+y) = \left(\left\{1 - \frac{u}{1-u}\theta(x)\right\}\left\{1 - \frac{u}{1-u}\theta(y)\right\} - \frac{u}{(1-u)^2}\theta(x)\theta(y)\right)^{-1}$$

$$= \sum_{k\geqslant 0} u^k(1-u)^{-2k}\left(\theta^k(x)\left\{1 - \frac{u}{1-u}\theta(x)\right\}^{-(k+1)}\right)$$

$$\times\left(\theta^k(y)\left\{1 - \frac{u}{1-u}\theta(y)\right\}^{-(k+1)}\right)$$

and

$$\frac{\theta^k(x)}{k!(1-u)^k}\left\{1 - \frac{u}{1-u}\theta(x)\right\}^{-(k+1)}$$

$$= \frac{x^k}{k!} + \frac{1}{2}\frac{x^{k+1}}{(k+1)!}\{k(1-u) + 2u(1+k)\}(k+1) + \mathcal{O}(x^{k+2}).$$

The result follows from the Stieltjes–Rogers J-fraction theorem.

5.2.7. (a) Let $f(x) = \Sigma_{k\geqslant 0} a_k x^k/k!$. Now $\int_0^\infty t^k e^{-tx^{-1}}\,dt = k!x^{k+1}$, and the result follows.

(b) Integrating by parts gives

$$t_n = x^{-1}\left[-e^{-tx^{-1}}\tan^n t\right]_0^\infty + \int_0^\infty n(\tan^{n-1}t\sec^2 t)e^{-tx^{-1}}\,dt$$

$$= xn(t_{n-1} + t_{n+1}),$$

whence $xn\,t_{n-1} - t_n + xn\,t_{n+1} = 0$. Thus

$$\frac{t_n}{t_{n-1}} = \frac{nx}{1 - nx\dfrac{t_{n+1}}{t_n}}.$$

But $t_0 = 1$, so $t_1 = xS_{x^2}[(k+1)(k+2):(0,\infty)]$ and the result follows from (a).

5.2.8. (a) When $x = 0$, then $\phi = 0$. The result follows.

(b) (i) $(\mathrm{cn}\,x)' = -\sin\phi(d\phi/dx) = -\mathrm{sn}\,x(d\phi/dx)$. But

$$\frac{dx}{d\phi} = (1 - m^2\sin^2\phi)^{-1/2}, \quad \text{so } \frac{d\phi}{dx} = (1 - m^2\sin^2\phi)^{1/2} = \mathrm{dn}\,x.$$

Thus $(\mathrm{cn}\,x)' = -\mathrm{sn}\,x\,\mathrm{dn}\,x$.

(ii) Similar to (i).

(iii) $(\mathrm{dn}\,x)' = -m^2\sin\phi\cos\phi\{1 - m^2\sin^2\phi\}^{-1/2}d\phi/dx$, and the result follows.

(c) Let $n > 2$. Integrating g_n by parts,

$$g_n = \left[-e^{-tx^{-1}}\mathrm{sn}^n t\right]_0^\infty + \int_0^\infty e^{-tx^{-1}}\left(\frac{d}{dt}\,\mathrm{sn}^n t\right)dt.$$

But we know from (a) that $\mathrm{sn}\,0 = 0$ so, integrating by parts again,

$$g_n = x\int_0^\infty e^{-tx^{-1}}\left(\frac{d^2}{dt^2}\,\mathrm{sn}^n t\right)dt.$$

But from (b),

$$\frac{d^2}{dt^2}\,\mathrm{sn}^n t = n(n-1)\mathrm{sn}^{n-2}t\,\mathrm{cn}^2 t\,\mathrm{dn}^2 t - n\,\mathrm{dn}^2 t\,\mathrm{sn}^n t - nm^2\mathrm{cn}^2 t\,\mathrm{sn}^n t.$$

However, from the definitions $\mathrm{cn}^2 t = 1 - \mathrm{sn}^2 t$ and $\mathrm{dn}^2 t = 1 - m^2\mathrm{sn}^2 t$, so

$$g_n = n(n-1)x^2 g_{n-2} - n^2(1+m^2)x^2 g_n + m^2 n(n+1)x^2 g_{n+2},$$

whence

$$g_n/g_{n-2} = \frac{n(n-1)x^2}{1 + n^2(1+m^2)x^2 - m^2 n(1+n)x^2 g_{n+2}/g_n}, \qquad \text{for } n > 2.$$

But

$$\begin{aligned} g_1 &= x^{-1}\int_0^\infty \mathrm{sn}\,t\,e^{-tx^{-1}}\,dt \\ &= \int_0^\infty \mathrm{cn}\,t\,\mathrm{dn}\,t\,e^{-tx^{-1}}\,dt, \quad \text{integrating by parts} \\ &= \left[-x\,\mathrm{cn}\,t\,\mathrm{dn}\,t\,e^{-tx^{-1}}\right]_0^\infty - \int_0^\infty \{(1+m^2)\mathrm{sn}\,t - 2m^2\mathrm{sn}^3 t\}e^{-tx^{-1}}\,dt \\ &= x - (1+m^2)x^2 g_1 + 2m^2x^2 g_3. \end{aligned}$$

Thus

$$\hat{f}(x) = g_1 = \frac{x}{1 + (1+m^2)x^2 - 2m^2x^2 g_3/g_1},$$

and the result follows from the recurrence equation for g_n/g_{n-2}.

(d) (i) We follow the method of (c). Integrating g_n by parts twice gives

$$g_n = x\int_0^\infty e^{-tx^{-1}}\frac{d^2}{dt^2}(\mathrm{sn}^n t\,\mathrm{cn}\,t)\,dt, \qquad \text{for } n \geqslant 2.$$

But

$$\frac{d^2}{dt^2}(\mathrm{sn}^n t\,\mathrm{cn}\, t) = n(n-1)\mathrm{sn}^{n-2} t\,\mathrm{cn}\, t$$

$$- \{m^2 n^2 + (n+1)^2\}\mathrm{sn}^n t\,\mathrm{cn}\, t + m^2(n+1)(n+2)\mathrm{sn}^{n+2} t\,\mathrm{cn}\, t,$$

so

$$g_n = n(n-1)x^2 g_{n-2} - \left(m^2 n^2 + (1+n)^2\right)x^2 g_n + m^2(n+1)(n+2)x^2 g_{n+2}$$

whence

$$g_n/g_{n-2} = \frac{n(n-1)x^2}{1 + \{m^2 n^2 + (n+1)^2\}x^2 - m^2(n+1)(n+2)x^2 g_{n+2}/g_n},$$

for $n \geqslant 2$. Integrating g_0 by parts twice gives

$$g_0 = x + x\int_0^\infty e^{-tx^{-1}} \frac{d^2}{dt^2}(\mathrm{cn}\, t)\, dt.$$

But

$$\frac{d^2}{dt^2}\,\mathrm{cn}\, t = 2m^2\mathrm{sn}^2 t\,\mathrm{cn}\, t - \mathrm{cn}\, t,$$

so

$$\hat{f}(x) = g_0 = \frac{x}{1 + x^2 - 2m^2 x^2 g_2/g_0},$$

and the result follows from the recurrence equation for g_n/g_{n-2}.

(ii) The result follows immediately by contraction (Proposition 5.2.2) of (i).

5.2.9. (a) Let $\Phi(a, b, c) = {}_2\Phi_1\left(\begin{matrix} a, b \\ c\end{matrix} : q, x\right)$ and

$$\phi_j(a, b, c) = \binom{a-1+j}{j}_q \binom{b-1+j}{j}_q \bigg/ \binom{c-1+j}{j}_q.$$

Then

$$\phi_j(a, b, c) = \left\{1 - q^b \frac{(j)_q (c-b)_q}{(b+j)_q (c)_q}\right\}\phi_j(a, b+1, c+1)$$

$$= \phi_j(a, b+1, c+1) - \lambda_0 \phi_{j-1}(a+1, b+1, c+2),$$

so $\Phi(a, b, c) = \Phi(a, b+1, c+1) - \lambda_0 x\Phi(a+1, b+1, c+2)$, whence

$$\frac{\Phi(a, b+1, c+1)}{\Phi(a, b, c)} = \left\{1 - xq^b \frac{(a)_q (c-b)_q}{(c)_q (c+1)_q} \frac{\Phi(a+1, b+1, c+2)}{\Phi(a, b+1, c+1)}\right\}^{-1}.$$

Now exchange a and b, and then replace b by $b+1$ and c by $c+1$ to get

$$\frac{\Phi(a+1, b+1, c+2)}{\Phi(a, b+1, c+1)}$$

$$= \left\{1 - xq^a \frac{(b+1)_q (c-a+1)_q}{(c+1)_q (c+2)_q} \frac{\Phi(a+1, b+2, c+3)}{\Phi(a+1, b+1, c+2)}\right\}^{-1}$$

where the obvious symmetry $\Phi(a, b, c) = \Phi(b, a, c)$ has been used. The result follows by alternating these two expressions. This is a q-analogue of **Gauss' theorem** for the hypergeometric function ${}_2F_1$.

(b) (i) Replace q by q^2, then x by $x(1-q)^{-1}$, and set $a = \frac{1}{2}$, $b = 0$, $c = \infty$ in Heine's theorem. Then

$${}_2\Phi_1\left(\begin{matrix}\frac{1}{2}, 1\\ \infty\end{matrix} : q^2, x(1-q)^{-1}\right) = 1 + \sum_{j \geq 1} (1)_q (3)_q \cdots (2j-1)_q x^j,$$

$${}_2\Phi_1\left(\begin{matrix}\frac{1}{2}, 0\\ \infty\end{matrix} : q^2, x(1-q)^{-1}\right) = 1,$$

$$\lambda_{2j} = (1-q)q^{2j}(2j+1)_q, \qquad \lambda_{2j-1} = (1-q)q^{2j-1}(2j)_q,$$

so the result follows.

(ii) Replace x by $x(1-q)^{-1}$ and set $a = 1$, $b = 0$, $c = \infty$ in Heine's theorem. Then

$${}_2\Phi_1\left(\begin{matrix}1, 1\\ \infty\end{matrix} : q, x(1-q)^{-1}\right) = 1 + \sum_{j \geq 1} j!_q x^j, \; {}_2\Phi_1\left(\begin{matrix}1, 0\\ \infty\end{matrix} : q, x(1-q)^{-1}\right) = 1,$$

$\lambda_{2j} = (1-q)q^j(j+1)_q$, $\lambda_{2j-1} = (1-q)q^j(j)_q$, and the result follows.

5.2.10. (a) We have, from **2.7.3**,

$$C(x) = \sum_{n \geq 0} C_n x^n = \frac{1}{2x}\{1 - (1-4x)^{1/2}\}$$

so $xC^2(x) - C(x) + 1 = 0$ and $C(x) = \{1 - xC(x)\}^{-1}$, where the inverse exists. Thus $C(x) = S_x[1 : (0, \infty)]$ and C_n is the number of level-free paths of length $2n+1$ from altitude 0 to altitude 0, from the path lemma, since $S_{x^2}[1 : (0, \infty)] = J_x[0, 1 : (0, \infty)]$.

(b) Let $M(x) = \Sigma_{n \geq 0} M_n x^n$, so $x^2 M^2(x) - (1 - x)M(x) + 1 = 0$ and $M(x) = \{1 - x - x^2 M(x)\}^{-1}$. Thus $M(x) = J_x[1, 1 : (0, \infty)]$ and M_n is the number of paths of length $n + 1$ from altitude 0 to altitude 0 by the path lemma.

5.2.11. (a) Clearly lead codes are in one-to-one correspondence with level-free paths from altitude 0 to altitude 0. Under this correspondence the lead code is the sequence of altitudes of vertices in the path, reading from left to right.

From the path lemma, the generating function for such paths is $J_x[0, \alpha_k \beta_{k+1} : (0, \infty)]$. Since the corresponding sequence is obtained by recording altitudes, and since the final element is a 0, we set $\alpha_k = x_k$ and $\beta_k = x_k$ to obtain the generating function

$$xx_0 J_x\left[0, x_k x_{k+1} : (0, \infty)\right] = xx_0 S_{x^2}\left[x_k x_{k+1} : (0, \infty)\right]$$

for lead codes.

(b) From part (a), we want the number of level-free paths from altitude 0 to altitude 0, with $2n$ steps. This is $\frac{1}{n+1}\binom{2n}{n}$, directly from [**5.2.10**(a)].

(c) The required number, from part (a), is

$$[x^{2n+1}] x S_{x^2}[(k+1)(k+2) : (0, \infty)] = \left[\frac{x^{2n+1}}{(2n+1)!}\right] \tan x$$

from **5.2.17**.

5.2.12. (a) Steps at altitude i contribute i to the area of a path. Let q be an indeterminate marking unit area. Then the result follows from the path lemma with $\alpha_i = \beta_i = \gamma_i = q^i$.

(b) The result is obtained by setting $x_i = q^i$ in [**5.2.11**(a)].

(c) Now $S_x[q^k : (0, \infty)] = \{1 - x S_{qx}[q^k : (0, \infty)]\}^{-1}$, so

$$N(x, q)/D(x, q) = \{D(qx, q) - xN(qx, q)\}^{-1} D(qx, q).$$

Comparing numerators and denominators, N and D are suitable if $N(x, q) = D(xq, q)$ and D satisfies

$$D(x, q) = D(qx, q) - xD(q^2 x, q).$$

Now let $D(x, q) = \Sigma_{n \geq 0} d_n(q) x^n$, so comparing coefficients of x^n gives

$$d_n(q) = \frac{-q^{2n-2}}{1 - q^n} d_{n-1}(q)$$

so $d_n(q) = (-1)^n q^{n(n-1)} (1 - q)^{-1} \cdots (1 - q^n)^{-1}$ for $n \geq 0$, since $d_0(q) = D(0, q) = 1$. Thus

$$D(x, q) = 1 + \sum_{n \geq 1} (-1)^n x^n q^{n(n-1)} \prod_{i=1}^{n} (1 - q^i)^{-1},$$

and the result follows since

$$S_x\left[q^{2k+1}:(0,\infty)\right] = D\left(xq^3, q^2\right)/D\left(xq, q^2\right).$$

5.2.13. (a) Let $\sigma = \sigma_1 \cdots \sigma_{2n+3}$ be a lead code, and let $1 = i_1 < i_2 < \cdots < i_{n+1} \leqslant 2n+1$ be such that $\sigma_{i_j} < \sigma_{i_j+1}$ for $j = 1,\ldots,n+1$. Thus $i_1,\ldots,i_{n+1}$ are the positions at which the $n+1$ rises in the lead code occur, and the lead code is uniquely determined by this set of positions. Now we must have

$$0 \leqslant \sigma_{i_j} \leqslant j-1 \tag{1}$$

for $j = 1,\ldots,n+1$, since the jth rise must occur at altitude $j-1$ or less. Moreover,

$$\sigma_{i_{j+1}} \leqslant \sigma_{i_j} + 1 \tag{2}$$

for $j = 1,\ldots,n$, since two consecutive rises can only be separated by a (possibly empty) sequence of falls. Now define $a_0, a_1,\ldots, a_n$ by $a_{j-1} = j-1 - \sigma_{i_j}$ for $j = 1,\ldots,n+1$. Then from (1) we have $0 \leqslant a_{j-1} \leqslant j-1$ and from (2) we have $a_{j-1} = j-1-\sigma_{i_j} \leqslant j - \sigma_{i_{j+1}} = a_j$, so $0 = a_0 \leqslant a_1 \leqslant \cdots \leqslant a_n$ and $a_j \leqslant j$ for $j = 0,\ldots,n$. Now we can uniquely determine $i_1,\ldots,i_{n+1}$ from $a_1,\ldots,a_{n+1}$, since (1) and (2) are not only necessary but sufficient, giving the required correspondence.

(b) In the level-free path corresponding to the lead code mark a rise at altitude k by xq^k and a fall by 1. Then by the path lemma, the number of lead codes of length $2n+3$ with $\sigma_{i_1} + \cdots + \sigma_{i_{n+1}} = m$ is

$$\left[x^{n+1}q^m\right] S_x\left[q^k:(0,\infty)\right].$$

But we have a unique sequence $a_1,\ldots,a_n$ such that $0 \leqslant a_1 \leqslant \cdots \leqslant a_n$, $a_j \leqslant j$ for $j = 1,\ldots,n$ and

$$\sum_{j=1}^{n} a_j = \sum_{j=1}^{n}\left(j - \sigma_{i_{j+1}}\right) = \binom{n+1}{2} - m, \quad \text{since } \sigma_{i_1} = 0.$$

(c) From [**5.2.12**(c)],

$$S_{-q}\left[q^k:(0,\infty)\right] = \frac{\sum_{n\geqslant 0} q^{n(n+1)}(1-q)^{-1}\cdots(1-q^n)^{-1}}{\sum_{n\geqslant 0} q^{(n^2)}(1-q)^{-1}\cdots(1-q^n)^{-1}}$$

$$= \prod_{n\geqslant 0}\frac{\left(1-q^{5n+1}\right)\left(1-q^{5n+4}\right)}{\left(1-q^{5n+2}\right)\left(1-q^{5n+3}\right)}$$

by the Rogers–Ramanujan identities (see [**2.5.19**]), and the result follows.

5.2.14. The path that corresponds to such permutations in the Françon–Viennot decomposition contains a rise, right-level, left-level, or fall at (k, α_k) according to whether k belongs to π_1, π_2, π_3, or π_4, respectively, for $k = 1, \ldots, n-1$. Thus the required number is equal to $\prod_{k=1}^{n-1}(1+\alpha_k)$, the number of possible ways of weighting such a path. But $\alpha_1 = 0$, $\alpha_k = \sum_{j=1}^{k-1} i_j$, and the result follows.

5.2.15. (a) Consider the set M of permutations that are terminated by their largest element. Suppose that $\sigma \in$ M has length $n+1$. From Corollary 5.2.18, we force $n+1$ to be the terminal element of σ by not attaching an elementary fragment corresponding to a modified maximum or double fall to the right-most open edge at any stage prior to $n+1$ in the construction of $t(\sigma)$. Thus, as in **5.2.19**, $\psi_{-1}(k) = \psi_\lambda(k) = k$, and since modified minima and double rises are unaffected, $\psi_1(k) = \psi_\rho(k) = k+1$. From the weighted path lemma, the number of permutations in M of length $n+1$ is $[x^n]J_x[2k+1, (k+1)^2 : (0, \infty)]$, and the result follows since there are $n!$ such permutations.

(b) Suppose that ρ is a permutation on N_n with k cycles, for some $k \geqslant 1$, and that $\sigma = \sigma_1 \cdots \sigma_n$ is the permutation with k upper records that corresponds to ρ under the Foata–Schützenberger decomposition (see [**3.3.17**(c)]). If $\sigma' = (n+1)(n+1-\sigma_1)\cdots(n+1-\sigma_n)$, then a cycle of length 1 consisting of the element $n+1-i$ in ρ corresponds to placing the elementary fragment corresponding to a left-level on the left-most open edge at stage i in constructing $t(\sigma')$. Thus derangements ρ correspond to weighted paths for σ' with possibility functions $\psi_1(k) = k+1$, $\psi_0(k) = 2k$, $\psi_{-1}(k) = k$. This is because right-levels and falls cannot have weight equal to 0, ensuring that $n+1$ is the left-most element in σ' (see Corollary 5.2.18). There are only k remaining choices of weights for falls and right-levels. Moreover, from the above, left-levels cannot be attached to the left-most open edge, so there are only k choices of weights for left-levels. The result follows.

5.2.16. (a) We use the Françon–Viennot decomposition, in which the elementary fragments corresponding to modified maxima must be assigned to the left-most open edge, so $\psi_{-1}(k) = 1$. This ensures that the modified maxima appear in increasing order, and no restrictions are required for the other elementary fragments, so $\psi_0(k) = 2(k+1)$ and $\psi_1(k) = k+1$. Thus the required number is, from the weighted path lemma,

$$[x^{n-1}]J_x[2(k+1), (k+1) : (0, \infty)].$$

(b) In the Françon–Viennot decomposition we have $\psi_0(k) = 0$ since the permutation is alternating. We must assign the elementary fragments corresponding to modified maxima to the left-most open edge so that the modified maxima appear in increasing order. Thus $\psi_{-1}(k) = 1$. To ensure that the modified minima appear in increasing order, since there are no right- or left-levels, and since modified maxima are placed on the left-most open edge, we therefore place the elementary fragments corresponding to the modified minima on the right-most open edge. Thus $\psi_1(k) = 1$, and the required number is

$$[x^{2n}]J_x[0,1:(0,\infty)] = [x^n]S_x[1:(0,\infty)] = \frac{1}{n+1}\binom{2n}{n}$$

from [**5.2.10**(a)].

5.2.17. (a) Let $u = 0$ in **4.2.15**, to give the generating function for sequences with no π_1-strings of length 3 as

$$\left\{(1-x)(1-x^3)^{-1}\circ\gamma(\pi_1)\right\}^{-1} = \left\{\sum_{i\geqslant 0}\gamma_{3i}(\pi_1) - \gamma_{3i+1}(\pi_1)\right\}^{-1}.$$

Applying the <-transformation lemma gives

$$C(x) = \left\{\sum_{i\geqslant 0}\frac{x^{3i}}{(3i)!} - \frac{x^{3i+1}}{(3i+1)!}\right\}^{-1}.$$

(b) If σ is a permutation on N_n with no <-strings of length 3, then $\sigma' = (n+1)\sigma$ is a permutation on N_{n+1} that begins with $n+1$ and has no double rises. We force $n+1$ to be the left-most element as in the solution to [**5.2.15**(b)]. Thus the possibility functions for the permutations σ' are $\psi_{-1}(k) = k$, $\psi_0(k) = k+1$, $\psi_1(k) = k+1$, so

$$\hat{C}(x) = J_x\left[k+1,(k+1)^2:(0,\infty)\right]$$

by the weighted path lemma and the Françon–Viennot decomposition.

(c) Let

$$A(x) = \sum_{i\geqslant 0}\left\{\frac{x^{3i}}{(3i)!} - \frac{x^{3i+1}}{(3i+1)!}\right\}$$

and

$$B(x) = \sum_{i\geqslant 0}\left\{\frac{x^{3i+1}}{(3i+1)!} - \frac{x^{3i+2}}{(3i+2)!}\right\}.$$

Then by multisection of e^x,

$$A(x) = \tfrac{1}{3}\left(e^x + e^{\omega x} + e^{\omega^2 x}\right) - \tfrac{1}{3}\left(e^x + \omega^2 e^{\omega x} + \omega e^{\omega^2 x}\right)$$

$$= \tfrac{1}{3}(1-\omega^2)e^{\omega x} + \tfrac{1}{3}(1-\omega)e^{\omega^2 x}$$

$$B(x) = \tfrac{1}{3}\left(e^x + \omega^2 e^{\omega x} + \omega e^{\omega^2 x}\right) - \tfrac{1}{3}\left(e^x + \omega e^{\omega x} + \omega^2 e^{\omega^2 x}\right)$$

$$= \tfrac{1}{3}(\omega^2 - \omega)\left(e^{\omega x} - e^{\omega^2 x}\right),$$

where $\omega = e^{2\pi i/3}$. Thus

$$A(x)A(y) = -\tfrac{1}{3}\omega^2 e^{\omega x}e^{\omega y} - \tfrac{1}{3}\omega e^{\omega^2 x}e^{\omega^2 y} + \tfrac{1}{3}\left(e^{\omega x}e^{\omega^2 y} + e^{\omega y}e^{\omega^2 x}\right)$$

$$B(x)B(y) = -\tfrac{1}{3}\left(e^{\omega x} - e^{\omega^2 x}\right)\left(e^{\omega y} - e^{\omega^2 y}\right)$$

since $1 + \omega + \omega^2 = 0$. Combining these yields

$$\begin{aligned} A(x)A(y) - B(x)B(y) &= \tfrac{1}{3}(1 - \omega^2)e^{\omega x}e^{\omega y} + \tfrac{1}{3}(1 - \omega)e^{\omega^2 x}e^{\omega^2 y} \\ &= A(x + y), \end{aligned}$$

so

$$\begin{aligned} C(x + y) = \{A(x + y)\}^{-1} &= \frac{1}{A(x)A(y)}\left\{1 - \frac{B(x)B(y)}{A(x)A(y)}\right\}^{-1} \\ &= \sum_{k \geqslant 0} (k!)^2 \left\{\frac{B(x)^k}{k!A(x)^{k+1}}\right\}\left\{\frac{B(y)^k}{k!A(y)^{k+1}}\right\} \end{aligned}$$

and

$$\frac{1}{k!}\frac{B(x)^k}{A(x)^{k+1}} = \frac{x^k}{k!} + \tfrac{1}{2}(k + 1)(k + 2)\frac{x^{k+1}}{(k + 1)!} + \mathcal{O}(x^{k+2}).$$

The result follows from the Stieltjes–Rogers J-fraction theorem.

5.2.18. (a) We use the Françon–Viennot decomposition. Consider a doubled permutation of odd length $2n + 1$. Then this corresponds to a weighted path P with $2n$ steps. Since the permutation is a doubled permutation, then the $(2j + 1)$st. step, s_{2j+1}, and the $(2j + 2)$nd. step, s_{2j+2}, are both rises, or both falls, or both left-levels or both right-levels. Thus, corresponding uniquely to the weighted path P, we have the weighted path P' on n steps, where the jth step in P', s'_j, is the same as s_{2j-1} and s_{2j}. Furthermore, the weight of the step s'_j is the ordered pair of weights of s_{2j-1} and s_{2j}. Let m mark rises in the doubled permutation (not in the associated weighted path) and consider the possibility functions for the steps in P'.

Now each rise in a permutation immediately precedes a double rise or a modified maximum, and each double rise and modified maximum is immediately preceded by a rise, except for the initial element in the permutation, which is either a modified maximum or a double rise. Thus the number of rises in a permutation is equal to one less than the sum of the numbers of modified maxima and double rises in the permutation. This is equal to the sum of the numbers of right-levels and falls in the associated path P, since the maximum element $2n+1$ will always be a modified maximum, and has no corresponding edge in the path.

Now s_{2j-1} must be a step at an even altitude, say $2k$. Then s'_j is the same step at altitude k. If $s_{2j-1}=(1)_{2k}$, then $s_{2j}=(1)_{2k+1}$, so $\psi_1(k)=(2k+1)(2k+2)$ is the number of ordered pairs of weights that may be assigned to a rise at altitude k in P'. If $s_{2j-1}=(-1)_{2k}$, then $s_{2j}=(-1)_{2k-1}$ and $\psi_{-1}(k)=(2k+1)2k$. If $s_{2j-1}=(0)_{2k}$, then $s_{2j}=(0)_{2k}$ and $\psi_\rho(k)=\psi_\lambda(k)=(2k+1)^2$. To record the number of rises in the doubled permutation, mark falls and right-levels in P by m, and thus in P' by m^2. Finally by the path lemma the required number is

$$[x^n m^i]J_x\left[(2k+1)^2(m^2+1),(2k+1)(2k+2)^2(2k+3)m^2:(0,\infty)\right]$$

$$=(-1)^n[x^{2n+1}m^i]\mathrm{sn}(x,m),\qquad \text{from } [\mathbf{5.2.8}(\mathrm{c})].$$

(b) If σ is a doubled permutation on N_{2n} with i rises, then $(2n+1)\sigma$ is a doubled permutation on N_{2n+1} that has $2n+1$ as the left-most element and i rises. We enumerate the latter by modifying part (a) to make $2n+1$ the left-most element. But from the solution to [**5.2.15**(b)], we do this by reducing the possibility functions for falls and right-levels in P by one each, so the possibility functions for P' are

$$\psi_1(k)=(2k+1)(2k+2),\qquad \psi_\lambda(k)=(2k+1)^2,$$

$$\psi_\rho(k)=(2k)^2,\qquad \psi_{-1}(k)=2k(2k-1).$$

Now marking each right-level and fall in P' by m^2, the required number is

$$[x^n m^i]J_x\left[(2k+1)^2+(2k)^2m^2,(2k+1)^2(2k+2)^2m^2:(0,\infty)\right]$$

$$=(-1)^n[x^{2n}m^i]\mathrm{cn}(x,m),\qquad \text{from } [\mathbf{5.2.8}(\mathrm{d})].$$

5.2.19. (a) Let Π be a partition of N_n, each of whose blocks has size equal to 2. The elements of N_n are considered sequentially, from 1 to n, in the following algorithm, which constructs the line segments of the layer diagram from left to right and where placing the smallest element is regarded as "opening" a line segment and placing the largest element as "closing" the segment.

Suppose that there are m open line segments when $i\in\mathsf{N}_n$ is the next element to be considered. If i opens another line segment (that is, i is in a new block, not containing any element in N_{i-1}), then i corresponds to a rise at altitude m with weight equal to 0. If i closes a line segment (that is, i is in a block containing an element in N_{i-1}), then i corresponds to a fall at altitude m with weight j, where $0\leqslant j\leqslant m-1$, and the closed line segment has the

$(j+1)$st. smallest ordinate of the m open line segments. The resulting configuration is a level-free weighted path, and the construction is reversible, so there is a one-to-one correspondence between the partitions and weighted paths with $\psi_1(k) = 1$, $\psi_0(k) = 0$, $\psi_{-1}(k) = k$.

For example, the weighted path corresponding to the partition given in the exercise is

$$((1\,1\,1\,-1\,1\,-1\,-1\,-1),(0\,0\,0\,1\,0\,0\,1\,0))_0.$$

(b) In terms of the layer diagram, the new types of elements, namely, elements of sets of size 1 or internal elements in sets of size greater than 2, are either isolated vertices or nonterminal vertices of the line segments. We call such vertices *transient*, and modify the construction given in (a) to account for them.

Transient elements may be represented by weighted levels, where the weight for a nonterminal vertex is determined by the same means as for falls, and thus takes on values 0 to $k-1$ when there are k open edges. Isolated vertices are recognized by an assignment to a new line segment, an assignment that both opens and closes the line segment so that no other elements may belong to it. Thus for an isolated vertex there is only one choice of weight for the corresponding level and we let the weight be equal to its altitude, so $\psi_0(k) = k+1$. Moreover, $\psi_1(k) = 1$ and $\psi_{-1}(k) = k$, as in (a). This algorithm is called the **Flajolet decomposition.**

As an example, we give the layer diagram and the weighted path corresponding to the partition

$$\{\{1,7,11\},\{2,4,6,9\},\{3\},\{5,10\},\{8\},\{12,13\}\} \text{ of } \mathsf{N}_{13}.$$

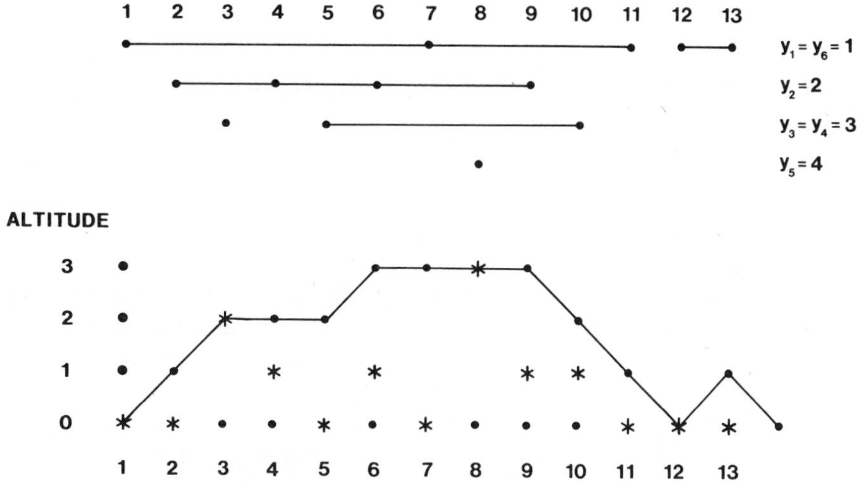

(c) The levels of weight k in part (b) correspond to subsets of size 1 and are marked by u. The rises are marked by v because they correspond to subsets of size at least 2.

(d) The levels of weight k in part (b) are marked by u, and other levels are forbidden because they correspond to internal elements of subsets of size greater than 2. The rises are marked by v.

5.2.20. (a) We use the correspondence in the solution to [**5.2.19**(a)] and let q record the intersection number. Then the number of intersection points in column i is equal to k, the number of open line segments, if i opens a new line segment. This corresponds to a rise at altitude k in the weighted path, which is thus marked by q^k. If i closes the line segment with weight j, then there are j points of intersection in column i, and a fall at altitude k is marked by $\sum_{j=0}^{k-1} q^j = (k)_q$. Thus the required number, from the path lemma, is

$$[x^{2n}q^l]J_x[0, q^k(k+1)_q:(0,\infty)] = [x^n q^l]S_x[q^k(k+1)_q:(0,\infty)].$$

(b) We construct such partitions directly in the following way. The first block consists of the smallest element, namely 1, and the (j_1+1)st. smallest element of the remaining $2n-1$ elements. The ith block consists of the smallest and (j_i+1)st. smallest of those elements that remain after the first $i-1$ blocks have been selected, for $i = 2,\ldots,n$. Thus $1 \leqslant j_i \leqslant 2(n-i)+1$ and the horizontal segment corresponding to the ith block contributes $j_i - 1$ points of intersection to the intersection number, for $i = 1,\ldots,n$. Thus the generating function for partitions of N_{2n} into blocks of size 2 with respect to intersection number is $\prod_{i=1}^{n}\prod_{j=1}^{2(n-i)+1} q^{j-1} = \prod_{i=1}^{n}(2i-1)_q$. The result follows from (a).

5.2.21. If P_k and Q_k are the numerator and denominator polynomials, then, for any $m \geqslant 0$,

$$\sum_{k\geqslant 0} J_k x^k = J^{\langle m\rangle}(x) + \sum_{i\geqslant m+1}\left(J^{\langle i\rangle}(x) - J^{\langle i-1\rangle}(x)\right)$$

$$= P_m Q_m^{-1} + \sum_{i\geqslant m+1} \lambda_0\cdots\lambda_{i-1}x^{2i}Q_i^{-1}Q_{i-1}^{-1}, \qquad \text{from } \mathbf{5.2.21(3)}$$

so

$$Q_m \sum_{k\geqslant 0} J_k x^k = P_m + Q_m \sum_{i\geqslant m+1} \lambda_0\cdots\lambda_{i-1}x^{2i}Q_i^{-1}Q_{i-1}^{-1}.$$

But P_m has degree at most m and $Q_m(0) = 1$ for $m \geqslant 0$. Thus comparing coefficients of x^n in the preceding equation, for $n > m$, gives (a) immediately, as well as the following system of linear equations for $c(l, m)$, $l = 0,\ldots, m+1$,

$$\sum_{l=0}^{m+1} c(l,m)J_{n-l} = \begin{cases} 0, & n = m+1,\ldots,2m+1,\\ \lambda_0\cdots\lambda_m, & n = 2m+2.\end{cases}$$

If the determinant of this system, $\|J_{m+1+i-j}\|_{(m+2)\times(m+2)}$, is denoted by D_m, then from Cramer's rule

$$(-1)^{m+1}\lambda_0 \cdots \lambda_m D_{m-1} D_m^{-1} = c(0, m) = 1.$$

Multiplying this equation for $m = 0, \ldots, k$ and replacing k by $m - 1$ yields (b), since $D_{-1} = 1$ and

$$\|J_{m+i-j}\|_{(m+1)\times(m+1)} = (-1)^{\binom{m+1}{2}} \|J_{i+j-2}\|_{(m+1)\times(m+1)},$$

by reversing the order of columns.

(c) Let $J^{\langle m\rangle}(x) = \sum_{k\geqslant 0} J_{k,m} x^k$. Then from the first equation above, $J_k \equiv J_{k,m} (\text{mod } \lambda_0 \cdots \lambda_m)$ for $k \geqslant 0$. But $Q_m J^{\langle m\rangle} = P_m$, so $\{J_{k,m} | k \geqslant 0\}$ satisfies a linear recurrence equation with characteristic polynomial Q_m. Let $\alpha_k = J_k \text{mod}(\lambda_0 \cdots \lambda_m)$. Then $\{\alpha_k | k \geqslant 0\}$ satisfies the same recurrence, and α_{n+1} is determined uniquely from $\alpha_{n-m}, \ldots, \alpha_n$, for $n \geqslant m$, since Q_m has degree $m + 1$. Thus $\{\alpha_n | n \geqslant 0\}$ is periodic if there exist two strings $\alpha_{i-m}, \ldots, \alpha_i$ and $\alpha_{j-m}, \ldots, \alpha_j$ of $m + 1$ consecutive elements, $i \neq j$. But there are only $(\lambda_0 \cdots \lambda_m)^{m+1}$ distinct such strings, so we must repeat a string by the pigeonhole principle, and the sequence $\{\alpha_n | n \geqslant 0\}$ is periodic past this first repetition.

5.2.22. Put $u = 1$ in [**5.2.6**(a)] to give $\sum_{n\geqslant 0} B_n x^n = J_x[k + 1, k + 1 : (0, \infty)]$. For this J-fraction, $\lambda_0 \cdots \lambda_m = (m + 1)!$ and $Q_3(x) = 1 - 10x + 29x^2 - 24x^3 + x^4$. The results follow from [**5.2.21**(a)] with $m = 3$, and from [**5.2.21**(c)] with $m = k - 1$, respectively.

5.2.23. (a) From **5.2.11**,

$$\sum_{n\geqslant 0} n! x^n = J_x\left[2k + 1, (k + 1)^2 : (0, \infty)\right],$$

so $\|(i + j - 2)!\|_{(m+1)\times(m+1)} = (1! \ldots m!)^2$. The result follows by dividing row i by $(i - 1)!$ and column j by $(j - 1)!$.

(b) From [**5.2.3**(c)], $\sum_{n\geqslant 0} E_{2n} x^{2n} = J_x[0, (k + 1)^2 : (0, \infty)]$. Thus from the proof of [**5.2.21**(c)] with $m = 2$,

$$\sum_{n\geqslant 0} E_{2n} x^{2n} \equiv \left\{1 - x^2(1 - 4x^2)^{-1}\right\}^{-1} \text{mod } 36$$

$$\equiv \left(1 + \sum_{n\geqslant 1} 5^{n-1} x^{2n}\right) \text{mod } 36.$$

5.2.24. Let $A(x, m)$, $B(x, m)$, $C(x, m)$ be the generating functions for the sets A, B, C of stratified, r-stratified, l-stratified permutations, respectively, with x an exponential marker for permutation length and m an ordinary marker for number of rises. If the minimum element in a permutation σ is removed by deleting the corresponding vertex in $t(\sigma)$, then an ordered pair of associated trees remain. This yields

$$\frac{d}{ds}\mathsf{A} \xrightarrow{\sim} \mathsf{C} * \mathsf{B}, \qquad \frac{d}{ds}(\mathsf{B} - \{\varepsilon\}) \xrightarrow{\sim} \mathsf{C} * \mathsf{A}, \qquad \frac{d}{ds}(\mathsf{C} - \{\varepsilon\}) \xrightarrow{\sim} \mathsf{A} * \mathsf{B},$$

where differentiation is realized **(3.2.20)** by deleting the minimum element. But the minimum element contributes a single rise when the right subtree is nonempty, so from the preceding decompositions

$$\frac{\partial}{\partial x}A = C\{m(B-1)+1\}, \qquad \frac{\partial}{\partial x}B = mCA, \qquad \frac{\partial}{\partial x}C = A\{m(B-1)+1\}.$$

Now let $D(x, m) = m(B(x, m) - 1) + 1$, giving

$$\frac{\partial}{\partial x}A = CD, \qquad \frac{\partial}{\partial x}D = m^2CA, \qquad \frac{\partial}{\partial x}C = AD,$$

where $A(0, m) = 0$, $C(0, m) = D(0, m) = 1$. By comparison with [**5.2.8**(a), (b)], we obtain from **1.1.5(3)**

$$A(x, m) = -i\,\mathrm{sn}(ix, m), \qquad C(x, m) = \mathrm{cn}(ix, m), \qquad D(x, m) = \mathrm{dn}(ix, m),$$

and the results follow.

SECTION 5.3

5.3.1. (a) We have $\Phi_{\mathsf{S}}(x, y, z) = zy^{-1}f(y)$ where $f(y) = x(1 + xy^3)$. Thus, from the lattice path theorem,

$$\begin{aligned}\Phi_{\mathsf{S}_-}(x, y, z) &= \{1 - y^{-1}w(z)\}^{-1}, \qquad \text{where } w = zf(w)\\ &= 1 + \sum_{m\geq 1} y^{-m}w^m\\ &= 1 + \sum_{m\geq 1} y^{-m}\sum_{n\geq 1}\frac{m}{n}z^n[\lambda^{n-m}]x^n(1+x\lambda^3)^n\end{aligned}$$

from the Lagrange theorem, and the result follows with $m = n - 3i$.

(b) From the lattice path theorem

$$\begin{aligned}\Phi_{\mathsf{S}_0}(x, y, z) &= z^{-1}w(z)f^{-1}(0) = x^{-1}z^{-1}w(z)\\ &= x^{-1}z^{-1}\sum_{m\geq 1}\frac{z^m}{m}[\lambda^{m-1}]x^m(1+x\lambda^3)^m, \qquad \text{from the Lagrange theorem}\\ &= x^{-1}z^{-1}\sum_{n\geq 0}\frac{z^{3n+1}}{3n+1}x^{3n+1}\binom{3n+1}{n}x^n,\end{aligned}$$

and the result follows.

(c) From the lattice path theorem

$$\Phi_{S_+}(x, y, z) = \{1 - y^{-1}zf(y)\}^{-1}\left(z^{-1}x^{-1}w(z)\right)^{-1}\left(1 - y^{-1}w(z)\right)$$

$$= \{1 - y^{-1}zf(y)\}^{-1}\left\{\left(z^{-1}x^{-1}w(z)\right)^{-1} - zxy^{-1}\right\}.$$

But

$$\{1 - y^{-1}zf(y)\}^{-1} = \sum_{m, i \geqslant 0} \binom{m}{i} x^{m+i} z^m y^{3i-m}$$

and

$$\left(z^{-1}x^{-1}w(z)\right)^{-1} = 1 - zx \sum_{n \geqslant 1} \frac{z^n}{n} [\lambda^{n+1}] x^n (1 + x\lambda^3)^n,$$

from the Lagrange theorem

$$= 1 - \sum_{k \geqslant 0} \frac{1}{3k+2} \binom{3k+2}{k+1} x^{4(k+1)} z^{3(k+1)},$$

and the result follows.

5.3.2. We have $\Phi_S(x, y, z) = zy^{-1}x(1 + x^a y^b)$ so that $f(y) = x(1 + x^a y^b)$ in the terminology of the lattice path theorem. Thus

$$\Phi_{S_0}(x, y, z) = z^{-1}w(z)f^{-1}(0) = z^{-1}x^{-1}w(z)$$

$$= z^{-1}x^{-1} \sum_{n \geqslant 1} \frac{z^n}{n} [\lambda^{n-1}] x^n (1 + x^a \lambda^b)^n$$

$$= \sum_{k \geqslant 0} \frac{1}{bk+1} \binom{bk+1}{k} x^{(a+b)k} z^{bk},$$

and the result follows.

5.3.3. We have $\Phi_S(1, y, z) = zy^{-1}(1 + y + y^2)$, so that $\Phi_{S_0}(1, y, z) = z^{-1}w(z)$, where $w = z(1 + w + w^2)$. Thus

$$\Phi_{S_0}(1, y, z) = \left\{1 - z - (1 - 2z - 3z^2)^{1/2}\right\}/2z^2,$$

where the negative sign has been selected since $\Phi_{S_0}(1, y, z) \in \mathbb{Q}[[z]]$. The coefficients in this series are Motzkin numbers. (See **[5.2.10]**.)

5.3.4. By considering a 1-shift and reverse of these paths, we obtain the zero-paths on step-set $\mathsf{S}^{(1)} = \{(i, i-j) \mid i, j \geqslant 0, i+j \neq 0\}$, for which

$$\Phi_{(\mathsf{S}^{(1)})^*}(x, y, 1) = \left\{1 - \sum_{\substack{i\geqslant 0 \\ i+j\neq 0}} \sum_{j \geqslant 0} x^i y^{j-i}\right\}^{-1}$$

$$= \left\{2 - \left(1 - xy^{-1}\right)^{-1}(1-y)^{-1}\right\}^{-1}$$

$$= \left(1 - xy^{-1}\right)(1-y)\left\{1 - y^{-1}(2x - 2xy + 2y^2)\right\}^{-1}.$$

Now let $\Phi_{\mathsf{T}^*}(x, y, 1) = \{1 - y^{-1}(2x - 2xy + 2y^2)\}^{-1}$. Then

$$\Phi_{(\mathsf{S}^{(1)})_0} = \Phi_{\mathsf{T}_0}, \qquad \Phi_{(\mathsf{S}^{(1)})_+} = (1-y)\Phi_{\mathsf{T}_+}, \qquad \Phi_{(\mathsf{S}^{(1)})_-} = \left(1 - xy^{-1}\right)\Phi_{\mathsf{T}_-}$$

since the factorization is unique. Thus we apply the lattice path theorem with $f(y) = 2(x - xy + y^2)$. Then

$$\Phi_{\mathsf{T}_0}(x, y, z) = z^{-1}w(z)f^{-1}(0) = \tfrac{1}{2}x^{-1}z^{-1}w(z)$$

where $w = 2z(x - xw + w^2)$. Solving this quadratic,

$$w(1) = \tfrac{1}{4}\left\{1 + 2x - \left(1 - 12x + 4x^2\right)^{1/2}\right\}$$

where the negative sign has been chosen so that Φ_{T_0} avoids negative powers of x. The result follows.

5.3.5. (a) By considering an $(m-1)$-shift of the paths in this problem, we obtain the paths from $(0,0)$ to $(n,0)$ on step-set $\{(0,1), (1, -(m-1))\}$, which never go above $y = 0$. If we reverse these paths to get zero-paths, then the step-set is not suitable for application of our main result unless $m = 2$. However, if we reflect these paths about $y = 0$, then we get the zero-paths from $(0,0)$ to $(n,0)$ with step-set $\mathsf{S} = \{(0, -1), (1, m-1)\}$, which can be enumerated by the lattice path theorem. We have $\Phi_{\mathsf{S}} = z(y^{-1} + xy^{m-1}) = zy^{-1}f(y)$, where $f(y) = 1 + xy^m$. Then

$$\Phi_{\mathsf{S}_0} = z^{-1}w(z)f^{-1}(0) = z^{-1}w(z), \qquad \text{where } w = zf(w)$$

$$= \sum_{n \geqslant 1} \frac{z^{n-1}}{n}[\lambda^{n-1}](1 + x\lambda^m)^n, \qquad \text{from the Lagrange theorem}$$

$$= \sum_{k \geqslant 0} \frac{1}{km+1}\binom{km+1}{k} z^{km} x^k,$$

and the result follows.

(b) If these paths are reflected about the line $y = x$, then μ-shifted and reflected about $y = 0$, then the reverses of the resulting paths are minus-paths on step-set $\{(0, -1), (1, \mu)\}$, from $(0,0)$ to $(n, \mu n - m)$. The required number is thus

$$[z^{m+n}x^n y^{\mu n-m}]\{1 - y^{-1}w(z)\}^{-1}$$

$$= [z^{m+n}x^n]w^{m-n\mu}, \qquad \text{where } w = z(1 + xw^{\mu+1})$$

$$= [x^n]\frac{m - n\mu}{m+n}[\lambda^{n(\mu+1)}](1 + x\lambda^{\mu+1})^{m+n}, \qquad \text{from the Lagrange theorem}$$

$$= \frac{m - \mu n}{m+n}\binom{m+n}{n}, \qquad \text{as required.}$$

(c) As in part (b), we wish to count minus-paths from $(0,0)$ to $(n, \mu n - m)$, on step-set $\{(0, -1), (1, \mu), (1, \mu - 1)\}$, that contain r occurrences of the step $(1, \mu - 1)$, and thus contain $m + n - r$ steps. Thus the required number is

$$[z^{m+n-r}x^n y^{\mu n-m}t^r]\{1 - y^{-1}w(z)\}^{-1}$$

$$= [z^{m+n-r}x^n t^r]w^{m-\mu n}, \qquad \text{where } w = z(1 + xtw^{\mu} + xw^{\mu+1})$$

$$= [x^n t^r]\frac{m - \mu n}{m+n-r}[\lambda^{n(\mu+1)-r}](1 + xt\lambda^{\mu} + x\lambda^{\mu+1})^{m+n-r},$$

from the Lagrange theorem, and the result follows.

Parts (a), (b), and (c) are generalized ballot problems.

5.3.6. Let a lead code be represented by a lattice path on step-set $\{(1,0), (0,1)\}$, beginning at the origin, where an increase of 1 between adjacent elements is represented by $(1,0)$ and a decrease of 1 by $(0,1)$. Then the paths corresponding to lead codes of length $2n + 1$ are the paths from $(0,0)$ to (n, n) that never go above the line $y = x$. This number is

$$\frac{1}{2n+1}\binom{2n+1}{n} = \frac{1}{n+1}\binom{2n}{n},$$

from [**5.3.5**(a)].

5.3.7. It is clear that the number of paths from $(0,0)$ to (n, n), but never above $y = x$ is $N_1 - N_2$, where N_1 is the total number of paths from $(0,0)$ to (n, n) and N_2 is the number of these that touch the line $y = x + 1$. Thus $N_1 = \binom{2n}{n}$, and the paths counted by N_2 have a first point at which they touch the line $y = x + 1$. If we reflect the initial portion of the path, to the first point on $y = x + 1$, about the line $y = x + 1$, then we obtain a path from $(-1, 1)$ to (n, n), and this procedure is reversible. Thus N_2 is the number of paths from

$(-1,1)$ to (n,n), so that $N_2 = \binom{2n}{n-1}$. This method of proof is called the **reflection principle**.

SECTION 5.4

5.4.1. Since the largest part is at most m, set $x_{m+1} = x_{m+2} = \cdots = 0$ in **5.4.7**. Then set $x_i = q^i$ for $1 \leqslant i \leqslant m$, and the required number is

$$[q^N]\|h_{\alpha_j+i-j}\|_{n\times n},$$

where

$$h_t = [x^t]\prod_{k=1}^{m}(1-xq^k)^{-1} = q^t\binom{m+t-1}{m-1}_q$$

from [**2.5.17**], and the result follows.

5.4.2. We use induction on $l = \alpha_1$. (i) When $l = 0$, the result is trivially true. (ii) Let $(\boldsymbol{\alpha}^*, \mathbf{1})$ be the shape of column-strict plane partitions obtained by deleting the first columns of column-strict plane partitions of shape $(\boldsymbol{\alpha}, \mathbf{1})$. We assume that the result is true for all column-strict plane partitions of shape $(\boldsymbol{\beta}, \mathbf{1})$, where $\boldsymbol{\beta} \leqslant \boldsymbol{\alpha}$, so it is true for those of shape $(\boldsymbol{\alpha}^*, \mathbf{1})$.

Now $\alpha_i^* = \alpha_i - 1$ when $\alpha_i > 0$. If the hook lengths of $\boldsymbol{\alpha}^*$ are $d_1,\ldots,d_{p-n}$, then the hook lengths of $\boldsymbol{\alpha}$ are $d_1,\ldots,d_{p-n}, \alpha_1+n-1, \alpha_2+n-2,\ldots,\alpha_n$, making p hook lengths in all. In addition, if the contents of $\boldsymbol{\alpha}^*$ are $c_1,\ldots,c_{p-n}$ then the contents of $\boldsymbol{\alpha}$ are $c_1+1,\ldots,c_{p-n}+1, 0, -1,\ldots,-n+1$, again making p in all.

We shall have proved the result by induction if we show that

$$H_m(\boldsymbol{\alpha}) = q^{\sum_{i=1}^n i\alpha_i}\frac{(m+1+c_1)_q\cdots(m+1+c_{p-n})_q(m)_q\cdots(m-n+1)_q}{(d_1)_q\cdots(d_{p-n})_q(\alpha_1+n-1)_q\cdots(\alpha_n)_q}$$

or, equivalently,

$$H_m(\boldsymbol{\alpha}) = q^{\binom{n+1}{2}}\frac{(m)_q(m-1)_q\cdots(m-n+1)_q}{(\alpha_1+n-1)_q\cdots(\alpha_n)_q}H_{m+1}(\boldsymbol{\alpha}^*)$$

where

$$H_{m+1}(\boldsymbol{\alpha}^*) = \left\| q^{\alpha_i-1-i+j}\binom{m+\alpha_i-i+j-1}{m}_q \right\|_{n\times n},$$

by the induction hypothesis.

To do this divide the jth column of the matrix appearing in $H_m(\boldsymbol{\alpha})$ by $q^{n-j+1}(m-n+j)_q$ and multiply the ith row by $(\alpha_i-i+n)_q$ for $1 \leqslant i, j \leqslant n$,

so

$$H_m(\boldsymbol{\alpha}) = q^{\binom{n+1}{2}} \frac{(m)_q \cdots (m-n+1)_q}{(\alpha_1+n-1)_q \cdots (\alpha_n)_q}$$

$$\times \left\| q^{\alpha_i - i + 2j - n - 1} \binom{m+\alpha_i - i + j - 1}{m-1}_q \frac{(\alpha_i - i + n)_q}{(m-n+j)_q} \right\|_{n\times n}.$$

Now denote the ith column of the preceding determinant by col_i, and apply the column operation

$$\mathrm{col}_{n-j} := \mathrm{col}_{n-j} - \frac{(j)_q}{q^{j+1}(m-j)_q}\mathrm{col}_{n+1-j}$$

for $j = 1, \ldots, n$ in order. Thus the preceding determinant is equal to

$$\left\| q^{\alpha_i - 1 - i + j} \binom{m+\alpha_i - i + j - 1}{m}_q \right\|_{n\times n} = H_{m+1}(\alpha^*)$$

by the induction hypothesis. The result follows.

5.4.3. From **5.4.8**, $f^{\boldsymbol{\alpha}} = p!\|1/(\alpha_i - i + j)!\|_{n\times n}$, where $p = \alpha_1 + \cdots + \alpha_n$ is the number of cells in Young tableaux of shape $\boldsymbol{\alpha}$. We evaluate this determinant indirectly as follows. First observe that

$$\left\| \frac{1}{(\alpha_i - i + j)!} \right\|_{n\times n} = \left\{ (1-q)^p \lim_{m\to\infty} H_m(\boldsymbol{\alpha}) \right\} \Big|_{q=1}$$

from [**5.4.1**], since

$$\lim_{m\to\infty} \binom{m+\alpha_j - j + i - 1}{m-1}_q = \frac{(1-q)^{-(\alpha_j - j + i)}}{(\alpha_j - j + i)!_q}$$

by Proposition 2.6.13. But, from [**5.4.2**],

$$\lim_{m\to\infty} \frac{q^a (m+c_1)_q \cdots (m+c_p)_q}{(d_1)_q \cdots (d_p)_q} = \frac{q^a (1-q)^{-p}}{(d_1)_q \cdots (d_p)_q}$$

so $\{(1-q)^p \lim_{m\to\infty} H_m(\alpha)\}|_{q=1} = 1/d_1 \cdots d_p$. The result follows immediately.

5.4.4. (a) Let $\mathsf{B}_{r,c}^{(m)}$ be the set of column-strict plane partitions of shape $(\boldsymbol{\alpha}, \mathbf{1})$, where $\boldsymbol{\alpha} = (\alpha_1, \ldots, \alpha_r) = (c, \ldots, c)$ and largest part at most m. Let $\mathsf{O}_{r,c}^{(m)}$ be the set of ordinary plane partitions with at most r rows, at most c columns, and largest part at most m. Then

$$\mathsf{B}_{r,c}^{(m)} \xrightarrow{\sim} \mathsf{O}_{r,c}^{m-r} : (p_{ij}) \mapsto (p_{ij} - r + i - 1)$$

is a decomposition for $\mathsf{B}_{r,c}^{(m)}$. Moreover, if $\mathbf{p}$ is a plane partition of n, then $\mathbf{p}'$ is a plane partition of $n - c\binom{r+1}{2}$, where $\mathbf{p}'$ is the image of $\mathbf{p}$ under this decomposition. Thus the required number is $[q^M]G_m(r, c)$, where

$$H_m(\boldsymbol{\alpha})q^{-c\binom{r+1}{2}} = G_{m-r}(r, c)$$

whence $G_m(r, c) = H_{m+r}(\boldsymbol{\alpha})q^{-c\binom{r+1}{2}}$.

(b) From **[5.4.2]**,

$$H_{m+r}(\boldsymbol{\alpha}) = q^a \frac{(m + c_1)_q \cdots (m + c_p)_q}{(d_1)_q \cdots (d_p)_q},$$

where $a = \sum_{i=1}^r i\alpha_i$ and c_i, d_i are the content and hook length associated with cell i for $1 \leqslant i \leqslant p$. The hook lengths and contents for each cell when $\boldsymbol{\alpha} = (c, \ldots, c)$ are as follows:

$$\begin{array}{llllll} c+r-1, & c+r-2, \ldots, & r+1, & r & 0, 1, 2, \ldots, & c-1 \\ c+r-2, & c+r-3, \ldots, & r, & r-1 & -1, 0, 1, \ldots, & c-2 \\ \vdots & \vdots & \vdots & \vdots & \vdots & \vdots \\ c, & c-1, \ldots, & 2, & 1 & -r+1, -r+2, \ldots, & c-r. \end{array}$$

Thus

$$q^{-c\binom{r+1}{2}} H_{m+r}(\boldsymbol{\alpha}) = \prod_{1 \leqslant j \leqslant c} \prod_{1 \leqslant i \leqslant r} (m + r + j - i)_q/(i + j - 1)_q$$

$$= \prod_{1 \leqslant j \leqslant c} \binom{m + r + j - 1}{r}_q \Big/ \binom{r + j - 1}{r}_q$$

and the result follows.

5.4.5. (a) The required number is $[q^M]F_r(q)$ where, from **[5.4.4**(b)**]**,

$$F_r(q) = \lim_{\substack{c \to \infty \\ m \to \infty}} G_m(r, c) = \prod_{j \geqslant 1} \prod_{i=1}^{r} (1 - q^{i+j-1})^{-1}$$

$$= \prod_{k \geqslant 1} (1 - q^k)^{-\min(k, r)}, \qquad \text{by Proposition 2.6.13.}$$

(b) The required number is $[q^M]\lim_{r\to\infty} F_r(q)$ and the result follows immediately from (a).

5.4.6. Let $\mathbf{b} = (b_1, \ldots, b_n)$ where $1 \leqslant b_i \leqslant m - k + 1$ for $i = 1, \ldots, n$. Let $\mathsf{B}_n(\mathbf{b})$ be the set of binary sequences $\sigma_1 \cdots \sigma_{mn}$ in which every substring of length m contains at least k 1's, and $\sigma_{(m-1)i+1} \cdots \sigma_{mi}$ contains exactly $b_i + k - 1$ 1's for $i = 1, \ldots, n$. Suppose $\mathsf{P} = (P_1, \ldots, P_n)$, $\mathsf{Q} = (Q_1, \ldots, Q_n)$ where $P_i = (s_{i-1}, s_{i-1})$, and $Q_i = (s_i + k - 1, s_i + k - 1 - m)$ for $i = 1, \ldots, n$. Now $P_1, \ldots, P_n$ and $Q_1, \ldots, Q_n$ are arranged in increasing order along the lines $y = x$ and $y = x - m$, respectively, so (P, Q) is proper. Moreover

$$\mathsf{B}_n(\mathbf{b}) \xrightarrow{\sim} \bar{\mathsf{L}}(\mathsf{P}, \mathsf{Q}) : \sigma_1 \cdots \sigma_{mn} \mapsto (w_1, \ldots, w_n)$$

where the $(1,0)$- and $(0,-1)$-steps in w_i correspond, respectively, to the 1's and 0's in $\sigma_{(m-1)i+1} \cdots \sigma_{mi}$, for $i = 1, \ldots, n$. But the number of paths in L_{ij} is

$$\Phi_{\mathsf{L}_{ij}}(1) = \binom{(s_j + k - 1 - s_{i-1}) + s_{i-1} - (s_j + k - 1 - m)}{s_j + k - 1 - s_{i-1}}$$

$$= \binom{m}{s_j + k - 1 - s_{i-1}}$$

where $\binom{m}{l} = 0$ for $l > m$ or $l < 0$. The result follows by the nonintersecting n-path theorem.

5.4.7. The terminus of such a lattice polygon (π_1, π_2) is $(k + 1, n - k + 1)$. Moreover, π_1 has initial step $(0, 1)$ and final step $(1, 0)$, and π_2 has initial step $(1, 0)$ and final step $(0, 1)$. Thus if these initial and final steps are deleted, the set of lattice polygons becomes the set of nonintersecting 2-paths on step-set $\{(0,1),(1,0)\}$, with end-type $(\{(0,1),(1,0)\},\{(k, n-k+1),(k+1, n-k)\})$. The nonintersecting n-path theorem clearly holds for this modified step-set, and the end-type is proper. But the number of paths on step-set $\{(0,1),(1,0)\}$ from (i, j) to (l, m) is $\binom{l-i+m-j}{l-i}$ if $l \geqslant i$, $m \geqslant j$, so the required number is

$$\begin{vmatrix} \binom{n}{k} & \binom{n}{k+1} \\ \binom{n}{k-1} & \binom{n}{k} \end{vmatrix} = \frac{1}{n+1}\binom{n+1}{k}\binom{n+1}{k+1}.$$

(b) From part (a), the required number is

$$\sum_{k=0}^{n} \frac{1}{n+1}\binom{n+1}{k}\binom{n+1}{k+1} = \frac{1}{n+1}\sum_{k=0}^{n}\binom{n+1}{k}\binom{n+1}{n-k}$$

$$= \frac{1}{n+1}\binom{2n+2}{n} = \frac{1}{n+2}\binom{2n+2}{n+1} \qquad \text{from } \mathbf{1.1.6(8)}.$$

5.4.8. We follow the proof of Theorem 5.4.5, but in this case the sets of paths L_{ij} have an additional restriction, that the altitude of the $(1,0)$ step at x-coordinate i is bounded above by ω_{i-n+1} and below by μ_i. The nonintersecting n-paths correspond precisely to column bounded plane partitions and $\Phi_{\mathsf{L}_{ij}} = \theta_{ij}$, so the result follows.

SECTION 5.5

5.5.1. (a) This is similar to Decomposition 5.3.3, in which we identify the first and last occurrence of vertices at maximum altitude, instead of at minimum altitude. The set of plus-paths is $(\varepsilon \cup \mathsf{P})$ and the set of zero-paths is B. The set of minus-paths is F^*, from the proof of Decomposition 5.5.1.

(b) From **5.5.2**, $f(z) = t\Psi_{\mathsf{F}}(z, 1, t)$ satisfies the functional equation $f = zg(f)$, where $g(y) = \sum_{n \geqslant 0} g_n y^n$. Now $\Psi_{\mathsf{M}}(z, 1, t) = zt^{-1}g(t)$ and $\Psi_{\mathsf{B}}(z, 1, t) = (zg_0)^{-1}f(z)$. It follows from (a) that

$$\{1 - zt^{-1}g(t)\}^{-1} = \{1 + S(z, t)\}(zg_0)^{-1}f(z)\{1 - t^{-1}f(z)\}^{-1},$$

where $S(z, t) = \Psi_{\mathsf{S}}(z, 1, t)$. Taking the logarithm gives

$$\sum_{n \geqslant 1} \frac{z^n}{n}\{t^{-1}g(t)\}^n = \log(1 + S) + \log\{(zg_0)^{-1}f(z)\} + \sum_{k \geqslant 1} \frac{1}{k}\{t^{-1}f(z)\}^k.$$

Now S has only strictly positive powers of t, and $(zg_0)^{-1}f(z)$ is independent of t. Thus, applying $[z^n t^{-k}]$, where $n, k \geqslant 0$, yields

$$\frac{1}{n}[t^{n-k}]g^n(t) = \frac{1}{k}[z^n]f^k(z).$$

This is the Lagrange theorem (in one variable).

We also note that, by applying $[z^n]$ for $n \geqslant 0$,

$$[z^n]\log\{(zg_0)^{-1}f(z)\} = \frac{1}{n}[t^0]\{g(t)/t\}^n = \frac{1}{n}[t^n]g(t).$$

This is a well-known result given by Lagrange.

5.5.2. Consider the lattice polygon

$$(u, v) = ((u_0 \cdots u_{n-1})_{(0,0)}, (v_0 \cdots v_{n-1})_{(0,0)})$$

with semiperimeter n. We associate with this a unique path $p = p_0 \cdots p_{n-1}$, where for $i = 0,\ldots, n-1$, if

$$(u_i, v_i) = \begin{cases} ((0,1),(1,0)) & \text{then } p_i \text{ is a rise} \\ ((1,0),(1,0)) & \text{then } p_i \text{ is a left-level} \\ ((0,1),(0,1)) & \text{then } p_i \text{ is a right-level} \\ ((1,0),(0,1)) & \text{then } p_i \text{ is a fall.} \end{cases}$$

The path $p' = p_1 \cdots p_{n-2}$ is a path from altitude 1 to altitude 1, as the term is used in Section 5.2. This can be seen by letting $k_i = \operatorname{alt}(p_i) = \operatorname{alt}(u_i) - \operatorname{alt}(v_i)$. Then $k_1 = k_{n-1} = 1$, $k_i \geqslant 1$, for $i = 1,\ldots, n-1$, and if u_i, v_i are parallel, divergent, or convergent steps, k_{i+1} equals k_i, $k_i + 1$, or $k_i - 1$, respectively. Moreover,

$$\begin{aligned} \operatorname{area}(u, v) &= \sum_{i=0}^{n-1} \operatorname{alt}(u_i) - \sum_{j=0}^{n-1} \operatorname{alt}(v_i) \\ &= \sum_{i=0}^{n-1} \operatorname{alt}(p_i) = 1 + \sum_{i=1}^{n-1} \operatorname{alt}(p_i), \end{aligned}$$

and the terminal abscissa of (u, v) is equal to the sum of the number of rises and left-levels in p', plus 1. The semiperimeter of (u, v) is equal to the length of p', plus 2. Thus, if z marks semiperimeter, q marks area, and s marks the terminal abscissa, then rises at altitude k in p' are marked by szq^k, falls at altitude k by zq^k, and levels at altitude k by $szq^k + zq^k = (s+1)zq^k$. It follows from the path lemma (**5.2.7**) that

$$P(z, q, s) = qsz^2 J_z\left[(1+s)q^k, sq^{2k+1} : (1, \infty)\right].$$

(b) In the terminology of Corollary 5.5.5, $g_1 = 1 + s$, $g_2 = s$, so $g(y, q) = (1+y)(1+sy)$. It follows that

$$G_k(z, q) = \sum_{n \geqslant 0} q^{\binom{n+k+1}{2}} z^{n+k} [t^n] \prod_{i=1}^{n+k} (1 + q^{-i}t)(1 + q^{-i}st)$$

$$= \sum_{n \geqslant 0} q^{\binom{n+k+1}{2}} z^{n+k} [t^n] \sum_{i,j=0}^{n+k} q^{\binom{i+1}{2} - \binom{j+1}{2}} \binom{n+k}{i}_{q^{-1}} \binom{n+k}{j}_{q^{-1}} s^j t^{i+j}$$

by **2.6.12(1)**. After routine manipulation

$$G_k(z, q) = q^{k(k+1)/2} \sum_{n \geqslant 0} \sum_{j=0}^{n} q^{-j(n-j)} \binom{n+k}{j}_q \binom{n+k}{j+k}_q s^j z^{n+k}.$$

Now $zqJ_z[(1+s)q^k, sq^{2k+1} : (1, \infty)] = G_1(z, q)/G_0(z, q)$ by Corollary 5.5.5(1), and the result follows.

(c) From Corollary 5.5.5(2),

$$(1+s)z + P(z, q^{-1}, s) + P(z, q, s) = 1 + G_0^{-1}(z, q)$$

$$= 1 - \left\{ \sum_{n \geqslant 0} \sum_{j=0}^{n} q^{-j(n-j)} \binom{n}{j}_q^2 s^j z^n \right\}^{-1},$$

from part (b), and the result follows.

5.5.3. From [**5.5.2**], the required number is

$$[z^{n+2} s^{k+1}] P(z, 1, s) = [z^{n+1} s^k] f,$$

where $f = zJ_z[s+1, s:(1, \infty)]$, so f satisfies the equation $f = z(1+sf)(1+f)$. It follows from the Lagrange theorem that

$$[z^{n+1} s^k] f = \frac{1}{n+1} [s^k t^n] \{(1+t)(1+st)\}^{n+1}$$

$$= \frac{1}{n+1} [s^k] \sum_{i+j=n} \binom{n+1}{i} \binom{n+1}{j} s^j,$$

and the result follows.

5.5.4. (a) It follows from [**3.3.48**(b)] that

$$J(x, q) = xE\left(\frac{xq}{q-1}, q\right) \Big/ E\left(\frac{x}{q-1}, q\right),$$

where

$$E(x, q) = \sum_{k \geqslant 0} q^{\binom{k}{2}} \frac{x^k}{k!}.$$

But

$$E(ux, q) = \sum_{k \geqslant 0} q^{\binom{k}{2}} \frac{x^k}{k!} \sum_{i=0}^{k} (u-1)^i \binom{k}{i}$$

$$= \sum_{i \geqslant 0} \frac{1}{i!} (u-1)^i x^i \sum_{k \geqslant i} q^{\binom{k}{2}} \frac{x^{k-i}}{(k-i)!}$$

$$= \sum_{i \geqslant 0} \frac{1}{i!} (u-1)^i x^i q^{\binom{i}{2}} \sum_{k \geqslant i} q^{\binom{k-i}{2}} \frac{1}{(k-i)!} (q^i x)^{k-i}$$

$$= \sum_{i \geqslant 0} \frac{1}{i!} (u-1)^i x^i q^{\binom{i}{2}} E(q^i x, q),$$

whence

$$E\left(\frac{ux}{u-1}, q\right) = \sum_{i \geqslant 0} \frac{1}{i!} x^i q^{\binom{i}{2}} E\left(\frac{q^i x}{u-1}, q\right).$$

Setting $u = q$ gives

$$J(x, q) = x \sum_{i \geqslant 0} \frac{1}{i!} x^i q^{\binom{i}{2}} E\left(\frac{q^i x}{q-1}, q\right) \Big/ E\left(\frac{x}{q-1}, q\right).$$

But

$$J^{(i)}(x, q) = x^i q^{\binom{i}{2}} E\left(\frac{q^i x}{q-1}, q\right) \Big/ E\left(\frac{x}{q-1}, q\right),$$

so $J(x, q)$ satisfies the functional equation

$$J(x, q) = x \sum_{i \geqslant 0} \frac{1}{i!} J^{(i)}(x, q).$$

(b) Part (a) gives a functional equation that may be solved by means of the q-Lagrange theorem with $g(t, q) = q^{-1}e^t$. Now

$$\begin{aligned}[t^n] g_{(n+k)}(q^{-1}t, q) &= [t^n] q^{-(n+k)} \exp\{t(q^{-1} + \cdots + q^{-(n+k)})\} \\ &= \frac{1}{n!} q^{-(n+k)} \{q^{-1} + \cdots + q^{-(n+k)}\}^n \\ &= \frac{1}{n!} q^{-(n+1)(n+k)} \{1 + q + \cdots + q^{n+k-1}\}^n\end{aligned}$$

and the result follows from Theorem 5.5.4(2).

References

Abramson, M. (1975). A note on permutations with fixed pattern, *J. Comb. Theory* (*A*) **19**, 237–239.

Abramson, M., and W. O. J. Moser (1967). Permutations without rising or falling *w*-sequences, *Ann. Math. Statist.* **38**, 1245–1254.

Aeppli, A. (1923). A propos de l'interprétation géometrique du problème du scrutin, *Enseignement Math.* **23**, 328–329.

Anand, H., V. C. Dumir, and H. Gupta (1966). A combinatorial distribution problem, *Duke Math. J.* **33**, 757–770.

André, D. (1881). Mémoire sur les permutations alternées, *J. Math.* **7**, 167–184.

André, D. (1887). Solution directe du problème résolu par M. Bertrand, *C. R. Acad. Sci. Paris* **105**, 436–437.

Andrews, G. E. (1967). A generalization of a partition theorem of MacMahon, *J. Comb. Theory* **3** 100–101.

Andrews, G. E. (1971). On the foundations of combinatorial theory, V: Eulerian differential operators, *Studies Appl. Math.* **50**, 345–375.

Andrews, G. E. (1975a). The theory of compositions, II: Simon Newcomb's problem, *Utilitas Math.* **7**, 33–54.

Andrews, G. E. (1975b). Identities in combinatorics, II: A q-analog of the Lagrange inversion theorem, *Proc. Amer. Math. Soc.* **53**, 240–245.

Andrews, G. E. (1976). *The theory of partitions*, in *Encyclopedia of Mathematics and Its Applications*, Vol. 2, G.-C. Rota, Ed., Addison-Wesley, Reading, Mass.

Bender, E. A. (1974). Partitions of multisets, *Discrete Math.* **9**, 301–312.

Bender, E. A., and J. R. Goldman (1971). Enumerative uses of generating functions, *Indiana Univ. Math. J.* **20**, 753–764.

Biggs, N. L., E. K. Lloyd, and R. J. Wilson (1976). *Graph Theory* 1736–1936, Oxford University Press, Oxford.

Blum, J. (1974). Enumeration of the square permutations in S_n, *J. Comb. Theory* (*A*) **17**, 156–161.

Bognár, J., J. Mogyoródi, A. Prékopa, A. Rényi, and D. Szász (1970). *Problem Book on Probability* (in Hungarian), Tankönyvkiadó, Budapest.

Breach, D. R., et al. (1976). Solution to problem 75-4, *SIAM Rev.* **18**, 303–304.

Bondy, J. A., and U. S. R. Murty (1976). *Graph Theory with Applications*, Macmillan, London.

Bressoud, D. M. (1980). Analytic and combinatorial generalizations of the Rogers–Ramanujan identities, *Amer. Math. Soc. Mem.* **227**.

Brooks, R. L., C. A. B. Smith, A. H. Stone, and W. T. Tutte (1940). The dissection of rectangles into squares, *Duke Math. J.* **7**, 312–340.

Brown, W. G. (1963). Enumeration of non-separable planar maps, *Canad. J. Math.* **15**, 526–545.

Brown, W. G. (1964). Enumeration of triangulations of the disk, *Proc. Lond. Math. Soc.* (3) **14**, 746–768.

Brown, W. G., and W. T. Tutte (1964). On the enumeration of rooted non-separable planar maps, *Canad. J. Math.* **16**, 572–577.

de Bruijn, N. G., and T. van Aardenne-Ehrenfest (1951). Circuits and trees in oriented linear graphs, *Simon Stevin* **28**, 203–217.

de Bruijn, N. G., and B. J. M. Morselt (1967). A note on plane trees, *J. Comb. Theory* **2**, 27–34.

Carlitz, L. (1956). Some polynomials related to theta functions, *Ann. Math. Pura Appl.* (4) **41**, 359–373.

Carlitz, L. (1962). The generating function for $\max(n_1, \ldots, n_k)$, *Portugaliae Math.* **21**, 201–207.

Carlitz, L. (1972). Enumeration of sequences by rises and falls: a refinement of the Simon Newcomb problem, *Duke Math. J.* **39**, 267–280.

Carlitz, L. (1973a). Enumeration of up-down permutations by number of rises, *Pacific J. Math.* **45**, 49–58.

Carlitz, L. (1973b). Enumeration of up-down sequences, *Discrete Math.* **4**, 273–286.

Carlitz, L. (1974). q-Analog of the Lagrange expansion, *Eulerian Series and Applications*, Pennsylvania State University, Middletown, Pennsylvania.

Carlitz, L. (1977). Enumeration of compositions by rises, falls and levels, *Math. Nachrichten* **77**, 361–371.

Carlitz, L. (1978). Permutations with prescribed pattern, II: Applications, *Math. Nachrichten* **83**, 101–126.

Carlitz, L. (1979). Restricted compositions, II, *Fib. Quart.* **17**, 321–328.

Carlitz, L., and R. Scoville (1972). Up-down sequences, *Duke Math. J.* **39**, 583–598.

Carlitz, L., and R. Scoville (1973). Enumeration of rises and falls by position, *Discrete Math.* **5**, 45–59.

Carlitz, L., and R. Scoville (1974). Generalized Eulerian numbers: combinatorial applications, *J. reine. angew. Math.* **265**, 110–137.

Carlitz, L., and R. Scoville (1975). Generating functions for certain types of permutations, *J. Comb. Theory* (*A*) **18**, 262–275.

Carlitz, L., R. Scoville, and T. Vaughan (1973). Enumeration of permutations and sequences with restrictions, *Duke Math. J.* **40**, 723–741.

Carlitz, L., R. Scoville, and T. Vaughan (1976). Enumeration of pairs of sequences by rises, falls and levels, *Manuscripta Math.* **19**, 211–243.

Cartier, P., and D. Foata (1969). Problemes combinatoires de commutation et rearrangements, *Lecture Notes in Mathematics* **85**, Springer-Verlag, Berlin.

Catalan, E. (1838). Note sur une equation aux différences finies, *J. M. Pures Appl.* **3**, 508–516.

Cauchy, A. (1893). *Oeuvres Ser. 1*, Vol. 8, Gauthier-Villars, Paris.

Cayley, A. (1889). A theorem on trees, *Quart. J. Math. Oxford* **23**, 376–378.

Cayley, A. (1890). On the partitions of a polygon, *Phil. Mag.* (4) **22**, 237–262.

Chihara, T. S. (1978). *An Introduction to Orthogonal Polynomials*, Gordon and Breach, New York.

Chowla, S., I. N. Herstein, and W. R. Scott (1952). The solutions of $X^d = 1$ in symmetric groups, *Norske Vid. Selsk.* **25**, 29–31.

Clarke, I. E. (1958). On Cayley's formula for counting trees, *J. Lond. Math. Soc.* **33**, 471–474.

Comtet, L. (1966). Recouvrements, bases de filtre et topologies d'un ensemble fini, *C. R. Acad. Sci. Paris* (*A*) **262**, 1091–1094.

Comtet, L. (1968). Birecouvrements et birevêtements d'un ensemble fini, *Studia Sci. Math. Hungar.* **3**, 137–152.

Comtet, L. (1974). *Advanced Combinatorics*, D. Reidel, Dordrecht, Holland.

David, F. N., and D. E. Barton (1962). *Combinatorial Chance*, Griffin, London.

Dershowitz, N., and S. Zaks (1980). Enumerations of ordered trees, *Discrete Math*. **31**, 9–28.

Devitt, J. S., and D. M. Jackson (1982). The enumeration of covers of a finite set, *J. Lond. Math. Soc.* (2) **25**, 1–6.

Dillon, J. F., and D. P. Roselle (1969). Simon Newcomb's problem, *SIAM J. Appl. Math*. **17**, 1086–1093.

Dixon, A. C. (1891). On the sum of the cubes of the coefficients in a certain expansion by the binomial theorem, *Manuscripta Math*. **20**, 79–80.

Doubilet, P. (1972). On the foundations of combinatorial theory, VII: Symmetric functions through the theory of distribution and occupancy, *Studies Appl. Math*. **51**, 377–396.

Doubilet, P., G.-C. Rota, and R. P. Stanley (1972). On the foundations of combinatorial theory, VI: The idea of generating function, *6th. Berkeley Symp. Math. Statist. Prob.* **2**, 267–318.

Dumont, D. (1979). A combinatorial interpretation for the Schett recurrence on the Jacobian elliptic functions, *Math. Comp*. **33**, 1293–1297.

Entringer, R. C. (1969). Enumeration of permutations of $(1, \ldots, n)$ by number of maxima, *Duke Math. J*. **36**, 575–579.

Etherington, I. M. H. (1937). Non-associate powers and a functional equation, *Math. Gaz*. **21**, 36–39.

Euler, L. (1748). *Introductio In Analysin Infinitorum*, Marcum-Michaelem Bousquet, Lausannae, Chap. 16.

Euler, L. (1758, 59). *Novi Commentarii Academiae Scientiarum Petropolitanae* **7**, 13–14.

Everett, C. J., and P. R. Stein (1973). The asymptotic number of $(0, 1)$-matrices with zero permanent, *Discrete Math*. **6**, 29–34.

Farrell, E. (1979). On a class of polynomials associated with the stars of a graph and its application to node-disjoint decompositions of complete graphs and complete bipartite graphs into stars, *Canad. B. Math*. **22**, 35–46.

Feller, W. (1950). *An Introduction to Probability Theory and Its Applications*, Vols. 1 and 2, Wiley, New York.

Fishburn, P. C. (1979). Transitivity, *Rev. Economic Studies* **46**, 163–173.

Flajolet, P. (1980). Combinatorial aspects of continued fractions, *Discrete Math*. **32**, 99–216.

Flajolet, P. (1982). On congruences and continued fractions for some classical combinatorial quantities, *Discrete Math*. **42**, 145–153.

Flajolet, P., and J. Françon (1981). Interprétation combinatoire des coéfficients des dévelopements en série de Taylor des fonctions elliptiques de Jacobi (preprint).

Foata, D. (1968). On the Netto inversion number of a sequence, *Proc. Amer. Math. Soc*. **19**, 236–240.

Foata, D. (1974). La série génératrice exponentielle dans les problèmes d'énumération, Les Presses de l'Université de Montréal.

Foata, D., and J. Riordan (1974). Mappings of acyclic and parking functions, *Aequationes Math*. **10**, 10–22.

Foata, D., and M. P. Schützenberger (1970). Théorie géométrique des polynomies Eulériens, *Lecture notes in Mathematics* **138**, Springer-Verlag, Berlin.

Frame, J. S., G. de B. Robinson, and R. M. Thrall (1954). The hook graphs of the symmetric group, *Canad. J. Math*. **6**, 316–324.

Françon, J., and G. Viennot (1979). Permutations selon les pics, creux, doubles montées, doubles descentes, nombres d'Euler et nombres de Genocchi, *Discrete Math*. **28**, 21–35.

Franklin, F. (1881). Sur le développement du produit infini $(1 - x)(1 - x^2)(1 - x^3)\ldots$, *C. R. Acad. Sci. Paris* **82**, 448–450.

Garsia, A. M. (1983). A q-analogue of the Lagrange inversion formula (preprint).

Garsia, A. M., and I. Gessel (1979). Permutation statistics and partitions, *Adv. Math.* **31**, 288–305.

Garsia, A. M., and S. A. Joni (1977). A new expression for umbral operators and power series inversion, *Proc. Amer. Math. Soc.* **64**, 179–185.

Gauss, C. F. (1813). Disquisitiones generales circa seriem infinitam

$$1 + \frac{\alpha\beta}{1 \cdot \gamma}x + \frac{\alpha(\alpha+1)\beta(\beta+1)}{1 \cdot 2 \cdot \gamma(\gamma+1)}xx + \frac{\alpha(\alpha+1)(\alpha+2)\beta(\beta+1)(\beta+2)}{1 \cdot 2 \cdot 3 \cdot \gamma(\gamma+1)(\gamma+2)}x^3 + \text{etc.},$$

Commentationes societatis regiae scientiarum Gottingensis recentiores.

Gauss, C. F. (1863). *Werke*, Vol. 2, Königliche Gesellschaft der Wissenschaften, Göttingen.

Gessel, I. M. (1977). "Generating Functions and Enumeration of Sequences," Doctoral thesis, Massachusetts Institute of Technology, Cambridge, Massachussetts.

Gessel, I. (1980a). A factorization for formal Laurent series and lattice path enumeration, *J. Comb. Theory (A)* **28**, 321–337.

Gessel, I. (1980b). A noncommutative generalization and q-analog of the Lagrange inversion formula, *Trans. Amer. Math. Soc.* **257**, 455–481.

Gessel, I. (1981). Congruences for Bell and tangent numbers, *Fib. Quart.* **19**, 137–143.

Gessel, I., and R. Stanley (1978). Stirling polynomials, *J. Comb. Theory (A)* **24**, 24–33.

Gessel, I., and G. Viennot (1983). Determinants and plane partitions (preprint).

Gessel, I., and D. Wang (1979). Depth-first search as a combinatorial correspondence, *J. Comb. Theory (A)* **26**, 308–313.

Gilbert, E. N. (1956). Enumeration of labelled graphs, *Canad. J. Math.* **8**, 405–411.

Glaisher, J. W. L. (1883). A theorem in partitions, *Messenger Math.* **12**, 158–170.

Glaz, J. (1979). The number of dense arrangements, *J. Comb. Theory (A)* **27**, 367–370.

Goldman, J. R., and G.-C. Rota (1970). On the foundations of combinatorial theory, IV: Finite vector spaces and Eulerian generating functions, *Studies Appl. Math.* **49**, 239–258.

Good, I. J. (1960). Generalizations to several variables of Lagrange's expansion, with applications to stochastic processes, *Proc. Cambridge Philos. Soc.* **56**, 367–380.

Good, I. J. (1963a). A short proof of MacMahon's "Master Theorem," *Proc. Cambridge Philos. Soc.* **58**, 160.

Good, I. J. (1963b). Proofs of some "binomial" identities by means of MacMahon's "Master Theorem," *Proc. Cambridge Philos. Soc.* **58**, 161–162.

Good, I. J. (1965). The generalization of Lagrange's expansion and the enumeration of trees, *Proc. Cambridge Philos. Soc.* **61**, 499–517. See Correction **64** (1968), 489.

Good, I. J. (1969). Legendre polynomials and trinomial random walks, *Proc. Cambridge Philos. Soc.* **54**, 39–42.

Good, I. J. (1976). The relationship of a formula of Carlitz to the generalized Lagrange expansion, *SIAM J. Appl. Math.* **30**, 103.

Gordon, M., and J. A. Torkington (1980). Enumeration of coloured plane trees with a given type partition, *Discrete Appl. Math.* **2**, 207–224.

Gould, H. W. (1972). *Combinatorial Identities. A Standardized Set of Tables Listing* 500 *Binomial Coefficient Summations*, West Virginia University, Morgantown, W.V.

Goulden, I. P., and D. M. Jackson (1978). The enumeration of generalized alternating subsets with congruences, *Discrete Math.* **22**, 99–104.

Goulden, I. P., and D. M. Jackson (1979). An inversion theorem for cluster decompositions of sequences with distinguished subsequences, *J. Lond. Math. Soc.* (2) **20**, 567–576.

Goulden, I. P., and D. M. Jackson (1981). The enumeration of directed closed Euler trails and directed Hamiltonian circuits by Lagrangian methods, *Eur. J. Comb.* **2**, 131–135.

Goulden, I. P., and D. M. Jackson (1982a). An inversion model for q-identities (preprint).

Goulden, I. P., and D. M. Jackson (1982b). A logarithmic connection for circular permutations, *Studies Appl. Math.* (to appear).

Goulden, I. P., and D. M. Jackson (1982c). The application of Lagrangian methods to the enumeration of labelled trees with respect to edge partition, *Canad. J. Math.* **34**, 513–518.

Goulden, I. P., D. M. Jackson, and J. W. Reilly (1983). The Hammond series of a symmetric function and its application to p-recursiveness, *SIAM J. Discrete and Algebraic Methods* (to appear).

Goulden, I. P., and S. A. Vanstone (1983). The number of solutions of an equation arising from a problem on latin squares, *J. Austral. Math. Soc.* **34**, 138–142.

Guibas, L. J., and A. M. Odlyzko (1981a). Periods in strings, *J. Comb. Theory (A)* **30**, 19–42.

Guibas, L. J., and A. M. Odlyzko (1981b). String overlaps, pattern matching and nontransitive games, *J. Comb. Theory (A)* **30**, 183–208.

Hadamard, J. S. (1892). Essai sur l'étude des fonctions données par leur développement de Taylor, *J. Mathé.* (4) **8**, 1–86.

Halphen, G. H. (1879). Sur un problème d'analyse, *Bull. S.M.F.* **8**, 62–64.

Hammond, J. (1883). On the use of certain differential operators in the theory of equations, *Proc. Lond. Math. Soc.* **14**, 119–129.

Harary, F., G. Prins, and W. T. Tutte (1964). The number of plane trees, *Indag. Math.* **26**, 319–329.

Harary, F., and E. M. Palmer (1973). *Graphical Enumeration*, Academic Press, New York.

Hardy, G. H., and E. M. Wright (1938). *An Introduction to the Theory of Numbers*, 4th ed. Clarendon Press, Oxford.

Harris, B. (1960). Probability distributions related to random mappings, *Ann. Math. Stat.* **31**, 1045–1062.

Heine, E. (1846, 1847). Über die Reihe $1 + \frac{(q^\alpha - 1)(q^\beta - 1)}{(q-1)(q^\gamma - 1)}x + \frac{(q^\alpha - 1)(q^{\alpha+1} - 1)(q^\beta - 1)(q^{\beta+1} - 1)}{(q-1)(q^2-1)(q^\gamma - 1)(q^{\gamma+1} - 1)}x^2 + \cdots$, *J. Math.* **32**, **34**.

Henle, M. (1972). Dissection of generating functions, *Studies Appl. Math.* **51**, 397–410.

Henrici, P. (1964). An algebraic proof of the Lagrange-Bürmann formula, *J. Math. Anal. Appl.* **8**, 218–224.

Henrici, P. (1974). *Applied and Computational Complex Analysis*, Vol. 1, Wiley, New York.

Hermite, C. (1891). *Oeuvres*, Vol. 2, Gauthier-Villars, Paris.

Hofbauer, J. (1979). A short proof of the Lagrange-Good formula, *Discrete Math.* **25**, 135–140.

Hutchinson, J. P., and H. S. Wilf (1975). On Eulerian circuits and words with prescribed adjacency patterns, *J. Comb. Theory (A)* **18**, 80–87.

Jabotinsky, E. (1953). Representation of functions by matrices: application to Faber polynomials, *Proc. Amer. Math. Soc.* **4**, 546–553.

Jackson, D. M. (1978). Some results on product-weighted lead codes, *J. Comb. Theory (A)* **25**, 181–187.

Jackson, D. M., and R. Aleliunas (1977). Decomposition based generating functions for sequences, *Canad. J. Math.* **29**, 971–1009.

Jackson, D. M., and I. P. Goulden (1979, 1980). A formal calculus for the enumerative system of sequences I, II, III, *Studies Appl. Math.* **61**, 141–178, 245–277; **62**, 113–142.

Jackson, D. M., and I. P. Goulden (1981a). The generalization of Tutte's result for chromatic trees, by Lagrangian methods, *Canad. J. Math.* **33**, 12–19.

Jackson, D. M., and I. P. Goulden (1981b). Algebraic methods for permutations with prescribed patterns, *Adv. Math.* **42**, 113–135.

Jackson, D. M., B. Jeffcott, and W. T. Spears (1980). Enumeration of sequences with respect to structures on a bipartition, *Discrete Math.* **30**, 133–149.

Jackson, D. M., and J. W. Reilly (1975). The enumeration of homeomorphically irreducible labelled graphs, *J. Comb. Theory (B)* **19**, 272–286.

Jacobi, C. G. J. (1830). De resolutione aequationum per series infinitas, *J. reine angew. Math.* **6**, 257–286.

Joyal, A. (1981). Une théorie combinatorie des séries formelles, *Adv. Math.* **42**, 1–82.

Kaplansky, I. (1943). Solution of the "problème des ménages," *Bull. Amer. Math. Soc.* **49**, 784–785.

Karlin, S., and G. McGregor (1959). Coincidence probabilities, *Pacific J. Math.* **9**, 1141–1164.

Kim, K. H., M. S. Putcha, and F. W. Roush (1977). Some combinatorial properties of free semigroups, *J. Lond. Math. Soc.* (2) **16**, 397–402.

Klarner, D. A. (1967). Cell growth problem, *Canad. J. Math.* **19**, 851–863.

Klarner, D. A. (1970). Correspondences between plane trees and binary sequences, *J. Comb. Theory* **9**, 401–411.

Knuth, D. E. (1968a). *The Art of Computer Programming, Vol. 1: Fundamental Algorithms*, Addison-Wesley, Reading, Mass.

Knuth, D. E. (1968b). Another enumeration of trees, *Canad. J. Math.* **20**, 1077–1086.

Kreweras, G. (1965). Sur une classe de problèmes de dénombrement liés au treillis de partitions d'entiers, *Cahiers Buro* **6**, 2–107.

Lagrange, R. (1963). Sur les combinaisons d'objets numérotés, *Bull. Sci. Math.* **87**, 29–42.

Levine, J. (1959). Note on the number of pairs of non-intersecting routes, *Scripta Math.* **24**, 335–338.

Littlewood, D. E. (1950). *The Theory of Group Characters*, Clarendon Press, Oxford.

Liu, C. L. (1968). *Introduction to Combinatorial Mathematics*, McGraw-Hill, New York.

Lovász, L. (1979). *Combinatorial Problems and Exercises*, North-Holland, New York.

Macdonald, I. G. (1979). *Symmetric Functions and Hall Polynomials*, Clarendon Press, Oxford.

MacMahon, P. A. (1886). Certain special partitions of numbers, *Quart. J. Math. Oxford* **21**, 367–373.

MacMahon, P. A. (1891). The theory of perfect partitions and the compositions of multipartite numbers, *Messenger Math.* **35**, 103–119.

MacMahon, P. A. (1915). *Combinatory Analysis*, 2 vols., Cambridge University Press, London, 1915–1916. Reprinted Chelsea, New York, 1960.

Mallows, C. L., and J. Riordan (1968). The inversion enumerator for labelled trees, *Bull. Amer. Math. Soc.* **74**, 92–94.

Meir, A., and J. W. Moon (1968). On nodes of degree two in random trees, *Mathematika* **15**, 188–192.

Mohanty, S. G. (1979). *Lattice Path Counting and Applications*, Academic Press, New York.

Montmort, P. R. (1708). *Essai d'Analyse sur les Jeux de Hazard*, Paris.

Moon, J. W. (1964). The second moment of the complexity of a graph, *Mathematika* **11**, 95–98.

Moon, J. W. (1970). Counting labelled trees, *Canad. Math. Monographs*, No. 1.

Moon, J. W. (1972). The variance of the number of spanning cycles in a random graph, *Stud. Sci. Math. Hungar.* **7**, 281–283.

Moser, W. O. J., and M. Abramson (1969a). Enumeration of combinations with restricted differences and cospan, *J. Comb. Theory* **7**, 162–170.

Moser, W. O. J., and M. Abramson (1969b). Generalizations of Terquem's problem, *J. Comb. Theory* **7**, 171–180.

Muir, T. (1960). *A Treatise on the Theory of Determinants* (revised by W. H. Metzler), Dover, New York.

Mullin, R. C. (1964a). A combinatorial proof of the existence of Galois fields, *Amer. Math. Monthly* **71**, 901–902.

Mullin, R. C. (1964b). Enumeration of rooted triangular maps, *Amer. Math. Monthly* **71**, 1007–1010.

Mullin, R. C. (1965). On counting rooted triangular maps, *Canad. J. Math.* **17**, 373–382.

Mullin, R. C., and G.-C. Rota (1970). On the foundations of combinatorial theory, III: Theory of binomial enumeration, in *Graph Theory and Its Applications* (B. Harris, Ed.) Academic Press, New York, 167–213.

Mullin, R. C., and R. G. Stanton (1969). Identities from graphs, *Duke Math. J.* **36**, 605–608.

Narayana, T. V. (1959). A partial order and its applications to probability theory, *Sankhyā* **21**, 91–98.

Narayana, T. V. (1979). *Lattice Path Combinatorics with Statistical Applications*, University of Toronto Press, Toronto.

Nemetz, T. (1970). On the number of Hamilton cycles having common edges on a given number with a fixed Hamilton cycle, *Mat. Lapok.* **21**, 65–81.

Netto, E. (1927). *Lehrbuch der Combinatorik*, Teubner (reprinted by Chelsea, New York, 1958).

Niven, I. (1969). Formal power series, *Amer. Math, Monthly* **76**, 871–889.

Ostrowski, A. (1929). Ueber einige Verallgemeinerungen des Eulerschen Produktes $(1 - x)^{-1} = \prod_{\nu=0}^{\infty}(1 + x^{2^\nu})$, *Verh. Naturf. Ges. Basel* **11**, 153–214.

Perron, O. (1929). *Die Lehre von den Kettenbruchen*, Leipzig and Berlin.

Pólya, G. (1937). Kombinatorische Anzahlbestimmungen für Gruppen, Graphen, und chemische Verbindungen, *Acta Math.* **68**, 145–254.

Pólya, G. (1969). On the number of certain lattice polygons, *J. Comb. Theory* **6**, 102–105.

Pólya, G. (1970). Gaussian binomial coefficients and the enumeration of inversions, *Proc. 2nd. Chapel Hill Conf. Combinatorial Mathematics and Its Applications*, Univ. of North Carolina at Chapel Hill, 381–384.

Pólya, G., and G. Szegö (1964). *Aufgaben und Lehrsätze aus der Analysis*, 3rd ed., Vols. 1 and 2, Springer-Verlag, New York.

Prabhu, N. U. (1980). *Stochastic Storage Processes*, Springer-Verlag, New York.

Ramanujan, S. (1927). *Collected Papers of S. Ramanujan*, Cambridge University Press, London (reprinted by Chelsea, New York).

Raney, G. N. (1960). Functional composition patterns and power series reversion, *Trans. Amer. Math. Soc.* **94**, 441–451.

Read, R. C. (1960). The enumeration of locally restricted graphs (II), *J. Lond. Math. Soc.* **35**, 344–351.

Read, R. C. (1979). The chord intersection problem, *Ann. New York Acad. Sci.* **319**, 444–454.

Read, R. C. (1983). The dimer problem for narrow rectangular arrays: a unified method of solution, and some extensions, *Aeq. Math.* (to appear).

Read, R. C., and N. C. Wormald (1980). Number of labelled 4-regular graphs, *J. Graph Theory* **4**, 203–212.

Reid, W. T. (1972). *Riccati Differential Equations*, Academic Press, London.

Reilly, J. W. (1977). "An Enumerative Combinatorial Theory of Formal Power Series," Doctoral thesis, University of Waterloo, Waterloo, Ontario.

Reilly, J. W., and S. M. Tanny (1980). Counting permutations by successions and other figures, *Discrete Math.* **32**, 69–76.

Rényi, A. (1962). Théorie des éléments saillants d'une suite d'observations, in *Colloquium Aarhus*, 104–117.

Rényi, A. (1970). On the enumeration of trees, in *Combinatorial structures and their application*, Gordon Breach, New York, 355–360.

Rényi, A., and G. Szekeres (1967). On the height of trees, *J. Austral. Math. Soc.* **7**, 497–507.

Riordan, J. (1944). Three-line latin rectangles, *Amer. Math. Monthly* **51**, 450–452.

Riordan, J. (1958). *An Introduction to Combinatorial Analysis*, Wiley, New York.

Riordan, J. (1968). *Combinatorial Identities*, Wiley, New York.

Rodrigues, O. (1839). Notes sur les inversions ou dérangement produit dans les permutations, *J. Math. Pures Appl.* **4**, 236–246.

Rogers, L. J. (1893a). On a three-fold symmetry in the elements of Heine's series, *Proc. Lond. Math. Soc.* **24**, 171–179.

Rogers, L. J. (1893b). On the expansion of certain infinite products, *Proc. Lond. Math. Soc.* **24**, 337–352.

Rogers, L. J. (1907). On the representation of certain asymptotic series as continued fractions, *Proc. Lond. Math. Soc.* (2) **4**, 72–89.

Rogers, L. J., and S. Ramanujan (1919). Proof of certain identities in combinatory analysis, *Proc. Cambridge Philos. Soc.* **19**, 211–216.

Roselle, D. P. (1968). Permutations by number of rises and successions, *Proc. Amer. Math. Soc.* **19**, 8–16.

Rosen, J. (1976). The number of product-weighted lead codes for ballots and its relation to the Ursell functions of the linear Ising model, *J. Comb. Theory (A)* **19**, 377–384.

Rota, G.-C. (1964a). On the foundations of combinatorial theory, I: Theory of Mobius functions, *Z. Wahrscheinlichkeitstheorie und verw. Gebiete* **2**, 340–368.

Rota, G.-C. (1964b). The number of partitions of a set, *Amer. Math. Monthly* **71**, 498–504.

Rubin, H., and R. Sitgreaves (1954). "Probability Distributions Related to Random Transformations on a Finite Set," Tech. Report #19A, Applied Math. and Stat. Lab., Stanford University, Cal.

Salié, H. (1963). Arithmetische Eigenschaften der Koeffizienten einer spezieller Hurwitzschen Potenzreihe, *Wiss. Z. der Karl-Marx-Univ. Leipzig Math.-Natur. Reihe* **12**, 617–618.

Schröder, E. (1870). Vier kombinatorische probleme, *Z. Math. Phys.* **15**, 361–376.

Sedláček, J. (1969). On the number of spanning trees of finite graphs, *Časopis pro pěstování matematiky* **94**, 217–221.

Sherman, J., and W. J. Morrison (1949). Adjustments of an inverse matrix corresponding to changes in the elements of a given row or a given column of the original matrix, *Ann. Math. Stat.* **20**, 621.

Smirnov, N. V., O. V. Sarmanov, and V. K. Zaharov (1966). A local limit theorem for transition numbers in a Markov chain, and its applications, *Sovi. Math. Dokl.* **7**, 563–566 (= *Dokl. Akad. Nauk. SSR* **167**, 1238–1241).

Spears, W. T., B. Jeffcott, and D. M. Jackson (1980). An algebraic theory of sequence enumeration, *J. Comb. Theory (A)* **28**, 191–218.

Stieltjes, T. J. (1889). Sur la réduction en fraction continue d'une série procédant suivant les puissances descendantes d'une variable, *Ann. Fac. Sci. Toulouse* **3**, 1–17.

Stanley, R. P. (1971). Theory and applications of plane partitions: Part I, II, *Studies Appl. Math.* **50**, 167–188, 259–279.

Stanley, R. P. (1972). Ordered structures and partitions, *Mem. Amer. Math. Soc.* **119**, 1–102.

Stanley, R. P. (1976). Binomial posets, Möbius inversion, and permutation enumeration, *J. Comb. Theory (A)* **20**, 336–356.

Stanley, R. P. (1980). Differentiably finite power series, *Eur. J. Comb.* **1**, 175–188.

Stanton, R. G., and D. A. Sprott (1962). Some finite inversion formulae, *Math. Gazette* **46**, 197–202.

Subbarao, M. V. (1971). On a partition theorem of MacMahon-Andrews, *Proc. Amer. Math. Soc.* **27**, 449–450.

Sylvester, J. J. (1882, 1884). A constructive theory of partitions in three acts, an interact, and an exodion, *Amer. J. Math.* **5**, 251–330; **6**, 334–336 (or pp. 1–83 of the *Collected Papers of J. J. Sylvester, Vol.* 4, *Cambridge Univ. Press, London*, 1912; *reprinted by Chelsea, New York*, 1974).

Szegö, G. (1926). Ein Beitrag zur Theorie der Thetafunktionen, *S.B. Preuss, Akad. Wiss. Phys.-Math. Kl*, 242–252.

Takács, L. (1967). *Combinatorial Methods in the Theory of Stochastic Processes*, Wiley, New York.

Tanny, S. M. (1975). Generating functions and generalized alternating subsets, *Discrete Math.* **13**, 55–65.

Tanny, S. M. (1976). Permutations and successions, *J. Comb. Theory* (*A*) **21**, 196–202.

Terquem, O. (1839). Sur un symbole combinatoire d'Euler et son utilité dans l'analyse, *J. Math. Pures Appl.* **4**, 177–184.

Tomescu, I. (1975). *Introduction to Combinatorics*, Collet's, Wellingborough, England.

Touchard, J. (1934). Sur un problème de permutations, *C. R. Acad. Sci. Paris* **198**, 631–633.

Touchard, J. (1952). Sur un problème de configurations et sur les fractions continues, *Canad. J. Math.* **4**, 2–25.

Touchard, J. (1953). Permutations discordant with two given permutations, *Scripta Math.* **19**, 109–119.

Tutte, W. T. (1948). The dissection of equilateral triangles into equilateral triangles, *Proc. Cambridge Philos. Soc.* **44**, 463–482.

Tutte, W. T. (1962). A census of planar triangulations, *Canad. J. Math.* **14**, 21–38.

Tutte, W. T. (1963). A census of planar maps, *Canad. J. Math.* **15**, 249–271.

Tutte, W. T. (1964). The number of planted plane trees with a given partition, *Amer. Math. Monthly* **71**, 272–277.

Tutte, W. T. (1973a). The enumerative theory of planar maps, in *A Survey of Combinatorial Theory*, (J. N. Srivastava, Ed.) North-Holland, New York, 437–448.

Tutte, W. T. (1973b). Chromatic sums for rooted planar triangulations, IV: The case $\lambda = \infty$, *Canad. J. Math.* **25**, 929–940.

Tutte, W. T. (1975). On elementary calculus and the Good formula, *J. Comb. Theory* (*B*) **18**, 97–137.

Viennot, G. (1978). Algèbres de Lie libres et monoides libres, *Lecture Notes in Mathematics*, **691**, Springer-Verlag, Berlin.

Viennot, G. (1980). Une interprétation combinatoire des développements en série entière des fonctions elliptiques de Jacobi, *J. Comb. Theory* (*A*) **29**, 121–133.

Wall, H. S. (1967). *Analytic Theory of Continued Fractions*, Chelsea, New York.

Walsh, T. R. S., and A. B. Lehman (1972a). Counting rooted maps by genus, I., *J. Comb. Theory* (*B*) **13**, 192–218.

Walsh, T. R. S., and A. B. Lehman (1972b). Counting rooted maps by genus, II, *J. Comb. Theory* (*B*) **13**, 122–141.

Walsh, T. R. S., and A. B. Lehman (1975). Counting rooted maps by genus, III: Nonseparable maps, *J. Comb. Theory* (*B*) **18**, 222–259.

Watson, G. N. (1952). *Bessell Functions*, Cambridge University Press, London.

Weinberg, L. (1958). Number of trees in a graph, *Proc. IRE* **46**, 1954–1955.

Whittaker, E. T., and G. N. Watson (1927). *A Course in Modern Analysis*, Cambridge University Press, New York.

Wormald, N. (1981a). On the number of planar maps, *Canad. J. Math.* **33**, 1–11.

Wormald, N. (1981b). Counting unrooted planar maps, *Discrete Math.* **36**, 205–226.

Wright, E. M. (1973). For how many edges is a digraph almost certainly Hamiltonian? *Proc. Amer. Math. Soc.* **41**, 384–388.

Zeilberger, D. (1981). Enumeration of words by their number of mistakes, *Discrete Math.* **34**, 89–92.

Index

Note: "a.b.c." denotes section number, "[a.b.c]" denotes exercise number; "*b*" denotes location immediately before, "*f*" denotes location immediately following; "*N & R*" denotes Notes and References, these being found at end of indicated section. *See* "Problems" for list of types of problems treated.